Lecture Notes in Computer Science　14461

Founding Editors

Gerhard Goos

Juris Hartmanis

The series Lecture Notes in Computer Science (LNCS), including its subseries Lecture Notes in Artificial Intelligence (LNAI) and Lecture Notes in Bioinformatics (LNBI), has established itself as a medium for the publication of new developments in computer science and information technology research, teaching, and education.

LNCS enjoys close cooperation with the computer science R & D community, the series counts many renowned academics among its volume editors and paper authors, and collaborates with prestigious societies. Its mission is to serve this international community by providing an invaluable service, mainly focused on the publication of conference and workshop proceedings and postproceedings. LNCS commenced publication in 1973.

Weili Wu · Jianxiong Guo
Editors

Combinatorial Optimization and Applications

16th International Conference, COCOA 2023
Hawaii, HI, USA, December 15–17, 2023
Proceedings, Part I

 Springer

Editors
Weili Wu
University of Texas at Dallas
Richardson, TX, USA

Jianxiong Guo
Beijing Normal University
Zhuhai, China

ISSN 0302-9743   ISSN 1611-3349 (electronic)
Lecture Notes in Computer Science
ISBN 978-3-031-49610-3   ISBN 978-3-031-49611-0 (eBook)
https://doi.org/10.1007/978-3-031-49611-0

This Springer imprint is published by the registered company Springer Nature Switzerland AG
The registered company address is: Gewerbestrasse 11, 6330 Cham, Switzerland

Paper in this product is recyclable.

Preface

The papers in these proceedings, which consist of two volumes, were presented at the 16th Annual International Conference on Combinatorial Optimization and Applications (COCOA 2023), December 15–17, 2023, in Honolulu, Hawaii, USA. The topics cover most aspects of combinatorial optimization and applications pertaining to computing.

All accepted papers were selected by an international program committee consisting of a large number of scholars from various countries and regions, distributed over the world, including Asia, North America, Europe, and Australia. Each paper was required to submit in double-blind style and was evaluated by at least three reviewers. The decision was made based on those evaluations through a process containing a discussion period.

Authors of selected papers come from the following countries and regions: Canada, China (including Hong Kong and Macau), Romania, Brazil, UK, India, Belgium, Japan, Germany, Israel, and USA. Many of these papers represent reports of continuing research, and it is expected that most of them will appear in a more polished and complete form in scientific journals.

We wish to thank all who have made this meeting possible and successful, the authors for submitting papers, the program committee members for their excellent work in reviewing papers, the sponsors, the local organizers, and Springer for their support and assistance, We are especially grateful to Yi Zhu and Xiao Li who made tremendous efforts in local arrangements and set-up.

December 2023

Weili Wu
Jianxiong Guo

Organization

General Co-chair

Ding-Zhu Du   University of Texas at Dallas, USA

PC Co-chairs

Weili Wu   University of Texas at Dallas, USA
Jianxiong Guo   Beijing Normal University, China

Web Co-chairs

Xiao Li   University of Texas at Dallas, USA
Ke Su   University of Texas at Dallas, USA

Finance Co-chairs

Jing Yuan   University of Texas at Dallas, USA
Smita Ghosh   Santa Clara University, USA

Registration Co-chairs

Xiao Li   University of Texas at Dallas, USA
Garg Priyanshi   University of Texas at Dallas, USA

Local Chair

Yi Zhu   Hawaii Pacific University, USA

Program Committee Members

An Zhang	Hangzhou Dianzi University, China
Andras Farago	University of Texas at Dallas, USA
Annalisa De Bonis	Università degli Studi di Salerno, Italy
Arash Rafiey	Indiana State University, USA
Bin Liu	Ocean University of China, China
Binhai Zhu	Montana State University, USA
Bo Li	Hong Kong Polytechnic University, China
Chenchen Wu	Tianjin University of Technology, China
Chuanwen Luo	Beijing Forestry University, China
Dachuan Xu	Beijing University of Technology, China
Donglei Du	University of New Brunswick, Canada
Fay Zhong	Shanghai Jiao Tong University, China
Guochuan Zhang	Zhejiang University, China
Guohui Lin	University of Alberta, Canada
Habib Ammari	Texas A&M University-Kingsville, USA
Ho-Lin Chen	National Taiwan University, Taiwan
Huaming Zhang	University of Alabama in Huntsville, USA
Jing Yuan	North Texas University, USA
Joong-Lyul Lee	University of North Carolina at Pembroke, USA
Juraj Hromkovic	ETH Zurich, Switzerland
Kazuo Iwama	Kyoto University, Japan
Lidong Wu	University of Texas at Dallas, USA
Ling-Ju Hung	National Taipei University of Business, Taiwan
Louxin Zhang	National University of Singapore, Singapore
Lu Han	Beijing University of Posts and Telecommunications, China
Meghana Satpute	University of Texas at Dallas, USA
Michael Khachay	Krasovsky Institute of Mathematics and Mechanics, Russia
Mihaela Cardei	Florida Atlantic University, USA
Qianping Gu	Simon Fraser University, Canada
Sergey Bereg	University of Texas at Dallas
Sergiy Butenko	Texas A&M University, USA
Shaojie Tang	University of Texas at Dallas, USA
Shengxin Liu	Harbin Institute of Technology (Shenzhen), China
Shuyang Gu	Texas A&M University - Central Texas, USA
Smita Ghosh	Santa Clara University, USA
Ueverton Souza	Universidade Federal Fluminense, Brazil
Viet Hung Nguyen	Université Pierre & Marie Curie, France
Wei Wang	Xian Jiaotong University, China

Contents – Part I

Set-Related Optimization

Applied Optimization and Algorithm

Graph Planer and Others

Contents – Part II

Optimization and Algorithms

Extreme Graph and Others

Machine Learning, Blockchain and Others

Optimization in Graphs

An Efficient Local Search Algorithm for Correlation Clustering on Large Graphs

Nathan Cordner[1]([✉])(iD) and George Kollios[2]

[1] Utah Valley University, Orem, UT, USA
ncordner@uvu.edu
[2] Boston University, Boston, MA, USA

Abstract. Correlation clustering (CC) is a widely-used clustering paradigm, with many applications to problems such as classification, database deduplication, and community detection. CC instances represent objects as graph nodes, and clustering is performed based on relationships between objects (positive or negative edges between pairs of nodes). The CC objective is to obtain a graph clustering that minimizes the number of incorrectly assigned edges (negative edges within clusters, and positive edges between clusters).

For large CC instances, lightweight algorithms like the Pivot method have been preferred due to their scalability. Because these algorithms do not have state-of-the-art approximation guarantees, LocalSearch (LS) methods have often then been applied to refine their clustering results. Unfortunately, LS does not enjoy the same ability to scale since it is inherently sequential and has the potential to converge slowly.

We propose a lightweight, parallelizable LS method called Inner-LocalSearch (ILS) to use in conjunction with the Pivot algorithm. We show that ILS still provides a significant improvement to clustering quality while dramatically reducing the additional running time costs incurred by LS. We demonstrate our algorithm's effectiveness against several LS benchmarks and other popular CC methods on real and synthetic data sets.

Keywords: data mining · correlation clustering · local search

1 Introduction

The "min disagreement" correlation clustering (CC) problem, as originally defined by Bansal et al. [6], inputs a complete graph $G = (V, E)$ where every pair of nodes is assigned a positive $(+)$ or negative $(-)$ relationship. The objective is to cluster together positively related nodes and separate negatively related ones, minimizing the total number of clustering "mistakes" (negatively related pairs within clusters, and positively related pairs separated between clusters). This clustering paradigm has been used in many applications, such has its original motivation of classification [6], database deduplication [20], and community detection in social networks [30,33]. This formulation of graph clustering has

W. Wu and J. Guo (Eds.): COCOA 2023, LNCS 14461, pp. 3–15, 2024.
https://doi.org/10.1007/978-3-031-49611-0_1

been especially useful, since a specific number of clusters does not need to be specified beforehand and the only information needed as input concerns the relationship between objects—not about the objects themselves.

One of the most popular CC algorithms is Pivot, presented by Ailon et al. [3], which gives an expected 3-approximation result for correlation clustering. It runs by choosing a "pivot" node at random, adding it and all other unclustered nodes with an edge to it into a cluster, and repeating until all nodes are clustered (think of choosing a random person in a social media network and clustering together all "friends" of the pivot person). Currently the best known approximation factor is $1.994 + \epsilon$, from a linear program rounding method due to Cohen et al. [13]. Unfortunately, linear programs for correlation clustering are quite large (the number of constraints is at least cubic in the number of graph nodes). Some work has been done to reduce the size of these linear programs [21], but even algorithms based on these methods become intractable once input graphs have millions of nodes. Instead, the Pivot algorithm has been revisited many times [1,2,4,11,12,16,19,22–24,26–28,32] because of its ease of implementation and scalability for large graphs.

A popular method of improving Pivot and other CC results is LocalSearch (LS)[1] [1,10,14,15,17,18,25,26,29,31] which moves nodes one at a time to improve clustering costs until no more improvements can be made or a self-imposed limit is reached. Each LS pass through the node set V examines all nodes and all positive edges E^+ (yielding a time complexity of $\Theta(|V| + |E^+|)$), and LS has the potential to make multiple passes while improvements slowly accrue. A LS pass that yields only a small number of improvements can trigger another full pass through the node and positive edge sets. For larger graphs, even running a small number of LS rounds can be less practical. The purpose of this paper is to develop a new LS technique that still yields significant improvement to clusterings produced by the Pivot algorithm, without the exorbitant time cost imposed by running a full LS algorithm.

One weakness in the design of the Pivot algorithm is that it only considers immediate connections of chosen pivot nodes. In many real-world settings, it is reasonable to assume that not all "friends-of-friends" edges of pivot nodes are present. We propose a new LS method called InnerLocalSearch (ILS), which runs LS *inside* clusters only. The ILS algorithm still starts out with an $O(|V| + |E^+|)$ pass through the node set, but now has the ability to ignore positive graph edges that go between clusters. Convergence within clusters tends to be much quicker since smaller sets of nodes are being compared against each other, and once individual clusters are converged the algorithm does not need to consider the nodes inside them in future iterations. ILS is also easily parallelizable, making it possible to run LS within multiple clusters at the same time. And though it necessarily will not yield the same level of clustering improvement as the full LS algorithm, we show experimentally that ILS still lowers objective values significantly while drastically reducing the running time needed for convergence. We compare Pivot with ILS against Pivot, several versions of Pivot with LS, and another popular CC algorithm called Vote [15].

---

[1] Also called "Best One Element Move" (BOEM).

1.1 Related Work

The NP-hard correlation clustering problem was introduced by Bansal et al. [6], who also provided its first constant approximation algorithm in the min disagreement setting. The best known approximation factor is $1.994 + \epsilon$, from a linear program rounding method due to Cohen et al. [13]. Correlation clustering remains an active area of research, and many variations of the problem have arisen over time; a general introduction to the correlation clustering problem and some of its early variants is given by Bonchi et al. [8]. Recent research has gone into developing sublinear time [5] and better parallel [7] CC algorithms.

The Pivot algorithm was first introduced by Ailon et al. [3]. Its efficient run time and ease of implementation have made it very popular, and it has been applied to many variants of correlation clustering that have arisen since. Recently, it has been used for uncertain graphs [26], query-constrained CC [16], online CC [24], chromatic CC [22], and fair CC [1]. It has also been shown how to run the Pivot algorithm in parallel in various settings [11,27]. Zuylen and Williamson [32] developed a deterministic version of Pivot that picks a best pivot at each round, though at the cost of an increased running time complexity. The most efficient non-parallel implementation of Pivot uses a neighborhood oracle, where a hash table stores lists of neighbors for each node [4].

Various authors have employed LocalSearch as post-processing to improve clustering results. LS refinements and LS-based algorithms have been used in many CC variants and applications [1,10,14,15,17,18,25,26,29,31]. Bonchi et al. [9] experimented with a heuristic method for running LocalSearch in parallel, sacrificing monotone decreasing objective values for potentially faster runtimes. Levinkov et al. [25] studied and compared other LocalSearch-based algorithms for correlation clustering.

2 Previous Algorithms

Let G be a complete graph on node set $V = \{1, \ldots, n\}$. A *clustering* of the graph G is a partition $\mathcal{C}$ of the node set V. For a given clustering $\mathcal{C} = \{C_1, ..., C_k\}$, define *intra-cluster edges* to be edges between nodes within the same cluster; define *inter-cluster edges* to be edges between nodes in distinct clusters.

For correlation clustering we assume the edge set E is partitioned into a set of *positive edges* E^+ and *negative edges* E^-. The objective of min disagreement correlation clustering is to find a clustering $\mathcal{C}$ of V that minimizes the number of negative intra-cluster edges and positive inter-cluster edges. Let similarity function $s(u,v) = 1$ if $(u,v) \in E^+$, and 0 otherwise. We write the *cost* (or *objective value*) of clustering $\mathcal{C}$ as

$$\text{Cost}(\mathcal{C}, V) = \sum_{\substack{u,v \in V,\ u \neq v \\ (u,v) \text{ is intra-cluster}}} (1 - s(u,v)) + \sum_{\substack{u,v \in V,\ u \neq v \\ (u,v) \text{ is inter-cluster}}} s(u,v).$$

For a given clustering $\mathcal{C}$, we define the *precision* of $\mathcal{C}$ to be the average number of positive edges inside clusters of $\mathcal{C}$. Let $\text{Intra}(\mathcal{C}) = \{(u,v) \mid (u,v) \text{ is}$

Algorithm 1. Pivot Clustering

1: **function** PIVOT($G = (V, E, s)$)
2:    Initialize empty clustering $\mathcal{C}$
3:    **while** $V \neq \emptyset$ **do**
4:       Choose random $u \in V$
5:       Let $C = \{u\} \cup \{v \in V \mid (u, v) \in E^+\}$ ▷ u and its unclustered neighbors
6:       $\mathcal{C} = \mathcal{C} \cup \{C\}$ ▷ Add cluster C to clustering $\mathcal{C}$
7:       $V = V \setminus C$ ▷ Remove clustered nodes from V
8:    **return** the finished clustering $\mathcal{C}$

---

intra-cluster}. We write

$$\text{Precision}(\mathcal{C}) = \frac{1}{|\text{Intra}(\mathcal{C})|} \cdot \sum_{(u,v) \in \text{Intra}(\mathcal{C})} s(u, v).$$

We also define the *recall* of $\mathcal{C}$ to be the ratio of the number of positive edges within clusters of $\mathcal{C}$ to the total number of positive edges $|E^+|$. We write

$$\text{Recall}(\mathcal{C}) = \frac{1}{|E^+|} \cdot \sum_{(u,v) \in \text{Intra}(\mathcal{C})} s(u, v).$$

2.1 Pivot

Ailon et al. [3] proposed the randomized Pivot algorithm (Algorithm 1) for unweighted correlation clustering. A cluster C is formed by picking a *pivot node* u at random from V, then adding u and all other nodes v in V to C that are connected by a positive edge to u (that is, $(u, v) \in E^+$). If $V \setminus C$ is not empty, the algorithm continues on the subgraph induced by $V \setminus C$.

The Pivot algorithm yields a 3-approximation clustering result. Following a common implementation method [4], we assume that every node $u \in V$ has access to a *neighborhood oracle* $N(u)$ that contains all nodes v with a positive relationship to u. We write

$$N(u) = \{v \in V \mid (u, v) \in E^+\}.$$

The time complexity of Pivot is thus $O(|V| + |E^+|)$. However, Pivot often runs much quicker than its worst-case time bound since many nodes can be removed from V with each choice of pivot (see lines 5–7 of Algorithm 1).

2.2 LocalSearch

LocalSearch (Algorithm 2) has been a popular technique for improving the clusterings output by CC algorithms. LS takes a current CC instance G and a current clustering $\mathcal{C}$ of the node set V. LS chooses a random permutation of nodes, and

---

Algorithm 2. Local Search Improvements

1: **function** LOCALSEARCH($G = (V, E, s)$, $\mathcal{C}$, π) ▷ π is a permutation of V (line 10)
2:     **for** $i \in \{1, \ldots, |V|\}$ **do**
3:         Let $u = \pi(i)$, with current cluster C
4:         Let $C' = \arg\min_{C'' \in \mathcal{C} \cup \{\emptyset\}} \{\text{Cost}(C'', u)\}$
5:         **if** $C' \neq C$ **then**
6:             $\mathcal{C} = (\mathcal{C} \setminus \{C\}) \cup \{C \setminus \{u\}\}$ ▷ Remove u from C
7:             $\mathcal{C} = (\mathcal{C} \setminus \{C'\}) \cup \{C' \cup \{u\}\}$ ▷ Add u to C'
8:     **return** the augmented clustering $\mathcal{C}$
9: **function** LOCALSEARCHLOOP($G = (V, E, s)$, $\mathcal{C}$)
10:     Choose a random permutation π of V
11:     $\mathcal{C}' = $ LOCALSEARCH(G, $\mathcal{C}$, π)
12:     **while** $\text{Cost}(\mathcal{C}', V) < \text{Cost}(\mathcal{C}, V)$ **do**
13:         $\mathcal{C} = \mathcal{C}'$
14:         $\mathcal{C}' = $ LOCALSEARCH(G, $\mathcal{C}$, π)
15:     **return** $\mathcal{C}$

---

iteratively makes improvements to the given clustering. A current node u considers all current clusters $C \in \mathcal{C}$, and the possibility of forming a new singleton cluster, and chooses to move to whatever cluster minimizes its own contribution to the overall clustering cost. When some pass through the node set yields no new improvements, the LS algorithm halts and returns the modified clustering $\mathcal{C}'$. Limits can be imposed on LS, such as the max number of allowable iterations through the node set or a time limit on how long LS can run before returning its improved clustering.

For a given cluster $C \subseteq V$, and current node $u \in V$, we define the cost of node u to be

$$\text{Cost}(C, u) = \sum_{v \in C \setminus \{u\}} (1 - s(u, v)) + \sum_{v \in V \setminus (C \cup \{u\})} s(u, v).$$

The cost of opening a new cluster is $\text{Cost}(\emptyset, u) = \sum_{v \in V \setminus \{u\}} s(u, v)$. The algorithm greedily minimizes the cost of each node relative to the current clustering.

With a neighborhood oracle and a hash table containing the current node-to-cluster assignment, line 4 of Algorithm 2 can be computed in $O(|N(u)|)$ time. We do this by iterating over $N(u)$, tracking how many neighbors of u lie in each cluster. We then consider only the clusters C that contain neighbors of u (as well as the possibility of opening a new cluster), and compute $\text{Cost}(C, u) = |C| - |\{\text{neighbors of } u \text{ within } C\}| + (|N(u)| - |\{\text{neighbors of } u \text{ within } C\}|)$; $\text{Cost}(\emptyset, u) = |N(u)|$, so we can safely ignore clusters that do not contain neighbors of u since in that case $\text{Cost}(C, u) = |C| + |N(u)| > \text{Cost}(\emptyset, u)$. At most $|N(u)| + 1$ clusters are considered in this process, so finding the minimum is still $O(|N(u)|)$. Looping over every node in V, the time complexity of a single LS iteration is thus $\Theta(|V| + |E^+|)$. The overall running time of LocalSearch is $\Theta((|V| + |E^+|)I)$, where I is the total number of iterations made through the node set V.

Algorithm 3. InnerLocalSearch Improvements

1: **function** INNERLOCALSEARCH($G = (V, E, s)$, $\mathcal{C} = \{C_1, \ldots, C_k\}$)
2:     **for** each cluster $C_i \in \mathcal{C}$ **do**   ▷ Form subgraph induced by C_i
3:         Let $E_i^+ = \{(u, v) \mid u, v \in C_i, u \neq v, s(u, v) = 1\}$
4:         **for** $v \in C_i$ **do**
5:             Let $N_i(v) = \{u \in C_i \mid (u, v) \in E_i^+\}$ ▷ new neighborhood oracle
6:         Let $G_i = (C_i, E_i^+, s)$ be the subgraph of G induced by C_i
7:         Let $\mathcal{C}_i = $ LOCALSEARCHLOOP($G_i, \{C_i\}$)
8:     **return** $\mathcal{C}_1 \cup \cdots \cup \mathcal{C}_k$

---

3 InnerLocalSearch

Though LocalSearch has been used effectively in several applications, its has a few drawbacks that make it less practical to run on larger instances. LS is inherently sequential, since the decision of where to place a current node depends on the decision of the previous nodes in the ordering, making it difficult to run in parallel (e.g. [9]). LS decisions are also slow, since comparisons are made across the entire vertex set (or at least all the neighbors of any given node).

To make up for these shortcomings, we propose an InnerLocalSearch algorithm (Algorithm 3). On a given clustering $\mathcal{C}$, we run LocalSearch to convergence inside each cluster $C \in \mathcal{C}$ and return the updated clustering.

We note that ILS runs LS inside each cluster, with a runtime of $O((|C_i| + |E_i^+|)I_i)$ per cluster C_i (line 7 of Algorithm 3, where I_i is the number of LS iterations needed for cluster C_i to converge). Forming the subgraph (lines 3 to 6) across all clusters can be done in $O(|V| + |E^+|)$ time. Thus the overall time complexity of ILS is $O(|V| + |E^+| + \sum_{i=1}^{k}(|C_i| + |E_i^+|)I_i)$, where k is the size of the input clustering $\mathcal{C}$. We note that ILS tends to converge much more quickly than LS since it greatly reduces the number of comparisons needed between nodes across the entire node set V. ILS cluster improvement can also be done in parallel by running the For loop in line 2 of Algorithm 3 on multiple threads.

3.1 InnerLocalSearch and Pivot

Though InnerLocalSearch (and LocalSearch too) can be run on any clustering result, we will focus on how ILS improves the Pivot algorithm. On any given clustering round, the Pivot algorithm only checks to see if nodes have positive edges to the chosen pivot node before deciding whether to cluster them together. As such it is possible for many negative edges to exist within clusters formed by the Pivot algorithm, making for a low precision clustering.

Consider Fig. 1, which shows a sample cluster from the Pivot algorithm (only positive edges are drawn between nodes). Here node A was chosen as pivot and all 10 nodes from A to J are put into one cluster (drawn in purple). The cost of this cluster alone is 27, whereas the optimal partition into 3 clusters (drawn in red) reduces the cost to just 6. By putting all nodes into one cluster, the Pivot

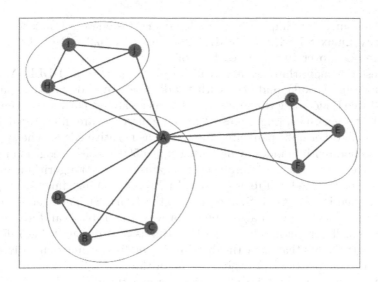

Fig. 1. Pivot Cluster Example

algorithm incurred a higher cost and ignored smaller tight-knit communities. In the worst case a Pivot cluster C might only contain $|C| - 1$ positive edges and $(|C|^2 - 3|C| + 2)/2$ negative edges, with a ratio of $(|C| - 2)/2$ negative edges for every positive.

The goal of InnerLocalSearch is thus to split larger Pivot clusters containing many negative edges into smaller clusters that contain a high number of positive edges. In other words, InnerLocalSearch seeks to quickly boost the precision of a Pivot clustering without greatly reducing its recall.

4   Experiments

We test the following algorithms: Pivot, Pivot with InnerLocalSearch (ILS), Pivot with Timed LocalSearch (Timed), Pivot with full LocalSearch (Full). For comparison, we include the Vote algorithm [15] which chooses a random node to start a cluster and adds new nodes by greedily minimizing increase of clustering cost. We present two implementations of ILS—sequential (clusters are improved one at a time), and parallel (multiple clusters being improved at once). The parallel implementation is done via Java parallel streams. For objective values, we include one benchmark (Outer) that lists the inter-cluster cost from the Pivot clustering; this is the maximum level of improvement that ILS can obtain if every misclassified edge within Pivot clusters is resolved. The Timed LocalSearch method allows full LocalSearch to run for the same amount of time used by InnerLocalSearch. For running times we include another benchmark (Match) that records the time full LocalSearch takes to match the same level of clustering improvement obtained by InnerLocalSearch.

All algorithms were implemented in Java[2] and tested on a Linux server running Rocky Linux 8.7 with a 2.9 GHz processor and 16.2 GB of RAM. Mean results are taken over 10 runs of each algorithm.

We test our algorithms on five real data sets[3] (Amazon, DBLP, Youtube, Livejournal, and Orkut), and one synthetic data set. Brief descriptions are provided in Table 1, including the value of U (the mean largest cluster size generated by the Pivot algorithm). Edges present in these data sets are interpreted as positive edges; all other node pairs are interpreted as negative edges. The synthetic data was generated by randomly assigning up to 5000 positive edges per node.

Objective values, relative improvements against the Pivot algorithm, and running times are reported in Tables 2 and 3. In Table 3 running times are reported for both sequential ILS (SeqILS) and parallel ILS (ParILS). The average number of clusters produced by each algorithm is provided in Table 4, and average precision and recall percentages for each clustering is provided in Table 5. Figure 2 contains scatter plots that show the distribution of disagreements for Pivot, ILS, Full, and Vote across the 10 runs of each algorithm.

We first note that InnerLocalSearch improves the Pivot clustering results significantly across all data sets. The Amazon data set has the smallest objective value decrease at just over 20% lower than Pivot, with all others decreasing over 25%. Some (Orkut and Synthetic) even decrease by at least 30% from the Pivot baseline.

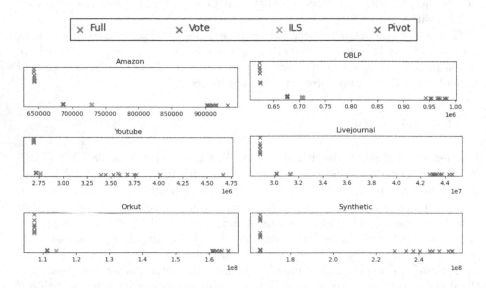

Fig. 2. Disagreement Plots for Pivot, ILS, Full, and Vote

---

[2] Code available at github.com/cc-conf-sub/ils-improvement.

[3] Available at snap.stanford.edu/data/#communities.

Table 1. Data Set Descriptions

Data Set	$|V|$	$|E^+|$	U	Description
Amazon	334863	925872	113	joint-purchase network
DBLP	317080	1049866	149	co-author network
YouTube	1134890	2987624	500	friend network
LiveJournal	3997962	34681189	926	friend network
Orkut	3072441	117185083	1762	friend network
Synthetic	100000	165987514	4852	random edges

Table 2. Mean Objective Values

Data Set	Pivot	Outer	ILS	Timed	Full	Vote
Amazon	912658	0.667	0.799	0.966	0.706	0.751
DBLP	964983	0.664	0.731	0.973	0.647	0.701
YouTube	3738286	0.7	0.739	1.0	0.722	0.728
LiveJournal	43455676	0.666	0.721	0.984	0.664	0.694
Orkut	162654949	0.67	0.7	0.995	0.661	0.684
Synthetic	244469396	0.665	0.679	0.888	0.678	0.679

Table 3. Mean Running Times (s)

Data Set	Pivot	SeqILS	ParILS	Match	Full	Vote
Amazon	0.12	0.24	0.13	1.2	14.73	0.52
DBLP	0.11	0.22	0.12	1.12	9.65	0.62
Youtube	0.73	0.3	0.32	4.92	24.8	1.92
Livejournal	2.01	6.2	3.5	47.8	1076	20.5
Orkut	1.73	14.6	7.93	134	4380	61.5
Synthetic	0.13	13.6	13.6	63.7	1370	53.4

Table 4. Mean Number of Clusters

Data Set	Pivot	ILS	Full	Vote
Amazon	143675	193028	155423	139842
DBLP	133627	158448	145191	134345
Youtube	801928	894597	827679	7851778
Livejournal	1817576	2556263	1995222	1837134
Orkut	651394	1771069	874592	818587
Synthetic	4327	37709	18669	14581

Table 5. Mean Precision/Recall (%)

Data Set	Pivot	ILS	Full	Vote
Amazon	51.4/34.2	88.5/24.4	85.4/36.7	76.8/37.2
DBLP	55.8/39.0	97.0/33.8	95.7/42.5	87.2/41.7
Youtube	25.9/12.4	87.3/8.7	84.0/12.0	76.2/13.0
Livejournal	28.4/16.6	88.7/11.1	85.4/20.3	78.0/18.1
Orkut	13.3/7.0	83.3/3.5	81.1/10.7	72.7/8.1
Synthetic	3.9/2.0	80.3/0.06	82.2/0.12	61.0/0.12

As expected, we see that in all cases Pivot with the full LS yields the lowest objective values. On data sets where the Pivot clusterings have higher precision (like Amazon and DBLP) the benefit of the full LocalSearch is greater. For example, LS decreases by nearly 30% on Amazon compared with the just-over 20% decrease yielded by ILS. However, on data sets where Pivot precision is quite low (like Orkut and Synthetic), the benefit of the full LS is negligible compared to ILS. We also note that ILS improvements approach their theoretical maximum (measured by Outer) on these low-precision data sets.

We see that the running times of full LocalSearch becomes a significant obstacle on the largest data sets. For Livejournal, full LocalSearch takes an average of 20 min to complete, and on Orkut the average is over an hour and 15 min. By contrast, ILS takes just under 15 s to finish on Orkut and provides nearly the same level of improvement as the full LocalSearch.

The Timed LocalSearch is unable to keep up with ILS in most cases; it is unable to improve even 5% over Pivot within the time limit across all of the real data sets, and on Youtube it is unable to make any gains whatsoever. The Match benchmark also slows down considerably on the larger examples. On Livejournal sequential ILS takes about 6 s to finish, whereas LocalSearch takes 47.8 s on average to match the same level of improvement. Again on Orkut sequential ILS takes under 15 s to finish, while LocalSearch takes over 2 min to match. Using parallel threads we see that ILS is able to halve its sequential running time on four out of the six data sets, and produce about the same running time as sequential ILS on the other two (Youtube and Synthetic).

As we have already noted, the Pivot algorithm has the potential to yield low-precision clusterings. On the other hand, for all the real data sets ILS reports the highest precision (with a close second on the synthetic data set). ILS necessarily reduces the recall from the Pivot algorithm, but it is not significantly lower than the recall of other clustering results across all data sets. In the worst case, ILS recall is still under 10 points lower than the Pivot recall.

5   Conclusion

In this paper we presented the InnerLocalSearch method, a viable alternative to running a full LocalSearch process to improve clustering results from the Pivot algorithm. We showed experimentally that ILS requires only a fraction of the amount of time spent by LS, while still yielding significant decreases in objective value. In many cases ILS yields nearly the same level of improvement as LS, especially when Pivot produces low-precision clusterings. We also showed that ILS greatly boosts Pivot precision without too much sacrifice to its clustering recall, and that it has the advantage of easily being run in parallel.

References

1. Ahmadian, S., Epasto, A., Kumar, R., Mahdian, M.: Fair correlation clustering. In: International Conference on Artificial Intelligence and Statistics, pp. 4195–4205. PMLR (2020)
2. Ahn, K., Cormode, G., Guha, S., McGregor, A., Wirth, A.: Correlation clustering in data streams. In: International Conference on Machine Learning, pp. 2237–2246. PMLR (2015)
3. Ailon, N., Charikar, M., Newman, A.: Aggregating inconsistent information: ranking and clustering. J. ACM (JACM) **55**(5), 1–27 (2008)
4. Ailon, N., Liberty, E.: Correlation clustering revisited: the 'True' cost of error minimization problems. In: Albers, S., Marchetti-Spaccamela, A., Matias, Y., Nikoletseas, S., Thomas, W. (eds.) ICALP 2009. LNCS, vol. 5555, pp. 24–36. Springer, Heidelberg (2009). https://doi.org/10.1007/978-3-642-02927-1_4
5. Assadi, S., Wang, C.: Sublinear time and space algorithms for correlation clustering via sparse-dense decompositions. In: 13th Innovations in Theoretical Computer Science Conference (ITCS 2022) (2022)
6. Bansal, N., Blum, A., Chawla, S.: Correlation clustering. Mach. Learn. **56**(1–3), 89–113 (2004)
7. Behnezhad, S., Charikar, M., Ma, W., Tan, L.Y.: Almost 3-approximate correlation clustering in constant rounds. In: 2022 IEEE 63rd Annual Symposium on Foundations of Computer Science (FOCS), pp. 720–731. IEEE (2022)
8. Bonchi, F., Garcia-Soriano, D., Liberty, E.: Correlation clustering: from theory to practice. In: KDD, p. 1972 (2014)
9. Bonchi, F., Gionis, A., Ukkonen, A.: Overlapping correlation clustering. Knowl. Inf. Syst. **35**, 1–32 (2013)
10. Chehreghani, M.H.: Clustering by shift. In: 2017 IEEE International Conference on Data Mining (ICDM), pp. 793–798. IEEE (2017)
11. Chierichetti, F., Dalvi, N., Kumar, R.: Correlation clustering in mapreduce. In: Proceedings of the 20th ACM SIGKDD International Conference on Knowledge Discovery and Data Mining, pp. 641–650 (2014)
12. Christiansen, L., Mobasher, B., Burke, R.: Using uncertain graphs to automatically generate event flows from news stories. In: Proceedings of Workshop on Social Media World Sensors at ACM Hypertext 2017 (SIDEWAYS, HT'17) (2017)
13. Cohen-Addad, V., Lee, E., Newman, A.: Correlation clustering with sherali-adams. In: 2022 IEEE 63rd Annual Symposium on Foundations of Computer Science (FOCS), pp. 651–661. IEEE (2022)

14. Coleman, T., Saunderson, J., Wirth, A.: A local-search 2-approximation for 2-correlation-clustering. In: Halperin, D., Mehlhorn, K. (eds.) ESA 2008. LNCS, vol. 5193, pp. 308–319. Springer, Heidelberg (2008). https://doi.org/10.1007/978-3-540-87744-8_26
15. Elsner, M., Schudy, W.: Bounding and comparing methods for correlation clustering beyond ilp. In: Proceedings of the Workshop on Integer Linear Programming for Natural Language Processing, pp. 19–27 (2009)
16. García-Soriano, D., Kutzkov, K., Bonchi, F., Tsourakakis, C.: Query-efficient correlation clustering. In: Proceedings of The Web Conference 2020, pp. 1468–1478 (2020)
17. Gionis, A., Mannila, H., Tsaparas, P.: Clustering aggregation. ACM Trans. Knowl. Discovery Data (TKDD) 1(1), 4-es (2007)
18. Goder, A., Filkov, V.: Consensus clustering algorithms: comparison and refinement. In: 2008 Proceedings of the Tenth Workshop on Algorithm Engineering and Experiments (ALENEX), pp. 109–117. SIAM (2008)
19. Halim, Z., Waqas, M., Hussain, S.F.: Clustering large probabilistic graphs using multi-population evolutionary algorithm. Inf. Sci. 317, 78–95 (2015)
20. Haruna, C.R., Hou, M., Eghan, M.J., Kpiebaareh, M.Y., Tandoh, L.: A hybrid data deduplication approach in entity resolution using chromatic correlation clustering. In: Li, F., Takagi, T., Xu, C., Zhang, X. (eds.) FCS 2018. CCIS, vol. 879, pp. 153–167. Springer, Singapore (2018). https://doi.org/10.1007/978-981-13-3095-7_12
21. Hua, J., Yu, J., Yang, M.S.: Star-based learning correlation clustering. Pattern Recogn. 116, 107966 (2021)
22. Klodt, N., Seifert, L., Zahn, A., Casel, K., Issac, D., Friedrich, T.: A color-blind 3-approximation for chromatic correlation clustering and improved heuristics. In: Proceedings of the 27th ACM SIGKDD Conference on Knowledge Discovery and Data Mining, pp. 882–891 (2021)
23. Kollios, G., Potamias, M., Terzi, E.: Clustering large probabilistic graphs. IEEE Trans. Knowl. Data Eng. 25(2), 325–336 (2011)
24. Lattanzi, S., Moseley, B., Vassilvitskii, S., Wang, Y., Zhou, R.: Robust online correlation clustering. In: Advances in Neural Information Processing Systems 34 (2021)
25. Levinkov, E., Kirillov, A., Andres, B.: A comparative study of local search algorithms for correlation clustering. In: Roth, V., Vetter, T. (eds.) GCPR 2017. LNCS, vol. 10496, pp. 103–114. Springer, Cham (2017). https://doi.org/10.1007/978-3-319-66709-6_9
26. Mandaglio, D., Tagarelli, A., Gullo, F.: In and out: Optimizing overall interaction in probabilistic graphs under clustering constraints. In: Proceedings of the 26th ACM SIGKDD International Conference on Knowledge Discovery and Data Mining, pp. 1371–1381 (2020)
27. Pan, X., Papailiopoulos, D., Oymak, S., Recht, B., Ramchandran, K., Jordan, M.I.: Parallel correlation clustering on big graphs. In: Advances in Neural Information Processing Systems, pp. 82–90 (2015)
28. Puleo, G.J., Milenkovic, O.: Correlation clustering with constrained cluster sizes and extended weights bounds. SIAM J. Optim. 25(3), 1857–1872 (2015)
29. Queiroga, E., Subramanian, A., Figueiredo, R., Frota, Y.: Integer programming formulations and efficient local search for relaxed correlation clustering. J. Global Optim. 81, 919–966 (2021)
30. Shi, J., Dhulipala, L., Eisenstat, D., Lacki, J., Mirrokni, V.: Scalable community detection via parallel correlation clustering. Proc. VLDB Endowment 14(11), 2305–2313 (2021)

31. Thiel, E., Chehreghani, M.H., Dubhashi, D.: A non-convex optimization approach to correlation clustering. In: Proceedings of the AAAI Conference on Artificial Intelligence, vol. 33, pp. 5159–5166 (2019)
32. Van Zuylen, A., Williamson, D.P.: Deterministic pivoting algorithms for constrained ranking and clustering problems. Math. Oper. Res. **34**(3), 594–620 (2009)
33. Veldt, N., Gleich, D.F., Wirth, A.: A correlation clustering framework for community detection. In: Proceedings of the 2018 World Wide Web Conference, pp. 439–448 (2018)

Algorithms on a Path Covering Problem with Applications in Transportation

Ruxandra Marinescu-Ghemeci, Alexandru Popa$^{(\boxtimes)}$, and Tiberiu Sîrbu

Department of Computer Science, University of Bucharest, Bucharest, Romania
{verman,alexandru.popa}@fmi.unibuc.com

Abstract. Given a road network, a source, a destination and K platoon paths, our aim is to reach the destination in a maximum given time using the benefits of traveling along platoons. We consider two objective functions (maximize the total time spent as a member of a platoon or minimize the time traveled without platoons) and for each such objective function, we have two scenarios: in the first one we are given the moments when platoons start to travel, while in second version we can decide these moments. We show several NP-hardness results as well as a dynamic-programming polynomial time algorithm (for the scenario in which the starting times are given). Then we describe an approximation algorithm with a factor $1/c$ that has running time exponential in K/c, with $1 \le c \le K$.

1 Introduction

There are many problems in literature involving resource constrains that are motivated by applications in transportation and scheduling. Simultaneously with the development of autonomous cars, platoon based transportation gained more interest in recent years. Based on the assumption that vehicles are capable to communicate with each other, a platoon is formed using multiple vehicles that move in a line, trying to maintain a fixed distance between them. Several works show the advantages of this method of transportation with emphasize on fuel consumption reduction by a careful planning of the route of the platoons, e.g., [4, 5]. New studies are developed for the optimization and stability of platoons, as well as studies that explore the possibility of forming platoons from existing vehicles on a road [8]. Moreover, there are experiments made in this field of platoons formation, like the PATH project [7] and SARTRE project [1]. Due to the dynamics of evolution in the transportation area, it is necessary to study new scenarios related to transportation using platoons.

OUR RESULTS. The models proposed in this article have as main objective the possibility to travel between two cities with a certain autonomy level. The input consists of an undirected graph where each node represents a city, and each

This work was supported by a grant of the Ministry of Research, Innovation and Digitization, CNCS - UEFISCDI, project number PN-III-P1-1.1-TE-2021-0253, within PNCDI III.

edge represents a road between two adjacent cities. We assume that the graph is undirected since usually platoons travel on larger roads (e.g. highways) which have typically lanes in both directions. Nevertheless, we stress the fact that with small modifications, our results work also for directed graphs. Each edge has assigned a weight which represents the time necessary to travel between the two nodes which are connected with that particular edge. On the input graph, the platoons are described as paths between nodes. We assume that each platoon follows a path (does not visit the same vertex twice) and that the speed of the platoons and the speed of our vehicle are equal and constant. We aim to find a route for our vehicle from a starting vertex to a destination that maximizes the total time travelled along with a platoon (we say that we travel along a platoon if we travel on a road at the same moment when a platoon traverses it), or to minimize the distance travelled without platoons. We consider two scenarios for these problems: in the first we are given the time when each platoons start to travel, in second we can choose the starting time for each platoon.

In Sect. 3 we prove several NP-hardness results (mostly for the case when we are not allowed to visit a vertex twice) and in Sect. 4 we give a dynamic-programming polynomial time algorithm (for the scenario in which the starting times are given), using time-expanded graphs. Then, in Sect. 5 we describe an $1/c$-approximation algorithm that has running time exponential in K/c, with $c \in [K]$.

These problems can be used to model other applications, for example in planning a travel where we aim to use public transportation with a given schedule or private transportation for which we can choose the time we use it, but not the route. Instead of transportation we can think of public events with given paths and schedules, like street festivals or of routes for controlling existing traffic. RELATED WORK. Our work is related with the one of Van De Hoef, Johansson and Dimarogonas [9, 10], where the focus is on fuel consumption reduction. The above mentioned studies use similar initial conditions as the one presented in this paper, and the objective is mainly to find a path that intersect existing platoons and minimize a function that captures the reduction in fuel consumption. In a certain way, we extend the work in [6, 9, 10] by having the same initial conditions, but different objectives.

As we mention later in the paper, our problems are also related to problems like constrained shortest/longest path problems or disjoint path problems.

2 Problem Formulation

First we introduce some notations. For a positive integer n, let $[n] = \{1, 2, \ldots, n\}$. For an array $v = (v_1, \ldots, v_n)$ of numbers denote by $v[i : j]$ the sub-array $(v_i, \ldots, v_j)$ and $sum(v) = \sum_{i=1}^{n} v_i$. Similarly, for a function $v : [k] \to \mathbb{R}_+$, we denote $sum(v) = \sum_{i \in [k]} v(i)$ and also use vectorial notation $v = (v(1), \ldots, v(k))$.

We model a road network using an undirected graph $G(V, E)$, where each city is represented by a vertex and each road between two cities v_i and v_j is represented by an edge $v_i v_j \in E$. Each edge has a certain time necessary to traverse it; more precisely we are given a function $d : E \to \mathbb{R}_+$ where, for an edge $v_i v_j \in E$, $d(v_i v_j)$ is the time necessary to travel between nodes v_i and v_j (also called cost or distance, since we can assume that our speed equals 1).

For a walk $P = [x_1, \ldots, x_p]$ denote by $d(P) = \sum_{i=1}^{p-1} d(x_i x_{i+1})$ its cost, that is the total time to travel P and by $l(P) = p - 1$ its length (number of edges). Recall that in a walk vertices and edges may repeat, but in a path vertices are distinct. Denote by $P[i : j]$ the subwalk $[x_i, \ldots, x_j]$.

We are given a set of K paths $P_1, P_2, \ldots, P_K$, that we call *platoon paths*; each path P_i is traversed by platoon i (platoon i travels along path P_i). We also have a *starting time function* $s : [K] \to \mathbb{R}_+$, that represents the starting time of each platoon ($s(i)$ is the moment when the platoon i is at the beginning of path P_i and starts to travel along it). Instead of notation $s(i)$ we also use notation $s(P_i)$ for the starting time of platoon on path P_i, if platoon paths have no associated indices.

Then we can determine for each vertex x of a platoon path P_z the moment when the platoon z arrives in vertex x, called *arrival-time* and denoted by $t(z, x)$: if x is the i-th vertex of platoon path P_z, then $t(z, x) = s(z) + d(P_z[1 : i])$. Denote by $\delta(z, x) = d(P_z[1 : i])$ the time platoon z needs to travel to reach vertex x.

In this context, with platoon paths and starting times fixed, we aim to find walks in G from *source* to *dest* on which we can travel (with same speed as platoons, assumed 1) with two distinct objectives: first to spent a maximum amount of time along platoons, second to travel a minimum distance without platoons, in a given time. To model that, consider t the total time we can travel and $s : [K] \to \mathbb{R}_+$ the starting time function.

A *t-time-travel* from *source* to *dest* is a pair (P, w) where $P = [x_1, \ldots, x_p]$ is a *(source, dest)*-walk and $w : V(P) \to \mathbb{R}_+$ is a waiting time function ($w(i)$ is the time we wait in vertex x_i) such that the total time of the traveling $d(P) + sum(w)$ is at most t (called simpler *t-travel* or *travel* if t can be deduced from context).

Let $P = [x_1, \ldots, x_p]$ be a walk and (P, w) a travel. We say that the i-th edge $x_i x_{i+1}$ of P is *covered by travel* (P, w) *along a platoon* z (relative to starting time s) or that we *travel on edge* $x_i x_{i+1}$ *along platoon* z if $x_i x_{i+1} \in P_z$ and the time we arrive in x_i traveling on P with waiting times w is equal to the arrival-time of the platoon z in x_i, that is: $d(P[1 : i]) + sum(w[1 : i]) = t(z, x_i)$. If there is no platoon z such that $x_i x_{i+1}$ is covered along platoon z we call the edge *uncovered* by travel (P, w). Denote by $Q(P, w)$ the set of edges covered by (P, w).

The *covering* of travel (P, w), denoted $cov_s(P, w)$, is the sum of the costs of all the edges covered by (P, w): $cov_s(P, w) = \displaystyle\sum_{xy \in Q(P,w)} d(xy)$. The *autonomy* of travel (P, w), denoted $auto_s(P, w)$, is the sum of the costs of all uncovered edges of P: $auto_s(P, w) = d(P) - cov_s(P, w) = \displaystyle\sum_{xy \in E(P) - Q(P,w)} d(xy)$.

For example, consider the input from Fig. 1(a) with the total time to travel $t = 30$. Let $P = [v_1, v_2, v_3, v_4, v_5]$. Depending on the time we wait in vertices (that is on the travel associated to P) different edges can be covered. Assume first that the travel has waiting times $w_1 = (0, 2, 0, 1, 0)$: we arrive in v_2 at

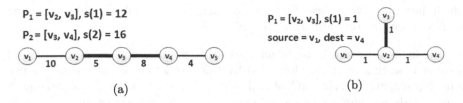

Fig. 1. Examples of input graphs with platoons.

moment 10, wait 2 moments and then travel on edge v_2v_3 along platoon 1; then we arrive in v_3 at moment 17, after platoon 2 already left; thus the only edge covered by the travel (P, w_1) is v_2v_3, hence $cov(P, w_1) = d(v_2v_3) = 5$ and $auto(P, w_1) = d(v_1v_2)+d(v_3v_4)+d(v_4v_5) = 22$. Consider now travel (P, w_2) with $w_2 = (0, 0, 1, 0, 0)$; we are in v_3 at moment 15, wait 1 and then travel on edge v_3v_4 along platoon 2. Now $Q(P, w_2) = \{v_3v_4\}$, $cov(P, w_2) = 8$ and $auto(P, w_2) = 19$.

A *t-travel from source to dest relative to s* is a t-travel (P, w) relative to s where P is a walk from *source* to *dest*. We use the following notations:

$$maxcov_s(source, dest, t) \qquad = \max\{cov_s(P, w)|(P, w)t\text{-travel from}$$
$$source \text{ to } dest \text{ relative to } s\}$$
$$maxpathcov_s(source, dest, t) = \max\{cov_s(P, w)|(P, w)t\text{-travel from}$$
$$source \text{ to } dest \text{ relative to } s, \text{with } P \text{ path}\}$$

Similarly define *minauto* and *minpathauto* for minimum autonomy. In the rest of the paper, if s is given and fixed, we can omit index s from the notations; the same holds for index t.

Remark 1. With above notations, the pair $(P[1 : i], w[1 : i])$ is an t_i-travel from $x_1 = source$ to x_i for $t_i = d(P[1 : i]) + sum(w[1 : i])$.

Denote by D the distance matrix of G. We consider two versions for our problems, given the maximum travel time T_{max} and for each version two different objectives, as already mention. In the first version we are also given the starting time of each platoon, in the second version we can decide the starting times. In both version we seek to find a travel from *source* to *dest* that optimize one of the criteria: maximizes the time spent as member of a platoon (determine $maxcov(source, dest, T_{max})$) or minimize the time (distance) traveled without platoons ($minauto(source, dest, T_{max})$). The formal definitions for our problems follow.

Problem 1. [MAXCOV] Given a graph $G(V, E)$, a distance function $d : E \to \mathbb{R}_+$, two vertices $source, dest \in V$, a total time to travel $T_{max} \in \mathbb{R}_+$, the platoon paths $P_1, P_2, \ldots, P_K$ and the starting-time function $s : [K] \to \mathbb{R}_+$, find an T_{max}−travel (P, w) from *source* to *dest* relative to s with maximum covering and its covering value $maxcov(source, dest, T_{max})$.

Problem 2. [s-MAXCOV] Given a graph $G(V, E)$, a distance function $d : E \to \mathbb{R}_+$, two vertices $source, dest \in V$, a total time to travel $T_{max} \in \mathbb{R}_+$ and the platoon paths $P_1, P_2, \ldots, P_K$, find a starting-time function $s : [K] \to \mathbb{R}$ for which $maxcov_s(source, dest, T_{max})$ is maximum and the value of this maximum: $\max\{maxcov_s(source, dest, T_{max})|s : [K] \to \mathbb{R}\}$.

But in some traveling problems and other applications we need to travel without repeating vertices. Motivated by this restriction, we also consider two similar problems: MAXPATHCOV and s-MAXPATHCOV, where we constrain P to be a path, not just a walk, so we replace $maxcov$ with $maxpathcov$.

As for our second objective - minimize the time we travel without platoons, that is the autonomy, we can consider also four similar problems: MINAUTO, MINPATHAUTO, s-MINAUTO and s-MINPATHAUTO.

To see the difference between these problems and their applications, we give some examples. First consider input from Fig. 1 (b). The solution both for MAXPATHCOV and MINPATHAUTO for $T_{max} = 4$ is the path $P = [v_1, v_2, v_4]$ with the travel having all waiting times equal to 0 ($w \equiv 0$). But the solution for MAXCOV is walk $P = [v_1, v_2, v_3, v_2, v_4]$ with $w \equiv 0$.. In this case we travel along a platoon on edge v_2v_3 and then travel back on v_2 just to maximize the covering. The effort to go back from v_3 to v_2 is not always worthing in some transportation problems, but is very useful in others, like travelling problems where what we want is actually to travel along platoons (for example clients or touristic events) as much as possible. Another example is the graph in Fig. 3. The solution for MAXCOV is $([v_1, v_2, v_5], (0, 1, 0))$ and for MINAUTO is $([v_1, v_3, v_4, v_5], 0)$.

We prove s-MAXCOV, MAXPATHCOV and s-MAXPATHCOV problems are NP-complete and propose a polynomial time dynamic programming algorithm for MAXCOV (by reducing the problem to a longest path problem in an associated time-expanded DAG) and an approximation algorithm for s-MAXCOV.

Remark 2. Let (P, w) be a t-travel with minimum autonomy, where P is a walk. If we successively replace a subwalk from P between two equal vertices x with x and add the cost of this subwalk to the waiting time in x we still obtain a t-travel (P', w') with minimum autonomy, where the walk P' is a path (after removing a subwalk we have $auto(P, w) \leq auto(P', w')$.

By Remark 2 we have $minauto_s(source, dest, t) = minpathauto_s(source, dest, t)$, so it suffices to consider only problems MINPATHAUTO and s-MINPATHAUTO. We prove that s-MINPATHAUTO is NP-complete and propose a polynomial time algorithm for MINPATHAUTO, similar to the one for MAXCOV.

The following remark also holds. This allows us to travel on shortest path when we do not travel along platoons.

Remark 3. Let $(P = [x_1, \ldots, x_p], w)$ be a t-travel. Let $x_i, x_j \in P$ such that all edges of subwalk $P[i : j]$ (from x_i to x_j) are uncovered by (P, w). Let (P', w') be the travel constructed as follows: P' is obtained from P by replacing the subwalk from x_i to x_j with a shortest path from x_i to x_j, and w' is obtained from w by setting the waiting times of the vertices from this shortest path to 0, except for x_j which has waiting time $sum(w[i : j]) + d(P[i : j]) - D(x_i, x_j)$. Then (P', w') is also an t-travel with $cov(P, w) \geq cov(P', w')$ and $auto(P', w') \leq auto(P, w)$.

3   NP-Hardness Result for MaxPathCov, MinPathAuto and s-MaxPathCov

To prove that our problems are NP-complete we used the following well known NP-Complete Problems, some of Karp's original NP-complete problems [3].

Problem 3. [Longest Path] Let G be an unweighted and undirected graph, *source* and *dest* two vertices of G and c a natural number. Decide if there exists a path P from *source* to *dest* in G with $l(P) \geq c$.

Problem 4. [Subset Sum] Let $B = \{b_1, b_2, \ldots, b_n\}$ a set of numbers and let c a number. Decide if there exists a set $Y \subseteq B$ with $\sum_{b \in Y} b = c$.

Problem 5. [Directed Edge-Disjoint - DEDP] Let G be a directed graph and $(s_1, t_1), \ldots, (s_k, t_k)$ be a set of k pairs of vertices. Decide if there are k edge-disjoint paths $D_1, \ldots, D_k$ such that D_i is a path from s_i to t_i for $i \in [k]$.

Theorem 1. *MaxPathCov and s-MaxPathCov are NP-complete and as hard to approximate as* Longest Path.

Proof. Consider G, *source* and *dest* an instance of Longest Path. We construct an instance of MaxPathCov as follows. Define $d(e_i) = 1, \forall i$. For every edge $xy \in G$ consider n platoon paths $[x, y]$ with starting time $0, 1 \ldots, n-1$ - denoted $P_{x,y,t}$ for $t \in \{0, 1 \ldots, n-1\}$ - and n platoon paths $[y, x]$ with starting times also $0, 1 \ldots, n-1$, denoted $P_{x,y,t}$ (that is there is a platoon traversing every edge e_i on both directions of the graph at any moment t from 0 to $n-1$). Let c be a fixed number. We prove that there is a path from *source* to *dest* of length at least c in G if and only if for $T_{max} = n-1$ we have $maxpathcov(sorce, dest, T_{max}) \geq c$.

Assume first that P is a path from *source* to *dest* with $l(P) \geq c$. Consider the T_{max} travel (P, w) with $w \equiv 0$. Then every edge of P is covered by this travel. Indeed, let xy be the t-th edge of P. We arrive in x at moment $t-1$ so this edge is covered along platoon on path $P_{x,y,t-1}$. It follows that each $cov(P, w) = l(P) \geq c$.

Conversely, if exists a travel (P, w) with $cov(P, w) \geq c$, then $d(P) = l(P) \geq cov(P, w) \geq c$.

The reduction from Longest Path to s-MaxPathCov is similar to the previous one, with the following differences. This time we consider two platoon paths for each edge xy of the graph: $P_{x,y} = [x, y]$ and $P_{y,x} = [y, x]$. If P is a path with $l(P) \geq c$ then let xy be the t-th edge of P; we can define the starting time of platoon $P_{x,y}$ as $t-1$ and then edge xy is covered by travel (P, w) with $w \equiv 0$. Thus, all edges of P are covered and we have $maxpathcov_s(source, dest, T_{max}) \geq cov(P, w) = l(P) \geq c$.   □

Theorem 2. *s-MaxCov is NP-complete (even for trees).*

Proof. We prove that Subset Sum is polynomial time reducible to s-MaxCov. Given an instance of the Subset Sum problem, we construct an instance of s-MaxCov as follows. Consider graph G as the star graph with center x_0 and terminal vertices $x_1, \ldots, x_{n+2}$. Consider *source* $= x_{n+1}$ and *dest* $= x_{n+2}$ and distance function defined as follows: $d(x_0x_i) = b_i$ for $i \in [n]$ and $d(x_0x_{n+1}) = d(x_0x_{n+2}) = \epsilon$ for some $\epsilon \ll b_i, \forall b_i \in B$. Consider platoon paths $P_j = [x_0, x_j]$ for $j \in [n]$. Let c be a fixed number. We will prove that there exist a subset of B with sum c if and only if there exists a starting time function s such that for time $T_{max} = 2c + 2\epsilon$ we have $maxcov_s(source, dest, T_{max}) \geq c$.

Assume first that $b_{i_1}, \ldots, b_{i_k}$ is a subset of B ($i_1 \leq \ldots \leq i_k$) with sum c. Consider walk $P = [x_{n+1} = source, x_0, x_{i_1}, x_0, x_{i_2}, \ldots, x_0, x_{i_k}, x_0, x_{n+2} = dest]$ and starting time function s defined as follows $s(i_j) = \epsilon + 2 * b_{i_1} + \ldots + 2 * b_{i_{j-1}}$ and $s(i) = T_{max}$ for $i \in [n] - B$. Consider also a travel (P, w) with $w \equiv 0$. Each edge $x_0x_{i_j}$ is covered since for every $j \in \{i_1, \ldots, i_k\}$ we arrive in x_0 at moment $\epsilon + 2b_{i_1} + \ldots + 2b_{i_{j-1}}$ and then travel the edge $x_0x_{i_j}$ along platoon i_j

Hence $maxcov_s(source, dest, T_{max}) \geq cov(P, w) = b_{i_1} + \ldots + b_{i_k} = c$.

Conversely, assume there exists s such that $maxcov_s(source, dest, T_{max}) \geq c$. Then there exists travel (P, w) such that $cov(P, w) \geq c$. When (P, w) covers an edge x_0x_i from a platoon path, it must traverse the edge back since x_i is a terminal vertex. Also, P must begin with edge $x_{n+1}x_0$ and end with x_0x_{n+2}

It follows that $d(P) = 2\epsilon + 2 * cov(P, w) \geq 2\epsilon + 2c$. Since we must have $d(P) \leq T_{max}$ it follows that $2\epsilon + 2 * cov(P, w) \leq T_{max} = 2c + 2\epsilon$ hence $cov(P, w) \leq c$.

Since the reverse inequality also holds, we have $cov(P, w) = c$ and the distances of the covered edges $Q(P, w)$ correspond to a subset of sum c in B.   □

Theorem 3. *Problem s-MINPATHAUTO is NP-complete (it is hard to decide even if the solution is 0).*

Proof. (sketch) We prove DEDP is reducible to s-MINPATHAUTO. Let $G(V, E)$ be a directed graph with $V = \{v_1, \ldots, v_n\}$ and let $(s_1, t_1), \ldots, (s_k, t_k)$ be a set of k pairs of vertices in G. Let $m = |E(G)|$. Consider ϵ such that $(3k - 1)\epsilon < 1$.

We build in polynomial time an instance for s-MINPATHAUTO as follows.

For every pair $(s_i = v_a, t_i = v_b)$ consider a gadget G^i with vertex set $V(G^i) = \{v_1^i, \ldots, v_n^i\} \cup \{x^i, y^i\}$ (a copy of vertices in G plus two new vertices) and edge set $E(G^i) = \{v_p^i v_q^i | v_p v_q \in\} \cup \{x^i v_a^i, y^i v_b^i, x^i y^i\}$ (corresponding to the edges of G but ignoring orientation, plus 3 new edges); we say that vertex v_j^i corresponds to vertex v_j from G (is a copy of it). Let $d(x^i v_a^i) = d(y^i v_b^i) = \epsilon$, $d(x^i y^i) = \epsilon$ and $d(e) = 1$ for every other edge in $E(G^i)$ (that correspond to edges in G).

Then for every $i \in [k - 1]$ add the following edges between gadgets: $y^i x^{i+1}$ with cost ϵ and edges $v_j^i v_k^{i+1}$ for j, k such that $v_k v_j \in E$ with cost $m + 1$. Then add a new vertex y^0 and edge $y^0 x^1$ with distance $A = (k + 1)(m + 2)$. Denote G' the obtained graph. An example for this construction is in Fig. 2.

For each edge $v_p v_q$ of G consider the platoon $P_{p,q} = [v_p^1, v_q^1, v_p^2, v_q^2, \ldots, v_p^k, v_q^k]$. Consider also for every $i \in [k]$ platoons: $P_i = [y^{i-1}, x^i]$ and $P_a^i = [x^i, v_a^i]$ and $P_b^i = [v_b^i, y^i]$ corresponding to pair $(s_i = v_a, t_i = v_b)$ from G.

For the example from Fig. 2 platoons are indicated by arrows: platoon path $P_{1,2} = [v_1^1, v_2^1, v_1^2, v_2^2]$ corresponding to edge $v_1 v_2$ with blue arrow, $P_{2,1}$ corresponding to edge $v_2 v_1$ with green, $P_{2,3}$ corresponding to $v_2 v_3$ with mauve arrow. With black arrow are platoons with one edge: $P_1 = [y^0, x^1]$, $P_2 = [y^1, x^2]$, plus platoon paths in gadget 1 for pair (v_1, v_3): $P_1^1 = [x^1, v_1^1]$, $P_3^1 = [y^1, v_3^1]$, plus platoon paths in gadget 2, for pair (v_2, v_1): $P_2^2 = [x^1, v_2^2]$, $P_1^2 = [y^1, v_1^2]$.

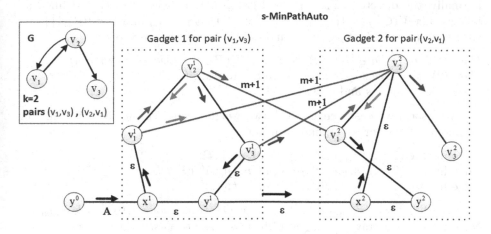

Fig. 2. Example of reduction from DEDP to s-MinPathAuto

Let $source = y^0$, $dest = y^k$, $T_{max} = A + (3k - 1)\epsilon + m$. Then DEDP has a solution in G for pairs (s_i, t_i), $i \in [k]$ if and only if in G' there is a starting time function s such that $minpathauto_s(source, dest, T_{max}) = 0$. $\square$

4   Polynomial Time Algorithm for MaxCov and MinPathAuto

Next we will describe how MaxCov and MinPathAuto can be reduced to longest path, respectively shortest path problem in an associated time-expanded graph. Since this graph is a DAG, polynomial time dynamic programming algorithms can be used to solve the two problems.

The idea of time-expanded graph was first considered for dynamic flow problems [2]. Informally, the time-expanded version of a dynamic network is a static network build as follows: for each vertex v and each time step t in time horizon a vertex (v, t) is added in the time-expanded graph; an arc between (u, t) and (v, t') correspond to traversal in time from u to v.

In our case, assume that we have the input MaxCov (which is the same as for MinPathAuto). The relevant moments for time horizon are the arrival-times of platoons in vertices, plus moments 0 and T_{max}. For uniformity, we define two more platoon paths with one vertex, corresponding to source, respectively destination: $P_{K+1} = [p_{K+1}^1 = source]$ and $P_{K+2} = [p_{K+2}^1 = dest]$ and extend function s to $[K + 2]$ by defining $s(K + 1) = 0$ and $s(K + 2) = T_{max}$.

Then a time-expanded graph for our problems has as vertices pairs (u, t) with t being the arrival time of a platoon in vertex u. An edge between two vertices

(u,t) and (v,t') correspond to a travel on edge uv along a platoon or to a travel using a shortest path from u to v (see Remark 3), having the possibility to wait in v. We can also wait in the *source*, so we have edges from $(source, 0)$ to $(source, t)$. Formally, the directed time-expanded graph G^T associated to G is defined as follows. Let $V(G^T) = \{(v,t) | \exists a \in [K] \text{ such that } v \in P_a \text{ and } t = t(a,v)\}\}$ An edge from (u,t) to (v,t') exists if and only if one of the cases occurs:

Case (1): $u \neq v \; \exists a \in [K]$ such that $uv \in E(P_a)$, $t = t(u,a)$, $t' \geq t(v,a)$
Case (2): $t' \geq D(u,v) + t$ and we are not in Case (1)
Case (3): $u = v = s$ and $t' > t$

We associate to each edge in G^T two weights representing the covering and autonomy of the corresponding path in G:

Edge in case (1): $c((u,t)(v,t')) = d(uv)$, $a((u,t)(v,t')) = 0$
Edge in case (2): $c((u,t)(v,t')) = 0$, $a((u,t)(v,t')) = D(u,v)$
Edge in case (3): $c((u,t)(v,t')) = 0$, $a((u,t)(v,t')) = 0$

A path $P^T = [(x_1, t_1), \ldots, (x_p, t_p)]$ in G^T from $(source, 0)$ to $(dest, T_{max})$ corresponds to a travel (P, w) in G from *source* to *dest*, where covered edges are actually the edges corresponding to the edges of G^T in Case (1), and edges in case (2) are replaced by shortest paths; Waiting time can be calculated for each vertex $w(x_i) = t_i - d(x_{i-1}x_i) - t_{i-1}$ if edge $x_{i-1}x_i$ is in Case (1), $w(x_i) = t_i - D(x_{i-1}, x_i) - t_{i-1}$ in Case (2), and $w(s) = \max\{t | (s,t) \in T\}$ (see Fig. 3). Conversely, a solution $(P = [x_1, \ldots, x_p], w)$ for MaxCov or MinAuto that satisfies property from Remark 3 corresponds to a path in G^T from $(source, 0)$ to $(dest, T_{max})$ with vertices (x_i, t_i) where $t_i = sum(d(P[1:i]) + sum(w[1:i]))$. We have $cov(P, w) = c(P^T)$ and $auto(P, w) = a(P^T)$

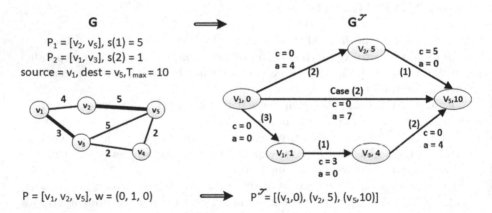

Fig. 3. Time-expanded graph example

Thus, MaxCov is equivalent to the longest path problem in G^T for weights c and MinAuto is equivalent to the shortest path problem in G^T for weights a. These problems can be solved using dynamic programming.

Theorem 4. MAxCov *and* MINPATHAUTO *are polynomial time solvable in* $O(n^2 K^2 + n^3)$.

Proof. Using the equivalence with longest path problem in the DAG G^T for weight c we obtain the following algorithm:

1. Consider two new platoon paths $P_{K+1} = [p^1_{K+1} = source]$ and $P_{K+2} = [p^1_{K+2} = dest]$, with $s(K+1) = 0$ and $s(K+2) = T_{max}$ and let $K = K+2$.
2. $N = \{(v,t)|\exists a \in [K]$ such that $v \in P_a$ and $t = t(a,v) \leq T_{max}\} = V(G^T)$
3. Sort N ascending by time arrival (this is a topological sorting for G^T).
4. Calculate distance matrix D of G using Floyd-Warshall algorithm
5. Set $DP(source, 0) = 0$
6. For every $(v,t) \in N - \{(source, 0)\}$ calculate $DP(v,t)$ using recurrence :
   $DP(v,t) = max\{R1, R2\}$ where:
   $R1 = max\{DP(u,t') + d(uv)|(u,t') \in D_1\}, R2 = max\{DP(u,t')|(u,t') \in D_2\}$,
   $D_1 = \{(u,t') \in N|\exists a \in [k]uv \in E(P_a), t' = t(a,u), t' + d(uv) \leq t\}$,
   $D_2 = \{(u,t') \in N|t' < t, t' + D(u,v) \leq t\}$
   ($D_1 \cup D_2$ are actually the predecessors of (v,t) in G^T)
7. return $DP(dest, T_{max})$.

Since $|N| \leq nK + 2$ and $|E(P_1)| + \ldots + |E(P_K)| \leq K(n-1)$, the complexity $O(n^2 K^2 + n^3)$ easily results. Using the classical way of following back the recursion we can also build an optimum T_{max}-travel in the same complexity.

The algorithm can be easily modified to solve MINAUTO. Moreover, by Remark 2 we can obtain a solution where the walk is actually a path, hence a solution for MINPATHAUTO.   □

Note that using this approach more general scenarios can be considered, for example when the profit of travelling on an edge uv along a platoon is given (so it is not necessary equal to $d(uv)$).

5   Approximation Algorithm for S-MAxCov

As we proved in Sect. 3, s-MAxCov is NP-complete. In this section we describe an approximation algorithm for this problem with a factor $1/c$ that has running time exponential in K/c, with $1 \leq c \leq K$.

Remark 4. When searching for a maximum covering we can consider only travels (P, w) such that the edges covered by (P, w) along a platoon a induce a subpath of P. Indeed, once we meet a platoon a there is no need to stop traveling along it and then intersect it again, since in this time platoon a travels with the same speed as us, so we cannot cover a bigger distance along another platoon before meeting a again. Then it suffices to consider only travels (P, w) with the property:

$$\begin{gathered} \forall i < j \text{ (the } i\text{-th and } j\text{-th edges are traveled along platoon } a) \neq 0 \Longrightarrow \\ \text{(the } k\text{-th edge is traveled along platoon } a, \forall i \leq k \leq j) \end{gathered} \qquad (1)$$

Remark 5. Consider a walk $P = [x_1, \ldots, x_p]$ and assume we defined the waiting times for $x_1, \ldots, x_{i-1}$. If a platoon path P_a contains $v = x_i$, we can always choose the starting time for a and waiting time in x_i such that we meet platoon a in x_i using travel $(P[1:i], w[1:i])$. More exactly, since using (P, w) we are in vertex v at time $t_v = d(P[1:i]) + w([1:i-1])$, we can choose the starting time for a such that we meet platoon a in v as follows: if $t_v > \delta(a, v)$ set $s(a) = t_v - \delta(a, v)$, otherwise set $s(a) = 0$ and set the waiting time in v to $\delta(a, v) - t_v$.

Remark 6. In finding an optimal solution for s-MAxCov, Remarks 3, 4, 5 allows us to consider only platoon paths with property (1) and starting-time function and waiting times defined according to the strategy given in Remark 5, such that the edges that are not covered induce shortest paths.

The main idea of our algorithm is to arbitrarily partition the set of platoon paths in c subsets of dimension $\left\lceil \frac{K}{c} \right\rceil$ or $\left\lceil \frac{K}{c} \right\rceil + 1$, where $c \in [K]$. For each subset we determine an exact solution for s-MAXCOV if the platoon paths are only the paths in this subset. For that, we build all walks P formed by subpaths of platoon paths joined by shortest paths (see Remark 4). Then we calculate starting times s and waiting times w as in Remark 5, if possible (if $d(P) + sum(w) \leq T_{max}$) and return the pair (P, w) with maximum covering obtained, together with the starting function s associated. The algorithm is fully described in Algorithm 1.

Theorem 5. *Denote by $l_{max} = max\{l_1, \ldots, l_K\}$. Algorithm 1 has running time $O\left(cn \cdot \left\lceil \frac{K}{c} \right\rceil! \, (l_{max})^{2\left\lceil \frac{K}{c} \right\rceil + 2}\right)$. The approximation factor is $1/c$.*

Proof. Denote by $k_c = \left\lceil \frac{K}{c} \right\rceil + 1$. Substeps from step 4 are repeated at most $c \cdot k_c!$ times. Since $|IO_i| = l_i(l_i - 1)/2 \leq l_{max}^2$ and $g \leq k_c$, there are at most $(l_{max})^{2k_c}$ combinations at step 4.3. For each such combination, P can be computed in $O(nk_c)$, since is composed from at most g shortest paths and g subpaths of platoon paths. So overall complexity is $O(c \cdot k_c! \, (l_{max})^{2k_c} nk_c)$.

Denote by $mcov_I$ the maximum value of a covering $cov(P, w)$ found by algorithm for group I. Since we build all the pairs (P, w) that satisfies the conditions from Remark 6, $mcov_I$ is the optimum solution if the set of platoon paths is restricted to I. Let OPT be the optimal solution. Denote by opt_I the total cost of edges covered by P belonging to platoon paths from the group I. The value returned by the algorithm is $ALG = mcov = max\{mcov_1, \ldots, mcov_c\}$. We then have that $OPT = opt_1 + \ldots + opt_c$ and $mcov_I \geq opt_I$ and hence: $ALG = max\{mcov_1, \ldots, mcov_c\} \geq max\{opt_1, \ldots, opt_c\} \geq OPT/c$. $\qquad \square$

Algorithm 1: Approximation Algorithm for s-MAXCOV Problem

Input: Graph G, paths $\mathcal{P} = \{P_1, \ldots, P_K\}$, $d : E \to \mathbb{R}_+$, $source, dest, T_{max}$
Output: starting function $s : [K] \to \mathbb{R}_+$, an T_{max})-travel relative to s
(P_{max}, w_{max}) and its covering $mcov = cov(P_{max}, w_{max})$

1. Set $mcov = 0$, $P_{max} = \emptyset$, $w_{max} = \emptyset$.
2. Compute distance matrix D and predecessor matrix F using Floyd-Warshall
3. Partition set P arbitrarily into c groups, each having $\left\lceil \frac{K}{c} \right\rceil$ or $\left\lceil \frac{K}{c} \right\rceil + 1$ elements.
4. For each group I let g be its dimension. Iterate over all $g!$ possible orderings of the paths in group I. For each such ordering repeat:
   4.1. Let $o = (i_1, \ldots, i_g)$ be the indices of the platoon paths in the current ordering
   4.2. For each $i \in o$ consider the set of pairs $IO_i = \{(in_i, out_i) | 1 \leq in_i < out_i \leq l_i\}$; each pair represents the vertex where a walk meets, respectively leaves platoon i when following the walk and travel build at next step
   4.3. For each combination $((in_{i_1}, out_{i_1}), \ldots, (in_{i_g}, out_{i_g}))$ with $(in_{i_j}, out_{i_j}) \in IO_{i_j}$:
       4.3.1. consider the walk P and define s and w as follows:
         - set $v = start$, $P = \emptyset$
         - for every $i \in o$:
             add to P a shortest path from v to in_i (determined using matrix F) and the subpath of platoon path P_i from in_i to out_i and set waiting time 0 to added vertices; set s, w for in_i according to Remark 5 (such that we meet platoon i in in_i).
         - add to P a shortest path from out_{i_g} to $dest$ and set waiting time 0 to added vertices
       4.3.2. If $d(P) + sum(w) \leq T_{max}$ and $mcov < cov(P, w)$ then set $mcov = cov(P, w)$, $(P_{max}, w_{max}) = (P, w)$, $s_{max} = s$.

---

6 Conclusions

This paper proposed viable models for transportation problems. We define more scenarios for each problem and provide algorithms and NP-hardness results. Tractable cases for the proposed problems are interesting to find. Notice that if we have just one platoon path $P_{x,y} = [x, y]$ and all weights are 1, MAXPATHCOV reduces to finding two vertex-disjoint paths D_1, D_2, with D_1 from *source* to x and D_2 from y to *dest* such that $d(D_1) \leq s(P_{x,y})$ and $d(D_2) \leq T_{max} - 1 - s(P_{x,y})$ or such that $d(D_1) + d(D_2) \leq T_{max} - 1$ in s-MAXPATHCOV case. These are well known problems, NP-complete: Length-bounded disjoint paths and min-sum disjoint path problems [11,12].

Various problems can be derived from our problems if we aim, for example, to maximize the percentage of time spent as a member of a platoon with respect to the total time traveled. Thus, our results may lead to future papers on path covering problems and on efficient algorithms for using platoons as a solution for transportation.

References

1. Coelingh, E., Solyom, S.: All aboard the robotic road train. IEEE Spectr. **49**(11), 34–39 (2012)
2. Ford Jr., L.R., Fulkerson, D.R.: Flows in networks, vol. 54. Princeton university press (1962 2015)
3. Garey, M.R.: Computers and intractability: A guide to the theory of np-completeness, freeman. Fundamental (1997)
4. Larson, J., Liang, K.-Y., Johansson, K.H.: A distributed framework for coordinated heavy-duty vehicle platooning. IEEE Trans. Intell. Transp. Syst. **16**(1), 419–429 (2014)
5. Larsson, E., Sennton, G., Larson, J.: The vehicle platooning problem: computational complexity and heuristics. Transp. Res. Part C Emerg. Technol. **60**, 258–277 (2015)
6. Meisen, P., Seidl, T., Henning, K.: A data-mining technique for the planning and organization of truck platoons. In: Proceedings of the International Conference on Heavy Vehicles, pp. 19–22 (2008)
7. Milanés, V., Shladover, S.E., Spring, J., Nowakowski, C., Kawazoe, H., Nakamura, M.: Cooperative adaptive cruise control in real traffic situations. IEEE Trans. Intell. Transp. Syst. **15**(1), 296–305 (2013)
8. Rajamani, R., Choi, S.B., Law, B., Hedrick, J., Prohaska, R., Kretz, P.: Design and experimental implementation of longitudinal control for a platoon of automated vehicles. J. Dyn. Sys., Meas., Control **122**(3), 470–476 (2000)
9. Van De Hoef, S., Johansson, K.H., Dimarogonas, D.V.: Fuel-optimal centralized coordination of truck platooning based on shortest paths. In: 2015 American Control Conference (ACC), pp. 3740–3745. IEEE (2015)
10. Van De Hoef, S., Johansson, K.H., Dimarogonas, D.V.: Fuel-efficient EN route formation of truck platoons. IEEE Trans. Intell. Transp. Syst. **19**(1), 102–112 (2017)
11. van der Holst, H., de Pina, J.C.: Length-bounded disjoint paths in planar graphs. Discret. Appl. Math. **120**(1–3), 251–261 (2002)
12. Zhang, P., Zhao, W.: On the complexity and approximation of the min-sum and min-max disjoint paths problems. In: Chen, B., Paterson, M., Zhang, G. (eds.) ESCAPE 2007. LNCS, vol. 4614, pp. 70–81. Springer, Heidelberg (2007). https://doi.org/10.1007/978-3-540-74450-4_7

Faster Algorithms for Evacuation Problems in Networks with a Single Sink of Small Degree and Bounded Capacitated Edges

Yuya Higashikawa, Naoki Katoh, Junichi Teruyama, and Yuki Tokuni[✉]

University of Hyogo, Kobe, Japan
{higashikawa,naoki.katoh,junichi.teruyama,af23v006}@gsis.u-hyogo.ac.jp

Abstract. In this paper, we propose new algorithms for *evacuation problems* defined on *dynamic flow networks*. A dynamic flow network is a directed graph in which *source* nodes are given supplies (i.e., the number of evacuees) and a single *sink* node is given a demand (i.e., the maximum number of acceptable evacuees). The evacuation problem seeks a dynamic flow that sends all supplies from sources to the sink such that its demand is satisfied in the minimum feasible time horizon. For this problem, the current best algorithms are developed by Schlöter (2018) and Kamiyama (2019), which run in strongly polynomial time but with high-order polynomial time complexity because they use submodular function minimization as a subroutine. In this paper, we propose new algorithms that do not explicitly execute submodular function minimization, and we prove that they are faster than the current best algorithms when an input network is restricted such that the sink has a small in-degree and every edge has the same capacity.

Keywords: dynamic flow · evacuation problem · quickest transshipment problem · polynomial-time algorithm · base polytope

1 Introduction

The network routing problem taking into account the movement of commodities over time is important from the viewpoint of evacuation planning. To model such movement over time, Ford and Fulkerson [8] introduced a *dynamic flow network*, which is a directed graph in which *source* nodes are given supplies (i.e., the number of evacuees), *sink* nodes are given a demand (i.e., the maximum number of acceptable evacuees), and edges are given capacities and transit times. Here, the capacity of an edge bounds the rate at which flow can enter the edge per unit time, and the transit time represents the time required for evacuees to travel across the edge.

Supported by JSPS KAKENHI Grant Numbers 19H04068, 20H05794, 22K11910, 23H03349.

The full version of this paper is available at the following link:
https://arxiv.org/abs/2301.06857.

The *evacuation problem* is one of the most basic problems defined on a dynamic flow network. In this problem, given a dynamic flow network with multiple sources and a single sink, the goal is to find a dynamic flow that sends all supplies from sources to the sink such that its demand is satisfied in the minimum feasible time horizon. For the evacuation problem, Kamiyama [16] and Schlöter [25] independently proposed the current best $\tilde{O}(m^2k^5 + m^2nk)$ time algorithms, where $\tilde{O}$-notation means computational complexity omitting all poly-logarithmic terms, n is the number of nodes, m is the number of edges, and k is the number of sources. Schlöter [25] showed that the minimum feasible time horizon can be calculated in $\tilde{O}(m^2k^4 + m^2nk)$ time. For a variant of the evacuation problem where a network has multiple sinks, so-called the *quickest transshipment problem*, Schlöter et al. [24] gave an $\tilde{O}(m^2\bar{k}^5 + m^3\bar{k}^3 + m^3n)$ time algorithm, where $\bar{k}$ denotes the total number of sources and sinks. All the algorithms by [16,24,25] call submodular function minimization (SFM for short) algorithms multiple times, which results in high-order polynomial time complexities.

On the other hand, for some restricted classes of networks, it is known that the problems can be solved without using SFM algorithms [9,10,12,17–19,21]. For the evacuation problem, Mamada et al. [19] and Higashikawa et al. [12] proposed $O(n\log^2 n)$ and $O(n\log n)$ time algorithms for tree networks with general and uniform edge capacity, respectively. Kamiyama et al. [17] proposed an $O(n\log n)$ time algorithm for restricted grid networks in which the capacity and transit times on edges are uniform, and the edges are oriented so that supplies can take only the shortest paths to the single sink. Kamiyama et al. [18] extended the grid networks considered in [17] to networks with a single sink and a uniform edge capacity where all paths from each node to the sink have the same length, and proposed an $O(m + n^4\log n)$ time algorithm. Chen and Golin [10] considered a variant of the quickest transshipment problem which has a restriction that evacuees from the same node have to go to the same sink, and proposed $O(\max\{k', \log n\}k'n\log^4 n)$ and $O(\max\{k', \log n\}k'n\log^3 n)$ time algorithms in tree networks with general edge capacity and uniform edge capacity, respectively, where k' denotes the number of sinks. For a special case of the evacuation problem where the number of sources is one, so-called the *quickest flow problem*, Saho and Shigeno [21] proposed an $O(nm^2\log^2 n)$ time algorithm in general networks.

Our Contribution: In this paper, we investigate new classes of networks in which the evacuation problem can be solved without using SFM algorithm. Specifically, for networks in which the sink has a small in-degree and every edge has the same capacity, we present an efficient algorithm. Note that our networks can model structures which cannot be represented by the networks in [10,18], i.e., in our network cycles can exist and paths from a node to the sink can have different lengths. Our main theorem is stated below.

Theorem 1. *Given a dynamic flow network $\mathcal{N}$ with a uniform edge capacity and a supply/demand function, the evacuation problem can be solved in $\tilde{O}(mndk^d + m^2k^2)$ time, where n is the number of nodes, m is the number of edges, d is the in-degree of the sink, and k is the number of sources.*

Note that when $d \leq 5$ and all edge capacities are uniform, our algorithm runs in $\tilde{O}(mnk^5 + m^2k^2)$ time, which is faster than the algorithms by [16,25] which run in $O(m^2k^5 + m^2nk)$.

Furthermore, our algorithm can be applied to a more general case, where each edge has any positive integer capacity. Let C, D be the maximum value of the all edge capacities and the sum of the capacities of the edges whose terminals are the sink, respectively. Basically, by replacing each edge with multiple edges of unit capacity, the network is constructed such that all edge capacities are uniform. In the modified network, the number of edges is $O(mC)$, and the in-degree of the sink is D. Therefore, Theorem 1 implies the following corollary.

Corollary 1. *Given a dynamic flow network $\mathcal{N}$ with a positive integer-valued edge capacity function, and a supply/demand function, the evacuation problem can be solved in $\tilde{O}(mnCDk^D + m^2k^2)$ time.*

For the quickest flow problem, in a network where the capacities of all edges are integer values, the running time of our algorithm is $\tilde{O}(m^2)$ if C and D are constant values. This is faster than the existing $O(nm^2 \log^2 n)$ algorithm [21].

In addition, we propose an even faster algorithm for square grid networks in settings more general than those in [17], where $k = \Theta(n)$ and for every pair of adjacent nodes u and v, there are two edges (u, v) and (v, u) with the same transit time and capacity. While we immediately have an $\tilde{O}(n^6)$ time algorithm for this case by applying Theorem 1 (because $m = O(n)$ and $d \leq 4$ in a grid network), we show the following theorem based on special properties of square grid networks (see Sect. 5 in the full version [13] for the proof).

Theorem 2. *Given a bidirected square grid network $\mathcal{G}$ and a supply/demand function, the evacuation problems can be solved in $\tilde{O}(n^4)$ time.*

Comparison with Existing Approaches: To solve the evacuation problem, the basic approach that we also employ is to compute the minimum feasible time horizon first and then find the corresponding dynamic flow, called a *quickest flow*. For the former phase, Fleischer and Hoppe [6] showed that a trivial solution space is $O(2^k)$ in size. To tackle this exponential solution space, the current best methods [16,25] use SFM algorithms in a discrete Newton method and solve the problem in polynomial time. For the latter phase, Schlöter and Skutella [23] proved that a quickest flow is obtained via a convex combination of at most k vertices of the base polytope with $O(2^k)$ facets defined on a submodular function. To obtain such a convex combination, Schlöter [25] utilizes an intermediate result of an execution of SFM algorithms relying on Cunningham's framework [5] (see Sect. 4 for more details). In contrast, our technical advancement is in proving a solution space of $O(k^d)$ size for computing the minimum feasible time horizon, and the base polytope for a quickest flow also has $O(k^d)$ facets. Thanks to the small number of facets of the base polytope, we also propose a geometric method for efficiently determining vertices and corresponding coefficients one by one for a convex combination representation of a quickest flow.

Organization: This paper is organized as follows. In Sect. 2, we introduce notations and important concepts used throughout the paper. In Sect. 3, we propose an algorithm to compute the minimum feasible time horizon. In Sect. 4, we propose an algorithm to find the dynamic flow that sends all supplies from sources to the sink within the minimum feasible time horizon.

2   Preliminaries

2.1   Notations and Problem Definition

Let $\mathbb{R}$ and $\mathbb{R}_+$ be the set of real values and the set of non-negative values, respectively. A dynamic flow network $\mathcal{N}$ is given as a 5-tuple $\mathcal{N} = (D = (V, E), u, \tau, S^+, S^-)$, where $D = (V, E)$ is a directed graph with node set V and edge set E, u is a capacity function $u\colon E \to \mathbb{R}_+$, τ is a transit time function $\tau\colon E \to \mathbb{R}_+$, and $S^+ \subseteq V$ and $S^- \subseteq V$ are sets of sources and sinks, respectively. For a path P on $\mathcal{N}$, let $|P|$ be the total transit time along P, i.e., $|P| = \sum_{e \in P} \tau(e)$. Throughout the paper, u is a constant function, so we abuse u as the constant capacity of every edge. Furthermore, in the evacuation problem, $\mathcal{N}$ contains a single sink denoted by s^- (i.e., $S^- = \{s^-\}$). For each node $v \in V$, let $\delta^+(v)$ and $\delta^-(v)$ denote the set of out-going edges from v and in-coming edges to v, respectively. Let $|\delta^+(v)|$ and $|\delta^-(v)|$ be the out-degree and in-degree of node v, respectively. We use n, m, and k as the cardinalities of V, E, and S^+, respectively.

The inputs of the evacuation problem are a dynamic flow network $\mathcal{N}$ and a supply/demand function $w\colon V \to \mathbb{R}$ that represents the amount of supply/demand at the sources/sinks. The value of the function w is $w(v) > 0$ for $v \in S^+$, $w(v) < 0$ for $v \in S^-$ (i.e., for $v = s^-$), $w(v) = 0$ for $v \in V \setminus (S^+ \cup S^-)$, and $\sum_{v \in V} w(v) = 0$. For any of sources/sinks subset $A \subseteq S^+ \cup S^-$, we define $w(A) := \sum_{s \in A} w(s)$.

On a dynamic flow network $\mathcal{N}$, a (continuous) *dynamic flow* f is defined as a function $f\colon E \times \mathbb{R} \to \mathbb{R}_+$, where $f(e, \theta)$ represents the flow rate entering edge e at time θ (≥ 0), and $f(e, \theta) = 0$ holds for any $\theta < 0$. We say that a dynamic flow f has *time horizon* T if no flow remains in the network after time T, i.e., $f(e, \theta) = 0$ holds for all $e \in E$ and $\theta \geq T - \tau(e)$. Let us consider the following constraints for a dynamic flow f:

$$0 \leq f(e, \theta) \leq u \quad \text{for each } e \in E, \text{ for each } \theta \in [0, \infty), \tag{1}$$

$$\int_0^\theta \left(\sum_{e \in \delta^+(v)} f(e, t) - \sum_{e \in \delta^-(v)} f(e, t - \tau(e)) \right) dt \leq \max\{w(v), 0\} \tag{2}$$

$$\text{for any } v \in V, \text{ for any } \theta \in [0, \infty).$$

The constraints (1) and (2) are called the *capacity constraint* and the *conserve constraint*, respectively. The conserve constraint (2) means that for any time θ and any node v, the amount of flow out of v within time θ is at most the amount of flow entering v within θ and the supply at v. Furthermore, for a time horizon $T \in \mathbb{R}_+$, consider the following *supply/demand constraint*:

$$\int_0^T \left(\sum_{e \in \delta^+(v)} f(e,t) - \sum_{e \in \delta^-(v)} f(e, t - \tau(e)) \right) dt = w(v) \text{ for any } v \in V. \quad (3)$$

The supply/demand constraint (3) implies that for each node $v \in V$, the net amount of flow accumulated at v within time horizon T equals its supply or demand. If a dynamic flow f satisfies the above constraints (1), (2), and (3), then f is said to be *feasible* w.r.t. (w, T), and such time horizon T is called a *feasible time horizon* w.r.t. w. Throughout the paper, T^* denotes the minimum feasible time horizon w.r.t. w. We call f^* a *quickest flow* if f^* is a feasible dynamic flow on $\mathcal{N}$ w.r.t. (w, T^*). Then the evacuation problem requires computing T^* and finding a corresponding dynamic flow, which is described precisely as follows.

EVACUATION PROBLEM
Input: A dynamic flow network $\mathcal{N} = (D = (V, E), u, \tau, S^+, S^- = \{s^-\})$ and a supply/demand function w.
Goal: Compute the minimum feasible time horizon T^* w.r.t. w and a quickest flow f^* w.r.t. w.

2.2 Maximum Dynamic Flows

Given a dynamic flow network $\mathcal{N}$, for a subset of the sources and sinks $A \subseteq S^+ \cup S^-$ and a time $\theta \in \mathbb{R}_+$, let $o^\theta(A)$ be the maximum amount of flow that can reach the sinks in $S^- \setminus A$ from the sources in A within time θ, where we assume that there is no restriction on the amount of flow out of each source of A. Note that $o^\theta(S^+ \cup S^-) = 0$ holds for any θ. We call a flow corresponding to $o^\theta(A)$ a *maximum dynamic flow* from A within time θ.

For a subset of sources $A \subseteq S^+$, we define the *minimum required time* $\theta(A)$ as

$$\theta(A) := \min\{\theta \mid o^\theta(A) - w(A) \geq 0\}. \quad (4)$$

Note that as shown later in (9), $o^\theta(A)$ is a nondecreasing continuous function in θ of range $[0, \infty)$. Thus, $\theta(A)$ is θ satisfying $o^\theta(A) = w(A)$, that is,

$$o^{\theta(A)}(A) = w(A). \quad (5)$$

Moreover, by Theorem 7.1 in [6] about the feasibility of a time horizon T, we immediately have the following corollary.

Corollary 2. *For a dynamic flow network $\mathcal{N}$ and a supply/demand function w, we have*

$$T^* = \max\{\theta(A) \mid A \subseteq S^+\}.$$

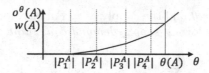

Fig. 1. $o^\theta(A)$ in θ and the minimum required time $\theta(A)$.

In the rest of this section, we give the properties of $o^\theta(A)$ and $\theta(A)$. For this purpose, we need to deal with a static flow network, which corresponds to an input dynamic flow network $\mathcal{N}$. A *static flow network* $\overline{\mathcal{N}}$ is a directed graph $D = (V, E)$ with a capacity $u(e)$ and a cost $c(e)$ for each edge $e \in E$. On a static flow network $\overline{\mathcal{N}}$, a *static flow* $\bar{f}$ is defined as a function $\bar{f} : E \to \mathbb{R}_+$, where $\bar{f}(e)$ represents the amount of flow on edge e. Given a source set $S^+ \subset V$ and a sink set $S^- \subset V$, a static flow $\bar{f}$ is said to be feasible if it holds that

$$0 \le \bar{f}(e) \le u(e) \qquad \text{for each } e \in E, \tag{6}$$

$$\sum_{e \in \delta^+(v)} \bar{f}(e) - \sum_{e \in \delta^-(v)} \bar{f}(e) = 0 \qquad \text{for each } v \in V \setminus (S^+ \cup S^-),$$
$$\tag{7}$$

$$\sum_{e \in \delta^+(s):\, s \in S^-} \bar{f}(e) - \sum_{e \in \delta^-(s):\, s \in S^+} \bar{f}(e) = 0. \tag{8}$$

The *minimum-cost maximum-flow* is a feasible static flow $\bar{f}$ that minimizes $\sum_{e \in E} c(e)\bar{f}(e)$ subject to $\sum_{e \in \delta^+(s):\, s \in S^-} \bar{f}(e)$ is maximized. Given a static flow network $\overline{\mathcal{N}}$ and a feasible static flow $\bar{f}$ on $\overline{\mathcal{N}}$, the residual network $\overline{\mathcal{N}}_{\bar{f}}$ is constructed as follows. For every edge $e = (v_1, v_2) \in E$ such that $\bar{f}(e) > 0$, reduce its capacity to $u(e) - \bar{f}(e)$ and add new edge $e' = (v_2, v_1)$ of capacity $\bar{f}(e)$ and cost $-c(e)$. In the following, given a dynamic flow network $\mathcal{N}$, $\overline{\mathcal{N}}$ means the static flow network comprising the same underlying graph as $\mathcal{N}$ with uniform capacity u and cost $\tau(e)$ for every edge $e \in E$.

According to Anderson and Philpott [1], given $A \subseteq S^+ \cup S^-$ and $\theta \in \mathbb{R}_+$, $o^\theta(A)$ can be obtained by applying the successively shortest path algorithm [3, 14, 15] for the minimum-cost maximum-flow problem in the following manner. Initially, set $\bar{f}$ as a zero static flow, that is, $\bar{f}(e) = 0$ for all $e \in E$. At each step $i\ (\ge 1)$, execute the following two procedures.
(i) Find the shortest (i.e., minimum-cost) path P_i^A from A to s^- in the current residual network $\overline{\mathcal{N}}_{\bar{f}}$. If there is no such path, then break the iteration.
(ii) Add to $\bar{f}$ a static flow of amount u along path P_i^A, denoted by $\bar{f}_i^A$.
Let p^A denote the number of paths obtained when the above iteration halts. We will refer to this algorithm as SSP from now on.

By the operation of the algorithm SSP, we see that $o^\theta(A)$ is a piecewise linear function in θ (see Fig. 1) represented as

$$o^\theta(A) = \max_{h=1,\ldots,p^A} \left\{ \sum_{i=1}^{h} (\theta - |P_i^A|)u \right\}. \tag{9}$$

As for $\theta(A)$, because $o^\theta(A)$ is a non-decreasing continuous function in θ of range $[0, \infty)$ as shown in (9), $\theta(A)$ is θ satisfying $o^\theta(A) = w(A)$ by definition (4). We thus have

$$\theta(A) = \min_{h=1,\ldots,p^A} \left\{ \frac{\sum_{i=1}^{h} |P_i^A|}{h} + \frac{w(A)}{hu} \right\}. \tag{10}$$

3 Computing the Minimum Feasible Time Horizon

In this section, we propose an algorithm that computes the minimum feasible time horizon T^* given the dynamic network $\mathcal{N}$ and a supply/demand function w. Throughout Sects. 3 and 4, let d denote the in-degree of the sink s^-.

Corollary 2 implies that T^* can be obtained by computing $O(2^k)$ values of $\theta(A)$, one for each $A \subseteq S^+$. We will define a family of $O(k^d)$ subsets of S^+, denoted by $\hat{\mathcal{A}}$, and show that $\hat{\mathcal{A}}$ contains $A^* \subseteq S^+$ that maximizes $\theta(A)$, that is, $T^* = \theta(A^*)$ holds.

3.1 Definition and Property of $\hat{\mathcal{A}}$

We now consider a tuple of p sources $(v_1, \ldots, v_p) \in (S^+)^p$ with an integer $p \in \{1, \ldots, d\}$. We say that a source subset $A \in S^+$ admits $(v_1, \ldots, v_p)$ if the following are satisfied: (i) $p^A = p$; (ii) the origins of paths $P_1^A, \ldots, P_p^A$ are $v_1, \ldots, v_p$, respectively. We notice that there may exist $(v_1, \ldots, v_p)$ such that any subset does not admit it. For each $(v_1, \ldots, v_p) \in (S^+)^p$, let $\hat{A}_{(v_1,\ldots,v_p)}$ be the subset of sources $A \subseteq S^+$ such that $|A|$ is the largest among all A admitting $(v_1, \ldots, v_p)$:

$$\hat{A}_{(v_1,\ldots,v_p)} := \operatorname{argmax}\{|A| \mid A \subseteq S^+, A \text{ admits } (v_1, \ldots, v_p)\}. \tag{11}$$

Note that $\hat{A}_{(v_1,\ldots,v_p)} = \emptyset$ if there is no subset A that admits $(v_1, \ldots, v_p)$. For each $p \in \{1, \ldots, d\}$, let $\hat{\mathcal{A}}_p$ denote the set of $\hat{A}_{(v_1,\ldots,v_p)}$ for all $(v_1, \ldots, v_p) \in (S^+)^p$, and let $\hat{\mathcal{A}}$ denote the union of all $\hat{\mathcal{A}}_p$:

$$\hat{\mathcal{A}} := \bigcup_{p \in \{1,\ldots,d\}} \hat{\mathcal{A}}_p, \text{ where } \hat{\mathcal{A}}_p := \{\hat{A}_{(v_1,\ldots,v_p)} \mid (v_1, \ldots, v_p) \in (S^+)^p\}. \tag{12}$$

We see that $\hat{\mathcal{A}}$ consists of $O(k^d)$ subsets of S^+ (because $\sum_{p=1}^{d} k^p < 2k^d$ for $k > 1$). Note that for any $A \subseteq S^+$ admitting the same $(v_1, \ldots, v_p)$, $o^\theta(A)$ is the same non-decreasing continuous function in θ by the definition (9), and $\hat{A}_{(v_1,\ldots,v_p)}$ maximizes $w(A)$ over such A. Therefore, $\hat{A}_{(v_1,\ldots,v_p)}$ maximizes $\theta(A)$ over source subsets A admitting $(v_1, \ldots, v_p)$. We then have the following lemma.

Lemma 1. *For a dynamic flow network $\mathcal{N}$ and a supply/demand function w, we have $T^* = \max\{\theta(A) \mid A \in \hat{\mathcal{A}}\}$.*

3.2   Algorithms

First of all, we describe an algorithm to compute $\hat{A}_{(v_1,\dots,v_p)}$ for given $(v_1,\dots,v_p) \in (S^+)^p$. We use the following lemma to check whether there exists a source subset A admitting $(v_1,\dots,v_p)$. See Sect. 3 in the full version [13] for its proof.

Lemma 2. *For each $p \in \{1,\dots,d\}$ and $(v_1,\dots,v_p) \in (S^+)^p$, there exists a source subset A that admits $(v_1,\dots,v_p)$ if and only if set $\{v_1,\dots,v_p\}$ admits $(v_1,\dots,v_p)$.*

In order to compute $\hat{A}_{(v_1,\dots,v_p)}$ for given $(v_1,\dots,v_p)$, let $A' = \{v_1,\dots,v_p\}$ and we successively compute the paths $P_1^{A'}, P_2^{A'}, \dots$ by the algorithm SSP and check if A' admits $(v_1,\dots,v_p)$. If so, $\hat{A}_{(v_1,\dots,v_p)}$ exists and $P_i^{\hat{A}_{(v_1,\dots,v_p)}} = P_i^{A'}$ and $\bar{f}_i^{\hat{A}_{(v_1,\dots,v_p)}} = \bar{f}_i^{A'}$ holds for $i \in \{1,\dots,p\}$. Let us see which source $\hat{A}_{(v_1,\dots,v_p)}$ contains. Consider static flows $\bar{f}_1^{A'},\dots,\bar{f}_p^{A'}$ corresponding to paths $P_1^{A'},\dots,P_p^{A'}$. Looking at the residual network $\mathcal{N}_{\bar{f}}$ for $\bar{f} = \sum_{h=1}^{i-1} \bar{f}_h^{A'}$, $\hat{A}_{(v_1,\dots,v_p)}$ does not contain any source from which s^- is closer than from v_i in $\mathcal{N}_{\bar{f}}$; otherwise, the origin of any minimum-cost path from $\hat{A}_{(v_1,\dots,v_p)}$ to s^- in $\mathcal{N}_{\bar{f}}$ is never v_i, contradiction. Furthermore, after taking all flows $\bar{f}_1^{A'},\dots,\bar{f}_p^{A'}$, there may remain source nodes which can reach s^- in $\mathcal{N}_{\bar{f}}$ for $\bar{f} = \sum_{h=1}^{p} \bar{f}_h^{A'}$ if $p < d$. Then $\hat{A}_{(v_1,\dots,v_p)}$ does not contain such nodes either. See Sect. 3 in the full version [13] for the correctness of this approach.

We then show the overall algorithm to compute the minimum feasible time horizon T^*. The algorithm follows Lemma 1 in a straightforward manner:

Step 1. Obtain $\hat{A}$ by computing $\hat{A}_{(v_1,\dots,v_p)}$ for all $p \in \{1,\dots,d\}$ and for all $(v_1,\dots,v_p) \in (S^+)^p$.

Step 2. Calculate $\theta(A)$ for all $A \in \hat{A}$ and take the maximum value among them.

The bottleneck of this algorithm is that at most d minimum-cost paths are computed $O(k^d)$ times in Step 1. Computing a minimum-cost path takes $O(nm)$ by applying the Moore-Bellman-Ford algorithm [2,7,20]. Therefore, we have the following theorem for the running time (see Sect. 3 in the full version [13] for its detailed proof).

Theorem 3. *Given a dynamic flow network $\mathcal{N}$ with a uniform edge capacity and a supply/demand function w, the minimum feasible time horizon T^* w.r.t. w can be computed in $O(mndk^d)$ time.*

4   An Algorithm for Finding a Quickest Flow

In this section, we propose an algorithm to find a quickest flow for the minimum feasible time horizon already given. Schlöter and Skutella [23] showed that an SFM algorithm relying on Cunningham's framework [5] directly gives a quickest flow, and the current fastest algorithms [16,25] use this approach explicitly.

(a) The gray face is a base polytope on $U = \{u_1, u_2, u_3\}$

(b) Example of a base polytope on $U = \{u_1, u_2, u_3, u_4\}$

Fig. 2. Base polytopes for submodular functions on U with three or four elements.

While our proposed algorithm solves the problem without using an SFM algorithm, it is also based on the relationship shown by [23] between a quickest flow and a base polytope. In the following, we give fundamental definitions for a base polytope, describe the relationship between a quickest flow and a base polytope, and then present our new algorithm for finding a quickest flow.

4.1 Base Polytope

Let U be a nonempty finite set, which can be considered as the set of numbers $1, 2, \ldots, |U|$. A set function $g \colon 2^U \to \mathbb{R}$ is said to be a *submodular function* if it satisfies $g(X) + g(Y) \geq g(X \cup Y) + g(X \cap Y)$ for every pair of subsets $X, Y \subseteq U$. For a vector $x \in \mathbb{R}^U$ and a subset $A \subseteq U$, let $x(s)$ denote the s-th component of x, and $x(A) := \sum_{s \in A} x(s)$. Given a submodular function g defined on U, a *base polytope* $B(g)$ is a convex polyhedron on $\mathbb{R}^U$ defined as

$$B(g) := \left\{ x \in \mathbb{R}^U \mid x(A) \leq g(A) \text{ for all } A \subseteq U, \text{ and } x(U) = g(U) \right\}.$$

Let us observe a property of $B(g)$. Let a vertex of $B(g)$ be the solution of simultaneous equations consisting of $x(U) = g(U)$ and $x(A) = g(A)$ for other $|U| - 1$ subsets $A \subset U$, which is contained in the base polytope $B(g)$. It is known that each vertex of $B(g)$ corresponds to a total order on U [4] (see Fig. 2). More precisely, given a total order $\prec = (u_{i_1}, \ldots, u_{i_{|U|}})$ on U, let $b^{(\prec, g)}$ denote the solution x of the following simultaneous equations:

$$x(\{u_{i_1}, \ldots, u_{i_\ell}\}) = g(\{u_{i_1}, \ldots, u_{i_\ell}\}) \text{ for each } \ell \in \{1, \ldots, |U|\}. \tag{13}$$

4.2 Relationship Between Quickest Flows and Base Polytopes

Hoppe and Tardos [11] showed that $o^\theta \colon 2^{S^+ \cup S^-} \to \mathbb{R}$ is a submodular function for any fixed time $\theta \in \mathbb{R}_+$. Note that in our problem, $B(o^\theta)$ is defined on $\mathbb{R}^{k+1}$ because $|S^+| = k$ and $|S^-| = 1$. To see the relationship between a quickest flow and the vertices of the base polytope, we need the concept of *lexicographically maximal dynamic flow* (*lex-max dynamic flow* for short), which is introduced by Hoppe and Tardos [11]. Given a dynamic flow network $\mathcal{N}$, a total order $\prec$ on

$S^+ \cup S^-$, and a time horizon $T \in \mathbb{R}_+$, the lex-max dynamic flow w.r.t. $(\prec, T)$, denoted by $f^{(\prec,T)}$, is the dynamic flow on $\mathcal{N}$ satisfying the capacity constraint (1) and the conserve constraint (2) that maximizes the amount of flow leaving the sources/sinks in order of $\prec$ within time horizon T. Note that the amount of flow entering the source/sink is treated as a negative value.

Schlöter and Skutella [23] showed that for any total order $\prec$ on $S^+ \cup S^-$, the vertex $b^{(\prec, o^T)}$ of $B(o^T)$ corresponds to $f^{(\prec,T)}$; in other words, for any source/sink $s \in S^+ \cup S^-$, $b^{(\prec, o^T)}(s)$ is the amount of flow leaving s by lex-max dynamic flow $f^{(\prec,T)}$. Moreover, considering a supply/demand function w as a vector in $\mathbb{R}^{k+1}$, there are an integer $h \le k$ and some h total orders $\prec_1, \ldots, \prec_h$ on $S^+ \cup S^-$ such that w can be a convex combination of $b^{(\prec_1, o^{T^*})}, \ldots, b^{(\prec_h, o^{T^*})}$. Furthermore, the quickest flow f^* can be expressed as the convex combination of lex-max dynamic flows $f^{(\prec_1, o^{T^*})}, \ldots, f^{(\prec_h, o^{T^*})}$ corresponding to $b^{(\prec_1, o^{T^*})}, \ldots, b^{(\prec_h, o^{T^*})}$, using the same coefficients. Summarizing these arguments, there are some h ($\le k$) total orders $\prec_1, \ldots, \prec_h$ and non-negative real values $\lambda_1, \ldots, \lambda_h$ with $\sum_{i=1}^h \lambda_i = 1$ such that

$$w = \sum_{i=1}^h \lambda_i b^{(\prec_i, o^{T^*})} \text{ and } f^* = \sum_{i=1}^h \lambda_i f^{(\prec_i, T^*)}. \tag{14}$$

In the rest of this subsection, we give the property of the base polytope $B(o^{T^*})$. Let $\hat{B}(o^{T^*})$ denote a convex polyhedron on $\mathbb{R}^{k+1}$ defined as

$$\hat{B}(o^{T^*}) := \left\{ x \in \mathbb{R}^{k+1} \mid x(A) \le o^{T^*}(A) \text{ for all } A \in \hat{\mathcal{A}}, \text{ and } x(S^+ \cup S^-) = 0 \right\}.$$

By the definition, it clearly holds that $B(o^{T^*}) \subseteq \hat{B}(o^{T^*})$. Not only that, we can show that $B(o^{T^*}) = \hat{B}(o^{T^*})$ holds.

Lemma 3. *For a dynamic flow network $\mathcal{N}$, a supply/demand function w, and the minimum feasible time horizon T^* w.r.t. w, it holds $B(o^{T^*}) = \hat{B}(o^{T^*})$.*

Proof. It is enough to show that $\hat{B}(o^{T^*}) \subseteq B(o^{T^*})$ holds. Let us consider any vector $x' \in \mathbb{R}^{k+1}$ in $\hat{B}(o^{T^*})$ and treat vector x' as a supply/demand function. Because $\hat{\mathcal{A}}$ does not depend on a supply/demand function by definition, we can replace w in the arguments in Sect. 3.1 with x'. Therefore, for any $A \subseteq S^+ \cup S^-$, there exists a source subset $\hat{A} \in \hat{\mathcal{A}}$ such that $x'(A) \le x'(\hat{A})$ and $o^{T^*}(A) = o^{T^*}(\hat{A})$ holds. Thus, we have $x'(A) \le o^{T^*}(A)$ for any $A \subseteq S^+ \cup S^-$ since $x'(\hat{A}) \le o^{T^*}(\hat{A})$ holds for any $\hat{A} \in \hat{\mathcal{A}}$ by the definition of $\hat{B}(o^{T^*})$. This means that vector x' is in $B(o^{T^*})$. Therefore, $\hat{B}(o^{T^*}) \subseteq B(o^{T^*})$ holds. $\qquad\square$

4.3   Algorithms

According to (14), our main task is to determine total orders $\prec_1, \ldots, \prec_h$ on $S^+ \cup S^-$ and non-negative real values $\lambda_1, \ldots, \lambda_h$ satisfying $w = \sum_{i=1}^h \lambda_i b^{(\prec_i, o^{T^*})}$. By Lemma 3, we see that the number of constraint equations that define $B(o^{T^*}) = \hat{B}(o^{T^*})$ is reduced from $O(2^k)$ to $|\hat{\mathcal{A}}| = O(k^d)$.

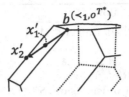

Fig. 3. Illustration of points x_1' and x_2'.

The algorithm inductively computes $\prec_i, f^{(\prec_i, T^*)}, b^{(\prec_i, o^{T^*})}, \lambda_i$ for $i = 1, \ldots, h$ in this order. Below, we detail the first iteration. Let $x_1' = w \in \mathbb{R}^{k+1}$ and $A_1 = A^*$. Note that we obtain A^* during the computation for the minimum feasible time horizon T^* described in Sect. 3. By (5) and $\theta(A_1) = T^*$, for $x = x_1'$, it holds that

$$x(A_1) = o^{T^*}(A_1). \tag{15}$$

By the definition of the supply/demand function and $o^{T^*}(S^+ \cup S^-)$, we have

$$x(S^+ \cup S^-) = o^{T^*}(S^+ \cup S^-) = 0. \tag{16}$$

In other words, x_1' is located on the facet of $\hat{B}(o^{T^*})$ determined by the above two equations in (15) and (16), denoted by B_1. This implies that all the vertices $b^{(\prec_1, o^{T^*})}, \ldots, b^{(\prec_h, o^{T^*})}$ satisfying (14) are located around B_1. We thus arbitrarily choose a total order $\prec_1 = (s_{i_{1,1}}, \ldots, s_{i_{1,k+1}})$ on $S^+ \cup S^-$ so that $b^{(\prec_1, o^{T^*})}$ is one of the vertices of B_1, that is, by (13), $\{s_{i_{1,1}}, \ldots, s_{i_{1,|A_1|}}\}$ and $\{s_{i_{1,|A_1|+1}}, \ldots, s_{i_{1,k+1}}\}$ coincide with A^* and $S^+ \cup S^- \setminus A^*$, respectively. We then compute the lex-max dynamic flow $f^{(\prec_1, T^*)}$ using an algorithm by Hoppe and Tardos [11], which immediately gives $b^{(\prec_1, o^{T^*})}$ as mentioned in Sect. 4.2. Next, to find the vertex $b^{(\prec_2, o^{T^*})}$, we determine the facet of B_1 denoted by B_2 with which the half line from $b^{(\prec_1, o^{T^*})}$ to x_1' in $\mathbb{R}^{k+1}$ intersects. Recall that B_2 is determined by the two equations in (15), (16) and one more equation $x(A') = o^{T^*}(A')$ for some $A' \in \hat{A} \setminus \{A_1, S^+ \cup S^-\}$. We thus calculate intersection points on all $O(|\hat{A}|)$ candidate hyperplanes with the half line, and choose one that minimizes the Euclidean distance from $b^{(\prec_1, o^{T^*})}$ to the intersection point. Let A_2 be an element of $\hat{A}$ corresponding to the chosen hyperplane, and let x_2' be the intersection point on B_2. Figure 3 shows the relationship between x_1' and x_2'. Once we obtain x_2', we can have x_1' as a convex combination of $b^{(\prec_1, o^{T^*})}$ and x_2' and then employ the coefficient of $b^{(\prec_1, o^{T^*})}$ as λ_1. Subsequently, we will represent x_2' as a convex combination of vertices around B_2 in an inductive manner. Algorithm 1 gives a formal description.

Here, we discuss the running time of Algorithm 1. Line 3 takes $O(k^2)$ time. Using an algorithm by Hoppe and Tardos [11] for computing a lex-max dynamic flow (together with Orlin's min-cost flow algorithm [22] and Thorup's shortest path algorithm [26]), Line 4 takes $O(mk(m + n \log \log n) \log n)$ time. Line 5 takes $O(k|\hat{A}|) = O(k^{d+1})$ time. Line 6 takes $O(1)$ time by reusing the results

---

Algorithm 1. Find a quickest flow f^*

Input: $\mathcal{N}, w, T^*, \hat{\mathcal{A}}, (P_1^A, \ldots, P_{p_A}^A)$ for all $A \in \hat{\mathcal{A}}, A^*$
Output: f^*

1: Set $i \leftarrow 1$, $x_1' \leftarrow w$, $\alpha_1 \leftarrow 1$, and $\bar{A}_1 \leftarrow A^*$
2: **while** x_i' is not a vertex of $\hat{B}(o^{T^*})$ **do**
3:     Arbitrarily choose a total order $\prec_i$ on $S^+ \cup S^-$ so that the first $|\bar{A}_h|$ nodes coincide with the elements of $\bar{A}_h$ for all $h \in \{1, \ldots, i\}$
4:     Compute $f^{(\prec_i, T^*)}$ and calculate $b^{(\prec_i, o^{T^*})}$
5:     Calculate the intersection points on hyperplanes $x(A) = o^{T^*}(A)$ for all $A \in \hat{\mathcal{A}}$ with the half line from $b^{(\prec_i, o^{T^*})}$ to x_i', set A_{i+1} as the one minimizing the distance from $b^{(\prec_i, o^{T^*})}$ to the intersection point, and set x_{i+1}' as that intersection point
6:     Calculate positive real values β_i and γ_i satisfying $x_i' = \beta_i b^{(\prec_i, o^{T^*})} + \gamma_i x_{i+1}'$, and set $\lambda_i \leftarrow \alpha_i \beta_i$ and $\alpha_{i+1} \leftarrow \alpha_i \gamma_i$
7:     Search $j \in \{0, \ldots, i\}$ such that $\bar{A}_j \subseteq A_{i+1} \subseteq \bar{A}_{j+1}$ (where $\bar{A}_0 = \emptyset$ and $\bar{A}_{i+1} = S^+ \cup S^-$), and set $\bar{A}_{j+1} \leftarrow A_{i+1}$ and $\bar{A}_{h+1} \leftarrow \bar{A}_h$ for $h \in \{j+1, \ldots, i\}$
8:     Set $i \leftarrow i + 1$
9: **return** $\lambda_1 f^{(\prec_1, T^*)} + \cdots + \lambda_i f^{(\prec_i, T^*)}$

---

calculated in Line 5. Line 7 takes $O(k)$ time. We repeat Lines 3–8 at most k times. Combining the above argument and Theorem 3, we have the following main theorem, which repeats Theorem 1 but with logarithmic factors.

Theorem 4. *Given a dynamic flow network $\mathcal{N}$ with a uniform edge capacity and a supply/demand function w, the evacuation problem can be solved in $O(mndk^d + mk^2(m + n \log \log n) \log n)$ time.*

5   Conclusion

In this paper, we proposed efficient algorithms for the evacuation problem without using SFM algorithms. Specifically, for networks in which the sink has a small in-degree and every edge has the same capacity, we presented an efficient time algorithm Note that our networks can model structures which cannot be represented by the networks in [10, 18], i.e., in our network cycles can exist and paths from a node to the sink can have different lengths.

In our proposed algorithm, the minimum feasible time horizon T^* can be computed without using an SFM algorithm by finding a dominant family of source subset $\hat{\mathcal{A}}$ such that $T^* = \max\{\theta(A) \mid A \in \hat{\mathcal{A}}\}$ and $|\hat{\mathcal{A}}| = O(k^d)$ holds. Furthermore, our proposed algorithm obtains a quickest flow without using an SFM algorithm by using the relationship shown by [23] between a quickest flow and a base polytope. As a result, given a dynamic flow network $\mathcal{N}$ with a uniform edge capacity and a supply/demand function w, the evacuation problem can be solved in $O(mndk^d + mk^2(m + n \log \log n) \log n)$ time, where n is the number of nodes, m is the number of edges, d is the in-degree of the sink, and k is the number of sources.

References

1. Anderson, E.J., Philpott, A.B.: Optimisation of flows in networks over time. In: Kelly, F.P. (ed.) Probability, Statistics and Optimisation, 27, pp. 369–382 (1994)
2. Bellman, R.E.: On a routing problem. Quarterly Appl. Math. **16**, 87–90 (1958)
3. Busacker, R.G., Gowen, P.J.: A procedure for determining a family of minimum-cost network flow patterns. ORO Tech. 15. Operational Research Office, Johns Hopkins University, Baltimore, 11 (1960)
4. Edmonds, J.: Submodular functions, matroids and certain polyhedra. Combinatorial Structures and Their Applications, pp. 69–87 (1970)
5. Cunningham, W.H.: On submodular function minimization. Combinatorica **5**(3), 185–192 (1985)
6. Fleischer, L., Tardos, É.: Efficient continuous-time dynamic network flow algorithms. Oper. Res. Lett. **23**(3–5), 71–80 (1998)
7. Ford Jr., L.R.: Network flow theory. Paper P-923, The Rand Corporation, Santa Monica 1956
8. Ford, L.R., Jr., Fulkerson, D.R.: Flows in networks. Princeton Univ. Press, Princeton, NJ (1962)
9. Chen, D., Golin, M.: Sink evacuation on trees with dynamic confluent flows. In: Proc. 27th Sympos. Algorithms and Compu. (ISAAC), pp. 25:1–25:13 (2016)
10. Chen, D., Golin, M.: Minmax centered k-partitioning of trees and applications to sink evacuation with dynamic confluent flows. Algorithmica, pp. 1–53 (2022)
11. Hoppe, B., Tardos, É.: The quickest transshipment problem. Math. Oper. Res. **25**(1), 36–62 (2000)
12. Higashikawa, Y., Golin, M.J., Katoh, N.: Minimax regret sink location problem in a dynamic tree network with uniform capacity. J. Graph Algorithms Appl. **18**(4), 539–555 (2014)
13. Higashikawa, Y., Katoh, N., Teruyama, J., Tokuni Y.: Faster Algorithms for Evacuation Problems in Networks with the Single Sink of Small Degree https://arxiv.org/abs/2301.06857 (2023)
14. Iri, M.: A new method of solving transportation-network problems. Oper. Res. Soc. Japan **3**(1) (1960)
15. Jewell, W.S.: Optimal flow through networks. Operat. Res. **6**(4), 633–633 (1958)
16. Kamiyama, N.: Discrete newton methods for the evacuation problem. Theoret. Comput. Sci. **795**, 510–519 (2019)
17. Kamiyama, N., Katoh, N., Takizawa, A.: An efficient algorithm for evacuation problem in dynamic network flows with uniform arc capacity. IEICE Trans. Inf. Syst. **89**(8), 2372–2379 (2006)
18. Kamiyama, N., Katoh, N., Takizawa, A.: An efficient algorithm for the evacuation problem in a certain class of networks with uniform path-lengths. Discret. Appl. Math. **157**, 3665–3677 (2009)
19. Mamada, S., Uno, T., Makino, K., Fujishige, S.: An $O(n \log^2 n)$ algorithm for the optimal sink location problem in dynamic tree networks. Discret. Appl. Math. **154**(16), 2387–2401 (2006)
20. Moore, E.F.: The shortest path through a maze. In: Proc. Sympos. Theory of Switching, Part II, pp. 285–292 (1959)
21. Saho, M., Shigeno, M.: Cancel-and-tighten algorithm for quickest flow problems. Networks **69**(2), 179–188 (2017)
22. Orlin, J.B.: A faster strongly polynomial minimum cost flow algorithm. Oper. Res. **41**(2), 338–350 (1993)

23. Schlöter, M., Skutella, M.: Fast and memory-efficient algorithms for evacuation problems. In: Proc. 28th Annu. ACM Sympos. Discrete Algo. (SODA), pp. 821–840 (2017)
24. Schlöter, M., Skutella, M., Van Tran, K.: A faster algorithm for quickest transshipments via an extended discrete newton method. In: Proceedings of 2022 Annual ACM Symposium Discrete Algorithm (SODA), pp. 90–102 (2022)
25. Schlöter, M.: Flows over time and submodular function minimization. Ph.D. thesis, Technische Universität Berlin (2018)
26. Thorup, M.: Integer priority queues with decrease key in constant time and the single source shortest paths problem. Comput. Syst. Sci. **69**(3), 330–353 (2004)

An $O(\log n)$-Competitive Posted-Price Algorithm for Online Matching on the Line

Stephen Arndt$^{(\boxtimes)}$, Josh Ascher, and Kirk Pruhs

Computer Science Department, University of Pittsburgh, Pittsburgh, PA 15260, USA
{sda19,joa71}@pitt.edu, kirk@cs.pitt.edu

Abstract. Motivated by demand-responsive parking pricing systems, we consider posted-price algorithms for the online metric matching problem. We give an $O(\log n)$-competitive posted-price randomized algorithm in the case that the metric space is a line. In particular, in this setting we show how to implement the ubiquitous guess-and-double technique using prices.

Keywords: Online Algorithms · Metric Matching · Posted-Price

1 Introduction

In this paper we are generally interested in addressing a particular difficulty that arises in the design of posted-price algorithms, which is a type of online algorithm that uses prices to incentive clients to take actions that increase the social good. Namely, we are interested in the "guess and double" technique that is ubiquitous in the online algorithms literature [11], but is challenging to implement with prices. In particular we will address this difficulty within the context of the problem of online metrical matching on a line metric, with the hope that the algorithmic techniques that we develop will be of use in addressing this difficulty in the setting of other online problems. Before giving more details, we need to give some background information.

As a motivating application for online metric matching, and for posted-price algorithms, let us consider SFpark, which is San Francisco's system for managing the availability of on-street parking [2,3,13]. The goal of SFpark is to reduce the time and fuel wasted by drivers searching for an open parking spot. The system monitors parking usages using sensors embedded in the pavement and distributes this information in real-time to drivers via SFpark.org and phone apps. SFpark periodically adjusts parking meter pricing to manage demand, to lower prices in under-utilized areas, and to raise prices in over-utilized areas. Several other cities in the world have similar demand-responsive parking pricing systems. For example, Calgary has had the ParkPlus system since 2008 [1].

Supported in part by NSF grants CCF-1907673, CCF-2036077, CCF-2209654 and an IBM Faculty Award.

The problem of centrally assigning drivers to parking spots to minimize time and fuel usage may be reasonably modeled by the online metric matching problem. The setting for this problem is a collection of servers $S = \{s_1, \ldots, s_n\}$ (the parking spots) located at various locations in a metric space. In the case that the metric space is a line, we name the servers so that $s_1 \leq s_2 \ldots \leq s_n$. Over time a sequence $R = \{r_1, \ldots, r_n\}$ of requests (the cars) arrive at various locations in the metric space. Upon the arrival of each request (car) r_i, the online algorithm must irrevocably be assigned r_i to an available server (parking spot) $s_{\sigma(i)}$, which results in $s_{\sigma(i)}$ being unavailable going forward. Conceptually think of the request (car) r_i moving to server (parking spot) $s_{\sigma(i)}$. Thus the cost incurred by such an assignment is the distance $d(s_{\sigma(i)}, r_i)$ between the location of $s_{\sigma(i)}$ and the location where r_i arrived. The objective is to minimize the total cost of matching the requests (cars) to the servers (parking spots).

However, in order to be implementable within the context of SFpark, online algorithms must be posted-price algorithms. In this setting, posted-price means that before each car arrives, the algorithm sets a price on each available parking spot without knowing the next car's arrival location. We assume each car is driven by a selfish agent who moves to the available parking spot that minimizes the sum of the price of that parking spot and the distance to that parking spot. The objective remains to minimize the aggregate distance traveled by the cars. It is important to note that conceptually the objective of the parking pricing agency is minimizing social cost (or equivalently maximizing social good), not maximizing revenue.

Research into posted-price algorithms for online metrical matching was initiated in [12], as part of a line of research to study the use of posted-price algorithms to minimize social cost in online optimization problems. As a posted-price algorithm is a valid online algorithm, one cannot expect to obtain a better competitive ratio for posted-price algorithms than what is achievable by online algorithms. So this research line has primarily focused on problems where the optimal competitive ratio achievable by an online algorithm is (perhaps approximately) known, and seeks to determine whether a similar competitive ratio can be (again perhaps approximately) achieved by a posted-price algorithm. The higher-level goal is to determine the increase in social cost that is necessitated by the restriction that an algorithm has to use posted prices to incentivize selfish agents, instead of being able to mandate agent behavior.

Essentially all results in the posted-price online algorithms literature use one of two algorithmic design techniques. The simpler algorithmic design paradigm is called *mimicry*. A posted-price algorithm $\mathcal{A}$ *mimics* an online algorithm $\mathcal{B}$ if the probability that $\mathcal{B}$ will take a particular action is equal to the probability that a self-interested agent will choose this same action when the prices of actions are set using $\mathcal{A}$. However, many online algorithms are not mimickable. So another algorithmic design paradigm, called *monotonization*, first seeks to identify a sufficient property for an online algorithm to be mimickable, and then seeks to design an online algorithm with this property. In all the examples in the literature, the identified property involves some sort of monotonicity in the behavior of the algorithm. In particular, for online metric matching on a tree metric (which includes a line as a special case), an online algorithm $\mathcal{A}$ is mimickable if and only if it is monotone in the sense that as the request location moves closer to the location of an available server the probability that the request is matched to that server cannot decrease [9].

There are three online algorithms for online metric matching on a line that interest us here:

- The Robust Matching (RM) algorithm is a deterministic primal-dual algorithm that is $\Theta(\log n)$-competitive [22]. The Robust Matching algorithm is not mimickable [8], and intuitively seems far from being mimickable.
- The Harmonic (H) algorithm is a randomized algorithm that is $\Theta(\log \Delta)$-competitive, where Δ is the ratio of the distance between the furthest pair of servers and the distance between the closest pair of servers [15]. The Harmonic algorithm chooses between the first available server to the left of the request and the first available server to the right of the request with probability inversely proportional to the distance from the request to these servers. [12] showed that the Harmonic algorithm is mimickable, thus obtaining an $O(\log \Delta)$-competitive posted-price algorithm.
- The Doubled Harmonic (DH) algorithm is a randomized algorithm that is $O(\log n)$-competitive. Doubled Harmonic combines a variation of Harmonic that uses an estimation Z of the optimal cost (between the requests and the servers), with a standard guess-and-double technique for maintaining a good estimate of the current optimal cost to date [15]. We show in Appendix B that Doubled Harmonic is not mimickable.

Thus the specific research question that we address is whether we can design a monotone variation of Doubled Harmonic that is $O(\log n)$-competitive, thus leading to an $O(\log n)$-competitive posted-price algorithm. But, even though it is the title of the paper, obtaining a better competitive ratio is only a secondary motivation for this research. Our primary motivation is to determine whether in this setting we can implement guess-and-double monotonically, with the hope that this will provide insights into designing posted-price algorithms in other settings where the standard online algorithms use the ubiquitous guess-and-double technique. To understand why answering this research question isn't completely straightforward, we need to first understand the Doubled Harmonic algorithm.

Firstly, for ease of presentation, we will make some simplifying assumptions, namely:

- No pair of servers is closer than 1 unit of distance from each other. We show that this is without loss of generality in Appendix A.1.
- All requests arrive at the location of some server. We show that this is without loss of generality in Appendix A.2.

Intuitively Doubled Harmonic modifies Harmonic in following ways[1]. Firstly, if the distance between consecutive servers is small (less than Z/n^2), where Z is the estimate of optimal maintained by the algorithm, then this distance is artificially inflated (to Z/n^2). Secondly, if the actual optimal cost between the requests and servers becomes at least the estimate Z, then the estimate Z is increased geometrically until it exceeds the current optimal cost, and the algorithm conceptually reruns itself on all the requests to date with this new estimate to compute which servers it would ideally like to be available now. The algorithm then continues forward imagining these servers are available, and then correcting to the actually available servers using some optimal matching between the imaginary available servers and the actually available servers. Unfortunately the full algorithm, with corner cases, is a bit more complicated.

Definition 1. *We define the pseudo-distance* $pd\,(s_i, s_{i+1})$ *between two adjacent servers* s_i *and* s_{i+1} *to be* ∞ *if* $s_{i+1} - s_i \geq Z$, *to be* Z/n^2 *if* $s_{i+1} - s_i \leq Z/n^2$, *and* $s_{i+1} - s_i$ *otherwise; here* Z *will be a parameter in the algorithms. We then define the pseudo-distance between two arbitrary servers* s_i *and* s_j, *where* $i < j$ *to be* $\sum_{h=i}^{j-1} pd\,(s_h, s_{h+1})$.

Definition 2 (Doubled Harmonic Algorithm Description).
Until a request arrives at a location where there is not an available server, the request is assigned to the available server where it arrives. When the first request r_t *arrives at a location where there isn't an available server, the Doubled Harmonic algorithm maintains the following invariants:*

- *An estimate* $Z = 10^j$, *for some integer* j, *such that optimal cost to date is at least* $Z/10$ *and is strictly less than* Z.
- *A set of imaginary servers* $S_\iota = \{s_{\iota(1)}, \ldots s_{\iota(k)}\}$ *that in some sense the algorithm imagines are available (but which may or may not actually be available).* S_ι *is initialized to* $S - \{s_{\sigma(1)}, \ldots, s_{\sigma(t-1)}\}$.
- *The set* $S_\rho = \{s_{\rho(1)}, \ldots s_{\rho(k)}\}$ *of servers that are really available.*
- *An arbitrary optimal matching* M *between* S_ι *and* S_ρ.

Then it responds to the arrival of a request r_t *in the following way:*

- *If* r_t *is triggering, meaning that it causes the optimal cost to date to be at least* Z, *then the estimate* Z *is set to* 10^j *where* j *is the minimum integer that will reestablish the invariant on* Z, *and the algorithm then performs what we call an adjustment operation (which we define below) up through request* r_{t-1}.
- *If there is an imaginary server* $s_{\iota(i)}$ *at the location of* r_t *then no action is taken (later we will think of this as an imaginary move of length 0).*
- *If there is no imaginary server to the left of* r_t *then it moves to the first imaginary server to its right. This is called an imaginary move.*

---

[1] Technically our description of Doubled Harmonic differs in some ways from how it is described in [15], but we believe that our description is a bit simpler, and the same analysis holds.

- Else if there is no imaginary server to the right of r_t then it moves to the first imaginary server to its left. This is called an imaginary move.
- Else let $s_{\iota(h)}$ and $s_{\iota(h+1)}$ be the first imaginary servers to the left and right of r_t, respectively. Then r_t moves to $s_{\iota(h)}$ with probability

$$L(s_{\iota(h)}, r_t, s_{\iota(h+1)}) = \frac{pd(r_t, s_{\iota(h+1)})}{pd(r_t, s_{\iota(h)}) + pd(r_t, s_{\iota(h+1)})}$$

and r_t moves to $s_{\iota(h+1)}$ with probability

$$R(s_{\iota(h)}, r_t, s_{\iota(h+1)}) = \frac{pd(r_t, s_{\iota(h)})}{pd(r_t, s_{\iota(h)}) + pd(r_t, s_{\iota(h+1)})}$$

So the algorithm chooses between the imaginary server to the left and the imaginary server to the right with probability inversely proportional to the pseudo-distance. Let us call this movement imaginary movement.
- After the imaginary movement of the request to a server in $s_{\iota(j)} \in S_\iota$, the request continues moving to the server in $s_{\rho(h)} \in S_\rho$ that $s_{\iota(j)}$ is matched to in M, which we call a corrective move, and $s_{\iota(j)}$ is removed from S_ι.

Definition 3 (Adjustment Operation Description).
This algorithm takes as input a request r_t. The algorithm simulates Doubled Harmonic on all requests up to r_t, sets S_ι to be the servers that would be available at the end of this simulation, and recomputes an optimal matching M.

There are two reasons why modifying Doubled Harmonic to be monotone isn't straightforward (and presumably why this wasn't done in [12]):

1. The first is that the behavior of the algorithm is quite different depending on whether the new request is triggering or not, which is challenging to implement with prices because the prices have to be set before the location of the request is known.
2. The correction moves used by Doubled Harmonic are intuitively not coordinated with the imaginary moves.

Our main contribution is an algorithm that we call Modified Doubled Harmonic (MDH) that circumvents these issues by modifying Doubled Harmonic in the following way:

1. Triggering requests r_t are just assigned as though they had appeared at a location x near r_t where r_t would not have been triggering had it arrived at location x. Intuitively because triggering requests are rare, it's not particularly critical that they be handled cheaply.
2. During the correction step the request moves in same direction as it would in Doubled Harmonic, but stops at the first available server. Note that this correction step cannot be implemented by any fixed matching, as Doubled Harmonic does.

One big hurdle in naturally extending poly-log competitiveness results on posted-price algorithms for online metric matching on a spider metric [9, 10] to tree metrics is the seeming need to be able to implement guess-and-double in a monotonic way on a tree, which was the main motivation for considering how to accomplish this on a line [8]. So our takeaway is that this result suggests trying to design the correction step for a tree to be as flexible as possible, so as to make it as easy as possible to monotonically blend with the imaginary movement.

Due to space limitations some proofs have been moved to appendix.

1.1 Additional Related Work

Online metric matching was first studied in [17, 18], and each showed independently that $(2n-1)$-competitive is the optimal competitive ratio for deterministic algorithms in a general metric space. The best known competitive ratio for a randomized algorithm against an oblivious adversary is $O\left(\log^2 n\right)$ [7, 20], and the best known lower bound is $\Omega(\log n)$.

In this paper, we focus on matching on the line, which is perhaps the most interesting case. [4] gave the first deterministic, $o(n)$-competitive algorithm for this problem. [19] showed that the Generalized Work Function algorithm is $\Omega(\log n)$ and $O(n)$ competitive. [21] showed that no randomized algorithm can achieve a competitive ratio of $o\left(\sqrt{\log n}\right)$ for online matching on the line.

[14] shows how to set prices to mimic the $O(1)$-competitive algorithm Slow-Fit from [5, 6] for the problem of minimizing makespan on related machines. Monotonization is used in [16] to obtain an $O(1)$-competitive posted-price algorithm for minimizing maximum flow time on related machines.

2   Modified Doubled Harmonic Description

We explain the Modified Doubled Harmonic algorithm mainly in terms of how it differs from Doubled Harmonic. Modified Doubled Harmonic makes the same initial assumptions about the instance, and maintains the same invariants, as does Doubled Harmonic. Intuitively Modified Doubled Harmonic modifies Doubled Harmonic in the following ways. Firstly, it handles a triggering request (by pretending it arrived at a nearby point where the request wouldn't have been triggering if it arrived there) before doing the double step of guess-and-double. Secondly, during the correction step the request moves in same direction as it would in Doubled Harmonic, but stops at the first available server. Unfortunately the details of both of these two modifications are a bit complicated.

Note that the optimal matching M between S_ι and S_ρ partitions the real line into subintervals of three different types:

Left Islands are maximal subintervals that contain points x where an $s_{\iota(j)} \in S_\iota$ to the right of x is matched to a $s_{\rho(h)} \in S_\rho$ to the left of x in M.

Right Islands are maximal subintervals that contain points x where an $s_{\iota(j)} \in S_\iota$ to the left of x is matched to a $s_{\rho(h)} \in S_\rho$ to the right of x in M.

Stationary Islands are maximal subintervals that are disjoint from left and right islands.

Note that this partitioning will be the same for all choices of M [22].

Definition 4 (Modified Doubled Harmonic).
The algorithm behaves the same way as Doubled Harmonic up until the first request that arrives at the location of an unavailable server. The algorithm responds to the arrival of a subsequent request r_t in the following manner:

1. *If r_t appears at the location of a available server $s_{\rho(j)}$, then it is assigned to $s_{\rho(j)}$.*
2. *Else if r_t appears to the left of the leftmost available server $s_{\rho(1)}$, then it is assigned to $s_{\rho(1)}$.*
3. *Else if r_t appears to the right of the rightmost available server $s_{\rho(k)}$, then it is assigned to $s_{\rho(k)}$.*
4. *Else if r_t is not triggering,*
   (a) *If r_t appears in a left island, it is assigned to the first available server to its left.*
   (b) *Else if r_t appears in a right island, it is assigned to the first available server to its right.*
   (c) *Else let $s_{\iota(h)}$ and $s_{\iota(h+1)}$ be the first imaginary servers to the left and right of r_t, respectively. Then r_t moves to the first available server to its left with probability*

$$L(s_{\iota(h)}, r_t, s_{\iota(h+1)}) = \frac{pd(r_t, s_{\iota(h+1)})}{pd(r_t, s_{\iota(h)}) + pd(r_t, s_{\iota(h+1)})}$$

   and r_t moves to the first available server to its right with probability

$$R(s_{\iota(h)}, r_t, s_{\iota(h+1)}) = \frac{pd(r_t, s_{\iota(h)})}{pd(r_t, s_{\iota(h)}) + pd(r_t, s_{\iota(h+1)})}$$

   So the algorithm chooses between the imaginary server to the left and the imaginary server to the right with probability inversely proportional to the pseudo-distance, and then moves to the nearest available server in that direction.

5. *Else (Comment: r_t is triggering)*
   (a) *Let $s_{\rho(h)}$ and $s_{\rho(h+1)}$ be the first available servers to the left and right of r_t, respectively.*
   (b) *Let y_ℓ be defined in the following way: If one moves from r_t to the left, let y_ℓ be the first point x that one comes to where either r_t would not have been triggering if it had arrived at x, or x is the location of $s_{\rho(h)}$.*
   (c) *Let y_r be defined in the following way: If one moves from r_t to the right, let y_r be the first point x that one comes to where either r_t would not have been triggering if it had arrived at x, or x is the location of $s_{\rho(h+1)}$.*

(d) *Let m be the midpoint between $s_{\rho(h)}$ and $s_{\rho(h+1)}$.*

(e) *If $R(s_{\rho(h)}, y_r, s_{\rho(h+1)}) < \frac{1}{2}$ then mimic the assignment of a request appearing at y_r.*

(f) *Else if $R(s_{\rho(h)}, y_\ell, s_{\rho(h+1)}) > \frac{1}{2}$ then mimic the assignment of a request appearing at y_ℓ.*

(g) *Else if $r_t < m$ then mimic the assignment of a request appearing at y_ℓ.*

(h) *Else $r_t \geq m$, and mimic the assignment of a request appearing at y_r.*

6. *If r_t was triggering (this could happen in Cases 1, 2, 3, or 5), the algorithm updates the estimate Z and calls the adjustment operation up through request r_t (note the adjustment operation was defined when we defined Doubled Harmonic).*

To show Modified Doubled Harmonic is well-defined, we make the following observations.

Observation 1. *The following hold for Case 4 of the definition of Modified Doubled Harmonic.*

(a) *If r_t appears in a left island, then it has an available server to its left.*

(b) *If r_t appears in a right island, then it has an available server to its right.*

(c) *If r_t appears in a stationary island, then there are imaginary servers on each side of r_t.*

Proof. The first two observations follow directly from the definitions of Left Island and Right Island. The third observation follows from the fact that r_t has available servers on each side, and so it must have imaginary servers on each side.

3 Monotonicity Analysis

Note that Modified Doubled Harmonic is a *neighbor* algorithm, that is it always assigns requests to a neighboring server. In Lemma 1 we show that if a neighbor algorithm is monotone on intervals between adjacent available servers $(s_{\rho(i)}, s_{\rho(i+1)})$ then it is monotone. In Lemma 2 we analyze the probability of a non-triggering request in $(s_{\rho(i)}, s_{\rho(i+1)})$ being assigned to $s_{\rho(i+1)}$. In Lemma 3 we analyze the probability of a triggering request in $(s_{\rho(i)}, s_{\rho(i+1)})$ being assigned to $s_{\rho(i+1)}$. Then we conclude in Theorem 1 that Modified Doubled Harmonic is monotone on each interval $(s_{\rho(i)}, s_{\rho(i+1)})$.

Let $r_t \to s_{\rho(j)}$ denote the event that request r_t is matched to $s_{\rho(j)}$. We will use the notation $r_t = x$ as shorthand for r_t arrived at location x. We say a point x on the line is a trigger point if a request arriving at location x would be a triggering request, and otherwise we say x is a non-trigger point.

Lemma 1. *A neighbor algorithm $\mathcal{A}$ is monotone if, for all intervals of adjacent available servers $(s_{\rho(i)}, s_{\rho(i+1)})$, $\Pr\left[r_t \xrightarrow{\mathcal{A}} s_{\rho(i+1)} \mid r_t = x\right]$ is non-decreasing across $(s_{\rho(i)}, s_{\rho(i+1)})$.*

Proof. Suppose for all intervals of adjacent available servers $\left(s_{\rho(i)}, s_{\rho(i+1)}\right)$, $\Pr\left[r_t \to s_{\rho(i+1)} \mid r_t = x\right]$ is non-decreasing across $\left(s_{\rho(i)}, s_{\rho(i+1)}\right)$. Let $u, v, s_{\rho(j+1)} \in \mathbb{R}^1$ be arbitrary such that $v \in [u, s_{\rho(j+1)}]$ and $\mathcal{A}$ has an available server at $s_{\rho(j+1)}$. We want to show the following monotonicity condition holds:

$$\Pr[r_t \to s_{\rho(j+1)} \mid r = u] \leq \Pr[r_t \to s_{\rho(j+1)} \mid r = v]$$

We proceed by simple casework. If $u = v$, then we have equality; and if $v = s_{\rho(j+1)}$, then $\Pr[r_t \to s_{\rho(j+1)} \mid r_t = v] = 1$. Thus it remains to consider $v \in \left(u, s_{\rho(j+1)}\right)$. If $s_{\rho(j)} \in [u, s_{\rho(j+1)})$, then $\Pr[r_t \to s_{\rho(j+1)} \mid r_t = u] = 0$. Otherwise, if there does not exist an available server to the left of u, then $\Pr[r_t \to s_{\rho(j+1)} \mid r_t = v] = 1$. Thus, it remains to consider the case where $u, v \in \left(s_{\rho(j)}, s_{\rho(j+1)}\right)$ for adjacent available servers at $s_{\rho(j)}, s_{\rho(j+1)}$. We know $\Pr\left[r_t \to s_{\rho(j+1)} \mid r_t = x\right]$ is non-decreasing across this interval, and so we must have $\Pr[r_t \to s_{\rho(j+1)} \mid r_t = u] \leq \Pr[r \to s_{\rho(j+1)} \mid r_t = v]$. Thus in all cases, the monotonicity condition holds. If instead we pick $u, v, s_{\rho(j+1)} \in \mathbb{R}^1$ arbitrary with $v \in [s_{\rho(j+1)}, u]$, the same reasoning holds. Thus the described condition implies $\mathcal{A}$ is monotone, and so it is equivalent to monotonicity for neighbor algorithms. $\square$

Let $r_t \xrightarrow{\text{MDH}} s_{\rho(j)}$ denote the event that request r_t is matched to available server $s_{\rho(j)}$ using Modified Doubled Harmonic. We now fix an arbitrary interval of adjacent available servers $\left(s_{\rho(i)}, s_{\rho(i+1)}\right)$.

Lemma 2. $\Pr\left[r \xrightarrow{\text{MDH}} s_{\rho(i+1)} \mid r_t = x\right]$ *is non-decreasing across the non-trigger points in* $\left(s_{\rho(i)}, s_{\rho(i+1)}\right)$.

Proof. Note that the interval $\left(s_{\rho(i)}, s_{\rho(i+1)}\right)$ can be expressed as the union of a left island, a stationary island, and a right island (any two of which could possibly be empty). Since [22] guarantees they must appear in this order, the fact that MDH assigns a request r_t in a stationary island to $s_{\rho(i+1)}$ with probability inversely proportional to its pseudodistance from $s_{\rho(i+1)}$ yields the result. $\square$

Lemma 3. *For all subintervals* $(x_L, x_R) \subseteq \left(s_{\rho(i)}, m\right) \cup \left(m, s_{\rho(i+1)}\right)$ *containing only trigger points, where m is the midpoint of $\left(s_{\rho(i)}, s_{\rho(i+1)}\right)$, we have* $\Pr\left[r_t \xrightarrow{\text{MDH}} s_{\rho(i+1)} \mid r_t = x\right]$ *is constant across* (x_L, x_R).

Proof. Let $(x_L, x_R) \subseteq \left(s_{\rho(i)}, m\right) \cup \left(m, s_{\rho(i+1)}\right)$ containing only trigger points be arbitrary. Note that the only information used to make the assignments of triggering requests are the adjacent non-trigger points (or endpoints of the interval) and the arrival location of the triggering requests relative to the midpoint. Since (x_L, x_R) contains no non-trigger points and is entirely contained on one side of m, all of this information is identical. Thus, all requests in (x_L, x_R) have the same probability of being assigned to $s_{\rho(i+1)}$. $\square$

Theorem 1. *Modified Doubled Harmonic is monotone.*

Proof. The non-trigger points in $\left(s_{\rho(i)}, s_{\rho(i+1)}\right)$, along with m, partition the interval into subintervals for which $\Pr\left[r_t \xrightarrow{\text{MDH}} s_{\rho(i+1)} \mid r_t = x\right]$ is constant via Lemma 3. Further, Lemma 2 shows that $\Pr\left[r_t \xrightarrow{\text{MDH}} s_{\rho(i+1)} \mid r_t = x\right]$ is non-decreasing across non-trigger points, and Case 5 of Definition 4 ensures that the probability of assigning a triggering request to $s_{\rho(i+1)}$ is sandwiched between the probability of assigning its neighboring non-trigger points to $s_{\rho(i+1)}$. So, Lemma 1 implies that MDH is monotone. □

4   Cost Analysis

In this section we prove Theorem 2, which states that Modified Doubled Harmonic is $O(\log n)$-competitive.

Theorem 2. *MDH is $O(\log n)$-competitive for online matching on the line.*

We first break the execution of Modified Doubled Harmonic into phases, where each phase terminates with a triggering request. We show in Lemma 4 that the aggregate cost of the nontriggering requests during a phase is at most $O(\log n)$ times the current estimate of the optimal cost plus the imaginary cost that Doubled Harmonic would have incurred during that phase. We accomplish this by showing that for each nontriggering request, the cost of the optimal matching between the imaginary and available servers decreases by at least the amount that the cost for Modified Doubled Harmonic exceeds the imaginary cost that Doubled Harmonic would have incurred on that request. In Lemma 5 we bound the cost to Modified Doubled Harmonic for a triggering request by twice the greedy cost (which is can be seen to be $O(\log n)$ times OPT via the traingle inequality) and the cost to Modified Doubled Harmonic if the request had arrived at a nearby non-trigger point. Once we have established Lemma 4 and Lemma 5, the bounding of Modified Doubled Harmonic's cost proceeds as in [15]. Details can be found in the Appendix.

We first need some definitions. Let $S_\iota(t)$ be the set of imaginary servers before the arrival of r_t, and let $S_\rho(t)$ be the set of available servers before the arrival of r_t. Let $D\left(S_\iota(t), S_\rho(t)\right)$ be the optimal cost of matching $S_\iota(t)$ and $S_\rho(t)$. Let $s_{\sigma(t)}$ be the available server that Modified Doubled Harmonic used for request r_t. For a nontriggering request r_t, if r_t appeared in a left island or a right island, let $s_{\gamma(t)}$ be the imaginary server that would be selected if one selected a neighboring imaginary server to either the left or right of r_t with probability inversely proportional to the pseudo-distance. If instead r_t appeared in a stationary island, then if one moves from r_t in the direction of $s_{\sigma(t)}$, let $s_{\gamma(t)}$ be the first imaginary server one hits. Define a phase as the sequence of requests which appear while MDH has the same estimate Z on the optimal cost. Phases begin with a sequence of nontriggering requests, and terminate with a single triggering request, after which the estimate Z inflates.

Lemma 4. *Consider an arbitrary phase, and renumber the nontriggering requests in that phase to $r_1, r_2, \ldots, r_k$. With probability one the expression*

$$D(S_\rho(t), S_\iota(t)) + \sum_{j=1}^{t-1} \left(d(r_j, s_{\sigma(j)}) - d(r_j, s_{\gamma(j)}) \right)$$

is a non-increasing function of t.

Proof. Define $g(t)$ to be the above expression for the chosen phase, and let $t \in [1, k]$ be arbitrary. Then we have

$$g(t+1) - g(t) = D(S_\rho(t+1), S_\iota(t+1)) - D(S_\rho(t), S_\iota(t)) + d(r_t, s_{\sigma(t)}) - d(r_t, s_{\gamma(t)})$$

where $S_\rho(t+1) = S_\rho(t) \setminus \{s_{\sigma(t)}\}$ and $S_\iota(t+1) = S_\iota(t) \setminus \{s_{\gamma(t)}\}$. Write $S_\rho(t) = \{s_{\rho(1)}, s_{\rho(2)}, \ldots, s_{\rho(\ell)}\}$ and $S_\iota(t) = \{s_{\iota(1)}, s_{\iota(2)}, \ldots, s_{\iota(\ell)}\}$ where the servers in each set have been ordered left-to-right. Suppose $s_{\sigma(t)} = s_{\rho(a)}$ and $s_{\gamma(t)} = s_{\iota(b)}$.

Now, suppose $a < b$. Then because $s_{\rho(a)} < s_{\rho(b)}$ and MDH is a neighbor algorithm, we must have $r_t < s_{\rho(b)}$. Further, because $s_{\iota(a)} < s_{\iota(b)}$ and $s_{\iota(b)}$ is a neighboring imaginary server to r_t, we must have $r_t > s_{\iota(a)}$. The final observation is the trickiest to notice: $r_t \leq s_{\rho(a)}$, meaning that $a < b$ implies MDH *cannot* assign r_t leftwards. We can show this through simple casework on the description of Modified Doubled Harmonic. The only cases where leftward assignment is possible are Case 3, Case 4a, and Case 4c. However, in all of these cases, we must have $a \geq b$. Indeed, in Case 3, $a = \ell \geq b$. In Case 4a, r_t is in a left island, and so [22] shows r_t must have more available servers than imaginary servers on its left, forcing $a \geq b$. In Case 4c, r_t is in a stationary island, and so [22] shows there must be an equal number of available and imaginary servers to the left of (and including the location of) r_t. By definition of $s_{\gamma(t)}$ when r_t is in a stationary island, we must have $a = b$. Thus given $a < b$, leftward assignment of r_t is not possible, and so $r_t \leq s_{\rho(a)}$. Finally, we can deduce $s_{\iota(a)} < r_t \leq s_{\rho(a)} < s_{\rho(b)}$. Simple computation yields

$$
\begin{aligned}
g(t+1) - g(t) &= D(S_\rho(t+1), S_\iota(t+1)) - D(S_\rho(t), S_\iota(t)) + d(r_t, s_{\rho(a)}) - d(r_t, s_{\iota(b)}) \\
&\leq \left(s_{\rho(b)} - s_{\rho(a)} \right) - d(s_{\rho(b)}, s_{\iota(b)}) + \left(s_{\rho(a)} - r_t \right) - d(r_t, s_{\iota(b)}) \\
&= \left(s_{\rho(b)} - r_t \right) - \left(d(s_{\rho(b)}, s_{\iota(b)}) + d(r_t, s_{\iota(b)}) \right) \\
&= d(s_{\rho(b)}, r_t) - \left(d(s_{\rho(b)}, s_{\iota(b)}) + d(r_t, s_{\iota(b)}) \right) \\
&\leq 0
\end{aligned}
$$

The first inequality follows by simple computation and application of the triangle inequality, but for completeness, the proof is given in the Appendix (Lemma 6). If $a = b$, direct computation gives the same result. If $a > b$, applying the same reasoning as before gives $s_{\rho(b)} < s_{\rho(a)} \leq r_t < s_{\iota(a)}$, and the same result follows. Thus in all cases, $g(t+1) - g(t) \leq 0$ giving $g(t+1) \leq g(t)$. Thus $g(t)$ is a non-increasing function of t, completing the proof. □

Lemma 5. *Consider a triggering request r_t. Let s_j be the available server closest to r_t. Let y_ℓ and y_r be defined as in the Modified Doubled Harmonic algorithm, and let s_ℓ and s_r be the available servers that Modified Doubled Harmonic would have assigned a request arriving at y_ℓ and y_r, respectively. Then*

$$\mathbb{E}\left[d\left(r_t, s_{\sigma(t)}\right)\right] \le 2\max\left(\mathbb{E}\left[d\left(y_\ell, s_\ell\right)\right], \mathbb{E}\left[d\left(y_r, s_r\right)\right], d\left(r_t, s_j\right)\right)$$

Proof (Sketch). For brevity we give the proof sketch, and the full proof is given in the Appendix. We proceed by showing the claim holds in each potential trigger case of Definition 4. In Cases 1, 2, and 3, the claim trivially holds, because r_t is assigned greedily to s_j. It remains to consider Case 5. Suppose r_t appeared in between adjacent available servers $s_{\rho(h)}$ and $s_{\rho(h+1)}$, and let m be the midpoint of $\left(s_{\rho(h)}, s_{\rho(h+1)}\right)$.

(a) If $R(s_{\rho(h)}, y_r, s_{\rho(h+1)}) < \frac{1}{2}$, then r_t mimics the assignment of a request arriving at y_r. Thus $\mathbb{E}\left[d\left(r_t, s_{\sigma(t)}\right)\right] \le \mathbb{E}\left[d\left(y_r, s_r\right)\right]$.
(b) Else if $R(s_{\rho(h)}, y_\ell, s_{\rho(h+1)}) > \frac{1}{2}$, then r_t mimics the assignment of a request appearing at y_ℓ. Thus $\mathbb{E}\left[d\left(r_t, s_{\sigma(t)}\right)\right] \le \mathbb{E}\left[d\left(y_\ell, s_\ell\right)\right]$.
(c) Else if $r_t < m$, then r_t mimics the assignment of a request appearing at y_ℓ. Because $y_\ell \le r_t \le m$ and $R(s_{\rho(h)}, y_\ell, s_{\rho(h+1)}) \le \frac{1}{2}$, we have $\mathbb{E}\left[d\left(r_t, s_{\sigma(t)}\right)\right] \le 2\max\left(\mathbb{E}\left[d\left(y_\ell, s_\ell\right)\right], d\left(r_t, s_j\right)\right)$.
(d) Else $r_t \ge m$, and r_t mimics the assignment of a request appearing at y_r. Because $y_r \ge r_t \ge m$ and $R(s_{\rho(h)}, y_r, s_{\rho(h+1)}) \ge \frac{1}{2}$, we have $\mathbb{E}\left[d\left(r_t, s_{\sigma(t)}\right)\right] \le 2\max\left(\mathbb{E}\left[d\left(y_r, s_r\right)\right], d\left(r_t, s_j\right)\right)$.

Thus in all cases, the claim holds.

A   Remedying Some Assumptions

A.1   Minimum Distance 1 Between Servers

First, note that we may always assume the minimum distance between servers at *different locations* is 1, which can be easily remedied by a suitable scaling. Thus to resolve the assumption that the minimum distance between adjacent servers is 1, the important piece to resolve is that no two servers exist at the same location.

Suppose we have a monotone neighbor algorithm $\mathcal{A}$ which is α-competitive under the assumptions that all servers exist at different locations, and requests appear at server locations. We will construct a monotone neighbor algorithm $\mathcal{B}$ which is 2α-competitive and removes the first assumption.

Again we may assume the instance given to $\mathcal{B}$ has minimum distance 1 between adjacent servers at different locations, which can be easily remedied by a suitable scaling. We do this primarily for ease of analysis. Let $\epsilon = \frac{1}{5n}$. On the instance given to $\mathcal{B}$, construct an instance for $\mathcal{A}$ by first placing one server per server location; and then perturbing the extra servers at the same location by at most ϵ (so that all servers are now at distinct locations). $\mathcal{B}$ then services request r in the following way.

- If r appears at an available server s in the instance of $\mathcal{B}$, place a simulated request $\tilde{r}$ at an available server $\tilde{s}$ in the same "ϵ-window" in the instance of $\mathcal{A}$. Then $r \xrightarrow{\mathcal{B}} s$ and $\tilde{r} \xrightarrow{\mathcal{A}} \tilde{s}$.
- Otherwise, let t be the location of r's appearance. Place a simulated request $\tilde{r}$ at t in the instance of $\mathcal{A}$. Given $\tilde{r} \xrightarrow{\mathcal{A}} s$ for an available server s, then $r \xrightarrow{\mathcal{B}} s$.

It is easy to see that $\mathcal{B}$ is a monotone neighbor algorithm given $\mathcal{A}$ is a monotone neighbor algorithm. It remains to show $\mathcal{B}$ is 2α-competitive. Note that each assignment in $\text{ON}_{\mathcal{B}}$ and $\text{OPT}_{\mathcal{B}}$ can differ from the corresponding assignment in $\text{ON}_{\mathcal{A}}$ and $\text{OPT}_{\mathcal{A}}$ by at most ϵ. Thus

$$\text{ON}_{\mathcal{B}} \leq \text{ON}_{\mathcal{A}} + n\epsilon = \text{ON}_{\mathcal{A}} + \frac{1}{5}$$

and

$$\text{OPT}_{\mathcal{B}} \geq \text{OPT}_{\mathcal{A}} - n\epsilon = \text{OPT}_{\mathcal{A}} - \frac{1}{5}$$

If $\text{OPT}_{\mathcal{B}} = 0$, then $\text{ON}_{\mathcal{B}} = 0$, because all requests appeared at available servers. Otherwise, $\text{OPT}_{\mathcal{B}} > 0$, and so some request is forced to match to a server at a different location. Because the minimum distance between adjacent servers (at different locations) is 1, we must have $\text{ON}_{\mathcal{B}} \geq \text{OPT}_{\mathcal{B}} \geq 1$. The same property holds for the instance of $\mathcal{A}$ (where some request is forced to assign outside of its "ϵ-window"), and so $\text{ON}_{\mathcal{A}} \geq \text{OPT}_{\mathcal{A}} \geq 1 - 2\epsilon \geq \frac{3}{5}$. Thus

$$\frac{\mathbb{E}\left[\text{ON}_{\mathcal{B}}\right]}{\text{OPT}_{\mathcal{B}}} \leq \frac{\mathbb{E}\left[\text{ON}_{\mathcal{A}}\right] + \frac{1}{5}}{\text{OPT}_{\mathcal{A}} - \frac{1}{5}} \leq \left(\frac{1 + \frac{1}{3}}{1 - \frac{1}{3}}\right)\left(\frac{\mathbb{E}\left[\text{ON}_{\mathcal{A}}\right]}{\text{OPT}_{\mathcal{A}}}\right) \leq 2\alpha$$

and so $\mathcal{B}$ is 2α-competitive, as desired.

A.2   Requests Appear at Server Locations

Suppose we have a monotone neighbor algorithm $\mathcal{B}$ which is β-competitive under the assumption that requests appear at server locations. We will construct a monotone neighbor algorithm $\mathcal{C}$ which is $(2\beta + 1)$-competitive and makes no such assumption. Specifically, $\mathcal{C}$ services request r in the following way.

- Let t be the server closest to r, regardless of whether t is available or not.
- Place a simulated request $\tilde{r}$ at the location of t in the running instance of $\mathcal{B}$.
- Given $r \xrightarrow{\mathcal{B}} s$ for an available server s, then $r \xrightarrow{\mathcal{C}} s$.

Let $S = \{s_1, s_2, \ldots, s_n\}$ be the set of servers in the instance and $R = \{r_1, r_2, \ldots, r_n\}$ be the set of requests. Without loss of generality, assume the servers of S and the requests of R have been written, according to their locations, in increasing order of coordinate value. Let t_i be the server nearest to r_i, regardless of whether it is available or not upon appearance of r_i. Then the set $T = \{t_1, t_2, \ldots, t_n\}$ is written as "ordered" as well.

First, we show $\mathcal{C}$ is $(2\beta + 1)$-competitive. Suppose $\mathcal{B}$ assigns $\tilde{r}_i$ to $s_{\sigma(i)}$ for each i. Then $\mathrm{OPT}_\mathcal{B} = \sum_{i=1}^n d(t_i, s_i)$, $\mathrm{ON}_\mathcal{B} = \sum_{i=1}^n d(s_{\sigma(i)}, t_i)$, $\mathrm{OPT}_\mathcal{C} = \sum_{i=1}^n d(r_i, s_i)$, and $\mathrm{ON}_\mathcal{C} = \sum_{i=1}^n d(s_{\sigma(i)}, r_i)$, where the structure of $\mathrm{OPT}_\mathcal{B}$ and $\mathrm{OPT}_\mathcal{C}$ is given by [22]. Note $\mathrm{OPT}_\mathcal{C} \geq \sum_{i=1}^n d(r_i, t_i)$ since t_i is the nearest server to r_i for each request r_i. Then we have

$$\mathrm{OPT}_\mathcal{C} = \frac{1}{2}\left(\mathrm{OPT}_\mathcal{C} + \mathrm{OPT}_\mathcal{C}\right)$$

$$\geq \frac{1}{2}\left(\sum_{i=1}^n d(r_i, s_i) + \sum_{i=1}^n d(r_i, t_i)\right)$$

$$\geq \frac{1}{2}\left(\sum_{k=1}^n d(t_i, s_i)\right)$$

$$= \frac{1}{2}\mathrm{OPT}_\mathcal{B}$$

and

$$\mathrm{ON}_\mathcal{C} = \sum_{i=1}^n d(s_{\sigma(i)}, r_i) \leq \sum_{i=1}^n d(s_{\sigma(i)}, t_i) + \sum_{i=1}^n d(r_i, t_i) \leq \mathrm{ON}_\mathcal{B} + \mathrm{OPT}_\mathcal{C}$$

Thus

$$\frac{\mathbb{E}\left[\mathrm{ON}_\mathcal{C}\right]}{\mathrm{OPT}_\mathcal{C}} \leq \frac{\mathbb{E}\left[\mathrm{ON}_\mathcal{B}\right] + \mathrm{OPT}_\mathcal{C}}{\mathrm{OPT}_\mathcal{C}} \leq \frac{\mathbb{E}\left[\mathrm{ON}_\mathcal{B}\right]}{\frac{1}{2}\mathrm{OPT}_\mathcal{B}} + 1 = 2\left(\frac{\mathbb{E}\left[\mathrm{ON}_\mathcal{B}\right]}{\mathrm{OPT}_\mathcal{B}}\right) + 1 \leq 2\beta + 1$$

as desired. Further, it is easy to see that $\mathcal{C}$ is a neighbor algorithm given $\mathcal{B}$ is a neighbor algorithm. Lastly, we must show $\mathcal{C}$ is monotone. Indeed, the sets of points closest to s_i for each server s_i partition the real line into disjoint intervals (where all servers at the same location are understood to share the same interval). Any requests appearing within the same interval are treated identically in $\mathcal{C}$. This discretization ensures that because $\mathcal{B}$ is monotone and thus satisfies the condition in Lemma 1, $\mathcal{C}$ satisfies the same condition, and so it is also monotone.

B Proof that Doubled Harmonic is Not Monotone

Consider the following instance.

Suppose that r_1 arrives at s_2. Then, $r_1 \xrightarrow{\text{DH}} s_2$. Next, suppose r_2 arrives at s_2. Then the optimal matching of r_1 and r_2 has cost 4, the estimate Z is set to 10, the set of imaginary servers is set to $S_\iota = \{s_1, s_3, s_4\}$, and the set of available servers is set to $S_\rho = \{s_1, s_3, s_4\}$. Clearly the optimal matching M between S_ι and S_ρ just assigns each server to itself. Suppose DH then performs the imaginary move $r_2 \to s_1$ and the subsequent corrective move $s_1 \to s_1$. This leaves $S_\iota = \{s_3, s_4\}$ and $S_\rho = \{s_3, s_4\}$. Now, we show that the assignment of r_3 is not monotone.

Suppose that r_3 appears at s_1. Then, the optimal matching of the requests has cost 7. DH performs the imaginary move $r_3 \to s_3$ and the subsequent corrective move $s_3 \to s_3$, and so DH assigns r_3 to s_3 with probability 1.

Suppose that r_3 instead appears at s_2. Now, the optimal matching of the requests has cost 11. The estimate Z is then set to 100, and the adjustment operation is performed. With probability $\frac{4}{11}$, DH simulates assigning r_1 to s_2 and r_2 to s_3. The imaginary move of r_3 is then to s_1 with probability $\frac{27}{31}$, and the subsequent corrective move assigns r_3 to s_3. The imaginary move of r_3 to s_4 has probability $\frac{4}{31}$, and the subsequent corrective move assigns r_3 to s_4. Thus with nonzero probability, DH assigns r_3 to s_4 (and thus NOT to s_3) in this case.

Thus the probability that DH assigns r_3 to s_3 is *higher* for arrival at s_1 (probability 1) than for arrival at s_2 (probability < 1). Further note that this violation of monotonicity is induced by the fact that an adjustment operation will not occur if r_3 arrives at s_1, but it will occur if r_3 arrives at s_2. Thus DH is not monotone.

C Auxiliary Lemma for Lemma 4

Lemma 6. *Let P, Q be two finite sets of points in $\mathbb{R}^1$ with the same number of elements. Suppose $P = \{p_1, p_2, \ldots, p_m\}$ and $Q = \{q_1, q_2, \ldots, q_m\}$, where the points have been written in increasing order of location. Let $D(P, Q)$ be the optimal cost of matching P and Q. Further, let $P' = P \setminus \{p_g\}$ and $Q' = Q \setminus \{q_h\}$ for arbitrary $g, h \in [1, m]$. Then*

$$D(P', Q') - D(P, Q) \leq \begin{cases} (p_h - p_g) - |p_h - q_h| & g \leq h \\ (q_g - q_h) - |q_g - p_g| & g > h \end{cases}$$

Proof. We know via [22] that

$$D(P,Q) = \sum_{k=1}^{m} |p_k - q_k|$$

Suppose $g \leq h$. Then we have

$$D(P',Q') = \sum_{k=1}^{g-1} |p_k - q_k| + \sum_{k=g}^{h-1} |p_{k+1} - q_k| + \sum_{k=h+1}^{m} |p_k - q_k|$$

Thus

$$
\begin{aligned}
D(P',Q') - D(P,Q) &= \left(\sum_{k=g}^{h-1} |p_{k+1} - q_k| - |p_k - q_k| \right) - |p_h - q_h| \\
&\leq \left(\sum_{k=g}^{h-1} |p_{k+1} - p_k| \right) - |p_h - q_h| \\
&= \left(\sum_{k=g}^{h-1} (p_{k+1} - p_k) \right) - |p_h - q_h| \\
&= (p_h - p_g) - |p_h - q_h|
\end{aligned}
$$

For $g > h$, the proof follows identically, only with P, Q and g, h switched. □

D Cost Analysis Definitions

To explicitly prove our cost results, we first introduce many useful definitions. Let $\mathrm{OPT}(t)$ be the optimal cost of matching the first t requests to the servers, and suppose that $\mathrm{OPT}(n) \in [10^\ell, 10^{\ell+1})$. For ease of presentation, suppose that before the estimate Z is instantiated during execution of MDH, it holds a default value of 1. Then the estimate Z runs through $Z = 10^{k_i}$ for $0 = k_0 < k_1 < k_2 \cdots < k_m = \ell + 1$. Let $Z_i = 10^{k_i}$ for each $0 \leq i \leq m$. We now introduce some definitions which allow us to partition the requests according to Z_i. Let

- τ_i be the **maximum index** t such that $\mathrm{OPT}(t) < Z_i$ for each $0 \leq i \leq m$.
- ρ_i be the i'th **triggering request**, which upon appearance causes $\mathrm{OPT}(t)$ to increase from $< Z_{i-1}$ to $\geq Z_{i-1}$. Equivalently $\rho_i = r_{\tau_{i-1}+1}$.
- B_i be the **sequence** of requests r_t arriving after ρ_i and before ρ_{i+1}.

Let B_0 and B_m be the sequence of requests appearing before ρ_1 and after ρ_m, respectively. This allows us to decompose the full request sequence as $B_0, \rho_1, B_1, \rho_2, \ldots, B_{m-1}, \rho_m, B_m$. The phase of the algorithm associated with Z_i is given by the pair (B_i, ρ_{i+1}). However, we will no longer use this phase terminology, and rather reference the B_i's and ρ_i's directly. We now introduce some definitions which allow us to partition MDH's assignments and DH's underlying imaginary moves according to Z_i. Let

- $W_i = \bigcup_{r_t \in B_i} \{(r_t, s_{\sigma(t)})\}$ be the set of **assigned edges** for the requests in B_i.
- $X_i = \bigcup_{r_t \in B_i} \{(r_t, s_{\gamma(t)})\}$ conceptually be the set of chosen **imaginary moves** for the requests in B_i.

Conceptually, X_i is a set of possible imaginary moves of Doubled Harmonic. These imaginary moves are relevant for us because we bound the cost of Modified Doubled Harmonic's assignments against the cost of these imaginary moves. We are also interested in how MDH/DH simulates request assignments during an adjustment operation. For this reason, define $s_{\mu(i,t)}$ to be the imaginary server chosen for the request r_t during the adjustment operation triggered by ρ_i. Of course, $s_{\mu(i,t)}$ is only defined for $t \leq \tau_{i-1}$, because the adjustment operation which occurs after the estimate inflates to $Z = Z_i$ only simulates request assignments up to the triggering request $\rho_i = r_{\tau_{i-1}+1}$. Now, let

- $Y_i = \bigcup_{t=1}^{\tau_{i-1}} \{(r_t, s_{\mu(i,t)})\}$ be the set of **simulated assignments** of the requests for the adjustment operation triggered by ρ_i.
- $e_i = \{(r_{t'}, s_{\sigma(t')})\}$ be the **assigned edge** of ρ_i. Here $t' = \tau_{i-1} + 1$.
- $f_i = \{(r_{t'}, s_{\gamma(t')})\}$ conceptually be the chosen **imaginary move** for ρ_i. Here $t' = \tau_{i-1} + 1$, and $s_{\gamma(t')}$ is a neighboring imaginary server to ρ_i chosen with probability inversely proportional to the pseudodistance *after* the adjustment operation triggered by ρ_i is performed.
- $E_i = \{e_1, e_2, \ldots, e_i\}$ be the set of **assigned edges** for the triggering requests up through ρ_i.

This allows us to decompose the full assigned edge set $W = (\bigcup_{i=0}^{m} W_i) \cup E_m$ in the order $W = W_0, e_1, W_1, e_2, \ldots, W_{m-1}, e_m, W_m$. We can further decompose the chosen imaginary moves in the order $X = X_0, f_1, X_1, f_2, \ldots, X_{m-1}, f_m, X_m$. Lastly, for an edge set U, let $|U|$ be the sum of the lengths of the edges in U.

E   Bounding Non-Trigger Costs

Let $i \in [0, m]$ be arbitrary. We now pursue the goal of bounding $|W_i|$, the total cost of the non-trigger assignments while Modified Doubled Harmonic has estimate $Z = Z_i$. We start by recalling an important result from [15], which gives a cost bound on the imaginary moves and the simulated assignments from the adjustment operation.

Lemma 7. *[15]* $\mathbb{E}[|X_i| + |f_i| + |Y_i|] \leq C \cdot Z_i$ *for* $C = O(\log n)$.

Moving forward, we will use C to refer to the specific $O(\log n)$ function which is used in Lemma 7. Because $|X_i|$ is properly bounded by $O(Z_i \log n)$, our goal now becomes bounding $|W_i| - |X_i|$, the amount Modified Doubled Harmonic exceeds the imaginary cost that Doubled Harmonic would have incurred on the requests in B_i.

Let $\hat{t}_i = \tau_{i-1} + 2$ be the time of the first request in B_i. To bound $|W_i| - |X_i|$, we will bound $D\left(S_\rho\left(\hat{t}_i\right), S_\iota\left(\hat{t}_i\right)\right)$, which will be sufficient for our purposes upon application of Lemma 4. We do so by constructing a matching $M_i : S_\iota(\hat{t}_i) \rightarrow S_\rho(\hat{t}_i)$ whose cost is appropriately bounded.

Lemma 8. *[15] There exists a bijection* $M_i : S_\iota(\hat{t}_i) \rightarrow S_\rho(\hat{t}_i)$ *such that*

$$\text{cost}(M_i) \leq |f_i| + |Y_i| + |E_i| + \sum_{j=0}^{i-1} |W_j|$$

Proof. [15] Cover the line with $\{f_i\} \cup Y_i$ and $\left(\bigcup_{j=0}^{i-1} W_j\right) \cup E_i$. For all imaginary and available servers at the same location, match them together. Otherwise, for each remaining imaginary server in $S_\iota(\hat{t}_i)$, follow the edges of this covering until an available server in $S_\rho(\hat{t}_i)$ is reached, and match them together. Via the triangle inequality, the induced matching $M_i : S_\iota(\hat{t}_i) \rightarrow S_\rho(\hat{t}_i)$ has

$$\text{cost}(M_i) \leq |f_i| + |Y_i| + |E_i| + \sum_{j=0}^{i-1} |W_j|$$

□

Lemma 9. $\mathbb{E}[|W_i|] \leq C \cdot Z_i + \mathbb{E}\left[|E_i| + \sum_{j=0}^{i-1} |W_j|\right]$.

Proof. $\hat{t}_i$ is the time of the first request in B_i, and $\tau_i + 1$ is the time of the $(i+1)$'st triggering request ρ_{i+1}. Thus $\hat{t}_i \leq \tau_i + 1$ and so Lemma 4 implies $g(\tau_i + 1) \leq g(\hat{t}_i)$. Thus

$$D(S_\rho(\tau_i + 1), S_\iota(\tau_i + 1)) + (|W_i| - |X_i|) \leq D(S_\rho(\hat{t}_i), S_\iota(\hat{t}_i))$$

This gives

$$
\begin{aligned}
|W_i| &\leq |X_i| + D(S_\rho(\hat{t}_i), S_\iota(\hat{t}_i)) - D(S_\rho(\tau_i + 1), S_\iota(\tau_i + 1)) \\
&\leq |X_i| + D(S_\rho(\hat{t}_i), S_\iota(\hat{t}_i)) \\
&\leq |X_i| + \text{cost}(M_i) \\
&\leq |X_i| + |f_i| + |Y_i| + |E_i| + \sum_{j=0}^{i-1} |W_j|
\end{aligned}
$$

The third inequality follows from the fact that $D(S_\rho(\hat{t}_i), S_\iota(\hat{t}_i))$ is the optimal cost of matching $S_\rho(\hat{t}_i)$ and $S_\iota(\hat{t}_i)$. The last inequality follows from Lemma 8. Applying Lemma 7 gives the desired result. □

F   Bounding Trigger Costs

We now prove a sequence of lemmas with the eventual goal of proving Lemma 13. We begin by introducing some basic functions to compute assignment costs. In Lemma 5, we bound the cost to Modified Doubled Harmonic for a triggering request by twice the greedy cost (which is clearly $O(\log n)$ times OPT) and the cost to Modified Doubled Harmonic if the request had arrived at a nearby non-trigger point. We bound the greedy cost of ρ_i in Lemma 12, and the cost bound on the non-trigger points is a simple corollary from Lemma 9. Combining these results, we prove Lemma 13.

First, we introduce some basic functions for computing assignment costs. The function $L_h(x)$ is the linear transformation of $\big(s_{\rho(h)}, s_{\rho(h+1)}\big)$ onto $(0, 1)$ (which maps $s_{\rho(h)}$ to 0 and $s_{\rho(h+1)}$ to 1). The function $N(\alpha, \gamma) = \alpha(1 - \gamma) + (1 - \alpha)\gamma$ is a "normalized" assignment cost, where we assume the adjacent available servers exist at 0 and 1. The following lemma makes these ideas rigorous.

Lemma 10. *Suppose request r_t appears in between adjacent available servers $s_{\rho(h)}$ and $s_{\rho(h+1)}$. Further, suppose r_t assigns to $s_{\rho(h)}, s_{\rho(h+1)}$ with probabilities $1 - p, p$. Then the expected cost of r_t's assignment is $\big(s_{\rho(h+1)} - s_{\rho(h)}\big) N(L_h(r_t), p)$.*

Proof. The proof follows directly from simple computation. $\square$

The utility of decomposing r_t's assignment cost in this way comes from the fact that we may now concern ourselves with studying N, the normalized assignment cost, which simplifies much of the computation. Next, we establish some useful facts about the function N. Each fact will be used in bounding the cost in each subcase of Case 5 of Definition 4.

Lemma 11. *The following facts hold for all $\alpha, \beta, \gamma \in [0, 1]$.*

(a) If $\alpha \leq \beta$ and $\gamma \leq \frac{1}{2}$, then $N(\alpha, \gamma) \leq N(\beta, \gamma)$.
(b) If $\alpha \geq \beta$ and $\gamma \geq \frac{1}{2}$, then $N(\alpha, \gamma) \leq N(\beta, \gamma)$.
(c) If $\beta \leq \alpha \leq \frac{1}{2}$ and $\gamma \leq \frac{1}{2}$, then $N(\alpha, \gamma) \leq 2 \max(\alpha, N(\beta, \gamma))$.
(d) If $\beta \geq \alpha \geq \frac{1}{2}$ and $\gamma \geq \frac{1}{2}$, then $N(\alpha, \gamma) \leq 2 \max(1 - \alpha, N(\beta, \gamma))$.

Proof.

(a) This follows directly via simple computation.
(b) This follows directly via simple computation.
(c)

$$\frac{N(\alpha, \gamma)}{\alpha} = \frac{\gamma + \alpha - 2\gamma\alpha}{\alpha} \leq \frac{\gamma + \alpha}{\alpha} = 1 + \frac{\gamma}{\alpha}$$

$$\frac{N(\alpha, \gamma)}{N(\beta, \gamma)} = \frac{\gamma + \alpha - 2\gamma\alpha}{\gamma + \beta - 2\gamma\beta} \leq \frac{\gamma + \alpha}{\gamma + \beta(1 - 2\gamma)} \leq \frac{\gamma + \alpha}{\gamma} = 1 + \frac{\alpha}{\gamma}$$

Because $\min\left(\frac{\gamma}{\alpha}, \frac{\alpha}{\gamma}\right) \leq 1$, we know $N(\alpha, \gamma) \leq 2\alpha$ or $N(\alpha, \gamma) \leq 2N(\beta, \gamma)$. Thus $N(\alpha, \gamma) \leq 2 \max(\alpha, N(\beta, \gamma))$.

(d) Via direct computation, $N(\alpha, \gamma) = N(1 - \alpha, 1 - \gamma)$ and $N(\beta, \gamma) = N(1 - \beta, 1 - \gamma)$. Thus upon application of (c), we have

$$N(1 - \alpha, 1 - \gamma) \leq 2 \max(1 - \alpha, N(1 - \beta, 1 - \gamma))$$
$$N(\alpha, \gamma) \leq 2 \max(1 - \alpha, N(\beta, \gamma))$$

$\square$

Proof (of Lemma 5). We proceed by showing the claim holds in each potential trigger case of Definition 4. In Cases 1, 2, and 3, the claim trivially holds, because ρ_i is assigned greedily to s_j. It remains to consider Case 5. Suppose ρ_i appeared in between adjacent available servers $s_{\rho(h)}$ and $s_{\rho(h+1)}$, and let m be the midpoint of $(s_{\rho(h)}, s_{\rho(h+1)})$. Suppose that under the linear transformation $L_h : (s_{\rho(h)}, s_{\rho(h+1)}) \rightarrow (0, 1)$, ρ_i maps to α, y_ℓ maps to β_ℓ, and y_r maps to β_r. Further, m trivially maps to $\frac{1}{2}$. In comparing the costs of assignments in $(s_{\rho(h)}, s_{\rho(h+1)})$, it suffices to compare the costs of the normalized assignments in $(0, 1)$, given the normalization factor of $s_{\rho(h+1)} - s_{\rho(h)}$ is always the same.

Let $p_\ell = R(s_{\rho(h)}, y_\ell, s_{\rho(h+1)})$ and $p_r = R(s_{\rho(h)}, y_r, s_{\rho(h+1)})$. Lemma 11 cleanly handles each subcase of Case 5.

(a) If $p_r < \frac{1}{2}$, then ρ_i mimics the assignment of a request arriving at y_r. We know $\alpha \leq \beta_r$, and so $N(\alpha, p_r) \leq N(\beta_r, p_r)$.
(b) Else if $p_\ell > \frac{1}{2}$, then ρ_i mimics the assignment of a request appearing at y_ℓ. We know $\alpha \geq \beta_\ell$, and so $N(\alpha, p_\ell) \leq N(\beta_\ell, p_\ell)$.
(c) Else if $\rho_i < m$, then ρ_i mimics the assignment of a request appearing at y_ℓ. We know $\beta_\ell \leq \alpha \leq \frac{1}{2}$ and $p_\ell \leq \frac{1}{2}$, and so $N(\alpha, p_\ell) \leq 2 \max(\alpha, N(\beta_\ell, p_\ell))$. Note that α is simply the normalized greedy assignment of ρ_i.
(d) Else $\rho_i \geq m$, and ρ_i mimics the assignment of a request appearing at y_r. We know $\beta_r \geq \alpha \geq \frac{1}{2}$ and $p_r \geq \frac{1}{2}$, and so $N(\alpha, p_r) \leq 2 \max(1 - \alpha, N(\beta_r, p_r))$. Note that $1 - \alpha$ is simply the normalized greedy assignment of ρ_i.

Given ρ_i assigns rightwards to $s_{\rho(h+1)}$ with probability p, in all cases, we have

$$N(\alpha, p) \leq 2 \max(N(\beta_\ell, p_\ell), N(\beta_r, p_r), \min(\alpha, 1 - \alpha))$$

Multiplying both sides by the normalization factor of $s_{\rho(h+1)} - s_{\rho(h)}$ gives the desired result.

It remains to bound the cost of all individual non-trigger assignments and the greedy assignment. First, we obtain a bound on the greedy assignment of ρ_i.

Lemma 12. *For a triggering request ρ_i, let s be the available server nearest to ρ_i. Then $\mathbb{E}[d(\rho_i, s)] \leq C \cdot Z_i + \mathbb{E}\left[|E_{i-1}| + \sum_{j=0}^{i-1} |W_j|\right]$.*

Proof. Run the adjustment operation on all requests up to $\rho_i = r_{\tau_{i-1}+1}$ to generate simulated assigned servers $s'_{\mu(i,t)}$ and a set of simulated assignments $Y'_i = \bigcup_{t=1}^{\tau_{i-1}} \{(r_t, s_{\mu(i,t)'})\}$ solely for the purposes of our argumentation. This produces a set of imaginary servers S'_ι. Cover the line with the edges in Y'_i and the assigned edges $\left(\bigcup_{j=0}^{i-1} W_j\right) \cup E_{i-1}$. This covering partitions the line into disjoint intervals for which each interval has the same number of requests, servers in Y'_i, and previously assigned servers in $\{s_{\sigma(1)}, s_{\sigma(2)}, \ldots, s_{\sigma(\tau_{i-1})}\}$. By extension, each partition must have the same number of imaginary servers in S'_ι and available servers in S_ρ.

Now pick an imaginary server $s'_{\gamma(t')}$ for the triggering request ρ_i in the same way we picked $s_{\gamma(t')}$, where here $t' = \tau_{i-1} + 1$. This gives a generated imaginary move $f'_i = \left\{\left(r_{t'}, s'_{\gamma(t')}\right)\right\}$, and add f'_i to this covering. From ρ_i, follow f'_i, reaching a (previously) imaginary server $s'_{\iota(g)} \in S'_\iota$. Some available server must exist within the partition containing $s'_{\iota(g)}$, and so the triangle inequality ensures that some available server exists at most distance $|f'_i| + |Y'_i| + |E_{i-1}| + \sum_{j=0}^{i-1} |W_j|$ from ρ_i. Given s is the available server nearest to ρ_i, we must have

$$\mathbb{E}\left[d\left(\rho_i, s\right)\right] \le \mathbb{E}\left[|f'_i| + |Y'_i| + |E_{i-1}| + \sum_{j=0}^{i-1} |W_j|\right]$$

$$\le C \cdot Z_i + \mathbb{E}\left[|E_{i-1}| + \sum_{j=0}^{i-1} |W_j|\right]$$

where in the final step we apply Lemma 7. $\qquad\square$

Next, we obtain a bound on assignments of requests appearing at non-trigger points, which is a direct corollary from Lemma 9.

Corollary 1. *For a non-trigger point y, let s be the available server that Modified Doubled Harmonic would have assigned a request arriving at y, given the estimate is currently $Z = Z_{i-1}$. Then $\mathbb{E}\left[d(y, s)\right] \le C \cdot Z_{i-1} + \mathbb{E}\left[|E_{i-1}| + \sum_{j=0}^{i-2} |W_j|\right]$.*

With all of the pieces in place, we establish a cost bound on $\mathbb{E}\left[|e_i|\right]$.

Lemma 13. $\mathbb{E}\left[|e_i|\right] \le 2\left(C \cdot Z_i + \mathbb{E}\left[|E_{i-1}| + \sum_{j=0}^{i-1} |W_j|\right]\right)$.

Proof. The proof follows directly from application of Lemma 5, Corollary 1, and Lemma 12. Note that we apply Corollary 1 when the estimate is $Z = Z_{i-1}$ because ρ_i causes the estimate to inflate from $Z = Z_{i-1}$ to $Z = Z_i$. $\qquad\square$

G   Proving Theorem 2

We now make the recursive bounds on $\mathbb{E}\left[|W_i|\right]$, $\mathbb{E}\left[|e_i|\right]$ established in Lemma 9, Lemma 13 explicit through induction. The key idea is that although $\mathbb{E}\left[|W_i|\right]$ and $\mathbb{E}\left[|e_i|\right]$ are bounded in terms of all previous assignments and imaginary moves, the geometrically increasing nature of Z_i ensures their costs are simply on the order of $C \cdot Z_i$.

Lemma 14. $\mathbb{E}\left[|W_i|\right] \le 8C \cdot Z_i$ and $\mathbb{E}\left[|e_i|\right] \le 5C \cdot Z_i$ for all $i \in [0, m]$.

Proof. First, note that for all $i \in [0, m-1]$,

$$\sum_{j=0}^{i} Z_j = \sum_{j=0}^{i} 10^{k_j} \le \sum_{h=0}^{k_i} 10^h = \frac{1}{9} \cdot \left(10^{k_i+1} - 1\right) \le \frac{1}{9} \cdot 10^{k_i+1} = \frac{1}{9} \cdot Z_{i+1}$$

We now proceed by induction on i. The base case of $|W_0| = 0$ is trivial, and simply define $|e_0| = 0$. Let $i \in [0, m-1]$ be arbitrary, and assume the claim holds for all $j \in [0, i]$. Then

$$\mathbb{E}\left[|e_{i+1}|\right] \le 2 \left(C \cdot Z_{i+1} + \mathbb{E}\left[|E_i| + \sum_{j=0}^{i} |W_j|\right] \right)$$

$$= 2C \cdot Z_{i+1} + 2 \cdot \sum_{j=0}^{i} \mathbb{E}\left[|e_j|\right] + 2 \cdot \sum_{j=0}^{i} \mathbb{E}\left[|W_j|\right]$$

$$\le 2C \cdot Z_{i+1} + 10C \cdot \sum_{j=0}^{i} Z_j + 16C \cdot \sum_{j=0}^{i} Z_j$$

$$= 2C \cdot Z_{i+1} + 26C \cdot \sum_{j=0}^{i} Z_j$$

$$\le 2C \cdot Z_{i+1} + \frac{26C}{9} \cdot Z_{i+1}$$

$$\le 5C \cdot Z_{i+1}$$

and

$$\mathbb{E}\left[|W_{i+1}|\right] \leq C \cdot Z_{i+1} + \mathbb{E}\left[|E_{i+1}| + \sum_{j=0}^{i}|W_j|\right]$$

$$= C \cdot Z_{i+1} + \sum_{j=0}^{i+1}\mathbb{E}\left[|e_j|\right] + \sum_{j=0}^{i}\mathbb{E}\left[|W_j|\right]$$

$$\leq C \cdot Z_{i+1} + 5C \cdot \sum_{j=0}^{i+1}Z_j + 8C \cdot \sum_{j=0}^{i}Z_j$$

$$= 6C \cdot Z_{i+1} + 13C \cdot \sum_{j=0}^{i}Z_j$$

$$\leq 6C \cdot Z_{i+1} + \frac{13C}{9} \cdot Z_{i+1}$$

$$\leq 8C \cdot Z_{i+1}$$

completing the induction. $\qquad\square$

Finally, we now prove Theorem 2. The $O(\log n)$-competitiveness of Modified Doubled Harmonic is a direct consequence of Lemma 14 and the fact the geometric sums are asymtotically equal to their largest summand.

Proof (of Theorem 2). Aggregating the edges $W = \left(\bigcup_{i=0}^{m}W_i\right) \cup E_m$, we have

$$\mathbb{E}\left[|W|\right] = \mathbb{E}\left[\left(\sum_{i=0}^{m}|W_i|\right) + |E_m|\right] = \sum_{i=0}^{m}\mathbb{E}\left[|W_i|\right] + \sum_{i=1}^{m}\mathbb{E}\left[|e_i|\right]$$

Applying Lemma 14,

$$\sum_{i=0}^{m}\mathbb{E}\left[|W_i|\right] + \sum_{i=1}^{m}\mathbb{E}\left[|e_i|\right] \leq 8C \cdot \sum_{i=0}^{m}Z_i + 5C \cdot \sum_{i=0}^{m}Z_i = 13C \cdot \sum_{i=0}^{m}Z_i$$

Simplifying yields

$$13C \cdot \sum_{i=0}^{m}Z_i \leq \frac{13C}{9} \cdot 10^{k_m+1} \leq 1.5C \cdot 10^{\ell+2} = 150C \cdot 10^{\ell} \leq 150C \cdot \mathrm{OPT}(n)$$

Recalling $C = O(\log n)$ completes the proof. $\qquad\square$

References

1. Calgary ParkPlus Homepage. https://www.calgaryparking.com/parkplus
2. SFPark Wikipedia page (2022). https://en.wikipedia.org/wiki/SFpark
3. SFPark Homepage (2023). https://sfpark.org/

4. Antoniadis, A., Barcelo, N., Nugent, M., Pruhs, K., Scquizzato, M.: A o(n)-competitive deterministic algorithm for online matching on a line, vol. 8952, pp. 11–22 (2014)
5. Aspnes, J., Azar, Y., Fiat, A., Plotkin, S., Waarts, O.: On-line routing of virtual circuits with applications to load balancing and machine scheduling. J. ACM **44**(3), 486–504 (1997)
6. Azar, Y., Kalyanasundaram, B., Plotkin, S., Pruhs, K.R., Waarts, O.: On-line load balancing of temporary tasks. J. Algorithms **22**(1), 93–110 (1997)
7. Bansal, N., Buchbinder, N., Gupta, A., Naor, J.S.: An $O(\log^2 k)$-competitive algorithm for metric bipartite matching. In: Arge, L., Hoffmann, M., Welzl, E. (eds.) Algorithms - ESA 2007, pp. 522–533. Springer, Berlin Heidelberg, Berlin, Heidelberg (2007)
8. Bender, M.: Personal communication. To appear in his Ph.D. thesis
9. Bender, M., Gilbert, J., Krishnan, A., Pruhs, K.: Competitively pricing parking in a tree. In: Chen, X., Gravin, N., Hoefer, M., Mehta, R. (eds.) Web and Internet Economics, pp. 220–233. Springer International Publishing, Cham (2020). https://doi.org/10.1007/978-3-030-64946-3_16
10. Bender, M., Gilbert, J., Pruhs, K.: A poly-log competitive posted-price algorithm for online metrical matching on a spider. In: Bampis, E., Pagourtzis, A. (eds.) Fundamentals of Computation Theory, pp. 67–84. Springer International Publishing, Cham (2021). https://doi.org/10.1007/978-3-030-86593-1_5
11. Borodin, A., El-Yaniv, R.: Online Computation and Competitive Analysis. Cambridge University Press, Cambridge (2005)
12. Cohen, I.R., Eden, A., Fiat, A., Łukasz Jeż: Pricing online decisions: beyond auctions (2015)
13. Shoup, D.: SFpark: pricing parking by demand (2018). https://www.accessmagazine.org/fall-2013/sfpark-pricing-parking-demand/
14. Feldman, M., Fiat, A., Roytman, A.: Makespan minimization via posted prices (2017)
15. Gupta, A., Lewi, K.: The online metric matching problem for doubling metrics. In: Czumaj, A., Mehlhorn, K., Pitts, A., Wattenhofer, R. (eds.) Automata, Languages, and Programming, pp. 424–435. Springer, Berlin, Heidelberg (2012). https://doi.org/10.1007/978-3-642-31594-7_36
16. Im, S., Moseley, B., Pruhs, K., Stein, C.: Minimizing maximum flow time on related machines via dynamic posted pricing. In: Embedded Systems and Applications (2017)
17. Kalyanasundaram, B., Pruhs, K.: Online weighted matching. J. Algorithms **14**(3), 478–488 (1993)
18. Khuller, S., Mitchell, S.G., Vazirani, V.V.: On-line algorithms for weighted bipartite matching and stable marriages. In: Albert, J.L., Monien, B., Artalejo, M.R. (eds.) Automata, Languages and Programming, pp. 728–738. Springer, Berlin, Heidelberg (1991). https://doi.org/10.1007/3-540-54233-7_178
19. Koutsoupias, E., Nanavati, A.: The online matching problem on a line. In: Solis-Oba, R., Jansen, K. (eds.) Approximation and Online Algorithms, pp. 179–191. Springer, Berlin, Heidelberg (2004). https://doi.org/10.1007/978-3-540-24592-6_14
20. Meyerson, A., Nanavati, A., Poplawski, L.: Randomized online algorithms for minimum metric bipartite matching. In: Proceedings of the Seventeenth Annual ACM-SIAM Symposium on Discrete Algorithm, pp. 954–959. SODA '06, Society for Industrial and Applied Mathematics, USA (2006)

21. Peserico, E., Scquizzato, M.: Matching on the line admits no $o(\sqrt{\log n})$-competitive algorithm (2020)
22. Raghvendra, S.: Optimal analysis of an online algorithm for the bipartite matching problem on a line. CoRR abs/1803.07206 (2018)

Online Dominating Set and Coloring

Minati De[1] , Sambhav Khurana[2], and Satyam Singh[1(✉)]

[1] Department of Mathematics, Indian Institute of Technology Delhi, New Delhi, India
{minati,satyam.singh}@maths.iitd.ac.in
[2] Department of Computer Science and Engineering, Texas A&M University,
College Station, TX, USA
sambhav_khurana@tamu.edu

Abstract. In this paper, we present online deterministic algorithms for minimum coloring, minimum dominating set and its variants in the context of geometric intersection graphs. We consider a graph parameter: the independent kissing number ζ, which is a number equal to 'the size of the largest induced star in the graph -1'. For a graph with an independent kissing number at most ζ, we obtain an algorithm having an optimal competitive ratio of ζ, for the minimum dominating set and the minimum independent dominating set problems; however, for the minimum connected dominating set problem, we obtain a competitive ratio of at most 2ζ. In addition, we prove that for the minimum connected dominating set problem, any deterministic online algorithm has a competitive ratio of at least $2(\zeta-1)$ for the geometric intersection graph of translates of a convex object in $\mathbb{R}^2$. Next, for the minimum coloring problem, we present an algorithm having a competitive ratio of $O\left(\zeta'\log m\right)$ for geometric intersection graphs of bounded scaled α-fat objects in $\mathbb{R}^d$ having a width in between $[1, m]$, where ζ' is the independent kissing number of the geometric intersection graph of bounded scaled α-fat objects having a width in between $[1, 2]$. Finally, we investigate the value of ζ for geometric intersection graphs of various families of geometric objects.

Keywords: α-Fat objects · Coloring · Connected dominating set · Dominating set · Independent kissing number · t-relaxed coloring

1 Introduction

We consider online algorithms for some well-known NP-hard problems: the minimum dominating set problem and its variants and the minimum coloring problem. Dominating set and its variants have several applications in wireless ad-hoc networks, routing, etc. [3,4]; while coloring has diverse applications in frequency assignment, scheduling and many more [1,14,15].

Work on this paper by M. De has been partially supported by SERB MATRICS Grant MTR/2021/000584, and work by S. Singh has been supported by CSIR (File Number-09/086(1429)/2019-EMR-I).

Minimum Dominating Set and its Variants. For a graph $G = (V, E)$, a subset $D \subseteq V$ is a *dominating set* (DS) if for each vertex $v \in V$, either $v \in D$ (containment) or there exists an edge $\{u, v\} \in E$ such that $u \in D$ (dominance). A dominating set D is said to be a *connected dominating set* (CDS) if the induced subgraph $G[D]$ is connected (if G is not connected, then $G[D]$ must be connected for each connected component of G). A dominating set D is said to be an *independent dominating set* (IDS) if the induced subgraph $G[D]$ is an independent set. The *minimum dominating set* (MDS) problem involves finding a dominating set of the minimum cardinality. Similarly, the objectives of the *minimum connected dominating set* (MCDS) problem and the *minimum independent dominating set* (MIDS) problem are to find a CDS and IDS, respectively, with the minimum number of vertices.

Throughout the paper, we consider online algorithms for *vertex arrival model* of graphs where a new vertex is revealed with its edges incident to previously appeared vertices. The dominating set and its variants can be considered in various online models [4]. In *Classical-Online-Model* (also known as "Standard-Model"), upon the arrival of a new vertex, an online algorithm must either accept the vertex by adding it to the solution set or reject it. In *Relaxed-Online-Model* (also known as "Late-Accept-Model"), upon the arrival of a new vertex, in addition to the revealed vertex, an online algorithm may also include any of the previously arrived vertices to the solution set. Note that once a vertex is included in the solution set for either model, the decision cannot be reversed in the future. For the MCDS problem, if we cannot add previously arrived vertices in the solution set, the solution may result in a disconnected dominating set [3]. Therefore, in this paper, for the case of the MCDS problem, we use Relaxed-Online-Model; while for MDS and MIDS, we use Classical-Online-Model. In addition, the revealed induced subgraph must always be connected for the MCDS problem.

Minimum Coloring Problem. For a graph $G = (V, E)$, the *coloring* is to assign colors (positive integers) to the vertices of G. The *minimum coloring* (MC) problem is to find a coloring with the minimum number of distinct colors such that no two adjacent vertices (vertices connected by an edge) have the same color. In the online version, upon the arrival of each new vertex v, an algorithm needs to immediately assign to v a feasible color, i.e., one distinct from the colors assigned to the neighbours of v that have already arrived. The color of v cannot be changed in future.

We analyze the quality of our online algorithm by competitive analysis [2]. An online algorithm ALG for a minimization problem is said to be *c-competitive*, if there exists a constant d such that for any input sequence $\mathcal{I}$, we have $\mathcal{A}(\mathcal{I}) \leq c \times \mathcal{O}(\mathcal{I}) + d$, where $\mathcal{A}(\mathcal{I})$ and $\mathcal{O}(\mathcal{I})$ are the cost of solutions produced by ALG and an optimal offline algorithm, respectively, for the input $\mathcal{I}$. The smallest c for which ALG is c-competitive is known as an *asymptotic competitive ratio* of ALG [3]. The smallest c for which ALG is c-competitive with $d = 0$ is called an *absolute competitive ratio* (also known as strict-competitive ratio) of *ALG* [3]. If not explicitly specified, we use the term "competitive ratio" to mean absolute competitive ratio.

1.1   Preliminaries

We use $[n]$ to denote the set $\{1, 2, \ldots, n\}$, where n is a positive real number. In this paper, we focus on geometric intersection graphs due to their applications in wireless sensors, network routing, medical imaging, etc. For a family S of geometric objects in $\mathbb{R}^d$, the *geometric intersection graph* G of S is an undirected graph with set of vertices same as S, and the set of edges is defined as $E = \{\{u, v\} | u, v \in S, u \cap v \neq \emptyset\}$. Several researchers have used the kissing number as a parameter to give an upper or lower bound for geometric problems. For instance, Butenko et al. [6] used it to prove the upper bound of the MCDS problem in the offline setup for unit balls in $\mathbb{R}^3$, whereas Dumitrescu et al. [12] used it to prove the upper bound for the unit covering problem for balls in $\mathbb{R}^d$. Similar to kissing number, we use a graph parameter- independent kissing number ζ. Let $\varphi(G)$ denote the size of a maximum independent set of a graph $G = (V, E)$. For any vertex $v \in V$, let $N(v) = \{u (\neq v) \in V | \{u, v\} \in E\}$ be the neighbourhood of the vertex v. Now, we define the *independent kissing number* ζ for graphs.

Definition 1. (Independent Kissing Number) *The independent kissing number ζ of a graph $G = (V, E)$ is defined as* $\max_{v \in V} \{\varphi(G[N(v)])\}$.

Note that the independent kissing number equals 'the size of the largest induced star in the graph -1'. In other words, a graph with independent kissing number ζ is a $K_{1,\zeta+1}$-free graph. Moreover, the value of ζ may be very small compared to the number of vertices in a graph. For example, the value of ζ is a fixed constant for the geometric intersection graph of several families of geometric objects like translated and rotated copies of a convex object in $\mathbb{R}^2$. However, the use of this parameter is not new. For example, in the offline setup, Marathe et al. [20] obtained a $2(\zeta - 1)$-approximation algorithm for the MCDS problem for any graph having an independent kissing number at most ζ.

Two geometric objects are said to be *non-overlapping* if they have no common interior, whereas we call them *non-touching* if their intersection is empty. An equivalent definition of independent kissing number ζ for a family S of geometric objects in $\mathbb{R}^d$ is given below.

Definition 2. *Let S be a family of geometric objects, and let u be any object belonging to the family S. Let ζ_u be the maximum number of pairwise non-touching objects in S that we can arrange in and around u such that all of them are intersected by u. The independent kissing number ζ of S is defined to be* $\max_{u \in S} \zeta_u$.

A set K of objects belonging to the family S is said to form an *independent kissing configuration* if (i) there exists an object $u \in K$ that intersects all objects in $K \setminus \{u\}$, and (ii) all objects in $K \setminus \{u\}$ are mutually non-touching to each other.

Here u and $K \setminus \{u\}$ are said to be the *core* and *independent set*, respectively, of the independent kissing configuration. The configuration is considered *optimal* if $|K \setminus \{u\}| = \zeta$, where ζ is the independent kissing number of $\mathcal{S}$. The configuration is said to be *standard* if all objects in $K \setminus \{u\}$ are mutually non-overlapping with u, i.e., their common interior is empty but touches the boundary of u.

A number of different definitions of fatness (not extremely long and skinny) are available in the geometry literature. For our purpose, we define the following. Let σ be an object and x be any point in σ. Let $\alpha(x)$ be the ratio between the minimum and maximum distance (under Euclidean norm) from x to the boundary $\partial(\sigma)$ of the object σ. In other words, $\alpha(x) = \frac{\min_{y \in \partial(\sigma)} d(x,y)}{\max_{y \in \partial(\sigma)} d(x,y)}$, where $d(.,.)$ denotes the Euclidean distance. The *aspect ratio* α of an object σ is defined as the maximum value of $\alpha(x)$ for any point $x \in \sigma$, i.e., $\alpha = \max\{\alpha(x) : x \in \sigma\}$. An object is said to be α-*fat object* if its aspect ratio is α. Observe that α-fat objects are invariant under translation, rotation and scaling. The *aspect point* of σ is a point in σ where the aspect ratio of σ is attained. The minimum distance (respectively, maximum distance) from the aspect point to the boundary of the object is referred to as the *width* (respectively, *height*) of the object. Note that fat objects are invariant under translation, rotation and scaling. For more details on α-fat object, one may see [9].

1.2   Related Work

The dominating set and its variants are well-studied in the offline setup. Finding MDS is known to be NP-hard even for the unit disk graphs [8,16]. A polynomial-time approximation scheme (PTAS) is known when all objects are homothets of a convex object [10]. King and Tzeng [18] initiated the study of the online MDS problem in Classical-Online-Model. They showed that for a general graph, the greedy algorithm achieves a competitive ratio of $n - 1$, which is also a tight bound achievable by any online algorithm for the MDS problem, where n is the length of the input sequence. Even for the interval graph, the lower bound of the competitive ratio is $n - 1$ [18]. Eidenbenz [13] proved that the greedy algorithm achieves a tight bound of 5 for the MDS of the unit disk graph.

Boyar et al. [3] considered a variant of the Relaxed-Online-Model for MDS, MIDS and MCDS problem in which, in addition to the Relaxed-Online-Model the revealed graph should always be connected. In this setup, they studied these problems for specific graph classes such as trees, bipartite graphs, bounded degree graphs, and planar graphs. Their results are summarized in [3, Table 2]. They proposed a 3-competitive algorithm for the MDS problem in the above-mentioned model for a tree. Later, Kobayashi [19] proved that 3 is also the lower bound for the tree in this setting. In the same setup, Eidenbenz [13] showed that, for the MCDS problem of unit disk graph, the greedy algorithm achieves a competitive ratio of $8+\epsilon$, whereas no online algorithm can guarantee a strictly better competitive ratio than 10/3. We observe that the (asymptotic) competitive ratio of the MCDS problem for the unit disk graph could be improved to 6.798 (see Sect. 2.2).

The minimum coloring problem is known to be NP-hard, even for unit disk graphs [8]. In the offline coloring case, a 3-approximation algorithm for coloring unit disk graphs was presented by Gräf et al. [17] and Marathe et al. [20]. Marathe et al. [20] generalised the approach for unit disk graphs to disk graphs. They proved that the obtained approximation for coloring disk graphs is at most 6. For online coloring, Erlebach and Fiala [14] proved that ALGORITHM-FIRST-FIT achieves a competitive ratio of $O(\log n)$ for disk graphs. Recently, Albers and Schraink [1] proved that the best competitive ratio of both deterministic and randomized online algorithms for disk graphs is $\Theta(\log n)$. Capponi and Pilloto [7] proved that for any graph with an independent kissing number (see Defnition 1) at most ζ, popular ALGORITHM-FIRST-FIT achieves a competitive ratio at most ζ. The existence of $O(\log m)$-competitive coloring algorithm for disk graphs, whose radius is in between $[1, m]$, is known due to Erlebach and Fiala [14]. In this paper, we generalize this result for the geometric intersection graph of bounded scaled α-fat objects in $\mathbb{R}^d$.

1.3 Our Contributions

In this paper, we obtain the following results.

1. First, we prove that for MDS and MIDS problems, the natural greedy algorithm has an optimal competitive ratio of ζ for a graph with an independent kissing number at most ζ (Theorems 1 and 2).
2. For the MCDS problem, we prove that, for any graph with the independent kissing number at most ζ, a greedy algorithm achieves a competitive ratio of at most 2ζ (Theorem 3). To complement this, we prove that the lower bound of the competitive ratio is at least $2(\zeta - 1)$ which holds even for a geometric intersection graph of translates of a convex object in $\mathbb{R}^2$ (Theorem 4).
3. Next, we consider coloring geometric intersection graphs of bounded scaled α-fat objects in $\mathbb{R}^d$ having a width in between $[1, m]$. For this, due to [7], the best known competitive ratio is ζ, where ζ is the independent kissing number of bounded scaled α-fat objects having a width in between $[1, m]$. Inspired by Erlebach and Fiala [14], we present ALGORITHM-LAYER having a competitive ratio of at most $O(\zeta'\log m)$, where ζ' is the independent kissing number of bounded scaled α-fat objects having a width in between $[1, 2]$ (Theorem 5). Since the value of ζ could be very large compared to $\zeta'\log m$ (see Remark 2), it is a significant improvement.
4. All results obtained above for the MC problem, MDS problem and its variants depend on the graph parameter: the independent kissing number ζ. Therefore, the value of ζ becomes a crucial graph parameter to investigate. To estimate the value of ζ, we consider various families of geometric objects. We show that for congruent balls in $\mathbb{R}^3$ the value of ζ is 12. For translates of a regular k-gon ($k \in ([5, \infty) \cup \{3\}) \cap \mathbb{Z}$) in $\mathbb{R}^2$, we show that $5 \le \zeta \le 6$. While the value is 2^d for translates of a hypercube in $\mathbb{R}^d$, for congruent hypercubes in $\mathbb{R}^d$ the value is at least 2^{d+1}. We also give bounds on the value of ζ for α-fat objects in $\mathbb{R}^d$ having a width in between $[1, m]$. We feel that these results will find applications in many problems. We illustrate a few in Sect. 5.

Note that all of our algorithms are deterministic. In particular, algorithms in items 1 and 2 do not need to know object's representation; whereas, ALGORITHM-LAYER needs to know the object's width upon arrival.

2 Dominating Set and Its Variants

In this section, we discuss the well-known greedy online algorithms for MDS and its variants for graphs. We show how their performance depends on the independent kissing number ζ of the graph. Note that algorithms need not know the value of ζ in advance, and the object's representation is also unnecessary. All the missing proofs of this section will appear in the final version of the paper.

2.1   Minimum Dominating Set

The greedy algorithm, ALGORITHM-GDS, for finding a minimum dominating set is as follows. The algorithm maintains a feasible dominating set $\mathcal{A}$. Initially, $\mathcal{A} = \emptyset$. On receiving a new vertex v, if the vertex is not dominated by the existing dominating set $\mathcal{A}$, then update $\mathcal{A} \leftarrow \mathcal{A} \cup \{v\}$. Eidenbenz [13] showed that this algorithm achieves an optimal competitive ratio of 5 for the unit disk graph. It is easy to generalize this result for graphs with the fixed independent kissing number ζ.

Observation 1. *The vertices returned by the* ALGORITHM-GDS *are pairwise non-adjacent. In other words, the solution set is always an independent set.*

Theorem 1. ALGORITHM-GDS *has an optimal competitive ratio of ζ for the MDS problem of a graph having an independent kissing number at most ζ.*

As a result of Observation 1, the output produced by ALGORITHM-GDS is an independent dominating set. Thus, we have the following.

Theorem 2. ALGORITHM-GDS *has an optimal competitive ratio of ζ for the MIDS problem of a graph having an independent kissing number at most ζ.*

2.2   Minimum Connected Dominating Set

Recall that if we cannot add previously arrived vertices in the solution set for the minimum connected dominating set problem, the solution may result in a disconnected dominating set. Therefore, we use Relaxed-Online-Model for this problem. Here, in addition to the Relaxed-Online-Model, the revealed induced subgraph must always be connected for the MCDS problem. Eidenbenz [13] proposed a greedy algorithm for unit disk graph in the aforementioned setup and showed that the algorithm achieves a competitive ratio of at most $8 + \epsilon$. We analyse the same algorithm for graphs with the fixed independent kissing number ζ.

Description of ALGORITHM-GCDS: Let V be the set of vertices presented to the algorithm and $\mathcal{A} \subseteq V$ be the set of vertices chosen by our algorithm such that $\mathcal{A}$ is a connected dominating set for the vertices in V. The algorithm maintains two disjoint sets $\mathcal{A}_1$ and $\mathcal{A}_2$ such that $\mathcal{A} = \mathcal{A}_1 \cup \mathcal{A}_2$. Initially, both $\mathcal{A}_1, \mathcal{A}_2 = \emptyset$. Let v be a new vertex presented to the algorithm. The algorithm first updates $V \leftarrow V \cup \{v\}$ and then does the following.

- If v is dominated by the set $\mathcal{A}$, do nothing.
- Otherwise, first, add v to $\mathcal{A}_1$. If v has at least one neighbour in V, choose any one neighbour, say u, of v from V, and add u to $\mathcal{A}_2$. In other words, update $\mathcal{A}_1 \leftarrow \mathcal{A}_1 \cup \{v\}$ and if necessary update $\mathcal{A}_2 \leftarrow \mathcal{A}_2 \cup \{u\}$. Note that u is already dominated by the existing dominating set $\mathcal{A}$. As a result, if we add u to the dominating set, it will result in a connected dominating set.

Note that the algorithm produces a feasible connected dominating set. The addition of vertex in $\mathcal{A}_1$ assures that $\mathcal{A}$ is a dominating set, and the addition of vertices in $\mathcal{A}_2$ ensures that $\mathcal{A}$ is a connected dominating set. Now, using induction, it is easy to prove the following.

Lemma 1. ALGORITHM-GCDS *maintains the following two invariants: (i) $\mathcal{A}_1$ is an independent set, and (ii) $|\mathcal{A}_1| \geq |\mathcal{A}_2|$.*

The next lemma is a generalization of a result by Wan et al. [23, Lemma 9].

Lemma 2. *Let $\mathcal{I}$ be an independent set, and $\mathcal{O}$ be a minimum connected dominating set of a graph with the independent kissing number ζ. Then $|\mathcal{I}| \leq (\zeta - 1)|\mathcal{O}| + 1$.*

Theorem 3. ALGORITHM-GCDS *has an asymptotic competitive ratio of at most $2(\zeta - 1)$ and an absolute competitive ratio of at most 2ζ for the MCDS problem for a graph having an independent kissing number at most ζ.*

Remark 1. Note that due to the result of Du and Du [11, Thm 1], for unit disk graphs, we have $|\mathcal{I}| \leq 3.399|\mathcal{O}| + 4.874$. As a result, similar to Theorem 3, one can prove that $|\mathcal{A}| \leq 6.798|\mathcal{O}| + 9.748$. Hence, for unit disk graphs, ALGORITHM-GCDS has an asymptotic competitive ratio of at most 6.798.

2.3   Lower Bound of the MCDS Problem

In this section, first, we propose a lower bound of the MCDS problem for a wheel graph. Then, using that, we propose a lower bound for the geometric intersection graph of translated copies of a convex object in $\mathbb{R}^2$.

Consider a *wheel graph* $W_k = (V, E)$ of order k, where $V = \{v_0, v_1, \ldots, v_k\}$ and $E = \big\{\{v_i, v_k\} \mid i \in [k-1]\big\} \cup \big\{\{v_i, v_{(i+1) \mod k}\} \mid i \in [k-1]\big\}$. In other words, in W_k, the vertices $v_0, v_1, \ldots, v_{k-1}$ form a cycle C_k and a single core vertex v_k is adjacent to each vertex of C_k. Now, we define a cyclone-order of vertices in W_k.

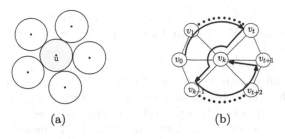

(a)   (b)

Fig. 1. (a) Optimal independent kissing configurations for unit disks; (b) Cyclone-order of vertices in a wheel graph. The bold arrow indicates a cyclone-order.

Definition 3. (Cyclone-order of vertices in a wheel graph) *For an integer t $(0 < t < k-1)$, in the the cyclone order of W_k, first, we enumerate $t+1$ vertices $v_0, v_1, \ldots, v_t$ of C_k, followed by an enumeration of the remaining $k-t-1$ vertices of C_k, i.e., $v_{k-1}, v_{k-2}, \ldots, v_{t+2}, v_{t+1}$. Finally, the core vertex v_k is appended. We denote the first $t+1$ length sequence of C_k as a cw-part and the remaining $k-t-1$ length sequence as an acw-part of the cyclone-order.*

Now, it is easy to obtain the following lemma.

Lemma 3. *If the vertices of a wheel graph W_k are enumerated in a cyclone-order, then any deterministic online algorithm reports a CDS of size at least $k-2$, where the size of an offline optimum is 1.*

Now, we give an explicit construction of a wheel graph $W_{2\zeta}$ using translates of a convex object having independent kissing number ζ.

Lemma 4. *For a family of translates of a convex object having independent kissing number ζ, there exists a geometric intersection graph $W_{2\zeta}$.*

Combining Lemma 3 and 4, we have the following result.

Theorem 4. *Let ζ be the independent kissing number of a family S of translated copies of a convex object in $\mathbb{R}^2$. Then the competitive ratio of every deterministic online algorithm for MCDS of S is at least $2(\zeta - 1)$.*

3   Algorithm for the Minimum Coloring Problem

Here, we present ALGORITHM-LAYER to find coloring for the geometric intersection graph of bounded scaled α-fat objects in $\mathbb{R}^d$ having a width in between $[1, m]$, where $m \geq 2$. First, we describe the well-known ALGORITHM-FIRST-FIT as follows. Upon the arrival of the object σ, the algorithm assigns the smallest color available, i.e., the smallest color that has not yet been assigned to an adjacent vertex of σ.

Lemma 5. *[7, Lemma 4]* ALGORITHM-FIRST-FIT *has a competitive ratio of ζ for the MC problem for a graph having an independent kissing number at most ζ.*

Now, we present a deterministic algorithm, ALGORITHM-LAYER, that is similar to the algorithm of Erlebach and Fiala [14] originally defined for bounded scaled disks.

Description of ALGORITHM-LAYER. For each $j \in \mathbb{Z}^+ \cup \{0\}$, let L_j be the jth layer containing all objects with a width in between $[2^j, 2^{j+1})$. Observe that the width of each layer's objects falls within a factor of two. For each layer L_j, we use ALGORITHM-FIRST-FIT separately to color the objects. When an object σ_i having width w_i arrives, our algorithm, first, determines the layer number $j = \lfloor \log w_i \rfloor$. Then we color σ_i using ALGORITHM-FIRST-FIT considering already arrived objects in L_j, and also we use the fact that a color that is used in any other layer cannot be used for σ_i. A pseudo-code of ALGORITHM-LAYER is given in Algorithm 1.

Algorithm 1. ALGORITHM-LAYER

1: $L_j \leftarrow \emptyset$, for all $j \in \mathbb{Z}^+$
2: **for** $i = 1$ to n; **do**   ▷ arrival of an object σ_i having a width w_i
3:   **begin**
4:     $j = \lfloor \log_2 w_i \rfloor$;   ▷ Identifying the index of the layer to which σ_i belongs, where r_i is the width of σ_i.
5:     $L_j \leftarrow L_j \cup \{\sigma_i\}$;   ▷ The layer containing σ_i
6:     $F = \{c(\sigma_k) : 1 \leq k < i, \sigma_k \in L_j, \sigma_k \cap \sigma_i \neq \emptyset\} \cup \{c(\sigma_k) : 1 \leq k < i, \sigma_k \notin L_j\}$;
       ▷ The set of forbidden colors
7:     $c(\sigma_i) = \min\{\mathbb{Z}^+ \setminus F\}$;   ▷ color assigned to σ_i
8:   **end**
9: **end for**

Theorem 5. *Let ζ' be the independent kissing number of bounded scaled α-fat objects having a width in between $[1, 2]$. ALGORITHM-LAYER has a competitive ratio of at most $O(\zeta' \log m)$ for MC of geometric intersection graph of bounded scaled α-fat objects in $\mathbb{R}^d$ having width in between $[1, m]$.*

Proof. Let $\mathcal{A}$ and $\mathcal{O}$ be the set of colors used by the ALGORITHM-LAYER and an offline optimum for an input sequence $\mathcal{I}$. For $i \in \{0, 1, \ldots, \lceil \log m \rceil\}$, let the layer L_i be the collection of all α-fat objects in $\mathcal{I}$ having a width in $[2^i, 2^{i+1})$. Let $\mathcal{O}_i$ be a set of colors used by an offline optimum algorithm for the layer L_i. Let $\mathcal{O}'_i \subseteq \mathcal{O}$ be the set of colors used for the layer L_i. Note that the set of colors in $\mathcal{O}'_i$ is a valid coloring for objects in L_i. Thus, we have $|\mathcal{O}_i| \leq |\mathcal{O}'_i|$. Let $\mathcal{A}_i$ be the set of colors used by ALGORITHM-LAYER to color layer L_i. Note that $\mathcal{A} = \cup_{i=0}^{\lceil \log m \rceil} \mathcal{A}_i$ and $\mathcal{A}_i \cap \mathcal{A}_j = \emptyset$, where $i \neq j \in [\lceil \log m \rceil]$. Due to Lemma 5, for all $i \in [\lceil \log m \rceil]$ we have $|\mathcal{A}_i| \leq \zeta_i |\mathcal{O}_i|$, where ζ_i is the independent kissing number of bounded scaled α-fat objects having width in between $[2^i, 2^{i+1})$. Since the width of objects in each layer is within a factor of two, for each i, the value of ζ_i is the same as ζ'. Since $|\mathcal{O}_i| \leq |\mathcal{O}'_i| \leq |\mathcal{O}|$, we have $|\mathcal{A}_i| \leq \zeta' |\mathcal{O}'_i| \leq \zeta' |\mathcal{O}|$. Then, we have $|\mathcal{A}| = \sum_{i=0}^{\lceil \log m \rceil} |\mathcal{A}_i| \leq \sum_{i=0}^{\lceil \log m \rceil} \zeta' |\mathcal{O}_i| = O(\zeta' \log m) |\mathcal{O}|$. Hence, the theorem follows.   □

4   Value of ζ for Families of Geometric Objects

Note that the value of ζ for unit disk graphs is already known to be 5 [13]. Here, we study the value of ζ for other geometric intersection graphs.

Theorem 6. *The independent kissing number for the family of*

(a) *congruent balls in $\mathbb{R}^3$ is 12;*
(b) *translated copies of a hypercube in $\mathbb{R}^d$ is 2^d, where $d \in \mathbb{Z}^+$;*
(c) *translated copies of an equilateral triangle is at least 5 and at most 6;*
(d) *translated copies of a regular k-gon $(k \geq 5)$ is at least 5 and at most 6;*
(e) *congruent hypercubes in $\mathbb{R}^d$ is at least 2^{d+1}, where $d \geq 2$ is an integer;*
(f) *bounded-scaled α-fat objects having a width in between $[1, m]$ is at least $\left(\frac{\alpha}{2} \left(\frac{m+2}{1+\epsilon} \right) \right)^d$ and at most $\left(\frac{m}{\alpha} + 2 \right)^d$, where $\epsilon > 0$ is a very small constant.*

For each item (except (e)) of the above theorem, we prove both the upper and lower bounds of the value of the independent kissing number.

Proof of Theorem 6(a).
First, we present a lower bound. Let C be a regular icosahedron whose each edge is of length $\ell = 2 + \epsilon$, where $0 < \epsilon < 1$ (in particular, one can choose $\epsilon = 0.001$). Let corner points of C be the centers of unit (radius) balls $\sigma_1, \sigma_2, \ldots, \sigma_{12}$. Since the edge length of the icosahedron C is greater than 2, all these balls are mutually non-touching. Let B be the circumscribed ball of the icosahedron C. It is a well-known fact that if a regular icosahedron has edge length ℓ, then the radius r of the circumscribed ball is $r = \ell \sin(\frac{2\pi}{5})$ [21]. In our case, it is easy to see that $r < 2$. In other words, the distance from the center of B to each corner point of C is less than two units. Thus, each of these unit balls $\sigma_1, \sigma_2, \ldots, \sigma_{12}$ is intersected by a unit ball σ_{13} whose center coincides with the center of B. This implies that the value of ζ for congruent balls in $\mathbb{R}^3$ is at least 12. The upper bound follows from the fact that the independent kissing number for balls in $\mathbb{R}^3$ is 12 [5,22]. Hence, the theorem follows.   □

Proof of Theorem 6(b).
Upper Bound Let K be an optimal independent kissing configuration for translates of an axis-parallel unit hypercube in $\mathbb{R}^d$. Let the core of the configuration be u. It is easy to observe that an axis-parallel hypercube R, with a side length of 2 units, contains all the centers of hypercubes in $K \setminus \{u\}$. Let us partition R into 2^d smaller axis-parallel hypercubes, each having unit side length. Note that each of these smaller hypercubes can contain at most one center of a hypercube in $K \setminus \{u\}$. As a result, we have $|K \setminus \{u\}| \leq 2^d$. Therefore, the independent kissing number for translates of a hypercube in $\mathbb{R}^d$ is at most 2^d.

Lower Bound We give an explicit construction of an independent kissing configuration K where the size of the independent set is 2^d. Let $\sigma_1, \sigma_2, \ldots, \sigma_{2^d}$ and σ_{2^d+1} be the d-dimensional axis-parallel unit hypercubes of K. We use c_i to denote the center of σ_i, for $i \in [2^d + 1]$. Let the center $c_{2^d+1} = (\frac{1}{2}, \frac{1}{2}, \ldots, \frac{1}{2})$, and $p_1, p_2, \ldots, p_{2^d} \in \mathbb{R}^d$ be corner points of the hypercube σ_{2^d+1}. It is easy to observe that each coordinate of $p_i, i \in [2^d]$ is either 0 or 1. Let ϵ be a positive constant satisfying $0 < \epsilon < \frac{1}{2\sqrt{d}}$. For $i \in [2^d]$ and $j \in [d]$, let us define the jth coordinates of c_i as follows:

$$c_i(x_j) = \begin{cases} -\epsilon, & \text{if } p_i(x_j) = 0 \\ 1+\epsilon, & \text{if } p_i(x_j) = 1, \end{cases} \tag{1}$$

where $c_i(x_j)$ and $p_i(x_j)$ are the jth coordinate value of c_i and p_i, respectively.

To complete the proof, here, we argue that the hypercubes $\sigma_1, \sigma_2, \ldots, \sigma_{2^d}$ are mutually non-touching, and each intersected by the hypercube σ_{2^d+1}. To see this, first, note that, for any $i \in [2^d]$, the Euclidean distance $d(p_i, c_i)$ between p_i and c_i is $\sqrt{d}\epsilon$ (follows from Eq. 1). Since $\epsilon < \frac{1}{2\sqrt{d}}$, we have $d(p_i, c_i) < \frac{1}{2}$. As a result, the corner point p_i of σ_{2^d+1} is contained in the hypercube σ_i. Thus, σ_{2^d+1} intersects $\sigma_i, \forall\ i \in [2^d]$. Now, consider any $i, j \in [2^d]$ such that $i \neq j$. Since p_i and p_j are distinct, it is easy to note that they will differ in at least one coordinate. As a result, the distance between c_i and c_j is $(1 + 2\epsilon)$ under L_∞-norm. So, σ_i and σ_j are non-touching. Hence, the theorem follows.   □

The proof of the remaining parts of Theorem 6 will appear in the final version of the paper.

Remark 2. Let ζ and ζ' be the independent kissing number for the family of bounded-scaled α-fat objects having a width in between $[1, m]$ and $[1, 2]$, respectively. Now, we have the following remarks.

(i) $\zeta' \leq \left(\frac{2}{\alpha} + 2\right)^d$: It follows from Theorem 6(f) by putting the value of $m = 2$.

(ii) $\zeta \geq \left(\frac{(m+2)\alpha^2}{4(1+\epsilon)(1+\alpha)}\right)^d \zeta'$: Due to Theorem 6(f), we have $\zeta \geq \left(\frac{\alpha}{2}\left(\frac{m+2}{1+\epsilon}\right)\right)^d$. Note

that $\zeta \geq \left(\frac{\alpha}{2}\left(\frac{m+2}{1+\epsilon}\right)\right)^d = \left(\frac{\alpha}{2}\left(\frac{m+2}{1+\epsilon}\right)\right)^d \left(\frac{1}{\frac{2}{\alpha}+2}\right)^d \left(\frac{2}{\alpha}+2\right)^d$. Now, using the

fact $\zeta' \leq \left(\frac{2}{\alpha} + 2\right)^d$ in the above expression, we have $\zeta \geq \left(\frac{(m+2)\alpha^2}{4(1+\epsilon)(1+\alpha)}\right)^d \zeta'$.

5   Applications

In this section, we mention some of the implications of Theorem 6. Combining Theorem 6 with Theorem 1, Theorem 2 and Lemma 5, respectively, we obtain the following results for the online MDS, MIDS and MC problems, respectively.

Theorem 7. *For each of the MDS, MIDS and MC problems, there exists a deterministic online algorithm that achieves a competitive ratio of*

(a) 12 for congruent balls in $\mathbb{R}^3$;
(b) 2^d for translated copies of a hypercube in $\mathbb{R}^d$, where $d \in \mathbb{Z}^+$;
(c) at most 6 for translated copies of a regular k-gon (for $k = 3$ and $k \geq 5$);
(d) at most $\left(\frac{m}{\alpha} + 2\right)^d$ for bounded-scaled α-fat objects having width in between $[1, m]$.

Similarly, combining Theorems 3 and 6, we have the following.

Theorem 8. *For the MCDS problems, there exists a deterministic online algorithm that achieves a competitive ratio of*

(a) 22 for congruent balls in $\mathbb{R}^3$;
(b) $2(2^d - 1)$ for translated copies of a hypercube in $\mathbb{R}^d$, where $d \in \mathbb{Z}^+$;
(c) at most 10 for translated copies of a regular k-gon (for $k = 3$ and $k \geq 5$);
(d) at most $2\left(\left(\frac{m}{\alpha} + 2\right)^d - 1\right)$ for bounded-scaled α-fat objects having width in $[1, m]$.

Now, we present the implication of Theorem 6 on the online t-relaxed coloring problem. For a nonnegative integer t, in the *online t-relaxed coloring* problem, upon the arrival of a new vertex, the algorithm must immediately assign a color to it, ensuring that the maximum degree of the subgraph induced by the vertices of this color class does not exceed t. The objective of the problem is to minimize the number of distinct colors. Combining the result of Capponi and Pilotto [7, Thm 5] with Theorem 6, we have the following.

Theorem 9. *For online t-relaxed coloring problem, there exists an online algorithm that achieves an asymptotic competitive ratio of*

(a) 288 for congruent balls in $\mathbb{R}^3$;
(b) 2^{2d+1} for translated copies of a hypercube in $\mathbb{R}^d$, where $d \in \mathbb{Z}^+$;
(c) at most 72 for translated copies of a regular k-gon (for $k = 3$ and $k \geq 5$);
(d) at most $2\left(\frac{m}{\alpha} + 2\right)^{2d}$ for bounded-scaled α-fat objects having width in between $[1, m]$.

6 Conclusion

We conclude by mentioning some open problems. The results obtained in this paper, as well as the results obtained in [7,20] for other graph problems, are dependent on the independent kissing number ζ. Consequently, the value of ζ becomes an intriguing graph parameter to investigate. For congruent balls in $\mathbb{R}^3$ and translates of a hypercube in $\mathbb{R}^d$, we prove that the value of ζ is tight and equals 12 and 2^d, respectively. In contrast, the value of ζ for translates of a regular k-gon (for $k \in ([5, \infty) \cup \{3\}) \cap \mathbb{Z}$) is either 5 or 6. We propose to settle the value ζ for this case as an open problem. For congruent hypercubes in $\mathbb{R}^d$, we prove that the value of ζ is at least 2^{d+1}; on the other hand, since congruent hypercubes are $\frac{1}{\sqrt{d}}$-fat objects, due to Theorem 6(f), it follows that ζ is at most $(2 + \sqrt{d})^d$. Bridging this gap would be an open question. It would be of independent interest to see parametrized algorithms for graphs considering ζ as a parameter.

Acknowledgments. The authors would like to thank anonymous reviewers for bringing to their attention articles [6,11,20] that they were unaware of.

References

1. Albers, S., Schraink, S.: Tight bounds for online coloring of basic graph classes. Algorithmica **83**(1), 337–360 (2021)
2. Borodin, A., El-Yaniv, R.: Online Computation and Competitive Analysis. Cambridge University Press (1998)
3. Boyar, J., Eidenbenz, S.J., Favrholdt, L.M., Kotrbčík, M., Larsen, K.S.: Online dominating set. Algorithmica **81**(5), 1938–1964 (2019)
4. Boyar, J., Favrholdt, L.M., Kotrbčík, M., Larsen, K.S.: Relaxing the irrevocability requirement for online graph algorithms. Algorithmica **84**(7), 1916–1951 (2022)
5. Brass, P., Moser, W.O.J., Pach, J.: Research Problems in Discrete Geometry. Springer (2005). https://doi.org/10.1007/0-387-29929-7
6. Butenko, S., Kahruman-Anderoglu, S., Ursulenko, O.: On connected domination in unit ball graphs. Optim. Lett. **5**(2), 195–205 (2011)
7. Capponi, A., Pilotto, C.: Bounded families for the on-line t-relaxed coloring. Inf. Process. Lett. **96**(4), 141–145 (2005)
8. Clark, B.N., Colbourn, C.J., Johnson, D.S.: Unit disk graphs. Discret. Math. **86**(1–3), 165–177 (1990)
9. De, M., Jain, S., Kallepalli, S.V., Singh, S.: Online geometric covering and piercing. CoRR, abs/2305.02445, (2023)
10. De, M., Lahiri, A.: Geometric dominating-set and set-cover via local-search. Comput. Geom. **113**, 102007 (2023)
11. Du, Y.L., Du, H.W.: A new bound on maximum independent set and minimum connected dominating set in unit disk graphs. J. Comb. Optim. **30**(4), 1173–1179 (2015)
12. Dumitrescu, A., Ghosh, A., Tóth, C.D.: Online unit covering in Euclidean space. Theor. Comput. Sci. **809**, 218–230 (2020)
13. Eidenbenz, S.: Online dominating set and variations on restricted graph classes, Technical Report No 380, ETH Library (2002)
14. Erlebach, T., Fiala, J.: On-line coloring of geometric intersection graphs. Comput. Geom. **23**(2), 243–255 (2002)
15. Erlebach, T., Fiala, J.: Independence and coloring problems on intersection graphs of disks. In: Bampis, E., Jansen, K., Kenyon, C. (eds.) Efficient Approximation and Online Algorithms. LNCS, vol. 3484, pp. 135–155. Springer, Heidelberg (2006). https://doi.org/10.1007/11671541_5
16. Garey, M.R., Johnson, D.S.: Computers and Intractability; A Guide to the Theory of NP-Completeness. W.H. Freeman & Co., USA (1990)
17. Gräf, A., Stumpf, M., Weißenfels, G.: On coloring unit disk graphs. Algorithmica **20**(3), 277–293 (1998)
18. King, G., Tzeng, W.: On-line algorithms for the dominating set problem. Inf. Process. Lett. **61**(1), 11–14 (1997)
19. Kobayashi, K.M.: Improved bounds for online dominating sets of trees. In Okamoto, Y., Tokuyama, T. (eds.) 28th International Symposium on Algorithms and Computation, ISAAC 2017, December 9–12, 2017, Phuket, Thailand, vol. 92 of LIPIcs, pp. 52:1–52:13. Schloss Dagstuhl - Leibniz-Zentrum für Informatik (2017)
20. Marathe, M.V., Breu, H., III, H.B.H., Ravi, S.S., Rosenkrantz, D.J.: Simple heuristics for unit disk graphs. Networks **25**(2), 59–68 (1995)

21. Noviyanti, D., Lestari, H.: The study of circumsphere and insphere of a regular polyhedron. In: Journal of Physics: Conference Series, vol. 1581–1, p. 012054. IOP Publishing (2020)
22. Schütte, K., van der Waerden, B.: Das problem der dreizehn kugeln. Mathematische Annalen **125**, 325–334 (1952)
23. Wan, P., Alzoubi, K.M., Frieder, O.: Distributed construction of connected dominating set in wireless ad hoc networks. Mob. Networks Appl. **9**(2), 141–149 (2004)

Near-Bipartiteness, Connected Near-Bipartiteness, Independent Feedback Vertex Set and Acyclic Vertex Cover on Graphs Having Small Dominating Sets

Maria Luíza L. da Cruz, Raquel S. F. Bravo, Rodolfo A. Oliveira, and Uéverton S. Souza(⊠)

Fluminense Federal University, Niterói, Brazil
{marialopez,rodolfooliveira}@id.uff.br, {raquel,ueverton}@ic.uff.br

Abstract. In the NEAR-BIPARTITENESS problem, we are given a simple graph $G = (V, E)$ and asked whether $V(G)$ can be partitioned into two sets $\mathcal{S}$ and $\mathcal{F}$ such that $\mathcal{S}$ is a stable set and $\mathcal{F}$ induces a forest. Alternatively, NEAR-BIPARTITENESS can be seen as the problem of determining whether G admits an independent feedback vertex set $\mathcal{S}$ or an acyclic vertex cover $\mathcal{F}$. Since such a problem is NP-complete even for graphs with diameter three, we first study the property of being near-bipartite on graphs having a dominating edge, a natural subclass of diameter-three graphs. Concerning graphs having a dominating edge, we present a polynomial-time algorithm for NEAR-BIPARTITENESS and prove that CONNECTED NEAR-BIPARTITENESS, the variant where the forest must be connected, is NP-complete. In addition, we show that INDEPENDENT FEEDBACK VERTEX SET, the problem of finding a near-bipartition $(\mathcal{S}, \mathcal{F})$ minimizing $|\mathcal{S}|$, and ACYCLIC VERTEX COVER, the problem of finding a near-bipartition $(\mathcal{S}, \mathcal{F})$ minimizing $|\mathcal{F}|$, are both NP-hard when restricted to such a class of graphs. Extending our polynomial-time approach to deal with NEAR-BIPARTITENESS on graphs having bounded dominating sets, we obtain a $O(n^2 \cdot m)$-time algorithm to solved NEAR-BIPARTITENESS on P_5-free graphs, improving the current $O(n^{16})$-time state of the art due to Bonamy, Dabrowski, Feghali, Johnson, and Paulusma [Algorithmica, 2019].

Keywords: Near-bipartite · independent feedback vertex set · acyclic vertex cover · stable set · dominating edge

1 Introduction

In 1972, Richard Karp presented the NP-completeness proof of 21 fundamental problems for Computer Science [18]. FEEDBACK VERTEX SET, INDEPENDENT SET and VERTEX COVER are three of these classical problems. FEEDBACK VERTEX SET consists of finding a minimum set of vertices such that its removal eliminates all cycles of the input graph, INDEPENDENT SET consists of determining a maximum set of pairwise nonadjacent vertices (also known as a stable set), and VERTEX COVER is the problem of determining a minimum set of vertices intersecting all edges (called vertex cover) of the input graph. Note that if S is a stable set of $G = (V, E)$ then $\mathcal{F} = V(G) \setminus S$ is a vertex cover of G.

An independent feedback vertex set (IFVS) of a graph G is a set of vertices that is independent/stable and also a feedback vertex set of G. Defined by Yang and Yuan in [25], a graph $G = (V, E)$ has a *near-bipartition* $(S, \mathcal{F})$ if there exist $S \subseteq V$ and $\mathcal{F} = V \setminus S$ such that S is a stable set, and $\mathcal{F}$ induces a forest (S and $\mathcal{F}$ may be empty sets). A graph that admits a near-bipartition is a *near-bipartite graph*. Note that the class of near-bipartite graphs is exactly the class of graphs having independent feedback vertex sets. Also, a graph G has an independent feedback vertex set S if and only if it has an acyclic vertex cover $\mathcal{F}$, i.e., a vertex cover $\mathcal{F}$ such that $G[\mathcal{F}]$ is acyclic (a vertex cover inducing a forest).

The problem of recognizing near-bipartite graphs, so-called NEAR-BIPARTITENESS, is NP-complete even when restricted to graphs with maximum degree four [25], graphs with diameter three [7], line graphs [8], and planar graphs [6,15]. On the other hand, Brandstädt et al. [10] showed that NEAR-BIPARTITENESS is polynomial-time solvable on cographs. Yang and Yuan [25] showed that NEAR-BIPARTITENESS is polynomial-time solvable for graphs of diameter at most two and that every connected graph of maximum degree at most three is near-bipartite except for the complete graph on four vertices (K_4). Also, Bravo, Oliveira, Silva Junior, and Souza [12] showed that near-bipartite P_4-tidy graphs admit finite forbidden induced subgraph characterization. Besides, Bonamy et al. [8] proved that NEAR-BIPARTITENESS on P_5-free graphs can be solved in $O(n^{16})$ time. FPT algorithms parameterized by k for finding an independent feedback vertex set of size at most k can be found in [2,19,22].

A *coloring* for a graph G is an assignment of colors (labels) to all vertices of G. A *proper coloring* for G is an assignment of color $c(u)$, for each vertex $u \in V$, such that $c(u) \neq c(v)$ if $uv \in E(G)$. A graph G is k-colorable if there exists a proper coloring for G with at most k colors. The *chromatic number* of G, $\chi(G)$, is the smallest number k for G being k-colorable. A clear necessary condition for a graph to be near-bipartite is the following.

Proposition 1. *If a graph G is near-bipartite then G is 3-colorable.*

However, the complexity of 3-COLORING and NEAR-BIPARTITENESS are not necessarily the same, depending on the graph class being explored. Grötschel, Lovász and Schrijver [17] proved that COLORING is solved in polynomial time for perfect graphs, while Brandstädt et al. [10] proved that NEAR-BIPARTITENESS is NP-complete in the same graph class. NEAR-BIPARTITENESS can also be seen as a variant of 2-COLORING. For an input graph G, the question is whether its vertex set can be colored with two colors (not necessarily properly coloring) such that one color class is K_2-free (a stable set), and the other is cycle-free (i.e., induces a forest). Other 2-COLORING variants have already received attention in the literature. In [1], Achlioptas studied the problem of determining if there exists a bipartition of $V(G)$ where each part (color class) is H-free for some fixed graph H. He showed that for any graph H on more than two vertices, the problem is NP-complete. Another variant was considered by Schaefer [24], who asked whether a given graph G admits a 2-coloring of the vertices such that each vertex has exactly one neighbor with the same color as itself. Schaefer proved that such a problem is NP-complete even for planar cubic graphs. The problem studied by Schaefer [24] is a particular case of a defective coloring called $(2,1)$-coloring. A (k,d)-coloring of a graph G is a k-coloring of $V(G)$ such that each vertex has at most d neighbors with the same color. Some studies on $(2,1)$-coloring include [9,14,20]. In addition, the problem of finding a bipartition where each part induces a subgraph of minimum degree at least k (for a given integer k) was studied in [5]. Also, the problem of partitioning the edge set of a graph into a stable set of edges (matching) and a forest has been studied in [21,23].

Motivated by the studies of 2-coloring variants and the natural relevance of feedback vertex sets that are independent/stable as well as vertex covers that are acyclic, we focus on the NEAR-BIPARTITENESS problem and its variants.

Since a near-bipartition $(\mathcal{S},\mathcal{F})$ of a graph G is a partition of $V(G)$ into a stable set $\mathcal{S}$ and an induced forest $\mathcal{F}$, we consider the following problems.

NEAR-BIPARTITENESS

Instance: A simple undirected graph $G = (V,E)$.
Question: Does G have a near-bipartition $(\mathcal{S},\mathcal{F})$?

INDEPENDENT FEEDBACK VERTEX SET

Instance: A simple undirected graph $G = (V,E)$.
Goal: Find (if any) a minimum independent feedback vertex set of G, i.e., a near-bipartition $(\mathcal{S},\mathcal{F})$ that minimizes $|\mathcal{S}|$.

ACYCLIC VERTEX COVER

Instance: A simple undirected graph $G = (V,E)$.
Goal: Find (if any) a minimum acyclic vertex cover of G, i.e., a near-bipartition $(\mathcal{S},\mathcal{F})$ of G that minimizes the size of $\mathcal{F}$.

Recall that the complement of an acyclic vertex cover is an independent feedback vertex set. So, the reader can assume that we are also dealing with the maximization version of both problems. Besides, we consider the problem of determining whether a graph G can have its set of vertices partitioned into a stable set and a *tree*, called CONNECTED NEAR-BIPARTITENESS, which was shown to be NP-complete even on bipartite graphs of maximum degree four [11].

CONNECTED NEAR-BIPARTITENESS

Instance: A simple undirected graph $G = (V,E)$.
Question: Does G have a near-bipartition $(\mathcal{S},\mathcal{F})$ such that $G[\mathcal{F}]$ is connected?

Motivated by the fact that NEAR-BIPARTITENESS remains NP-complete on graphs with diameter three [7], we first analyse the problem on graphs having a dominating edge, a natural subclass of graphs with diameter 3. In such a case, we show that NEAR-BIPARTITENESS can be solved in polynomial time, but CONNECTED NEAR-BIPARTITENESS is NP-complete. We also prove the NP-hardness of finding a *minimum* independent feedback vertex set or a *minimum* acyclic vertex cover on graphs having a dominating edge. Finally, we present a $O(n^2 \cdot m)$-time algorithm to solved NEAR-BIPARTITENESS on P_5-free graphs, improving the current $O(n^{16})$-time state of the art [8].

2   On Graphs Having a Dominating Edge

Recall that the class of graphs having a dominating edge is a natural subclass of graphs with diameter three, a class for which NEAR-BIPARTITENESS remains NP-complete [7]. In this section, we consider the problem of partitioning a graph having a dominant edge into a stable set and a *tree* (CONNECTED NEAR-BIPARTITENESS), as well as the problem of partitioning it into a stable set and a forest (NEAR-BIPARTITENESS).

Next, we show that CONNECTED NEAR-BIPARTITENESS is NP-complete on graphs having a dominating edge, while NEAR-BIPARTITENESS becomes polynomial-time solvable in the same class.

Theorem 1. CONNECTED NEAR-BIPARTITENESS *is NP-complete even when restricted to graphs having a dominating edge.*

Proof. The proof is based on a reduction from 1-IN-3SAT, a well-known NP-complete problem [16]. In such a problem we are given a formula φ in conjunctive normal form where each clause is limited to at most three literals, and asked whether there exists a satisfying assignment so that exactly one literal in each clause is set to true.

Given an instance φ of 1-IN-3SAT, we construct a graph G such that φ has a truth assignment such that each clause has exactly one literal set to true if and only if G is partitionable into a stable set and a tree.

From φ we construct G as follows:

1. first consider $G = (\{u, v\}, \{uv\})$;
2. add a chordless cycle C of size 4 in G induced by $\{k_1, k_2, k_3, k_4\}$, and add edges from u for all vertices in C;
3. add a chordless cycle $C' = l_1, m, l_2, n_1, n_2$;
4. add the edges ul_1, ul_2, vm, vn_1 and vn_2;

At this point, notice that every $(\mathcal{S}, \mathcal{T})$-partition of G has $v \in \mathcal{S}$ and $u \in \mathcal{T}$.

5. for each variable x_i of φ create vertices v_{x_i} and $v_{\overline{x}_i}$ and add edges $v_{x_i} v_{\overline{x}_i}$, uv_{x_i} and $uv_{\overline{x}_i}$;
6. for each clause C_j of φ create a vertex c_j in G and add the edge vc_j;
7. Finally, add an edge $c_j v_{x_i}$ if the clause C_j contains the literal x_i, and add an edge $c_j v_{\overline{x}_i}$ if the clause C_j contains the literal $\overline{x}_i$.

If φ is a 3-CNF formula having a truth assignment A such that each clause has exactly one literal set as true, then we can construct an $(\mathcal{S}, \mathcal{T})$-partition of G by setting $\mathcal{S} = \{k_1, k_3, l_1, v\} \cup \{v_{x_i} : x_i = false \in A\} \cup \{v_{\overline{x}_i} : x_i = true \in A\}$ (clearly $\mathcal{S}$ is a stable set). Since A defines a 1-in-3 truth assignment then each vertex c_j has exactly one neighbor in $G[V \setminus \mathcal{S}]$ then $\mathcal{T} = V \setminus \mathcal{S}$ induces a tree.

Conversely, if G admits an $(\mathcal{S}, \mathcal{T})$-partition then, by construction, it holds that $v \in \mathcal{S}$ and $u \in \mathcal{T}$. This implies that every vertex c_j belongs to $\mathcal{T}$, and that for each pair $v_{x_i}, v_{\overline{x}_i}$ exactly one of these vertices belongs to $\mathcal{T}$. Also, since $\mathcal{T}$ is connected each c_j has at least one neighbor in $\mathcal{T}$, thus as $\mathcal{T}$ is acyclic each vertex c_j has exactly one neighbor in $\mathcal{T}$ (each c_j must be a leaf in $\mathcal{T}$). Therefore, we can construct a 1-in-3 truth assignment by setting $x_i = true$ iff $v_{x_i} \in \mathcal{T}$. $\square$

Contrasting with Theorem 1, we show that when we remove the connectivity constraint, i.e., we look for a forest instead of a tree, the problem becomes polynomial-time solvable.

Theorem 2. *Given a graph G and a dominating edge of G, one can determine in $O(n^2)$ time whether G is a near-bipartite graph.*

Proof. Let $u, v \in V(G)$ be two vertices of G such that uv is a dominant edge of G. Suppose that G has a near-bipartition $(\mathcal{S}, \mathcal{F})$. Without loss of generality, we may assume that G does not have vertices with degree one. At this point, we may consider just two cases:

Case 1. Suppose that $u, v \in \mathcal{F}$.

As $uv \in E(F)$, then $N(u) \cap N(v) \subseteq \mathcal{S}$, otherwise $\mathcal{F}$ has cycles. Thus, $N(u) \cap N(v)$ must be a stable set. For a remaining vertex w belonging to either $N(u) \setminus N(v)$ or $N(v) \setminus N(u)$: if it has a neighbor in $\mathcal{S}$ then it must belong to $\mathcal{F}$; if it has a neighbor z ($z \neq u$ and $z \neq v$) that must be in $\mathcal{F}$, then w must belong to $\mathcal{S}$, otherwise, the edge wz together with uv induces a cycle in $\mathcal{F}$. Thus, by checking if $N(u) \cap N(v)$ is stable and then successively applying the operations previously described according to a Breadth-First Search from $N(u) \cap N(v)$, in linear time, we can either conclude that such a near-bipartition with $u, v \in \mathcal{F}$ does not exist, or build a partition (S', F', U) of $V(G)$ such that S' is stable, $F' \supseteq \{u, v\}$ induces a forest, and U is the set of unclassified vertices. Note that, by construction, no vertex in U has neighbors in $S' \cup F' \setminus \{u, v\}$. Since any pair of adjacent vertices together with u and v induces a cycle, G has a near-bipartition $(\mathcal{S}, \mathcal{F})$ with $\{u, v\} \subseteq \mathcal{F}$ if and only if $G[U]$ has an independent vertex cover, which is equivalent to U inducing a bipartite graph.

Case 2. Suppose that $u \in \mathcal{S}$ and $v \in \mathcal{F}$.

If $u \in \mathcal{S}$ and $v \in \mathcal{F}$ then $N(u) \subseteq \mathcal{F}$. Thus, $N(u)$ must induce a forest and $N(u) \cap N(v)$ must be a stable set. At this point, only the vertices belonging to $N(v) \setminus N[u]$ are unclassified.

Let $B = N(v) \setminus \{u\}$.

If G has a near-bipartition $(\mathcal{S}, \mathcal{F})$ then $G[B]$ must be bipartite, so that its vertices can be partitioned into two sets (B_1, B_2) such that $B_1 \subseteq \mathcal{S}$ and $B_2 \subseteq \mathcal{F}$. Thus, we must find a bipartition of B that satisfies the following conditions:

- $N(u) \cap N(v) \subseteq B_2$;
- for each component T of $G[N(u) \setminus N[v]]$ (which is a tree) it holds that:
  - For each $w \in B_2$, $|N_T(w)| \leq 1$ (otherwise $\{w\} \cup V(T)$ induces a cycle);
  - T has at most one neighbor in B_2 (otherwise $\mathcal{F}$ has cycles).

Note that any bipartition (B_1, B_2) satisfying the above restrictions is sufficient to form a near-bipartition such that $B_1 \cup \{u\} = \mathcal{S}$. Now, we can reduce the
problem of finding such a bipartition of $G[B]$ to the 2SAT problem by building
a 2-CNF formula φ as follows:

1. for each vertex $w \in B$ create a variable x_w;
2. for each vertex $w \in N(u) \cap N(v)$ create a clause (x_w);
3. for each edge $w_1 w_2 \in E(G[B])$ create the clauses $(x_{w_1} + x_{w_2})$ and $(\overline{x}_{w_1} + \overline{x}_{w_2})$;
4. for each vertex $w \in B$ with at least two neighbors in the same component T
   of $G[N(u) \setminus N[v]]$, create a clause $(\overline{x}_w)$;
5. For each component T of $G[N(u) \setminus N[v]]$, and for each pair of vertices w_1, w_2
   in the neighborhood of T, create a clause $(\overline{x}_{w_1} + \overline{x}_{w_2})$.

At this point, it is easy to see that φ is satisfied if and only if $G[B]$ has a
partition (B_1, B_2) as requested (variables equal to true correspond to the vertices
of B_2). Since φ can be built in $O(n^2)$ time with respect to the size of $G[B]$ and
2SAT can be solved in linear time [3], a near-bipartition $(\mathcal{S}, \mathcal{F})$ of G can be
found in $O(n^2)$ time (if any). $\qquad\square$

Recall that NEAR-BIPARTITENESS can be seen as the problem of determining
whether G admits an independent feedback vertex set $\mathcal{S}$ or an acyclic vertex cover
$\mathcal{F}$. In contrast to the previous theorem, we show that the problems of finding a
minimum independent feedback vertex set and a minimum acyclic vertex cover
are both NP-complete on graphs with a dominating edge.

Theorem 3. INDEPENDENT FEEDBACK VERTEX SET *is NP-hard when
restricted to graphs having a dominating edge.*

Proof. In POSITIVE MIN-ONES-2SAT we are given a 2SAT formula φ having
only positive literals and asked to decide whether there exists a satisfying assignment for φ with at most k variables set to true. Note that POSITIVE MIN-ONES-
2SAT is equivalent to MINIMUM VERTEX COVER, a well-known NP-complete
problem.

Given an instance φ of POSITIVE MIN-ONES-2SAT, we can construct a graph
G by applying the same construction as in Theorem 1 (disregarding negative
literals). At this point, variables x_i set to *true* are equivalent to the vertices v_{x_i}
assigned to $\mathcal{S}$. Therefore, φ has a satisfying truth assignment with at most k
trues if and only if G is partitionable into a stable set $\mathcal{S}$ and a forest $\mathcal{F}$ such
that $|\mathcal{S}| \leq k + 4$. $\qquad\square$

Theorem 4. ACYCLIC VERTEX COVER *is NP-hard when restricted to graphs
having a dominating edge.*

Proof. First, we define a construction algorithm f that receives as input a CNF formula φ and outputs a graph having a dominating edge, similar to that presented in Theorem 1. The first six steps of this construction are the same as in Theorem 1. (to make it easier for the reader, we repeat their description here).

From φ we construct G as follows:

1. first consider $G = (\{u, v\}, \{uv\})$;
2. add a cycle C of size 4 in G induced by $\{k_1, k_2, k_3, k_4\}$, and add edges from u for all vertices in C;
3. add a cycle $C' = l_1, m, l_2, n_1, n_2$;
4. add the edges ul_1, ul_2, vm, vn_1 and vn_2;
5. for each variable x_i of φ create vertices v_{x_i} and $v_{\overline{x}_i}$ and add edges $v_{x_i}v_{\overline{x}_i}$, uv_{x_i} and $uv_{\overline{x}_i}$;
6. for each clause C_j of φ create a vertex c_j in G and add the edge vc_j;

Now, in Step 7, we present the necessary change for our reduction.

7. add an edge $c_j v_{\overline{x}_i}$ if the clause C_j contains the literal x_i, and add an edge $c_j v_{x_i}$ if the clause C_j contains the literal $\overline{x}_i$; (Note that an edge between c_j and $v_{\overline{x}_i}$ is added when C_j contains the literal x_i)
8. Finally, for each vertex v_{x_i} representing a positive literal, we replace it with n copies $v_{x_1}^1, v_{x_1}^2, \ldots, v_{x_1}^n$, having the same neighbors as v_{x_1}.

Now, given an instance φ with n variables and m clauses of POSITIVE MIN-ONES-2SAT, we denote by $G = f(\varphi)$ the graph obtained by applying f from φ.

Suppose that φ has a satisfying assignment A with weight k. From A we construct a near-bipartition of G as follows: for each variable x_i of φ, if x_i equals *true* in A, then the vertices associated with the literal x_i are in $\mathcal{F}$, and if x_i equals *false* in A, then the vertex associated with the literal $\overline{x}_i$ is in $\mathcal{F}$. So, we can construct an near-bipartition $(\mathcal{S}, \mathcal{F})$ of G by setting

$$
\begin{aligned}
\mathcal{F} = \{ & \{k_2, k_4, l_2, m, n_1, n_2, u\} \\
& \cup \{v_{x_i}^1 \ldots v_{x_i}^n : \text{``}x_i = true\text{''} \in A\} \\
& \cup \{v_{\overline{x}_i} : \text{``}x_i = false\text{''} \in A\} \\
& \cup \{c_1, c_2, \ldots, c_m\}\}, \\
\mathcal{S} = & V \setminus \mathcal{F}.
\end{aligned}
$$

Note that $|\mathcal{F}| = m + kn + (n-k) + 7$, where m is the number of clause vertices, kn is the number of vertices that represent positive literals ("$x_i = true$" in A), $n - k$ is the number of vertices that represent negative literals ("$x_i = false$" in A), and 7 is the number of auxiliary vertices (k_2, k_4, l_2, m, n_1, n_2, u) in $\mathcal{F}$.

Now, let's analyze the graph induced by the vertices in $\mathcal{F}$. Every clause vertex is in $\mathcal{F}$. Also, if a clause vertex has degree two in the graph induced by $\mathcal{F}$, by construction, in the instance φ of POSITIVE MIN-ONES-2SAT, the corresponding clause has 2 false literals, which contradicts the fact that A is an assignment that satisfies φ. So, the clause vertices have degree at most 1, and do not belong to any cycle. At this point, it is easy to see that $\mathcal{F}$ induces a forest with $m + 7 + k \cdot n + (n - k)$ vertices, and $\mathcal{S} = V \setminus \mathcal{F}$ is a stable set.

Conversely, suppose that G has a near-bipartition $(\mathcal{S}, \mathcal{F})$ where $|\mathcal{F}| = m+7+k\cdot n+(n-k)$. First, let's argue that at most k gadgets g_{x_i} have vertices associated with true (x_i) in $\mathcal{F}$. By construction, each gadget g_{x_i} contains either one vertex representing $\overline{x}_i$ in $\mathcal{F}$ or n vertices representing x_i in $\mathcal{F}$. So let's suppose that $k + 1$ gadgets have $n \cdot (k + 1)$ vertices associated with true in $\mathcal{F}$. So this implies that $\mathcal{F}$ has at least $m + 7 + (k + 1) \cdot n$ vertices, which is bigger than the initially defined budget, $|\mathcal{F}| = m + 7 + k \cdot n + (n - k)$. So, at most k gadgets g_{x_i} have vertices associated with true in $\mathcal{F}$. With this, we will construct an assignment A to φ as follows:

- for each gadget g_{x_i} set $x_i = false$ if the vertex associated with $\overline{x}_i$ (i.e., $v_{\overline{x}_i}$), belongs to $\mathcal{F}$ and set $x_i = true$ otherwise.

Note that A has weight at most k. By the construction of G, each vertex associated with the clause has degree at most 1 in $\mathcal{F}$, given the property of the forest to be acyclic. So, for each clause at least one of its literals is $true$. Therefore, A satisfies φ with weight at most k. □

3   On P_5-Free Graphs

In 2019, Bonamy, Dabrowski, Feghali, Johnson, and Paulusma showed that NEAR-BIPARTITENESS and INDEPENDENT FEEDBACK VERTEX SET can be solved in $O(n^{16})$ time. In 1990, Bacsó and Tuza [4] showed that any connected P_5-free graph has a dominating clique or a dominating P_3. In 2016, Camby and Schaudt [13] generalized this result and showed that such a dominating set can be computed in polynomial time.

In this section, using the same approach presented in Theorem 2, we show how to handle NEAR-BIPARTITENESS on graphs having a dominating clique or a dominating P_3. Our results imply a faster algorithm to solve NEAR-BIPARTITENESS on P_5-free graphs in time $O(n^4)$. Interestingly, we can observe that the same technique combined with Bacsó and Tuza's result is not very useful to get a more efficient algorithm for INDEPENDENT FEEDBACK VERTEX SET on P_5-free graphs, due our Theorem 3 showing that this problem remains NP-complete on graphs having a dominating edge.

Theorem 5. *Given a graph G and a dominating triangle of G, one can determine in $O(n^2)$ time whether G is a near-bipartite graph.*

Proof. Let $\{u, v, z\} \in V(G)$ be a dominating set of G that induces a triangle. Suppose that G has a near-bipartition $(\mathcal{S}, \mathcal{F})$. Assume that $z \in \mathcal{S}$ and $u, v \subseteq \mathcal{F}$.

Since $z \in \mathcal{S}$, then $N(z) \subseteq F$ and must induce a forest. Observe that $N(u) \cap N(v) \cap N(z) = \emptyset$, because near-bipartite graphs have no K_4 (see Proposition 1).

At this point, only the vertices belonging to $N(v) \cup N(u) \setminus N[z]$ are unclassified. Let $B = N(v) \cup N(u) \setminus \{u, v, z\}$. Note that $G[B]$ must be bipartite. So, we have that the vertices of $G[B]$ can be partitioned into two sets (B_1, B_2) such that $B_1 \subseteq S$ and $B_2 \subseteq F$. Thus, we must find a bipartition of B that satisfies the following conditions:

- $N(u) \cap N(v) \subseteq B_1$, otherwise F will have cycles.
- $N(z) \cap (N(u) \cup N(v)) \subseteq B_2$, otherwise S will have edges.
- For each component T of $G[N(z) \cap N[u] \cup N[v]]$, which is a tree, holds that:
  - For each $w \in B_2$, $|N_T(w)| \leq 1$ (otherwise $\{w\} \cup V(T)$ induces a cycle);
  - T has at most one neighbor in B_2 (otherwise F has cycles);

Note that any bipartition (B_1, B_2) satisfying the above restrictions is sufficient to form a near-bipartition such that $B_1 \cup \{z\} = S$. At this point, we can reduce the problem of finding such a bipartition of $G[B]$ to the 2SAT problem, similarly as in Theorem 2, this concludes the proof. $\qquad\square$

Theorem 6. *Given a graph G and a dominating induced P_3 of G, one can determine in $O(m \cdot n^2)$ time whether G is a near-bipartite graph.*

Proof. Let uvz be an induced P_3 of G such that $\{u, v, z\}$ is a dominating set in G. Suppose that G has a near-bipartition (S, F). At this point, we may analyze four cases:

Case 1. Suppose that the vertices $\{u, v\} \in F$ and the vertex $\{z\} \in S$.

In this case the proof is similar to that of Theorem 5. The same holds if $\{v, z\} \in F$ and $\{u\} \in S$.

Case 2: Suppose that only the vertex $v \in F$ and the vertices $\{u, z\} \in S$.

In this case, $N(u) \cup N(z) \subseteq F$, otherwise S has edges. Thus, $N(u) \cup N(z)$ must induce a forest. Also, $G[N(u) \cup N(z)]$ must contain no path between two vertices of $N(v) \cap (N(u) \cup N(z))$, otherwise F has cycles.

Furthermore, $N(v) \setminus (N(u) \cup N(z))$ must induce a bipartite graph. Also, for a vertex w belonging to $N(v) \setminus (N(u) \cup N(z))$ if it has a neighbor $p \in N(u) \cup N(z)$ such that $p \neq v$ and it reaches v in $G[N(u) \cup N(z)]$ then p must be in S. At this point, similarly as in Theorem 2, we can use a 2SAT formula to decide which unclassified vertices of $N(v) \setminus (N(u) \cup N(z))$ must be in F and S. It is not hard to see that this case can be performed with linear searches in addition to the construction and resolution of a 2SAT instance in $O(n^2)$ time.

Case 3: Suppose that $\{u, v, z\} \subseteq F$.

In this case, any vertex with at least two neighbors in $\{u, v, z\}$ must be in S. Note that

$$(N(v) \setminus (N(u) \cup N(z))) \cup (N(z) \setminus (N(u) \cup N(v))) \cup (N(u) \setminus (N(v) \cup N(z)))$$

must induce a bipartite graph B.

Furthermore, for a remaining vertex w belonging to B: If it has a neighbor that must be in S then it must belong to F; On the other hand, if w has a neighbor $p \notin \{u, v, z\}$ that must be in $\mathcal{F}$, then w must belong to $\mathcal{S}$, otherwise, there is a cycle in $\mathcal{F}$.

Thus, by checking if $(N(v) \cap N(z)) \cup (N(u) \cap N(v)) \cup (N(u) \cap N(z))$ is a stable set and then successively applying the classification process previously described (a 2-coloring into $\mathcal{S}$ and $\mathcal{F}$) from $(N(v) \cap N(z)) \cup (N(u) \cap N(v)) \cup (N(u) \cap N(z))$, in linear time, we can either conclude that such a near-bipartition of G with $\{v, u, z\} \in \mathcal{F}$ does not exist, or we construct a partition (S', F', U) of $V(G)$ such that S' is stable, $F' \supseteq \{v, u, z\}$ induces a forest, and U is the set of unclassified vertices. Note that, by construction, no vertex in U has neighbors in $S' \cup F' \setminus \{v, u, z\}$.

Since any pair of adjacent vertices together with u, v and z induces a cycle, G has a near-bipartition $(\mathcal{S}, \mathcal{F})$ with $\{v, u, z\} \subseteq \mathcal{F}$ if and only if $G[U]$ has an independent vertex cover, which is equivalent to U inducing a bipartite graph, which can also be checked in linear time.

Next, we discuss the most intriguing case where the vertices of the dominating set that must be in the forest do not induce a connected component.

Case 4: Suppose that $\{u, z\} \subseteq \mathcal{F}$ and $v \in \mathcal{S}$.
Recall that $N(v) \subseteq \mathcal{F}$.

In the following, we consider two cases to be analyzed, either u and z are in the same tree of $\mathcal{F}$ or they are in distinct trees.

Case A - u and z are in the same tree of $\mathcal{F}$. Thus, there is a path $\mathcal{P}$ between them in $\mathcal{F}$. Such a path contains exactly one neighbor of u and exactly one neighbor of z; otherwise, $V(\mathcal{P})$ induces a cycle in $\mathcal{F}$.

Therefore, we enumerate each pair a_u, a_z ($a_u = a_z$ is allowed) such that $a_u \neq v$ and it is neighbor of u, $a_z \neq v$ and it is neighbor of z, and $\{a_u, a_z\} \cup N(v)$ induces a forest having a tree containing u and z.

Observe that we have $O(n^2)$ pairs a_u, a_z, and in $O(m)$ time we can check if $\{a_u, a_z\} \cup N(v)$ induces a forest having a tree containing u and z.

Now, for each enumerated pair a_u, a_z, we check if there is a near-bipartition having $\{a_u, a_z\} \cup N(v) \subseteq \mathcal{F}$ as follows.

- $(N(u) \cup N(z)) \setminus (\{a_u, a_z\} \cup N[v])$ must induce a bipartite graph B, otherwise there is no near-bipartition having $\{a_u, a_z\} \cup N(v) \subseteq \mathcal{F}$.
  Thus, it is enough to decide if B has a bipartition $V(B) = B_1 \cup B_2$ satisfying the following:
  Let T be the tree of $G[\{a_u, a_z\} \cup N(v)]$ containing u and z.
  - Each vertex of B having at least two neighbors in a tree of $G[\{a_u, a_z\} \cup N(v)]$ must be in B_1.
  - Each tree of $G[\{a_u, a_z\} \cup N(v)]$ distinct from T has at most one neighbor in B_2.

If B has such a bipartition then $B_1 \cup \{v\} = \mathcal{S}$ and $B_2 \cup \{a_u, a_z\} \cup N(v) = \mathcal{F}$ form a near-bipartition of G. Again, such a bipartition, if any, can be found using a 2SAT formula.

The overall running time for this case is $O(n^4)$, because we consider $O(n^2)$ pairs and for each one the described procedure can be performed in $O(n^2)$ time.

Case B - u and z are disconnected in $\mathcal{F}$.

In this situation there is no path between vertices u and z in $\mathcal{F}$.

Therefore, $N(u) \cap N(z) \subseteq \mathcal{S}$, $G[N(u) \cup N(z)]$ is bipartite, and the vertices of $(N(u) \cup N(z)) \cap V(\mathcal{F})$ is a stable set, otherwise the vertices u and z are connected in $\mathcal{F}$. Besides that, $N(v) \subseteq \mathcal{F}$.

At this point, analogously to the previous cases, we can use 2SAT to find an appropriated classification of the vertices of $N(z) \cup N(u)$ into $\mathcal{S}$ and $\mathcal{F}$ (if any).

Since all the cases are performed in $O(n^4)$ time, we conclude the proof. $\square$

Next, we improve the Bonamy, Dabrowski, Feghali, Johnson, and Paulusma's result [8] concerning NEAR-BIPARTITENESS on P_5-free graphs.

Corollary 1. NEAR-BIPARTITENESS *on* P_5-*free graphs can be solved in* $O(n^2 \cdot m)$ *time.*

Proof. Near-bipartite graphs are K_4-free and K_4's can be found in $O(m^2)$ time. Also, near-bipartite P_5-free graphs have either a dominating triangle or a dominating P_3 due to Bacsó and Tuza's result [4]. Hence, it is enough to apply Theorem 5 and Theorem 6. Since G is P_5-free then Case 4A of Theorem 6 can be performed in $O(n^2 \cdot m)$ time, since either $a_u = a_z$ or $a_u a_z$ is an edge of G. $\square$

In the light of previous demonstrations, the reader may be realizing that we are able to extend our approach to deal with NEAR-BIPARTITENESS parameterized the domination number. Actually, we can proceed as follows.

Theorem 7. *Given a graph G and a dominating set D of G with size k, one can determine whether G is near-bipartite in $O(2^k \cdot n^{2k})$ time.*

Proof. Given G and D we can "guess" the vertices of D in $\mathcal{S}$ and in $\mathcal{F}$ in $O(2^k)$ time. For the vertices of $D \cap V(\mathcal{F})$ there are at most $2k - 2$ neighbors used to connect some of them in the forest $\mathcal{F}$ (at most one pair a_u, a_z for a pair $u, z \in D$). We can "guess" such a vertices in $O(n^{2k-2})$ time. Then we can proceed in $O(n^2)$ time using 2SAT as in the previous results. $\square$

References

1. Achlioptas, D.: The complexity of G-free colourability. Discret. Math. **165–166**, 21–30 (1997)
2. Agrawal, A., Gupta, S., Saurabh, S., Sharma, R.: Improved algorithms and combinatorial bounds for independent feedback vertex set. In: 11th International Symposium on Parameterized and Exact Computation (IPEC 2016). Schloss Dagstuhl-Leibniz-Zentrum fuer Informatik (2017)
3. Aspvall, B., Plass, M.F., Tarjan, R.E.: A linear-time algorithm for testing the truth of certain quantified Boolean formulas. Inf. Process. Lett. **14**(4) (1982)

4. Bacsó, G., Tuza, Z.: Dominating cliques in P_5-free graphs. Period. Math. Hung. **21**(4), 303–308 (1990)
5. Bang-Jensen, J., Bessy, S.: Degree-constrained 2-partitions of graphs. Theoret. Comput. Sci. **776**, 64–74 (2019)
6. Bonamy, M., Dabrowski, K.K., Feghali, C., Johnson, M., Paulusma, D.: Recognizing graphs close to bipartite graphs. In: 42nd International Symposium on Mathematical Foundations of Computer Science (MFCS 2017). Schloss Dagstuhl-Leibniz-Zentrum fuer Informatik (2017)
7. Bonamy, M., Dabrowski, K.K., Feghali, C., Johnson, M., Paulusma, D.: Independent feedback vertex sets for graphs of bounded diameter. Inf. Process. Lett. **131**, 26–32 (2018)
8. Bonamy, M., Dabrowski, K.K., Feghali, C., Johnson, M., Paulusma, D.: Independent feedback vertex set for P_5-free graphs. Algorithmica **81**(4), 1342–1369 (2019)
9. Borodin, O., Kostochka, A., Yancey, M.: On 1-improper 2-coloring of sparse graphs. Discret. Math. **313**(22), 2638–2649 (2013)
10. Brandstädt, A., Brito, S., Klein, S., Nogueira, L.T., Protti, F.: Cycle transversals in perfect graphs and cographs. Theoret. Comput. Sci. **469**, 15–23 (2013)
11. Brandstädt, A., Le, V.B., Szymczak, T.: The complexity of some problems related to graph 3-colorability. Discret. Appl. Math. **89**(1), 59–73 (1998)
12. Bravo, R., Oliveira, R., da Silva, F., Souza, U.S.: Partitioning p4-tidy graphs into a stable set and a forest. Discret. Appl. Math. **338**, 22–29 (2023)
13. Camby, E., Schaudt, O.: A new characterization of p_k-free graphs. Algorithmica **75**(1), 205–217 (2016)
14. Cowen, L., Goddard, W., Jesurum, C.E.: Defective coloring revisited. J. Graph Theory **24**(3), 205–219 (1997)
15. Dross, F., Montassier, M., Pinlou, A.: Partitioning a triangle-free planar graph into a forest and a forest of bounded degree. Eur. J. Comb. **66**, 81–94 (2017)
16. Garey, M.R., Johnson, D.S.: Computers and Intractability, vol. 174. Freeman, San Francisco (1979)
17. Grötschel, M., Lovász, L., Schrijver, A.: Polynomial algorithms for perfect graphs. Ann. Discret. Math. **21**, 325–356 (1984)
18. Karp, R.M.: Reducibility among combinatorial problems. In: Miller, R.E., Thatcher, J.W., Bohlinger, J.D. (eds.) Complexity of Computer Computations. The IBM Research Symposia Series, pp. 85–103. Springer, Boston (1972). https://doi.org/10.1007/978-1-4684-2001-2_9
19. Li, S., Pilipczuk, M.: An improved FPT algorithm for independent feedback vertex set. Theory Comput. Syst. **64**(8), 1317–1330 (2020)
20. Lima, C.V., Rautenbach, D., Souza, U.S., Szwarcfiter, J.L.: On the computational complexity of the bipartizing matching problem. Ann. Oper. Res. (2021)
21. Lima, C.V., Rautenbach, D., Souza, U.S., Szwarcfiter, J.L.: Decycling with a matching. Infor. Proc. Lett. **124**, 26–29 (2017)
22. Misra, N., Philip, G., Raman, V., Saurabh, S.: On parameterized independent feedback vertex set. Theoret. Comput. Sci. **461**, 65–75 (2012)
23. Protti, F., Souza, U.S.: Decycling a graph by the removal of a matching: new algorithmic and structural aspects in some classes of graphs. Discret. Math. Theoret. Comput. Sci. **20**(2) (2018)
24. Schaefer, T.J.: The complexity of satisfiability problems. In: Proceedings of the Tenth Annual ACM Symposium on Theory of Computing, pp. 216–226 (1978)
25. Yang, A., Yuan, J.: Partition the vertices of a graph into one independent set and one acyclic set. Discret. Math. **306**(12), 1207–1216 (2006)

Exactly k MSTs: How Many Vertices Suffice?

Apratim Dutta, Rahul Muthu, Anuj Tawari[✉] [iD], and V. Sunitha[iD]

Dhirubhai Ambani Institute of Information and Communication Technology,
Gandhinagar, India
{201821001,rahul_muthu,anuj_tawari,v_suni}@daiict.ac.in

Abstract. In this work, we study the problem of finding a weighted graph with exactly k minimum spanning trees (MSTs, in short) while minimizing the number of vertices. While finding a graph with k MSTs is easy, finding such a graph with the minimum number of vertices remains an interesting open problem. Recently, Stong [15] proved an upper bound within $\log k$ multiplicative factor of the minimum. In this work, we prove the following results which make further progress on this problem:

1. Large weights do not help in constructing a minimal weighted graph with prime number of spanning trees.
2. For $n \geq 6$ and $1 \leq k \leq n^2$, n vertices suffice for constructing a graph with k spanning trees.

Keywords: weighted complete graphs · minimum spanning trees · elementary symmetric polynomials

1 Introduction

Spanning trees and their weighted counterparts, especially minimum weight ones, have been of central interest and importance to the field of graph theory and algorithmic graph theory. The work straddles computation, enumeration and counting, among others. One of the cornerstone results on spanning trees is Cayley's Theorem, which places the number of spanning trees on an n vertex complete graph at n^{n-2}. Complete graphs are edge maximal simple graphs on n vertices, and contain all n vertex graphs as subgraphs. In an unweighted graph, every spanning tree is a minimum spanning tree and thus the number of MSTs of a weighted graph is upper bounded by the number of spanning trees of the equivalent unweighted graph (meaning the edge weights are all altered to 1).

Motivated by Cayley's Theorem, and the universal nature of complete graphs (in the sense they contain all graphs on the corresponding number of vertices, as subgraphs) on which Cayley's theorem is focussed, we decided to investigate whether it is possible to assign edge weights to an n vertex complete graph to get a prespecified number k of MSTs. Clearly, we limit the range of k to the set of integers $S = \{1, 2, \ldots, n^{n-2}\}$. While the motivation for the problem was sparked by Cayley's theorem, we found that this generalised problem is rich in combinatorics, algorithms and complexity.

W. Wu and J. Guo (Eds.): COCOA 2023, LNCS 14461, pp. 94–106, 2024.
https://doi.org/10.1007/978-3-031-49611-0_7

Problem 1. Given a pair of integers n and k, such that $1 \le k \le n^{n-2}$, is it possible to assign weights to the edges of K_n, complete graph on n vertices such that the resulting weighted graph has exactly k MSTs.

This is a decision problem, and upon further investigation, one observes that for any fixed n, for certain values of k, the question is easy to answer, while for others it appears more difficult. We are unaware of the status of the problem with respect to NP-completeness.

Problem 2. Given k, what is the smallest n, for which we can assign weights to the edges of K_n such that the resulting weighted graph has exactly k MSTs.

Although the two problems are equivalent, and rewordings of each other, we address them both. When we focus on a fixed, but arbitrary, n, we try to break the set $S = \{1, 2, \ldots, n^{n-2}\}$ into subsets and obtain results on the distribution of yes and no instances in different ranges of the spectrum. We also consider specific elements in the spectrum and classify them as yes or no. This results in a detailed study of the structure of the spectrum from various interesting angles. This includes techniques involving prime and composite numbers and polynomials arising thereof.

Both directly, and using results from the study of the spectrum, we obtain results that try to get the optimal n for which one can assign weights to the edges of K_n to achieve exactly k MSTs. At the time of writing, these results are not guaranteed to be optimal, and the complexity of obtaining the optimal n for any given k is unknown. We could attempt to obtain approximation algorithms using known lower bounds on the optimum as a benchmark for the approximation factor. In this connection, it is in order to mention a fairly simple result: When we can achieve exactly k MSTs on weighted K_n, we can also do so on weighted K_{n+1}. Equivalently, when we cannot achieve exactly k MSTs on weighted K_n we cannot, on K_{n-1} either.

In general, we can try to find the smallest n such that there exists an unweighted graph on n vertices that has exactly k spanning trees. This can be extended to a weighted complete graph on the same vertices by giving the edges of the graph weight 1 and the absent edges weight 2, resulting in the same number of MSTs as the number of spanning trees of the original graph. We must add the caveat that the original unweighted graph must be connected. This scenario is a special one, and allowing edge weights other than $\{1, 2\}$ might allow us to get a better optimum for some values of k. Thus as a simpler version of the problem, we could consider finding the smallest n such that there exists an unweighted graph on n vertices with exactly k spanning trees. The optimal answers to this restricted problem are lower bounded by the optimal answers to our original formulation. It would be interesting to investigate the possible extent of the gap as also the values for which the two optima coincide.

Related Work

The notion of a graph with minimum number of vertices in a graph with exactly k spanning trees was introduced in [13]. This notion was also studied by [1,

9]. In particular, it was conjectured in [1] that there is an unweighted graph with $o(\log k)$ vertices with exactly k spanning trees. It is known that at least $\Omega(\frac{\log k}{\log \log k})$ vertices are needed to generate a graph with k spanning trees [15]. We also know that every integer up to k^k can be realized as the number of spanning arborescences (directed spanning trees) with a fixed root in a digraph on $3k - 3$ vertices without multiple arcs [12]. It follows that the lower bound of $\Omega(\frac{\log k}{\log \log k})$ is tight for at least directed graphs. For undirected graphs, we need to find a small symmetric, diagonally dominant matrix whose determinant equals k. Symmetric determinantal representations have also been well studied [5,6].

There has also been considerable work on enumerating [8] and counting [2] spanning trees in special classes of graphs and maximizing the number of spanning trees in graphs [3,4,7,10,11,14].

Our Results and Techniques

We now formally state our results and briefly comment on the techniques used for proving them:

Theorem 1. *For any prime p, there is a weighted graph with minimum number of vertices and p MSTs which uses edge weights only from the set* $S = \{1, 2, 3\}$.

The proof crucially uses the fact that minimum weight edges must appear in every minimum spanning tree. We show that connected components induced by weight 1 and weight 2 edges must have only one minimum spanning tree. Using these facts, we proceed to show that it suffices to only use weights from the set S.

Theorem 2. *For any integer $n \geq 6$ and any k such that $1 \leq k \leq n^2$, there is an n-vertex unweighted graph with exactly k spanning trees.*

The proof essentially employs a case analysis. In all the cases, the witnessing graphs comprise of three edge-disjoint paths with common start and end vertices. Note that this result immediately follows from [15]. We give an alternative proof here which may be of interest.

Theorem 3. *Every geometric range*

$$[1, n], [n + 1, n^2], \ldots, [n^{n-3} + 1, n^{n-2}]$$

has at least one integer k such that there is a n-vertex graph with exactly k MSTs.

The proof uses an inductive argument.

2   Preliminaries

2.1   Notation

We use $[n]$ to denote the set $\{1, 2, \ldots, n\}$. The n-variate degree k elementary symmetric polynomial, denoted S_n^k, is defined as follows:

$$S_n^k(x_1, \ldots, x_n) = \sum_{A \subseteq [n], |A| = k} \prod_{i \in A} x_i$$

We use the notation $G = (V, E)$ to denote a graph whose vertex set is V and edge set is E. A graph is said to be complete if there is an edge between every pair of vertices. We will denote the complete graph on n vertices by K_n. Given two disjoint subsets $S, T \subseteq V$, we will denote the edges between one vertex in S and another vertex in T as *cross-connections* between between S and T. A graph is said to be connected if for any pair of vertices u, v, there is at least one path from u to v.

A tree is a connected and acyclic graph. A tree T is said to be a spanning tree of G if T visits every vertex of G. Further, T is said to be a minimum spanning tree of G if the sum of weights of the constituent edges of T is the minimum possible among all spanning trees of G.

Below, we formally define our problem.

Definition 1. *Let $\kappa(k)$ denote the smallest integer k such that we can assign integer weights to the edges of K_n so that the resulting weighted graph has exactly k MSTs.*

Specifically, we are interested in knowing how small is $\kappa(k)$.

2.2   Known Lower Bounds

We begin with the following well-known result on the number of spanning trees in a complete graph.

Fact 4 (Cayley's formula). *The unweighted complete graph on n vertices, K_n has exactly n^{n-2} spanning trees.*

So any weight assignment to the edges of K_n will result in a weighted graph that can have at most n^{n-2} MSTs. Hence if a n-vertex graph has k MSTs, k is at most n^{n-2}. Using this fact, one can show that $\kappa(k)$ is at least $\frac{\log k}{\log \log k}(1 + o(1))$ as stated in [15]. We could not find a proof of this lower bound in [15]. For sake of completeness, we give a detailed proof below.

Let c_1, c_2 be parameters and n_0 be a constant such that for all $n \geq n_0$, $c_1 n \log n \leq \log n^{n-2} \leq c_2 n \log n$. For instance, $c_2 = 1 - \frac{2}{n}$.

$$\frac{\log k}{\log \log k} \leq \frac{\log n^{n-2}}{\log \log n^{n-2}} \leq \frac{c_2 n \log n}{\log(c_1 n \log n)} = c_2 \frac{n \log n}{\log c_1 + \log n + \log \log n}$$

Therefore,

$$n \geq \frac{1}{c_2} \cdot \frac{\log c_1 + \log n + \log \log n}{\log n} \cdot \frac{\log k}{\log \log k} \geq c_3 \frac{\log k}{\log \log k}$$

where c_3 is a positive constant (for example, $c_3 = 1/c_2$). Finally, we note that $c_3 = \frac{1}{c_2} = \frac{n}{n-2} = 1 + \frac{2}{n-2}$.

2.3   Connection to Arithmetic Expressions

There are a wide variety of algorithms for computing MSTs, but all of them have some central common properties which we crucially use in our proofs. The following fact is the main insight behind Kruskal's algorithm:

Fact 5. *Any algorithm that considers the edges sorted in non-decreasing order of weights and picks a considered edge if and only if it doesn't create a cycle with edges picked earlier, always results in a minimum spanning tree. By considering different linearisations (it is a partial order, since some edge weights may be equal) of the sorted lists of edge weights, we will get different MSTs.*

Fact 6. *Every minimum spanning tree has a sorted ordering of the graph's edges such that the generic algorithm described in Fact 5 will generate it as the output.*

Fact 7. *The actual weights of the edges of a graph give more information than necessary, to determine how many MSTs it has and which spanning trees are minimum. The only relevant information is the relative weights of the edges. By this we mean the classification of every pair of edges according to $>, <, =$, with regard to their edge weights.*

Thus, we follow the convention that the edge weights assigned are all positive integers beginning with 1 and following successive numbers with no gaps in between. Thus, throughout this paper, if the maximum edge weight is w_{MAX}, all weighted graphs have edge weights $1, 2, \ldots, w_{MAX}$ with at least one edge associated with each weight.

Further observations lead to the conclusion, that if an edge of weight $w > 1$ is a cut edge, in the subgraph induced by edges of weights $1, \ldots, w$, then it can be changed to $w - 1$, without any change in the number of MSTs. Although we study the counting of the number of MSTs, the idea of selecting edges in non-decreasing order of their weights (as occur in Kruskal's, Prim's and other algorithms for computing MSTs) is used extensively by us. Rather than thinking of one edge at a time, we employ a multi-round strategy where in each round we consider the edges in batches in increasing order of weights. All edges of the same weight constitute a batch.

In each round, the number of components in a minimum spanning forest, contained in a minimum spanning tree decreases. We maintain a sequence of integers, one for each component in each round, with the final value being the number of MSTs. To calculate the integer associated with a particular component in round i, we:

1. List the components from round $i - 1$ that merge into this component.
2. Multiply the integers associated with those components from round $i - 1$ and store it.
3. These components from round $i - 1$ are all linked together by edges of weight i. Treat each of these components as single vertices, and assign a weight to an edge linking two such components. The weight of any such edge is equal to the number of edges of weight i between those components. This is for the

purpose of calculation only, and the original graph structure is available for future rounds.

4. List all spanning trees of the graph constructed in step 3, ignoring the weights of the edges.
5. Compute the cost of each spanning tree obtained in step 4 as the product of the numbers associated with each edge of that spanning tree, as defined in step 3.
6. Add the numbers obtained in step 5 and multiply that total by the number stored in step 2. This is the weight of this component in round i.

We can clearly see that the weight assignment problem for prescribed number of MSTs maps directly to an arithmetic expression consisting of alternating levels of addition and multiplication of numbers. Thus, in several cases we may use arithmetic expressions, and calculate suitable values of the variables that lead to the target.

2.4 Upper Bounds

Recall the definition of $\kappa(k)$ (Definition 1. In this section, we present some results on upper bounds for $\kappa(k)$ where k is a positive integer. The following lemma shows that for every integer k less than or equal to n, $\kappa(k) \leq n$.

Lemma 1. For $1 \leq k \leq n$, we can assign weights to the edges of K_n such that the resulting weighted graph has exactly k MSTs.

Proof. For the value $k = 1$ we can assign the edges of some spanning tree weight 1 and all other edges weight 2.

For the value $k = 2$, create a triangle whose vertices are v_1, v_2, v_3 whose edges $(v_1, v_2), (v_2, v_3), (v_1, v_3)$ receive weights $1, 2, 2$ respectively. Connect each one of v_1, v_2, v_3 to all the remaining vertices, using $3(n - 3)$ edges of weight 1. All remaining edges are assigned weight 2. To cover the vertices other than v_1, v_2, v_3, some $n - 3$ edges of weight 1 described above need to be included. To cover v_1, v_2, v_3 we need to pick the edge (v_1, v_2) and any one of the edges (v_2, v_3) and (v_1, v_3) yielding 2 MSTs.

For $3 \leq k \leq n$, we use a similar argument to the one used for $k = 2$. We create a length k cycle C whose vertices are $v_1, v_2, \ldots, v_k$ and edges have weight 1, connect the remaining vertices to each one of $v_1, v_2, \ldots, v_k$ using $k(n - k)$ edges of weight 1 edges and assign weight 3 to all other edges. To cover the vertices other than $v_1, v_2, \ldots, v_k$, some $n - k$ edges of weight 1 described above need to be included. To cover $v_1, v_2, \ldots, v_k$, we need to pick $k - 1$ edges of the cycle C which can be chosen in k different ways, yielding k different MSTs.

The following proposition shows that for upper bounding $\kappa(k)$ for any positive integer k, it is sufficient to find a good enough upper bound for $\kappa(p)$ for any prime p.

Proposition 1. For any two positive integers r, s, $\kappa(r \cdot s) \leq \kappa(r) + \kappa(s) - 1$.

Proof. Let $G_1 = (V_1, E_1)$ be a graph with exactly r minimal spanning trees and $\kappa(r)$ vertices and $G_2 = (V_2, E_2)$ be another graph with exactly s minimal spanning trees and $\kappa(s)$ vertices and $|V_1 \cap V_2| = 1$. Then the graph $G_1 \vee G_2 = (V_1 \cup V_2, E_1 \cup E_2)$ has exactly $\kappa(r) + \kappa(s) - 1$ vertices and exactly $r \cdot s$ MSTs.

This suggests the following natural strategy for upper bounding $\kappa(m)$ where m is a composite number: Consider the unique prime factorization of m. Let $m = p_1^{\alpha_1} \cdot p_2^{\alpha_2} \cdots p_k^{\alpha_k}$ where $p_1, p_2, \ldots p_k$ are primes and $\alpha_1, \alpha_2, \ldots, \alpha_k \geq 0$. By Proposition 1, we get the following.

$$\kappa(m) \leq \sum_{i=1}^{k} \alpha_i \kappa(p_i)$$

The following proposition yields another upper bound for κ-value of composite numbers.

Proposition 2. *For any two positive integers r, s, $\kappa(r \cdot s) \leq \kappa(r) + \left\lceil \frac{s}{\kappa(r)} \right\rceil$.*

Proof. Consider a weighted complete graph $K_{\kappa(r)}$ with $\kappa(r)$ vertices having exactly r MSTs. Let S be an additional set of vertices with $|S| = \left\lceil \frac{s}{\kappa(r)} \right\rceil$, then we link the vertices in S together with a tree of weight 1 edges and all other edges with both endpoints in S are given weight 2. Exactly s of the cross connections between the original $K_{\kappa(r)}$ and the set S of vertices are given weight one more than the maximum weight occurring in the initial complete graph, and the remaining cross-connections are assigned the next higher integer.

This Proposition can be used to get some good upper bounds. Unlike Proposition 1 we will not provide an explicit upper bound here, but will rather give the method that can be adopted. We pick a factor and get as good a solution as possible using known methods. We incorporate the remaining factors iteratively. At each stage, when attempting to incorporate a factor, we pick the number of vertices to be added, by invoking Proposition 2. Note that the best results using this proposition will depend on carefully choosing the order in which the factors are incorporated and some strategic use of suboptimal number of vertices, when appropriate. Below is one example.

By Fact 4, n vertices are both sufficient as well as necessary to generate n^{n-2} spanning trees. In what follows, we call numbers of the form $m = n^{n-2}$ as Cayley numbers and n is called the inverse of the Cayley number m.

Definition 2. *We call a factorization of a positive integer m a cayley factorization when it is rendered into product of cayley numbers (numbers of the form $n_1^{n_1-2}$ for any positive integer n_1), covering as large a factor of the integer as possible, and the rest being residues. In particular m can be expressed as below:*

$$m = c \cdot n_1^{n_1-2} n_2^{n_2-2} \cdots n_k^{n_k-2}$$

where $c, n_1, n_2, \ldots, n_k$ are positive integers.

Note that a cayley factorization for a given number may or may not exists. Assuming it does, given a cayley factorization, we can use the inverse of the largest cayley factor and subsequently use a combination of Propositions 1 and 2 to get a good solution.

The problem now reduces to finding a good upper bound for $\kappa(p)$ for any prime p. We begin with observing that it suffices to find an unweighted graph with exactly p spanning trees.

Lemma 2. *If there is an unweighted graph $G = (V, E)$ with $|V| = n$ and k spanning trees, there is a weight assignment w to the edges of K_n so that the resulting weighted graph H has k MSTs.*

Proof. For any edge e of K_n we assign weight 1 if $e \in E$ and weight 2 otherwise.

By Lemma 1, it follows that p vertices suffice for constructing a graph with exactly p MSTs. The following result shows that one can obtain better bounds which hold for an infinite sequence of primes.

Theorem 8. *[13] Let $p \equiv 2 \mod 3, p > 5$. Then $\kappa(p) \leq \frac{p+4}{3}$*

The witnessing graph for the above result consists of a cycle of length $\frac{p+4}{3}$ with a single chord. We also know a near-optimal upper bound for a specific class of primes, namely, Fibonacci primes.

Theorem 9. *[2] The n-fan graph is formed by having a vertex adjacent to every vertex of a n-vertex path. n-fan has F_{2n} spanning trees where F_{2n} is the $2n^{th}$ Fibonacci number.*

Using induction, one can show that $F_{2n} \geq 2^n$. Let $p = F_{2n}$ be a Fibonacci prime. Then we get that $\kappa(p) = O(\log p)$ for Fibonacci primes. It is however unknown whether there are infinitely many Fibonacci primes.

The following lemma is useful for constructing unweighted graphs with prime number of spanning trees.

Lemma 3. *Let $v_1, v_2, \ldots, v_n$ be non-negative integers with at most one of them equal to 1. There is a graph with at most $\sum_{i \in [n]} v_i$ vertices and*

$$S_n^{n-1}(v_1, v_2, \ldots, v_n) = \sum_{A \subseteq [n], |A| = n-1} \prod_{i \in A} v_i$$

many spanning trees.

Proof. Consider the theta graph $\theta(v_1, v_2, \ldots, v_n)$ which consists of n edge disjoint paths of lengths $v_1, v_2, \ldots, v_n$ with common start and end vertices. Since we are interested only in simple graphs, there is at most one edge between any pair of vertices. Hence we require the condition that at most one among $\{v_1, v_2, \ldots, v_n$ is equal to one. Choosing $n-1$ of these n paths and deleting one edge from each path yields a spanning tree and conversely.

3   Minimal Weighted Graphs with Prime Number of MSTs

Let p be a prime. In this section, we focus our attention on minimal weighted graphs with p MSTs where minimality is with respect to number of vertices. Since we do not know of any minimal unweighted graphs with p MSTs, a natural question to ask is whether large edge weights can help. Here we prove Theorem 1 which shows that it is sufficient to consider edge weights from the set $S = \{1, 2, 3\}$.

Let G be a graph with $\kappa(p)$ vertices and exactly p MSTs. Due to Fact 6, Fact 5 and Fact 7, we assume that if the maximum edge weight is w_{MAX}, all weighted graphs have edge weights $1, 2, \ldots, w_{MAX}$ with at least one edge associated with each weight. In fact, for a vertex v, the minimum edge weight of an edge incident with v can be replaced by weight 1 without affecting the number of MSTs. Consider the components of the graph induced by edges of weight 1. If it is just one component then the other edges will never be used and can all be set to weight 2.

In case there is more than one component of the subgraph induced by weight 1 edges, if any of these components has more than one MST, then that number will be a divisor of the final number of MSTs. Since the final number is a prime, this factor has to be that same prime. This means we can do it in fewer number of vertices (restricted to the vertices of that component). Contradiction to the optimality. Now, since each component of the subgraph induced by edges of weight 1 have exactly one minimum spanning tree of the vertices they span, they are linked to each other by edges of higher weight. The claim is they must all be connected to each other using edges of weight 2 alone. The reason is that if after using linking weight 2 edges they do not span and connect, then again each component (1 and 2 together) must have only one spanning tree. The reasoning, as before is, if not, then the number of spanning trees of that component is a divisor of the prime which is a contradiction. And since the set of all edges of weight 1 and 2 in the forest built up so far have no cycles, the edges of weight 2 across components of weight 1 can all be relabelled to 1 without impacting the number of MSTs. Thus, it follows that in two rounds of weights, we span and connect the whole graph. Thus weights $1, 2, 3$ are all that are used.

4   Study of the Spectrum

We recall the spectrum problem below.

Problem 3. Given a positive integer n, for what values $k \in [1, n^{n-2}]$ is it possible to assign weights to the edges of K_n, such that the resulting weighted graph has exactly k MSTs.

In this section, as a first step towards solving the above problem, we prove Theorem 2.

The proof uses mathematical induction for composites and a direct approach for numbers which are $1, 3, 7, 9 \mod 10$ (which duplicates some composites too, but covers all primes). We choose to cover all prime integers which are $1, 3, 7, 9 \mod 10$.

For numbers $1 \mod 10$, we use the base case of 31. We consider three groups of vertices with the number of cross connections being 1,3,7. The number of MSTs is $1 \times 3 + 3 \times 7 + 7 \times 1 = 31$. If we replace the 1 with c in general the expression becomes $c \times 3 + 3 \times 7 + 7 \times c = 10c + 21$. This clearly covers all numbers of type $1 \mod 10$. In order to have the required number of cross connections, we can split the vertices into $1, \frac{n-1}{2}, \frac{n-1}{2}$ where $n \geq 15$ so as to enable 7 cross connections. The number of MSTs clearly covers all numbers which are $1 \mod 10$ upto n^2, asymptotically. For those numbers bypassing the clause "asymptotically", we have direct proofs, which are minor tinkering of specific values. We omit the details as they are expansive.

For numbers $3 \mod 10$ we cover it in two subcases: $13 \mod 30$ and $23 \mod 30$. The remaining case of $3 \mod 30$ need not be considered as we deal with integers greater than or equal to six and such numbers are always composite.

For numbers $13 \mod 30$, we take the base case as 43. For 43, we can create a 13 cycle with a chord that splits the cycle as 10, 3. This gives us $10+3+10 \times 3 = 43$ MSTs. When we want to get other numbers which are $13 \mod 30$, we need to add multiples of 30. Notice that the number of spanning trees using the chord is 10×3 and and thus if we replace the vertices that are the endpoints of the chord with a larger groups of vertices with just one spanning tree in each of the two groups, and make extra chord connections we can increase the number of spanning trees by 30 for every two additional chords. In order to have t chords we need approximately $\sqrt{t}$ vertices in each group. This results in a total of $11 + \sqrt{t+1}$ vertices to achieve $13 + 30t$ MSTs. From the relationship it is clear that the number of spanning trees achievable runs upto n^2 asymptotically, and for all small values not covered, we have done a direct verification.

For numbers $23 \mod 30$ we use the base case as 23. Here we need an eight cycle with a chord that splits the cycle as 3 and 5. This gives us $3+5+3 \times 5 = 23$ MSTs. For higher cases, the argument is as before, except that we need two chords for every additional 30 MSTs, since each chord contributes 15 MSTs.

For numbers $7 \mod 10$, we use 17 as the base case. To achieve 17, we need a 7 cycle with a chord that splits it as 5 and 2. This gives us $2 + 5 + 2 \times 5 = 17$ MSTs. Since the chord contributes 10 of the 17 spanning trees, every additional chord provides an extra 10 while the number of spanning trees not using the chords remains fixed at 7. Thus we cover all numbers $7 \mod 10$ in a range which is asymptotically upto n^2. For smaller instances, we have verified the result directly.

For numbers $9 \mod 10$, we create three groups of vertices containing $1, \frac{n-1}{2}, \frac{n-1}{2}$ vertices. This allows the number of cross connections to be $\frac{n-1}{2}, \frac{n-1}{2}, \frac{(n-1)^2}{4}$ respectively. For adequately large values of n, this enables connections numbering $1, 9, c$ where $c \leq \frac{(n-1)^2}{4}$. This allows us to get $k = 1 \times c + 9 \times c + 1 \times 9 = 10c + 9$

MSTs and for sufficiently large values of c this number exceeds n^2. This covers all numbers of the type $9 \mod 10$ upto n^2, for sufficiently large choices of n and c. For all values escaping the clause "sufficiently large", we have verified the result directly, using other methods.

For Composites: We first observe that for $n = 4, 5$ there are instances with $1 \leq k \leq n^2$ such that there is no n-vertex graph with exactly k MSTs. $n = 6$ serves as the base case for our proof and has been directly verified by us. We omit the details as they are expansive.

Inductive Hypothesis: Assume that for some $k \geq 6$ and for any l such that $1 \leq l \leq k^2$, there is a k-vertex weighted graph with exactly l MSTs. We wish to prove that for any l such that $1 \leq l \leq (k+1)^2$, there is a $k+1$-vertex weighted graph with exactly l MSTs. When we move from k to $k+1$, a composite number c_1 in the range $[k^2 + 1, (k+1)^2 - 1]$ will have a factor c_2 such that $2 \leq c_2 \leq k$. Suppose not . Then it has at least two factors of magnitude at least $k+1$. This implies that c_1 is at least $(k+1)^2$, contradicting the fact that c_1 is in the range $[k^2 + 1, (k+1)^2 - 1]$. Let $c_1 = c_2 \cdot c_3$ where $2 \leq c_2 \leq k$. Then c_3 is a number which is at most k^2, and hence there is a k-vertex weighted graph with exactly c_3 MSTs. Since the $(n+1)^{th}$ vertex can have upto n cross connections to the first n vertices, we can use c_2 number of cross connections of light weight and achieve the target.

For the value $(k+1)^2$, we use induction from two levels below. We use the assumption that there is a weighted graph with $k-1$ vertices and exactly $k+1$ spanning trees. We observe that this also true for the case when $k = 5$. The number of cross connections between the two extra vertices and the first $k-1$ vertices is $2 \times (k-1) > (k+1)$. Thus, using that many cross connections of light weight, we achieve $(k+1)^2$ MSTs.

The next result shows that one can partition the range $[1, n^{n-2}]$ into several subranges such that each of the subranges has at least one value k such that there is a weighted n-vertex graph with exactly k MSTs. Specifically, we prove Theorem 3.

The proof is by induction on n.

Base Case: let $n = 4$. Here, we can generate a graph with n vertices and k MSTs where k can be any number from 1 to 4 in $[1, 4]$, $\{5, 6, 8, 9, 16\}$ in $[5, 16]$. The base case is clearly true.

Inductive Step: For K_n: Assume for any integer $i \leq n - 2$, there is a n-vertex weighted graph with k MSTs where k belongs to the interval $[n^i + 1, n^{i+1}]$. Without loss of generality let it be $(n^i + x)$, where $x \in \mathbb{N}$ & $x \leq n^i(n-1)$.

When we move to K_{n+1}, we have an extra vertex that can have any number t of light weight cross connections where $1 \leq t \leq n$. This will result in a weighted K_{n+1} with value $t \cdot (n^i + x)$. We aim to show that for some choice of $t \in \{1, \ldots, n\}$, this gives a number in the range $[(n+1)^i + 1, (n+1)^{i+1}]$.

We do this by showing that:

1. The size (number of numbers) in the target range for K_{n+1} is at least as large as k. This establishes that there is a number in the range that is an integral multiple of k. The following formula establishes this:

$$(n+1)^{i+1} - (n+1)^i \geq n^i + x$$

This is equivalent to:
$$n \cdot (n+1)^i \geq n^i + x$$

This may be inferred from:
$$n \cdot (n+1)^i > n^{i+1} \geq n^i + x$$

2. n times the yes instance in K_n is either within or greater than the target range. This will establish that **some** positive integral multiple ($\leq n$) of the k is in the target range for K_{n+1}. This is established by the following inequality:
$$n \cdot (n^i + x) > (n+1)^i$$

To see this, observe that:
$$n \cdot (n^i + x) > n^{i+1} > (n+1)^i, \text{ for } n \geq 4, n - 3 \geq i \geq 1.$$

So, for K_{n+1}, every geometric range $[(n+1)^i + 1, (n+1)^{i+1}]$ has a value k_1 such that there is a weighted K_{n+1} with exactly k_1 MSTs for $0 \leq i \leq n - 3$. However, we have an extra range in K_{n+1}, i.e. $[(n+1)^{n-2} + 1, (n+1)^{n-1}]$. We know that, there is a weighted K_n with $n^{n-3}(n - 2)$ MSTs for $n \geq 5$ [13], and $n \cdot n^{n-3}(n - 2) = n^{n-2}(n - 2)$ is in the desired range.

The following is another result which can help in constructing graphs with prespecified number of MSTs.

Lemma 4. *If there is a weight assignment to the edges of K_n resulting in $a + b$ MSTs, such that some edge e lies in exactly a of them and missing from the remaining b. Then there exists a weight assignment to the edges of K_{n+t} such that it has exactly $a + t \cdot b$ MSTs.*

Proof. We subdivide the edge e into a path of length $t + 1$ via t new vertices. Each edge in the subdivision get the same weight as the original edge which was subdivided. In the newly constructed graph:

- Every MST in the original graph containing the edge e maps to a tree containing the new subdivided path. Thus there are a of them.
- Every MST missing the edge e in the original graph corresponds to t MSTs in the new graph where the missing edge e is supplanted by $t - 1$ of the t edges of the subdivided edge. There are thus t MSTs in the new graph associated with each MST of the original graph, missing the edge e. This constitutes bt MSTs.

Thus, the resulting weighted graph has exactly $a + tb$ MSTs.

5 Conclusion

In this work, we showed that n vertices suffice for generating k spanning trees where $1 \leq k \leq n^2$. A natural next question is to extend this result to higher powers of n. In particular, find the maximum possible α such that n vertices suffice for generating k spanning trees where $1 \leq k \leq n^\alpha$.

References

1. Azarija, J., Škrekovski, R.: Euler's idoneal numbers and an inequality concerning minimal graphs with a prescribed number of spanning trees. Mathematica Bohemica (2012)
2. Bogdanowicz, Z.R.: Formulas for the number of spanning trees in a fan. Appl. Math. Sci. **2**(16), 781–786 (2008)
3. Cheng, C.S.: Maximizing the total number of spanning trees in a graph: two related problems in graph theory and optimum design theory. J. Comb. Theor. Ser. B **31**(2), 240–248 (1981)
4. Gilbert, B., Myrvold, W.: Maximizing spanning trees in almost complete graphs. Netw. Int. J. **30**(1), 23–30 (1997)
5. Grenet, B., Kaltofen, E.L., Koiran, P., Portier, N.: Symmetric determinantal representation of weakly-skew circuits. In: Schwentick, T., Dürr, C. (eds.) 28th International Symposium on Theoretical Aspects of Computer Science, STACS 2011, March 10-12, 2011, Dortmund, Germany. LIPIcs, vol. 9, pp. 543–554. Schloss Dagstuhl - Leibniz-Zentrum für Informatik (2011). https://doi.org/10.4230/LIPIcs.STACS.2011.543
6. Grenet, B., Monteil, T., Thomassé, S.: Symmetric determinantal representations in characteristic 2. CoRR **abs/1210.5879** (2012). http://arxiv.org/abs/1210.5879
7. Kelmans, A.K.: On graphs with the maximum number of spanning trees. Random Struct. Algorithms **9**(1–2), 177–192 (1996)
8. Mokhlissi, R., El Marraki, M.: Enumeration of spanning trees in a closed chain of fan and wheel. Appl. Math. Sci. **8**(82), 4053–4061 (2014)
9. Nebeský, L.: On the minimum number of vertices and edges in a graph with a given number of spanning trees. Časopis pro pěstování matematiky **98**(1), 95–97 (1973)
10. Petingi, L., Boesch, F., Suffel, C.: On the characterization of graphs with maximum number of spanning trees. Discret. Math. **179**(1–3), 155–166 (1998)
11. Petingi, L., Rodriguez, J.: A new technique for the characterization of graphs with a maximum number of spanning trees. Discret. Math. **244**(1–3), 351–373 (2002)
12. Rote https://mathoverflow.net/users/30800/gwithaprescribednumberofspanningtrees.MathOverflow. https://mathoverflow.net/q/122093
13. Sedláček, J.: On the minimal graph with a given number of spanning trees. Can. Math. Bull. **13**(4), 515–517 (1970)
14. Shier, D.R.: Maximizing the number of spanning trees in a graph with n nodes and m edges. J. Res. National Bur. Stan. Sect. B **78**(193–196), 3 (1974)
15. Stong, R.: Minimal graphs with a prescribed number of spanning trees. Australas. J. Comb. **82**(2), 182–196 (2022)

Minimum Monotone Tree Decomposition of Density Functions Defined on Graphs

Lucas Magee[✉][iD] and Yusu Wang[iD]

University of California San Diego, La Jolla, CA 92093, USA
lmagee@ucsd.edu

Abstract. Monotone trees - trees with a function defined on their vertices that decreases the further away from a root node one travels, are a natural model for a process that weakens the further one gets from its source. Given an aggregation of monotone trees, one may wish to reconstruct the individual monotone components. A natural representation of such an aggregation would be a graph. While many methods have been developed for extracting hidden graph structure from datasets, which makes obtaining such an aggregation possible, decomposing such graphs into the original monotone trees is algorithmically challenging.

Recently, a polynomial time algorithm has been developed to extract a minimum cardinality collection of monotone trees (M-Tree Set) from a given density tree - but no such algorithm exists for density graphs that may contain cycles. In this work, we prove that extracting such minimum M-Tree Sets of density graphs is NP-Complete. We additionally prove three additional variations of the problem - such as the minimum M-Tree Set such that the intersection between any two monotone trees is either empty or contractible (SM-Tree Set) - are also NP-Complete. We conclude by providing some approximation algorithms, highlighted by a 3-approximation algorithm for computing the minimum SM-Tree Set for density cactus graphs.

1 Introduction

A common problem in modern data analysis is taking large, complex datasets and extracting simpler objects that capture the true nature and underlying structure. In this paper we are interested in the case when the input data is the aggregation of a collection of trees. In fact, each tree also has attributes over nodes (e.g., the strength of certain signal) which decreases monotonically from its root – we call such a tree a monotone tree. Such trees come naturally in modeling a process that dissipates as it moves away from the root. One such example is in the construction of neuronal cells: a single neuron has tree morphology, with the cell body (soma) serving as the root. In (tracer-injection based) imaging of brains, the signal often tails off as it moves away from the cell body and out of the injection region, naturally giving rise to a rooted monotone tree. See Fig. 1 (D) for one such example.

Generally, we are interested in the following: given input data that is the aggregation of a collection of monotone trees, we aim to reconstruct the individual monotone trees. The specific version of the problem we consider in this

© The Author(s), under exclusive license to Springer Nature Switzerland AG 2024
W. Wu and J. Guo (Eds.): COCOA 2023, LNCS 14461, pp. 107–125, 2024.
https://doi.org/10.1007/978-3-031-49611-0_8

paper is where the input data is a graph $G = (V, E)$ with a density function $f : V \to \mathbb{R}^{\geq 0}$ defined on its vertices. Our goal is to decompose (G, f) into a collection of monotone trees $(T_1, f_1), \ldots, (T_k, f_k)$ whose union sums to the original (G, f) at each $v \in V$. See Sect. 2 for precise definitions. A primary motivation for considering graphs to be the input is because graphs are flexible and versatile, and recently, a range of methods have been proposed to extract the hidden graph structure from a wide variety of datasets; see e.g., [1,4,6,8,12,13,15,16,18,20,22]. In the aforementioned example of neurons, the discrete Morse-based algorithm of [6] has been applied successfully to extract a graph representing the summary of a collection of neurons [2,21]. To extract the individual neurons from such a summary would be a significant achievement for the neuroscience community - which has developed many techniques to extract individual neuron skeletonizations from imaging datasets; see e.g., [5,9,19]. However, going from a graph to a collection of trees poses algorithmic challenges.

The monotone-tree decomposition problem has been studied in the work of [3], which develops a polynomial-time algorithm for computing the minimum cardinality set of monotone trees (M-Tree Set) of a density function defined on **a tree** (instead of a graph). However many applications for such a decomposition have graphs that may contain cycles, with the authors of [3] explicitly mentioning a need for algorithms that can handle such input domains.

New Work. We consider density functions defined on graphs, which we refer to as *density graphs*. Our goal is to decompose an input density graph (G, f) into as few monotone trees as possible, which we call the *minimum M-Tree Decomposition problem*. See Sect. 2 for formal definitions and problem setup. Unfortunately, while the minimum M-Tree Decomposition problem can be solved efficiently in polynomial time via an elegant greedy approach when the density graph is itself a tree [3], we show in Sect. 3 that the problem for graphs in general is NP-Complete. In fact, no polynomial time constant factor approximation algorithm exists for this problem under reasonable assumptions (see Sect. 3). Additionally, we show NP-Completeness for several variations of the problem (Sect. 3). We therefore focus on developing approximation algorithms for this problem. In Sect. 4, we first provide two natural approximation algorithms but with additive error. For the case of multiplicative error, we provide a polynomial time 3-approximation algorithm for computing the so called minimum SM-Tree Set of a density cactus graph.

2   Preliminaries

2.1   Problem Definition

We will now introduce definitions and notions in order to formally define what we wish to compute. Given a graph $G(V, E)$, a *density function* defined on G is a function $f : V \to \mathbb{R}^{\geq 0}$. A *density graph* (G, f) is a graph G paired with a density function f defined on its vertices. A *monotone tree* is a density tree with a *root* $v \in V$ such that the path from the root to every node $u \in V$ is non-increasing in density values. See Fig. 1 for explicit examples of density trees and monotone trees. While multiple nodes may have the global maximum value on the monotone tree, exactly one node is the root. For example, in Fig. 1 (B), either node with the global maximum value may be its root, but only one of them is the root.

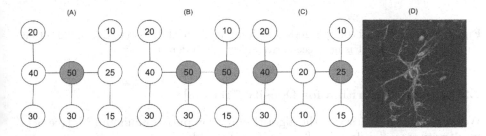

Fig. 1. (A)–(C) Contain examples of density trees with relative maxima colored red. (A) shows a monotone tree. (B) shows a monotone tree with multiple nodes having the global maximum density value. (C) shows an example of a density tree that is not a monotone tree. (D) A zoom in an individual neuron within a full mouse brain imaging dataset. The dataset is an fMOST imaging dataset that was created as part of the Brain Initiative Cell Census Network and is publicly available for download. (Color figure online)

Given a density graph $(G(V, E), f)$, we wish to build a set of monotone subtrees $(T_1, f_1), (T_2, f_2), \ldots, (T_n, f_n)$ such that $T_i \subseteq G$ for all i and $\sum_{i=1}^{n} f_i(v) = f(v)$ for all $v \in V$. Note that if a node $v \in V$ is not in a tree T_i then we say that $f_i(v) = 0$ and vice versa. We will refer to such a decomposition as a *monotone tree (M-tree) decomposition* of the density graph, and refer to the set as an *M-Tree Set* throughout the remainder of the paper. An M-Tree Set is a *minimum M-Tree Set* for a density graph if there does not exist an M-Tree Set of the density graph with smaller cardinality. An example of a density graph and a minimum M-Tree Set is shown in Fig. 2. Note that a density graph may have many different minimum M-Tree Sets. We abbreviate the cardinality of a minimum M-Tree Set for a density graph (G, f) as $|\mathsf{minMset}((G, f))|$.

There are different types of M-Tree Sets that may be relevant for different applications. A *complete M-Tree (CM-Tree) Set* is an M-Tree Set with the additional restriction that every edge in the density graph G must be in at least one tree in the set. A *strong M-Tree (SM-Tree) Set* is an M-Tree Set such that the intersection between any two trees in the set must be either empty or contractible. We similarly abbreviate the cardinality of a minimum SM-Tree Set of (G, f) as $|\mathsf{minSMset}((G, f))|$. A *full M-Tree (FM-Tree) Set* is an M-Tree Set such that for each element $(T_i(V_i, E_i), f_i)$, $f_i(v) = f(v)$ for the root node $v \in V_i$ of (T_i, f_i). The (minimum) M-Tree Set in Fig. 2 is also a (minimum) CM-Tree Set but is neither a SM-Tree Set nor a FM-Tree Set.

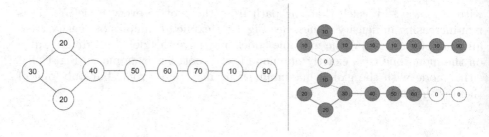

Fig. 2. A density graph (left) together with a minimum M-Tree Set (right). Note that a minimum M-Tree Set is not necessarily unique for a density graph.

2.2   Greedy Algorithm for Density Trees [3]

We will now briefly describe the algorithm for computing minimum M-Tree Sets for density trees developed in [3], as some of the ideas will be useful in our work. Please refer to [3] for more details. The approach of [3] relies on a so-called *monotone sweeping operation* to build individual elements of a minimum M-Tree Sets of density trees. Algorithm 1 explicitly defines a generalized version of this operation that we will need in a later proof. The operation takes a density tree $(T(V, E), f)$, a node $v \in V$, and a starting function value α such that $0 < \alpha \le f(v)$ as input. A monotone subtree $(T', h_{f,v,\alpha})$ and the remainder density tree $(T, R_{v,\alpha}f)$ where $R_{v,\alpha}f(u) = f(u) - h_{f,v,\alpha}(u)$ for all $u \in V$ is returned.

Algorithm 1: monotone-sweep$((T(V, E), f), v \in V, \alpha)$

Input: A density tree $(T(V, E), f)$, a starting node $v \in V$, and a staring value α
     such that $0 < \alpha \le f(v)$
Output: A monotone subtree $(T', h_{f,v,\alpha})$ and a remainder $(T, R_{v,\alpha}f)$
(Step 1) Initialize output density subtree T' to only contain the input vertex v,
    with corresponding density function $h_{f,v,\alpha}(v) = \alpha$
(Step 2) Perform DFS starting from v. For each edge $(u \to w)$ traversed:

$$h_{f,v,\alpha}(w) = \begin{cases} h_{f,v,\alpha}(u) & f(w) \ge f(u) \\ max(0, h_{f,v,\alpha}(u) - (f(u) - f(w))) & otherwise \end{cases}$$

Return monotone tree $(T', h_{f,v,\alpha})$ and remainder density tree $(T, R_{v,\alpha}f)$.

Algorithm 2, which outputs a minimum M-Tree Set of density trees, performs the monotone sweeping operation iteratively from certain nodes, called the *mode-forced* nodes of the density tree. To compute these mode-forced nodes, one iteratively remove leaves from the tree if their parent has greater or equal density. Such leaves are referred to as *insignificant vertices*. Once it is no longer possible to remove any additional nodes, the leaves of the remaining graph are the mode-forced nodes of the original density graph.

Algorithm 2: tree-algo$((T(V, E), f))$

Input: A density tree $(T(V, E), f)$

Output: A minimum M-Tree set of $(T(V, E), f)$

(Step 1) Find a mode forced vertex $v \in V$

(Step 2) Perform monotone-sweep$((T(V, E), f), v, f(v))$ to build a single element of a minimum M-Tree Set.

(Step 3) Repeat Steps 1 and 2 on remainder $(T, R_{v, f(v)}f)$ until no density remains.

An example of a single iteration of the tree algorithm is shown in Fig. 3. The running time complexity of Algorithm 2 is $O(n * |\text{minMset}((T, f))|)$ where n is the number of nodes in T. We note that all M-Tree Sets of a density tree are also SM-Tree Sets, so Algorithm 2 also outputs a minimum SM-Tree Set.

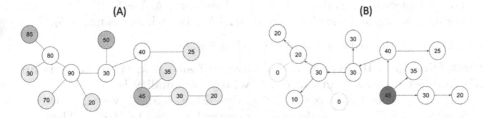

Fig. 3. (A) A density graph with mode-forced nodes colored green and insignificant vertices colored yellow. (B) A single element built by the monotone sweep operation from a mode forced node as performed in Algorithm 2 (Color figure online).

2.3 Additional Property of Monotone Sweeping Operation

Unfortunately, the previous work of [3] cannot be directly applied to density graphs. Nevertheless, we can show Claim 2.3 which will later be of use in developing approximation algorithms in Sect. 4. Its relatively simple proof is in Appendix A.

Claim. Given a density tree $(T(V, E), f)$, let $v \in V$. Let $a, b \in \mathbb{R}^+$ such that, without loss of generality, $0 < a < b \leq f(v)$. Let $(T, R_{v,a}f)$ be the remainder of monotone-sweep$((T, f), v, a)$. We can define a similar remainder $(T, R_{v,b}f)$. Then we have $|\text{minMset}((T, R_{v,b}f))| \leq |\text{minMset}((T, R_{v,a}f))|$.

3 Hardness Results

Given that there exists a polynomial time algorithm for computing minimum M-Tree Sets of density trees, it is natural to ask whether or not such an algorithm exists for density graphs. We prove Theorem 1, stating that the problem is NP-Complete.

Theorem 1. *Given a density graph* $(G(V, E), f)$ *and a parameter* k, *determining whether or not there exists an M-Tree set of size* $\leq k$ *is NP-Complete.*

Proof. It is easy to see that this problem is in NP, so we will now show it is also in NP-Hard. First we consider a variation of the Set Cover problem where the intersection between any two sets is at most 1. We refer to this problem as Set Cover Intersect 1 (SC-1). SC-1 is a generalization of the NP-Complete problem of covering points in a plane with as few lines as possible [17], and approximation bounds of SC-1 are well studied in [14]. Given an instance of SC-1 (m sets $S_1, S_2, \ldots, S_m$ covering a universe of n elements $e_1, e_2, \ldots, e_n$, and a number k), we reduce to an instance of the M-Tree Set decision problem as follows:

- Create a bipartite graph $G(V = A \cup B, E)$ equipped with a density function $f : V \to \mathbb{R}^{\geq 0}$ based on the input (SC-1) instance.
- In particular, for each set S_i, add a node a_{S_i} to A and set $f(a_{S_i}) = |S_i|$.
- For each element e_j, add a node b_{e_j} to B and set $f(b_{e_j}) = 1$
- For each set S_i, add edge between a_{S_i} and b_{e_j} for each element $e_j \in S_i$.

An example of this reduction is illustrated in Fig. 4.

First Direction: If there is a Set Cover of size $\leq k$, **then there is an M-Tree Set of density graph** (G, f) **whose cardinality is** $\leq k$.

Let S_{cover} be a set cover of size $n \leq k$. For each $S_i \in S_{cover}$, we will construct a monotone tree (T_i, f_i) rooted at a_{S_i}. In particular, $f_i(a_{S_i}) = f(a_{S_i})$. Then, for each element $e_j \in S_i$, T_i will include b_{e_j} and the edge (a_{S_i}, b_{e_j}), with $f_i(b_{e_j}) = 1$. Note that if e_j is an element in multiple sets in S_{cover}, simply pick one $S_i \in S_{cover}$ such that $e_j \in S_i$ to be the representative set of e_j. Finally, for each set $S_l \notin S_{cover}$, for each element $e_j \in S_l$, add the node a_{S_l} and the edge (b_{e_j}, a_{S_l}) to T_i with $f_i(a_{S_l}) = 1$, where (T_i, f_i) is the monotone tree rooted at the node a_{S_i} where $S_i \in S_{cover}$ is the representative set containing e_j.

Firstly, each element in the M-Tree Set is connected by construction. The only nodes in an element (T_i, f_i) are the root node a_{S_i}, where $S_i \in S_{cover}$, nodes of the form b_{e_j}, where $e_j \in S_i$, and nodes of the form a_{S_l}, where $S_l \notin S_{cover}$ and there exists e_j in both S_i and S_l. Edges of the form (a_{S_i}, b_{e_j}) are part of the domain by construction and are included in T_i. Similarly, edges of the form (a_{S_l}, b_{e_j}) are also part of the domain by construction and are included in T_i. For each edge $(a_{S_l}, b_{e_j}) \in T_i$, there must also exist an edge (a_{S_i}, b_{e_j}). Thus all nodes in T_i are connected to a_{S_i} - and in particular at most 2 edges away.

Secondly, each element in the M-Tree Set is a tree. Consider element (T_i, f_i). By construction, if a cycle were to exist in T_i it would have to be of the form $a_{S_i}, b_{e_p}, a_{S_l}, b_{e_q}, a_{S_i}$, where both e_p and e_q are in both S_i and S_l. However, such a cycle would imply that two sets have at least two elements in their intersection, which is not possible given we reduced from SC-1.

Next, each element in the M-Tree Set is a monotone tree. $f_i(v) = 1$ for all $v \in T_i$ that are not the root a_{S_i} of (T_i, f_i) and $f_i(a_{S_i}) \geq 1$.

Finally, $f(v) = \sum_{a=1}^{n} f_i(v)$ for all $v \in G$. Each node a_{S_i} such that $S_i \in S_{cover}$ is part of one monotone tree (T_i, f_i) and $f_i(a_{S_i}) = f(a_{S_i})$. Each node $b_{e_j} \in B$

is also part of only one monotone tree (T_i, f_i) and $f_i(e_j) = 1 = f(e_j)$. Finally, for a set $S_l \notin S_{cover}$, a_{S_l} is included in $m = |S_l|$ monotone trees. For each such monotone tree (T_i, f_i), $f_i(a_{S_l}) = 1$, thus $\sum_{a=1}^{n} f_i(a_{S_l}) = m = f(a_{S_l})$.

Thus, we have proven that there exists a M-Tree Set of (G, f) of size $\leq k$.

Second Direction: If there is an M-Tree Set of density graph (G, f) of size $\leq k$, then there is a Set Cover of size $\leq k$.

Let $\{(T_i, f_i)\}$ be an M-Tree set of density graph (G, f) of size k. Each monotone tree (T_i, f_i) in the set has a root node m_i. If multiple vertices in T_i have the maximum value of f_i (as seen in Fig. 1(B)) simply set one of them to be m_i. Each edge in T_i has implicit direction oriented away from m_i. First we prove Lemma 1.

Lemma 1. *Let $b_{e_j} \in B$. Either b_{e_j} is the root of a monotone tree in the M-Tree Set or at least one of its neighbors is the root of a monotone tree in the M-Tree Set.*

Proof. Assume b_{e_j} is not a root of any monotone tree. Consider a monotone tree (T_i, f_i) of the M-Tree Set containing b_{e_j}. This means that $f_i(b_{e_j}) > 0$. Consider node a_{S_l} that is the parent of b_{e_j} in (T_i, f_i). Assume a_{S_l} is not the root node of (T_i, f_i). Because a_{S_l} is not the root of the component, it must have a parent b_{e_d}. Consider the remaining density graph $(G, g = f - f_i)$. By definition of monotone tree, $0 < f_i(b_{e_j}) \leq f_i(a_{S_l}) \leq f_i(b_{e_d})$. By construction, we also know $f(a_{S_l}) = \sum_{e_j \in S_l} f(b_{e_j})$. Therefore, $g(a_{S_l}) > \sum_{e_j \in S_l} g(b_{e_j})$. Because a_{S_l} has more density than the sum of all of its neighbors in (G, g), it is impossible for a_{S_l} to not be the root of at least one monotone tree in any M-Tree Set of (G, g). Thus if b_{e_j} is not the root of any monotone tree in the M-Tree Set, a_{S_l} must be the root of a monotone tree in the M-Tree Set.

We now construct a set cover from the M-Tree Set with the help of Lemma 1. Initialize S_{cover} to be an empty set. For each $a_{S_i} \in A$ that is a root of a monotone tree in the M-Tree Set, add S_i to S_{cover}. Next for each $b_{e_j} \in B$ that is the root of a monotone tree in the M-Tree Set, if there is not already a set $S_i \in S_{cover}$ such that $e_j \in S_i$, choose a set S_l such that $e_j \in S_l$ to add to the Set Cover. Every element must now be covered by S_{cover}. A node e_j that is not the root in any monotone tree in the M-Tree Set must have a neighbor a_{S_l} that is a root in some monotone tree by Lemma 1. The corresponding set S_l was added to S_{cover} - thus e_j is covered. A node e_m such that b_{e_m} is the root of a monotone tree in the M-Tree Set must also be covered by S_{cover} - as a set was added explicitly to cover e_m if it was not already covered. We've added at most one set to the cover for every monotone tree in the M-Tree Set, therefore $|S_{cover}| \leq k$.

Combining both directions, we prove that, given a SC-1 instance, we can construct a density graph (G, f) such that there exists a set cover of size $\leq k$ if and only if the density graph has a M-tree Set of size $\leq k$. This proves the problem is NP-Hard, and thus the problem is NP-Complete.

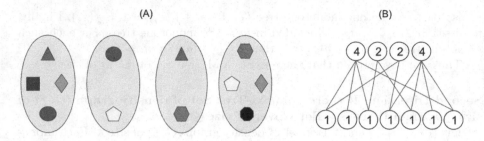

Fig. 4. (A) SC-1 instance with 4 sets and seven elements (B) M-Tree decision problem instance created by following reduction outlined in proof of Theorem 1. The top row consists of nodes in $A \subset V$ in the bipartite graph, which are nodes representing sets, while the bottom row consists of nodes in $B \subset V$ in the bipartite graph, which are nodes representing elements.

3.1   Approximation Hardness

From the proof of Theorem 1, it is easy to see that given an instance of SC-1, the size of its optimal set cover is equivalent to the cardinality of the minimum M-Tree Set of the density graph constructed in the reduction. Hence the hardness of approximation results for SC-1 translate to the minimum M-Tree Set problem too. We therefore obtain the following result, stated in Corollary 1 which easily follows from a similar result for SC-1. The SC-1 result from [14] is stated in Appendix C. We note that while Corollary 1 states the bound in terms of $n =$ of number of relative maxima, a similar bound can be obtained where $n =$ number of vertices.

Corollary 1. *There exists a constant $c > 0$ such that approximating the minimum M-Tree Decomposition problem within a factor of $c \frac{log(n)}{log(log(n))}$, where n is the number of relative maxima on the given density graph, in deterministic polynomial time is possible only if $NP \subset DTIME(2^{n^{1-\epsilon}})$ where ϵ is any positive constant less than $\frac{1}{2}$.*

Proof. Under the assumptions mentioned above, there exists a $c > 0$ such that SC-1 cannot be approximated within a factor of $c \frac{log(n)}{log(log(n))}$, where n is the number of elements in the universe [14]. We note that for a given SC-1 instance, performing the reduction to the M-Tree Set decision problem seen in the proof of Theorem 1 results in a density graph with at most $\frac{n(n-1)}{2} + n$ relative maxima - the upper bound on the number of sets in the SC-1 instance. Thus, the number of relative maxima on the density graph is $O(n^2)$.

For sufficiently large n, we have the following:
$$c \frac{log(n^2)}{log(log(n^2))} = 2c \frac{log(n)}{log(2log(n))} = 2c \frac{log(n)}{log(log(n))+1} < 2c \frac{log(n)}{log(log(n))}$$

Thus there exists a $c > 0$ such that minimum M-Tree Decomposition problem cannot be approximated within a factor of $c \frac{log(n^2)}{log(log(n^2))}$ under the same assumptions mentioned previously. Because the number of relative maxima on the density graph is $O(n^2)$, we can substitute the number of relative maxima for n^2 to establish our final bound.

3.2 Variations of Minimum M-Tree Sets are Also NP-Complete

In addition to proving that computing minimum M-Tree Sets of density graphs is NP-Complete, we have also proven Theorem 5 in Appendix B. The theorem states that computing the minimum CM-Tree Sets, minimum SM-Tree Sets, and minimum FM-Tree Sets of density graphs is also NP-Complete. It should be noted that Corollary 1 can be extended to CM-Tree Sets and FM-Tree Sets. In contrast, SC-1 is not used in the NP-Complete proof for SM-Tree Sets. Thus, Corollary 1 does not apply to minimum SM-Tree Set and there is hope we can develop tighter bounded approximation algorithms for this problem than for the other variations.

4 Algorithms

4.1 Additive Error Approximation Algorithms

Now that we have shown that computing minimum M-Tree Sets of density graphs, as well as several additional variations, is NP-Complete, we focus on developing approximation algorithms. We define two algorithms with different additive error terms. Firstly, we note that a naive upper bound for a given density graph is the number of relative maxima on the graph. We include Algorithm 6 in Appendix F to establish this naive upper bound.

Shifting focus to nontrivial approaches, Algorithm 3 computes the minimum M-Tree Set of a density graph restricted to a spanning tree $T \subseteq G$. We prove that $|\mathsf{minMset}((T,f))| \leq |\mathsf{minMset}((G,f))| + 2g$, where g the *genus* of G, denoted as $\beta_1 G = g$. For a connected graph, $G(V,E)$, $\beta_1 G = |E| - |V| + 1$, which is the number of independent cycles on the graph. This approximation error bound for Algorithm 3 is stated in Theorem 2 and proven in Appendix D.

Algorithm 3: additive-error-algo(($G(V, E)$, f))

Input: A density graph $(G(V,E), f)$ such that $\beta_1 G = g$
Output: An (S)M-Tree set of G, f
(Step 1) Compute g edges that if removed leave a spanning tree T of G
(Step 2) Compute minimum (S)M-Tree set of density tree (T, f) via Algorithm 2

Theorem 2. *Let $(G(V,E), f)$ be a density graph with $\beta_1 G = g$. Let k^* be the size of a minimum (S)M-Tree Set of (G, f). Algorithm 3 outputs an (S)M-Tree Set of size at most $k^* + 2g$.*

4.2 Approximation Algorithm for Minimum SM-Tree Sets of Density Cactus Graphs

A cactus graph is a graph such that no edge is part of more than one simple cycle [10]. See Fig. 5 (A) for an example. Many problems that are NP-hard on graphs belong to P when restricted to cacti - such as vertex cover and independent set [11]. While we do not yet know whether or not computing a minimum M-Tree Set (or any variations) of density cactus graphs is NP-hard, we have developed a 3-approximation algorithm for computing the minimum SM-Tree Set of a density cactus graph.

(A)   (B)

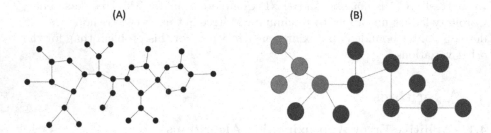

Fig. 5. (A) An example of a cactus graph where no edge is part of more than a single simple cycle. (B) An example of an input for Algorithm 4. The density tree is broken into two subtrees (green and blue) that have a single node as intersection (red). Monotone sweeping is performed iteratively at mode-forced nodes only in one of the subtrees. Once the only remaining mode-forced nodes lie on the other tree, the output tuple containing the number of monotone sweeps performed and the remaining density at the intersection node is returned. (Color figure online)

We first prove Theorem 3, which states that for any density cactus graph (G, f), there exists a spanning tree $T \subseteq G$ such that $|\mathsf{minSMset}(T, f)|$ is at most 3 times $|\mathsf{minSMset}(G, f)|$. The proof can be found in Appendix E.

Algorithm 4: split-tree-algo$((\mathrm{T(V, E)}, \mathrm{f}), T_1, T_2)$

Input: A density tree $(T(V, E), f)$ and two subtrees T_1, T_2 of T that share a
           single node v as intersection

Output: A tuple (a, b) representing the number of monotone sweeps a from
            mode-forced nodes on T_1 to make all mode-forced nodes on T be part
            of T_2, and the remaining function value b at v after the monotone
            sweeps.

While there exists mode-forced node $u \in V$ off of T_2:
- monotone-sweep$((T, f), u, f(u))$
Set $a =$ number of monotone sweeps performed
Set $b =$ remaining density on v
Return (a, b)

Theorem 3. *Let $(G(V,E), f)$ be a density cactus graph. There exists a spanning tree T of G such that $|\text{minSMset}(T, f)| \leq 3|\text{minSMset}(G, f)|$.*

With Theorem 3 proven, we aim to compute the optimal density spanning tree of a density cactus graph. To help compute such an optimal density spanning tree, we first define Algorithm 4. Given a density tree $(T(V, E), f)$ divided into two subtrees T_1 and T_2 that share a single node $v \in V$ as intersection, Algorithm 4 performs monotone sweeping operations on the mode-forced nodes of T_1 until all mode-forced nodes of (T, f) are on T_2. An example of a valid input is seen in Fig. 5 (B). The output is a tuple (a, b), where a is the number of monotone sweeps performed and b is the remaining density on v after performing the monotone sweeps. The tuple will essentially capture how helpful monotone sweeping from T_1 is for building a minimum (S)M-Tree Set on T_2.

Algorithm 4 can be used to help compute the desired density spanning tree. In particular, it is used in Algorithm 5, to cut the optimal edge from each cycle, one cycle at a time. We prove Theorem 4 which states that Algorithm 5 returns an SM-Tree Set at most 3 times larger than a minimum SM-Tree Set of a density cactus graph. The running time of Algorithm 5 is $O(n^3)$ where n is the number of nodes in the input cactus graph. Algorithm 2 ($O(n^2)$) is performed once for each edge ($O(n)$) that is part of a simple cycle.

Algorithm 5: opt-spanning-tree-algo$((G(V, E), f))$

Input: Density cactus graph $(G(V, E), f)$
Output: SM-Tree Set of (G, f)
If G is a tree
- Compute optimal (S)M-Tree Set of (G, f) using Algorithm 2.
Else If G has only a single cycle
- compute optimal sized (S)M-Tree Set of each density spanning tree of G and return smallest cardinality (S)M-Tree Set.
Else (G has multiple simple cycles)
- Compute a leaf cycle $C = c_1, \ldots, c_m$ connected to rest of cycles at c_i
- Let G_C = the simple cycle C with all branches off of each node in the cycle - not including the branches off of c_i that do not lead to other cycles in the graph. Let $G_{\bar{C}} = T - G_C + c_i$.
- Fix a spanning tree $T_{G_{\bar{C}}}$ of $G_{\bar{C}}$.
- For each spanning tree T_i of G_C compute split-tree-algo$((G_C \cup G_{\bar{C}}, f), T_i, T_{G_{\bar{C}}})$
- Set $G = G(V, E - e^*)$ such that e^* is edge removed from C that results in spanning tree with smallest output of split-tree-algo.
- Iterate until basecase (single cycle graph) is achieved

Theorem 4. *Algorithm 5 outputs an SM-Tree Set that is at most 3 times the size of a minimum SM-Tree Set of the input density cactus graph.*

Proof. Theorem 4 clearly holds when G is a tree or contains a single cycle (Lemma 4). Therefore, we only need to prove Theorem 4 holds when G contains multiple simple cycles. Because G is a cactus, a leaf cycle $C = c_1, \ldots, c_m$ exists. Let G_C be the graph of all nodes in the cycle and branches off of those nodes, excluding branches off of c_i that do not lead to other cycles. Let $G_{\bar{C}}$ be the graph of G excluding C and all branches off of C, except for the node c_i itself. G_C has m spanning trees, $T_1, \ldots, T_m$ corresponding to the m edges of C. Fix a spanning tree $T_{G_{\bar{C}}}$ of $G_{\bar{C}}$. We next introduce Lemma 2, which we subsequently prove with two claims. It follows from Lemma 2 that Algorithm 5 outputs a minimum SM-Tree Set on the density spanning tree that has the smallest sized minimum SM-Tree Set of all density spanning trees. Using this with Theorem 3 finishes the proof.

Lemma 2. *Let* $T^* =$ *spanning tree of* G_C *such that the output of* split-tree-algo$(T^* \cup T_{G_{\bar{C}}}, T^*, T_{G_{\bar{C}}})$ *is minimized.* $|\mathsf{minSMset}((T^* \cup T_{G_{\bar{C}}}, f))| \leq |\mathsf{minSMset}((T_k \cup T_{G_{\bar{C}}}, f))|$ *for any spanning tree* $T_k \subseteq G_C$.

Claim. If split-tree-algo$(T_j \cup T_{G_{\bar{C}}}, T_j, T_{G_{\bar{C}}})[0] <$ split-tree-algo$(T_k \cup T_{G_{\bar{C}}}, T_k, T_{G_{\bar{C}}})[0]$ then $|\mathsf{minSMset}((T_j \cup G_{\bar{C}}, f))| \leq |\mathsf{minSMset}((T_k \cup G_{\bar{C}}, f))|$.

Proof. Let T_j, T_k be spanning trees of G_C such that $a_j < a_k$, where $a_j = $ split-tree-algo$(T_j \cup T_{G_{\bar{C}}}, T_j, T_{G_{\bar{C}}})[0]$ and $a_k = $ split-tree-algo$(T_k \cup T_{G_{\bar{C}}}, T_k, T_{G_{\bar{C}}})[0]$. Let $s^* = |\mathsf{minSMset}((T_{G_{\bar{C}}}, f))|$.

Algorithm 4 performs Algorithm 2 sweeping from mode-forced nodes on T_j, but stops once mode-forced nodes only remain on $T_{G_{\bar{C}}}$. Thus it is still constructing minimum SM-Tree Sets but stopping short of completion. The first element of the output of Algorithm 4 indicates the number of iterations required to have only mode-forced nodes on $T_{G_{\bar{C}}}$. $|\mathsf{minSMset}((T_j \cup T_{G_{\bar{C}}}, f))| \leq a_j + s^*$. Similarly, $|\mathsf{minSMset}((T_k \cup T_{G_{\bar{C}}}, f))| \geq a_k + s^* - 1$.

Claim. If split-tree-algo$(T_j \cup T_{G_{\bar{C}}}, T_j, T_{G_{\bar{C}}})[0] = $ split-tree-algo$(T_k \cup T_{G_{\bar{C}}}, T_k, T_{G_{\bar{C}}})[0]$ and split-tree-algo$(T_j \cup T_{G_{\bar{C}}}, T_j, T_{G_{\bar{C}}})[1] <$ split-tree-algo$(T_k \cup T_{G_{\bar{C}}}, T_k, T_{G_{\bar{C}}})[1]$ then $|\mathsf{minMset}((T_j \cup G_{\bar{C}}, f))| \leq |\mathsf{minMset}((T_k \cup G_{\bar{C}}, f))|$.

Proof. Let T_j, T_k be spanning trees of G_C such that $a_j = a_k$ and $b_j < b_k$ where $(a_j, b_j) = $ split-tree-algo$(T_j \cup T_{G_{\bar{C}}}, T_j, T_{G_{\bar{C}}})$ and $(a_k, b_k) = $ split-tree-algo$(T_k \cup T_{G_{\bar{C}}}, T_k, T_{G_{\bar{C}}})$.

$a_j = a_k$ indicates that both T_j and T_k require the same number of iteration of monotone sweeps to leave mode-forced nodes on $T_{G_{\bar{C}}}$. However, $b_j < b_k$ means that T_j is more helpful than T_k for reducing the minimum SM-Tree Set size on the remainder in the same number of monotone sweeps by Claim 2.3.

5   Conclusion

We have proven that decomposing density graphs into minimum M-Tree Sets, and many other variations, becomes NP-Complete when the input graph is not restricted to trees. We have also shown that, under reasonable assumptions, no polynomial time constant factor approximation exists for most variations. We provided additive error approximations algorithms for the minimum M-Tree Set problem, as well as developed a 3-approximation algorithm for minimum SM-Tree Sets for density cactus graphs. Future work will be to close the gap between the bounds of approximation we have established with the error bounds of the algorithms we have developed.

A   Proof of Claim 2.3

Proof. We will prove the claim by contradiction. Assume that $|\mathsf{minMset}((T, R_{v,b}f))| > |\mathsf{minMset}((T, R_{v,a}f))|$.

In particular, we will construct two new density trees, (T_a, f_a) and (T_b, f_b), as follows: T_a is equal to our starting tree with the addition of two nodes v_a and v_∞, with two additional edges connecting to v_a to both v_∞ and v. Set $f_a(v_a) = a$ and $f_a(v_\infty) = \infty$. Similarly define T_b and f_b.

Now imagine we run Algorithm 2 on (T_a, f_a). v_∞ is a mode-forced node, and thus we can perform the first iteration in Algorithm 2 on v_∞. Sweeping from v_∞ will leave remainder with a minimum M-Tree Set of size $|\mathsf{minMset}((T_a, f_a))| - 1$. The remainder is exactly the same as $(T, R_{v,a}f)$ at all nodes $v \in V$, and is zero at our newly added nodes. Hence, $|\mathsf{minMset}((T_a, f_a))| = |\mathsf{minMset}((T, R_{v,a}f))| + 1$. Similarly, by performing Algorithm 2 on (T_b, f_b), we have $|\mathsf{minMset}((T_b, f_b))| = |\mathsf{minMset}(T, R_{v,b}f)| + 1$. Now if our initial assumption is true, then by the above argument we have that

$$|\mathsf{minMSet}((T_b, f_b))| > |\mathsf{minMSet}((T_a, f_a))|. \tag{1}$$

However, we could construct an M-tree set of (T_b, f_b) as follows: First construct one monotone tree rooted at v_∞ that leaves no remainder at both v_∞ and v_b, then perform the monotone sweep operation starting at v with starting value a to build the rest of the component. Note that the remainder after removing this tree is in fact $(T_a, R_{v,a}f)$, which we can then decompose using the minimum M-tree set of $(T_a, R_{v,a}f)$. In other words, we can find a M-tree set for (T_b, f_b) with $|\mathsf{minMset}(T, R_{v,a}f)| + 1 = |\mathsf{minMset}(T_a, f_a)|$. This however contradicts with Eq. (1) (and the correctness of Algorithm 2). Hence our assumption cannot hold, and we must have that $|\mathsf{minMset}(T, R_{v,b}f)| \leq |\mathsf{minMset}(T, R_{v,a}f)|$. This proves the claim.

We note that while this proof is for M-Tree Sets specifically, the proof for SM-Tree Sets follows identical arguments.

B   Complexity

In this section, we prove Theorem 5, which states many variations of the minimum M-Tree set problem are also NP-Complete.

Theorem 5. *Given a density graph $(G(V, E), f)$ and a parameter k, determining whether or not there exists a CM-Tree Set, SM-Tree Set, or FM-Tree Set of size $\leq k$ are all NP-Complete.*

B.1   Proof of Theorem 5: CM-Tree Sets

Firstly, the problem is clearly in NP. We will follow the same reduction from SC-1 as seen in the proof of Theorem 1 to prove NP-Hardness, with one additional step.

The first direction we follow identical arguments to create an M-Tree Set of appropriate size, but do not yet have a CM-Tree Set. In particular, consider the M-Tree Set at the end of the proof - the only possible edges missing are edges (a_{S_j}, b_e) such that S_j is in the Set Cover and contains e, but another set S_i containing e is in the Set Cover and $f_i(b_e) = 1$. We will modify the M-Tree Set to ensure every such edge that is left out is included in a component. Consider an element e that is in n sets in the set cover, where $n > 1$. Let (T_i, f_i) be the monotone tree in the M-Tree Set such that $f_i(b_e) = 1$. Set $f_i(b_e) = \frac{1}{n}$. Additionally, for each set S_j in the set cover such that $b_e \in S_j$, add (a_{S_j}, b_e) to monotone tree (T_j, f_j) and set $f_j(b_e) = \frac{1}{n}$. Then, for each set S_k such that S_k is not in the set cover and $b_e \in S_k$, add (b_e, a_{S_k}) to each (T_j, f_j) and set $f_j(a_{s_k}) = \frac{1}{n}$. We still have an M-Tree Set, as each component clearly remains a monotone tree, and the sum of function values at each node is equal to what it was prior to the modification. Once this modification is performed for every element contained within multiple sets in the set cover, we have an M-Tree set with every edge in the input domain included in at least one monotone tree. The second direction is identical to the previous proof.

B.2   Proof of Theorem 5: SM-Tree Sets

Firstly, the problem is clearly in NP. In order to prove this decision problem is NP-Hard - we first show that a specific instance of Vertex Cover - where for the given input graph $G(V, E)$, for any two vertices $u, v \in V$, there is at most one vertex $w \in V$ that is adjacent to both u and v - is NP-Complete. This will limit the number of connected components in the intersection between two components to be at most one in our reduction to the M-Tree problem.

Lemma 3. *Given a graph $G(V, E)$ such that for any two vertices $u, v \in V$, there is at most one vertex $w \in V$ that is adjacent to both u and v and an integer k, determining whether or not there exists a vertex cover of size $\leq k$ is NP-Complete.*

Proof. This is a specific instance of Vertex Cover and is clearly in NP. To show it is in NP-Hard use the same reduction from 3-SAT to regular Vertex Cover as seen in [7], but use a "restricted" version of 3-SAT where we can assume the following:

- A clause has 3 unique literals
- A clause cannot have a literal and its negation

These assumptions are safe because we can transform any 3-SAT instance that has any such clauses to an equivalent 3-SAT instance with no such clauses in polynomial time. Thus this restricted version of 3SAT is also NP-Complete.

Consider the graph created in the reduction of this restricted 3-SAT to Vertex Cover. Two literal vertices will have no shared neighbors by design. Any literal vertex and vertex in a clause will only have 1 neighbor if the literal vertex is also in the clause or the clause vertex is the negation of the literal vertex. Two vertices in clauses will only have a single shared neighbor - the literal vertex if they are the same literal (and are thus in different clauses), or the final clause vertex if they are in the same clause. Thus, the reduction is also a reduction to our special Vertex Cover problem, and follows the exact same proof.

We will now reduce the special instance of Vertex Cover to the SM-Tree decision problem to prove NP-Hardness. We follow a very similar reduction as seen in the proof of Theorem 1. We construct a bipartite graph $G(V = A \cup B, E)$, with nodes in A corresponding to nodes in the instance of Vertex Cover, and nodes in B representing edges in the instance of Vertex Cover. We then build density function f on the nodes of V, setting $f(a_v) =$ degree of node v in Vertex Cover instance for each node $a_v \in A$, and $f(b_e) = 1$ for each edge e in the Vertex Cover instance. Prove both directions the exact same way as shown in the proof of Theorem 1, but note that for the first direction, because of the restriction on our input graph, any two components of the decomposition can have at most a single vertex in their intersection.

B.3   Proof of Theorem 5: FM-Tree Sets

Firstly, the problem is clearly in NP. To show the problem is in NP-Hard, we follow the exact same reduction from SC-1 as seen in proof of Theorem 1. For the first direction - we note that the M-Tree Set we have constructed is also an FM-Tree Set - as each set S_i in the set cover is a root of a component (T_i, f_i) such that $f_i(a_{S_i}) = f(a_{S_i})$. The second direction remains the same - though the argument that if a node b_e is not the maximum of any monotone tree then one of its neighbors must be is slightly different. In this case, the neighbor must be a maximum in the same monotone tree it is a parent of b_e in - not being so would contradict that the set is in fact a FM-Tree Set.

C   Set Cover Intersection 1 Approximation Bound [14]

Theorem 6. *There exists a constant $c > 0$ such that approximating the SC-1 problem within a factor of $c \frac{log(n)}{log(log(n))}$, where n is the number of elements in the universe, in deterministic polynomial time is possible only if $NP \subset DTIME(2^{n^{1-\epsilon}})$ where ϵ is any positive constant less than $\frac{1}{2}$.*

D   Proof of Theorem 2

Proof. We need to prove Lemma 4 to provide an upper bound on $|\text{minMset}(T, f)|$ for any spanning tree $T \subseteq G$. Algorithm 2 will then output an M-Tree Set of size at most equal to the upper bound, thus completing our proof. The proof is identical for SM-Tree Sets.

Lemma 4. *Let $(G(V, E), f)$ be a density graph with $\beta_1 G = g$ and $|\mathsf{minMset}(G, f)| = k^*$. For any spanning tree $T \subseteq G$, $|\mathsf{minMset}(T, f)| \leq k^* + 2g$*

Let $M = \{(T_i, f_i)\}$ be a minimum M-Tree set of (G, f). Let E_{cut} be the set of g edges that if removed from G leave spanning tree T. Firstly, we note that $|\mathsf{minMset}(G, f)| \leq |\mathsf{minMset}(T, f)|$, as any M-Tree Set of (T, f) is also an M-Tree Set of (G, f).

We will construct an M-Tree Set of (T, f) from a minimum M-Tree Set of G. For each monotone tree $(T_i, f_i) \in M$, consider an edge $e_j = (u, v) \in E_{cut}$ that is in T_i. There is implicit direction to e_j with respect to the root of (T_i, f_i), meaning either (1) $(u \to v)$ or (2) $(v \to u)$. If (1) is the case, we can cut the branch rooted at v off of (T_i, f_i) to create two non-intersecting monotone trees. See Fig. 6 for an example. We perform a similar operation if (2) is the case, but instead cut the branch rooted at u. Perform this cut for each edge in E_{cut} to divide (T_i, f_i) into, at most, $|E_{cut}| + 1$ non-intersecting monotone trees. After dividing each tree into at most $|E_{cut}| + 1$ non-intersecting monotone trees, we make 2 key observations - (1) we still have a M-Tree Set of (G, f) and (2) no edge in E_{cut} is in any monotone tree in the M-Tree Set. Thus the M-Tree Set is also an M-Tree Set of (T, f).

We can shrink the size of this M-Tree Set by summing the components that share the same root. In particular, consider an edge $e_j = (u, v) \in E_{cut}$. We have created as many as k^* additional monotone trees rooted at u and as many as k^* additional monotone trees rooted at v. Sum the monotone trees rooted at u to create a single monotone tree rooted at u. The sum would clearly still be a monotone tree because all monotone trees are subtrees of tree T, so no cycle or non-non-increasing path from u will be created. We can similarly do the same for v, and for all edges in E_{cut}. This we have a new M-Tree Set of (T, f), with (at most) an additional monotone tree rooted at each node of each edge in E_{cut} when compared to the original M-Tree Set of G. Thus $|\mathsf{minMset}(T, f)|$ is bounded above by $k^* + 2g$.

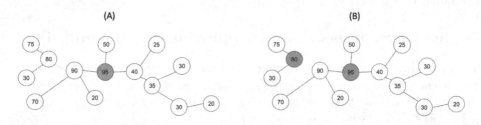

Fig. 6. (A) shows a single monotone tree with its root colored red and an edge colored green. Cutting the green edge leaves us with two non-intersecting monotone trees shown in (B). (Color figure online)

E   Proof of Theorem 3

Proof. Let $M = \{(T_i, f_i)\}$ be a minimum SM-Tree Set of (G, f), $k = |M|$, and $\beta_1 G = g$. Consider graph $G' = \bigcup_{i=1}^{k} T_i$. Let $\beta_1 G' = g'$. We note that $g' \leq g$. and that M is also a minimum SM-Tree Set of G'. We will use G' to help construct a spanning tree T of G with an SM-Tree Set with the desired cardinality. Note that if G' has no cycles then there obviously exists a spanning tree T of G such that $|\mathsf{minSMset}(T, f)| = |\mathsf{minSMset}(G, f)|$. Additionally, if $g' = 1$, then creating T by removing any edge from the simple cycle $|\mathsf{minSMset}(T, f)| \leq |\mathsf{minSMset}(G, f)| + 2$ (similar arguments to Lemma 4 and Theorem 2). Therefore assume $g' \geq 2$. Construct spanning tree T as follows:

- Add each edge in G that is not part of a simple cycle.
- For each simple cycle in G that is not in G', add all edges of cycle to T except for one missing in G' (it does not matter which if multiple such edges exist).
- For each simple cycle in G that is in G', add all edges of that cycle to T except for one (does not matter which).

Let k' be the $|\mathsf{minSMset}(T, f)|$. k' is bounded above by $k + 2g'$, because removing edges that aren't used in the any monotone tree in M from the domain will not change $|\mathsf{minSMset}(G', f)|$. Additionally, removing an edge from a simple cycle in g' will increase the $|\mathsf{minSMset}(G', f)|$ by at most 2 (again by Lemma 4).

k' is also bounded below by $2 + g'$. For each cycle in G', the number of monotone trees in M that contain nodes in a simple cycle must be at least 3 - otherwise the set cannot be an SM-Tree Set. So consider a leaf cycle C_0 in G'. We know that there are at least 3 monotones trees in M that cover C_0. For a cycle C_1 adjacent to C_0 in G' that there is a single path between the two cycles, and the monotone trees that cover C_0 cannot completely cover C_1, otherwise M would not be an SM-Tree Set. There must be at least one monotone tree with nodes on C_1 and no nodes on C_0. Continuing traversing the graph to all cycles and it is clear that for each cycle there must be an additional monotone tree added to the SM-Tree Set. Thus, we cannot have an SM-Tree Set of size less than $2 + g'$.

From above, we have $\frac{k'}{k} \leq \frac{k+2g'}{k} \leq \frac{2+g'+2g'}{2+g'} \leq \frac{3g'+2}{g'+2} < 3$.

F   Algorithms

F.1   Naive Approximation Algorithm

As stated in the main paper, given a density graph (G, f), a natural upper bound for $|\mathsf{minSMset}(G, f)|$ is the number of relative maxima on the density graph. Algorithm 6 constructs monotone trees rooted at each relative maxima on the input density graph. Starting at a root, depth-first search (DFS) is performed to reach every node that can be reached via a non-increasing path from the root. DFS stops once no nodes remain or all remaining nodes are not reachable from the root via a non-increasing path. We call this graph traversal algorithm *monotone DFS*. Perform monotone DFS from each relative maxima to build an M-Tree Set. The M-Tree Set will have size at most one less than the number of relative maxima more than the size of a minimum M-Tree Set. Figure 7 shows an example output of Algorithm 6.

Algorithm 6: naive-algo$((G(V, E), f))$

Input: A density graph $(G(V, E), f)$
Output: An (S)M-Tree set of $(G(V, E), f)$
(Step 1) Compute set M containing the relative maxima of f on G.
(Step 2) For each relative maxima $m_i \in M$, perform monotone DFS to build a component (T_i, f_i)
(Step 3) Return all (T_i, f_i)

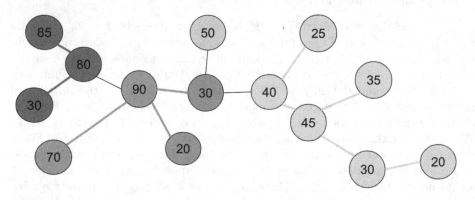

Fig. 7. An M-Tree Set with of a density tree with 4 monotone trees computed by Algorithm 6.

References

1. Aanjaneya, M., Chazal, F., Chen, D., Glisse, M., Guibas, L., Morozov, D.: Metric graph reconstruction from noisy data. In: Proceedings of 27th Symposium on Computational Geometry, pp. 37–46 (2011)
2. Banerjee, S., et al.: Semantic segmentation of microscopic neuroanatomical data by combining topological priors with encoder-decoder deep networks. Nat. Mach. Intell. **2**, 585–594 (2020)
3. Baryshnikov, Y., Ghrist, R.: Minimal unimodal decompositions on trees. J. Appl. Comput. Topol. **4**, 199–209 (2020)
4. Chazal, F., Huang, R., Sun, J.: Gromov-Hausdorff approximation of filamentary structures using reeb-type graphs. Discrete Comput. Geom. **53**(3), 621–649 (2015)

5. Chen, H., Xiao, H., Liu, T., Peng, H.: SmartTracing: self-learning-based neuron reconstruction. Brain Inform. **2**(3), 135–144 (2015)
6. Dey, T.K., Wang, J., Wang, Y.: Graph reconstruction by discrete Morse theory. In: Proceedings of International Symposoium on Computational Geometry, pp. 31:1–31:15 (2018)
7. Garey, M.R., Johnson, D.S.: Computers and Intractability, vol. 174. Freeman, San Francisco (1979)
8. Ge, X., Safa, I.I., Belkin, M., Wang, Y.: Data skeletonization via reeb graphs. In: Shawe-Taylor, J., Zemel, R.S., Bartlett, P.L., Pereira, F., Weinberger, K.Q. (eds.) Advances in Neural Information Processing Systems, vol. 24, pp. 837–845. Curran Associates, Inc. (2011)
9. Hang, Z., et al.: Dense reconstruction of brain-wide neuronal population close to the ground truth. bioRxiv (2018)
10. Harary, F., Uhlenbeck, G.E.: On the number of Husimi trees: I. Proc. Natl. Acad. Sci. **39**(4), 315–322 (1953)
11. Hare, E., Hedetniemi, S., Laskar, R., Peters, K., Wimer, T.: Linear-time computability of combinatorial problems on generalized-series-parallel graphs. In: Johnson, D.S., Nishizeki, T., Nozaki, A., Wilf, H.S. (eds.) Discrete Algorithms and Complexity, pp. 437–457. Academic Press (1987)
12. Hastie, T.J.: Principal curves and surfaces. Ph.D. thesis, Stanford University (1984)
13. Kégl, B., Krzyżak, A.: Piecewise linear skeletonization using principal curves. IEEE Trans. Pattern Anal. Machine Intell. **24**, 59–74 (2002)
14. Kumar, V.S.A., Arya, S., Ramesh, H.: Hardness of set cover with intersection 1. In: Montanari, U., Rolim, J.D.P., Welzl, E. (eds.) ICALP 2000. LNCS, vol. 1853, pp. 624–635. Springer, Heidelberg (2000). https://doi.org/10.1007/3-540-45022-X_53
15. Lecci, F., Rinaldo, A., Wasserman, L.: Statistical analysis of metric graph reconstruction. J. Mach. Learn. Res. **15**(1), 3425–3446 (2014)
16. Magee, L., Wang, Y.: Graph skeletonization of high-dimensional point cloud data via topological method. J. Comput. Geometry **13**, 429–470 (2022)
17. Megiddo, N., Tamir, A.: On the complexity of locating linear facilities in the plane. Oper. Res. Lett. **1**(5), 194–197 (1982)
18. Ozertem, U., Erdogmus, D.: Locally defined principal curves and surfaces. J. Mach. Learn. Res. **12**, 1249–1286 (2011)
19. Quan, T., et al.: Neurogps-tree: automatic reconstruction of large-scale neuronal populations with dense neurites. Nat. Methods **13**(1), 51–54 (2016)
20. Sousbie, T.: The persistent cosmic web and its filamentary structure i. Theory and implementation. Mon. Not. Roy. Astron. Soc. **414**, 350–383 (2011)
21. Wang, D., et al.: Detection and skeletonization of single neurons and tracer injections using topological methods. arXiv preprint arXiv:2004.02755 (2020)
22. Wang, S., Wang, Y., Li, Y.: Efficient map reconstruction and augmentation via topological methods. In: Proceedings of 23rd ACM SIGSPATIAL, p. 25. ACM (2015)

Scheduling

Exact and Approximation Algorithms for the Multi-depot Data Mule Scheduling with Handling Time and Time Span Constraints

Minqin Liu, Wei Yu$^{(\boxtimes)}$, Zhaohui Liu, and Xinmeng Guo

School of Mathematics, East China University of Science and Technology,
Shanghai 200237, China
{y30211273,y30221654}@mail.ecust.edu.cn, {yuwei,zhliu}@ecust.edu.cn

Abstract. In this paper, we investigate the data mule scheduling with handling time and time span constraints (DMSTC) in which the goal is to minimize the number of data mules dispatched from a depot that are used to serve target sensors located on a wireless sensor network. Each target sensor is associated with a handling time and each dispatched data mule must return to the original depot before time span D. We also study a variant of the DMSTC in which the objective is to minimize the total travel distance of the data mules dispatched.

We give exact and approximation algorithms for the DMSTC on a path and their multi-depot version. For the first objective, we show an $O(n^4)$ polynomial time algorithm for the uniform 2-depot DMSTC on a path where at least one depot is on the endpoint (n indicates the number of target sensors). And we present a new 2-approximation algorithm for the non-uniform DMSTC on a path. For the second objective, we derive an $O((n+k)^2)$-time algorithm for the uniform multi-depot DMSTC on a path, where k is the number of depots. For the non-uniform multi-depot DMSTC on a path or cycle, we give a 2-approximation algorithm.

Keywords: Data Mule Scheduling · Multi-Depot · Handling Time · Time Span Constraint · Approximation Algorithm

1 Introduction

Recently, Chen et al. [3] introduced the data mule scheduling with handling time and time span constraints (DMSTC). In this paper, we consider the following multi-depot extension of the DMSTC.

Let $G = (V, E)$ be an undirected graph, which represents a stationary sensor network with vertex set V and edge set E. The vertex set V is partitioned into a depot set $R = \{d_1, d_2, \ldots, d_k\}$ and a set $N = \{u_1, u_2, \ldots, u_n\}$ of target sensors. Each edge $e = (u, v) \in E$ is associated with a nonnegative length, denoted by $l(e)$ (or $l(u, v)$ if $e = (u, v)$), indicating the travel time (or travel distance) of the

© The Author(s), under exclusive license to Springer Nature Switzerland AG 2024
W. Wu and J. Guo (Eds.): COCOA 2023, LNCS 14461, pp. 129–140, 2024.
https://doi.org/10.1007/978-3-031-49611-0_9

data mules along the edge e. Here data mules are mobile devices that can move around the network to perform maintenance operations, including data collection and battery charging/replacement, etc. All the data mules are initially located at the depots and each data mule must return to the same depot where it is initially situated. Each vertex $u \in N$, i.e., a target sensor, is associated with a handling time (or service time) $c(u)$. Let $c(N') = \sum_{u \in N'} c(u)$ for any $N' \subseteq N$. The route of a data mule corresponds to a closed walk starting from and ending at some depot $d \in R$. The length $l(C)$ of a closed walk C is defined as the total length of the edges traversed by it. We are asked to find a series of closed walks $C_1, C_2, \ldots, C_m$ as the routes of m data mules and a partition $(N_1, N_2, \ldots, N_m)$ of N such that for $j = 1, 2, \ldots, m$,

(i) C_j traverses each target sensor in N_j at least once;
(ii) C_j is feasible, i.e. the load of C_j, defined as $L(C_j) = l(C_j) + c(N_j)$, is no more than a prescribed time span constraint D.

The goal of the problem, which we call the multi-depot DMSTC, is to minimize the number m of data mules dispatched.

When $k = 1$, the multi-depot DMSTC turns into the DMSTC investigated in [3]. We use (multi-depot) DMSTC^{tree} (DMSTC^{path}, DMSTC^{cycle}) to represent the special case of the (multi-depot) DMSTC where the sensor network is a tree (path, cycle).

In a variant of the DMSTC, called the DMSTC$_l$, the objective is to minimize the total travel distance of the data mules dispatched, i.e. $\sum_{j=1}^{m} l(C_j)$. The special cases of the DMSTC$_l$, i.e. DMSTC$_l^{tree}$, DMSTC$_l^{path}$, DMSTC$_l^{cycle}$, are defined similarly. Furthermore, we also have the multi-depot DMSTC$_l$ where there are multiple depots available.

The uniform (multi-depot) DMSTC/DMSTC$_l$ is a particular case of the (multi-depot) DMSTC/DMSTC$_l$ where all the handling times are identical.

The DMSTC is a fundamental optimization problem arised naturally in wireless sensor networks [7]. In the environment where the energy consumption of sensors is huge for transmitting straightforwardly data between the sensors and the depots, using data mules to move between sensors to collect data can greatly save the energy of the sensors and hence prolong the lifetime of the whole network. Apart from the wireless sensor networks, the DMSTC and its variants have been widely used in many other practical applications as well as in our daily life, including real-time surveillance of the battlefield [10], monitoring of environmental conditions [12], healthcare [1] and so on.

Since the DMSTC and the DMSTC$_l$ generalize the well-known Metric Traveling Salesman Problem (Metric TSP), they are both NP-hard. Consequently, we focus on approximability results as well as their polynomially solvable special cases.

1.1   Related Work

Chen et al. [3] proved the NP-hardness of a restricted case of the DMSTC^{path}, which we denote by DMSTC$_0^{path}$, where the depot is located at one of the two endpoints of the path network. However, they showed that the uniform DMSTC$_0^{path}$ can be solved in $O(n)$ time by a greedy algorithm. The authors also generalized the greedy algorithm to the (non-uniform) DMSTC$_0^{path}$ and achieved an approximation ratio of $\frac{2D-2l_{\max}}{D-2l_{\max}}$, where $l_{\max}$ is defined for a general network as the length of the shortest path between the depot and the farthest target sensor[1]. For the path network with the depot being one of the two endpoints, $l_{\max}$ is simply the total edge length of the network.

Based on the structure properties given by Perez-Escalona et al. [11], Yu and Liu [13] were able to show that the uniform DMSTC^{path} can be solved in $O(n^3)$ time. They also gave an $O(n^5)$-time algorithm to solve the uniform DMSTC^{cycle} to optimality. In addition, they described an approximation preserving reduction from the DMSTC to the Distance Constrained Vehicle Routing Problem (DVRP), a special case of the DMSTC with all the handling times being zero, and this reduction may be extended to deal with the multi-depot problems. As a result, a 2-approximation algorithm for the DMSTC^{path} can be derived from a 2-approximation algorithm for the DVRP defined on tree metrics given by Nagarajan and Ravi [9] and an approximation algorithm with the same ratio for the DMSTC can be obtained from a $O\left(\min\{\log n, \frac{\log D}{\log\log D}\}\right)$-approximation algorithm developed by Friggstad and Swamy [5]. Very recently, Li and Zhang [7] presented an $O(\log \frac{D}{\mu})$-approximation algorithm for the DVRP, which they called the Rooted Budgeted Cycle Cover Problem. Here μ indicates the minimum difference of any two distances between V and the depot. For the multi-depot DVRP, they provided an approximation algorithm with ratio $O(\log n)$ using the s-t orienteering problem as a subroutine. Liang et al. [8] proved that any γ-approximation algorithm for the DVRP can be transformed into a $k\gamma$-approximation algorithm for the multi-depot DVRP. Actually, they derived this result in a more general context of distance constrained sweep coverage (see also [2]).

As for the DMSTC$_l$, all the previous results are with respect to its special case with all the handling times being zero, which we term as the DVRP$_l$ for consistency (Li et al. [6] and Liang et al. [8] used other names.) Li et al. [6] demonstrated a close relation between the DVRP and the DVRP$_l$ by showing that any α-approximation for one of these two problems yields a 2α-approximation for the other problem. They also gave $\left(1 + \frac{\alpha D}{D-2l_{\max}}\right)$-approximation algorithms for both problems based on a tour-partitioning approach, where α indicates the best available approximation ratio for the Metric TSP. Liang et al. [8] pointed out that this approximation ratio only holds for the DVRP$_l$ with the assumption that D is no more than the length of the optimal traveling salesman tour. They showed how to remove this assumption while achieving a better approximation ratio of $\frac{\alpha D}{D-2l_{\max}}$. Moreover, the authors proved that the algorithm by Nagarajan

---

[1] One can see that if $D < 2l_{\max} + c(u_i)$ for some i then the DMSTC/DMSTC$_l$ has no feasible solution and any target sensor u_i with $D = 2l_{\max} + c(u_i)$ has to be served by a private data mule. Therefore, we assume that $D > \max_{1 \le i \le n}\{2l_i + c(u_i)\}$ and hence $D > 2l_{\max}$.

and Ravi [9] is versatile enough to achieve an approximation ratio of 2 for the $DVRP_l$ defined on a tree.

1.2 Our Results

In this paper, we give exact and approximation algorithms for the $DMSTC/DMSTC_l$ and their multi-depot version.

First, we present a polynomial algorithm with time complexity $O(n^4)$ for the uniform 2-depot $DMSTC^{path}$ where at least one depot is an endpoint of the path network. For the general uniform 2-depot $DMSTC^{path}$, we give an example to demonstrate the complicated structure of the optimal solution. Second, we show that the algorithm for the (non-uniform) $DMSTC_0^{path}$ by Chen et al. [3] has a tight approximation ratio of 2. Previously, Chen et al. only showed a super-constant upper bound on the approximation ratio. We also propose a new 2-approximation algorithm for the $DMSTC_0^{path}$ motivated by the First-Fit algorithm for the classical bin packing problem. In addition, we extend both algorithms to the $DMSTC^{path}$. Finally, we derive an $O((n+k)^2)$-time exact algorithm for the uniform multi-depot $DMSTC_l^{path}$ based on which we present a 2-approximation algorithm for the multi-depot $DMSTC_l^{path}$. We also extend the results on the (uniform) multi-depot $DMSTC_l^{path}$ to the (uniform) multi-depot $DMSTC_l^{cycle}$.

The rest of the paper is organized as follows. Section 2 describes the preliminaries and notations. In Sect. 3, we deal with the polynomial algorithms for the uniform 2-depot $DMSTC^{path}$. In Sect. 4, we discuss the approximation algorithms for the $DMSTC^{path}$. Approximation and polynomial algorithms for the multi-depot $DMSTC_l^{path}$ and the multi-depot $DMSTC_l^{cycle}$ are presented in Sects. 5 and 6, respectively. Finally, we conclude the paper in Sect. 7.

2 Preliminaries

Throughout the paper, we analyze algorithms for different versions of the $DMSTC/DMSTC_l$ formulated in the introduction. When discussing a particular problem, we denote by OPT both the optimal solution and the optimal value. We use SOL to represent the feasible solution obtained by the algorithm and its objective value.

For the multi-depot $DMSTC/DMSTC_l$, the input consists of an undirected graph $G = (V, E)$, depot set $R = \{d_1, d_2, \ldots, d_k\}$, a set $N = \{u_1, u_2, \ldots, u_n\}$ of target sensors, length function $l(\cdot)$ on E and a time span constraint $D \geq 0$. For any $N' \subseteq N$, let $c(N') = \sum_{u \in N'} c(u)$. For a subgraph G' of G, $V(G')$ and $E(G')$ indicates the vertex set and edge set of G', respectively. Given a feasible solution consisting of a series of feasible routes (or closed walks, tours) $C_1, C_2, \ldots, C_m$ and a partition $(N_1, N_2, \ldots, N_m)$ of N such that $N_j \subseteq V(C_j)$ for each j, we say that $u \in N_j$ is served by the jth data mule (or C_j). The vertices in $V(C_j) \setminus N_j$ are passed by the jth data mule (or C_j) but not served by it. Here each C_j

has a length $l(C_j) = \sum_{e \in E(C_j)} l(e)^2$ and C_j is feasible means that the load of C_j, denoted by $L(C_j) = l(C_j) + c(N_j)$, is at most D. When the partition $(N_1, N_2, \ldots, N_m)$ is clear in the context, a feasible solution is simply represented by the feasible routes $C_1, C_2, \ldots, C_m$. The region of a depot d (with respect to a feasible solution) is the set of targeted sensors served by the data mules departing from depot d. Note that the regions of distinct depots are disjoint.

For the uniform multi-depot DMSTC/DMSTC$_l$, we denote by c the common handling time of all the target sensors. The following two lemmas describe some situations where we can modify a given feasible solution to derive a new feasible solution of no worse objective value.

Lemma 1. *Given a feasible solution S for the uniform multi-depot DMSTC/DMSTC$_l$, if there are two routes C_1 and C_2 in S such that C_1 (C_2) serves a target sensor u (v) and both routes pass by u and v, then we can derive a new feasible solution $\hat{S}$ by switching the service of u and v (between C_1 and C_2) and maintaining the service of all the vertices in $N \setminus \{u, v\}$. Moreover, the objective value of $\hat{S}$ is no more than that of S.*

Proof. Suppose that after switching the two routes C_1 and C_2 turn into $\hat{C}_1$ and $\hat{C}_2$, respectively. Then u (v) is served by $\hat{C}_2$ ($\hat{C}_1$) in $\hat{S}$. Since $c(u) = c(v) = c$ and each of C_1 and C_2 pass by both u and v, the load of $\hat{C}_1$ ($\hat{C}_2$) is no more than that of C_1 (C_2), which can not exceed D by the feasibility of S. Because the other routes in S are unchanged, $\hat{S}$ is indeed a feasible solution. Moreover, $l(\hat{C}_1)$ ($l(\hat{C}_2)$) is not greater than $l(C_1)$ ($l(C_2)$), which implies that the objective value of $\hat{S}$ can not exceed that of S. $\qquad\square$

Lemma 2. *Given a feasible solution S for the uniform multi-depot DMSTC/DMSTC$_l$, if there is a route C departing from some depot d in S such that C passes by another depot d', then we can derive a new feasible solution $\hat{S}$ by replacing C with another route C' departing from d' and maintaining the other routes in S. Moreover, the objective value of $\hat{S}$ is no more than that of S.*

Proof. Since C is a closed walk departing from d and containing d', this walk can also be seen as a closed walk C' that departs from d' and passes by d. Clearly, $l(C') = l(C)$ and $L(C') = L(C)$. In addition, all the target sensors served by C are now served by C'. Thus, $\hat{S}$ is a desired feasible solution. $\qquad\square$

For the DMSTC^{path}/DMSTC$_l^{path}$, $G = (V, E)$ is a path network where the depots in $R = \{d_1, d_2, \ldots, d_k\}$ and the target sensors in $N = \{u_1, u_2, \ldots, u_n\}$ are aligned from left to right on a line. Without loss of generality, we assume that d_i is on the left of d_j and u_i is on the left of u_j for any $i < j$. For convenience, we also rename the vertices from left to right as $v_1, v_2, \ldots, v_{n+k}$. Then the edge set E consists of $n + k - 1$ edges $(v_1, v_2), (v_2, v_3), \ldots, (v_{n+k-1}, v_{n+k})$. In particular, $d_1 = v_1$ means the depot d_1 is the left endpoint of the path and $d_k = v_{n+k}$ implies

---

² $E(C_j)$ may be a multiset because an edge may be traversed multiple times by the closed walk C_j. In that case, an edge e appearing t times in C_j will contribute $t \cdot l(e)$ to $l(C_j)$.

that d_k is the right endpoint of the path. A route departing from a depot $d \in R$ is called a one-side route if it only serves target sensors located on one side (left or right) of d. For $1 \leq i \leq j \leq n$, we define $[u_i, u_j] = \{u_i, u_{i+1}, \ldots, u_j\}$. For $1 \leq i \leq j \leq n + k$, we use $[v_i, v_j]$ to denote the set $\{v_i, v_{i+1}, \ldots, v_j\}$ of consecutive vertices. For any $V' \subseteq V$, we use $lf(V')$ $(rt(V'))$ to indicate the leftmost (rightmost) vertex of V'. The region of a depot (in a given feasible solution) is called continuous if it takes the form $[u_i, u_j]$ for some $1 \leq i \leq j \leq n$. For $1 \leq p < q \leq k$, if both the region of d_p, say S_p, and the region of d_q, say S_q, are continuous, we have that $rt(S_p)$ is on the left of $lf(S_q)$.

3   Uniform 2-Depot DMSTC on a Path

In this section, we deal with the uniform 2-depot DMSTC^{path}. For the special case where at least one depot is an endpoint of the path network, we show a polynomial algorithm with time complexity $O(n^4)$. In addition, if both depots are endpoints, the running time can be improved to $O(n^2)$. For the general uniform 2-depot DMSTC^{path}, we give an example to demonstrate the complicated structure of the optimal solution. As a result, the approach for the special case with at least one depot being an endpoint can not be extended to the general case.

For the uniform 2-depot DMSTC^{path} where at least one depot is an endpoint, we assume without loss of generality that d_1 is located at the left endpoint, i.e. $v_1 = d_1$, and d_1 is on the left of d_2.

Lemma 3. *For the uniform 2-depot DMSTC^{path} with at least one depot being an endpoint, there exists an optimal solution where both the region of d_1 and the region of d_2 are continuous and the rightmost vertex of the region of d_1 is on the left of d_2.*

Based on the above lemma, the uniform 2-depot DMSTC^{path} with at least one depot being an endpoint can be simplified to optimally partition the original path into two disjoint subpaths by deleting a proper edge between d_1 and d_2 and then solve two single-depot subproblems defined on these two subpaths. One subproblem with depot d_1 being an endpoint can be solved in $O(n)$ time by the Greedy Algorithm (GA) of Chen et al. [3]. The other subproblem with d_2 being the depot can be solved in $O(n^3)$ time by the algorithm of Yu and Liu [13]. Since there are at most $O(n)$ possible choices for the partition edge[3], we have obtained the following result.

Theorem 1. *The uniform 2-depot DMSTC^{path} where at least one depot is an endpoint can be solved in $O(n^4)$ time.*

When d_2 is also an endpoint of the original path network, the above subproblem with d_2 being the depot can also be solved in $O(n)$ time by the algorithm in [3]. This yields the following result.

---

[3] It is possible that one of the two subproblems derived by some partition edge e is infeasible. Then we will never choose e as the partition edge.

Theorem 2. *The uniform 2-depot DMSTC^{path} with both depots being endpoints can be solved in $O(n^2)$ time.*

Next we treat the general case where none of the two depots is located at an endpoint. It turns out that for the general case, the optimal solution may not have a continuous structure as described in Lemma 3. We will give an example to illustrate this in the full version of the paper (The omitted proofs of the lemmas and theorems are also provided in the full version.)

4   Non-uniform DMSTC on a Path

In this section, we treat the (non-uniform) DMSTC^{path}. We first show that the algorithm for the DMSTC$_0^{path}$ by Chen et al. [3], which we call the Non-uniform Greedy Algorithm (NGA), has a tight approximation ratio of 2. As mentioned in the introduction, Chen et al. [3] only proved a super-constant bound on the approximation ratio of the NGA. Our analysis of the NGA is based on a stronger lower bound on the optimal value given by Nagarajan and Ravi [9]. Moreover, we propose a new 2-approximation algorithm for the DMSTC$_0^{path}$. Finally, we generalize both the NGA and our new algorithm to the DMSTC^{path} while preserving the approximation ratio 2.

For convenience, we introduce some more notations. Let $G = (V, E)$ be the path network where $V = \{v_1, v_2, \ldots, v_{n+1}\}$ and the single depot $d = v_h$ for some h with $1 \leq h \leq n + 1$. Given a subpath $L' = (V(L'), E(L'))$ containing d, we see L' as a tree rooted at d. For any vertex $v \in V(L')$, the depth of v in L' is measured by the number of edges between d and v. The total length of the edges between d and v is called the distance between d and v in L' and is denoted by l_v. We say that v' is a descendant of v if v lies between d and v' (it is possible that $v' = v$.) L_v denotes the subpath of L' induced by the set of all descendants of v and v is called the root of L_v. Note that $L_d = L'$.

For a subset of vertices $U \subseteq [v_i, v_j]$ $(1 \leq i \leq j \leq n+1)$ on the path $G = (V, E)$ with $v_i = lf(U)$ and $v_j = rt(U)$, we use $R(U)$ to represent the route that departs from d, serves precisely the target sensors in U and return to d. $R(U)$ is also called a d-tour. One can see that the load of $R(U)$ is given by

$$L(R(U)) = \begin{cases} 2\max\{l_{v_i}, l_{v_j}\} + c(U), & \text{if } d \notin [v_i, v_j], \\ 2(l_{v_i} + l_{v_j}) + c(U \setminus \{d\}), & \text{if } d \in [v_i, v_j]. \end{cases}$$

If $L(R(U)) \leq D$, then $R(U)$ is a feasible route. If $U = [v_i, v_j]$ and $L(R(U)) > D$, we say that U is a heavy cluster (this definition of heavy cluster is equivalent to the definition given by Nagarajan and Ravi [9] for the DVRP on a tree.) For simplicity, we write $R(L')$ for $R(V(L'))$ if L' is a subpath.

Lemma 4. *(Nagarajan and Ravi 2012) If there are h disjoint heavy clusters in G, then the number of d-tours in an optimal solution is at least $h + 1$.*

Algorithm 1: Non-uniform Greedy Algorithm (NGA) by Chen et al. [6]

Input: A path network $G = (V, E)$ with $V = \{v_1, v_2, \ldots, v_{n+1}\}$ and
$E = \{(v_i, v_{i+1}) \mid i = 1, 2, \ldots, n\}$, V consists of a single depot $d = v_1$
and a set $N = V \setminus \{d\}$ of n target sensors, handling time $c(u)$ for each
$u \in N$, length function $l(\cdot)$ on E, and time constraint D.

Output: A set of feasible routes departing from d and serving all the target sensors.

1 Initialize $L' = G$, $SOL = \emptyset$.

2 **while** $V(L') \neq \{d\}$ **do**

3   (a) Pick a deepest vertex $v_i \in V(L')$ s.t. $R(L_{v_i})$ is not feasible (hence
    $V(L_{v_i})$ is a heavy cluster). Add the d-tour $R(L_{v_{i+1}})$, which is feasible by
    the definition of v_i, to SOL. Update L' as the subpath induced by
    $V(L') \setminus V(L_{v_{i+1}}) \cup \{d\}$.

4   (b) If no such v_i exists, add the d-tour $R(L_d)$, which must be feasible, to
    SOL and END.

5 **end**

6 Return SOL.

Fig. 1. Description of the NGA.

Now we state in Fig. 1 the NGA for the DMSTC_0^{path} where the depot d is the left endpoint of the path. The NGA uses a greedy idea to serve the vertices in decreasing order of the distances from the depot. Once a data mule can not serve one vertex, say v_i, due to the time constraint, it returns directly to the depot without considering whether the vertices between d and v_i can be served.

One can see that the above algorithm will execute a series of consecutive Step (a) of the while loop and end at a execution of Step (b). The algorithm can be easily implemented in $O(n)$ time. Let $i_1 > i_2 > \cdots > i_m$ be the indices of the vertex v_i picked in the executions of Step (a) in this order. Set $i_0 = n + 1$ and $i_{m+1} = 1$. Then the algorithm adds to SOL $m + 1$ feasible routes $R(L_{v_{i_1+1}}), R(L_{v_{i_2+1}}), \ldots, R(L_{v_{i_m+1}}), R(L_d) = R(L_{v_{i_{m+1}}})$, where $R(L_{v_{i_t}})$ serves exactly the target sensors in $[v_{i_t+1}, v_{i_{t-1}}]$ $(t = 1, \ldots, m + 1)$. The corresponding heavy clusters generated in Step (a) are $[v_{i_t}, v_{i_{t-1}}]$ $(t = 1, \ldots, m)$. Note that $\{[v_{i_{2t-1}}, v_{i_{2t-2}}] \mid t = 1, \ldots, \lceil \frac{m}{2} \rceil\}$ are $\lceil \frac{m}{2} \rceil$ disjoint heavy clusters. Applying Lemma 4, we have $OPT \geq \lceil \frac{m}{2} \rceil + 1$. This implies that $|SOL| = m + 1 \leq 2OPT$. That is, the NGA is a 2-approximation algorithm.

On the other hand, the following example shows that the ratio 2 is tight. There are $4n$ vertices to be served. The handling time list for all vertices from left to right is $[\frac{1}{2}, \frac{1}{2n}, \frac{1}{2}, \frac{1}{2n}, \ldots, \frac{1}{2}, \frac{1}{2n}]$. The distance from all vertices to the depot is 0. Time span constraint $D = 1$. It can be verified that $SOL = 2n$ and $OPT = n + 1$ (the optimal solution consists of one route serving $2n$ vertices of handling time $\frac{1}{2n}$ and n routes each of which serves two vertices of handling time $\frac{1}{2}$. So $\frac{SOL}{OPT} = \frac{2n}{n+1} \to 2$ $(n \to +\infty)$.

To sum up, we have the following result.

Algorithm 2: Modified NGA

    Input: Same as Algorithm 1
    Output: Same as Algorithm 1
1 Initialize $L' = G$, $SOL' = \emptyset$.
2 **while** $V(L') \neq \{d\}$ **do**
3   (a) Pick a deepest vertex $v_i \in V(L')$ s.t. $R(L_{v_i})$ is not feasible. Update L'
       as the subpath induced by $V(L') \setminus V(L_{v_{i+1}}) \cup \{d\}$. Construct a feasible
       d-tour $R := R(L_{v_{i+1}})$. Consider the unserved target sensors in $[v_1, v_i]$ in
       decreasing order of the indices. For an unserved target sensor $v_j \in [v_1, v_i]$,
       if $L(R) + c(v_j) \leq D$ then update R as a new feasible route serving one
       more target sensor v_j than the original route R. Correspondingly, remove
       v_j from $V(L')$ and replace two edges $(v_{j-1}, v_j), (v_j, v_{j+1})$ by a new edge
       (v_{j-1}, v_{j+1}) of length $l(v_{j-1}, v_j) + l(v_j, v_{j+1})$. After considering all the
       target sensors in $[v_1, v_i]$, add R to SOL'.
4   (b) If no such v_i exists, add the d-tour $R(L_d)$, which must be feasible, to
       SOL' and END.
5 **end**
6 Return SOL.

Fig. 2. Description of the Modified NGA.

Theorem 3. *The NGA is an $O(n)$-time 2-approximation algorithm for the $DMSTC_0^{path}$ and the ratio is tight.*

We observe that the NGA is actually an extension of the famous Next-Fit algorithm for the bin packing problem, which is a special case of the $DMSTC_0^{path}$ with all the distances from the target sensors to the depot being zero. The above tight example is exactly the folklore worst-case example for the Next-Fit algorithm. The connection between the $DMSTC_0^{path}$ and the bin packing problem motivates us to propose a modified version of the NGA, described in Fig. 2, which is a generalization of another well-studied bin packing algorithm, i.e. the First-Fit algorithm (see e.g. [4]).

Next we analyze the performance of the Modified NGA by comparing it with the NGA. Recall that $i_1 > i_2 > \cdots > i_m$ are the indices of the vertex v_i picked in the executions of Step (a) of the NGA in this order. Set $i_0 = n + 1$ and $i_{m+1} = 1$. By a simple induction argument, it is not hard to show that the first t $(t = 1, 2, \ldots, m + 1)$ feasible routes constructed in the Modified NGA must have served all the target sensors in $[v_{i_t+1}, v_{i_0}]$, which are exactly the target sensors served by the first t feasible routes obtained by the NGA. As a consequence, the number $|SOL'|$ of d-tours generated by the Modified NGA is no more than the number $|SOL|$ of d-tours generated by the NGA. Combining this with Theorem 3, we have $|SOL'| \leq |SOL| \leq 2OPT$. In addition, the Modified NGA can be implemented easily in $O(n^2)$ time. Therefore, we have obtained the following conclusion.

Theorem 4. *The Modified NGA is an $O(n^2)$-time 2-approximation algorithm for the $DMSTC_0^{path}$.*

Algorithm 3: Extended NGA for the DMSTC^{path}

Input: A path network $G = (V, E)$ with $V = \{v_1, v_2, \ldots, v_{n+1}\}$ and
$E = \{(v_i, v_{i+1}) \mid i = 1, \ldots, n\}$, V consists of a single depot $d = v_h$ for
some $1 \leq h \leq n + 1$ and a set $N = V \setminus \{d\}$ of n target sensors,
handling time $c(u)$ for each $u \in N$, length function $l(\cdot)$ on E, and time
constraint D.

Output: A set of feasible routes departing from d and serving all the target
sensors.

1 Use the NGA to find a feasible solution SOL_1 for G_L induced by $[v_1, v_h]$ and a
feasible solution SOL_2 for G_R induced by $[v_{h+1}, v_n]$;

2 Set $SOL = SOL_1 \cup SOL_2$;

3 Let R_1 be the last route in SOL_1 serving $[v_l, v_{h-1}]$ for some $l < h$ and R_2 be
the last route in SOL_2 serving $[v_{h+1}, v_r]$ for some $r > h$. If
$L(R_1) + L(R_2) \leq D$, merge the two routes R_1 and R_2 into one route in SOL;

4 Return SOL.

Fig. 3. Description of the Extended NGA.

Now we extend the NGA to solve the DMSTC^{path} where the depot is not on
the endpoint. Let G_L (G_R) be the subpath on the left (right) side of the depot,
including d. The extended algorithm is described in Fig. 3.

Theorem 5. *The Extended NGA is an $O(n)$-time 2-approximation algorithm
for the DMSTC^{path}.*

Proof. It is not hard to see that the Extended NGA has a time complexity of
$O(n)$ and indeed obtains a feasible solution.

Suppose that $|SOL_1| = m_1 + 1$ and $|SOL_2| = m_2 + 1$. By the above analysis
of the NGA, we know that there are $\lceil \frac{m_1}{2} \rceil$ disjoint heavy clusters in G_L and $\lceil \frac{m_2}{2} \rceil$
disjoint heavy clusters in G_R. By Lemma 4, this leads to $OPT \geq \lceil \frac{m_1}{2} \rceil + \lceil \frac{m_2}{2} \rceil + 1$.
Then we have $|SOL| \leq |SOL_1| + |SOL_2| = m_1 + m_2 + 2 \leq 2OPT$. This completes
the proof. □

By replacing the NGA with the Modified NGA in Step 1 of the Extended
NGA, we obtain another algorithm, called the Extended Modified NGA, for the
DMSTC^{path}. Using a similar analysis, we can show the following result.

Theorem 6. *The Extended Modified NGA is an $O(n^2)$-time 2-approximation
algorithm for the DMSTC^{path}.*

5   Multi-depot DMSTC$_l$ on a Path

In this section, we discuss the multi-depot DMSTC$_l^{path}$, where the objective is
to minimize the total travel distance of the data mules. For the uniform case,
we derive an $O((n + k)^2)$-time algorithm to solve it. For the non-uniform case,
we present a 2-approximation algorithm.

Theorem 7. *The uniform multi-depot $DMSTC_l^{path}$ can be solved in $O((n+k)^2)$ time.*

Theorem 8. *The multi-depot $DMSTC_l^{path}$ admits a 2-approximation algorithm.*

6 Multi-depot DMSTC$_l$ on a Cycle

The approach in the previous section can be extended straightforwardly to the multi-depot $DMSTC_l^{cycle}$ by noting that, this problem reduces to solving k 2-depot $DMSTC_l^{path}$ with both depots being endpoints (the corresponding 2-depot pairs are $(d_1, d_2), \ldots, (d_{k-1}, d_k), (d_k, d_1)$). Thus, we can obtain the following results.

Theorem 9. *The uniform multi-depot $DMSTC_l^{cycle}$ can be solved in $O((n+k)^2)$ time.*

Theorem 10. *The multi-depot $DMSTC_l^{cycle}$ admits a 2-approximation algorithm.*

7 Conclusions

We give an $O(n^4)$ polynomial algorithm for the uniform 2-depot DMSTC^{path} where at least one depot is an endpoint of the path network. However, the approach for the special case can not be extended to the general case, as shown by the example in Sect. 3. We leave it as an open question whether the general uniform 2-depot DMSTC^{path} is polynomially solvable or NP-hard. For the (non-uniform) $DMSTC_0^{path}$, we show that the Non-uniform Greedy Algorithm (NGA) has a tight approximation ratio of 2 and we also propose a new 2-approximation algorithm, i.e. the Modified NGA, which is an extension of the famous First-Fit algorithm for the bin packing problem. For future research, it would be interesting to give a tight analysis on the Modified NGA. We conjecture that the real approximation ratio of the Modified NGA is below 2. In other words, the Modified NGA may be a better-than-2 approximation algorithm for the non-uniform $DMSTC_0^{path}$.

Acknowledgments. This research is supported by the National Natural Science Foundation of China under grant number 12371317.

References

1. Canete, E., Chen, J., Diaz, M., Llopis, L., Rubio, B.: Wireless sensor networks and structural health monitoring: experiences with slab track infrastructures. Int. J. Distrib. Sens. Netw. **15**(3) (2019). https://doi.org/10.1177/1550147719826002
2. Chen, Q., Huang, X., Ran, Y.: Approximation algorithm for distance constraint sweep coverage without predetermined base stations. Discrete Math. Algorithms Appl. **10**(5), 1850064 (2018)

3. Chen, Z., Zhang, Z., Ran, Y., Shi, Y., Du, D.Z.: Data mule scheduling on a path with handling time and time span constraints. Optim. Lett. **14**, 1701–1710 (2020)
4. G. Dosa, J. Sgall, First Fit bin packing: A tight analysis. In: the Proceedings of the 30th International Symposium on Theoretical Aspects of Computer Science, pp. 538–549 (2013)
5. Friggstad, Z., Swamy, C.: Approximation algorithms for regret-bounded vehicle routing and applications to distance-constrained vehicle routing. In: the Proceedings of the 46th Annual ACM Symposium on Theory of Computing, pp. 744–753 (2014)
6. Li, C.L., Simchi-Levi, D., Desrochers, M.: On the distance constrained vehicle routing problem. Oper. Res. **40**(4), 790–799 (1992)
7. Li, J., Zhang, P.: New approximation algorithms for the rooted budgeted cycle cover problem. Theoret. Comput. Sci. **940**, 283–295 (2023)
8. Liang, J., Huang, X., Zhang, Z.: Approximation algorithms for distance constraint sweep coverage with base stations. J. Comb. Optim. **37**(4), 1111–1125 (2019)
9. Nagarajan, V., Ravi, R.: Approximation algorithms for distance constrained vehicle routing problems. Networks **59**(2), 209–214 (2012)
10. Nazib, R.A., Moh, S.: Energy-efficient and fast data collection in UAV-aided wireless sensor networks for hilly terrains. IEEE Access **9**, 23168–23190 (2021)
11. Pérez-Escalona, P., Rapaport, I., Soto, J., Vidal, I.: The multiple traveling salesman problem on spiders. In: Bureš, T., et al. (eds.) SOFSEM 2021. LNCS, vol. 12607, pp. 337–348. Springer, Cham (2021). https://doi.org/10.1007/978-3-030-67731-2_24
12. Popescu, D., Stoican, F., Stamatescu, G., Ichim, L.: Advanced UAV-WSN system for intelligent monitoring in precision agriculture. Sensors **20**(3), 817 (2020)
13. Yu, W., Liu, Z.: Approximation and polynomial algorithms for the data mule scheduling with handling time and time span constraints. Inf. Process. Lett. **178**, 106299 (2022)

Two Exact Algorithms for the Packet Scheduling Problem

Fei Li[1]([⊠])[iD] and Ningshi Yao[2][iD]

[1] Department of Computer Science, George Mason University,
Fairfax, VA 22030, USA
fli4@gmu.edu
[2] Electrical and Computer Engineering Department, George Mason University,
Fairfax, VA 22030, USA
nyao4@gmu.edu

Abstract. We consider a classic packet scheduling problem [7] and its variants. This packet scheduling problem has applications in the areas of logistics, road traffic, and more. There is a network and a set of unit-length packets are to be transmitted over the network from their respective sources to their respective destinations. Each packet is associated with a directed path on which it must travel along. Time is discrete. Initially, all the packets stay on the first edges of their respective paths. Packets are pending on the edges at any time. In each time step, a packet can move along its path by one edge, given that edge having no other packets move onto it in the same time step. The objective is to minimize *makespan* – the earliest time by which all the packets arrive at their respective destination edges. This problem was proved NP-hard [1] and it has been studied extensively in the past three decades. In this paper, we first provide a semi-online algorithm GRD and show that GRD is optimal for scheduling packets on arborescence and/or anti-arborescence forests. We then provide a parameterized algorithm PDP which finds an optimal makespan for the general case. PDP is a dynamic programming algorithm and its running time complexity depends on the *congestion* and *dilation* in the input instance. The algorithm PDP's idea is new and it is derived from an insightful lower bound construction for the general packet scheduling problem.

Keywords: Packet scheduling · exact algorithms · dynamic programming

1 Introduction

A packet scheduling problem [7] has been studied extensively in the past three decades. Consider a directed graph $G = (V, E)$ with a set of vertices V and a set of edges E, where $|V| = n$ and $|E| = m$. There are N packets that are

---

N. Yao's research is partially supported by NSF grant ECCS-2218517.

to be transmitted on G. Each packet $p \in \{1, 2, \ldots, N\}$ has a *source* $s_p \in V$, a *destination* $t_p \in V$, and a directed *path* P_p with a *length* l_p denoting the path P_p's number of edges $|P_p|$. A path P_p can be represented by an ordered set of edges $e_p(1), e_p(2), \ldots, e_p(l_p)$ and each $e_p(i) \in E$ is an edge on P_p, $\forall i = 1, 2, \ldots, l_p$. All the packet paths are simple ones and thus we have $l_p = |P_p| \leq |E| = m$, $\forall p$. The path P_p specifies the edges as well as the order that a packet p should travel along on G.

Time is discrete. Assume that any edge $e \in E$ represents a sufficiently-large-size buffer so that at any time, any number of packets can stay on the edge e. The vertices V act as switches so that in each time step, each edge e accepts at most one packet to be forwarded to the queue represented by e: In each time slot, for any outgoing edge $e = (v, w)$ incident to a vertex v, at most one packet staying on v's incoming edges (u, v) and having e as their next step's edges can be moved onto e. Multiple packets can be forwarded simultaneously to their next edges in one single time step as long as these packets are not getting into the same queue after being forwarded. For ease of notation, we assume that all the packets p stay on their respective paths P_p's first edges $e_p(1)$ at the end of time step 1. A packet p's *duration* d_p is the time slot by which p arrives at the last edge $e_p(l_p)$ of its path P_p. Clearly, we have $d_p \geq l_p$, $\forall p$. We design a scheduling algorithm with the objective of minimizing *makespan*, defined as the maximum duration $\max_p d_p$ for all the packets $p \in \{1, 2, \ldots, N\}$.

2   Related Work

This packet scheduling problem was proved NP-hard [1]. The authors in [13] showed that an optimal makespan cannot be approximated down to the ratio 1.2 unless P = NP, even for the case in which the graph is a tree (with bidirectional edges). Define the *dilation* D as the maximum path length, $D = \max_p l_p$, and the *congestion* C as the maximum number of paths having a single edge in common, $C = \max_{e \in E} |C(e)| = \max_{e \in E} |\{p | e \in P_p\}|$, where $C(e)$ is the set of packets having e in their paths. It is clear that any schedule's makespan has a lower bound of $\max(C, D) \geq \lceil \frac{C+D}{2} \rceil = \Omega(C + D)$ [7].

A class of scheduling algorithms are called *greedy*, if an algorithm in such a class never leaves an outgoing edge $e = (u, v)$ *idle* (an idle edge does not accept packets) as long as there are packets waiting in the incoming edges of v having e as their next-step edges in the corresponding paths [9]. Any simple randomized algorithm with the greedy strategy achieves a makespan $O(C \cdot D)$. In [8], the authors gave a schedule of length $O(C + D)$ in time $O(L(\log \log L) \log L)$ with a probability at least $1 - L^\beta$, where $\beta < 0$ is a constant and $L = \sum_p l_p$. This randomized algorithm can be derandomized using the method of conditional probabilities [11] and it became the first constant approximation algorithm, against the lower bound $\lceil \frac{C+D}{2} \rceil$ of makespan. In [16], the author gave a simpler proof of the algorithm in [8]. The algorithm is an *offline algorithm* which is given its input including a complete description of the graph G and the packets' paths P_p, $\forall p$ in designing a schedule. For the algorithm in [8], the hidden constant in

$O(C + D)$ is high. The best known approximation ratio is 24 [14], against $C + D$ as well.

Some variants of this packet scheduling problem have been studied as well. In an *online setting*, an algorithm only uses the information that is available locally to a vertex v in order to determine which packet to be forwarded to the edge (v, w), among the ones waiting on v's incoming edges. For the class of *layered networks*, [6] gave a simple online randomized algorithm with a makespan $O(C + B + \log N)$, where B ($\leq D$) is the number of layers of this network. For the case in which all the packets' paths P_p are assumed to be the shortest ones from the sources s_p to the destinations d_p (in terms of the number of edges), there was an online algorithm with a maximum duration bounded by $D + N - 1$ [10]. In [12], the authors improved the result to be $O(C + D + \log(N \cdot D))$, with a high probability, and in [18], the authors gave a simple online randomized algorithm with a duration $O(C + D + \log N)$, with a high probability. The most recent known work on online algorithms is [12], giving a universal deterministic $O(C + D + \log^{1+\epsilon} N)$ algorithm. This result is almost optimal. The problem whether there exists an online algorithm with competitive ratio bounded by $O(C + D)$ is still open. The algorithm in [10] is a greedy online algorithm. Some other variants in which the edges are bufferless (i.e., at most one packet is on an edge at any time) or the packets are allowed to wait (i.e., staying on each of such edges for more than 1 time slots) only on some predefined edges were discussed in [17]. Another line of research is to consider *packet scheduling* and *packet routing* (packet routing algorithms allow packets to choose paths to get to the destinations) together in order to minimize the makespan. The *competitive packet scheduling* problem is also studied. In this problem, the packets select their paths rationally and the makespan is the social welfare to be optimized [3]. The paper [9] gave a brief survey. More recent related work can be found from the work following [9].

Our Contributions. In this paper, we study exact algorithms for the packet scheduling problem. We design two algorithms. One is named GRD. GRD is a simple, fast semi-online algorithm and it optimizes the makespan in scheduling packets on arborescence and/or anti-arborescence forests. The other one is an exact algorithm, named PDP, for the general packet scheduling case and its running-time complexity depends on the parameters (congestion and dilation) of the input instances. In Sect. 3 and Sect. 4, we describe the algorithms GRD and PDP, along with their running-time analysis and performance analysis, respectively.

3   GRD: Scheduling Packets on Arborescence and Anti-Arborescence Forests

In the *semi-online setting*, an algorithm has no complete knowledge of the graph G and a packet has no information regarding to the other packets' status at any time. A semi-online algorithm may allow packets carry some information on themselves — normally, such information is a constant value that cannot embed the whole graph information nor any information on the other packets. The values carried by the packets waiting on the incoming edges of a vertex can be used to make the decision of transmitting them.

In this section, we design a *semi-online* algorithms and name it GRD (standing for greedy). Each packet p has its path information P_p.

3.1   The Ideas

GRD is based on the following two greedy ideas: Consider a packet p at the beginning of a time step t.

1. The packet p greedily moves onto the next edge e as long as no other packets are competing for e in the same time step. Such a best-effort movement of p will not increase p's duration and will not increase any delays to other packets.
2. Consider the case when there are more than one packets including p competing for an edge e. If p is not chosen to move onto e, then p's duration is increased by 1. Therefore, in this time step t, the idea is to forward the packet whose duration's increase affects the algorithm's makespan the most. Recall that a semi-online algorithm has no global information on the graph G nor the information of the packets not-competing for e in the time step t, thus, a packet p's duration is *estimated* as the sum of the length of its remaining path $(e_p(i+1), e_p(i+2), \ldots, e_p(l_p))$ and the current time t, assuming $e = e_p(i+1)$. The packet with the largest number of time steps to reach its destination under the assumption of no future delays, among those pending packets for the edge e, is moved onto the edge e.

3.2   The Algorithm

We use c_p to denote the number of *remaining edges that the packet p should take in order to reach its destination*, assuming there are no delays along p's remaining path. At the beginning of a time step t, p's duration is estimated as $t + c_p$. When $t = 1$, c_p is initialized as l_p, and p should take the path P_p with l_p edges to its destination. In the algorithm GRD, the value c_p is updated by the packet p using a counter. At a time, given an edge e, GRD uses the value c_p to select the largest-value packet p to send to e. Since the value c_p may be updated over time, we use a function $c_p(t)$ to denote the value c_p at the end of time step t. The algorithm GRD is described in Algorithm 1.

Note that for each edge $e = (u, v)$, the decision of accepting a packet p or not by the edge e depends on the local packets' $c_p(t - 1)$ values, hence GRD is a semi-online algorithm.

3.3   The Analysis

In the following, we analyze GRD. We first state two assumptions with which we do not lose generality. These two assumptions facilitate the analysis of GRD

Algorithm 1. GRD

1: For each packet p, associate p with a value $c_p(t-1)$ to denote its remaining time slots needed to get p to its destination t_p, starting from time t and assuming no delay incurred in the future for forwarding p. Initially, $c_p(0) = l_p$.

2: Forward a packet p to an edge e as long as no other packets are waiting for being forwarded to e or $c_p(t-1)$ is the largest value for all such packets competing for the edge e. Ties are broken arbitrarily.

3: Update $c_p(t) \leftarrow c_p(t-1) - 1$, for each time of forwarding a packet p. Update $c_q(t) \leftarrow c_q(t-1)$, for each time of not forwarding a packet q.

---

as well as the analysis of PDP which is introduced in Sect. 4. We then show GRD's running time analysis and prove that it is optimal for scheduling packets on *anti-arborescence forests*.

Assumption 1. *Any edge in the graph G must belong to a packet's path, say $\forall e \in E$, we have $e \in \bigcup_p P_p$.*

Assumption 1 holds since for any algorithm, it does not schedule a packet over an edge outside of the set of edges $\bigcup_p P_p$ and thus, such removals of edges do not hurt the algorithm in generating the makespan. A useful fact is that Assumption 1 implies $m \leq N$.

In the packet scheduling problem's statement, we assume that all the paths are simple ones. Given an input instance with some packet paths having cycles, we can always convert the input instance to be one with simple paths only. Such conversion does not introduce a larger makespan.

Assumption 2. *All packets' paths are simple ones.*

Consider a packet p and its path P_p that have cycles. We modify the path P_p and the graph G so that the modified path P_p has no cycles. Such cycles, if any, are removed one by one from the input instance as below. Let

$$P_p = \{s_p, v_1, v_2, \ldots, v_{k-1}, v_k, v_{k+1}, \ldots, v_w, v_k, v_{w+1}, \ldots, t_p\}$$

and there is one simple cycle $v_k, v_{k+1}, \ldots, v_w, v_k$. We create a new graph: having two vertices v_k' and v_k'' so that all the edges having v_k as the heads originally now have v_k' as the heads. All the edges having v_k as the tails originally now have v_k'' as the tails. We create a subpath $v_k', v_{k+1}, \ldots, v_w, v_k''$ to replace the subpath $v_k, v_{k+1}, \ldots, v_w, v_k$. The vertex v_k is removed from the new graph and the new path is:

$$P_p = \{s_p, v_1, v_2, \ldots, v_{k-1}, v_k', v_{k+1}, \ldots, v_w, v_k'', v_{w+1}, \ldots, t_p\}$$

Recall here that though in the new graph we have two new edges (v_{k-1}, v_k') and (v_w, v_k''), these two new edges belong to the packet p's path only but not to any others. These two edges replaces the edges (v_{k-1}, v_k) and (v_w, v_k). Having these two edges does not increase p's duration, nor any other packet's duration. Any algorithm on the original graph G has the same makespan on the new graph.

Theorem 1. *GRD has a running time of $O(m \cdot D \log N)$.*

Proof. We are using a charging scheme to calculate GRD's running time complexity and will show that it is $O(\max(m, N)D \log N)$. With Assumption 1, we will have Theorem 1.

First, we show GRD's running time is $O(N \cdot D \log N)$. For each edge e in the graph G, we use a priority queue to maintain all the packets staying on the edge e at a time and the value c_p is used as the key. We charge GRD's running time on the priority queue operations on the packets during GRD's execution. For each edge e, it takes time $O(\log N)$ to get the packet with the largest c_p value. For each packet p, it incurs at most l_p times of getting into a new packet queue. Note that a packet queue is associated with each edge of the path P_p. For each such a packet transmission, it incurs queue-operation time $O(\log N)$. There are N packets. Thus, the total running time is $O(N \log N \max_p l_p) = O(N \log N \cdot D)$, where D is the dilation.

Second, we show that the running time can be calculated as $O(m \cdot D \log N)$ by charging the cost to each packet at a time. Label the edges as $e_1, e_2, \ldots, e_m$. Note that among the packets $S(e_i)$ waiting to be sent to an edge e_i, only the max-c_p-value packet p experiences $\log |S(e_i)|$ time while the other packets in $S(e_i) \setminus \{p\}$ experiences search time 0. Consider a time step t and let $S(e_1), S(e_2), \ldots, S(e_m)$ denote the m priority queues containing the N packets, with some queues being possibly empty. For this single time step t, the total search time incurred to those packets being sent is $\sum_i \log |S(e_i)|$ and the total search time incurred for those packets not being sent is 0. Note $S(e_i) \cap S(e_j) = \emptyset, \forall i \neq j$. We have the total search cost for a packet moving one step along its path (assuming $m \geq 2$):

$$\sum_i \log |S(e_i)| = \log \prod_i |S(e_i)|$$

$$\leq \log \left(\frac{\sum_i |S(e_i)|}{m} \right)^m$$

$$= m \log \frac{N}{m} \leq m \log N - m \tag{1}$$

Inequality 1 is based on Edwin Beckenbach and Richard Bellman's work presented in [2]. Recall that we only need to count the search time for a packet being sent in a time step, thus, the number of searches associated with a packet is its length, bounded by D. The total running cost of GRD is also bounded by $O(m \log N \cdot D)$. Theorem 1 is proved.

In the following, we analyze GRD's performance when the underlying graph G is an *arborescence and anti-arborescence forest*. An arborescence and anti-arborescence forest contains multiple arborescences and anti-arborescence. An *arborescence* [4] is a directed graph having a root so that there is exactly one directed path from the root to any vertex of this graph. An *anti-arborescence* [5] is one created by reversing all the directed edges of an arborescence, i.e. making them all point to the root rather than away from it.

Theorem 2. *GRD is optimal in scheduling packets on arborescence and anti-arborescence forests.*

Proof. In order to prove Theorem 2, we only need to show that GRD is optimal for packet scheduling on one arborescence and one anti-arborescence since each packet is scheduled only on one arborescence or one anti-arborescence. In the following, we prove that GRD is optimal in scheduling packets on an anti-arborescence. The analysis for GRD on arborescence is similar but easier. We leave it in our full journal paper.

We inductively prove Theorem 2 using an exchange argument. Let ADV denote an adversary. At the beginning of time step 1, ADV is the same as an optimal algorithm with the minimum makespan d^*. Consider an anti-arborescence T with a root r and label the depth of the edges as $1, 2, \ldots$ based on the distances from the directed edges' tails to the root r. The 1-depth edges are the edges having r as their heads. An observation is that, given the graph being an anti-arborescence, a packet moves from an edge labelled as i to an edge labelled as $i - 1$ if the packet is transmitted in this time step. We are going to show that there exists an invariant maintained during the algorithm's execution. The invariant guarantees Theorem 2 since at the end of the schedule, we have $d^* = d$, where d is GRD's makespan.

- (Invariant): At the beginning of any time step t, ADV and GRD have the same configuration so that each edge holds the same set of packets.

At the beginning of time step 1, the invariant holds. Now, we consider the first time step t, in which, ADV and GRD sends different packets, say, q and p respectively, to an edge e. If such a time step t does not exist, then ADV and GRD are the same and therefore, $d^* = d$.

Consider the time step t. Recall that G is an anti-arborescence, thus, the fact that e is the edge that p and q plan to step onto in time step t implies that p and q have their paths overlap from time t till one packet reaches to its destination. GRD chooses p instead of q because of $c_p(t - 1) \geq c_q(t - 1)$, which implies that the remaining path for q is embedded in the remaining path for p. The modification on ADV is as below:

1. In time step t, we modify ADV so that ADV sends p instead of q in t.
2. In the remaining schedule, ADV switches the orders of scheduling packets p and q. In each time step that ADV originally schedules q, the packet p is available (considering that p is ahead of q on q's remaining path and q's remaining path is embedded in p's remaining path) and p scheduled.
3. Similarly, in each time step ADV originally schedules p, q can be scheduled until q reaches to its destination.
4. The order and time slots of sending other packets than p and q are not changed.
5. For the possible case in which at some point t' in the future, the original ADV sends q instead of p making q is again before p on their shared subpath, then the modified ADV switches back and follows the original schedule starting from time t'.

Realize that such modification on ADV at time t does not increase the duration for p since p moves ahead of where it was in its original schedule. This modification does not make q's duration more than p's original duration, which is no more than the makespan d^*. This modification does not change any other packet's duration as well. Thus, d^* keeps the same after we modify ADV and the invariant holds.

For each edge that ADV and GRD schedule different packets, we apply the above procedure to modify ADV. The above modifications make sure that the modified ADV is with the same configuration as GRD. At the end of time step t, ADV and GRD are with the same configuration again. Inductively, the invariant is proved. Thus Theorem 2 holds.

4   An Optimal Algorithm for the General Case

In this section, we use the dynamic programming technique to design a parameterized optimal algorithm, named PDP, for the packet scheduling problem in the general case. This algorithm's idea is different from the ones in [14], which were based on the integer linear programming technique. Our algorithm catches some properties of the makespan lower bound construction for the packet scheduling problem and we hope that such properties can be used to design better approximation algorithms.

4.1   The Ideas

We introduce some concepts that will be used to describe our ideas. Consider a packet p in a given schedule. If p moves onto an edge e at time t, then we say that the edge e is *busy* at time t. Otherwise, we say that the edge e is *idle* at time t. The maximal interval $[t, t']$ in which an edge e is continuously busy (to accept different packets) is called a *busy interval*, and thus, the edge e is idle in time step $t - 1$ and time step $t' + 1$, if any. If at the beginning of a time step t, a packet p and a packet q have e as their next edges in their respective paths, then we call p and q the *competing packets for* e in time step t. The edge e is called a *congested edge*.

Consider a packet p at the beginning of a time step t. Assume p is on the edge $e_p(i)$ where $i \neq l_p$. The lower bound of time steps needed for p to arrive at its destination t_p is $l_p - i$ — In the lower bound case, all the edges in $e_p(i+1), e_p(i+2), \ldots, e_p(l_p)$ should be busy for p. Assume p has the maximum duration d^* in an optimal algorithm. Our ideas in PDP is to make sure that p experiences not many delays along its path to its destination.

The first idea is as below: the packet p greedily moves onto the next edge e as long as no other packets are competing for e in the same time step. This idea is identical to one used by GRD. Let OPT denote an optimal algorithm. OPT forwards a packet as long as it can. Based on the above observation, we have the following lemma.

Lemma 1. *Assume that at the beginning of a time step t, there are k packets competing for an edge e. Then the edge e must be continuously busy from time t to time $t + k - 1$ in OPT.*

Proof. As OPT is a greedy algorithm, it schedules a packet onto e as long as e is not busy. The edge e accepts one packet at a time, then there are at least k packets available for e to accept in the time interval $[t, t + k - 1]$.

Another observation based on the first idea is as below: Let us category all the packets $1, 2, \ldots, N$ into different groups, based on the needed number of transmissions to their destinations. We use $G(t, i)$ to denote the group containing the packets p which need i transmissions to their respectively destinations, starting from the beginning of a time step t. Initially, we have at most D groups:

$$G(1,1), G(1,2), \ldots, G(1,D),$$

where D is the dilation of the input instance. Initially, a packet p has its path length l_p and thus, it belongs to the group $G(1, l_p)$. From the best-effort manner of forwarding packets, we have the following observation.

Lemma 2. *In OPT, we have*

$$G(t, i) = G^1(t + 1, i - 1) \bigcup G^2(t + 1, i)$$

$$\emptyset = G^1(t + 1, i - 1) \bigcap G^2(t + 1, i)$$

$$G(t, i + 1) = G^1(t + 1, i) \bigcup G^2(t + 1, i + 1)$$

$$G(t + 1, i) = G^1(t + 1, i) \bigcup G^2(t + 1, i)$$

where $G^1(\cdot, \cdot)$ denotes the set of packets forwarded in the time step, $G^2(\cdot, \cdot)$ denotes the set of packets not being forwarded, and any one of them can be an empty set.

Proof. Lemma 2 holds due to the facts that a packet is either forwarded or kept stay in a single time step and a packet can be moved at most one step in one time slot. Consider the beginning of a time step t. For a packet $p \in G(t, i)$, if p is forwarded to the next edge of its path, then p is added to $G^1(t + 1, i - 1)$. If p stays on the edge in the time slot t, then p is added to $G^2(t + 1, i)$.

Though Lemma 2 is obvious, it provides us a way of constructing the dynamic program using the index i in $G(t, i)$. Lemma 2 implies that when t is increased by 1, the number of groups is not strictly increased.

In the following, we introduce some new observations and ideas that our algorithm needs.

Consider a packet p. For each edge e in the path P_p, the packet experiences at least one time step on an edge. We define

$$b(p, e) = \text{delayed time slots for the packet } p \text{ on the edge } e,$$

where $b(p,e) \geq 1$ and the packet p is on the edge $e_p(i+1)$ at time $t + b(p,e)$ given $e = e_p(i)$.

A packet p's duration is the sum of its delays on the edges, say, $d_p = \sum_{e \in P_p} b(p,e)$. In order to calculate the values $b(p,e)$, we introduce the time slots to calculate $b(p,e)$. For any edge e, the packet p arrives at e at time

$$t_{in}(p,e) := \sum_{e'} b(p,e'), \tag{2}$$

where $e' \in e_p(1), e_p(2), \ldots, e_p(i-1)$ given $e_p(i) = e$. Also, the packet p leaves the edge e at time

$$t_{out}(p,e) = \sum_{e'} b(p,e'), \tag{3}$$

where $e' \in e_p(1), e_p(2), \ldots, e_p(i)$ given $e_p(i) = e$. $b(p,e)$ is calculated as below:

$$b(p,e) = t_{out}(p,e) - t_{in}(p,e). \tag{4}$$

Instead of assigning integer variables to the values $t_{in}(p,e)$ in Eq. (2) and $t_{out}(p,e)$ in Eq. (3), we come up with a new idea. We regard the packet p as a unit-length job, $t_{in}(p,e)$ as p's *release time*, $t_{out}(p,e)$ as p's *deadline*, and e as the machine processing p. The machine e processes at most one job at a time. In the following, we introduce the way of tuning up the values in Eq. (4) to make the job p successfully processed. The range $[t_{in}(p,e), t_{out}(p,e)]$ is the *interval* to schedule the packet p on e.

We want to guarantee that for each edge e, in the time ranges that the packets p are ready/competing to move on e, there are sufficient number of time steps to do so. Lemma 1 indicates that starting from a time t, the edge e is busy for at least k time slots given k competing packets for the edge e. In order to specify the interval to schedule a packet p on the edge e, we must guarantee that the *work load density* (the ratio of the number of packets and the number of time slots in any continuous range) for the edge e cannot exceed 1 [15]. That is, for any time range $[t, t']$, we have

$$\frac{|\{p | t \leq t_{in}(p,e) < t_{out}(p,e) \leq t'\}|}{t' - t + 1} \leq 1, \qquad \forall e \tag{5}$$

Note that Inequality (5) is a lower bound construction for the general case's makespan t' for the edge e. When this inequality is tight, it is feasible to schedule all the packets successfully using the EDF (earliest-deadline-first) policy, where $t_{out}(p,e)$ denotes the deadline. Consider the maximal interval in which the edge e is busy. We have the following observation.

Our dynamic programming algorithm is based on the formulation in Inequality (5). Given a makespan d for an edge e (for example, $d = t'$ in Inequality 5), we are looking at the earliest release time $t_{in}(p,e)$ for a packet p so that a schedule on an edge e ending at time d is feasible.

4.2   The Algorithm and Its Analysis

Along the way of describing our algorithm, we give the running time analysis as well as the correctness analysis. Some part of the correctness analysis has been given when we introduced the algorithm's ideas.

Denote $C(e)$ the set of packets having their paths P_p covering an edge e, $C(e) = \{p | e \in P_p\}$. Note that $\max_e C(e) = C$ where C denotes the congestion. Consider a packet p. Define $\psi(i,j)$, $\forall i,j$, as the state of a packet i arrives *at least* the j-th edge $e_p(j)$ on its path P_i. Recall that $j \leq D$, where D is the dilation. Therefore, we have in total as most D^N different *configurations* to show the states of all the packets at a time. We index these configurations as $\Psi(1), \Psi(2), \ldots, \Psi(Z)$, where $Z \leq D^N$. Our algorithmic contribution is to reduces the number of configurations needed. Our analysis below shows that the total number of configurations needed in the algorithm PDP is 2^N, much less than D^N where $D = \max_p l_p$.

Now, define

$$OPT(\Psi(k), t) = \begin{cases} 1, & \text{the configration } \Psi(k) \text{ happens at the end of time step} t \\ 0, & \text{otherwise} \end{cases}$$

The Objective. The objective of minimizing the makespan d^* is to return the smallest value t so that $OPT(\Psi(k), t) = 1$ for the configuration $\Psi(k)$ when all the packets i arriving at least their destination edges with indexes j $(= l_i)$ in G.

The Base Case. The base case happens at the end of time step 1. We calculate $OPT(\Psi(k), 1)$ for all the indexes k. For each movement of a packet p, we list all the configurations that a greedy schedule moves packets. Consider each edge e and the competing packets $C(e)$. Define $C(e, 1) := \bigcup_p e_p(1)$ and $P(e, 1) := \{p | e_p(1) = e\}$. The total running time of the base case is thus to enumerate all the configurations $\Psi(k)$ and get the value $OPT(\Psi(k), 1)$.

$$\prod_{e \in C(e,1)} |P(e, 1)| \leq \left(\frac{N}{|C(e,1)|} \right)^{|C(e,1)|} < \binom{N}{|C(e,1)|} \leq \binom{N}{(N/2)} \approx \frac{2^N}{\sqrt{\pi \cdot N}}$$

(6)

since $|\bigcup_e P(e, 1)| = N$ and $P(e, 1) \cap P(e', 1) = \emptyset$, for all $e \neq e'$. This inequality holds due to Edwin Benckenbach and Richard Bellman's formula, as well as the Stirling's approximation. We remark here that the parameterized running time $\prod_{e \in C(e,1)} |P(e, 1)|$ can be much less than the upper bound.

The Recursive Step. We consider the ways of calculating $OPT(\Psi(k), t)$. Due to the ideas introduced above, this configuration $\Psi(k)$ comes from the one step move for some packets and being idle for the remaining packets. For these N packets, we consider to partition them into two groups, the group of packets moving forward in a time step and the group of packets staying in the same time step. For each of such a partition, we transform from one configuration to another configuration. These two configurations are called *neighboring configurations*. We have the following recursion:

$$OPT(\Psi(k), t) = \max_{k'} OPT(\Psi(k'), t - 1) \tag{7}$$

where $\Psi(k')$ at time $t - 1$ is a neighboring configuration of $\Psi(k)$ at time t. The correctness of the recurrence in Eq. 7 is based on the recursion discussed in Sect. 4.1.

In the following, we calculate the running time of the recursive step. Though the total configuration number is up to D^N, in this recursion, we only consider the neighboring configurations so that if p is in $\Psi(k)$ given p being on at least the j-th edge of its path P_p, then p is on at least the $(j - 1)$th edge in the configuration $\Psi(k)$ if j is forwarded in time step t, otherwise, p should be on at least the jth edge at the beginning of time step t. The total running time in this recursive step is therefore bounded by 2^N. As t is bounded by $O(C + D)$ [8], we have the following result.

Theorem 3. *PDP is an optimal algorithm for scheduling packets on a graph with a total running time $O\left(2^N(C + D)\right)$.*

The instance-dependent running time has been provided above, as $\prod_{e \in C(e,1)} |P(e, 1)|$ in Inequality 6.

5   Conclusions

In this paper, we present two exact algorithms for the packet scheduling problem. The solution to the general problem brings more insights on designing approximation algorithms. We expect these algorithmic techniques help with solving packet routing problems.

References

1. Clementi, A.E.F., Ianni, M.D.: On the hardness of approximating optimum schedule problems in store and forward networks. IEEE/ACM Trans. Network. **4**(2), 272–280 (1996)
2. Graham, D.L., Knuth, D.E., Patashnik, O.: Concrete Mathematics, A Foundation for Computer Science, 2nd edn. Addison-Wesley, Boston (1994)
3. Harks, T., Peis, B., Schmand, D., Tauer, B., Koch, L.V.: Competitive packet routing with priority lists. ACM Trans. Econ. Comput. (TEAC) (2018)
4. Kleinberg, J., va Tardos: Algorithm Design. Pearson, New York (2006)
5. Korte, B., Vygen, J.: Combinatorial Optimization, Theory and Algorithms. Springer, Heidelberg (2018). https://doi.org/10.1007/978-3-662-56039-6
6. Leighton, F.T., Maggs, B.M., Ranade, A.G., Rao, S.B.: Randomized routing and sorting on fixed-connection networks. J. Algorithms **14**(2), 167–180 (1994)
7. Leighton, F.T., Maggs, B.M., Rao, S.B.: Packet routing and job-shop scheduling in O(congestion + dilation) steps. Combinatorica **14**(2), 167–186 (1994)

8. Leighton, F.T., Maggs, B.M., Richa, A.W.: Fast algorithms for finding O(congestion+dilation) packet routing schedules. Combinatorica **19**(2), 375–401 (1999)
9. Maggs, B.M.: A survey of congestion + dilation results for packet scheduling. In: Proceedings of the 40th Annual Conference on Information Sciences and Systems (CISS) (2006)
10. Mansour, Y., Patt-Shamir, B.: Greedy packet scheduling on shortest paths. J. Algorithms **14**, 449–465 (1993)
11. Mitzenmacher, M., Upfal, E.: Probability and Computing, 2nd edn. Cambridge University Press, Cambridge (2017)
12. Ostrovsky, R., Rabani, Y.: Universal O(congestion + dilation + $\log^{1+\epsilon} n$) local control packet switching algorithms. In: Proceedings of the 29th Annual ACM Symposium on Theory of Computing (STOC), pp. 644–653 (1997)
13. Peis, B., Skutella, M., Wiese, A.: Packet routing: complexity and algorithms. In: Bampis, E., Jansen, K. (eds.) WAOA 2009. LNCS, vol. 5893, pp. 217–228. Springer, Heidelberg (2010). https://doi.org/10.1007/978-3-642-12450-1_20
14. Peis, B., Wiese, A.: Universal packet routing with arbitrary bandwidths and transit times. In: Proceedings of the International Conference on Integer Programming and Combinatorial Optimization (IPCO), pp. 362–375 (2011)
15. Pinedo, M.L.: Scheduling: Theory, Algorithms, and Systems, 6th edn. Springer, New York (2022). https://doi.org/10.1007/978-1-4614-2361-4
16. Rothvob, T.: A simpler proof for O(congestion + dilation) packet routing. In: Proceedings of the 16th Annual Conference on Integer Programming and Combinatorial Optimization (IPCO), pp. 336–348 (2013)
17. Tauer, B., Fischer, D., Fuchs, J., Koch, L.V., Zieger, S.: Waiting for trains: complexity results. In: Proceedings of the 6th Annual International Conference on Algorithms and Discrete Applied Mathematics (CALDAM), pp. 282–303 (2020)
18. Vcking, F.M.B.: A packet routing protocol for arbitrary networks. In: Proceedings of the 28th Annual Symposium on Theoretical Aspects of Computer Science (STACS), pp. 291–302 (1995)

Improved Scheduling with a Shared Resource

Christoph Damerius[✉], Peter Kling, and Florian Schneider

University of Hamburg, 22527 Hamburg, Germany
{damerius,fschneider}@informatik.uni-hamburg.de,
peter.kling@uni-hamburg.de

Abstract. We consider the following shared-resource scheduling problem: Given a set of jobs J, for each $j \in J$ we must schedule a job-specific processing volume of $v_j > 0$. A total resource of 1 is available at any time. Jobs have a resource requirement $r_j \in [0, 1]$, and the resources assigned to them may vary over time. However, assigning them less will cause a proportional slowdown.

We consider two settings. In the first, we seek to minimize the makespan in an online setting: The resource assignment of a job must be fixed before the next job arrives. Here we give an optimal $e/(e-1)$-competitive algorithm with runtime $O(n \log n)$. In the second, we aim to minimize the total completion time. We use a continuous linear programming (CLP) formulation for the fractional total completion time and combine it with a previously known dominance property from malleable job scheduling to obtain a lower bound on the total completion time. We extract structural properties by considering a geometrical representation of a CLP's primal-dual pair. We combine the CLP schedule with a greedy schedule to obtain a $(3/2 + \varepsilon)$-approximation for this setting. This improves upon the so far best-known approximation factor of 2.

Keywords: Approximation Algorithm · Malleable Job Scheduling · Makespan · List Scheduling · Completion Time · Continuous Linear Program

1 Introduction

Efficient allocation of scarce resources is a versatile task lying at the core of many optimization problems. One of the most well-studied resource allocation problems is parallel processor scheduling, where a number of *jobs* need (typically at least temporarily exclusive) access to one or multiple *machines* to be completed. The problem variety is huge and might depend on additional constraints, parameters, available knowledge, or the optimization objective (see [14]).

In the context of computing systems, recent years demonstrated a bottleneck shift from *processing power* (number of machines) towards *data throughput*. Indeed, thanks to cloud services like AWS and Azure, machine power is available in

Full Version (Preprint): https://arxiv.org/abs/2310.05732.

abundance while data-intensive tasks (e.g., training LLMs like ChatGPT) rely on a high data throughput. If the bandwidth of such data-intensive tasks is, say halved, they may experience a serious performance drop, while computation-heavy tasks care less about their assigned bandwidth. In contrast to the number of machines, throughput is (effectively) a *continuously* divisible resource whose distribution may be easily changed *at runtime*. This opens an opportunity for adaptive redistribution of the available resource as jobs come and go. Other examples of similarly flexible resources include power supply or the heat flow in cooling systems.

This work adapts formal models from a recent line of work on such flexible resources [1,5,13] and considers them under new objectives and settings. Classical *resource constrained scheduling* [8,11,15,16] assumes an "all-or-nothing" mentality (a job can be processed if it receives its required resource but is not further affected). One key aspect of the model we consider is the impact of the amount of received resource on the jobs' performance (sometimes referred to as *resource-dependent processing times* [9–12]). The second central aspect is that we allow a job's resource assignment to change while the job is running.

1.1 Model Description and Preliminaries

We consider a scheduling setting where a set $J = [n] := \{1, 2, \ldots, n\}$ of $n \in \mathbb{N}$ *jobs* compete for a finite, shared resource in order to be processed. A *schedule* $R = (R_j)_{j \in J}$ consists of an (integrable) function $R_j \colon \mathbb{R}_{\geq 0} \to [0, 1]$ for each $j \in J$ (the job's *resource assignment*) that returns what fraction of the resource is assigned to j at time $t \in \mathbb{R}_{\geq 0}$. We use $R(t) = (R_j(t))_{j \in J}$ to refer to j's *resource distribution* at time t and $\bar{R}(t) := \sum_{j \in J} R_j(t)$ for the *total resource usage* at time t. Each $j \in J$ comes with a (*processing*) *volume* $v_j \in \mathbb{R}_{\geq 0}$ (the total amount of resource the job needs to receive over time in order to be completed) and a *resource requirement* $r_j \in [0, 1]$ (the maximum fraction of the resource the job can be assigned). We say a schedule $R = (R_j)_{j \in J}$ is *feasible* if:

– the resource is never overused: $\forall t \in \mathbb{R}_{\geq 0} \colon \bar{R}(t) \leq 1$,
– a job never receives more than its resource requirement: $\forall t \in \mathbb{R}_{\geq 0} \colon R_j(t) \leq r_j$, and
– all jobs are completed: $\forall j \in J \colon \int_0^\infty R_j(t)\, \mathrm{d}t \geq v_j$.

For $j \in J$ we define its *processing time* $p_j := v_j/r_j$ as the minimum time that j requires to be completed. See Fig. 1a for an illustration of these notions.

For a schedule $R = (R_j)_{j \in J}$ we define $C_j(R) := \sup\{t \geq 0 | R_j(t) > 0\}$ as the *completion time* of job $j \in J$. We measure the quality of a schedule R via its *makespan* $M(R) := \max\{C_j(R) | j \in J\}$ and its *total completion time* $C(R) := \sum_{j \in J} C_j(R)$. Our analysis additionally considers the *total fractional completion time* $C^F(R) := \sum_{j \in J} C_j^F(R)$, where $C_j^F(R) := \int_0^\infty R_j(t) \cdot t/v_j\, \mathrm{d}t$ is job j's *fractional completion time*.

Relation to Malleable Tasks with Linear Speedup. Our problem assumes an arbitrarily divisible resource, as for example the bandwidth shared by jobs running

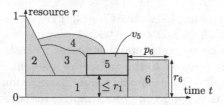

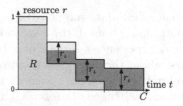

(a) A schedule for six jobs. The resource assignment of job j is given by the height of j's area at time t and must never exceed r_j. The total area of a job is equal to its volume.

(b) Augmenting a schedule R by a job ι via WFSTEP(R,ι,C). The volume of ι is „poured" into the fat-outlined area. The blue area indicates where it is eventually scheduled.

Fig. 1. Illustration of model notions (left) and a WATERFILL step.

on the same host. Another common case are jobs that compete for a *discrete* set of resources, like a number of available processing units. This is typically modeled by a scheduling problem where a set J of n *malleable* jobs of different sizes s_j (length when run on a single machine) must be scheduled on m machines. Each machine can process at most one job per time, but jobs j can be processed on up to $\delta_j \in [m]$ machines in parallel with a linear speedup. Jobs are preemptable, i.e., they can be paused and continued later on, possibly on a different number of machines. See [14, Ch. 25] for a more detailed problem description.

This formulation readily maps to our problem by setting j's processing volume to $v_j = s_j/m$ and its resource requirement to $r_j = \delta_j/m \in (0,1]$. The only difference is that our schedules allow for arbitrary resource assignments, while malleable job scheduling requires that each job j gets an *integral* number δ_j of machines (i.e., resource assignments must be multiples of $1/m$). However, as observed by Beaumont et al. [3], fractional schedules can be easily transformed to adhere to this constraint:

Observation 1 ([3, Theorem 3, reformulated]). Consider a feasible schedule R for a job set J in which $j \in J$ completes at C_j. Let $m := 1/\min\{r_j | j \in J\}$. We can transform each R_j without changing C_j to get $R_j(t) \in \{i/m | i \in [m] \cup \{0\}\}$ for any $t \in \mathbb{R}_{\geq 0}$ and such that each R_j changes at most once between consecutive completion times.

We first consider *online makespan minimization* (Sect. 2), where the scheduler must commit to future resource assignments as jobs arrive (as in list-scheduling). Afterwards, we consider *offline total completion time minimization* (Sect. 3).

1.2 Related Work

Our model falls into the class of continuous shared-resource job scheduling as introduced in [1] and its variants [5,13]. These models have the same relation between a job's resource requirement, the assigned resource, and the resulting processing time as we but only consider makespan minimization as objective. The two main differences are that they assumed an additional constraint on the number of machines and considered discrete time slots in which resource assignments may not change.

Another closely related model is *malleable* job scheduling, where the number of machines assigned to a job can be dynamically adjusted over time. If each job j has its own upper limit δ_j on the number of processors it can be assigned, the model becomes basically equivalent to our shared-resource job scheduling problem (as discussed at the end of Sect. 1.1). Drozdowski [7] gave a simple greedy algorithm for minimizing the makespan in the offline setting (see also Sect. 2). Decker et al. [6] considered total completion time minimization for *identical* malleable jobs for an otherwise rather general (possibly non-linear) speed-up function. They gave a 5/4-approximation for this setting. Beaumont et al. [3] is closest to our model. In particular, they assumed job-dependent resource limits δ_j that correspond to our resource requirements. For minimizing weighted total completion time, they used a water-fill approach to prove the existence of structurally nice solutions (cf. to the our water-filling approach in Sect. 2). Their main result is a (non-clairvoyant) 2-approximation algorithm for the weighted case. Their algorithm WDEQ assigns each job a number of processors according to their relative weight, but no more than the limit imposed by δ_j. Our results in Sect. 3 yield an improved approximation ratio of $3/2 + \varepsilon$ at the cost of clairvoyance (i.e., we must know the job's volumes and resource requirements). Also, our algorithm only handles the unweighted case.

Other related models, such as rigid and moldable scheduling, disallow the resource assignment of a job to be adjusted after it has been started (see [14] for details).

1.3 Our Contribution and Methods

For our model, makespan minimization is known to be offline solvable (see Sect. 2). We thus concentrate on an online (list-scheduling) setting where jobs are given sequentially and we must commit to a resource assignment without knowing the number of jobs and future jobs' properties. We use a water-filling approach that is known to produce "flattest" schedules [3]. We derive properties that are necessary and sufficient for any c-competitive algorithm by providing conditions on c-*extendable* schedules (c-competitive schedules to which we can add any job while remaining c-competitive). From this, we derive slightly weaker *universal schedules* that are just barely c-extendable and show that schedules derived via water-fill are always flatter than universal schedules. Optimizing the value of c yields $e/(e-1)$-competitiveness. We then show that no algorithm can have a lower competitive ratio than $e/(e-1)$.

Our main result considers *offline total completion time minimization*. We improve upon the so far best result for this variant (a 2-approximation [3]) by providing a $(3/2 + \varepsilon)$-approximation running polynomial time in $n, 1/\varepsilon$. The result relies on a continuous linear programming (CLP) formulation for the fractional total completion time, for which we consider primal-dual pairs. The primal solution represents the resource assignments over time, while the dual represents the *priority* of jobs over time. We then extract additional properties about the primal/dual pair. Roughly, our method is as follows. We draw both the primal and dual solutions into a two-dimensional coordinate system. See Fig. 3b for an example. We then merge both solutions into a single 3D coordinate system by sharing the time axis and use the these solutions as a blueprint for shapes in this coordinate system (see Fig. 4). The volume of these shapes then correspond to parts of the primal and dual objective. We use a second algorithm called GREEDY that attempts to schedule jobs as early as possible. Choosing the better one of the CLP and the greedy solution gives us the desired approximation.

2   Makespan Minimization

This section considers our resource-aware scheduling problem under the makespan objective. For the offline problem, it is well-known that the optimal makespan $M^*(J)$ for a job set $J = [n]$ with total volume $V(J) = \sum_{j \in J} v_j$ is $M^*(J) = \max\{V(J)\} \cup \{p_j | j \in J\}$ and that a corresponding schedule can be computed in time On [14, Section 25.6]. The idea is to start with a (possibly infeasible) schedule R that finishes all jobs at time $p_{\max} := \max\{p_j | j \in J\}$ by setting $R_j(t) = v_j/p_{\max}$ for $t \in [0, p_{\max})$ and $R_j(t) = 0$ for $t > p_{\max}$. This schedule uses a constant total resource of $\bar{R} := V(J)/p_{\max}$ until all jobs are finished. If $\bar{R} \leq 1$ (the resource is not overused), this schedule is feasible and optimal (any schedule needs time at least $p_{\max}$ to finish the "longest" job). Otherwise we scale all jobs' resource assignments by $1/\bar{R}$ to get a new feasible schedule that uses a constant total resource of 1 until all jobs are finished at time $V(J)$. Again, this is optimal (any schedule needs time at least $V(J)$ to finish a total volume of $V(J)$).

List-Scheduling Setting. Given that the offline problem is easy, the remainder of this section considers the (online) list-scheduling setting. That is, an (online) algorithm $\mathcal{A}$ receives the jobs from $J = [n]$ one after another. Given job $j \in J$, $\mathcal{A}$ must fix j's resource assignment $R_j \colon \mathbb{R}_{\geq 0} \to [0, 1]$ without knowing n or the properties of future jobs. We refer to the resulting schedule by $\mathcal{A}(J)$. As usual in the online setting without full information, we seek to minimize the worst-case ratio between the costs of the computed and optimal schedules. More formally, we say a schedule R for a job set J is *c-competitive* if $M(R) \leq c \cdot M^*(J)$. Similarly, we say an algorithm $\mathcal{A}$ is *c-competitive* if for any job set J we have $M(\mathcal{A}(J)) \leq c \cdot M^*(J)$.

An Optimal List-Scheduling Algorithm. Water-filling algorithms are natural greedy algorithms for scheduling problems with a continuous, preemptive character. They often yield structurally nice schedules [2–4]. In this section, we show that water-filling (described below) yields a simple, optimal online algorithm for our problem.

Theorem 1. *Algorithm* WATERFILL *has competitive ratio $e/(e-1)$ for the makespan. No deterministic online algorithm can have a lower worst-case competitive ratio.*

We first describe a single step $\text{WFSTEP}(R, \iota, C)$ of WATERFILL (illustrated in Fig. 1b). It takes a schedule $R = (R_j)_{j \in J}$ for some job set J, a new job $\iota \notin J$, and a *target completion time* C. Its goal is to *augment R by ι with completion time C*, i.e., to feasibly complete ι by time C without altering the resource assignments R_j for any $j \in J$. To this end, define the h-*water-level* $\text{wl}_h(t) := \min\{r_\iota, \max\{h - \bar{R}(t), 0\}\}$ at time t (the resource that can be assigned to ι at time t without exceeding total resource h). Note that ι can be completed by time C iff $\int_0^C \text{wl}_1(t)\,dt \geq v_\iota$ (the total leftover resource suffices to complete ι's volume by time C). If ι cannot be completed by time C, $\text{WFSTEP}(R, \iota, C)$ *fails*. Otherwise, it *succeeds* and returns a schedule that augments R with the resource assignment $R_\iota = \text{wl}_{h^*}$ for job ι, where $h^* := \inf_{h \in [0,1]}\{h \mid \int_0^C \text{wl}_h(t)\,dt \geq v_\iota\}$ is the smallest water level at which ι can be scheduled.

WATERFILL is defined recursively via WFSTEP. Given a job set $J = [n]$, define $H_j := M^*([j]) \cdot e/(e-1)$ as the target completion time for job $j \in J$ (remember that $M^*([j])$ can be easily computed, as described at the beginning of this section). Assuming WATERFILL computed a feasible schedule $R^{(j-1)}$ for the first $j-1$ jobs (with $R^{(0)}(t) = 0\ \forall t \in \mathbb{R}_{\geq 0}$), we set $R^{(j)} := \text{WFSTEP}(R^{(j-1)}, j, H_j)$. If this step succeeds, the resulting schedule is clearly $e/(e-1)$-competitive by the choice of H_j. The key part of the analysis is to show that indeed these water-filling steps always succeed.

We start the observation that water-fill schedules always result in "staircase-like" schedules (see Fig. 1b), a fact also stated in [3] (using a slightly different wording).

Observation 2 ([3, Lemma 3]). Consider a schedule R whose total resource usage $\bar{R}$ is non-increasing (piecewise constant). If we $\text{WFSTEP}(R, \iota, C)$ successfully augments R by a job ι, the resulting total resource usage is also non-increasing (piecewise constant).

Next, we formalize that WFSTEP generates the "flattest" schedules: if there is *some* way to augment a schedule by a job that completes until time C, then the augmentation can be done via WFSTEP.

Definition 1. The *upper resource distribution* $A_R^C(y)$ of a schedule R is the total volume above height y before time C in R. Given schedules R, S (for possibly different job sets), we say R is *flatter* than S ($R \preceq S$) if $A_R^C(y) \leq A_S^C(y)$ $\forall C \in \mathbb{R}_{\geq 0}, y \in [0,1]$.

Lemma 1 ([3, Lemma 4, slightly generalized]). *Consider two schedules $R \preceq S$ for possibly different job sets. Let S' denote a valid schedule that augments S by a new job ι completed until time C. Then $\text{WFSTEP}(R, \iota, C)$ succeeds and $\text{WFSTEP}(R, \iota, C) \preceq S'$.*

Next, we characterize c-competitive schedules that can be augmented by *any* job while staying c-competitive.

Definition 2. A schedule R is *c-extendable* if it is c-competitive and if it can be feasibly augmented by *any* new job ι such that the resulting schedule is also c-competitive.

Lemma 2. *Consider a job set J of volume V and with maximal processing time $p_{\max}$. A c-competitive schedule R for J is c-extendable if and only if*

$$\forall y \text{ with } (c-1)/c < y \le 1: \quad A_R^\infty(y) \le (c-1) \cdot (1-y)/y \cdot \max\{V, p_{\max} \cdot y\}. \quad (1)$$

See the Full Version for the proof of Lemma 2. While Lemma 2 gives a strong characterization, the bound on the right hand side of Eq. (1) cannot be easily translated into a proper schedule for the given volume. Thus we introduce proper (idealized) schedules that adhere to a slightly weaker version of Eq. (1). These schedules are barely $e/(e-1)$-extendable. Our proof of Theorem 1 combines their existence with Lemma 1 to deduce that WATERFILL is $e/(e-1)$-competitive.

Definition 3. For any $V \in \mathbb{R}_{\ge 0}$ we define the *universal schedule*[1] $U_V \colon \mathbb{R}_{\ge 0} \to [0,1]$ via

$$U_V(t) := \begin{cases} 1 & \text{if } t < \frac{1}{e-1} \cdot V, \\ 1 - \ln\left(t \cdot \frac{e-1}{V}\right) & \text{if } \frac{1}{e-1} \cdot V \le t < \frac{e}{e-1} \cdot V, \text{ and} \\ 0 & \text{otherwise.} \end{cases} \quad (2)$$

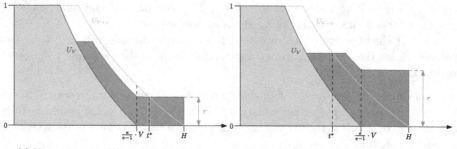

(a) U_{V+v} intersects W at the lower plateau.   (b) U_{V+v} intersects W at the upper plateau.

Fig. 2. Universal schedules U_V and U_{V+v}. The blue area indicates a new job ι with volume v and resource requirement r that is scheduled via WFSTEP(U_V, ι, H). Depending on the resource requirement r, the yellow line enters the blue area exactly once, either on the upper plateau (a) or on the lower plateau (b).

See Fig. 2 for an illustration of universal schedules. With $c = e/(e-1)$, one can easily check that $A_{U_V}^\infty(y) = \frac{e^{1-y}-1}{e-1} \cdot V \le (c-1) \cdot \frac{1-y}{y} \cdot V$. Thus, by Lemma 2, universal schedules (and any flatter schedules for the same volume) are $e/(e-1)$-extendable. Our final auxiliary lemma extends the optimality of WATERFILL from Lemma 1 to certain augmentations of universal schedules.[2] See the Full Version for the proof of Lemma 3.

---

[1] One can think of U_V as a schedule for a single job of volume V and resource requirement 1. Since there is only one job, we identify U_V with its total resource requirement function $\bar{U}_V$.

[2] Lemma 3 is not a special case of Lemma 1: the schedule S' from Lemma 1 must adhere to the new job's resource requirement, which is not the case for the universal schedule U_{V+v}.

Lemma 3. *Consider the universal schedule U_V, a new job ι of volume v and processing time p, as well as a target completion time $H \geq \frac{e}{e-1} \cdot \max\{V + v, p\}$. Then $\text{WFSTEP}(U_V, \iota, H) \preceq U_{V+v}$.*

The above enables us to prove the competitiveness of WATERFILL from Theorem 1: We show inductively that WATERFILL produces a feasible schedule $R^{(j)}$ for the first j jobs (using that $R^{(j-1)}$ is "flatter" than $U_{V([j-1])}$ together with Lemma 1) and use this to prove $R^{(j)} \preceq U_{V([j])}$ (via Lemma 3). By universality, this implies that all $R^{(j)}$ are $e/(e-1)$-extendable (and thus, in particular, $e/(e-1)$-competitive). The full proof of WATERFILL is given in the Full Version.

3   Total Completion Time Minimization

This section considers the total completion time minimization and represents our main contribution. In contrast to offline makespan minimization (Sect. 2), it remains unknown whether there is an efficient algorithm to compute an offline schedule with minimal total completion time. The so far best polynomial-time algorithm achieved a 2-approximation [3]. We improve upon this, as stated in the following theorem.

Theorem 2. *There is a $(3/2 + \varepsilon)$-approximation algorithm for total completion time minimization. Its running time in polynomial time in n and $1/\varepsilon$.*

For clarity of presentation, we analyze an idealized setting in the main part. The details for the actual result can be found in the Full Version.

Algorithm Description. Our algorithm computes two *candidate schedules* using the two sub-algorithms GREEDY and LSAPPROX (described below). It then returns the schedule with smallest total completion time among both candidates.

Sub-algorithm GREEDY processes the jobs in ascending order of their volume. To process a job, GREEDY assigns it as much resource as possible as early as possible in the schedule. Formally, for jobs $J = [n]$ ordered as $v_1 \leq \cdots \leq v_n$, the schedule R^G for GREEDY is calculated recursively using $R_j^G(t) = \mathbb{1}_{t < t_j} \cdot \min(r_j, 1 - \sum_{i=1}^{j-1} R_i^G(t))$, where the completion time t_j for job j is set such that j schedules exactly its volume v_j. See Fig. 3b for an example of a GREEDY schedule. Sub-algorithm LSAPPROX deals with solutions to following *continuous linear program* (CLP).

$$\text{minimize} \sum_{j \in J} \int_0^\infty \frac{t \cdot R_j(t)}{v_j} \, dt \qquad \int_0^\infty R_j(t)\, dt \geq v_j \; \forall j \in J$$

$$0 \leq R_j(t) \leq r_j \; \forall j \in J, t \in \mathbb{R}_{\geq 0} \qquad \sum_{j \in J} R_j(t) \leq 1 \; \forall t \in \mathbb{R}_{\geq 0}$$

Roughly, LSAPPROX first subdivides the job set into those jobs that produce a high completion time and the remaining jobs. For the former, an approximate solution is computed using the dual to the discretization (an LP) of above CLP. For the latter, is enough to reserve a small portion of the resource to schedule

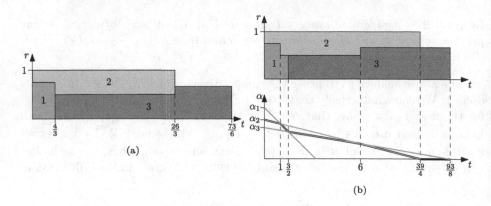

Fig. 3. Schedules for a job set $J = [3]$ with $(v_1, r_1) = (1, 3/4)$, $(v_2, r_2) = (4, 1/2)$ and $(v_3, r_3) = (6, 2/3)$. (a) GREEDY's schedule, (b) Above: A primal (resource) schedule. Below: A dual (priority) schedule. With the dual variables having values $\alpha_1 = 51/16, \alpha_2 = 39/16$ and $\alpha_3 = 31/16$, the volumes of the jobs are exactly scheduled. (See the Full Version.)

them with small completion times. For clarity of presentation, the main part will only do a simplified analysis using an idealization of LSAPPROX. For the details of this algorithm, the analysis using LSAPPROX and the analysis of GREEDY, we refer to the Full Version.

3.1   Analysis via a Bounded Fractionality Gap

Throughout the analysis, we use C^* to denote the optimal total completion time and C^{F*} for the optimal fractional total completion time. We require an algorithm that produces a schedule R with a small *fractionality gap* $\gamma(R) := C(R)/C^{F*}$, i.e., we compare the total completion time of R with the optimal fractional total completion time for the same job set. We show the following generalization of Theorem 2.

Theorem 3. *Assume that there is a polynomial-time algorithm A for total completion time minimization that produces a schedule R with $\gamma(R) \geq 1$. Then there exists a polynomial-time $(\gamma(R) + 1)/2$-approximation for total completion time minimization.*

The proof of Theorem 3 relies on Proposition 1 (three lower bounds on the optimal total completion time) and Proposition 2 (GREEDY's objective in relation to these bounds). Lower Bound (1) (*Squashed Area* Bound) and Bound (2) (*Length* or *Height* Bound) are due to Beaumont et al. [3, Def. 6,7]. Bound (3) is our novel lower bound. The proof can be found in the Full Version.

Proposition 1. *Assuming $v_1 \leq \cdots \leq v_n$, the following are lower bounds on C^*:*

(1) $C^L := \max_{j \in J} p_j$ *(2)* $C^A := \sum_{j=1}^{n} \sum_{i=1}^{j} v_i$ *(3)* $C^{F*} + 1/2 \cdot C^L$

Proposition 2. *The* GREEDY *schedule* R^G *satisfies* $C(R^G) \le C^A + C^L$.

Using them, we can give the proof of Theorem 3.

Proof of Theorem 3. We run both GREEDY and A in polynomial time to produce schedules R^G and R^A, respectively, and choose the schedule with the smaller total completion time. Using Proposition 1 and 2 and the fractionality gap $\gamma := \gamma(R^A)$, we can bound the cost $C := \min(C(R^A), C(R^G))$ of the resulting schedule in terms of C^*:

$$C \le \min(\gamma \cdot C^{F*}, C^A + C^L) \le \min(\gamma \cdot (C^* - 1/2 \cdot C^L), C^* + C^L)$$
$$= \frac{\gamma+1}{2} C^* - \frac{\gamma+2}{4} C^L + \min\left(\frac{\gamma-1}{2} C^* - \frac{\gamma+2}{4} C^L, \frac{\gamma+2}{4} C^L - \frac{\gamma-1}{2} C^*\right) \le \frac{\gamma+1}{2} C^*$$

$\square$

3.2   The Fractionality Gap of Line Schedules

For the remainder of this paper, we will introduce *line schedules* and their structural properties. Roughly, a line schedule is a certain primal-dual pair for the *CLP* defined in Sect. 3, and its dual, which we call *DCP*:

$$\text{maximize } \sum_{j \in J} \alpha_j v_j - \sum_{j \in J} r_j \int_0^\infty \beta_j(t)\, dt - \int_0^\infty \gamma(t)\, dt$$
$$\text{s.t. } \alpha_j, \beta_j(t), \gamma(t) \ge 0 \ \forall j \in J, t \in \mathbb{R}_{\ge 0} \qquad \gamma(t) + \beta_j(t) \ge \alpha_j - t/v_j \ \forall j \in J, t \in \mathbb{R}_{\ge 0}$$

It is obtained by dualizing the time-discretized version of the *CLP* (see the Full Version) and extending its constraints to the continuous time domain. *Line schedules* formalize the idea that, if we know the dual α-values, we can reconstruct all remaining primal/dual variables to obtain a primal-dual pair. If the α-values are chosen correctly, then the volumes scheduled in the primal are exactly the desired volumes $(v_j)_{j \in J}$.

To this end, we will *assume* that we have access to an algorithm called LS that produces such a line schedule R^F with $C^F(R^F) = C^{F*}$. We can then show that LS produces schedules with a fractionality gap of 2:

Proposition 3. *The* LS *schedule* R^F *satisfies* $\gamma(R^F) \le 2$.

In the following, we develop the details of line schedules. To this end, first define *primal-dual pair* as a tuple $(R, \alpha, \beta, \gamma, v)$ that fulfills the following continuous *slackness conditions (sc)*. Again, these are found by extending the time-discretized version of the *CLP* to the continuous time domain. These conditions hold for all $j \in J$ and $t \in \mathbb{R}_{\ge 0}$.

$$(\alpha\text{-sc}) : \alpha_j \left(\bar{v}_j - \int_0^\infty R_j(t)\, dt\right) = 0 \qquad (\beta\text{-sc}) : \beta_j(t)(r_j - R_j(t)) = 0$$

$$(\gamma\text{-sc}) : \gamma(t)\left(1 - \sum_{j \in J} R_j(t)\right) = 0 \qquad (R\text{-sc}) : R_j(t)(\alpha_j - t/v_j - \beta_j(t) - \gamma(t)) = 0$$

If we choose arbitrary α-values, then the corresponding line schedule is still a primal-dual pair, except that it possibly schedules a different set of volumes, i.e., the α-sc is only true if we replace v_j in the constraint by some other volume $\bar{v}_j$. This fact is used for the detailed proof of our $(3/2 + \varepsilon)$-approximation, see the Full Version.

To this end, define the *dual line* $d_j(t) := \alpha_j - t/v_j$ for each $j \in J$. The intuition behind a line schedule is now that the heights of the dual lines represent priorities: Jobs are scheduled (with maximum remaining schedulable resource) in decreasing order of the dual line heights at the respective time points. Jobs are not scheduled if their dual line lies below zero. This is formalized in the following definition. (In Fig. 3b, we supplement the example from Fig. 3a by a depiction of the dual lines.)

Definition 4. We call a job set J *non-degenerate* if all job volumes are pairwise distinct, i.e., $v_j \neq v_{j'}$ for all $j, j' \in J$.[3] Define a total order for each $t \geq 0$ as $j' \succ_t j :\Leftrightarrow d_{j'}(t) > d_j(t)$ or $d_{j'}(t) = d_j(t)$ and $v_{j'} > v_j$.[4] The *line schedule* of α is a tuple $(R, \alpha, \beta, \gamma, v)$ (recursively) defined as follows.

$$R_j(t) = \mathbb{1}_{d_j(t)>0} \cdot \min(r_j, 1 - \sum\nolimits_{j' \succ_t j} R_{j'}(t)) \qquad \beta_j(t) = \max(0, d_j(t) - \gamma(t))$$

$$\gamma(t) = \max(0, d_j(t)), \text{ where } j \text{ is the smallest job according to } \succ_t \text{ with } R_j(t) > 0$$

Equipped with the definition of a line schedule, we can now tackle the proof of Proposition 3. It requires the following two properties about the assumed algorithm LS. First, Lemma 4 allows us to bound the completion times of a fractional schedule in terms of the α-variables in the DCP:

Lemma 4. *Algorithm LS produces a schedule R^F with $C_j(R^F) \leq \alpha_j v_j$ for all $j \in J$.*

Second, we show the following lemma. Abbreviate $P = \sum_{j \in J} \int_0^\infty t \cdot R_j(t)/v_j \, dt$ (the primals objective), and $A = \sum_{j \in J} \alpha_j v_j$, $B = \sum_{j \in J} r_j \int_0^\infty \beta_j(t) \, dt$ and $\Gamma = \int_0^\infty \gamma(t) \, dt$ (the parts of the dual objective).

Lemma 5. *Algorithm LS produces a schedule R^F such that there exists a primal-dual pair $(R^F, \cdot, \cdot, \cdot)$ that fulfills strong duality ($A = B + \Gamma + P$) and balancedness ($P = B + \Gamma$).*

Using these lemmas, we can show Proposition 3.

Proof of Theorem 3. Using Lemmas 4 and 5, we show the statement as follows:

$$C(R^F) = \sum_{j \in J} C_j(R^F) \leq \sum_{j \in J} \alpha_j v_j = A = A - B - \Gamma + P = 2P = 2C^F(R^F) = 2C^{F*}$$

---

[3] While not strictly required, this makes line schedules unique and simplifies the analysis.

[4] The second part of the definition ($d_{j'}(t) = d_j(t)$ and $v_{j'} > v_j$) only exists for disambiguation of the line schedule, but is not further relevant.

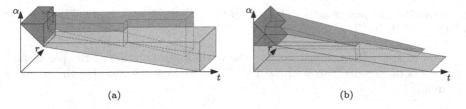

Fig. 4. (a) P-shapes for job set from Fig. 3b. P-shapes are delimited from below by $d_j(t)$ (extended into the resource axis), from above by α_j, and their top surface follows the primal schedule. (b) The shapes shown represent the union of B- and Γ-shapes. They are delimited from the left (right) by $t = 0$ $(d_j(t))$ (extended into the resource axis), and from top and bottom by the value of $d_j(t)$ at the starting and finishing time of some piece of j. See the Full Version for the formal definition of these shapes.

□

In the Full Version, we show the following Lemma 6, stating that line schedules are indeed primal-dual pairs. We then define LS to output a schedule R^F for a line schedule $(R^F, \alpha, \beta, \gamma, v)$ according to Lemma 6, i.e., for each $j \in J$, $\int_0^\infty R_j(t)\, dt = v_j$. Using this definition, we can show Lemma 4.

Lemma 6. *For any job set J there exists an α such that the line schedule of α is a primal-dual pair.*

Proof of Lemma 4. By definition, $R_j^F(t) = 0$ if $d_j(t) \le 0$. Hence, as d_j is monotonically decreasing, $C_j(R^F)$ is bounded by the zero of $d_j(t)$, which lies at $t = \alpha_j v_j$.

□

The remainder of this section will initiate the proof of Lemma 5. We first give a geometric understanding of the involved quantities (P, A, B, Γ). We build a 3D coordinate system from a line schedule. The time axis is shared, and the ordinates form the remaining two axes. We then draw 3D shapes into this coordinate system that correspond to parts of the above quantities and therefore of the CLP/DCP objectives. These shapes are described in detail in the Full Version. Generally, these shapes are constructed such that the primal and dual schedules can be "seen" from above or front. In our case, the primal schedule will be seen from the top, and the dual schedule from the front. Figure 4 illustrates the shapes in our construction. For each part of the objective $\Psi \in \{P, A, B, \Gamma\}$, we have a corresponding shape Ψ^{all}, which is subdivided into pieces $\Psi^{i,l}$, respectively.

We can show that certain pieces are pairwise non-overlapping (Lemma 7), that the A-pieces make up all other pieces (Lemma 8) and we can relate the volume of these pieces with one another and with the actual objective (Lemma 9).

Lemma 7. *Let V and W, $V \ne W$, be P-pieces, B-pieces or Γ-pieces (every combination allowed), or both be A-pieces. Then V and W do not overlap.*

Lemma 8. *A^{all} is composed of the other shapes, i.e., $A^{\text{all}} = P^{\text{all}} \cup B^{\text{all}} \cup \Gamma^{\text{all}}$.*

Lemma 9. *The pieces satisfy* $|P^{i,l}| = |B^{i,l}| + |\Gamma^{i,l}|$ *for all* i, l *and* $|\Psi^{\mathrm{all}}| = \Psi$ *for all* $\Psi \in \{P, A, B, \Gamma\}$.

Due to space limitations, we give the actual construction of the pieces and the proofs of Lemma 7 to 9 in the Full Version. Now we can give the proof of Lemma 5.

Proof of Lemma 5. Using Lemma 7 to 9, we get

$$A = |A^{\mathrm{all}}| = |P^{\mathrm{all}} \cup B^{\mathrm{all}} \cup \Gamma^{\mathrm{all}}| = |P^{\mathrm{all}}| + |B^{\mathrm{all}}| + |\Gamma^{\mathrm{all}}| = P + B + \Gamma$$

$$= |P^{\mathrm{all}}| + |B^{\mathrm{all}}| + |\Gamma^{\mathrm{all}}| = \sum\nolimits_{i,l} |P^{i,l}| + |B^{i,l}| + |\Gamma^{i,l}| = \sum\nolimits_{i,l} 2|P^{i,l}| = 2P.$$

$\square$

References

1. Althaus, E., et al.: Scheduling shared continuous resources on many-cores. J. Sched. **21**(1), 77–92 (2018). https://doi.org/10.1007/s10951-017-0518-0
2. Antoniadis, A., Kling, P., Ott, S., Riechers, S.: Continuous speed scaling with variability: a simple and direct approach. Theor. Comput. Sci. **678**, 1–13 (2017)
3. Beaumont, O., Bonichon, N., Eyraud-Dubois, L., Marchal, L.: Minimizing weighted mean completion time for malleable tasks scheduling. In: IEEE 26th IPDPS, pp. 273–284 (2012)
4. Chen, D., Wu, J., Liu, P.: Data-bandwidth-aware job scheduling in grid and cluster environments. In: 15th IEEE ICPADS, pp. 414–421 (2009)
5. Damerius, C., Kling, P., Li, M., Schneider, F., Zhang, R.: Improved scheduling with a shared resource via structural insights. In: COCOA, pp. 168–182 (2020)
6. Decker, T., Lücking, T., Monien, B.: A 5/4-approximation algorithm for scheduling identical malleable tasks. TCS **361**(2–3), 226–240 (2006)
7. Drozdowski, M.: New applications of the Muntz and Coffman algorithm. J. Sched. **4**(4), 209–223 (2001)
8. Garey, M.R., Johnson, D.S.: Complexity results for multiprocessor scheduling under resource constraints. SIAM J. Comput. **4**(4), 397–411 (1975)
9. Grigoriev, A., Sviridenko, M., Uetz, M.: Machine scheduling with resource dependent processing times. Math. Program. **110**(1), 209–228 (2007)
10. Grigoriev, A., Uetz, M.: Scheduling jobs with time-resource tradeoff via nonlinear programming. Discret. Optim. **6**(4), 414–419 (2009)
11. Jansen, K., Maack, M., Rau, M.: Approximation schemes for machine scheduling with resource (in-)dependent processing times. In: Proceedings of the Twenty-Seventh Annual ACM-SIAM SODA, pp. 1526–1542 (2016)
12. Kellerer, H.: An approximation algorithm for identical parallel machine scheduling with resource dependent processing times. Oper. Res. Lett. **36**(2), 157–159 (2008). https://doi.org/10.1016/j.orl.2007.08.001
13. Kling, P., Mäcker, A., Riechers, S., Skopalik, A.: Sharing is caring: multiprocessor scheduling with a sharable resource. In: Proceedings of the 29th ACM Symposium on Parallelism in Algorithms and Architectures, SPAA, pp. 123–132. ACM (2017)
14. Leung, J.Y. (ed.): Handbook of Scheduling - Algorithms, Models, and Performance Analysis. Chapman and Hall/CRC, Boca Raton (2004). ISBN 978-1-58488-397-5

15. Maack, M., Pukrop, S., Rasmussen, A.R.: (in-)approximability results for interval, resource restricted, and low rank scheduling. In: 30th Annual European Symposium on Algorithms, ESA, LIPIcs, vol. 244, pp. 77:1–77:13 (2022)
16. Niemeier, M., Wiese, A.: Scheduling with an orthogonal resource constraint. Algorithmica **71**(4), 837–858 (2015). https://doi.org/10.1007/s00453-013-9829-5

An Energy-Efficient Scheduling Method for Real-Time Multi-workflow in Container Cloud

Zaixing Sun, Zhikai Li, Chonglin Gu$^{(\boxtimes)}$, and Hejiao Huang

School of Computer Science and Technology, Harbin Institute of Technology
(Shenzhen), Shenzhen 518055, China
guchonglin@hit.edu.cn

Abstract. Cloud computing has a powerful ability to handle a large
number of tasks. Correspondingly, it also consumes a lot of energy.
Reducing the energy consumption of cloud service platforms while ensuring the quality of service has become a crucial issue. In this paper,
we propose a heuristic energy-saving scheduling algorithm named Real-
time Multi-workflow Energy-efficient Scheduling (RMES) with the aim
to minimize the total energy consumption in container cloud. RMES exe-
cutes tasks as parallel as possible to enhance the resource utilization of
the running machines in cluster, therefore reducing the time of the global
process, saving energy as a result. RMES takes advantage of the affinity
between containers and machines to meet the resource quantity and performance requirements of containers during scheduling. In order to follow
the change of the system state overtime, we introduce the re-scheduling
mechanism, which can automatically adjust the scheduling decisions of
the tasks that have not yet been executed in the scheduling scheme. The
experimental results show that RMES has obvious advantages over other
scheduling algorithms in terms of energy consumption and success ratio.

Keywords: Multi-workflow scheduling · Real time · Container cloud ·
Energy minimization

1 Introduction

Cloud service platform (CSP) have powerful ability to handle large-scale scientific applications. These applications are submitted to CSP in real time in the
form of workflow. The different users' workflow requests with various structures
are mixed into a multi-workflow for CSP to process. Each workflow has its Quality of Service (QoS, such as deadline) needs. Different workflows consist of tasks
with various resource requirements. CSP provides consumers with on-demand
compute and storage resources [2]. Container, a new virtualization technique, is
better suited for this multi-workflow scenario than classic virtualization technology [14]. It has three advantages including less memory, faster startup speed and
lower management overhead [17]. In container cloud, users can specify the affinity
between containers for applications, which facilitates the container orchestration
on clusters [11], such that the special resource requirements of the tasks can be
met.

© The Author(s), under exclusive license to Springer Nature Switzerland AG 2024
W. Wu and J. Guo (Eds.): COCOA 2023, LNCS 14461, pp. 168–181, 2024.
https://doi.org/10.1007/978-3-031-49611-0_12

CSP has a large number of physical machines, which consume massive energy. Data centres are reported [6] to spend around $13 billion a year on electricity. Massive energy usage not only increases the expense of the data center, but also causes some damage to the environment. However, the main challenge is that we should not only minimize the energy consumption of cloud service center, but also ensure all the tasks to be completed on time. At present, many researches have focused on the scheduling algorithm to reduce energy consumption in data centers. In [12,15], the authors proposed energy-saving scheduling algorithms on the traditional cloud. Since the complex affinity relationship between containers and physical machines brings more constraints to energy-saving scheduling decisions, these methods can not be directly used in container cloud scenarios. In [8], an energy-saving scheduling algorithm based on Q-Learning is proposed. However, the algorithm does not consider the dependencies between tasks. The workflow scheduling problem is an NP-hard problem [16]. These algorithms require a lot of computation to make decisions, which may not be suitable for real-time scheduling scenarios that require rapid response.

The heuristic method is to set some scheduling rules to make the task scheduling results on the cluster reach an approximately optimal state. Heuristic approaches are faster than other methods at producing scheduling decisions because they are based on empirical design rules. Therefore, this method is more suitable for real-time scheduling scenarios. In [9], principles from generational garbage collection (GC) reduce energy consumption in homogeneous clusters and ensure that all requests do not violate deadline constraints as much as possible. In [10], the authors adjust the task scheduling decision by balancing energy consumption and task execution time in the real-time scenario. However, the above methods either do not consider the real-time constraints, or ignore some special conditions of resource constraints in container cloud, such as their affinity.

In view of the above shortcomings, we propose a real-time multi-workflow energy-efficient scheduling (RMES) algorithm to solve real-time multi-workflow scheduling. The objective is to minimize the energy consumption of the cluster while completing as many workflows on time as possible. The main contributions of this paper are as follows:

- We build a real-time multi-workflow scheduling model on heterogeneous clusters considering the affinity constraints between tasks and machines.
- We propose a heuristic scheduling algorithm called RMES, which decreases the base energy consumption of cluster by compressing the time of the global process through executing tasks in parallel.
- RMES evaluates the current running status of the physical machines in cluster, and shuts down the physical machines with low utilization in time, thus reducing unnecessary energy consumption of the cluster.
- The performance of scheduling algorithm is verified by using workflow in the real world. Compared with the existing algorithms, RMES reduces more energy consumption for CSP while meeting the affinity constrains between container and physical machine.

The rest of this paper is organized as follows. Section 2 introduces the real-time multi-workflow scheduling and energy consumption model. In Sect. 3, a heuristic energy-saving scheduling algorithm is proposed. Section 4 presents the experimentation and evaluation. Finally, Sect. 5 concludes the paper.

2　Problem Formulation

2.1　Workflow Modeling

In cloud, the system needs to schedule the workflow applications submitted dynamically by users in real time. These workflows are composed of many requests which can be denoted as $\mathcal{W} = \{w_1, w_2, ..., w_m\}$. A single request can be described as a directed acyclic graph (DAG). We model a request $w_m \in \mathcal{W}$ as $w_m = \{w_m^{at}, w_m^{trt}, w_m^d, G_m\}$, where w_m^{at}, w_m^{trt}, w_m^d and G_m represent the arrival time, tolerable running time (the maximum time a user can tolerate for a request to be execute), deadline and structure of w_m, respectively. w_m^d can be calculated as $w_m^d = w_m^{at} + w_m^{trt}$. $G_m = (\mathcal{T}_m, E_m)$, where $\mathcal{T}_m$ is the set of tasks and E_m represents the dependency between tasks. $\mathcal{T}_m = \{t_{m1}, t_{m2}, \cdots, t_{m|\mathcal{T}_m|}\}$, where $T_{mi}(0 < i \leq |\mathcal{T}_m|)$ represents the ith task of w_m and $|\mathcal{T}_m|$ is the total number of tasks contained in w_m. E_m is the 0–1 matrix of $T_m \times T_m$. $e_{u,v}^m = 1$ means that there is a data dependence between t_{mu} and t_{mv}, where t_{mu} is the immediate predecessor of t_{mv}. For the t_{mi}, we further model it as $t_{mi} = \{t_{mi}^I, t_{mi}^{image}, t_{mi}^{type}, t_{mi^d}\}$. t_{mi}^I is the number of instructions contained in t_{mi}; t_{mi}^{image} is the image of container executing t_{mi}; t_{mi}^{type} is the t_{mi}'s type; t_{mi}^d is the sub-deadline of t_{mi}.

A task is executed within a container and then deployed on physical machine (PM). Container bundles the software configuration of a specific workflow into a container image and is the smallest execution unit in the resource scheduling system. We model container c_j as $c_j = \{c_j^{type}, c_j^{cpu}, c_j^{mem}, c_j^{cache}, c_j^{taints}\}$, where $c_j \in \mathcal{C}$ $(0 < j \leq |\mathcal{C}|)$. $\mathcal{C}$ is the set of all containers in the cluster. c_j^{type} is the type of task that container c_j runs; c_j^{cpu} is the number of CPU cores required by container c_j; c_j^{mem} is memory size required by container c_j; c_j^{cache} is the set of tasks that have been assigned to container c_j; c_j^{taints} is taint nodes (the set of PMs that cannot run container c_j).

Since a single container can only process one task at a time, the tasks in the cache can be divided into two types: waiting and executing. The taint node is a set of PMs that cannot run the container. This set is defined by the user. The reasons why a PM cannot run the container include that there are no specific devices required by the container, the performance of some devices cannot meet the minimum requirements for container operation, and so on. There is a one-to-one correspondence between containers and tasks. The same type container can only run the same type task, and tasks of the same type can only run on the same type of container. A container is created based on the images contained in the corresponding task.

2.2　Service Instance Modeling

A cloud service provider can provide a variety of cloud service instances, such as virtual machines and PMs. In this paper, we only consider the case of PM $\mathcal{P} =$

$\{p_1, p_2, \ldots, p_{|\mathcal{P}|}\}$. We use $p_k (0 < k \le |\mathcal{P}|)$ to represent the kth PM and $\mathcal{P}$ is the set of PMs in this CSP. We model a PM as $p_k = \{p_k^{t\_cpu}, p_k^{t\_mem}, p_k^{\bar{t},u\_cpu}, p_k^{\bar{t},u\_mem}, p_k^{c}, p_k^{e\_total}, p_k^{e\_base}, p_k^{ips}\}$. $p_k^{t\_cpu}$ is the Number of CPU cores of the p_k; $p_k^{t\_mem}$ is the memory resources of the p_k; $p_k^{\bar{t},u\_cpu}$ is the Number of CPU cores used in p_k at $\bar{t}$ moment; $p_k^{\bar{t},u\_mem}$ is the memory resources used in p_k at $\bar{t}$ moment; p_k^{c} is the container set which contains container runs in p_k; $p_k^{e\_total}$ is the power of full load operation of p_k; $p_k^{e\_base}$ is the basic power of no-load operation of p_k; p_k^{ips} is the number of instructions that a single core of p_k can process per second. $p_k^{\bar{t},u\_cpu}, p_k^{\bar{t},u\_mem}$ can be calculate by Eqs.(1–2), where $x_{j,k}^{\bar{t}} \in \{0,1\}$, $x_{j,k}^{\bar{t}} = 1$ means that the container c_j is running in p_k at $\bar{t}$ moment.

$$p_k^{\bar{t},u\_cpu} = \sum_{c_j \in \mathcal{C}} x_{j,k}^{\bar{t}} c_j^{cpu}, \tag{1}$$

$$p_k^{\bar{t},u\_mem} = \sum_{c_j \in \mathcal{C}} x_{j,k}^{\bar{t}} c_j^{mem}, \tag{2}$$

Based on the model widely used [4,8] in cloud computing energy analysis, we model the relationship between the power of the p_k and the CPU utilization at $\bar{t}$ moment by Eq. (3), where $\mathbf{P}_k^{\bar{t}}$ represent the power of p_k at $\bar{t}$ moment.

$$\mathbf{P}_k^{\bar{t}} = \begin{cases} 0, & p_k \ is \ off, \\ \frac{p_k^{\bar{t},u\_cpu}}{p_k^{t\_cpu}}(p_k^{e\_total} - p_k^{e\_base}) + p_k^{e\_base}, & p_k \ is \ on. \end{cases} \tag{3}$$

2.3 Workflow Scheduling Model

In this paper, workflow scheduling aims to minimize the total energy consumption for executing workflows. The total energy consumption for a CSP can be expressed by Eq. (4), where T is total running time of the system.

$$\mathbb{E} = \sum_{\bar{t}=0}^{T} \sum_{k=1}^{|\mathcal{P}|} \mathbf{P}_k^{\bar{t}}, \tag{4}$$

Task Dependency Constraints. Due to the dependency between tasks, all tasks can be executed only when their predecessors are completed or there are no predecessors. t_{mu}^{EST} and t_{mv}^{FT} mean earliest start time of t_{mu} and finish time of t_{mv}, respectively. $Pred(t_{mu})$ is a set of immediate predecessors of t_{mu}, where $Pred(t_{mu}) = \{t_{mv} | e_{v,u}^{m} = 1, \forall \ t_{mv} \in \mathcal{T}_m\}$.

$$\max_{t_{mv} \in Pred(t_{mu})} t_{mv}^{FT} \le t_{mu}^{EST}, \tag{5}$$

Task Completion Time Constraint. When the system makes scheduling decisions, we should ensure that the real-time tasks submitted by users can be completed on time.

$$\max_{t_{mv} \in \mathcal{T}_m} t_{mv}^{FT} \le w_m^{d}. \tag{6}$$

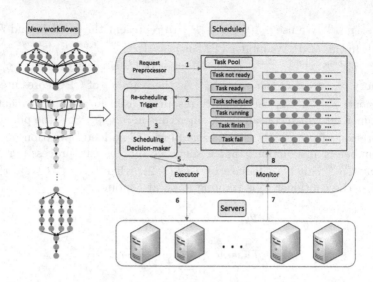

Fig. 1. Schedule System

Task and Container Placement Constraint. A task can only be deployed on a single container, and its type should be the same as the task type that the container can handle. In Eq. (7), $y^{\bar{t}}_{mi,j} \in \{0,1\}$ and $y^{\bar{t}}_{mi,j} = 1$ means t_{mi} is deployed on c_j, otherwise we set $y^{\bar{t}}_{mi,j}$ to 0. When we deploy containers on PMs, we need to meet some resource level constraints. Equation (9) and Eq. (10) mean the resources occupied by the deployed container on the PM cannot exceed the total resources of the PM. Equation (11) means a container can only be deployed on one PM and this PM can't be taint node of c_j.

$$\sum_{j=1}^{|\mathcal{C}|} y^{\bar{t}}_{mi,j} = 1. \tag{7}$$

$$t^{type}_{mi} = c^{type}_j, y^{\bar{t}}_{mi,j} = 1. \tag{8}$$

$$\sum_{c_j \in \mathcal{C}} x^{\bar{t}}_{j,k} c^{cpu}_j \le p^{t\_cpu}_k, \tag{9}$$

$$\sum_{c_j \in \mathcal{C}} x^{\bar{t}}_{j,k} c^{men}_j \le p^{t\_mem}_k, \tag{10}$$

$$\sum_{j=1}^{|\mathcal{C}|} x^{\bar{t}}_{j,k} = 1, \ p_k \notin c^{taints}_j. \tag{11}$$

3 Real-Time Multi-workflow Energy-Efficient Scheduling Algorithm

3.1 Scheduling Architecture

The real-time multi-workflow scheduling architecture is shown in Fig. 1. The architecture can be divided into three parts: end user, instance cluster and scheduler. End users can submit workflow requests to the system at any time. The cloud service platform provides instance clusters to handle the requests submitted by users. The scheduler is to arrange the workflow submitted by users into the instance cluster reasonably, so that the whole system can operate efficiently. Schedulers are mainly divided into several components: request preprocessor, task pool, rescheduling trigger, scheduling decision maker, executor and monitor. After the workflow request is accepted by the scheduler, the request preprocessor first decomposes the workflow submitted by the user into tasks and sets the deadline and priority for each task. These tasks will be placed in the task pool (Step 1). The rescheduling trigger will receive the status in the task pool (Step 2) and inform the scheduling decision-maker whether to perform general scheduling or rescheduling (Step 3). After receiving instructions, the scheduler extracts the task information (Step 4) to be dispatched from the task pool and sends the generated scheduling decisions to the executor (Step 5). During the execution period, adjust the running state of the instance cluster according to the received instructions (Step 6). The monitor will constantly monitor the status of cluster (Step 7) and update the information in the task pool (Step 8).

3.2 Request Preprocessor

This component sets the sub-deadline for the tasks contained in the request submitted by the user, and sorts them according to the priority. The sub-deadline setting of each task is related to the topology level of the task in request. The topological level of a t_{mi} is defined as Eq. (12). For each level, we calculate the task with the largest number of instructions in the level as the critical task of the level. We take the duration of the task on the fastest machine in the system as the execution time of this level ($level_l^{time}$).

$$Lev(t_{mi}) = \begin{cases} 1, & Pred(t_{mi}) = \emptyset \\ \max\limits_{t_{mj} \in Pred(t_{mi})} Lev(t_{mj}) + 1, other. \end{cases} \tag{12}$$

$$Level_l^t = \{t_{mi} | Lev(t_{mi}) = l\} \tag{13}$$

$$level_l^{time} = c_j^{load} + \frac{\hat{t}_l}{\hat{p}^{ips} \cdot c_j^{cpu}}, c_j^{type} = \hat{t}_l^{type}, \hat{p} = \arg\max\limits_{p_k \in \mathcal{P}}(p_k^{ips}), \hat{t}_l = \arg\max\limits_{t_{mi} \in Level_l^t}(t_{mi}^I),$$

$$\tag{14}$$

where $\hat{t}_l$, $\hat{p}$ and c_j^{load} represent the task with maximum number of instructions in level l, the fastest single core PM and the preparation time before the container is able to handle tasks after deploying, respectively. The estimated processing time for w_m can be model as Eq. (15), where L denotes the maximum level contained in the w_m. After that, we can set the sub-deadline of t_{mi} as Eq. (16).

$$w_m^{et} = \sum_{l=1}^{L} level_l^{time}, \tag{15}$$

$$t_{mi}^d = \frac{level_l^{time}}{w_m^{et}} \cdot w_m^{trt} + w_m^{at}, \ Lev(t_{mi}). \tag{16}$$

Our priority ranking of tasks is mainly calculated according to the number of subsequent tasks related to the task which include all immediate and mediate successors. We defined $d(t_{mu})$ as the set of tasks which are dependent on t_{mu}.

$$d(t_{mu}) = (\bigcup_{t_{mv} \in Sub(t_{mu})} d(t_{mv})) \cup t_{mu}, \tag{17}$$

$$Rank(t_{mu}) = |d(t_{mu})|, \tag{18}$$

$Sub(t_{mv}) = \{t_{mu}|e_{u,v}^m = 1, \forall \ t_{mu} \in \mathcal{T}_m\}$ represents the set of immediate successors of t_{mv}. A task with higher rank means the task has higher scheduling priority than other tasks at the same topology level.

3.3   Task Pool

Task pool is a mapping of task states in the system. In the cluster, tasks are mainly divided into the following types: *Task not ready:* Tasks whose predecessors have not been completed and have not entered the executable state; *Task ready:* Tasks whose predecessors have completed but are not scheduled by the system; *Task scheduled:* Tasks that have been scheduled to the container but have not started to execute; *Task running:* Tasks being processed by the container; *Task finished:* Tasks that have been completed on time; *Task fail:* Tasks that have timed out.

3.4   Re-scheduling Trigger

In most scheduling strategies, the system only schedules tasks once, which often falls into local optimization in real-time scenarios. In the real-time system, the request will arrive at the cloud platform at any time, thus the state of the task in the system will fluctuate with time. For tasks that have been scheduled before but the container has not started to execute, there may be a better scheduling decision in the current new system state. However, if every new task arrives, rescheduling all the tasks in the system will greatly increase the scheduling cost of the system and affect the quality of service. To trade off this decision, we use $\theta_t = \frac{|Newtask_t|}{|Alltask_t|}$ to describe the state of unprocessed tasks in the system at t moment. $Newtask_t$ represents the set of new executable tasks at t moment and $Alltask_t$ represents the set of tasks that can be executed but not started. $|Newtask_t|$ and $|Alltask_t|$ mean the number of elements in corresponding set. α is re-schedule factor, $(0 < \alpha < 1)$. If $\theta_t > \alpha$, this means that the task state in the system has changed greatly, and rescheduling decision will be a better choice.

3.5   Scheduling Decision-Maker

After the above stages, the system has completed the screening and sorting of the tasks to be scheduled. We will schedule all tasks once according to the priority of the tasks. The selection of target container and machine should consider the scheme with the lowest energy consumption as far as possible on the basis of ensuring that the task can be completed on time, and also consider the universality of the machine. If too many containers are deployed on a machine with higher versatility (the machine is used as taint with fewer tasks), more picky tasks (tasks with many Taints) may not have enough resources to deploy in the cluster. Therefore, the selection of scheduling objectives should be comprehensively determined by weighing the new energy consumption caused by the deployment of the machine and the universality of the target machine.

The calculation of new energy consumption after deployment can be divided into the following situations: $\mathcal{A}$: There is a deployed container, and the running time of the container after deployment will not exceed the maximum running time of the container deployed on the PM to which the container is deployed. $\mathcal{B}$: There is a deployed container, and the running time of the container after deployment will exceed the maximum running time of the container deployed on the PM to which the container is deployed. $\mathcal{C}$: A new container needs to be deployed on the PM that has been powered on, and the running time of the PM will not be extended. $\mathcal{D}$: A new container needs to be deployed on the PM that has been powered on, which will cause the increment of running time of the PM. $\mathcal{E}$: Need to open a new PM to deploy the container. $\mathcal{F}$: The cluster does not have enough resources to complete the task on time. Considering the above situations, the new energy consumption caused by the deployment tasks can be calculated by Eq. (19).

$$\Delta E = \begin{cases} \frac{c_j^{cpu}}{p_k^{cpu}} \cdot (p_k^{e\_total} - p_k^{e\_base}) \cdot \bar{t}_r, & \mathcal{A} \\ \frac{c_j^{cpu}}{p_k^{cpu}} \cdot (p_k^{e\_total} - p_k^{e\_base}) \cdot \bar{t}_r + \bar{t}_e \cdot p_k^{e\_base}, & \mathcal{B} \\ \frac{c_j^{cpu}}{p_k^{cpu}} \cdot (p_k^{e\_total} - p_k^{e\_base}) \cdot (\bar{t}_r + \bar{t}_p), & \mathcal{C} \\ \frac{c_j^{cpu}}{p_k^{cpu}} \cdot (p_k^{e\_total} - p_k^{e\_base}) \cdot (\bar{t}_r + \bar{t}_p) + \bar{t}_e \cdot p_k^{e\_base}, & \mathcal{D} \\ \frac{c_j^{cpu}}{p_k^{cpu}} \cdot (p_k^{e\_total} - p_k^{e\_base}) \cdot (\bar{t}_r + \bar{t}_p) + (\bar{t}_r + \bar{t}_p + \bar{t}_s) \cdot p_k^{e\_base}, & \mathcal{E} \end{cases} \quad (19)$$

$$t_r = \frac{t_{mi}^I}{p_k^{ips} \cdot c_j^{cpu}}, \quad (20)$$

where c_j, p_k are the container and PM for task t_{mi} plan to deployment, respectively. $\bar{t}_r$, $\bar{t}_e$, $\bar{t}_p$ and $\bar{t}_s$ are task execution time, extended execution time of PM, start time of container and start time of PM. For each PM in the platform, the system calculates the universality of each PM (p_k^u, (21)) according to the taints node information submitted by the user. $\mathcal{C}_{avail}$ is a set of containers that can be deployed to p_k. The higher p_k^u means that p_k has higher versatility.

$$p_k^u = \frac{|\mathcal{C}_{avail}|}{|\mathcal{C}|}, \quad (21)$$

In order to meet the scheduling objectives, the system needs to allocate tasks to lower energy consumption and more "exclusive" PMs. $Q_{i,j,k} \in \mathcal{Q}$ is calculated by Eq. (22), which denote the energy consumption of deploying t_i to container c_j and PM p_k. The system will deploy the task with a higher Q scheme.

$$Q_{i,j,k} = \beta \Delta E_{i,j,k} + (1 - \beta)p_k^u, \tag{22}$$

where β is the weight of energy consumption in scheduling decision, $(0 < \beta < 1)$.

The detailed pseudocode of the our algorithm can be found in Appendix A.

4   Performance Evaluation

4.1   Experimental Setup

We use five well-known workflows widely used in previous work to evaluate the algorithm: Montage, LIGO, Epigenomics, CyberShake and SIPHT[1]. Montage is I/O intensive, LIGO and CyberShake are CPU intensive containing task with high memory requirements. Epigenomics and SIPHT are CPU intensive. Details of these workflows are described in [13]. Similar to [7], we use the rule that the arrival interval of any two requests in the actual scenario obeys the Poisson distribution to generate a real-time workflow.

We establish a simulation platform in Python which generates the request workflow according DAX format file. The platform runs on a Ubuntu 20.04.2 LTS a 64bit PC with i5-9500 3.0 GHz CPU and 32 GB RAM, python 3.8.5. In the experiment, we use eight types of PMs (see Table 1.), and the relevant configuration parameters of these PMs are obtained by [1]. We set re-schedule factor α to 0.05. We set w_m^{min} as the time it takes for the longest critical path in the workflow to run on the fastest machine in the cluster. We select deadline factor γ according to the uniform distribution $[2, 8]$ and then assign the deadline calculated by $\gamma \cdot w_m^{min}$ to workflows.

Table 1. Real-world PM types

Type	CPU cores	Mem (GB)	basic power (w)	full load power (w)
PM1	4	4	43	115
PM2	4	8	63	115
PM3	8	8	89.4	173
PM4	8	16	155	269
PM5	8	18	173	334
PM6	8	32	226	294
PM7	16	16	299	521
PM8	32	32	260	748

---

[1] https://confluence.pegasus.isi.edu/display/pegasus/WorkflowGenerator.

We constructed five homogeneous scenarios (with only one workflow type) and one heterogeneous scenario (with all workflow types) using the above workflows. In each scenario, the experimental parameters are composed of three parameters, including arrival rate λ, workflow scale *scale* and compatibility δ. Similar to previous works [3,7], λ includes the arrival rate of four poisson distribution of 1 workflows/s, 5 workflows/s, 10 workflows/s and 15 workflows/s. Workflow scale represents the workflow intensity of three requests, *small*, *medium* and *large*. They are composed of a mixture of multiple workflows with a total of 1000, 2000 and 3000 tasks. The workflow contains more tasks, if it is in large workflow scale. We set compatibility to 0.2, which represent 20% of the PMs in the cluster as taints of the task. This represents the degree to which the task is picky about the PMs in the cloud service provider cluster. For each workflow structure, we conducted experiments on the different values of the above three parameters, and a total of 24 groups of experiments were conducted.

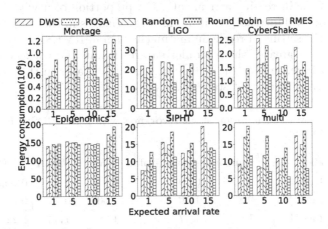

Fig. 2. The energy consumption of each workflow with DWS, ROSA, Round-Robin, Random and RMES

4.2   Comparison Algorithm

To verify the effectiveness of our proposed algorithm, we compare it with four existing algorithms: Random and Round-Robin, ROSA [5] and DWS [3]. Random is a random scheduling strategy. After disassembling the arriving workflow into sub tasks, the system randomly assigns the tasks to the container and schedules the container to run on the random machine. Round-Robin is one of the default scheduling strategies of Kubernetes. The algorithm schedules tasks to appropriate containers and PMs according to the polling rules. ROSA is an uncertainty-aware online scheduling algorithm to schedule dynamic and multiple workflows with deadlines. The algorithm first estimates the completion time of the task, and then schedules the task to minimize the cost. DWS is an online heuristic algorithm, which aims to minimize the cost of renting service instances under the deadline. When the new workflow arrives, the system sets heuristic rules according to the cost deadline to schedule the task to a more reasonable instance. To ensure the fairness of the experiment, we equivalently replace the optimization objective function in ROSA and DWS with the same energy consumption objective function as RMES.

4.3   Simulation Results

Arrival Rate. Figure 2 shows the energy consumption result of the algorithms under different workflow structures. The arrival rate is increased from 1 to 15 in the case of medium workflow scale. We can see that in most cases, the experimental results of RMES are excellent, which is significantly improved compared with other algorithms. When the arrival rate increases, our algorithm performs better. When the arrival rate is 1, our algorithm improves -1%, -10%, 18.45% and 34.11% respectively compared with DWS, ROSA, Random and Round-Robin. When arrival rate reaches 15, RMES increases to 40.57%, 19.42%, 27.89% and 39.40%. With the increasing arrival rate, the proportion of newly arrived tasks in the task pool will increase, resulting in large changes in the status of the task pool. Due to the setting of rescheduling mechanism in RMES, when the state of task pool changes greatly, the scheduler can reschedule the unexecuted tasks in the system in time, so that the scheduling results of tasks in the task pool are more in line with the current state of task pool.

Table 2. The energy consumption of each workflow with DWS, ROSA, Round-Robin, Random and RMES

Workflow, size	Random	Round-robin	DWS	ROSA	RMES
Montage, S	0.47	0.79	0.64	0.52	**0.37**
Montage, M	0.83	1.10	1.06	0.73	**0.55**
Montage, L	1.62	1.77	1.72	**1.17**	1.22
LIGO, S	13.12	18.96	9.95	**9.16**	9.64
LIGO, M	19.86	22.81	21.68	14.71	**11.98**
LIGO, L	23.17	24.69	27.85	19.74	**17.46**
CyberShake, S	0.85	0.95	0.80	0.75	**0.67**
CyberShake, M	1.51	2.55	1.84	1.25	**0.94**
CyberShake, L	1.85	2.44	2.17	1.60	**1.30**
Epigenomics, S	208.48	**205.63**	216.23	216.22	206.24
Epigenomics, M	**141.98**	144.15	145.72	147.13	146.19
Epigenomics, L	163.81	162.89	**160.84**	172.21	162.00
SIPHT, S	12.92	13.14	17.49	12.08	**11.18**
SIPHT, M	13.09	15.20	12.37	**10.02**	10.08
SIPHT, L	30.89	46.99	44.93	28.90	**23.99**
multi, S	6.21	8.06	6.01	4.44	**3.86**
multi, M	10.95	13.91	10.85	10.95	**5.51**
multi, L	27.49	34.94	24.80	17.12	**12.44**

Workflow Scale. Table 2 shows the energy consumption result of the algorithm under different sizes and different workflow structures when the arrival rate is 10 workflows/s. In Table 2, 'Montage, S' means Montage workflow small-scale test example. In addition to the two workflow structures with fluctuating completion rates, RMES has obvious advantages over other algorithms. Under the three workflow structures of multi, SIPHT and CyberShake, the improvement of RMES and the comparison algorithm with the best performance increases from 13.06%, 7.45% and 10.66% on a small scale to 27.33%, 16.98% and 18.75% on a large scale. We find that RMES algorithm has a greater improvement under the condition of large-scale workflow.

5 Conclusion

In this paper, we focus on the real-time energy-saving multi-workflow scheduling on container cloud. Firstly, we establish a cloud-based workflow scheduling model, which considers resource quantity and performance constraints of container deployment. Then we propose an real-time multi-workflow energy-efficient scheduling (RMES) algorithm. By executing tasks in parallel on the running PM, RMES can compress the time of the global process to reduce the base energy consumption. Furthermore, RMES introduces rescheduling mechanism, so that the task scheduling decision can be adjusted with the change of system state. Finally, we conduct several groups of experiments under the actual workflow conditions. Compared with other algorithms, RMES significantly reduces the energy consumption generated by CSP.

Acknowledgements. This work is supported by the Shenzhen Science and Technology Program under Grant No. GXWD20220817124827001, and No. JCYJ20210324132406016.

Appendix for "An Energy-Efficient Scheduling Method for Real-Time Multi-workflow in Container Cloud"

A   Detailed Pseudocode of the Proposed Algorithm

The detailed procedure is given in Algorithm 1. $task_{ready}$ and $task_{scheduled}$ represent task sets of type task ready and task scheduled in the task pool, respectively. $<Q_{n,i,j}, c_i, p_j>$ means assign $task_n$ to c_i running in p_j. Firstly, we evaluate the task state in the system and judge whether to reschedule the scheduled tasks according to the current task state of the system, as shown in lines 2–10 of Algorithm 1. Then, for the task to be scheduled, the system calculates the Q value of the task deployed on the existing container, and deploys the task to the container with the lowest Q value, as shown in lines 11–24 of Algorithm 1. For a task without a suitable container to run, the system will create a new container for it, calculate the Q value of the container deployed to each PM in the cluster, and select the PM with the lowest Q value to run the container, as shown in lines 25–41 of Algorithm 1.

Algorithm 1: RMES

Input: $task_{ready}, task_{scheduled}, \mathcal{C}, \mathcal{P}$

Output: $\{x_{n,j}\}, \{y_{i,j}\}$

1   $scheduletask \leftarrow \emptyset$;

2   Calculate θ_t;

3   **if** $\theta_t > \alpha$ **then**

4   **foreach** $t_n \in task_{scheduled}$ **do**

5 | $scheduletask \leftarrow scheduletask \cup \{t_n\}$;

6   **end**

7   **end**

8   **foreach** $t_n \in task_{ready}$ **do**

9 | $scheduletask \leftarrow scheduletask \cup \{t_n\}$;

10   **end**

11   **foreach** $t_n \in scheduletask$ **do**

12     $target \leftarrow \emptyset$;

13   **foreach** $c_i \in \mathcal{C}$ **do**

14   **if** $c_i^{type} = task_n^{type}$ **then**

15   Calculate $Q_{n,i,j}$ according Eq. (22);

16        $target \leftarrow <Q_{n,i,j}, c_i, p_j>$;

17   **end**

18   **end**

19   **if** $target \neq \emptyset$ **then**

20   select c_i with minimum Q;

21       $x_{n,j} \leftarrow 1$;

22       $scheduletask \leftarrow scheduletask - \{task_n\}$;

23   **end**

24   **end**

25   **if** $scheduletask \neq \emptyset$ **then**

26   **foreach** $t_n \in scheduletask$ **do**

27       $target \leftarrow \emptyset$;

28   create container c_i;

29   **foreach** $pm_j \in \mathcal{P}$ **do**

30   Calculate $Q_{n,i,j}$ according Eq. (22);

31         $target \leftarrow <Q_{n,i,j}, c_i, p_j>$;

32   **end**

33   **if** $target \neq \emptyset$ **then**

34   select c_i with minimum Q;

35         $x_{n,j} \leftarrow 1$;

36         $y_{i,j} \leftarrow 1$;

37         $scheduletask \leftarrow scheduletask - \{task_n\}$;

38         $\mathcal{C} \leftarrow \mathcal{C} \cup \{c_i\}$;

39   **end**

40   **end**

41   **end**

References

1. Third quarter 2021 specpower_ssj2008 results (2021). www.spec.org/power_ssj2008/results/res2021q3/
2. Al-Dulaimy, A., Taheri, J., Kassler, A., HoseinyFarahabady, M.R., Deng, S., Zomaya, A.: Multiscaler: a multi-loop auto-scaling approach for cloud-based applications. IEEE Trans. Cloud Comput. **10**(4), 2769–2786 (2022)
3. Arabnejad, V., Bubendorfer, K., Ng, B.: Dynamic multi-workflow scheduling: a deadline and cost-aware approach for commercial clouds. Futur. Gener. Comput. Syst. **100**, 98–108 (2019)
4. Beloglazov, A., Buyya, R., Lee, Y.C., Zomaya, A.: A taxonomy and survey of energy-efficient data centers and cloud computing systems. In: Advances in Computers, vol. 82, pp. 47–111 (2011)
5. Chen, H., Zhu, X., Liu, G., Pedrycz, W.: Uncertainty-aware online scheduling for real-time workflows in cloud service environment. IEEE Trans. Serv. Comput. **14**(4), 1167–1178 (2018)
6. Cheng, M., Li, J., Nazarian, S.: DRL-cloud: deep reinforcement learning-based resource provisioning and task scheduling for cloud service providers. In: 2018 23rd Asia and South Pacific Design Automation Conference, pp. 129–134 (2018)
7. Deng, F., Lai, M., Geng, J.: Multi-workflow scheduling based on genetic algorithm. In: 2019 IEEE 4th International Conference on Cloud Computing and Big Data Analysis (ICCCBDA), pp. 300–305. IEEE (2019)
8. Ding, D., Fan, X., Zhao, Y., Kang, K., Yin, Q., Zeng, J.: Q-learning based dynamic task scheduling for energy-efficient cloud computing. Futur. Gener. Comput. Syst. **108**, 361–371 (2020)
9. Havet, A., Schiavoni, V., Felber, P., Colmant, M., Rouvoy, R., Fetzer, C.: Genpack: a generational scheduler for cloud data centers. In: 2017 IEEE International Conference on Cloud Engineering (IC2E), pp. 95–104 (2017)
10. Hu, B., Cao, Z., Zhou, M.: Scheduling real-time parallel applications in cloud to minimize energy consumption. IEEE Trans. Cloud Comput. **10**(1), 662–674 (2022)
11. Hu, Y., Zhou, H., de Laat, C., Zhao, Z.: Concurrent container scheduling on heterogeneous clusters with multi-resource constraints. Futur. Gener. Comput. Syst. **102**, 562–573 (2020)
12. Hussain, M., Wei, L.F., Lakhan, A., Wali, S., Ali, S., Hussain, A.: Energy and performance-efficient task scheduling in heterogeneous virtualized cloud computing. Sustain. Comput. Inform. Syst. **30**, 100517 (2021)
13. Juve, G., Chervenak, A., Deelman, E., Bharathi, S., Mehta, G., Vahi, K.: Characterizing and profiling scientific workflows. Futur. Gener. Comput. Syst. **29**(3), 682–692 (2013)
14. Merkel, D., et al.: Docker: lightweight Linux containers for consistent development and deployment. Linux J. **2014**(239), 2 (2014)
15. Sun, Z., Huang, H., Li, Z., Gu, C., Xie, R., Qian, B.: Efficient, economical and energy-saving multi-workflow scheduling in hybrid cloud. Expert Syst. Appl. **228**, 120401 (2023)
16. Sun, Z., Zhang, B., Gu, C., Xie, R., Qian, B., Huang, H.: ET2FA: a hybrid heuristic algorithm for deadline-constrained workflow scheduling in cloud. IEEE Trans. Serv. Comput. **16**(3), 1807–1821 (2023)
17. Zhang, F., Tang, X., Li, X., Khan, S.U., Li, Z.: Quantifying cloud elasticity with container-based autoscaling. Futur. Gener. Comput. Syst. **98**, 672–681 (2019)

Set-Related Optimization

Weakly Nondominated Solutions of Set-Valued Optimization Problems with Variable Ordering Structures in Linear Spaces

Zhiang Zhou[1(✉)], Wenbin Wei[1], and Kequan Zhao[2]

[1] College of Sciences, Chongqing University of Technology, Chongqing 400054, China
zhi_ang@163.com
[2] School of Mathematical Sciences, Chongqing Normal University, Chongqing 401331, China

Abstract. In this paper, weakly nondominated solutions of set-valued optimization problems with variable ordering structures are investigated in linear spaces. Firstly, the notion of weakly nondominated element of a set with a variable ordering structure is introduced in linear spaces, and the relationship between weakly nondominated element and nondominated element is also given. Secondly, under the assumption of nearly $\mathcal{C}(y)$-subconvexlikeness of set-valued maps, scalarization theorems of weakly nondominated solutions for unconstrained set-valued optimization problems are established. Finally, two duality theorems of constrained set-valued optimization problems are obtained. Some examples are given to illustrate our results. The results obtained in this paper improve and generalize some known results in the literatures.

Keywords: Set-valued maps · Variable ordering structures · Weakly nondominated solution · Scalarization · Duality

1 Introduction

Set-valued analysis has become an important branch of nonlinear analysis since it is widely applied in various areas of the human real life. For example, Debreu [5] used the fixed point theorem of the set-valued map, which is an important mathematical tool, to prove the existence of the Walrasian equilibrium theorem. Some works about set-valued analysis can be founded in [2, 3, 10]. Recently, many researchers have paid attention to the set-valued optimization problem which is a kind of optimization problem with the objective map being a set-valued map. Yang et al. [16] introduced the nearly cone subconvexlike set-valued map and

Supported by the National Nature Science Foundation of China (12171061), the Science and Technology Research Program of Chongqing Education Commission (KJZD-K202001104) and Chongqing Graduate Innovation Project (CYS22671).

established optimality conditions involving Lagrangian multiplier and scalarization of set-valued optimization problems. Zhao et al. [19] used the improvement set E, which was introduced by Chicco et al. [4], to define nearly E-subconvexlike set-valued map and investigated the weak E-optimal solution of set-valued optimization problems. Zhou et al. [20] studied scalarizations and optimality of constrained set-valued optimization using improvement sets and image space analysis.

On the other hand, in order to compare different objective values of the optimization problem, we need to establish a partial ordering relation which is induced by a pointed closed convex cone. Generally speaking, the ordering relation involving optimization problems is determined by a fixed convex cone. However, in actual situations, different decision-makers have different preferences in different environments. Therefore, the partial ordering relation involved in optimization problems is no longer determined by a fixed convex cone. Instead, it is determined by a variable ordering cone related to environment, times, economy and other factors. This kind of optimization problems are called optimization problems with variable ordering structures. More general concepts of ordering structures were introduced by Yu [18] in terms of domination structures. Eichfelder and Kasimbeyli [6,7] studied optimal elements and proper optimal elements in vector optimization with variable ordering structures. Shahbeyk [13] investigated Hartley properly and super nondominated solutions in vector optimization with variable ordering structures. Further, approximate solutions of vector optimization problems with variable ordering structures were also studied in [12,14,17].

Recently, some researchers have studied optimization problems in linear spaces without any topology structure. Li [11] used the separation theorem of convex sets in a real linear space to establish a theorem of the alternative for cone subconvexlike set-valued maps and obtained optimality conditions for vector optimization of set-valued maps. In linear spaces, properly efficient solutions of set-valued optimization problems, including Benson properly efficient solution [8] and super efficient solution [21], also were introduced.

However, to the best of our knowledge, there are few literatures involving set-valued optimization with variable ordering structures in linear spaces. Therefore, how to generalize some results obtained by the above references from topological spaces to linear spaces is interesting.

Inspired by [6,11,16,18], we will research weakly nondominated solution of set-valued optimization problems with variable ordering structures in linear spaces. This paper is organized as follows. In Sect. 3, we give some preliminaries including some basic notions and lemmas. In Sects. 4, we establish scalarization characterizations of weakly nondominated solution of unconstrained set-valued optimization problems with variable ordering structures in linear spaces. In Sects. 5, we obtain two duality theorems of unconstrained set-valued optimization problems, including a weak dual theorem and a strong dual theorem.

2   Preliminaries and Lemmas

Throughout this paper, we suppose that X and Y are two real linear spaces. Let A and M be two nonempty sets in X and Y, respectively. 0 stands for the zero element in every space. The generated cone of M is defined as $\mathrm{cone} M := \{\lambda m \mid m \in M, \ \lambda \geq 0\}$. M is called a convex cone iff

$$\lambda_1 m_1 + \lambda_2 m_2 \in M, \ \forall \lambda_1, \lambda_2 \geq 0, \ \forall m_1, m_2 \in M.$$

M is said to be pointed iff $M \cap (-M) = \{0\}$. M is said to be nontrivial iff $M \neq \{0\}$ and $M \neq Y$. The algebraic dual of Y is denoted by Y^*. Let C be a nontrivial, pointed and convex cone in Y. The algebraic dual cone C^+ of C is defined as $C^+ := \{y^* \in Y^* \mid \langle y, \ y^* \rangle \geq 0, \ \forall y \in C\}$, where $\langle y, \ y^* \rangle$ denotes the value of the linear functional y^* at the point y.

Definition 2.1 [9]. Let M be a nonempty subset in Y. The algebraic interior of M is the set $\mathrm{cor} M := \{m \in M \mid \forall h \in Y, \ \exists \epsilon > 0, \ \forall \lambda \in [0, \epsilon], m + \lambda h \in M\}$.

Definition 2.2 [1]. Let M be a nonempty subset in Y. The vector closure of M is the set $\mathrm{vcl} M := \{m \in Y \mid \exists h \in Y, \ \forall \epsilon > 0, \ \exists \lambda \in [0, \epsilon], m + \lambda h \in M\}$.

In this paper, we assume that the variable ordering structure is given by the set-valued map $\mathcal{C} : Y \rightrightarrows Y$ with $\mathcal{C}(y)$ being a nontrivial pointed convex cone and $\mathrm{cor} \mathcal{C}(y) \neq \emptyset$ for any $y \in Y$. Let $F : X \rightrightarrows Y$ be a set-valued map on A. We write

$$\langle F(x), \ y^* \rangle := \{\langle y, \ y^* \rangle \mid y \in F(x)\}$$

and

$$F(A) := \bigcup_{x \in A} F(x).$$

Now, we give a new notion of generalized convexity with the variable ordering structure.

Definition 2.3. Let $F : X \rightrightarrows Y$ be a set-valued map on A, and $\mathcal{C} : Y \rightrightarrows Y$ be a set-valued map with $\mathcal{C}(y)$ being a nontrivial pointed convex cone for all $y \in Y$. F is called near $\mathcal{C}(y)$-subconvexlike on A iff, for any $y \in Y$, $\mathrm{vcl}(\mathrm{cone}(F(A) + \mathcal{C}(y)))$ is a convex set in Y.

Remark 2.1. When Y is a topological space and $\mathcal{C}(y) = C$ for any $y \in Y$, Definition 2.3 reduces to Definition 2.2 in [16].

Definition 2.4. [18]. Let M be a nonempty subset of Y, and $\mathcal{C} : Y \rightrightarrows Y$ be a set-valued map with $\mathcal{C}(y)$ being a nontrivial pointed convex cone and $\mathrm{cor} \mathcal{C}(y) \neq \emptyset$ for all $y \in M$. $\overline{y} \in M$ is called a nondominated element of M w.r.t. $\mathcal{C}$ (denoted by $\overline{y} \in \mathrm{N}(M, \mathcal{C}(\cdot))$) iff there does not exist $y \in M$ such that $\overline{y} \in y + \mathcal{C}(y) \backslash \{0\}$. Equivalently, $\overline{y} \notin M + \mathcal{C}(y) \backslash \{0\}$ for any $y \in M$.

Remark 2.2. It follows from Definition 2.4 that $\overline{y} \in M$ is a nondominated element of M w.r.t. $\mathcal{C}$ iff there exists $\mathcal{C} : Y \rightrightarrows Y$ such that $M \cap (\overline{y} - \mathcal{C}(y)) = \{\overline{y}\}$ for any $y \in M$.

Definition 2.5 [6]. Let M be a nonempty subset of Y, and $\mathcal{C} : Y \rightrightarrows Y$ be a set-valued map with $\mathcal{C}(y)$ being a nontrivial pointed convex cone and $\mathrm{cor}\mathcal{C}(y) \neq \emptyset$ for all $y \in M$. $\overline{y} \in M$ is called a weakly nondominated element of M w.r.t. $\mathcal{C}$ (denoted by $\overline{y} \in \mathrm{WN}(M, \mathcal{C}(\cdot))$) iff there does not exist $y \in M$ such that $\overline{y} \in y + \mathrm{cor}\mathcal{C}(y)$.

Remark 2.3. It follows from Definition 2.5 that $\overline{y} \in \mathrm{WN}(M, \mathcal{C}(\cdot))$ iff there exists $\mathcal{C} : Y \rightrightarrows Y$ such that $(M - \overline{y}) \cap (-\mathrm{cor}\mathcal{C}(y)) = \emptyset$ for any $y \in M$.

Remark 2.4. Clearly, $\mathrm{N}(M, \mathcal{C}(\cdot)) \subseteq \mathrm{WN}(M, \mathcal{C}(\cdot))$. However, the following example shows that $\mathrm{WN}(M, \mathcal{C}(\cdot)) \not\subseteq \mathrm{N}(M, \mathcal{C}(\cdot))$.

Example 2.1. Let $Y = \mathbb{R}^2$, $M = \{(y_1, y_2) \in \mathbb{R}^2 | (y_1 - 1)^2 + (y_2 - 1)^2 \leq 1\} \cup \{(0,0), (0,-1)\}$ and $\overline{y} = (0,0)$. The set-valued map $\mathcal{C} : Y \rightrightarrows Y$ is defined as

$$\mathcal{C}(y) := \begin{cases} \{(y_1, y_2) \in \mathbb{R}^2 | \; y_2 - y_1 \geq 0, \; y_2 \geq 0, \; y_1 \geq 0\}, & y \in Y \setminus \{(1, \frac{1}{2})\} \\ \{(y_1, y_2) \in \mathbb{R}^2 | \; y_2 - y_1 \leq 0, \; y_2 \geq 0, \; y_1 \geq 0\}, & y = (1, \frac{1}{2}). \end{cases}$$

It is easy to check

$$(M - \overline{y}) \cap (-\mathrm{cor}\mathcal{C}(y)) = \emptyset, \forall y \in M.$$

However, there exists $\tilde{y} = (0, -1) \in M$ such that $M \cap (\overline{y} - \mathcal{C}(\tilde{y})) = \{(0, -1), (0, 0)\} \neq \{(0, 0)\}$. Therefore, $\overline{y} \in \mathrm{WN}(M, \mathcal{C}(\cdot))$ and $\overline{y} \notin \mathrm{N}(M, \mathcal{C}(\cdot))$. Thus, $\mathrm{WN}(M, \mathcal{C}(\cdot)) \not\subseteq \mathrm{N}(M, \mathcal{C}(\cdot))$.

Definition 2.6 [7]. let M be a nonempty subset of Y, and $\mathcal{C} : Y \rightrightarrows Y$ be a set-valued map with $\mathcal{C}(y)$ being a nontrivial pointed convex cone and $\mathrm{cor}\mathcal{C}(y) \neq \emptyset$ for all $y \in M$. $\overline{y} \in M$ is called a weakly max-nondominated element of M w.r.t. $\mathcal{C}$ (denoted by $\mathrm{WMN}(M, \mathcal{C}(\cdot))$) iff there does not exist $y \in M$ such that $\overline{y} \in y - \mathrm{cor}\mathcal{C}(y)$.

Let $F : A \rightrightarrows Y$ be a set-valued map with nonempty value. Consider the following unconstrained set-valued optimization problem:

$$(\mathrm{SVOP}) \begin{cases} \mathrm{Min} \; F(x) \\ x \in A, \end{cases}$$

where $A \subseteq X$.

Based on Definition 2.5, we introduce the concept of the weakly nondominated solution of (SVOP).

Definition 2.7. $\overline{x} \in A$ is called a weakly nondominated solution of (SVOP) w.r.t $\mathcal{C}$ iff there exist $\overline{x} \in A$, $\overline{y} \in F(\overline{x})$ and $\mathcal{C} : Y \rightrightarrows Y$ with $\mathcal{C}(y)$ being a nontrivial pointed convex cone and $\mathrm{cor}\mathcal{C}(y) \neq \emptyset$ for all $y \in F(A)$ such that $\overline{y} \in \mathrm{WN}(F(A), \mathcal{C}(\cdot))$. $(\overline{x}, \overline{y})$ is called a weakly nondominated element of (SVOP) w.r.t. $\mathcal{C}$.

Lemma 2.1 [15]. Let $P, Q \subseteq Y$ be two convex sets such that $P \neq \emptyset$, cor$Q \neq \emptyset$ and $P \cap \text{cor}Q = \emptyset$. Then, there exists a hyperplane separating P and Q in Y.

Similarly to Lemma 2.1 [11] and Lemma 3.21(b) [9], we have the following lemmas.

Lemma 2.2. Let $\mathcal{C} : Y \rightrightarrows Y$ be a set-valued map with $\mathcal{C}(y)$ being a nontrivial pointed convex cone and cor$\mathcal{C}(y) \neq \emptyset$ for all $y \in Y$. Then, $\mathcal{C}(y) + \text{cor}\mathcal{C}(y) = \text{cor}\mathcal{C}(y)$ for $y \in Y$.

Lemma 2.3. Let $\mathcal{C} : Y \rightrightarrows Y$ be a set-valued map with $\mathcal{C}(y)$ being a nontrivial pointed convex cone and cor$\mathcal{C}(y) \neq \emptyset$ for all $y \in Y$. Then,

$$\text{cor}\mathcal{C}(y) \subseteq \{b \in Y | \langle b, b^* \rangle > 0, \forall b^* \in \mathcal{C}(y)^+ \setminus \{0\}\}, \forall y \in Y.$$

3   Scalarization

In this section, we will establish scalarization theorems of an unconstrained set-valued optimization problem in the sense of weakly nondominated element. Now, we consider the following scalar problem of (SVOP):

$$(\text{SVOP})_\varphi \begin{cases} \text{Min } \langle F(x), \varphi \rangle \\ x \in A, \end{cases}$$

where $\varphi \in Y^* \setminus \{0\}$.

Definition 3.1 [11]. Let $\overline{x} \in A$ and $\overline{y} \in F(\overline{x})$. $\overline{x}$ is called an optimal solution of $(\text{SVOP})_\varphi$ iff

$$\langle \overline{y}, \varphi \rangle \leq \langle y, \varphi \rangle, \forall y \in F(A).$$

$(\overline{x}, \overline{y})$ is called an optimal element of $(\text{SVOP})_\varphi$.

Now, we give an optimality necessary condition of weakly nondominated element of (SVOP) under the suitable assumptions.

Theorem 3.1. Let $\mathcal{C} : Y \rightrightarrows Y$ be a set-valued map with $\mathcal{C}(y)$ being a nontrivial pointed convex cone and cor$\mathcal{C}(y) \neq \emptyset$ for any $y \in F(A)$. Suppose that the following conditions hold.

(i) $(\overline{x}, \overline{y})$ is a weakly nondominated element of (SVOP) w.r.t. $\mathcal{C}$;

(ii) $F - \overline{y}$ is nearly $\mathcal{C}(\cdot)$-subconvexlike on A.

Then, for any $y \in F(A)$, there exists $\varphi \in (\mathcal{C}(y))^+ \setminus \{0\}$ such that $(\overline{x}, \overline{y})$ is an optimal element of $(\text{SVOP})_\varphi$.

Proof. Since $(\overline{x}, \overline{y})$ is a weakly nondominated element of (SVOP) w.r.t. $\mathcal{C}$, we have

$$(F(A) - \overline{y}) \cap (-\text{cor}\mathcal{C}(y)) = \emptyset, \forall y \in F(A). \tag{1}$$

We assert that

$$\text{cone}(F(A) + \mathcal{C}(y) - \overline{y}) \cap (-\text{cor}\mathcal{C}(y)) = \emptyset, \forall y \in F(A). \tag{2}$$

Otherwise, there exsists $y_0 \in F(A)$ such that

$$\text{cone}(F(A) + \mathcal{C}(y_0) - \overline{y}) \cap (-\text{cor}\mathcal{C}(y_0)) \neq \emptyset. \tag{3}$$

By (3), there exist $d > 0$, $y_1 \in F(A)$ and $c \in \mathcal{C}(y_0)$ such that $d(y_1 + c - \overline{y}) \in -\text{cor}\mathcal{C}(y_0)$. Hence,

$$y_1 + c - \overline{y} \in -\text{cor}\mathcal{C}(y_0). \tag{4}$$

It follows from (4) and Lemma 2.2 that

$$y_1 - \overline{y} \in -c - \text{cor}\mathcal{C}(y_0) \subseteq -\mathcal{C}(y_0) - \text{cor}\mathcal{C}(y_0) = -\text{cor}\mathcal{C}(y_0),$$

which contradicts (1). Hence, (2) holds. We again assert that

$$\text{vcl}(\text{cone}(F(A) + \mathcal{C}(y) - \overline{y})) \cap (-\text{cor}\mathcal{C}(y)) = \emptyset, \forall y \in F(A). \tag{5}$$

Otherwise, there exist $y_1 \in F(A)$ and $a \in \text{vcl}(\text{cone}(F(A) + \mathcal{C}(y_1) - \overline{y}))$ such that

$$a \in -\text{cor}\mathcal{C}(y_1). \tag{6}$$

Since $a \in \text{vcl}(\text{cone}(F(A) + \mathcal{C}(y_1) - \overline{y}))$, there exist $h \in Y$ and $\lambda_n > 0$ with $\lim_{n \to \infty} \lambda_n = 0$ such that

$$a + \lambda_n h \in \text{cone}(F(A) + \mathcal{C}(y_1) - \overline{y}), \quad \forall n \in \mathbb{N}, \tag{7}$$

where $\mathbb{N}$ is the set of the natural numbers. It follows from (6) that $a \in \text{cor}(-\text{cor}\mathcal{C}(y_1))$. Therefore, for the above h, there exists $\lambda' > 0$ such that

$$a + \lambda h \in -\text{cor}\mathcal{C}(y_1), \forall \lambda \in [0, \lambda'].$$

Taking a sufficiently big $n' \in \mathbb{N}$ such that $\lambda_{n'} \in [0, \lambda']$, we have

$$a + \lambda_{n'} h \in -\text{cor}\mathcal{C}(y_1). \tag{8}$$

It follows from (7) and (8) that $a + \lambda_{n'} h \in \text{cone}(F(A) + \mathcal{C}(y_1) - \overline{y}) \cap (-\text{cor}\mathcal{C}(y_1))$, which contradicts (2). Therefore, (5) holds.

By Condition (ii), $\text{vcl}(\text{cone}(F(A) + \mathcal{C}(y) - \overline{y}))$ is a convex set for any $y \in F(A)$. Clearly, $\text{vcl}(\text{cone}(F(A) + \mathcal{C}(y) - \overline{y})) \neq \emptyset$ and $\text{cor}(\mathcal{C}(y)) \neq \emptyset$ for any $y \in F(A)$. Hence, it follows from Lemma 2.1 that there exists $\varphi \in Y^* \setminus \{0\}$ such that

$$\langle y_2, \varphi \rangle \geq \langle y_3, \varphi \rangle, \forall y \in F(A), \forall y_2 \in \text{vcl}(\text{cone}(F(A) + \mathcal{C}(y) - \overline{y})), \forall y_3 \in -\mathcal{C}(y). \tag{9}$$

By (9), we obtain

$$\langle y_2, \varphi \rangle \geq 0, \forall y \in F(A), \forall y_2 \in F(A) + \mathcal{C}(y) - \overline{y}. \tag{10}$$

Since $0 \in \mathcal{C}(y)$ for any $y \in F(A)$, it follows from (10) that

$$\langle \overline{y}, \varphi \rangle \leq \langle y, \varphi \rangle, \forall y_2 \in F(A). \tag{11}$$

We assert that $\varphi \in (\mathcal{C}(y))^+ \setminus \{0\}$ for all $y \in F(A)$. Otherwise, there exists $\tilde{y} \in F(A)$ such that $\varphi \notin (\mathcal{C}(\tilde{y}))^+ \setminus \{0\}$. Thus, there exists $c' \in \mathcal{C}(\tilde{y})$ such that

$$\langle c', \varphi \rangle < 0. \tag{12}$$

Since $\bar{y} \in F(A), \tilde{y} \in F(A)$ and $c' \in \mathcal{C}(\tilde{y})$, it follows from (10) that

$$\langle c', \varphi \rangle \geq 0,$$

which contradicts (12). Therefore, $\varphi \in (\mathcal{C}(y))^+ \setminus \{0\}$ for all $y \in F(A)$. (11) shows that $(\overline{x}, \overline{y})$ is an optimal element of $(\text{SVOP})_\varphi$. $\qquad\square$

Remark 3.1. Theorem 3.1 improves the necessity of Theorem 3.1 [11] in the following two aspects. Firstly, the fixed ordering cone C in Theorem 3.1 [11] has been replaced by the variable ordering cone $\mathcal{C}(.)$ in Theorem 3.1. Secondly, the C-subconvexlikeness of F in Theorem 3.1 [11] has been replaced by the near $\mathcal{C}(\cdot)$-subconvexlikeness of F in Theorem 3.1.

The following example is used to illustrate to Theorem 3.1.

Example 3.1. Let $X = Y = \mathbb{R}^2$ and $A = [0, 2] \times \{0\} \subseteq \mathbb{R}^2$. The set-valued map $F : X \rightrightarrows Y$ on A is defined as follows:

$$F(x_1, x_2) := \{(y_1, y_2) \in \mathbb{R}^2 \mid y_1 = x_1, 1 - \sqrt{1 - (x_1 - 1)^2} \leq y_2 \leq 1 + \sqrt{1 - (x_1 - 1)^2}\} \cup \{(0,0)\},$$

where $(x_1, x_2) \in A$. Let $\overline{x} = (0, 0)$ and $\overline{y} = (0, 0)$. The set-valued map $\mathcal{C} : Y \rightrightarrows Y$ is defined as

$$\mathcal{C}(y) := \begin{cases} \{(y_1, y_2) \in \mathbb{R}^2 \mid y_2 - y_1 \geq 0, \ y_2 \geq 0, \ y_1 \geq 0\}, & y \in Y \setminus \{(1, \frac{1}{2})\} \\ \{(y_1, y_2) \in \mathbb{R}^2 \mid y_2 - y_1 \leq 0, \ y_2 \geq 0, \ y_1 \geq 0\}, & y = (1, \frac{1}{2}). \end{cases}$$

It is easy to check that Conditions (i) and (ii) in Theorem 3.1 are satisfied. Therefore, for any $y \in F(A)$, there exists $\varphi = (1, 1) \in (\mathcal{C}(y))^+ \setminus \{(0, 0)\} = \{(y_1, y_2) \in \mathbb{R}^2 \mid y_2 \geq 0, \ y_1 \geq 0\} \setminus \{(0, 0)\}$ such that

$$\langle (0, 0), \varphi \rangle = 0 \leq \langle y, \varphi \rangle = y_1 + y_2, \forall (y_1, y_2) \in F(A).$$

Hence, $((0, 0), (0, 0))$ is an optimal element of $(\text{SVOP})_\varphi$.

Theorem 3.2. Let $\mathcal{C} : Y \rightrightarrows Y$ be a set-valued map with $\mathcal{C}(y)$ being a nontrivial pointed convex cone and $\text{cor}\mathcal{C}(y) \neq \emptyset$ for any $y \in F(A)$. Let $\overline{x} \in A$ and $\overline{y} \in F(\overline{x})$. Suppose that the following conditions hold.
(i) $\varphi \in (\mathcal{C}(y))^+ \setminus \{0\}$ for any $y \in F(A)$;
(ii) $(\overline{x}, \overline{y})$ is an optimal element of $(\text{SVOP})_\varphi$.
Then, $(\overline{x}, \overline{y})$ is a weakly nondominated element of (SVOP) w.r.t. $\mathcal{C}$.

Proof. By Condition (ii), we have

$$\langle \overline{y}, \varphi \rangle \leq \langle y, \varphi \rangle, \forall y \in F(A). \tag{13}$$

Suppose that $(\overline{x}, \overline{y})$ is not a weakly nondominated element of (SVOP) w.r.t. $\mathcal{C}$. Then, there exists $\tilde{y} \in F(A)$ such that $(F(A) - \overline{y}) \cap (-\text{cor}\mathcal{C}(\tilde{y})) \neq \emptyset$. Let

$$a \in (F(A) - \overline{y}) \cap (-\text{cor}\mathcal{C}(\tilde{y})). \tag{14}$$

It follows from (14) that there exists $y_1 \in F(A)$ such that

$$a = y_1 - \overline{y} \in -\mathrm{cor}(\mathcal{C}(\tilde{y})). \tag{15}$$

By (15) and Lemma 2.3, we have

$$\langle y_1 - \overline{y}, \varphi \rangle < 0. \tag{16}$$

On the other hand, it follows from (13) that $\langle y_1 - \overline{y}, \varphi \rangle \geq 0$, which contradicts (16). Therefore, $(\overline{x}, \overline{y})$ is a weakly nondominated element of (SVOP) w.r.t. $\mathcal{C}$. $\square$

Remark 3.2. When $\mathcal{C}(y) = C$ for any $y \in F(A)$, Theorem 3.2 reduces to the sufficiency of Theorem 3.1 in [11].

The following example is used to illustrate to Theorem 3.2.

Example 3.2. In Example 3.1, let $\overline{x} = (0,0)$ and $\overline{y} = (0,0) \in F(0,0)$. There exists $\varphi = (1,1) \in (\mathcal{C}(y))^+ \setminus \{0\} = \{(y_1, y_2) \in \mathbb{R}^2 | \ y_2 \geq 0, \ y_1 \geq 0\} \setminus \{(0,0)\}$. Hence, Condition (i) in Theorem 3.2 holds. Clearly,

$$\langle (0,0), (1,1) \rangle = 0 \leq \langle (y_1, y_2), (1,1) \rangle = y_1 + y_2, \forall (y_1, y_2) \in F(A).$$

Therefore, Condition (ii) in Theorem 3.2 holds. It is easy to check

$$(F(A) - \overline{y}) \cap (-\mathrm{cor}\mathcal{C}(y)) = \emptyset, \forall y \in F(A).$$

Thus, $(\overline{x}, \overline{y})$ is a weakly nondominated element of (SVOP) w.r.t. $\mathcal{C}$.

4   Duality

In this section, we will consider the duality problem of the constrained set-valued optimization problem and present a weak and a strong duality theorem in sense of weakly nondominated element.

Let $\widehat{S} \neq \emptyset$ be a nonempty subset of X. Let $D \subseteq Z$ be a nontrivial pointed convex cone in Z. Let $F : X \rightrightarrows Y$ and $G : X \rightrightarrows Z$ be two set-valued map on $\widehat{S}$. We consider the following constrainted set-valued optimization problem:

$$(\mathrm{CSVOP}) \begin{cases} \mathrm{Min} \ F(x) \\ G(x) \cap (-D) \neq \emptyset \\ x \in \widehat{S}. \end{cases}$$

The feasible set of (CSVOP) is denoted by $S := \{x \in \widehat{S} | \ G(x) \cap (-D) \neq \emptyset\}$.

Let $\mathcal{C} : Y \rightrightarrows Y$ be a set-valued maps with $\mathcal{C}(y)$ being a nontrivial pointed convex cone and $\mathrm{cor}\mathcal{C}(y) \neq \emptyset$ for any $y \in Y$. We write

$$C_1 := \{y \in Y | \exists (\varphi, \mu) \in (\mathcal{C}(y)^+ \setminus \{0\}) \times D^+, \forall d \in \bigcup_{x \in \widehat{S}} (\langle F(x), \varphi \rangle + \langle G(x), \mu \rangle), d \geq \langle y, \varphi \rangle$$

Now, we give the definition of the weakly nondominated element of (CSVOP).

Definition 4.1. Let $\bar{x} \in S$ and $\bar{y} \in F(\bar{x})$. $\bar{x} \in S$ is called a weakly nondominated solution of (CSVOP) w.r.t $\mathcal{C}$ iff there exists $\mathcal{C} : Y \rightrightarrows Y$ with $\mathcal{C}(y)$ being a nontrivial pointed convex cone and cor$\mathcal{C}(y) \neq \emptyset$ for all $y \in F(S)$ such that $\bar{y} \in \mathrm{WN}(F(S), \mathcal{C}(\cdot))$. $(\bar{x}, \bar{y})$ is called a weakly nondominated element of (CSVOP) w.r.t. $\mathcal{C}$.

Firstly, we present a weak duality theorem.

Theorem 4.1. For any $\bar{y} \in C_1$, there exists $\varphi \in (\mathcal{C}(\bar{y}))^+ \backslash \{0\}$ such that

$$\langle \bar{y}, \varphi \rangle \leq \langle y, \varphi \rangle, \forall y \in F(S). \tag{17}$$

Proof. Since $\bar{y} \in C_1$, there exists $(\varphi, \mu) \in (\mathcal{C}(\bar{y})^+ \backslash \{0\}) \times D^+$ such that

$$d \geq \langle \bar{y}, \varphi \rangle, \forall d \in \bigcup_{x \in \widehat{S}} (\langle F(x), \varphi \rangle + \langle G(x), \mu \rangle). \tag{18}$$

By (18), we have

$$\langle y, \varphi \rangle + \langle z, \mu \rangle \geq \langle \bar{y}, \varphi \rangle, \forall x \in S, \forall y \in F(x), \forall z \in G(x). \tag{19}$$

According to $\mu \in D^+$, we have

$$\langle z, \mu \rangle \leq 0, \forall z \in G(x) \cap (-D). \tag{20}$$

It follows from (19) and (20) that (17) holds. □

Remark 4.1. When set-valued maps $F : X \rightrightarrows Y$ and $G : X \rightrightarrows Z$ become vector-valued maps $f : X \to Y$ and $g : X \to G$, Theorem 4.1 reduces Theorem 4.5 in [6].

Next, we state the following strong duality theorem.

Theorem 4.2. $\mathcal{C} : Y \rightrightarrows Y$ with $\mathcal{C}(y)$ being a nontrivial pointed convex cone and cor$\mathcal{C}(y) \neq \emptyset$ for all $y \in C_1$. Suppose that the following conditions hold:
   (i) $(\bar{x}, \bar{y})$ is a weakly nondominated element of (CSVOP);
   (ii) $F - \bar{y}$ is nearly $\mathcal{C}(\cdot)$-subconvexlike on S;
   (iii) There exists $\varphi \in (\mathcal{C}(\bar{y}))^+ \backslash \{0\}$ such that

$$\langle \bar{y}, \varphi \rangle \leq \langle y, \varphi \rangle, \forall y \in F(S); \tag{21}$$

(iv) For the above φ,

$$\inf \bigcup_{x \in S} \langle F(x), \varphi \rangle = \sup \{\inf \bigcup_{x \in \widehat{S}} (\langle F(x), \varphi \rangle + \langle G(x), \mu \rangle) | \mu \in D^+\}, \tag{22}$$

and

$$\sup \{\inf \bigcup_{x \in \widehat{S}} (\langle F(x), \varphi \rangle + \langle G(x), \mu \rangle) | \mu \in D^+\}$$

has at least one solution.
Then, $\bar{y}$ is a weakly max-nondominated element of C_1 w.r.t. $\mathcal{C}$.

Proof. Since $\sup\{\inf \bigcup_{x\in\widehat{S}}(\langle F(x),\varphi\rangle + \langle G(x),\mu\rangle)|\mu \in D^+\}$ has at least one solution, it follows from (22) that there exists $\overline{\mu} \in D^+$ such that

$$\inf \bigcup_{x\in S} \langle F(x),\varphi\rangle = \inf \bigcup_{x\in\widehat{S}} (\langle F(x),\varphi\rangle + \langle G(x),\overline{\mu}\rangle). \tag{23}$$

By (21) and (23), we obtain

$$d \geq \inf \bigcup_{x\in S} \langle F(x),\varphi\rangle = \langle \overline{y},\varphi\rangle, \forall d \in \bigcup_{x\in\widehat{S}} (\langle F(x),\varphi\rangle + \langle G(x),\overline{\mu}\rangle). \tag{24}$$

(24) shows that $\overline{y} \in C_1$. Therefore, $\overline{y} \in F(S) \cap C_1$.

We assert that

$$c - \overline{y} \notin \mathrm{cor}\mathcal{C}(c), \forall c \in C_1. \tag{25}$$

Otherwise, there exists $\overline{c} \in C_1$ such that

$$\overline{c} - \overline{y} \in \mathrm{cor}\mathcal{C}(\overline{c}). \tag{26}$$

According to (26) and Lemma 2.3, we have

$$\langle \overline{c} - \overline{y}, \varphi'\rangle > 0, \forall\varphi' \in (\mathcal{C}(\overline{c}))^+ \setminus \{0\}. \tag{27}$$

It follows from (27) that

$$\langle \overline{c}, \varphi'\rangle > \langle \overline{y}, \varphi'\rangle, \forall\varphi' \in (\mathcal{C}(\overline{c}))^+ \setminus \{0\}. \tag{28}$$

Since $\overline{y} \in F(S)$, (28) contradicts Theorem 4.1. Therefore, (25) holds. Thus, $\overline{y}$ is a weakly max-nondominated element of C_1 w.r.t. $\mathcal{C}$. $\qquad\square$

Remark 4.2. It follows from Theorem 3.1 that Conditions (i) and (ii) ensure the existence of φ in Condition (iii).

Remark 4.3. Theorem 4.2 improves Theorem 4.6 [6] in the following two aspects. First, the $\mathcal{C}(\overline{y})$-convexity of f in Theorem 4.6 [6] has been replaced by the nearly $\mathcal{C}(\cdot)$-subconvexlikeness of $F - \overline{y}$ in Theorem 4.2 which is much weaker than the $\mathcal{C}(\overline{y})$-convexity of f. Secondly, we delete the convexity of $\widehat{S}$ and D-convexity of G which is need in Assumption 4.1 of Theorem 4.6 [6].

5   Conclusions

In this paper, we studied weakly nondominated solutions of set-valued optimization problems with variable ordering structures. We obtain some scalarization characterizations and dual theorems. Our results are obtained in linear spaces without any topological structure. In the future, we will investigate properly nondominated solutions of set optimization problems with variable ordering structures in linear spaces.

Acknowledgements. The authors would like to express their thanks to two anonymous referees for their valuable comments and suggestions.

References

1. Adán, M., Novo, V.: Weak efficiency in vector optimization using a closure of algebraic type under cone-convexlikeness. Eur. J. Oper. Res. **149**, 641–653 (2003)
2. Aubin, J.P., Frankowska, H.: Set-Valued Analysis. Birkhauser, Basel (1990)
3. Chen, G.Y., Huang, X.X., Yang, X.Q.: Vector Optimization, Set-Valued and Variational Analysis. Springer, Heidelberg (2005). https://doi.org/10.1007/3-540-28445-1
4. Chicco, M., Mignanego, F., Pusillo, L., Tijs, S.: Vector optimization problems via improvement sets. J. Optim. Theory Appl. **150**, 516–529 (2011)
5. Debreu, G.: A social equilibrium existence theorem. Proc. Natl. Acad. Sci. U.S.A. **38**, 803–886 (1952)
6. Eichfelder, G.: Optimal elements in vector optimization with a variable ordering structure. J. Optim. Theory Appl. **151**, 217–240 (2011)
7. Eichfelder, G., Kasimbeyli, R.: Properly optimal elements in vector optimization with variable ordering structure. J. Global Optim. **60**, 689–712 (2014)
8. Hernández, E., Jiménez, B., Novo, V.: Weak and proper efficiency in set-valued optimization on real linear spaces. J. Convex Anal. **14**, 275–296 (2007)
9. Jahn, J.: Vector Optimization-Theory, Applications and Extensions, 2nd edn. Springer, Heidelberg (2011). https://doi.org/10.1007/978-3-642-17005-8
10. Khan, A.A., Tammer, C., Zălinescu, C.: Set-Valued Optimization: An Introduction with Applications. Springer, Heidelberg (2015). https://doi.org/10.1007/978-3-642-54265-7
11. Li, Z.M.: The optimality conditions for vector optimization of set-valued maps. J. Math. Anal. Appl. **237**, 413–424 (1999)
12. Sayadi-Bander, A., Kasimbeyli, R., Pourkarimi, L.A.: A coradiant based scalarization to characterize approximate solutions of vector optimization problems with variable ordering structures. Oper. Res. Lett. **45**, 93–97 (2017)
13. Shahbeyk, S., Soleimani-damaneh, M., Kasimbeyli, R.: Hartley properly and super nondominated solutions in vector optimization with a variable ordering structure. J. Global Optim. **71**, 383–405 (2018)
14. Soleimani, B.: Characterization of approximate solutions of vector optimization problems with variable order structure. J. Optim. Theory Appl. **162**, 605–632 (2014)
15. Tiel, J.V.: Convex Analysis. Wiley, New York (1984)
16. Yang, X.M., Li, D., Wang, S.Y.: Near-subconvexlikeness in vector optimization with set-valued functions. J. Optim. Theory Appl. **110**, 413–427 (2001)
17. You, M.X., Li, G.H.: Optimality characterizations for approximate nondominated and Fermat rules for nondominated solutions. Optimization **71**, 2865–2889 (2022)
18. Yu, P.L.: Cone convexity, cone extreme points, and nondominated solutions in decision problems with multiobjectives. J. Optim. Theory Appl. **14**, 319–377 (1974)
19. Zhao, K.Q., Yang, X.M., Peng, J.W.: Weak E-optimal solution in vector optimization. Taiwan. J. Math. **17**, 1287–1302 (2013)
20. Zhou, Z.A., Chen, W., Yang, X.M.: Scalarizations and optimality of constrained set-valued optimization using improvement sets and image space analysis. J. Optim. Theory Appl. **183**, 944–962 (2019)
21. Zhou, Z.A., Yang, X.M.: Scalarization of ε-super efficient solutions of set-valued optimization problems in real ordered linear spaces. J. Optim. Theory Appl. **162**, 680–693 (2014)

The MaxIS-Shapley Value in Perfect Graphs

Junqi Tan, Dongjing Miao[✉], and Pengyu Chen

Harbin Institute of Technology, Harbin, China
miaodongjing@hit.edu.cn

Abstract. We investigate the application of the Shapley value to quantifying the contribution of vertices to the maximum independent set (MaxIS) in perfect graphs. The MaxIS problem in perfect graphs can be computed in polynomial time. Many well-studied families of graphs are perfect, for example, bipartite graphs, chordal graphs, forests, etc. The Shapley value is a widely known numerical measure for assessing the contribution of individuals. We study this measure in the context of MaxIS by redefining corresponding concepts. We show that computing the Shapley value with respect to MaxIS in perfect graphs, bipartite graphs, line graphs of bipartite graphs, chordal graphs is #P-complete. We present parameterized algorithm and polynomial-time algorithm for some special cases: perfect graphs whose vertices have a small number of types, and graphs with maximum degree two. We also propose a fully polynomial-time randomized approximation scheme (FPRAS) for general perfect graphs.

Keywords: Shapley value · Maximum independent set · Perfect graphs · Computational complexity · Algorithms

1 Introduction

1.1 Background

The independent set is a classic concept in graph theory. An independent set in a graph is a vertex set that any two vertices in it are not connected. The maximum independent set (MaxIS) is an independent set which has the largest number of vertices among all independent sets and there may be more than one MaxIS in a graph. Finding a MaxIS is a traditional research problem and has a wide range of applications, for example, the MaxIS problem has been applied in databases to handle inconsistency [17].

An inconsistent database is a database which violates some given integrity constraints, for example, a database is inconsistent if there are two tuples which

This work was supported by National Natural Science Foundation of China 62372138, 61972110 and Heilongjiang Provincial Natural Science Foundation of China YQ2023F008.

have the same credit card but different name. In most scenarios, each tuple has the same weight. Therefore, we can view a database as a graph. A tuple in a database can be viewed as a vertex in a graph. Two vertices are connected if the tuples corresponded to them conflict with each other. A subset repair of the database corresponds to a unique independent set [1]. An optimal subset repair is the subset repair with minimum cost, where the cost is usually defined as the number of the tuples deleted to make the database consistent. So, an optimal subset repair corresponds to a unique MaxIS.

The computation of optimal subset repairs has attracted a lot of attention. However, there are some scenarios that are overlooked: a database may have some important tuples that should not be deleted although it is necessary to remove these tuples to make the database consistent, and the number of the tuples deleted should be limited to guarantee the reliability of the database. Existing research results do not deal well with these situations. If the contribution of tuples to the optimal subset repair can be quantified, we can give tuples priorities and decide the amount or the order of tuples to be deleted. This problem equals to quantifying the contribution of vertices to the MaxIS. A conventional approach to dividing the contribution for a quantitative property (here the cardinality of a MaxIS) among players (here the vertices) is the Shapley value, which is a highly desirable solution concept for wealth distribution in cooperative games.

The Shapley value was named in honor of Lloyd Shapley [20], who introduced it in 1951 and won the Nobel Memorial Prize in Economic Sciences for it in 2012. It was proposed for the aim of allocating profits or wealth between players based on a wealth function. The wealth function v accepts a set of players as its input and outputs the profits or wealth these players can produce. To compute the Shapley value of a player p, we consider all the permutations of all the players. For each permutation, we view the set of players which appear in front of p as S. The Shapley value of p is the average value of $v(S \cup \{p\}) - v(S)$.

With perfect properties (efficiency, symmetry, linearity, and dummy player property), the Shapley value has been applied in many fields, such as answering queries in database [15], assigning profits in economics [12], pollution responsibility in ecology [19]. The Shapley value with respect to MaxIS can not only work on optimal subset repair, but also deal with problems in other areas which adopt MaxIS as the core technique, such as collusion detection [2], automated map labeling [9], and social network analysis [10]. The Shapley value provides a priority for each vertex and it works well when the vertices we can operate is limited. The Shapley value with respect to MaxIS can be easily extended to the Shapley value with respect to minimum vertex cover or maximum clique as well. What's terrible is that computing a MaxIS is a NP-hard problem [13], so it is intractable to compute the Shapley value with respect to MaxIS in general graphs as its formula involves computing MaxISs.

In this paper, we apply the Shapley value to quantifying the contribution of vertices to MaxIS in perfect graphs. Perfect graph is a class of graph in which the chromatic number of every induced subgraph equals the order of the largest clique of that subgraph (clique number). The MaxIS problem in perfect graphs can be solved in polynomial time [11]. Perfect graphs contain a variety of famous subclasses, such as bipartite graphs, chordal graphs, comparability graphs, forests. These graphs correspond to a lot of real-world scenarios. The Shapley value with respect to MaxIS in perfect graphs will have many real-world applications.

1.2 Related Work

The complexity and algorithmic issues of computing the Shapley value in cooperative games have been the topic of detailed studies, varieties of results are presented. Computing the Shapley value in a number of cooperative games has been proved to be intractable, such as weighted voting games [16], weighted majority games [8]. On the other hand, Deng and Papadimitriou [8] and Ieong and Shoham [14] presented polynomial-time algorithms to compute the Shapley value in weighted subgraph games and marginal contribution nets respectively.

In recent years, computing the Shapley value in matching games has attracted much attention, which is established on optimal matching problems. Deng et al. [7] showed that the core was characterized efficiently by the dual theorem and problems related to the core and the least core of matching games can be easily solved. Aziz and Keijzer [3] showed that computing the Shapley value of matching games can be computed in polynomial time when restricted on paths or graphs with a constant number of clique or coclique modules. Based on the results of Aziz and Keijzer, Zhao et al. [23] showed that the Shapley value of threshold cardinality matching games can be computed in polynomial time when graphs are restricted to some special graphs, such as linear graphs and complete k-partite graphs.

1.3 Contribution

We study the algorithmic aspects and computational complexity of computing the MaxIS-Shapley value in perfect graphs. For better application of Shapley value, we redefine some corresponding concepts. We show that computing the MaxIS-Shapley value in perfect graphs, bipartite graphs, line graphs of bipartite graphs, chordal graphs is #P-complete. The proofs are by Turing reductions from three classic #P-complete problems. We present a parameterized algorithm for perfect graphs whose vertices have a small number of types, a polynomial-time algorithm for graphs of maximum degree two, a fully polynomial time randomized approximation scheme (FPRAS) for general perfect graphs.

2 Preliminaries

The Shapley value is a solution concept in cooperative game theory. A cooperative game consists of a set A of n players and a characteristic function v that maps subsets of players to the real numbers: $2^A \to \mathbb{R}$. The Shapley value measures the share of each individual player $a \in A$ in $v(A)$. Let σ be a permutation over the players in A and σ_a be the set of players that appear before a in σ. We refer to the value $v(\sigma_a \cup \{a\}) - v(\sigma_a)$ as the marginal contribution of a to

σ. The Shapley value of a player $a \in A$ in a cooperative game is denoted by Shapley(A, v, a) and is defined as follows, where Π_A is the set of all possible permutations over the players in A.

$$\text{Shapley}(A, v, a) \overset{\text{def}}{=} \frac{1}{|A|!} \sum_{\sigma \in \Pi_A} (v(\sigma_a \cup \{a\}) - v(\sigma_a)) \tag{1}$$

Let $G = (V, E)$ be a perfect graph and v_a be a vertex of G. For $S \subseteq V$, we denote by $G(S)$ the subgraph of G induced by S, i.e., the graph $(S, \{e \subseteq E : e \subseteq S \times S\})$. The size of a MaxIS in G is called the independence number of G and is usually denoted by $\alpha(G)$. We denote the Shapley value with respect to MaxIS as MaxIS-Shapley value. The MaxIS-Shapley value of v_a in G is denoted by Shapley(G, α, v_a) and is defined as follows.

$$\text{Shapley}(G, \alpha, v_a) \overset{\text{def}}{=} \sum_{S \subseteq V \setminus \{v_a\}} \frac{|S|! \cdot (|V| - |S| - 1)!}{|V|!} (\alpha(G(S \cup \{v_a\})) - \alpha(G(S)))$$

$$\tag{2}$$

In this paper, we view G as consisting of two types of vertices: general vertices and stable vertices. Stable vertices stand for the vertices that we do not have permission to operate and are assumed not to claim any contribution or responsibility to the MaxIS of G. We denote by X the set of all stable vertices. For a perfect graph $G = (V, E)$ with stable vertices, the formula for MaxIS-Shapley value is the following, where $\alpha(G, X, S, v_a)$ stands for $\alpha(G(X \cup S \cup \{v_a\})) - \alpha(G(X \cup S))$.

$$\text{Shapley}(G, \alpha, v_a) = \sum_{S \subseteq V \setminus (X \cup \{v_a\})} \frac{|S|! \cdot (|V \setminus X| - |S| - 1)!}{|V \setminus X|!} \alpha(G, X, S, v_a) \tag{3}$$

Lemma 1. *Adding a vertex to a graph can only change the size of a MaxIS by 0 or 1. The MaxIS-Shapley value of a vertex is 1 if and only if this vertex is not adjacent to any other vertices.*

Therefore, computing the MaxIS-Shapley value is a counting problem.

3   Computational Complexity

In this section, we examine the computational complexity of computing the MaxIS-Shapley value in perfect graphs. As computing the MaxIS-Shapley value is a counting problem, it is easy to see that this problem is a member of the complexity class #P. We want to prove this problem is also a member of the complexity class #P-*complete*. Instead of investigating the computational complexity of computing the MaxIS-Shapley value in perfect graphs, we study it in three common subclasses of perfect graphs, bipartite graphs, line graphs of bipartite graphs and chordal graphs.

Theorem 1. *Computing the MaxIS-Shapley Value in bipartite graphs is #P-complete.*

Proof. The proof is by a Turing reduction from the problem of computing the Tutte polynomial $T(\mathcal{U}, 2, 1)$ on transversal matroids. The input of this problem is an undirected and unweighted bipartite graph $G = (I \cup S, E)$, the output is the number of $B \subseteq S$, such that the size of a maximum cardinality matching in $G(I \cup B)$ equals the size of B. Colbourn et al. proved this problem is #P-*complete* [6].

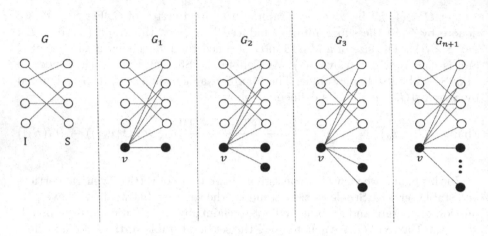

Fig. 1. Constructions in the reduction of the proof of Theorem 1.

The reduction is illustrated in Fig. 1. Given an undirected and unweighted bipartite graph $G = (I \cup S, E)$, we construct $n+1$ graphs by adding new vertices and edges to G, where $n = |S|$. For the i-th graph G_i, we add i independent vertices to S. Then we add an vertex v to I and make it connected to all vertices in S.

Next, we compute the MaxIS-Shapley value of v in G_i. We assume that all the vertices in $I \setminus \{v\}$ are stable vertices. According to the formula of MaxIS-Shapley value, the core part is computing the number of $B \subseteq S$, such that $\alpha(G_i(I \cup B)) - \alpha(G_i(I \setminus \{v\} \cup B)) = 1$. To make $\alpha(G_i(I \cup B)) - \alpha(G_i(I \setminus \{v\} \cup B)) = 1$, all the MaxISs in $G_i(I \cup B)$ should contain v. Then, I must be a MaxIS in $G_i(I \cup B)$. $I \setminus \{v\}$ is a MaxIS in $G_i(I \setminus \{v\} \cup B)$. According to König's theorem, the size of a maximum cardinality matching in $G_i(I \setminus \{v\} \cup B)$ equals $|B|$. Thus, the number of B is the output of $T(\mathcal{U}, 2, 1)$.

We split $T(\mathcal{U}, 2, 1)$ into n terms: $T(\mathcal{U}, 2, 1) = \sum_{k=0}^{n} T^k(\mathcal{U}, 2, 1)$. Each term $T^k(\mathcal{U}, 2, 1)$ computes the number of $B \subseteq S$, such that the size of a maximum cardinality matching in $G(I \cup B)$ equals $|B|$ and $|B| = k$.

The MaxIS-Shapley value of v in G_i can be computed as following.

$$\text{Shapley}(G_i, \alpha, v) = \frac{1}{(n+i+1)!} \sum_{k=0}^{n} T^k(\mathcal{U}, 2, 1) \cdot k! \cdot (n - k + i)! \qquad (4)$$

Since we construct $n+1$ bipartite graphs based on G, we can get the following system of equations, where each S_i stands for Shapley(G_i, α, v):

$$
\begin{pmatrix}
0!(n+1)! & 1!n! & \cdots & n!1! \\
0!(n+2)! & 1!(n+1)! & \cdots & n!2! \\
\vdots & \vdots & \vdots & \vdots \\
0!(2n+1)! & 1!(2n)! & \cdots & n!(n+1)!
\end{pmatrix}
\begin{pmatrix}
T^0(\mathcal{U}, 2, 1) \\
T^1(\mathcal{U}, 2, 1) \\
\vdots \\
T^n(\mathcal{U}, 2, 1)
\end{pmatrix}
=
\begin{pmatrix}
(n+2)!S_1 \\
(n+3)!S_2 \\
\vdots \\
(2n+2)!S_{n+1}
\end{pmatrix}
$$
(5)

We denote the $(n+1) \times (n+1)$ matrix in the above systems of equations as M. We divide each column j of M by $j!$ and reverse the order of the columns. Bacher [4] proved this matrix is non-singular. We can use Gaussian elimination to obtain each unique $T^k(\mathcal{U}, 2, 1)$ in $\mathcal{O}(n^3)$ time, and then, compute $T(\mathcal{U}, 2, 1)$ in linear time.

Lemma 2. *Given an undirected and unweighted bipartite graph G, let $L(G)$ be the line graph of G, v be a vertex in $L(G)$, and e be an edge corresponding to v in G. Computing the MaxIS-Shapley value of v in $L(G)$ is equivalent to computing the MaxCM-Shapley value of e in G.*

Proof. We denote the Shapley value of an edge with respect to maximum cardinality matching (MaxCM) as MaxCM-Shapley value. There is a one-to-one correspondence between the vertex sets of $L(G)$ and the edge sets of G. Let S_1 be a vertex set of $L(G)$ and S_2 be an edge set corresponding to S_1 in G. If S_1 is a independent set of $L(G)$, S_2 must be a matching of G, and they have same size. As computing the MaxIS-Shapley value (MaxCM-Shapley value) can be reduced to computing the number of permutations in which a vertex's(edge's) marginal contribution doesn't equal 0, the lemma is proved.

Theorem 2. *Computing the MaxCM-Shapley Value in bipartite graphs is #P-complete.*

Theorem 3. *Computing the MaxIS-Shapley Value in chordal graphs is #P-complete.*

The proofs of Theorem 2 and Theorem 3 can be found in the Appendix.

4   Algorithms

As computing the MaxIS-Shapley value in perfect graphs has been proved to be $\#P-complete$ in Sect. 3, it is difficult to propose polynomial-time algorithms. In this section, we propose efficient algorithms for special graphs and approximate algorithm for general perfect graphs.

For the sake of concise representation of the algorithms, we assume that graphs are connected. So we need to prove that computing the MaxIS-Shapley value in a disconnected graph can be reduced to computing the MaxIS-Shapley value in a connected component.

Lemma 3. *Let $G = (V, E)$ be a disconnected graph with k connected components: $G_1 = (V_1, E_1)$, $G_2 = (V_2, E_2)$, ..., and $G_k = (V_k, E_k)$. Let v be a vertex in G_i. Then, $\text{Shapley}(G, \alpha, v) = \text{Shapley}(G_i, \alpha, v)$.*

Proof. The MaxIS-Shapley value is equivalent to the problem of computing the average marginal contribution of v in all permutations. We can generate the permutations of V based on the permutations of V_i. For each permutation π of V_i, we can generate $(|V_i| + 1) \cdot (|V_i| + 2) \dots |V|$ permutations by inserting each vertex in $V \setminus \{V_i\}$ into π one by one. Let τ be a permutation generated by π. We denote by τ_v the set of vertices that appear before v in τ, and π_v the set of vertices that appear before v in π. Then, the following holds.

$$
\begin{aligned}
&\alpha(G(\tau_v \cup \{v\})) - \alpha(G(\tau_v)) \\
&= (\alpha(G(\pi_v \cup \{v\})) + \alpha(G(\tau_v \setminus \pi_v))) \\
&\quad - (\alpha(G(\pi_v)) + \alpha(G(\tau_v \setminus \pi_v))) \\
&= \alpha(G(\pi_v \cup \{v\})) - \alpha(G(\pi_v)) \\
&= \alpha(G_i(\pi_v \cup \{v\})) - \alpha(G_i(\pi_v))
\end{aligned}
\tag{6}
$$

Therefore, the following holds.

$$
\begin{aligned}
&\text{Shapley}(G_i, \alpha, v) \\
&= \frac{1}{|V_i|!} \sum_{\sigma \in \Pi_{V_i}} (\alpha(G_i(\sigma_v \cup \{v\})) - \alpha(G_i(\sigma_v))) \\
&= \frac{(|V_i| + 1) \dots |V|}{|V_i|! \cdot (|V_i| + 1) \dots |V|} \sum_{\sigma \in \Pi_{V_i}} (\alpha(G_i(\sigma_v \cup \{v\})) - \alpha(G_i(\sigma_v))) \\
&= \frac{1}{|V|!} \sum_{\sigma \in \Pi_V} (\alpha(G(\sigma_v \cup \{v\})) - \alpha(G(\sigma_v))) \\
&= \text{Shapley}(G, \alpha, v)
\end{aligned}
\tag{7}
$$

4.1 Parameterized Algorithm

Ueda et al. [21] showed that computing the Shapley value of any player can be done in polynomial time by using a type-based characteristic function representation. Our algorithm is based on this technique. At first, we need to know how to partition vertices into different types.

Lemma 4. *Let v_i, v_j be two vertices in a perfect graph $G = (V, E)$. If $N(v_i) \setminus \{v_j\} = N(v_j) \setminus \{v_i\}$ where $N(v) = \{u \in V : (u, v) \in E\}$, then $\alpha(G(S \cup \{v_i\})) = \alpha(G(S \cup \{v_j\}))$ for every $S \subseteq V \setminus \{v_i, v_j\}$.*

Proof. For any independent set of $G(S \cup \{v_i\})$ which contains v_i, if we replace v_i with v_j, it is still an independent set. Therefore, $\alpha(G(S \cup \{v_i\})) = \alpha(G(S \cup \{v_j\}))$.

Algorithm 1: Partition Algorithm 1

Input: A perfect graph $G = (V, E)$ and $V = \{v_1, v_2, \ldots, v_n\}$
Output: A size k partition

1   $k = 1$;
2   $S_1 = \{v_1\}$;
3   **for** $i = 2$ *to* n **do**
4   *find* =False;
5   **for** $j = 1$ *to* k **do**
6   Let v_x be a vertex in S_j;
7   **if** v_x *and* v_i *are adjacent and* $N(v_i) \setminus \{v_x\} = N(v_x) \setminus \{v_i\}$ **then**
8   *find* =True;
9            $S_j = S_j \cup \{v_i\}$;
10   **break**;

11   **if** *find* == False **then**
12         $k = k + 1$;
13         $S_k = \{v_i\}$;

14 **return** $\{S_1, S_2, \ldots, S_k\}$.

According to Lemma 4, there are two partition algorithms. The first is shown in Algorithm 1. We can obtain the second partition algorithm by replacing the condition in line 7 with $N(v_i) = N(v_x)$.

Both algorithms take only polynomial time. The following figure shows the results of two algorithms executed on the same perfect graph. Vertices of same type have same color. We choose the partition with a smaller size of the two algorithms, here is 3 (Fig. 2).

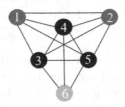 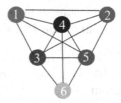

Fig. 2. The results of two partition algorithms executed on a perfect graph $G = (V, E)$. (Color figure online)

We denote each $S \subseteq V$ as a vector $\langle t_1, t_2, t_3 \rangle$. For example, $S = \{v_1, v_3\}$ can be denoted as $\langle 1, 1, 0 \rangle$ because S contains one vertex of type 1, one vertex of type 2, and no vertex of type 3. Thus, 64 possible S can be represented by 24 vectors. For a perfect graph G with a size k partition, we denote each set of vertices of G as a vector $\langle t_1, t_2, \ldots, t_k \rangle$. Let n_i be the number of vertices of type i and v_a be a vertex of type k in G, the function $c(\langle t_1, t_2, \ldots, t_k \rangle)$ is defined as follows where t stands for $t_1 + t_2 + \ldots + t_k$ and n stands for $n_1 + n_2 + \ldots + n_k$.

$$c(\langle t_1, t_2, \ldots, t_k \rangle) = \binom{n_1}{t_1} \binom{n_2}{t_2} \cdots \binom{n_k - 1}{t_k} \frac{t!(n - t - 1)!}{n!} \tag{8}$$

Let $G(\langle t_1, t_2, \ldots, t_k \rangle)$ be an induced subgraph of G which contains t_1 vertices of type 1, t_2 vertices of type 2, ..., t_k vertices of type k. $G(\langle t_1, t_2, \ldots, t_k \rangle)$ is not unique. The function $\varphi(\langle t_1, t_2, \ldots, t_k \rangle)$ is defined as follows.

$$\varphi(\langle t_1, t_2, \ldots, t_k \rangle) = \alpha(G(\langle t_1, t_2, \ldots, t_k \rangle \cup \{v_a\})) - \alpha(G(\langle t_1, t_2, \ldots, t_k \rangle)) \tag{9}$$

The MaxIS-Shapley value of v_a can be computed by the following formula.

$$\mathrm{Shapley}(G, \alpha, v_a) = \sum_{t_1=0}^{n_1} \sum_{t_2=0}^{n_2} \cdots \sum_{t_k=0}^{n_k - 1} c(\langle t_1, t_2, \ldots, t_k \rangle) \varphi(\langle t_1, t_2, \ldots, t_k \rangle) \tag{10}$$

If we denote the time of computing the MaxIS of a perfect graph as $\mathcal{O}(poly(n))$ and assume that k is a constant, the MaxIS-Shapley value can be computed in $\mathcal{O}(poly(n) \cdot n^k)$ time.

4.2 $\mathcal{O}(n^4)$ Algorithm for Graphs of Degree at Most Two

A graph with degree at most two is a disjoint union of cycles and linear graphs. Our algorithm takes the approach that Aziz and Keijzer [3] used for computing the Shapley value of a player in a matching game on an unweighted linear graph: subdivide the problem in separate cases and take the sum of them. For the sake of brevity, our algorithm takes a special case, a linear graph for example.

Let $G = (V, E)$ be a linear graph, where $V = \{v_j : 1 \leq j \leq n\}$ and $E = \{\{v_j, v_{j+1}\} : 1 \leq j \leq n-1\}$. Let $[n]$ be the set of $\{j : 1 \leq j \leq n\}$. For any vertex $v_i \in V$, let P_i^s be the number of $B \in V \setminus \{v_i\}$ of size s, for which $\alpha(G(B \cup \{v_i\})) - \alpha(G(B)) = 1$. We subdivide P_i^s as $p_i^{s,left} + p_i^{s,right} + p_i^{s,connect} + p_i^{s,isolated}$.

- Define $P_i^{s,left}(k)$ to be the number of $B \in V \setminus \{v_i\}$ of size s for which $\alpha(G(B \cup \{v_i\})) - \alpha(G(B)) = 1$ and B contains the line segment $\{v_{i+1}, v_{i+k+1}\}$, but does not contain $\{v_{i-1}, \ldots, v_{i+k+2}\}$. As adding a vertex to the left of a (non-empty) line segment L increases the size of a MaxIS if and only if L has an even number of vertices, we can compute $P_i^{s,left}$ as follows.

$$P_i^{s,left}(k) = \begin{cases} 0 & \text{if } k \text{ is even,} \\ \binom{|[n] \setminus \{i - 1, \ldots, i + k + 2\}|}{s - |\{i + 1, \ldots, i + k + 1\} \cap [n]|} & \text{otherwise.} \end{cases} \tag{11}$$

Define $P_i^{s,left}$ to be $\sum_{k=1}^{max(n-i-1,s-1)} p_i^{s,left}(k)$.
- $P_i^{s,right}(k)$ and $P_i^{s,right}$ can be computed in a similar way.

– Define $P_i^{s,connect}(k_1, k_2)$ to be the number of $B \in V \setminus \{v_i\}$ of size s for which $\alpha(G(B \cup \{v_i\})) - \alpha(G(B)) = 1$ and B contains the line segments $\{v_{i-k_1-1}, \ldots, v_{i-1}\}$ and $\{v_{i+1}, \ldots, v_{i+k_2+1}\}$, but does not contain $\{v_{i-k_1-2}, v_{i+k_2+2}\}$. As a vertex increases the size of a MaxIS of the union of two line segments L_1 and L_2 by connecting L_1 and L_2 if and only if L_1 and L_2 both have an even number of vertices, we can compute $P_i^{s,connect}(k_1, k_2)$ as follows.

$$P_i^{s,connect}(k_1, k_2) = \begin{cases} 0 & \text{if } k_1 \text{ is even or } k_2 \text{ is even,} \\ \binom{|[n] \setminus \{i - k_1 - 2, \ldots, i + k_2 + 2\}|}{s + 1 - |\{i - k_1 - 1, \ldots, i + k_2 + 1\} \cap [n]|} & \text{otherwise.} \end{cases}$$
(12)

We can then define $P_i^{s,connect}$ as the sum of all possible $P_i^{s,connect}(k_1, k_2)$:

$$P_i^{s,connect} = \sum_{k_1=1}^{max(i-2,s-1)} \sum_{k_2=1}^{max(n-i-1,s-k_1-2)} P_i^{s,connect}(k_1, k_2) \quad (13)$$

– Define $P_i^{s,isolated}$ to be the number of $B \in V \setminus \{v_{i-1}, v_i, v_{i+1}\}$ of size s. $P_i^{s,isolated}$ is easy to determine:

$$P_i^{s,isolated} = \left(|[n] \setminus \{i - 1, i, i + 1\}| \right) \quad (14)$$

Hence, we have that:

$$\text{Shapley}(G, \alpha, v_i) = \sum_{s=0}^{n-1} \frac{s! \cdot (n - s - 1)!}{n!} p_i^s \quad (15)$$

It is easy to see that computing $\text{Shapley}(G, \alpha, v_i)$ takes $\mathcal{O}(n^4)$ time. Our algorithm is easily extended to a cycle. It is worth noting that a cycle of odd length greater than three is not a perfect graph according to the strong perfect graph theorem [5]. But our algorithm still works. From Lemma 3, the MaxIS-Shapley value in graphs with maximum degree 2 can be computed in polynomial time.

4.3 Approximation Algorithm

Our approximation algorithm works as a *Fully Polynomial-Time Randomized Approximation Scheme*(FPRAS). Let f be a numeric function and x be the input of f, a FPRAS is an algorithm for f that returns an ϵ-approximation of $f(x)$ with probability at least $1 - \delta$ where $\epsilon, \delta \in (0, 1)$. The running time of a FPRAS is required to be polynomial in x, $1/\epsilon$, and $\log(1/\delta)$. Our FPRAS is shown in Algorithm 2.

Using the Chebyshev's inequality, we easily prove that it is an FPRAS. The proof can be found in the Appendix.

Algorithm 2: Approximation Algorithm

 Input: A perfect graph $G = (V, E)(|V| = n)$ and a specified vertex $v_i \in V$
 Output: Shapley(G, α, v_i)

1   $sv = 0$;
2   **if** $N(v_i) == \emptyset$ **then**
3      $sv = 1$;
4   **else if** $N(v_i) == V \setminus \{v_i\}$ **then**
5      $sv = 1/n$;
6   **else**
7   **for** $i = 1$ *to* $\lceil 6n^2(n-1)^2/\epsilon^2 \rceil$ **do**
8   Select a random permutation π over V.
            $sv = sv + \alpha(G(\pi_{v_i} \cup \{v_i\})) - \alpha(G(\pi_{v_i})$;
9      $sv = sv / \lceil 6n^2(n-1)^2/\epsilon^2 \rceil$
10 **return** sv;

---

5   Conclusion

In this paper, we apply the Shapley value to quantify the contribution of vertices to the MaxIS in perfect graphs. We prove that computing the Shapley value with respect to MaxIS in perfect graphs, bipartite graphs, line graphs of bipartite graphs, chordal graphs is #P-complete. We propose a parameterized algorithm and a polynomial-time algorithm for some special cases: perfect graphs whose vertices have a small number of types, graphs with maximum degree two. Finally, we present a fully polynomial time randomized approximation scheme (FPRAS) for general perfect graphs.

In the future, we plan to obtain the computational complexity of the MaxIS-Shapley value in other subclasses of perfect graphs, like forests.

Appendix A

Proof of Theorem 2. The proof is by a Turing reduction from #PERFECT MATCHINGS. The input of this problem is a bipartite graph G with $2n$ nodes, the output is the number of perfect matchings. Valiant [22] proved this problem is #P-*complete*.

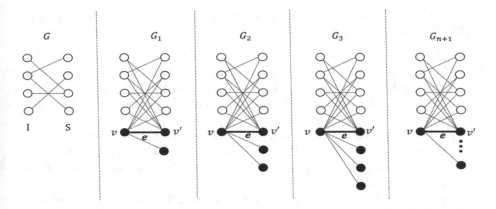

Fig. 3. Constructions in the reduction of the proof of Theorem 2.

The reduction is illustrated in Fig. 3. Given a bipartite graph $G = (I \cup S, E)(|I| = |S|)$, we construct $n + 1$ graphs by adding new vertices and edges to G, where $n = |E|$. For the i-th graph G_i, we add i independent vertices to S. Then we add an vertex v to I and make it connected to all vertices in S. Finally, we add an vertex v' to S and make it connected to all vertices in I.

Next, we compute the MaxCM-Shapley value of e in G_i. We assume that all the edges which connect only one black vertex are stable edges. We need to compute the number of $B \subseteq E$, such that there exists a MaxCM in the spanning subgraph of G_i with edge set $B \cup X$ that doesn't have commom vertex with e. Then, B should be a subset of E in G and contain a perfect matching of G. Otherwise, every MaxCM in the spanning subgraph of G_i with edge set $B \cup X$ that would have commom vertex with e.

We define a counting problem #CPM: Given a bipartite graph $G = (I \cup S, E)$ $(|I| = |S|)$, compute the number of $B \subseteq E$, such that B contains a perfect matching of G. We split #CPM into n terms: $\#\text{CPM} = \sum_{k=0}^{n} \#\text{CPM}_k$. Every item $\#\text{CPM}_k$ computing the number of $B \subseteq E$, such that B contains a perfect matching of G and $|B| = k$. Thus, $\#\text{CPM}_{|I|}$ is equivalent to #PERFECT MATCHINGS. The MaxCM-Shapley value of e in G_i can be computed as follows, where $\text{Shapley}(G_i, |MaxCM|, e)$ stands for the MaxCM-Shapley value of e in G_i.

$$\text{Shapley}(G_i, |MaxCM|, e) = \frac{1}{(n + i + 1)!} \sum_{k=0}^{n} \#\text{CPM}_k \cdot k! \cdot (n - k + i)! \quad (16)$$

Since we construct $n+1$ bipartite graphs based on G, we can get the following system of equations, where each S_i stands for $\text{Shapley}(G_i, |MaxCM|, e)$.

$$\begin{pmatrix} 0!(n+1)! & 1!n! & \cdots & n!1! \\ 0!(n+2)! & 1!(n+1)! & \cdots & n!2! \\ \vdots & \vdots & \vdots & \vdots \\ 0!(2n+1)! & 1!(2n)! & \cdots & n!(n+1)! \end{pmatrix} \begin{pmatrix} \#\text{CPM}_0 \\ \#\text{CPM}_1 \\ \vdots \\ \#\text{CPM}_n \end{pmatrix} = \begin{pmatrix} (n+2)!S_1 \\ (n+3)!S_2 \\ \vdots \\ (2n+2)!S_{n+1} \end{pmatrix}$$

$$(17)$$

The $(n+1) \times (n+1)$ matrix in the above equation is the same as the matrix in (5). We can obtain a unique $\#\text{CPM}_{|I|}$ in $\mathcal{O}(n^3)$ time by Gaussian elimination.

Proof of Theorem 3. The proof is by a Turing reduction from #SET COVERS. Let X be a finite set, and $\mathcal{S} \subseteq 2^X$ be a family of subsets of X. A set cover of X is a subfamily $\mathcal{F} \subseteq \mathcal{S}$ such that $\bigcup \mathcal{F} = X$. The output of #SET COVERS is the number of set covers of X. Provan and Ball [18] proved this probelem is *#P-complete*.

The idea is similar with Theorem 2 and Theorem 3. Given a finite set X and a family of subsets $\mathcal{S}$ for which we wish to compute #SET COVERS, we construct $n + 1$ chordal graphs, where $n = |\mathcal{S}|$. The construction process of the i-th graph is as follows:

Step 1. For each element in X, we build a vertex. We build an edge between every two vertices.

Step 2. For each subset of X in $\mathcal{S}$, we build a vertex and let it connected to the vertices in Step 1 that it contains.

Step 3. We build a vertex v and make it connected to all vertices in Step 1.

Step 4. We build i independent vertices and make them connected to v.

Next, we compute the MaxIS-Shapley value of v in G_i. We assume that all the vertices in Step 1 are stable vertices. We need to compute the number of B, such that B doesn't contain v and stable vertices, and there exists a MaxIS in $G_i(B)$ in which every vertex isn't connected to v. According to our constructions, the number of B equals the output of #SET COVERS. We regard #SET COVERS as the sum of n terms: #SET COVERS $= \sum_{k=0}^{n} \#SC_k$. Each term $\#SC_k$ computes the number of set covers with size k.

The MaxIS-Shapley value of v in G_i can be computed as follows.

$$\text{Shapley}(G_i, \alpha, v) = \frac{1}{(n+i+1)!} \sum_{k=0}^{n} \#SC_k \cdot k! \cdot (n - k + i)! \qquad (18)$$

Since we construct $n + 1$ chordal graphs based on the input of #SET COVERS, we can get a system of equations like (5). We can obtain each $\#SC_k$ in polynomial time (e.g., via Gaussian elimination), and then, compute #SET COVERS in linear time.

Theorem 4. *Algorithm 2 is an FPRAS.*

Proof. If we denote $\alpha(G(\pi_{v_i} \cup \{v_i\})) - \alpha(G(\pi_{v_i})$ as X, the following holds.

$$\mathbf{E}[X] = \text{Shapley}(G, \alpha, v_i) \qquad (19)$$

If v_i has $n - 1$ neighbors, Algorithm 2 is certainly an FPRAS. We assume that v_i has at most $n - 2$ neighbors. Let v' be not a neighbor of v_i. There are $(n - 2)!$ permutations which satisfy the conditions that v' is in the first position and v_i is in the second position. Therefore:

$$\frac{(n-2)!}{n!} \leq \text{Shapley}(G, \alpha, v_i) \qquad (20)$$

Therefore, the following holds.

$$\mathbf{Var}[X] = \mathbf{E}[X^2] - \mathbf{E}[X]^2$$

$$\leq \mathbf{E}[X^2] \leq 1 = \frac{(n!)^2}{(n!)^2} \tag{21}$$

$$\leq n^2(n-1)^2 \, \text{Shapley}(G, \alpha, v_i)^2$$

Let $Y = \frac{\sum_{j=1}^{\lceil 6n^2(n-1)^2/\epsilon^2 \rceil} X_j}{\lceil 6n^2(n-1)^2/\epsilon^2 \rceil}$, where X_j are independent random variables with the same distribution as X. Hence, we have that:

$$\mathbf{E}[Y] = \mathbf{E}[X] = \text{Shapley}(G, \alpha, v_i) \tag{22}$$

Using the Chebyshev's inequality, the following holds, where Shapley stands for $\text{Shapley}(G, \alpha, v_i)$.

$$\Pr[|Y - \mathbf{E}[Y]| \geq \epsilon \mathbf{E}[Y]]$$
$$= \Pr[|Y - \text{Shapley}| \geq \epsilon \, \text{Shapley}]$$
$$\leq \frac{\mathbf{Var}[Y]}{\epsilon^2 \, \text{Shapley}^2} = \frac{\mathbf{Var}\left[\frac{1}{\lceil 6n^2(n-1)^2/\epsilon^2 \rceil} \sum_{j=1}^{\lceil 6n^2(n-1)^2/\epsilon^2 \rceil} X_j \right]}{\epsilon^2 \text{Shapley}^2}$$
$$= \frac{\left(\frac{\mathbf{Var}[X]}{\lceil 6n^2(n-1)^2/\epsilon^2 \rceil} \right)}{\epsilon^2 \, \text{Shapley}^2} \tag{23}$$
$$\leq \frac{n^2(n-1)^2 \, \text{Shapley}^2}{\lceil 6n^2(n-1)^2/\epsilon^2 \rceil \cdot \epsilon^2 \, \text{Shapley}^2}$$
$$= \frac{1}{6}$$

It is easy to see that Algorithm 2 is an FPRAS.

References

1. Afrati, F.N., Kolaitis, P.G.: Repair checking in inconsistent databases: algorithms and complexity. In: Proceedings of the 12th International Conference on Database Theory, pp. 31–41 (2009)
2. Araujo, F., Farinha, J., Domingues, P., Silaghi, G.C., Kondo, D.: A maximum independent set approach for collusion detection in voting pools. J. Parallel Distrib. Comput. **71**(10), 1356–1366 (2011)
3. Aziz, H., de Keijzer, B.: Shapley meets Shapley. In: 31st International Symposium on Theoretical Aspects of Computer Science, p. 99 (2014)
4. Bacher, R.: Determinants of matrices related to the Pascal triangle. J. de théorie des nombres de Bordeaux **14**(1), 19–41 (2002)
5. Chudnovsky, M., Robertson, N., Seymour, P., Thomas, R.: The strong perfect graph theorem. Ann. Math. **164**, 51–229 (2006)

6. Colbourn, C.J., Provan, J.S., Vertigan, D.: The complexity of computing the Tutte polynomial on transversal matroids. Combinatorica **15**, 1–10 (1995)
7. Deng, X., Ibaraki, T., Nagamochi, H.: Algorithmic aspects of the core of combinatorial optimization games. Math. Oper. Res. **24**(3), 751–766 (1999)
8. Deng, X., Papadimitriou, C.H.: On the complexity of cooperative solution concepts. Math. Oper. Res. **19**(2), 257–266 (1994)
9. Gemsa, A., Nöllenburg, M., Rutter, I.: Evaluation of labeling strategies for rotating maps. J. Exp. Algorithmics (JEA) **21**, 1–21 (2016)
10. Goldberg, M., Hollinger, D., Magdon-Ismail, M.: Experimental evaluation of the greedy and random algorithms for finding independent sets in random graphs. In: Nikoletseas, S.E. (eds.) Experimental and Efficient Algorithms: 4th International Workshop, WEA 2005, Santorini Island, Greece, 10–13 May 2005, Proceedings 4, pp. 513–523. Springer, Cham (2005). https://doi.org/10.1007/11427186_44
11. Grötschel, M., Lovász, L., Schrijver, A.: Polynomial algorithms for perfect graphs. In: North-Holland Mathematics Studies, vol. 88, pp. 325–356. Elsevier (1984)
12. Gul, F.: Bargaining foundations of Shapley value. Econometrica J. Econometric Soc. **57**, 81–95 (1989)
13. Hartmanis, J.: Computers and intractability: a guide to the theory of np-completeness (Michael R. Garey and David S. Johnson). SIAM Rev. **24**(1), 90 (1982)
14. Ieong, S., Shoham, Y.: Marginal contribution nets: a compact representation scheme for coalitional games. In: Proceedings of the 6th ACM Conference on Electronic Commerce, pp. 193–202 (2005)
15. Livshits, E., Bertossi, L., Kimelfeld, B., Sebag, M.: The Shapley value of tuples in query answering. arXiv preprint arXiv:1904.08679 (2019)
16. Matsui, T., Matsui, Y.: A survey of algorithms for calculating power indices of weighted majority games. J. Oper. Res. Soc. Japan **43**(1), 71–86 (2000)
17. Miao, D., Cai, Z., Li, J., Gao, X., Liu, X.: The computation of optimal subset repairs. Proc. VLDB Endow. **13**(12), 2061–2074 (2020)
18. Okamoto, Y., Uno, T., Uehara, R.: Counting the number of independent sets in chordal graphs. J. Discrete Algorithms **6**(2), 229–242 (2008)
19. Petrosjan, L., Zaccour, G.: Time-consistent Shapley value allocation of pollution cost reduction. J. Econ. Dyn. Control **27**(3), 381–398 (2003)
20. Roth, A.E.: The Shapley Value: Essays in Honor of Lloyd S. Shapley. Cambridge University Press, Cambridge (1988)
21. Ueda, S., Kitaki, M., Iwasaki, A., Yokoo, M.: Concise characteristic function representations in coalitional games based on agent types. In: Twenty-Second International Joint Conference on Artificial Intelligence (2011)
22. Valiant, L.G.: The complexity of computing the permanent. Theoret. Comput. Sci. **8**(2), 189–201 (1979)
23. Zhao, L., Chen, X., Fang, Q.: Computing the Shapley value of threshold cardinality matching games. In: Li, D.-F., Yang, X.-G., Uetz, M., Xu, G.-J. (eds.) China-Dutch GTA/China GTA -2016. CCIS, vol. 758, pp. 174–185. Springer, Singapore (2017). https://doi.org/10.1007/978-981-10-6753-2_13

Asteroidal Sets and Dominating Paths

Oleksiy Al-saadi[1]([⊠]) and Jamie Radcliffe[2]

[1] School of Computing, University of Nebraska, Lincoln, NE 68588, USA
`oal-saadi2@unl.edu`
[2] Department of Mathematics, University of Nebraska, Lincoln, NE 68588, USA
`jamie.radcliffe@unl.edu`

Abstract. An independent set of three vertices is called an asteroidal triple (AT) if there exists a path between any two of them that avoids the neighborhood of the third. Asteroidal triple-free (AT-free) graphs are very well-studied, but some of their various superclasses are not. We study two of these superclasses: hereditary dominating pair (HDP) graphs and diametral path graphs. We correct a mistake that has appeared in the literature claiming that the class of diametral path graphs are a superclass of HDP. More specifically, we show that a graph with a dominating shortest path does not necessarily contain a dominating diametral path. We say a graph is a strict dominating pair graph if it contains a dominating pair but has no dominating diametral path, and we show structural and algorithmic properties of these graphs. To study properties of HDP graphs, we introduce the notion of spread in asteroidal triples. Given a dominating pair, we show that all paths between this pair meet the common neighborhood of some pair from each asteroidal triple. We use these results to improve the best known time complexity for the recognition of chordal HDP graphs.

Keywords: asteroidal triple · hereditary dominating pair · diametral path · chordal · distance

1 Introduction

Asteroidal triple-free (AT-free) graphs capture a common property which imposes the linearity we see in a multitude of classic graph classes. For example, asteroidal triples are forbidden in interval, permutation, and trapezoidal graphs. A famous result by Lekkerkerker and Boland [12] states that interval graphs are exactly the class of chordal AT-free graphs. Several problems that are NP-Complete in general have been shown to have polynomial-time solutions on AT-free graphs, including INDEPENDENT SET and FEEDBACK VERTEX SET [2,9]. The recognition of AT-free graphs is known to be polynomial [10]. Corneil, Olariu and Stewart [3] showed that AT-free graphs have two key properties: First, all AT-free graphs have a *dominating pair*, a set of two vertices such that any path between them dominates the vertex set of the graph. Second, all AT-free graphs have a *dominating diametral path*. Because the class of AT-free graphs is

© The Author(s), under exclusive license to Springer Nature Switzerland AG 2024
W. Wu and J. Guo (Eds.): COCOA 2023, LNCS 14461, pp. 211–225, 2024.
https://doi.org/10.1007/978-3-031-49611-0_15

hereditary, these two properties hold true for all connected, induced subgraphs of any AT-free graph. Later, these two properties were generalized to their own graph classes by Deogun and Kratsch [4,5].

The class of dominating pair graphs is the largest graph class for which the dominating pair property is hereditary. Dominating pair graphs that have asteroidal triples were investigated by Pržulj, Corneil, and Köhler [13], where the name *hereditary dominating pair* (HDP) was introduced. We will use the term HDP when referring to dominating pair graphs in order to better distinguish them from the weaker class of graphs that *have* a dominating pair but do not necessarily have a dominating pair in every connected, induced subgraph. Polynomial-time algorithms for STEINER SETS, MINIMUM CONNECTED DOMINATING SETS exist for HDP graphs [1]. DOMINATING SET and TOTAL DOMINATING SET have also been solved in polynomial-time [11] on graphs with dominating shortest path. In a general sense, HDP graphs may find an intuitive application as a topology for wireless, ad hoc networks or critical systems [5] because communication may be less susceptible to disruption.

A graph is *diametral path* if it contains a dominating diametral path in every connected, induced subgraph. Recently, RAINBOW VERTEX COLORING was studied on the class of diametral path graphs, drawing parallels to encryption and data security [7]. Although it has been claimed that a graph with a dominating shortest path also has a dominating diametral path [4,14] and this has been referred to in other literature [6], we show a counterexample. If a graph contains a dominating pair but has no dominating diametral path, then the graph is *strict dominating pair*.

In Sect. 3, we analyze strict dominating pair graphs. We prove that the diameter of these graphs is close to the distance of a dominating pair, a result that can be used to quickly compute the diameter if given a dominating pair. We give a necessary condition (Theorem 1) to be in strict dominating pair. We show that strict dominating pair graphs have diameter 3, 4, or have an asteroidal number of at least 4.

In Sect. 4, we introduce the notion of spread in asteroidal triples. We show that HDP graphs may not contain asteroidal triples with 3-spread. Intuitively speaking, an asteroidal triple without 3-spread has a pair of vertices that remain "close" in every connected, induced subgraph that contains the given asteroidal triple. So, 3-spread describes a hereditary structure that can be exploited to design algorithms, such as identifying cut sets or points of weakness of a network. This allows us to improve the run-time complexity for the recognition of chordal HDP graphs.

2   Preliminaries

Our graph theory notation basically follows that of Golumbic [8]. By G we denote a simple, undirected, and finite graph with n vertices. The vertex set of G is denoted by V and the edge set of G is denoted by E. For any $x \in G$ let $N(x) = \{y : \{x, y\} \in E\}$ be the (open) neighborhood of x and $N[x] = N(x) \cup \{x\}$ the closed neighborhood of x.

A set $S \subseteq V$ *dominates* a set $T \subseteq S$ if every vertex in T is contained in the closed neighborhood of S. We write $u \sim v$ to indicate that vertices u and v are adjacent. More generally, we write $u \sim S$ if there exists $v \in S$ such that $u \sim v$. We denote by $C_N(S)$ the *common neighborhood* of S. i.e.

$$C_N(S) = \bigcap_{v \in S} N(v).$$

A sequence of vertices $P = \langle u = x_0, x_1, \ldots, x_{k-1}, x_k = v \rangle$ is called a *walk* if $x_i \sim x_{i+1}$ for all $i \in \{0, 1, \ldots, k-1\}$. We say P is a path if the vertices $x_0, x_1, \ldots, x_k$ are all distinct. A walk with endpoints u and v may be called a u, v-walk. If $v \sim v'$ then we denote by $P-v'$ the walk $\langle u = x_0, x_1, \ldots, x_{k-1}, x_k = v, v' \rangle$. Similarly, if $P = \langle x_0, x_1, \ldots, x_k \rangle$ and $P' = \langle y_0, y_1, \ldots, y_q \rangle$ are walks and $x_k \sim y_0$, then we let $P-P' = \langle x_0, \ldots, x_k, y_0, \ldots, y_q \rangle$. We write $P[x_i, x_j]$ where $(0 \leq i \leq j \leq k)$ for the subwalk $\langle x_i, \ldots, x_j \rangle$ of P. It is well known that every walk contains a path. In other words, we can extract a u, v-path from the vertex set of every u, v-walk. The *length* of a path is the number of edges it contains.

We say a walk P *meets* some set S if $P \cap S \neq \emptyset$ and that P *avoids* S if $P \cap S = \emptyset$. The notation $d_G(u, v)$ denotes *geodesic distance*, the length of a shortest path, between the vertices u and v. The *diameter* of G, denoted also as $\mathrm{diam}(G)$, is the greatest geodesic distance between any two vertices in G. Formally, $\mathrm{diam}(G) = \max\{d_G(u, v) : u, v \in G\}$. A *diametral pair* is a pair of vertices (u, v) such that $d_G(u, v) = \mathrm{diam}(G)$. A *diametral vertex* is any vertex that belongs to a diametral pair, and a *diametral path* is any path whose endpoints form a diametral pair. We use the following generalization of asteroidal triples:

Definition 1. *An* asteroidal set S *is an independent set of at least three vertices such that, for every vertex $v \in S$, there exists a path between any two remaining vertices of $S \setminus \{v\}$ that avoids $N[v]$. We call such a path an* asteroidal path. *The cardinality of the largest asteroidal set in a graph is known as the* asteroidal number *of that graph, denoted by* $\mathrm{an}(G)$. *An* asteroidal triple *is an asteroidal set of size 3.*

Definition 2. *Two vertices a and b form a* dominating pair (a, b) *if every a, b-path dominates G. A* dominating pair path *is any path between a dominating pair of vertices. A graph is called a* hereditary dominating pair graph (HDP) *if every connected, induced subgraph has a dominating pair.*

It is simple to see that a graph with a dominating pair contains a dominating shortest path. We introduce the notion of *spread* as a characteristic of certain asteroidal triples.

Definition 3. *An asteroidal triple has k-spread if between any pair of vertices in the triple there is an induced path of length at least k that avoids the neighborhood of the third.*

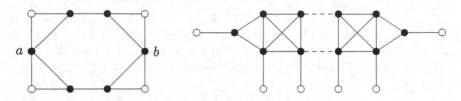

Fig. 1. On the left, a strict dominating pair graph where (a, b) is a dominating pair and the white vertices form an asteroidal set of size 4. On the right, a diametral path graph that does not contain a dominating pair

3   Dominating Pairs and Diametral Paths

In this section, we study the structural discrepancy between graphs with dominating pairs and graphs with dominating diametral paths. We state new results on the relationship between the two properties and expressly state conditions for when a graph is strict dominating pair. We begin with a correction to the assumption in past literature that any graph with a dominating shortest path contains a dominating diametral path.

Proposition 1. *A graph with a dominating shortest path does not necessarily contain a dominating diametral path.*

Proof. See the left graph in Fig. 1 for an example. The pair (a, b) is a dominating pair, and thus every shortest a, b-path is dominating. Notice that (a, b) is not diametral. It is easy to verify that no dominating diametral path exists.   □

With respect to the HDP and diametral path graph classes, both are superclasses of AT-free graphs, yet by Proposition 1 neither is a superclass of the other. To prepare for our main results, we present several results that relate the diameter of G to the distance between any dominating pair.

Lemma 1. *Let (a, b) be a dominating pair. There exists a diametral pair with one vertex contained in $N[a]$ and the other contained in $N[b]$.*

Proof. Let P be a shortest a, b-path of the form $\langle a = x_0, x_1, \ldots, x_k = b \rangle$, i.e. $d_G(a, b) = k$. Pair (a, b) is not diametral, otherwise we are done. Let (d_1, d_2) be a diametral pair. Certainly d_1, d_2 are adjacent to P because P is a dominating path by definition of dominating pair (a, b). If $d_1 \sim \langle x_1, \ldots, x_{k-1} \rangle$ and $d_2 \sim \langle x_1, \ldots, x_{k-1} \rangle$ then $d_G(d_1, d_2) \leq k$, a contradiction. Therefore, at least one of d_1 and d_2 is adjacent to a or b. If $d_1 \sim a$ and $d_2 \sim b$ then we are done; therefore, w.l.o.g. suppose $d_1 \sim x_1$ and $d_2 \sim b$. If $d_G(d_1, d_2) = d_G(a, d_2)$ then (a, d_2) is diametral s.t. $d_2 \in N(b)$, and we are done. Notice that $d_G(a, d_2) = k + 1$. So, there exists an a, d_2-path Q of length k. Let $Q = \langle a = u_0, u_1, \ldots, u_k = d_2 \rangle$. The path $Q - b$ must be dominating and $d_1 \not\sim b$, so $d_1 \sim Q$. We let $d_1 \sim u_i$ where $i > 0$. Now $d_G(d_1, d_2) \leq 1 + (k - i) < k$, a contradiction to the diameter of G. Thus, $d_1 \in N[a]$ and $d_2 \in N[b]$.   □

Immediately, we see the following:

Corollary 1. *Let G contain a dominating pair (a, b). The diameter of G is bounded by the following inequality:*

$$d_G(a, b) \leq \mathrm{diam}(G) \leq d_G(a, b) + 2.$$

Moreover, if G is strict dominating pair then $\mathrm{diam}(G) = d_G(a, b) + 1$ and there exists $d_1 \in N(a)$ and $d_2 \in N(b)$ s.t. (d_1, d_2) is a diametral pair.

Proof. By Lemma 1, there exists a diametral pair (d_1, d_2) s.t. $d_1 \in N[a]$ and $d_2 \in N[b]$. If $\mathrm{diam}(G) = d_G(a, b)$ then any a, b-path is a dominating diametral path, a contradiction. If $\mathrm{diam}(G) = d_G(a, b) + 2$, let P be any shortest a, b-path. We must have $d_1 \neq a$ and $d_2 \neq b$. The path $d_1 - P - d_2$ is a diametral path that dominates G because it contains P as a subpath.

In the remaining case, $\mathrm{diam}(G) = d_G(a, b) + 1$. If $d_1 = a$ or $d_2 = b$ then either $P - b$ or $a - P$ is a dominating diametral path, a contradiction.   □

Theorem 1 gives an interesting necessary condition, but not a characterization, of strict dominating pair graphs. The proof of Theorem 1 appears in the appendix.

Theorem 1. *Let G be a strict dominating pair graph. Either $\mathrm{diam}(G) \in \{3, 4\}$ or $\mathrm{an}(G) \geq 4$.*

We remark that the converse of the above theorem is not true. For example, consider the graph shown on the left in Fig. 1. If a pendant is added to the leftmost black vertex, we obtain a graph that is HDP, has a dominating diametral path, has asteroidal number 4, and has diameter 4.

As stated by Corollary 1, strict dominating pair graphs have a diameter more tightly constrained to the distance between any two dominating pair vertices than HDP graphs in general. It is also easy to check that these graphs may not have diameter less than or equal to 2. No linear-time method to compute the diameter is known to exist for AT-free, HDP, or diametral path graphs even when a dominating pair is given. Hence, we have proven the following corollary concerning the complexity of diameter on this class of graphs.

Corollary 2. *Let G be a strict dominating pair graph with a given dominating pair. The diameter can be computed in linear-time.*

Proof. Let (a, b) be the given dominating pair. We perform a breadth-first-search to calculate $d_G(a, b)$. By Corollary 1, we have that $\mathrm{diam}(G) = d_G(a, b) + 1$.   □

4   Cut Sets in HDP Graphs

In this section, we explore the structure of graphs that have dominating pairs and asteroidal triples, and show that dominating pair paths are necessarily "funnelled" through the common neighborhood of some pair of each asteroidal triple. Our intuition is that as asteroidal number increases, the placement of a dominating pair path is restricted such that there are never more than two distinct points of contact required to dominate a asteroidal set. We complete this section with an improved algorithm for the recognition of chordal HDP graphs.

4.1   Asteroidal Paths and Dominating Pairs

If a graph contains a dominating pair (a, b) and an asteroidal set S, then any a, b-path P dominates every vertex in S. We define notation in order to more easily refer to the outermost vertices in P that dominate vertices in S.

Definition 4. *Let (a, b) be a dominating pair, let S be an asteroidal set, and let P be an a, b-path. We denote by f^P (resp. ℓ^P) the first (resp. last) vertex of P that is adjacent to any vertex in S. These exist since P dominates S. We let $F^P = N(f^P) \cap S$ and $L^P = N(\ell^P) \cap S$. When necessary to distinguish, we will write F_S^P and L_S^P.*

Certainly $f^P \in C_N(F^P)$ and $\ell^P \in C_N(L^P)$. It is possible that $f^P = \ell^P$. With the following, we show that no such vertex is on any asteroidal path.

Proposition 2. *Let G contain an dominating pair (a, b) and an asteroidal set S, then if P is an a, b-path with $f^P = \ell^P$ then S is an asteroidal set in $G \setminus \{f^P\}$.*

Proof. Since there is only one vertex in P, viz. $f^P = \ell^P$, that is adjacent to any vertex in S, we have that f^P dominates S, and in particular is not on any asteroidal path for S.   □

Next, we discuss the cardinality of F^P and L^P, for a given dominating pair path P.

Lemma 2. *Let G be HDP. Given a dominating pair (a, b), an a, b-path P and an asteroidal set S, then $|F^P| \geq 2$ or $|L^P| \geq 2$.*

Proof. If F^P or L^P have size at least 2, we are done. Otherwise, suppose $F^P = \{x\}$. If $L^P = \{x\}$ then $P[a, f^P] - x - P[\ell^P, b]$ is an a, b-path not dominating any vertex in $S \setminus \{x\}$. Suppose then w.l.o.g. that $L^P = \{y\}$, with $y \neq x$. Pick $z \in S \setminus \{x, y\}$. Since S is an asteroidal set, there exists an x, y-path P' that does not dominate z. Now $P[a, f^P] - P' - P[\ell^P, b]$ is an a, b-walk from which we can extract an a, b-path that does not dominate z, a contradiction.   □

Pržulj, Corneil, and Köhler [13] investigated a subclass of HDP graphs called *frame HDP*. Below, we give a definition of a frame HDP graph.

Definition 5. *A frame hereditary dominating pair (frame HDP) graph G is a hereditary dominating pair graph with an asteroidal triple T such that all vertices of G are on some asteroidal path with endpoints in T.*

In particular, Pržulj, Corneil, and Köhler explored the location of dominating pairs in frame HDP graphs and proved that such graphs have $\mathrm{diam}(G) \leq 5$. They showed that every dominating pair satisfies strong constraints on the location of the endpoints relative to the fixed asteroidal triple. Lemma 2 generalizes this by putting constraints on all paths between any dominating pair of vertices.

An important result about the structure of HDP graphs is directly implied by Lemma 2:

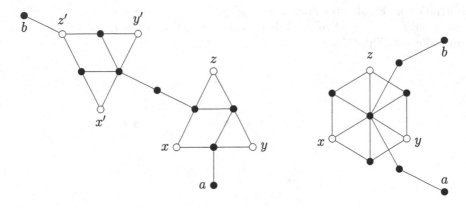

Fig. 2. Some HDP graphs that are not frame HDP. Note that in each case vertices a, b are not on any asteroidal path for $T = \{x, y, z\}$ or $T = \{x', y', z'\}$

Corollary 3. *HDP graphs do not contain an asteroidal triple with 3-spread.*

Proof. Let G be HDP. Given an asteroidal triple T with 3-spread, let $H = G \backslash (C_N(x, y) \cup C_N(y, z) \cup C_N(x, z))$. H has a dominating pair (a, b) and contains T. Let P be an arbitrary a, b-path. Then, by Lemma 2 one of f^P, ℓ^P is in the common neighborhood of some pair from T, a contradiction. □

Diametral path graphs may contain asteroidal triples with arbitrarily large spread. See, for instance, the right-hand graph in Fig. 1.

In Lemma 2, we proved that given a dominating pair (a, b) and an asteroidal triple T, every a, b-path passes through the common neighborhood of some pair from T. In fact, we will show in Theorem 2 that such a pair can be chosen uniformly for all a, b-paths. Before we can prove such a theorem, we prove two useful lemmas. For now, we apply Definition 4 with respect to asteroidal sets of cardinality 3 (i.e., asteroidal triples). Later, we will generalize these lemmas to graphs with greater asteroidal number.

Lemma 3. *Let G be HDP. Given a dominating pair (a, b), an AT $\{x, y, z\} = T$, and a, b-paths P and P', we have $F^P \cup L^{P'} = F^{P'} \cup L^P = T$.*

Proof. By Lemma 2, either $|F^P| \geq 2$ or $|L^P| \geq 2$. W.l.o.g., suppose that $F^P = \{x, y\}$. If $L^{P'}$ contains z, then we are done. First we will show that $F^P \cup L^{P'} = T$. Otherwise, since $L^{P'}$ is not empty, we may suppose that $L^{P'}$ contains x. The walk $P[a, f^P] - x - P'[\ell^{P'}, b]$ contains an a, b-path that does not dominate z, a contradiction.

Now we will show that $F^{P'} \cup L^P = T$. By the first paragraph, applied with $P = P'$, we know that $z \in L^P$. If $\{x, y\} \subseteq F^{P'} \cup L^P$ we are done. Otherwise, w.l.o.g. $x \notin F^{P'} \cup L^P$. If $y \in F^{P'}$ then let R be an asteroidal y, z-path. The walk $P'[a, f^{P'}] - R - P[\ell^P, b]$ contains an a, b-path that does not dominate x, a contradiction. Finally, the only remaining possibility is that $F^{P'} = \{z\}$, in which case $P'[a, f^{P'}]-z-P[\ell^P, b]$ contains an a, b-path that does not dominate x, a contradiction (Fig. 3). □

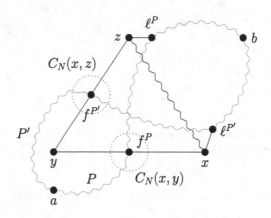

Fig. 3. Setup of the first part of Lemma 3. The a, b-walk $P[a, f^P]-x-P[\ell^{P'}, b]$ does not dominate z

At this stage, Lemmas 2 and 3 give us a strong understanding of how a dominating pair path dominates asteroidal triples. Next, we give a lemma that describes the orientation of two or more a, b-paths that meet the same common neighborhood of some pair in an asteroidal triple.

Lemma 4. *Let G be HDP. Given a dominating pair (a, b), an AT $\{x, y, z\} = T$, and two a, b-paths P and P', it cannot be that $F^P = L^{P'}$ unless $F^P = L^{P'} = T$.*

Proof. Suppose that $F^P = L^{P'} \neq T$. W.l.o.g. F^P contains x but not z. Then let $Q = P[a, f^P]-x-P'[\ell^{P'}, b]$. From the walk Q, we can extract a path that does not dominate z, a contradiction. □

We prove an important lemma that will be useful for proving the proceeding theorem.

Lemma 5. *Let G be HDP. Given dominating pair (a, b), an asteroidal triple $T = \{x, y, z\}$, and two a, b-paths P, P', there exists a pair in T s.t. P and P' meet its common neighborhood.*

Proof. First note that if either path meets $C_N(T)$, then by Lemma 2 the result holds.

Case 1. Suppose that one of the paths, w.l.o.g. P, has F^P and L^P disjoint, so in particular one of them is a singleton. W.l.o.g. $F^P = \{x, y\}$ and $L^P = \{z\}$. By Lemma 3, $F^{P'}$ contains $\{x, y\}$ and we are done.

Case 2. Otherwise, each of $F^P, L^P, F^{P'}$ and $L^{P'}$ have size 2. Then two of them and equal and by Lemma 2 they cannot be contained on the same path; therefore, we are done.   □

We are prepared to prove a major theorem that describes the structure of dominating pair paths in asteroidal triples in HDP graphs.

Theorem 2. *Let G be HDP. Given a dominating pair (a,b) and an asteroidal triple $T = \{x,y,z\}$, there exists a pair $D_{a,b} \subseteq T$ s.t. all a,b-paths meet $C_N(D_{a,b})$.*

Proof. Suppose for the sake of contradiction that P, P', and P'' are a,b-paths avoiding $C_N(x,y), C_N(x,z)$, and $C_N(y,z)$, respectively. Consider $I = F^P \cap F^{P'} \cap F^{P''}$. Everything in $J = T \setminus I$ is contained, by Lemma 3, in $J = L^P \cap L^{P'} \cap L^{P''}$. Thus one of I, J has size at least 2, a contradiction.   □

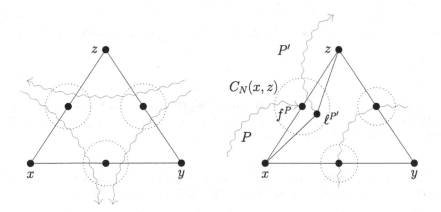

Fig. 4. On the left, any pair of a,b-paths shown satisfy Lemma 5. But, Theorem 2 is contradicted. Dotted circles represent the the common neighborhoods of pairs in the AT $\{x,y,z\}$. On the right, we detail that the left graph contains an a,b-walk $P[a, f^P]-z-P'[\ell^{P'}, b]$ that does not dominate y

Theorem 2 is important because it shows that an example like the one shown on the left of Fig. 4 may not occur. The unique structure of asteroidal triples that are allowed in HDP graphs shows that if (a,b) is a dominating pair in such a graph and $D_{a,b}$ corresponds to some asteroidal triple T then either $C_N(D_{a,b}) \cap (a,b)$ is non-empty or $C_N(D_{a,b})$ is a cut set in G.

We have strengthened Lemma 2 with Theorem 2. Recall that they apply to every asteroidal triple in an asteroidal set S. The next lemma generalizes Lemma 3 to asteroidal sets of arbitrary size.

Lemma 6. *Let G be HDP. Given an asteroidal set S, a dominating pair (a,b), and a,b-paths P and P', we have $F^P \cup L^{P'} = F^{P'} \cup L^P = S$.*

Proof. Suppose that $F_S^P \cup L_S^{P'} \neq S$. By Lemma 2, one of F_S^P and $L_S^{P'}$ has size at least 2. W.l.o.g. we assume $|F_S^P|$ is at least 2. Therefore, we can pick $x \in F_S^P$ and $y \in L_S^{P'}$ with $x \neq y$. By assumption, there exists $z \notin F_S^P \cup L_S^{P'}$. Now we will consider $F_{\{x,y,z\}}^P$ and $L_{\{x,y,z\}}^{P'}$. Since $x \in F_S^P$ and $y \in L_S^{P'}$, we have $f_S^P = f_{\{x,y,z\}}^P$ and $\ell_S^{P'} = \ell_{\{x,y,z\}}^{P'}$. By assumption, z belongs to neither of $F_{\{x,y,z\}}^P$ nor $L_{\{x,y,z\}}^{P'}$. This contradicts Lemma 3. □

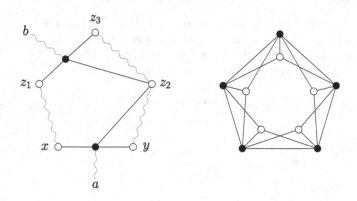

Fig. 5. Given a dominating pair (a, b) and letting P be an a, b-path on the left, we have an HDP graph with $F^P \cup L^P = S$. On the right, we show an example with asteroidal number 5. White vertices denote the asteroidal set

Next, we make a more general statement regarding all dominating pair paths between a given dominating pair and asteroidal sets of any size in HDP graphs.

Theorem 3. *Let G be HDP. Given an asteroidal set S and a dominating pair (a, b), let $F = \bigcap_P F_S^P$ and $L = \bigcap_P L_S^P$ for all a, b-paths P. Then $F \cup L = S$.*

Proof. Suppose otherwise. There exists $x \in S$ where $x \notin F$ and $x \notin L$. Then, there exist a, b-paths P', P'' such that $x \notin F_S^{P'}$ and $x \notin L_S^{P''}$. This contradicts Lemma 6. □

Consider Fig. 5. In both graphs, the white vertices denote the asteroidal set S. On the left graph we depict Theorem 3 where $F = \{x, y, z_2\}$ and $L = \{z_1, z_2, z_3\}$. On the right graph, we show a specific example of the more symbolic depiction shown on the left.

In general, HDP graphs become inherently more dense as asteroidal number increases. An interesting and extreme case of Theorem 3 is where $F = S$ (or, $L = S$). Then, it trivially holds that $F \cup L = S$. Consequently, $C_N(F)$ dominates S despite $C_N(F)$ being a cut set in G. Observe that $C_N(F)$ does not establish S in this case, so the removal of $C_N(F)$ leaves a connected, induced subgraph that contains the asteroidal set S, and this follows by Proposition 2. We demonstrate this case in the right graph of Fig. 2.

4.2   On Networks and Faster Recognition of Chordal HDP Graphs

With respect to application of HDP graphs in critical systems or ad hoc networks, Theorem 3 poses a problem. A cut vertex or cut set is naturally a point of weakness in a network. Therefore, if one is interested in utilizing dominating pairs, the inclusion of an asteroidal triple may necessitate the reinforcement of articulation vertices in some manner. Certainly, such a restriction gives us greater control in regards to the algorithmic complexity of certain problems in HDP graphs. The notion of spread, in particular, is a useful algorithmic tool.

We present a method for faster recognition of chordal HDP graphs from a previous best run-time of $O(n^7)$ in [13]. A complete set of forbidden subgraphs for the class of chordal HDP graphs is shown in their paper and reproduced in Fig. 6. The forbidden subgraphs have an asteroidal triple with 3-spread. Therefore, a faster algorithm for recognition is apparent.

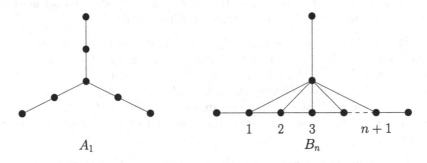

Fig. 6. Forbidden induced subgraphs for Chordal HDP graphs

Theorem 4. *Chordal HDP graphs are exactly the graphs that are chordal and have no asteroidal triples with 3-spread. In particular, chordal HDP graphs can be recognized in* $O(n^{3.82})$.

Proof. If G contains an asteroidal triple with 3-spread, then by Corollary 3, G is not HDP. On the other hand, if G contains no asteroidal triple with 3-spread, then in particular it does not contain any of the forbidden subgraphs that characterize chordal HDP graphs (see Fig. 6), and hence is chordal HDP.

Iterating through all asteroidal triples in a graph G requires time $O(n^{2.82})$ [10]. Additionally, checking whether an asteroidal triple has 3-spread requires time $O(n)$ and can be accomplished as follows. Let $T = \{x, y, z\}$ be any asteroidal triple in G. We remove the sets $C_N(x, y)$, $C_N(y, z)$ and $C_N(x, z)$ from G to form subgraph H. We check that T is an asteroidal triple in H, which requires linear time. If so, then T has 3-spread. Lastly, it is well-known that checking that a graph is chordal is linear [8]. Thus, the total time-complexity is $O(n^{3.82})$.   □

5 Appendix

In this appendix, we prove Theorem 1. To simplify the proofs, we introduce the notion of *corner vertices*, vertices that bear witness to the fact that a diametral path is not dominating but a given pair (a, b) is a dominating pair.

Definition 6. *If Q is a non-dominating diametral path in a graph G and (a, b) is a dominating pair, then vertex c is a corner w.r.t. (Q, a) if c is not dominated by Q, an endpoint of Q is adjacent to a, and $c \in N(a)$.*

Note that corners are adjacent to elements of the dominating pair, rather than to intermediate vertices on dominating pair paths. Next, we prove that corner vertices are inevitable in strict dominating pair graphs.

Lemma 7. *Let G have a dominating pair (a, b) and let $k = d_G(a, b)$, but no dominating diametral path (and thus in particular $\mathrm{diam}(G) = k+1$). Let (d_1, d_2) be a diametral pair s.t. $d_1 \in N(a)$ and $d_2 \in N(b)$. There exists a diametral d_1, d_2-path Q that contains b and a corner vertex c w.r.t. (Q, a).*

Proof. By Corollary 1 we have $d_G(d_1, d_2) = k + 1$. We claim that $d_G(d_1, b) = k$. By Corollary 1, $d_G(d_1, b) \leq k+1$. If $d_G(d_1, b) < k$ then $d_G(d_1, d_2) < k+1$ because $d_2 \sim b$, a contradiction. Thus, $d_G(d_1, b) = k$. Similarly, $d_G(a, d_2) = k$. Let M be a shortest d_1, b-path. By assumption, diametral path $M-d_2$ is not dominating while dominating pair path $a-M$ is dominating. Since $(a-M) \setminus (M-d_2) = \{a\}$, there exists $c \in N(a)$ s.t. $c \not\sim (M-d_2)$. Thus, we can set $Q = M-d_2$ and we are done (Fig. 7). □

The remaining proofs will utilize the existence of corner vertices to resolve various properties of strict dominating pair graphs.

Lemma 8. *Under the hypotheses of Lemma 7, let P_1, P_2 be diametral d_1, d_2-paths containing b and a, respectively. Let c_1 and c_2 be corner vertices w.r.t. (P_1, a) and (P_2, b), respectively. If $c_1 \sim c_2$ then $\mathrm{diam}(G) \in \{3, 4\}$.*

Proof. It is easy to check that $k \geq 2$. Notice that there exists a path $R = \langle a, c_1, c_2, b \rangle$ of length 3. If R is induced, then $k = 3$ and thus $\mathrm{diam}(G) = k+1 = 4$. If R is not induced, then $k < 3$. □

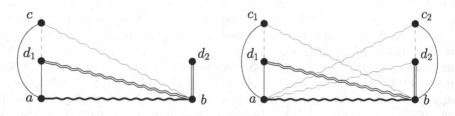

Fig. 7. On the left we depict a single corner vertex c. On the right, Lemma 7 is applied to both dominating pair vertices a and b. The thick path is a shortest a, b-path of length k. The double-lined path is Q within the proof. Dotted edges may not exist

Next, we consider the effect that corner vertices have on the asteroidal number of the graph. To prepare for the theorem, we resolve general consequences of having a corner vertex that does *not* belong to a diametral pair.

Lemma 9. *Under the hypotheses of Lemma 7, let c be a corner vertex w.r.t. (Q, a) where Q is a diametral d_1, d_2-path that contains b. The following hold:*

1. *If R is a shortest c, b-path and $d_G(c, b) < k$, then $N[d_1] \cap R = \emptyset$.*
2. *If $G \setminus N[d_1]$ disconnects c from b, then (c, d_2) is a diametral pair. Moreover, if R' is any shortest c, d_2-path in G, then d_1 is adjacent to the first vertex on R' following c.*

Proof. Assume for the sake of contradiction that $v \in N[d_1] \cap R$. Since $v \neq c$ the walk $d_1 - R[v, b] - d_2$ has length strictly less than $k + 1$, a contradiction.

For the second half, let R' be a shortest c, d_2-path and suppose that $d_G(c, d_2) \leq k$. Since $R' - b$ is a c, b-walk, there exists w on $R' - b$ such that $w \in N[d_1]$. Note that $w \neq c$ by definition and $w \neq b$ since $k \geq 2$. The path $R'' = d_1 - R'[w, d_2]$ has length strictly less than $k + 1$, a contradiction. Thus (c, d_2) is diametral and R' has length $k + 1$. Note that the same argument shows that w cannot occur later than the first vertex on R' following c. $\square$

We can see that Lemma 9 is symmetric with respect to a and b for a given dominating pair (a, b).

To simplify the following proof, we define new relationship notation. Let (a, b) be a dominating pair and let (d_1, d_2) be a diametral pair such that $d_1 \in N(a)$ and $d_2 \in N(b)$. Let c be a corner vertex with respect to (P, a) where P is a diametral d_1, d_2-path that contains b. In that case, we say that $d_1 \prec^P c$ (Fig. 8).

Lemma 10. *Let G satisfy the hypotheses of Lemma 7 and let $\operatorname{diam}(G) > 4$. There exist $c, c' \in N(a)$ with a shortest c, b-path P and a shortest c', b-path P' s.t. $N[c] \cap P' = N[c'] \cap P = \emptyset$.*

Proof. By Corollary 1, we let (d_1, d_2) be a diametral pair s.t. $d_1 \in N(a)$ and $d_2 \in N(b)$. We let $c_0 = d_1$ and P_0 be a diametral d_1, d_2-path that contains b. We will construct a sequence of distinct corner vertices $c_1, c_2, \ldots$ in $N(a)$, and diametral paths $P_1, P_2, \ldots$ such that P_i is a c_i, d_2-path and c_{i+1} is a corner vertex w.r.t. (P_i, a). Moreover, our sequence will satisfy the condition that for each c_i, set $N[c_i]$ meets every shortest c_{i+1}, b-path.

To be precise, given $c_0 \prec^{P_0} c_1 \prec^{P_1} \cdots \prec^{P_{i-1}} c_i \prec^{P_i}$ we let c_{i+1} be a corner vertex w.r.t. (P_i, a), as promised by Lemma 7, distinct from $c_0, \ldots, c_i$. If there is no corner vertex distinct from the earlier ones, the process terminates at the path P_i.

The step above alternates with the one we describe now, that of finding a path to add to $c_0 \prec^{P_0} c_1 \prec^{P_1} \cdots \prec^{P_{i-1}} c_i \prec^{P_i} c_{i+1}$. At this point there are two ways at which we might be done. If $d_G(c_{i+1}, b) < k$ then by the definition of a corner vertex and Lemma 9 we have $N[c_{i+1}] \cap P_i = \emptyset$ and $N[c_i] \cap P_{i+1} = \emptyset$, so we are done, letting $c = c_{i+1}$, $c' = c_i$, $P' = P_i \setminus \{d_2\}$, and $P = P_{i+1} \setminus \{d_2\}$. Also, if $N[c_i]$ does not meet every shortest c_{i+1}, b-path, then there exists a c_{i+1}, b-path

R of length k in $G \setminus N[c_i]$. Now $c = c_{i+1}$, $c' = c_i$, $P' = P_i \setminus \{d_2\}$, and $P = R$ satisfy the conclusion of the lemma.

On the other hand if $d_G(c_{i+1}, b) = k$ and $N[c_i]$ does meet every shortest c_i, b-path, then by Lemma 7 there exists a diametral c_{i+1}, d_2-path P_{i+1} that contains b which we add to the end of our sequence.

If in our construction we never succeeded in producing the required c, c', P, P' then the construction must have terminated because we could not find a corner vertex c_{f+1} distinct from all the earlier c_i. Thus we have constructed

$$c_0 \prec^{P_0} c_1 \prec^{P_1} c_2 \prec^{P_2} \cdots \prec^{P_{f-1}} c_f \prec^{P_f} .$$

By Lemma 7 there *is* a corner vertex c_{f+1} w.r.t. (P_f, a) and so, by assumption, there exists h in the set $\{0, 1, \ldots, f-1\}$ such that $c_{f+1} = c_h$. Suppose that $h = f - 1$. Then we have $(c_{f-1} = c_h) \sim P_f$, a contradiction. Otherwise, we have $h < f - 1$. We will prove that this leads to a contradiction.

Let $P'_h = c_h - (P_{h+1} \setminus \{c_{h+1}\})$ and let s, s' be the first vertices in P_h, P'_h following c_h. We will prove by reverse-induction that for all $h + 1 < j \le f$ that $c_j \sim s$ and $c_j \sim s'$. This holds for $j = f$ since $c_{f+1} = c_h$ and by construction $N[c_f]$ meets every shortest c_{f+1}, b-path.

Now suppose that $h+1 < j < f$. By the inductive hypothesis $c_{j+1} - (P_h \setminus \{c_h\})$ is a diametral c_{j+1}, d_2-path, and thus, since $N[c_j]$ meets every diametral c_{j+1}, d_2-path at its second vertex, $c_j \sim s$. Similarly, since $c_{j+1} - (P'_h \setminus \{c_h\})$ is diametral, $c_j \sim s'$ (Fig. 8)

Finally, when $j = h + 2$ we reach a contradiction. We must have $c_{h+2} \not\sim s'$ since c_{h+1} is a corner w.r.t. (P_{h+1}, a) and $s' \in P_{h+1}$. This contradiction establishes that the construction must have terminated through one of the conditions that give us appropriate c, c', P, P'. $\qquad\square$

Lemma 10 is also symmetric with respect to either vertex in a given dominating pair. Applying the lemma twice, we find there are four vertices such that

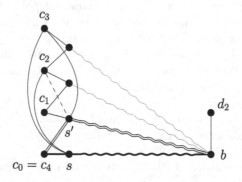

Fig. 8. Depicting Lemma 10 where $c_3 \prec^{P_3} c_0$, so $h = 0$. The thick path is $P_h \setminus \{d_2\}$ and the double-lined path is $P_{h'} \setminus \{d_2\}$

the removal of the closed neighborhood of any one of them does not disconnect the remaining three. These four vertices form an asteroidal set of size 4. Thus, we complete the proof of Theorem 1.

References

1. Aggarwal, D., Dubey, C., Mehta, S.: Algorithms on graphs with small dominating targets. In: International Symposium on Algorithms and Computation, pp. 141–152 (2006)
2. Broersma, H., Kloks, T., Kratsch, D., Müller, H.: Independent sets in asteroidal triple-free graphs. SIAM J. Discrete Math **12**(2), 276–287 (1999)
3. Corneil, D.G., Olariu, S., Stewart, L.: Asteroidal triple-free graphs. In: van Leeuwen, J. (ed.) WG 1993. LNCS, vol. 790, pp. 211–224. Springer, Heidelberg (1994). https://doi.org/10.1007/3-540-57899-4_54
4. Deogun, J.S., Kratsch, D.: Diametral path graphs. In: Nagl, M. (ed.) WG 1995. LNCS, vol. 1017, pp. 344–357. Springer, Heidelberg (1995). https://doi.org/10.1007/3-540-60618-1_87
5. Deogun, J., Kratsch, D.: Dominating pair graphs. SIAM J. Discrete Math. **15**(3), 353–366 (2002)
6. Ducoffe, G.: The diameter of AT-free graphs. J. Graph Theor. **99**, 594–614 (2021). https://doi.org/10.1002/jgt.22754
7. Dyrseth, J., Lima, P.: On the complexity of rainbow vertex colouring diametral path graphs. In: 33rd International Symposium on Algorithms and Computation (2022)
8. Golumbic, M.: Algorithmic Graph Theory and Perfect Graphs. Elsevier (2004)
9. Kratsch, D., Müller, H., Todinca, I.: Feedback vertex set on AT-free graphs. Discrete Appl. Math. **156**(10), 1936–1947 (2008)
10. Kratsch, D., Spinrad, J.: Between O(nm) and O(n^{alpha}). SIAM J. Comput. **36**(2) (2003). https://doi.org/10.1137/S0097539704441435
11. Kratsch, D.: Domination and total domination on asteroidal triple-free graphs. Discrete Appl. Math. **99**, 111–123 (2000). https://doi.org/10.1016/S0166-218X(99)00128-6
12. Lekkerkerker, C., Boland, J.C.: Representation of a finite graph by a set of intervals on the real line. Fund. Math. **51**, 45–64 (1962)
13. Pržulj, N., Corneil, D., Köhler, E.: Hereditary dominating pair graphs. Discrete Appl. Math. **134**(1–3), 239–261 (2004). https://doi.org/10.1016/S0166-218X(03)00304-4
14. Pržulj, N.: Minimal Hereditary Dominating Pair Graphs. Master's thesis, University of Toronto (2000)

A Novel Approximation Algorithm for Max-Covering Circle Problem

Kaiqi Zhang[1]([✉]), Siyuan Zhang[1], Jirun Gao[1], Hongzhi Wang[1], Hong Gao[2], and Jianzhong Li[1,3]

[1] School of Computer Science and Technology, Harbin Institute of Technology, Harbin, China
{zhangkaiqi,wangzh,lijzh}@hit.edu.cn
[2] School of Computer Science and Technology, Zhejiang Normal University, Jinhua, China
honggao@zjnu.edu.cn
[3] Faculty of Computer Science and Control Engineering, Shenzhen Institute of Advanced Technology, Chinese Academy of Sciences, Shenzhen, China

Abstract. We study the efficient approximation algorithm for max-covering circle problem. Given a set of weighted points in the plane and a circle with specified size, max-covering circle problem is to find the proper place where the center of the circle is located so that the total weight of the points covered by the circle is maximized. Our core approach is to approximate the circle with a symmetrical rectilinear polygon (SRP). We first present a method to construct the circumscribed SRP of a given circle and disclose their area relationship. Then, we convert max-covering SRP problem to SRP intersection problem, which can be efficiently solved with simple partition and modification based on the existing method. Finally, the optimal solution returned from max-covering SRP problem can be used to produce an approximate answer to max-covering circle problem. We prove that for most of the inputs, our algorithm can give a $(1 - \varepsilon)$ approximation to the optimal solution, which only needs $O\left(n\varepsilon^{-1}\log n + n\varepsilon^{-1}\log\left(\frac{1}{\varepsilon}\right)\right)$ time for unit points and $o\left(n\varepsilon^{-2}\log n\right)$ time for weighted points.

Keywords: Max-covering problem · Symmetrical rectilinear polygon · $(1 - \varepsilon)$ approximation

1 Introduction

Max-covering problem is a fundamental operation in computational geometry and database community. Given a set of weighted points Q in the plane $\mathbb{R}^2$ and a planar geometry G with specified size, max-covering problem is to find the proper place where the center of G is located so that the total weight of the points covered by G is maximized. In a word, max-covering problem aims to cover the points with maximum weight in a fixed-size region. There are many studies on this problem that we only present some important results on rectangle and circle.

For max-covering rectangle problem, two efficient exact algorithms are presented with $O(n \log n)$ time complexity where n is the number of points. Imai et al. [14] solve max-covering rectangle problem by finding a maximum clique of an intersection graph of rectangles in the plane. An alternative algorithm is proposed by Nandy and Bhattacharya in [19]. They transform max-covering rectangle problem into rectangle intersection problem and employ plane-sweep technique and interval tree data structure [20] to find the intersection region with maximum weight. Up to now, the time complexity has not been improved. Hemmer et al. [12] solve this algorithm regardless of whether the boundaries of the rectangles are open or closed. Tao et al. [21] design a grid-sampling-based approximation algorithm, which obtains a $(1 - \varepsilon)$-approximate answer with extremely high probability in $O\left(n \log \frac{1}{\varepsilon} + n \log \log n\right)$ time. Due to the weak scalability of the above in-memory algorithms, Choi et al. [5] propose an I/O-optimal external-memory algorithm. Later on, this problem is extended to trajectories and exact and approximation algorithms are proposed in [22].

For max-covering circle problem, Chazelle and Lee [4] substantially give an exact algorithm that runs in $O(n^2)$ time. It is widely acknowledged that the time complexity of any exact algorithm for this problem could not be improved any further, because it has been proved to be a 3SUM-HARD problem [2], which means a lower bound of $\Omega\left(n^2\right)$ running time must be held by any exact algorithms. Mark De Berg et al. [8] present a method that gives a $(1-\varepsilon)$-approximate solution in a deterministic $O\left(n \log n + n\varepsilon^{-3}\right)$ time. Although Choi et al. [5] give a $\frac{1}{4}$-approximate solution in $O(n \log n)$ time, the approximation accuracy cannot be specified by users. A special case of this problem argues that points have unit weight. Thus this problem changes to compute the maximum number of covered points by a circle. Aronov and Har-Peled [2] propose a random sampling method that returns a $(1 - \varepsilon)$- approximate solution in $O\left(n\varepsilon^{-2} \log n\right)$ time with high probability.

Besides max-covering problem, another type of geometric covering problem is k-covering problem. It aims to find a smallest geometry G covering at least k $(k \leq n)$ points of Q. For k-covering circle problem, Efrat et al. [9] give an exact algorithm for $O(nk\log^2 n)$ time. Subsequently, Matouvsek et al. [17] present a stochastic algorithm that runs in $O(n \log n + nk)$ expected time. All the above exact algorithms degenerate to $O(n^2)$ when $k = O(n)$. Har-Peled et al. [11] improve the exact result to $O(nk)$ and give a stochastic approximation algorithm that runs in $O\left(n + n \cdot \min\left(\frac{1}{k\varepsilon^3} \log^2 \frac{1}{\varepsilon}, k\right)\right)$ expected time. The case where G is an axis-aligned rectangle or square has also been studied somewhat for k-covering problem. When G is an axis-aligned square, the best result belongs to Mahapatra [16]. He gives the algorithm with time complexity $O\left(n + (n - k) \log^2(n - k)\right)$. The idea is to transform k-covering problem into iteratively solving max-covering problem by using the characteristics of the square. When G is an axis-aligned rectangle, the objective of solving k-covering problem usually falls into two categories: minimizing the perimeter of the rectangle and minimizing the area of the rectangle. For the former, the first two algorithms with $O\left(n^3\right)$ time complexity and $O\left(k^2 n \log n\right)$ time complexity are given by Aggarwal *et al.* [1]. The second result is subsequently improved by Eppstein & Erickson [10] and Datta *et al.* [6] to $O\left(n \log n + k^2 n\right)$ running time and again by Kaplan *et al.* [15] to $O\left(n \log n + n k^{3/2} \log^2 k\right)$ time complexity. For the latter, there is an $O\left(n^{5/2} \log^2 n\right)$ algorithm given by Kaplan *et al.* [15]. Besides, de Berg *et al.* [7] gives a k-sensitive algorithm with $O\left(n \log^2 n + n k^2 \log n\right)$ time complexity. The optimal result for k-covering problem of rectangular perimeter or area minimization belongs to Timothy M. Chan and Sariel Har-Peled [3]. They propose a general algorithm with $O\left(n^2 \log n\right)$ running time (for both perimeter and area), and, at the same time, a k-sensitive algorithm with $O(n \log n + n k \log k)$ time (perimeter) and a k-sensitive algorithm with $O\left(n k \log \frac{n}{k} \log k\right)$ time (area).

Our Contributions. In this paper, we concentrate on studying the efficient approximation algorithm for max-covering circle problem. From the previous work, we have the following observations. (1) For existing algorithms, solving max-covering rectangle problem has a better time complexity than solving max-covering circle problem. (2) A symmetrical rectilinear polygon (SRP) can approximate the circle in terms of area. Thus, the solution of max-covering SRP problem can be used to approximatively solve max-covering circle problem. These give birth to our core idea. We first present a method to construct the circumscribed SRP of a given circle and disclose their area relationship. The area difference is determined by the number of edges in the circumscribed SRP. Then, we convert max-covering SRP problem to SRP intersection problem, which can be efficiently solved with simple partition and modification based on the existing method. Finally, the optimal solution returned from max-covering SRP problem can be used to produce an approximate answer to max-covering circle problem. We prove that for most of the inputs, our algorithm can give a $(1 - \varepsilon)$ approximation to the optimal solution, which only needs $O\left(n\varepsilon^{-1}\log n + n\varepsilon^{-1}\log\left(\frac{1}{\varepsilon}\right)\right)$ time for unit points and $o\left(n\varepsilon^{-2}\log n\right)$ time for weighted points.

2   Preliminaries

Let us consider a set of points Q in 2-dimensional space $\mathbb{R}^2$. Each point $q \in Q$ has a non-negative weight $w(q)$.

Definition 1. *(Covering Weight). Given a set of points Q and a geometry G, the covering weight of G is:*

$$covering\text{-}weight(G, Q) = \sum_{q \in Q \cap G} w(q)$$

Definition 2. *(Max-Covering Problem). Given a set of weighted points Q and a geometry G with specified size, max-covering problem is to find the proper place of G to maximize the covering weight of G.*

The geometry can be arbitrary shape, including circle, rectangle, rectilinear polygon, etc. Similarly, let G be a circle with a given diameter, this problem can be renamed as max-covering circle (MaxC-C) problem. In the following, we formally define $(1 - \varepsilon)$-approximate max-covering circle problem.

Definition 3. *($(1 - \varepsilon)$-Approximate Max-Covering Circle Problem). Given a set of points Q and a circle C with specified diameter, for any ε $(0 < \varepsilon < 1)$, $(1 - \varepsilon)$-approximate max-covering circle problem finds a place in $\mathbb{R}^2$ to place C that satisfies*

$$covering\text{-}weight(C) \geq (1 - \varepsilon) \times covering\text{-}weight(C^*)$$

where C^* is an optimal circle of the original problem.

In this paper, we propose an approximation algorithm for solving MaxC-C problem with a $(1 - \varepsilon)$ accuracy to the optimal solution. In other words, our algorithm can solve $(1-\varepsilon)$-approximate MaxC-C problem. We achieve this result by approximatively converting MaxC-C problem to Max-covering symmetrical rectilinear polygon problem. Here, symmetrical rectilinear polygon is a special polygon which will be introduced below.

3 Symmetrical Rectilinear Polygon Construction

We first formulate several definitions of rectilinear polygon and symmetrical rectilinear polygon.

Definition 4. *(Rectilinear Polygon). A polygon is said to be a rectilinear polygon if the following conditions hold in a two-dimensional rectangular coordinate system: (1) For each of the given x-axis and y-axis, each side of this polygon is either perpendicular to the given coordinate axis or parallel to the given coordinate axis. (2) Any two sides of this polygon do not intersect except at the endpoints.*

Definition 5. *(Symmetrical Rectilinear Polygon). A rectilinear polygon RP, assuming that its center is at the origin, is symmetrical if $\forall q(x,y) \in RP$ such that $q_1(-x,y) \in RP$, $q_2(x,-y) \in RP$ and $q_3(-x,-y) \in RP$.*

In a word, symmetrical rectilinear polygon is both centrosymmetric and axisymmetric. To simplify the presentation, symmetrical rectilinear polygon can be abbreviated to SRP. Max-covering symmetrical rectilinear polygon problem can thus be called as MaxC-SRP problem.

Definition 6. *(Circumscribing and Inscribing). Given a circle C and a positive number k, take the diameter (Y-axis-parallel) of the circle and divide it into k segments equally. Then, make $k - 1$ vertical lines of this diameter through these k-bisected points respectively and intersect the circle at $2k-2$ points axisymmetrically. Moreover, connect adjacent points on the circumference with line segments perpendicular to and parallel to this diameter. The resulting polygon is said to be the circumscribed SRP of the circle. Similarly, circle C is said to be the inscribed circle of this circumscribed SRP.*

For a given circle, the circumscribed SRP depends only on k. Once k is determined, let S_k be the circumscribed SRP of the circle.

Given a circle, the construction methods of every circumscribed SRP have been presented above. As far as our work goes, that is not the exact point. What we need to do is, for a given parameter ε, choose an appropriate k such that for most of the inputs, the optimal solution of MaxC-SRP problem with respect to S_k can be used to produce a $(1-\varepsilon)$ approximation answer to MaxC-C problem. We denote the part that belongs to S_k but not C as $S_k - C$. It is a simple fact that, the area of $S_k - C$ can be a useful measure of how well the circumscribed SRP S_k approximates the circle C.

Let A_G be the area of any closed geometry G. We can infer the following relationship between A_{S_k-C} and A_{S_k}.

Theorem 1. *e is the base of natural logarithm and ε is in range $(0,1)$, there is an even number $k = O\left(\varepsilon^{-1}\right)$ such that $\frac{A_{S_k-C}}{A_{S_k}} < \frac{\varepsilon}{e}$.*

Proof. We consider the length from the center of the circle to each corner of S_k. Since k is even, we can only consider the part of S_k which is above the horizontal line that goes through the center of the circle. Let r be the radius. For the i-th corner $\left(1 \leq i \leq \frac{k}{2}\right)$, from bottom to top of this part, the length from the center of the circle to it is $r\sqrt{\left(1 + \frac{8i-4}{k^2}\right)}$, with simple geometric derivations.

Obviously, the following inequality holds

$$\pi r^2 < A_{S_k} < \pi r^2 (1 + \frac{4k-4}{k^2})$$

Then, we have

$$\frac{A_{S_k-C}}{A_{S_k}} < \frac{\frac{4k-4}{k^2}}{1 + \frac{4k-4}{k^2}} < \frac{4k-4}{k^2}$$

We just choose

$$k = 2\lfloor \frac{1 + \lceil 4e\varepsilon^{-1}\rceil}{2}\rfloor$$

This guarantees $4e\varepsilon^{-1} \leq k \leq 4e\varepsilon^{-1} + 1$ and $2|k$. Therefore, there is an even number k, which satisfies

$$k = O\left(\varepsilon^{-1}\right)$$
$$\frac{A_{S_k-C}}{A_{S_k}} < \frac{4k-4}{k^2} < \frac{4}{k} < \frac{\varepsilon}{e}$$

4   Algorithm for MaxC-SRP Problem

In this section, we discuss how to solve MaxC-SRP problem. In fact, this problem is not difficult and we can solve it completely with existing techniques.

Given a set of geometries $\mathcal{G}$ and a point q in $\mathbb{R}^2$, let $\mathcal{G}_q$ be the set of geometries covering q, denoted by $\mathcal{G}_q = \{G \mid \forall G \in \mathcal{G}, q \in G\}$. Each geometry $G \in \mathcal{G}$ has a non-negative weight $w(G)$.

Definition 7. *(Geometry Intersection Problem). Given a set of weighted geometries $\mathcal{G}$, the intersection problem of $\mathcal{G}$ is to find a point $q \in \mathbb{R}^2$ that maximizes*

$$covered\text{-}weight(q, \mathcal{G}) = \sum_{G \in \mathcal{G}_q} w(G)$$

Given a point q and a circumscribed SRP SRP, let SRP_q be the circumscribed SRP whose center point is at q and size is the same with SRP. For a set of points Q, we define $SRP_Q = \{SRP_q \mid \forall q \in Q\}$.

Lemma 1. *Given a set of weighted points Q, $\forall q \in \mathbb{R}^2$, covering-weight$(SRP_q, Q) = covered\text{-}weight(q, SRP_Q)$.*

Proof. We first prove the following two statements:

(1)$\forall q' \in Q$, if q' is covered by SRP_q, q is also covered by $SRP_{q'}$.

We let the coordinates of q be (x_q, y_q) and the coordinates of q' be $(x_{q'}, y_{q'})$. By the centrosymmetry of SRP in the definition, the fact that SRP_q contains $q'(x_{q'}, y_{q'})$ implies that it also contains $q''(2x_q - x_{q'}, 2y_q - y_{q'})$. Considering translating SRP_q so that its center becomes q', we obtain $SRP_{q'}$, which contains q, since q is the position of q'' after translation. Now, statement (1) is proved.

(2)$\forall q' \in Q$, if q' is not covered by SRP_q, q is not covered by $SRP_{q'}$ either.

We assume that $\exists q \in \mathbb{R}^2$, q' is not contained by SRP_q, but q is contained by $SRP_{q'}$. If so, then by centrosymmetry it is equally possible to obtain that $SRP_{q'}$ contains the point $q''(2x_{q'} - x_q, 2y_{q'} - y_q)$. Thus, after translating the center of $SRP_{q'}$ to q, the point corresponding to q'' is q'. This implies that SRP_q contains the point q', contradicting the assumption, and thus statement (2) holds.

Based on the above statements, we have

$covering\text{-}weight(SRP_q, Q) = \sum_{q' \in Q_{covered}} w(q') = \sum_{SRP \in SRP_Q\ covering\ q} w$
$(SRP) = covered\text{-}weight(q, SRP_Q)$.

Theorem 2. *Given a set of weighted points Q, let q^* is the answer to intersection problem of SRP_Q, then SRP_{q^*} is an optimal SRP to MaxC-SRP problem in Q.*

Proof. We assume that SRP_{q^*} is not an optimal SRP to MaxC-SRP problem in Q. Then, $\exists q \in \mathbb{R}^2$, $covering\text{-}weight(SRP_q, Q) > covering\text{-}weight(SRP_{q^*}, Q)$. With Lemma 1, we have both $covering\text{-}weight(SRP_q, Q) = covered\text{-}weight(q, SRP_Q)$ and $covering\text{-}weight(SRP_{q^*}, Q) = covered\text{-}weight(q^*, SRP_Q)$. This shows that $covered\text{-}weight(q, SRP_Q) > covered\text{-}weight(q^*, SRP_Q)$, which implies that q^* is not the answer to intersection problem of SRP_Q, contradicting our premise. Therefore, the theorem is proved.

Now we just need to give the method to solve the intersection problem of n SRPs. We use the division method to divide each SRP into several rectangles with parallel axes. We specify that for each SRP, the tangent along the line where its respective horizontal edges is located. Due to symmetry, let the number of horizontal edges to the left of its axis be l. Then each SRP is divided into

$l - 1$ rectangles. With a known input SRP, the time required to compute these rectangles is $O(l)$: we scan the SRP horizontally from top to bottom and group its horizontal boundaries two by two. For n SRPs, it takes $O(nl)$ time to compute all divided $n(l - 1)$ rectangles. Meanwhile, we want these $n(l - 1)$ rectangles to satisfy that the individual rectangles divided by the same SRP are disjoint, which requires determining the attribution of common edges between these rectangles at the time of division. We simply specify that the common edge of two adjacent rectangles divided by the same SRP belongs to the rectangle that is entirely above this edge.

Obviously, computing the intersection problem of these n SRPs is equivalent to settle the intersection problem of corresponding $n(l - 1)$ axis-parallel rectangles. This has been proved in [22]. The existing sweepline algorithm [19] solves this problem well by using an interval tree for updating and counting events, returning the region with the greatest total weight and the corresponding weight. In the original MaxC-SRP problem, the former of the returned results corresponds to the positions where the SRP can be placed and the latter corresponds to the total weight of points it covers at most. Therefore, our algorithm is to divide the SRP into rectangles and then use the sweepline algorithm for these $n(l - 1)$ rectangles. Considering that these rectangles may have open or closed boundaries (open boundaries may exist only in the horizontal direction), we use the improved algorithm [12], which is able to handle this case with no increase in time complexity and space complexity compared to the classical sweepline algorithm. When this algorithm is applied to solve our problem, it has a time complexity of $O\left(nl(\log n + \log l)\right)$ and a space complexity of $O\left(nl\right)$. The total time complexity of this algorithm is the same as it, for the time of dividing is $O\left(nl\right)$.

5   Approximation Algorithm for MaxC-C Problem

Now, we know that the MaxC-SRP algorithm, based on binary tree search, returns the correct result in deterministic $O\left(nl(\log n + \log l)\right)$ time and $O\left(nl\right)$ space when each SRP has $O\left(l\right)$ edges. Actually, this algorithm can be used to approximatively compute the MaxC-C problem. Given a set of points Q and a specified circle C, we first construct the circumscribed SRP S_k of C based on a given approximate accuracy ε. Then, the optimal SRP SRP^* can be returned by MaxC-SRP algorithm. Finally, the inscribed circle of SRP^* can be a desirable approximation to the optimal solution of MaxC-C problem for most of the inputs of Q. For a few specific inputs of Q, our result is not a satisfying approximation. Obviously, the correctness of the approximation degree depends on the distribution of points (*i.e.*, the inputs of Q). We would like to analyze whether the algorithm can give correct results for various different inputs and discuss its performance.

Assuming a random distribution of n points in the plane is clearly not a good choice. To discuss the various inputs without bias, we assume a random distribution of m points inside the SRP (the remaining $n - m$ points are outside the SRP). For practical applications with a large number of different inputs, we believe that this assumption is reasonable. For a given ε, we use our MaxC-SRP algorithm on S_k, where $k = O\left(\varepsilon^{-1}\right)$ and $\frac{A_{S_k - C}}{A_{S_k}} < \frac{\varepsilon}{e}$ are both satisfied. Then, for each point within S_k, let the probability that it falls within C be $1 - p$ and the probability that it falls within $S_k - C$ be p. By Theorem 1 and the assumptions, we thus easily know that $p < \frac{\varepsilon}{e}$. Let the total weight of these m points falling within $S_k - C$ be the random variable X.

5.1 Points with Unit Weight

If each point has unit weight, max-covering problem changes to compute the maximum number of covered points and X represents the number of these m points falling within $S_k - C$. Hence, X follows a binomial distribution, that is $X \sim B(m, p)$. The mathematical expectation $E(X)$ of X is mp. Based on Chernoff inequality [18], we have the following bound.

Theorem 3. *Let $Y_1, Y_2, ..., Y_m$ be m independent Bernoulli trials, where $Pr\{Y_i = 1\} = p_i$ and $Pr\{Y_i = 0\} = 1 - p_i$ for $i = 1, 2, ..., m$. Define $Y = \sum_{i=1}^{m} Y_i$. Then, for any $\varepsilon > 0$,*

$$Pr\left\{Y \geq (1 + \varepsilon)E(Y)\right\} \leq \left(\frac{e^{\varepsilon}}{(1 + \varepsilon)^{1+\varepsilon}}\right)^{E(Y)}$$

X coincides with Theorem 3, we thus have

$$Pr\left\{X \geq m\varepsilon\right\} < Pr\left\{X \geq emp\right\} = Pr\left\{X \geq (1 + e - 1)E(X)\right\} \leq e^{-E(X)}$$

For any $0 < p < \frac{\varepsilon}{e}$, if $p \geq \frac{\varepsilon}{3e}$, then

$$Pr\left\{X \geq m\varepsilon\right\} < e^{-E(X)} = e^{-mp} \leq e^{-\frac{m\varepsilon}{3e}}$$

For $0 < p < \frac{\varepsilon}{3e}$, let $\frac{\varepsilon}{3e} \leq p^* < \frac{\varepsilon}{e}$ and its corresponding random variable to be X^*, $X^* \sim B(m, p^*)$. It is obvious that

$$Pr\left\{X \geq m\varepsilon\right\} < Pr\left\{X^* \geq m\varepsilon\right\} < e^{-\frac{m\varepsilon}{3e}}$$

Therefore, for given m and ε, $Pr\left\{X \geq m\varepsilon\right\} < e^{-\frac{m\varepsilon}{3e}}$ always holds. Now we discuss the value of this probability. We consider two cases separately. When $m = \Omega\left(\varepsilon^{-1}\log n\right)$, we can easily say that our algorithm gives $(1 - \varepsilon)$ approximations for most inputs. This is because

$$Pr\left\{X \geq m\varepsilon\right\} < e^{-\frac{m\varepsilon}{3e}} < n^{-\frac{c}{3e}}$$

where c is a suitable constant. Then we have

$$Pr\left\{(m - X) \geq m(1 - \varepsilon)\right\} > 1 - n^{-\frac{c}{3e}}$$

We know that $m - X$ represents the number of points covered by the inscribed circle of SRP. Therefore, we know that under our assumption

$$Pr\left\{\frac{(m - X)}{m} \geq (1 - \varepsilon)\right\} > 1 - n^{-\frac{c}{3e}}$$

We note the fact that the maximum number of points that can be covered by the circumscribed SRP of a circle must be not less than the maximum number of points that can be covered by this circle. Let C^* be the optimal circle of the MaxC-C problem and $m^* = covering\text{-}weight(C^*)$. We have

$$m^* \leq m$$

$$Pr\left\{\frac{(m - X)}{m^*} \geq (1 - \varepsilon)\right\} \geq Pr\left\{\frac{(m - X)}{m} \geq (1 - \varepsilon)\right\} > 1 - n^{-\frac{c}{3e}}$$

The above equation tells us that, when $m = \Omega\left(\varepsilon^{-1}\log n\right)$, for most of the inputs (at least $1 - n^{-c_1}$ times of the total input ($c_1 = \frac{c}{3e}$)), the inscribed circle of the SRP returned by our algorithm can be used as a $(1 - \varepsilon)$ approximation to the optimal solution. From Sect. 4, we know that the MaxC-SRP algorithm takes $O\left(n\varepsilon^{-1}\log n + n\varepsilon^{-1}\log\left(\frac{1}{\varepsilon}\right)\right)$ running time when $l = O(k) = O\left(\varepsilon^{-1}\right)$.

For the other cases (i.e., $m = o\left(\varepsilon^{-1}\log n\right)$), the order of magnitude of m is not sufficient for our algorithm to return satisfactory solutions for most inputs. However, due to $m^* \leq m$, we know that the maximum number of points that can be covered by a given circle does not exceed m. This is equivalent to the fact that the depth of the deepest point in the arrangement of a set of circles does not exceed m. Another fact is that m is already computed when we run the algorithm. Now we simply use the conclusion [2] that, given any set X of n psedocircles, we can compute the deepest point in $\mathcal{A}(X)$ in $O(nd + n\log n)$, where $d = depth(X) = o\left(\varepsilon^{-1}\log n\right)$. Therefore, we just use the exact algorithm to compute the solution when m is not big enough. The time cost is $o\left(n\varepsilon^{-1}\log n\right)$.

Given n points Q, circle radius r and approximation accuracy ε, the basic procedure of the algorithm for $(1-\varepsilon)$-approximate MaxC-C problem is as follows.

Step 1. Construct S_k based on given r and ε.

Step 2. Compute the optimal SRP SRP^* of MaxC-SRP problem based on Q and S_k.

Step 3. Count the number of points inside SRP^*, denoted by m.

Step 4. Make a judgment based on the value of m: (1) If $m \geq c\varepsilon^{-1}\log n$ where c is an arbitrarily chosen positive real number, the center of the corresponding SRP^* is used as the center of the circle, and r is used as the radius to construct the circle. The circle is returned as the approximate solution of the circle position, and the number of points covered by this circle (which requires additional $O(n)$ time complexity to calculate it) is returned as the approximate solution of the maximum number of points covered by the circle. (2) Otherwise, we directly calculate max-covering circle problem.

In summary, our algorithm finishes running in $O\left(n\varepsilon^{-1}\log n + n\varepsilon^{-1}\log\left(\frac{1}{\varepsilon}\right)\right)$ time and returns a $(1 - \varepsilon)$ approximation for at least $1 - n^{-c_1}$ events for all possible values of m. Compared to almost all previous methods, our algorithm consumes surprisingly less time. It can be used either alone as an approximation algorithm (running when there is some tolerance for errors) or as a preprocessor for other approximation algorithms to reduce their time consumption.

5.2 Weighted Points

If points have different weights, X cannot follow a binomial distribution. The mathematical expectation $E(X)$ of X is Mp where M is the total weight of point inside the SRP. From Hoeffding inequality [13], we have the following bound.

Theorem 4. *Let* $Y_1, Y_2, ..., Y_m$ *be independent and identically distributed random variables where* $Y_i \in [a, b]$ *for* $i = 1, 2, .., m$. *Define* $Y = \sum_{i=1}^{m} Y_i$. *Then*

$$Pr(Y - E[Y] \geq \alpha) \leq e^{-\frac{2\alpha^2}{m(b-a)^2}}$$

X coincides with Theorem 4, we thus have

$$Pr\{X \geq M\varepsilon\} < Pr\{X \geq eMp\} = Pr\{X - E(X) \geq (e-1)E(X)\} \leq e^{-\frac{2(e-1)^2 E(X)^2}{mb^2}}$$

For any $0 < p < \frac{\varepsilon}{e}$, if $p \geq \frac{\varepsilon}{3e}$, then

$$Pr\{X \geq M\varepsilon\} \leq e^{-\frac{2(e-1)^2 E(X)^2}{mb^2}} = e^{-\frac{2(e-1)^2 M^2 p^2}{mb^2}} \leq e^{-\frac{2(e-1)^2 M^2 \varepsilon^2}{9e^2 mb^2}}$$

For $0 < p < \frac{\varepsilon}{3e}$, let $\frac{\varepsilon}{3e} \leq p^* < \frac{\varepsilon}{e}$ and its corresponding random variable to be X^*, X^* coincides with Theorem 4. It is obvious that

$$Pr\{X \geq M\varepsilon\} < Pr\{X^* \geq M\varepsilon\} < e^{-\frac{2(e-1)^2 M^2 \varepsilon^2}{9e^2 mb^2}}$$

Therefore, for given M and ε, $Pr\{X \geq M\varepsilon\} < e^{-\frac{2(e-1)^2 M^2 \varepsilon^2}{9e^2 mb^2}}$ always holds. In fact, $M = \sum_{i=1}^{m} w(q_i) = m\overline{w}$ where $\overline{w} = \frac{\sum_{i=1}^{m} w(q_i)}{m}$. Then

$$Pr\{X \geq M\varepsilon\} < e^{-\frac{2(e-1)^2 m\overline{w}^2 \varepsilon^2}{9e^2 b^2}}$$

We argue that $\overline{w}$ and b are stable and can be regarded as constants. When $m = \Omega(\varepsilon^{-2}\log n)$, we can easily say that our algorithm gives $(1 - \varepsilon)$ approximations for most inputs. Because

$$Pr\{X \geq M\varepsilon\} < n^{-c}$$

where c is a suitable constant. Then we have

$$Pr\{(M - X) \geq M(1 - \varepsilon)\} > 1 - n^{-c}$$

Similar to points with unit weight, $M - X$ is the total weight of points covered by the inscribed circle of SRP. Thus

$$Pr\left\{\frac{(M-X)}{M} \geq (1-\varepsilon)\right\} > 1 - n^{-c}$$

Let C^* be the optimal circle of the MaxC-C problem and $M^* = $ $covering\text{-}weight(C^*)$. We have

$$M^* \leq M$$

$$Pr\left\{\frac{(M-X)}{M^*} \geq (1-\varepsilon)\right\} \geq Pr\left\{\frac{(M-X)}{M} \geq (1-\varepsilon)\right\} > 1 - n^{-c}$$

The above equation tells us that, when $m = \Omega\left(\varepsilon^{-2}\log n\right)$, for most of the inputs (at least $1 - n^{-c}$ times of the total input), the inscribed circle of the SRP returned by our algorithm can be used as a $(1-\varepsilon)$ approximation to the optimal solution. Recall Sect. 4 again, MaxC-SRP algorithm takes $O\left(n\varepsilon^{-1}\log n + n\varepsilon^{-1}\log\left(\frac{1}{\varepsilon}\right)\right)$ time when $l = O(k) = O\left(\varepsilon^{-1}\right)$.

For the other cases (i.e., $m = o\left(\varepsilon^{-2}\log n\right)$), the order of magnitude of m is not sufficient for our algorithm to return satisfactory solutions for most inputs. However, due to $M^* \leq M$, that is the total weight of points that can be covered by a given circle does not exceed M. We know that $M^* \leq M \leq bm$ where b is the maximum weight of all the points. In fact, M can be computed when we run MaxC-SRP algorithm. Then, in the case of $m = o\left(\varepsilon^{-2}\log n\right)$, the griding method of Mark De Berg et al. [8] can achieve a time complexity of $o\left(n\varepsilon^{-2}\log n\right)$. It suffices to note that $M^* \leq M \leq bm$ holds, such that each circular region equal in size to the input circle has at most $\frac{b}{a}m$ points. And each grid in [8] can be covered by a constant number of circles, so the number of points in each grid is $o\left(\varepsilon^{-2}\log n\right)$ and the running time of their method is $o\left(n\varepsilon^{-2}\log n\right)$.

In summary, our algorithm finishes running in $o\left(n\varepsilon^{-2}\log n\right)$ time when $l = O(k) = O\left(\varepsilon^{-1}\right)$ time and returns a $(1-\varepsilon)$ approximation for at least $1 - n^{-c}$ events for all possible values of m.

6   Conclusion

We propose a novel approximation algorithm for max-covering circle problem. We first construct the circumscribed SRP of a given circle and disclose their area relationship. The area difference is determined by the number of edges in the circumscribed SRP. Then, we convert max-covering SRP problem to SRP intersection problem, which can be efficiently solved with simple partition and modification based on the existing method. Finally, the optimal solution returned from max-covering SRP problem can be used to produce an approximate answer to max-covering circle problem. We prove that for most of the inputs, our algorithm can give a $(1-\varepsilon)$ approximation to the optimal solution, which only needs $O\left(n\varepsilon^{-1}\log n + n\varepsilon^{-1}\log\left(\frac{1}{\varepsilon}\right)\right)$ time for unit points and $o\left(n\varepsilon^{-2}\log n\right)$ time for weighted points.

Acknowledgements. This work was supported by NSFC grant (Nos. 62102119, U19A2059, U22A2025 and 62232005).

References

1. Aggarwal, A., Imai, H., Katoh, N., Suri, S.: Finding k points with minimum diameter and related problems. J. Algorithms **12**(1), 38–56 (1991)
2. Aronov, B., Har-Peled, S.: On approximating the depth and related problems. SIAM J. Comput. **38**(3), 899–921 (2008)
3. Chan, T.M., Har-Peled, S.: Smallest k-enclosing rectangle revisited. Discrete Comput. Geomet. **66**(2), 769–791 (2021)
4. Chazelle, B.M., Lee, D.T.: On a circle placement problem. Computing **36**(1), 1–16 (1986)
5. Choi, D., Chung, C., Tao, Y.: A scalable algorithm for maximizing range sum in spatial databases. Proc. VLDB Endow **5**(11), 1088–1099 (2012)
6. Datta, A., Lenhof, H.P., Schwarz, C., Smid, M.: Static and dynamic algorithms for k-point clustering problems. J. Algorithms **19**(3), 474–503 (1995)
7. De Berg, M., Cabello, S., Cheong, O., Eppstein, D., Knauer, C.: Covering many points with a small-area box. arXiv preprint arXiv:1612.02149 (2016)
8. De Berg, M., Cabello, S., Har-Peled, S.: Covering many or few points with unit disks. Theory Comput. Syst. **45**(3) (2009)
9. Efrat, A., Sharir, M., Ziv, A.: Computing the smallest k-enclosing circle and related problems. Comput. Geom. **4**(3), 119–136 (1994)
10. Eppstein, D., Erickson, J.: Iterated nearest neighbors and finding minimal polytopes. Discrete Comput. Geomet. **11**(3), 321–350 (1994)
11. Har-Peled, S., Mazumdar, S.: Fast algorithms for computing the smallest k-enclosing circle. Algorithmica **41**(3), 147–157 (2005)
12. Hemmer, M., Kleinbort, M., Halperin, D.: Improved implementation of point location in general two-dimensional subdivisions. In: Epstein, L., Ferragina, P. (eds.) ESA 2012. LNCS, vol. 7501, pp. 611–623. Springer, Heidelberg (2012). https://doi.org/10.1007/978-3-642-33090-2_53
13. Hoeffding, W.: Probability inequalities for sums of bounded random variables. J. Am. Stat. Assoc. **58**, 13–30 (1963)
14. Imai, H., Asano, T.: Finding the connected components and a maximum clique of an intersection graph of rectangles in the plane. J. Algorithms **4**(4), 310–323 (1983)
15. Kaplan, H., Roy, S., Sharir, M.: Finding axis-parallel rectangles of fixed perimeter or area containing the largest number of points. Comput. Geom. **81**, 1–11 (2019)
16. Mahapatra, P.R.S., Karmakar, A., Das, S., Goswami, P.P.: k-enclosing axis-parallel square. In: Murgante, B., Gervasi, O., Iglesias, A., Taniar, D., Apduhan, B.O. (eds.) ICCSA 2011. LNCS, vol. 6784, pp. 84–93. Springer, Heidelberg (2011). https://doi.org/10.1007/978-3-642-21931-3_7
17. Matoušek, J.: On enclosing k points by a circle. Inf. Process. Lett. **53**(4), 217–221 (1995)
18. Motwani, R., Raghavan, P.: Randomized Algorithms. Cambridge University Press (1995)
19. Nandy, S.C., Bhattacharya, B.B.: A unified algorithm for finding maximum and minimum object enclosing rectangles and cuboids. Comput. Math. Appl. **29**(8), 45–61 (1995)

20. Preparata, F.P., Shamos, M.I.: Computational Geometry. Springer, New York (1985). https://doi.org/10.1007/978-1-4612-1098-6
21. Tao, Y., Hu, X., Choi, D., Chung, C.: Approximate Maxrs in spatial databases. Proc. VLDB Endow **6**(13), 1546–1557 (2013)
22. Zhang, K., Gao, H., Han, X., Chen, J., Li, J.: Maximizing range sum in trajectory data. In: IEEE International Conference on Data Engineering, pp. 755–766 (2022)

GAMA: Genetic Algorithm for k-Coverage and Connectivity with Minimum Sensor Activation in Wireless Sensor Networks

Syed F. Zaidi[1], Kevin W. Gutama[2], and Habib M. Ammari[3]([✉])

[1] Kean University, Union, NJ 07083, USA
zaidisye@kean.edu
[2] New Jersey Institute of Technology, Newark, NJ 07102, USA
kg567@njit.edu
[3] Texas A&M University - Kingsville, Kingsville, TX 78363, USA
habib.ammari@tamuk.edu

Abstract. In wireless sensor networks, ensuring k-coverage and connectivity is crucial in order to efficiently gather data and relay it back to the base station. We propose an algorithm to achieve k-coverage and connectivity in randomly deployed wireless sensor networks while minimizing the number of active sensors. It has been shown that selecting a minimum set of sensors to activate from an already deployed set of sensors is NP-hard. We address this by using a genetic algorithm that efficiently approximates a solution close to the optimal solution. The algorithm works by selecting random solutions and mutating them, retaining only the best solutions for the next generation until it converges to a near-optimal solution. We examine the time complexity of our approach and discuss possible optimizations. Our simulation results show that our approach works consistently across different types of wireless sensor networks and for different degrees of required coverage.

Keywords: Wireless sensor networks · k-coverage · Connectivity · Sensor selection · Genetic algorithm

1 Introduction

A wireless sensor network (WSN) is a network consisting of wireless sensors capable of measuring various environmental conditions. These sensors are deployed in predetermined patterns or placed randomly within a target region, enabling a comprehensive view of the environment and valuable data collection. WSNs face the challenge of achieving adequate coverage, connectivity, and energy efficiency. The coverage problem entails that the target region is covered by at least one sensor, while the k-coverage problem focuses on covering each point with at least k sensors, crucial for fault tolerance. Connectivity is vital for relaying information to the base station.

To optimize the network's lifetime, it is essential to manage sensor states actively or inactively. All sensors operating simultaneously lead to energy waste,

W. Wu and J. Guo (Eds.): COCOA 2023, LNCS 14461, pp. 239–251, 2024.
https://doi.org/10.1007/978-3-031-49611-0_17

and redundant data collection. Balancing active and inactive sensors is challenging while maintaining connectivity and coverage. Randomly deployed sensor networks pose additional challenges. Uniform deployment is often impractical, and environmental factors can lead to sensor movement. Applications may require random or near-random deployments, necessitating optimal sensor placement.

The problem of finding the minimum number of sensors for k-coverage in randomly deployed networks is NP-hard, making an exact solution computationally infeasible. Genetic algorithms (GA), inspired by natural selection, offer an approximation approach. GAs generate a population of solutions, evaluate their proximity to the optimal solution, and iteratively improve them through mutation. Our GA is designed for static sensors, as mobility consumes more energy. We identify optimal sensor locations for k-coverage and connectivity while retaining network lifetime. Inactive sensors can activate as active sensors deplete energy.

2   Related Work

Yang et al. [1] establishes that selecting the minimum set of active sensors in a randomly deployed network to achieve k-coverage is NP-hard. Previous research efforts have tackled the k-coverage problem by introducing mobile sensors capable of moving to areas within the network where coverage is lacking. In [2], the authors propose a GA approach that utilizes mobile sensors to optimize coverage. This differs from our GA as we only consider static, pre-deployed sensors.

Hurizan and Kuila [3] investigate the activation of a specific set of nodes instead of deploying all nodes within the network. They employ a GA approach to assess the minimum selection of nodes required for full coverage, connectivity, and energy optimization. The main distinction between our approach and [3] lies in the integration of mobile sensors, which prevents the genetic algorithm from frequent reactivation. This prolongs the network's lifespan and effectively addresses potential environmental challenges that may arise within the network.

3   Preliminaries

This section provides an introduction to some of the terminology and notation used in our explanation of the genetic algorithm. Additionally, we outline some assumptions relating to the coverage model.

3.1   Key Terminology and Notation

The following are key definitions:

Definition 1. Sensors and points - A sensor is denoted by S and a point is denoted by p. The total number of sensors in a network is denoted by N and the total number of points in a network is denoted by P.

Definition 2. Sensing and communication range - The sensing radius of a sensor S is denoted by r and the communication radius is denoted by c. The sensing range and communication range of a sensor is the area formed by the circular disk centered at the sensor with radius r and c, respectively. The sensing range of a sensor S is denoted by S_r and the communication range is denoted by S_c.

Definition 3. Distance - The distance between a sensor S and a point p is denoted by $d(S, p)$. Similarly, the distance between two sensors S_1 and S_2 is denoted by $d(S_1, S_2)$.

Definition 4. Communication neighbors - Sensors S_1 and S_2 are considered communication neighbors if the distance between them $d(S_1, S_2)$ is less than or equal to their communication range c (i.e. S_1 and S_2 are communication neighbors if $d(S_1, S_2) \leq c$).

Definition 5. Active sensors - The active sensor set consists of all sensors that are currently in the active state and is denoted by S_{active}.

Definition 6. Parent sensors - A sensor is considered a parent sensor if it is active and the base station is within its communication range. The set of all parent sensors is denoted by S_{parent}. The parent sensor set is a subset of the active sensor set ($S_{parent} \subseteq S_{active}$).

Definition 7. Connected sensors - The set of all active sensors that are connected to a parent sensor via communication neighbors is denoted by S_{conn}. The connected sensor set is a subset of the active sensor set ($S_{conn} \subseteq S_{active}$).

Definition 8. Covered points and k-covered points - The set of all points in the target region that are at least 1-covered is denoted by p_{cov}. The set of all points in the target region that are at least k-covered is denoted by p_{kcov}. The set of k-covered points is a subset of the set of covered points ($p_{kcov} \subseteq p_{cov}$).

3.2   Assumptions

The following are assumptions that our approach is based upon:

Assumption 1. All sensors in the network are homogeneous i.e. all sensors have the same sensing and communication range.

Assumption 2. Sensors communicate to the base station via neighboring sensors within their communication range. If a sensor is a parent sensor it relays information from neighboring nodes to base station.

Assumption 3. The target region is populated with mobile sensors, which are randomly deployed. Similarly, the base station is also positioned randomly within a predefined area located at the center of the target region.

Assumption 4. The deployed sensor network is dense enough to k-cover the region despite the random deployment. If the sensor network is not dense enough to k-cover the region, the algorithm will return the minimum set of active sensors that k-covers the region as much as possible.

Assumption 5. The algorithm operates under the assumption that the sensors remain stationary; however, the inclusion of mobile sensors serves the purpose of seamlessly replacing a failed or dying sensor within the network by utilizing a neighboring sensor. This enables continuous k-coverage of the area even in the event of sensor failure.

4   Genetic Algorithm

In this section, we present our genetic algorithm approach and go into detail about the design.

4.1   Coverage Model

The target region consists of an $m \times n$ grid with $m * n$ grid points. We utilize a point coverage model where a target region is considered k-covered if all grid points within that target region are k-covered. For example, a $50\,\mathrm{m} \times 50\,\mathrm{m}$ target region would have 2500 grid points. We consider this target region k-covered if all 2500 grid points are k-covered. A point p in the target region is considered covered by a sensor S if the distance $d(S,\ p)$ from S to p is less than or equal to the sensing radius r (i.e. p is covered if $d(S,p) \le r$.).

4.2   Genetic Algorithm

Before starting the algorithm, we deploy a given number of sensors in the target region and we deploy the base station randomly in a bounded region towards the center of the target region. A sensor determines its location using GPS technology which is communicated to the base station through a communication path. Since the algorithm is centralized, the base station takes charge of monitoring sensor locations and keeping track of potential solutions.

The algorithm randomly generates an initial population of potential solutions (see Fig. 1). Generating additional potential solutions for a larger sensor network is logical, however, scaling the population size linearly with the number of sensors in the network would result in excessive computational costs. On the other hand, scaling logarithmically with the number of sensors would result in a population size that is too small to properly explore solutions in a large network. Therefore, we compute the population size as a radical function of the total number of sensors N. We start with a base population size of 10 for small networks where $\sqrt{N}$ would not result in a sufficiently large population size. Then, we add to the population size $\frac{\sqrt{N}}{2}$ (we divide by 2 to reduce the population size further). Since the population size must be a whole number, we can apply a floor operation to the $\frac{\sqrt{N}}{2}$ term in case the result is a fraction. This calculation can be represented as the following function of N:

$$pop(N) = 10 + \lfloor \frac{\sqrt{N}}{2} \rfloor$$

To generate potential solutions, the algorithm picks a random number of sensors from 1 to N, denoted by $rand(N)$, and then randomly picks $rand(N)$ sensors to activate from the network. After activating these sensors, it determines the following metrics to evaluate the fitness of the solution: the rate of coverage of the target region $RoC(p_{cov},\ P)$, the rate of k-coverage of the target region $RokC(p_{kcov},\ P)$, the rate of connectivity among the set of active sensors $Conn(S_{conn},\ S_{active})$, and the rate of inactivity $RoI(S_{active},\ N)$. This process is repeated $pop(N)$ times to get $pop(N)$ possible solutions in a single generation.

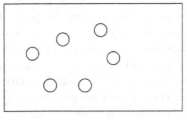
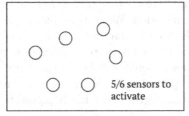

(a) Initial random sensor deployment. All sensors are inactive.

(b) The number of sensors to activate is randomly picked from 1 to N $(rand(N))$.

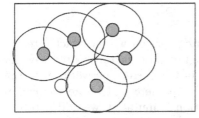
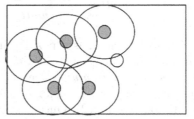

(c) $rand(N)$ sensors are randomly activated.

(d) Another possible formation of $rand(N)$ active sensors.

Fig. 1. Represents how a solutions is generated.

Fitness Metrics

$$RoC(p_{cov},\ P) = \frac{|p_{cov}|}{P}$$

$$RokC(p_{kcov},\ P) = \frac{|p_{kcov}|}{P}$$

$$Conn(S_{conn},\ S_{active}) = \frac{|S_{conn}|}{|S_{active}|}$$

$$RoI(S_{active},\ N) = \frac{N - |S_{active}|}{N}$$

After determining the metrics of each solution, the algorithm calculates the score of each solution using the fitness function and keeps track of the top 20% of the solutions as this provides the best results (determined through trial and error). We also keep track of the best solution for each generation and compare it to the universal best solution. If the best solution of the current generation has a higher score than the universal best solution, then the universal best solution is assigned the best solution of the generation (i.e. *bestSolUniv = max(bestSolGen, bestSolUniv)*). Storing the best universal solution is not only useful for determining the final result, but it also helps terminate the generation loop.

The next step is to create a new generation based on the best solutions of the current generation. We start by picking a random solution from the best solutions and apply a mutation factor to it in order to mimic random genetic mutation. This is an important step in order to explore different solution permutations. If the number of sensors N is less than or equal to 100, then the mutation factor is randomly picked from the range of integers from -3 to 3, inclusive. Otherwise, the mutation factor is randomly picked from the range of integers from $\frac{-N}{30}$ to $\frac{N}{30}$, inclusive. These ranges represent a 0–3% mutation. The function for computing the mutation factor can be written as:

$$mut(N) = \begin{cases} randInt(-3,3), & N \leq 100 \\ randInt(\frac{-N}{30}, \frac{N}{30}), & N > 100 \end{cases}$$

If the mutation factor is a positive integer, then we must activate $mut(N)$ more sensors in the current network. If the mutation factor is a negative integer, then we must deactivate $mut(N)$ sensors in the current network. Note that the mutation factor can also be 0, in which case there will be no changes to the current solution (see Fig. 2). Once a solution is mutated, we add it to our new generation. This step of randomly selecting a solution from the best solutions and mutating it is done $pop(N)$ times to produce a new generation. After creating a new generation, we can repeat the algorithm, starting from the fitness evaluation step, to create an even more fit generation. See *Algorithm 1* for the psuedo-code of the genetic algorithm.

4.3   Terminating Conditions

As we start to create better solutions in each generation and approach the optimal solution, we need a terminating condition to stop the generational loop. Since there is no way to verify the correctness of a solution in polynomial time, we must utilize a heuristic that assumes we have determined the optimal solution based on a terminating condition. One way is to set a constant limit to the number of generational loops. This limit can be a large upper bound to the number of generations required to compute an optimal solution (such as 1000 generations) to ensure that we arrive at the most optimal solution that the algorithm can generate before exiting the loop. This approach, however, introduces redundancy as some networks will arrive at the optimal solution significantly quicker than other networks despite having the same number of sensors. So, a

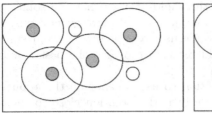
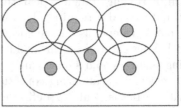

(a) Possible solution for 4/6 active sensors.

(b) Possible mutation activating 2 sensors $(mut(N) = 2)$.

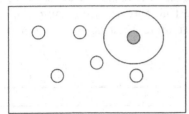
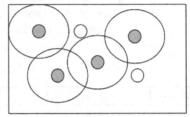

(c) Possible mutation deactivating 3 sensors $(mut(N) = -3)$

(d) No mutation occurs $mut(N) = 0$

Fig. 2. Represents how a solutions is generated.

given network may arrive at the optimal solution long before the generation limit is reached and continue to unnecessarily calculate new solutions, wasting computational resources.

Another way to terminate the generational loop is to compute a score threshold and stop the loop when that threshold has been exceeded. The score threshold can be computed by inputting the metrics of a desired solution into the fitness function. We can then use this threshold as our terminating condition for the generational loop, ensuring that the resulting solution meets the metrics requirement of our desired solution. In other words, the resulting solution is as good as or better than our desired solution. An issue that can arise with setting a score threshold is that an infinite loop can occur if the algorithm is unable to generate a solution that exceeds the threshold. To combat this, we can set a constant limit for the number of generations in the event that a solution with a score exceeding the threshold is unattainable. If this is the case, then the algorithm will continue computing more generations despite having already achieved the most optimal solution it can produce, resulting in the same issue of excess computation.

Perhaps the best terminating condition - if we are concerned with efficiently arriving at a solution reasonably close to the most optimal solution - is to keep track of the highest score across all generations, and if that score has not been exceeded after a set number of generations, we can assume that the algorithm has arrived at the optimal solution and terminate the generational loop. The number of generations after which we want to break the loop if we have not achieved a higher score can be referred to at the "repeat threshold" (since the highest score is repeating). Having a high repeat threshold ensures that the solution is the most optimal, but also requires the computation of more generations. Finding the right number depends on the specific use case; however, in our simulations we found having the repeat threshold set to 5 provided the best results when considering accuracy and saving time.

Deciding which terminating condition to use depends on the application of the algorithm. If the goal is to efficiently find a solution close to the optimal solution that the algorithm can generate, then we can use the repeat threshold. If we simply aim to achieve a desired solution and stop computation once that solution has been attained, we can set a score threshold. If the goal is to simply achieve the most optimal possible solution that the algorithm can come up with regardless of time and computational constraints, then we can set a large upper bound on the generational loop.

Algorithm 1: Genetic Algorithm

---

1 initialize number of sensors as N
2 initialize population size as *popSize*

3 initialize empty list *solutions*
4 **for** *(i = 0 to popSize)* **do**
5   create a solution object *sol*
6   *numActive* = random integer from 1 to N
7   activate *numActive* sensors in *sol*
8   append *sol* to *solutions*

9 initialize empty list *scoredSolutions*
10 **while** *repeatCounter < repeatThreshold* **do**
11   **for** *(j = 0 to popSize)* **do**
12   score = fitness(*solutions[j]*)
13   add current solution and score to *scoredSolutions*

14   sort *scoredSolutions* in ascending order
15   *bestSolutions* = top 20% of *scoredSolutions*

16   update *repeatCounter* and highest score

17   // create new generation
18   initialize empty list *newGen*
19   **for** *(j = 0 to popSize)* **do**
20   *sol* = random solution from *bestSolution*
21   **if** $mut(N) > 0$ **then**
22   activate $mut(N)$ more sensors in *sol*
23   **else**
24   deactivate $mut(N)$ sensors in *sol*
25   append *sol* to *newGen*

26   solutions = newGen

4.4   Fitness Function

The fitness function calculates a score for each generated solution based on the following parameters: coverage, k-coverage, connectivity, and inactivity. The goal is to maximize each of these metrics, but to do it with different priority. For example, we must prioritize connectivity over inactivity since a connected sensor network is preferable to a disconnected sensor network with less sensors. Therefore, we must multiply each metric by a weight that represents its priority. Connectivity takes the highest precedence as a sensor network must be connected to the base station in order to relay any information it gathers. As such, we assign the highest weight to connectivity. Next, we prioritize coverage, then k-coverage, and lastly inactivity. This order of precedence ensures that we first achieve a connected network, then achieve 1-coverage, after which we focus on achieving k-coverage, and lastly, once we have a connected and k-covered sensor network, we can focusing on reducing the number of active sensors.

An issue that can arise when calculating the fitness of a solution is that a solution can come close to achieving an optimal metric, but not be exactly optimal. For example, if the optimal achievable k-coverage in a randomly deployed sensor network is 100% but requires the activation of far more sensors than achieving 99% k-coverage, then the fitness function will give a higher score to a solution that achieves 99% k-coverage with fewer sensors than to a solution that achieves 100% k-coverage with more sensors. To prioritize achieving optimal k-coverage and connectivity before minimizing the number of active sensors, we enhance the scoring of solutions that reach these optimal metrics. Specifically, we multiply the weight of a metric by 10 when it is considered optimal, thereby assigning a significantly higher score to solutions that meet these criteria compared to those that do not achieve any optimal metrics. By adopting this approach, the genetic algorithm will experience notably faster convergence to the optimal solution, given that solutions with optimal metrics will consistently attain the highest scores. If a solution achieves optimal connectivity, coverage, and k-coverage, then its score will be 0.99. From there, the fewer the number of active sensors in a solution, the closer its score will be to 1. The score of a solution should never actually be 1 as this would require the solution to have optimal metrics with 0 active sensors, which is not possible. See *Algorithm 2* for the pseudo-code of the fitness function.

4.5   Time Complexity

The algorithm computes $pop(N)$ solutions in every iteration of the generational loop. The generational loop will typically terminate when an optimal solution has been achieved; however, we set some bounding constant C as the limit of the generational loop in the event that this does not occur. This operation reduces to $O(\sqrt{N})$.

The computation of every solution requires us to activate or deactivate sensors and update coverage. To update coverage, the algorithm maintains two sets, $p_c ov$ and $p_k cov$. It iterates through all active sensors, employing a depth-first

---

Algorithm 2: Fitness Function

Input : coverageRate, kCoverageRate, connectivity, inactivity
Ouput: solutionScore

1 // boost score if solution reaches optimal metrics
2 **if** *(coverageRate == optimalCoverageRate)* **then**
3 $\quad\lfloor$ coverageRate = coverageRate * 10

4 **if** *(kCoverageRate == optimalKCoverageRate)* **then**
5 $\quad\lfloor$ kCoverageRate = kCoverageRate * 10

6 **if** *(connectivity == optimalConnectivity)* **then**
7 $\quad\lfloor$ connectivity = connecitivity * 10

8 solutionScore = (0.045 * connectivity + 0.030 * coverageRate + 0.024 * kCoverageRate + 0.01 * inactivity)

9 return solutionScore

---

search starting at the sensor's location to update the coverage of only the points covered by the sensor. This approach has a time complexity of $O(a*r^2)$, where a is the number of active sensors, and r is the radius of the sensing disk, accounting for the quadratic scaling of the area. This process is also used when deactivating sensors to remove points from the respective sets if they are no longer 1-covered or k-covered.

Activating a sensor also requires that we initialize its communication neighbors to determine connectivity. In order to do this, we must iterate through every currently active sensor and check if it is communication neighbors with the newly activated sensor. Similarly, when we deactivate a sensor, we need to remove it from the communication neighbor set of every other active sensor, which also requires us to iterate through S_{active}. Doing this for every active sensor results in a time complexity of $O(a^2)$.

Once we have computed a potential solution, we need to determine its metrics in order to give it a score. Calculating coverage, k-coverage, and inactivity are constant time operations, but computing connectivity requires a depth-first search traversal of all currently active sensors. Since we need to traverse all active sensors at least once, the time complexity of this traversal is $O(a)$.

The time complexity of this algorithm stands at $O(\sqrt{N} * (a * r^2 + a^2))$ (the $O(a)$ step of computing connectivity reduces here). In the worst case, the number of total sensors N is equal to the number of active sensors a and taking this into account, we must write the time complexity as $O(\sqrt{(N)} * (N * r^2 + N^2))$. N^2 is greater than $N * r^2$ when the number of sensors N is greater than the sensing radius r squared which is the most likely case in a randomly deployed sensor network dense enough to k-cover a target region. Therefore, we can say that the overall time complexity of the genetic algorithm is $O(N^2\sqrt{N})$.

5   Simulation

In this section, we present our experimental results under different simulation conditions. Figures 3a, 3b and 3c show the minimum number of active sensors generated by the genetic algorithm with respect to different degrees of required coverage. Figure 3d hows the number of active sensors generated by the algorithm with respect to different sensing ranges.

5.1   Coverage Degree Vs. Number of Active Sensors

Simulation parameters for Fig. 3a: 50 m × 50 m target region (2500 points to cover), 100 deployed sensors, 10 m sensing range, and 20 m communication range ($N = 100$, $r = 10$, $c = 20$). Simulation parameters for Fig. 3b: 100 m × 100 m target region (10,000 points to cover), 300 deployed sensors, 15 m sensing range, and 30 m communication range ($N = 300$, $r = 15$, $c = 30$). Simulation parameters for Fig. 3c: 150 m × 150 m target region (22,500 points to cover), 500 deployed sensors, 20 m sensing range, and 40 m communication range ($N = 500$, $r = 20$, $c = 40$).

Note that the minimum number of active sensors for each value of k in Figs. 3a, 3b and 3c is the average of 10 simulations, each ran with a different randomly deployed sensor network. In Fig. 3a the algorithm computes an average of 40.7 active sensors for $k = 2$, 57 for for $k = 3$, and 72.9 for $k = 4$. We can see that the number of active sensors increases linearly with the required degree of coverage. The same trend can be observed in Figs. 3b and 3c, demonstrating that the algorithm performs similarly for sensor networks of different sizes and varying sensor density.

Furthermore, the average number of generations to compute a solution (which is averaged among 30 different simulations - 10 for each value of k) is roughly the same for each figure, and does not scale with any of the simulation parameters. This demonstrates that the algorithm computes solutions in roughly the same amount of generations regardless of the specific attributes of a wireless sensor network, and also explains why the number of generations is constant when computing the time complexity of the algorithm.

5.2   Sensing Range Vs. Number of Active Sensors

Following are the simulation parameters for Fig. 3d 100 m × 100 m target region (10,000 points to cover), 300 deployed sensors, the required degree of coverage is 3, and the communication range is twice the sensing range ($N = 300$, $k = 3$, $c = 2 * r$). Note that, as with the previous experimental results, the number of active sensors for each value of r is the average of the results of 10 simulations. We can see in Fig. 3d that as we increase the sensing range r, the number of active sensors that the algorithm computes decreases which is consistent with the expected result.

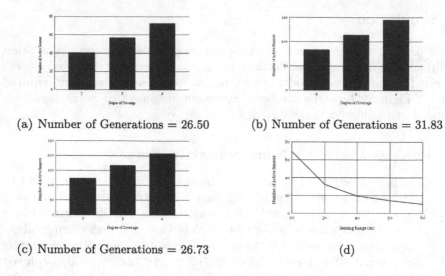

(a) Number of Generations = 26.50   (b) Number of Generations = 31.83

(c) Number of Generations = 26.73   (d)

Fig. 3. Simulation Results

6 Conclusion and Future Work

In this paper, we proposed a genetic algorithm approach to the k-coverage problem. Specifically, we focused on finding the minimum set of active sensors required to k-cover a region among a randomly deployed wireless sensor network while ensuring connectivity. We detailed how our genetic algorithm selects only the best solutions in each generation and mutates them, converging closer to the optimal solution in each iteration. Through our simulation results, we showed that the algorithm performs consistently across different types of wireless sensor networks and across varying degrees of required coverage.

Our future works consists of improving the time complexity of our algorithm by finding a faster way to verify area or point coverage. Furthermore, we plan to extend our approach by developing a scheme that allows inactive sensors to activate and move to the location of dying active sensors in order to prolong the lifetime of the network.

Acknowledgements. We would like to thank Dr. Habib M. Ammari, NSF REU Site PI, for his diligent support and review of our paper, which helped improve its overall quality. This work is funded by the US National Science Foundation under NSF grant 2338521.

References

1. Yang, S., Dai, F., Cardei, M., Wu, J.: On multiple point coverage in wireless sensor networks. In: IEEE International Conference on Mobile Adhoc and Sensor Systems Conference, Washington, DC, 2005, pp. 8–764 (2005). https://doi.org/10.1109/MAHSS.2005.1542868

2. Elhoseny, M., Tharwat, A., Yuan, X., Hassanien, A.E.: Optimizing K-coverage of mobile WSNs. Exp. Syst. Appl. **92**, 142–153 (2018). https://doi.org/10.1016/j.eswa. 2017.09.008. ISSN 0957–4174
3. Harizan, S., Kuila, P.: Coverage and connectivity aware energy efficient scheduling in target based wireless sensor networks: an improved genetic algorithm based approach. Wirel. Netw. **25**, 1995–2011 (2019). https://doi.org/10.1007/s11276-018-1792-2

Simple Heuristics for the Rooted Max Tree Coverage Problem

Jiang Zhou and Peng Zhang[✉]

School of Software, Shandong University, Jinan 250101, China
jiangz@mail.sdu.edu.cn, algzhang@sdu.edu.cn

Abstract. The Max Tree Coverage (MTC) problem is the dual of the classic k-MST problem and has wide applications in areas such as network design and vehicle routing. Given a graph G with nonnegative costs defined on edges, a vertex $r \in V(G)$, and a budget B, the rooted Max Tree Coverage problem asks to find a tree rooted at r having total cost at most B, so that the number of vertices included in the tree is maximized. This problem is NP-hard and has constant factor approximation algorithms. However, the existing approximation algorithms for rooted MTC is very complicated and hard to be implemented practically.

In this paper, we develop a simple CMSA heuristic for rooted MTC for the first time, where CMSA (Construct, Merge, Solve and Adapt) is a meta-heuristic proposed recently. We also formulate a polynomial size mixed integer linear program for rooted MTC for the first time. Experimental results show that CMSA has very good practical performance. For the small size instances of the problem, CMSA almost finds the optimal solutions. For the large size instances, CMSA finds solutions better than that of CPLEX within the same running time and two additional greedy algorithms. Note that within an admissible running time limit, CPLEX returns the best solutions ever found (not guarantee to be optimal).

Keywords: Max Tree Coverage · Meta-heuristic · CMSA · k-MST · Combinatorial Optimization

1 Introduction

The Minimum Spanning Tree (MST) problem is one of the most fundamental problems in algorithm theory and combinatorial optimization. Given a weighted graph G, we want to find a subgraph containing all the vertices of G so that the total weight of all edges of the subgraph is minimized. As an important variant of MST, the k-MST problem has also been extensively studied in the literature [1,2,6,12,16], which asks to find a minimum cost tree of the input graph spanning at least k vertices. In this paper, we consider the rooted Max Tree Coverage (MTC) problem, which is the dual of the k-MST problem.

Definition 1. The Rooted Max Tree Coverage Problem.

© The Author(s), under exclusive license to Springer Nature Switzerland AG 2024
W. Wu and J. Guo (Eds.): COCOA 2023, LNCS 14461, pp. 252–264, 2024.
https://doi.org/10.1007/978-3-031-49611-0_18

Instance: An undirected graph $G = (V, E)$ with edge costs $c : E \rightarrow \mathbb{R}^+$, a vertex $r \in V$ and a non-negative budget B.

Goal: Find a tree in G rooted at r with cost at most B such that the number of vertices spanned by the tree is maximized.

One can find that the goal of rooted k-MST is to minimize the total cost of the tree, which is just a constraint in rooted MTC (i.e., the total cost of the tree is at most B). The goal of rooted MTC is to maximize the number of vertices spanned by the tree, which is just a constraint in rooted k-MST (i.e., spanning at least k vertices). In this sense, we say that rooted MTC is a dual of rooted k-MST.

The rooted MTC problem finds wide applications in real world [10,18]. For example, a communication company wants to lay fiber optic cables between cities in an area, where a central city is designated to must be included. The cost of laying fiber optic cable varies from city to city. However, the communication company has a limited budget and needs to maximize the number of cities connected by the laid cables. This scenario is just captured by the rooted MTC problem.

Like k-MST, the rooted MTC problem is in fact a very basic problem and its algorithms can be used as sub-routines in network design and vehicle routing [5,13]. For example, finding a cycle (the trajectory of a vehicle) containing a depot that visits as many as possible nodes can be solved by using the rooted MTC problem.

The Max Tree Coverage problem with budget consists of two sub-problems, i.e., the (unrooted) MTC problem and the rooted MTC problem. In the MTC problem, there is merely no root given in the input. The problem is to find a tree of the input graph spanning as many as possible vertices, with the total cost of the tree not exceeding a given budget. The MTC problem is just the dual of the k-MST problem, while the rooted MTC problem is the dual of the rooted k-MST problem.

We will mainly consider the rooted version of the Max Tree Coverage problem, since the un-rooted version can be reduced easily to the rooted one by trying every vertex as the root and returning the cheapest one of the solutions obtained.

As usual, given a graph, we use n to denote the number of vertices in the graph and m the number of edges. Given a combinatorial optimization problem, we use OPT to denote the optimal value of its instance.

1.1 Related Work

The k-MST problem is NP-hard, and its best known approximation ratio is 2 [12]. As its dual, the rooted MTC problem is also NP-hard. Although k-MST has been paid much attention from researchers, only less research has been done for the MTC problem.

The (unrooted) MTC problem was first studied in [13]. Johnson et al. [13] gave a $5 + \epsilon$ approximation algorithm for the MTC problem. The proof for this result was only sketched in [13].

On the other hand, the rooted MTC problem was explicitly proposed first until the work of Blum et al. [5] (the conference version is in FOCS 2003). This problem is called the (rooted) Max Prize Tree problem in [5]. Blum et al. [5] observed that the Max Prize Tree problem can be reduced to the (rooted) Max Prize Path problem. Specifically, Blum et al. [5] proved that if the Max Prize Path problem can be approximated within α, then the Max Prize Tree problem can be approximated within 2α. In [5], the Max Prize Path problem (also called the (rooted) Orienteering problem in [5]) is approximated within 4 via still a reduction to the Min Excess Path problem. The Min Excess Path problem in [5] is approximately solved by using a dynamic programming approach, which in turn calls an approximation algorithm for the k-Path problem given in [8]. As a result, [5] showed that the rooted MTC problem (i.e., the Max Prize Tree problem) can be approximated within 8.

Later, Bansal et al. [3] gave an improved 3-approximation algorithm for the Orienteering problem. This means that rooted MTC can be approximated within 6.

Chekuri et al. [9] gave an improved approximation algorithm for the Orienteering problem. Based on the framework of [5], Chekuri et al. [9] reduced the Orienteering problem to the k-Stroll problem via the Min Excess Path problem. In [9], the k-Stroll problem was approximately solved using dynamic programming, and the solution was adapted to a solution to the Orienteering problem, resulting in a $(2 + \epsilon)$-approximation for Orienteering. Consequently, Chekuri et al. [9] showed that the rooted MTC problem can be approximated within $4 + \epsilon$, which is currently the best ratio for the problem.

1.2   Our Results

Both the work of [5,9] aim to design approximation algorithms with better ratios for the rooted MTC problem (and other problems therein). One can see that the algorithms given in [5,9] for rooted MTC, although running in polynomial time, are very complicated and difficult to be implemented. In this paper, we study the rooted MTC problem from the aspect of heuristics, giving simple heuristic algorithms for rooted MTC with good practical performance. Our contributions are summarized as follows.

(i) We develop a CMSA heuristic for the rooted MTC problem for the first time. The CMSA (Construct, Merge, Solve and Adapt) heuristic is a powerful meta-heuristic proposed recently by Blum et al. [7]. CMSA has been widely applied to solve many combinatorial optimization problems, e.g., [14,15,17].

(ii) Moreover, we provide a polynomial-size MILP (mixed integer linear program) model for the rooted MTC problem for the first time. The most common integer linear program (ILP) for the spanning tree-like problems such as MST and rooted MTC, usually contains a connectivity constraint which is defined on cuts. Since the number of cuts is exponential in the number of vertices, this way cannot lead to a polynomial-size ILP for rooted MTC. We overcome this difficulty via a flow approach.

(iii) We do experiments to test the performance of the CMSA heuristic. In order to make a comparison, we additionally provide two more polynomial time greedy algorithms for rooted MTC. They are Algorithm $\mathcal{G}$ and Algorithm $\mathcal{P}$. Algorithm $\mathcal{G}$ uses the approach of the Prim Algorithm for MST, while Algorithm $\mathcal{P}$ uses an idea of minimum average cost. Moreover, we also use CPLEX to solve the MILP for rooted MTC. So, in the experiments we test four algorithms for rooted MTC, namely, CMSA, Algorithm $\mathcal{G}$, Algorithm $\mathcal{P}$, and CPLEX. They all run on five data sets. The first two data sets come from a benchmark for a similar problem (since there is no direct benchmark for rooted MTC), and the last three data sets are randomly generated according to different graph models.

We set a running time limit of 1800 s for both CMSA and CPLEX. The experimental results show that for the small size instances of rooted MTC, CMSA almost always finds the optimal solutions. For the large size instances, within the same running time limit, CMSA performs best in the sense that it almost always finds solutions better than that of CPLEX, Algorithm $\mathcal{G}$, and Algorithm $\mathcal{P}$. We also find that in the experiments Algorithm $\mathcal{G}$ performs better than Algorithm $\mathcal{P}$. Our experimental findings show that CMSA can be a good heuristic for the rooted MTC problem in practice.

It is well-known that CPLEX can find optimal solutions by solving integer program for a combinatorial optimization problem. However, in our experiments, since the instances of MTC have large sizes, if one insists to use CPLEX to find the optimal solutions, the running time would be impractically high and cannot be tolerated. Therefore, we have to set a running time limit. To guarantee the fairness, the running time limit is used for both CPLEX and CMSA. Note that within the running time limit, CPLEX returns the best solution ever found. In other words, CPLEX cannot guarantee to find the optimal solutions within this time limit. So, it is possible that CMSA performs better than CPLEX on large size instances using a common running time limit.

Algorithm $\mathcal{G}$ and Algorithm $\mathcal{P}$ are fast and they normally finish within 1800 s in all of our experiments.

(iv) We provide a bi-criteria approximation algorithm for rooted MTC. We prove that when the input graph is restricted to be a tree, the rooted MTC problem is polynomial time solvable using a dynamic programming approach.

Organization of the Paper. The remainder of the paper is organized as follows. In Sect. 2, we show that rooted MTC is NP-hard and give a bi-criteria approximation algorithm for the problem. In Sect. 3, we describe the CMSA heuristic. In Sect. 4, we give two additional greedy algorithms for the rooted MTC problem. Next, we show the experimental results in Sect. 5. In Sect. 6, we show that rooted MTC in trees is polynomial time solvable using the dynamic programming approach. Finally, we give the conclusions of this paper in Sect. 7.

2   NP-Hardness and Bi-criteria Approximation

Theorem 1. *The rooted MTC problem is NP-hard.*

Proof. The k-MST problem is known to be NP-hard even in the graphs with edge costs only coming from the set $\{1, 2, 3\}$ [16]. Given such an instance (denoted by $\mathcal{I}$) of the k-MST problem (rooted version), its optimal solution cost clearly lies in the interval $[k, 3k]$. Now we assume that there exists an algorithm (denoted by $\mathcal{A}$) which optimally solves the rooted MTC problem. We then run algorithm $\mathcal{A}$ on instance $\mathcal{I}$ with budget B initialized to be k, the smallest possible optimum of the k-MST instance $\mathcal{I}$. If the solution obtained spans less than k vertices, we then increase the budget B by one and continue to run algorithm $\mathcal{A}$ on instance $\mathcal{I}$. The first solution found in this procedure that spans at least k vertices must be the optimal solution to the k-MST instance. The whole procedure obviously consists of polynomial number of iterations. In this way, we reduce k-MST to rooted MTC via a Turing reduction. Since k-MST is NP-hard, this means that rooted MTC is also NP-hard. □

We now use the search idea similar to the proof of Theorem 1 to give a bi-criteria approximation algorithm for the rooted MTC problem.

Theorem 2. *Given a polynomial time α-approximation algorithm for the rooted k-MST problem, we can find in polynomial time a solution to the rooted MTC problem whose total cost $\leq \alpha B$ and which spans $\geq OPT$ number of vertices, where OPT is the optimum of the rooted MTC problem.*

Proof. Let $\mathcal{A}$ be the α-approximation algorithm for k-MST. Starting from $k = 1$ and each time increasing k by one, we repeatedly run algorithm $\mathcal{A}$ on the rooted MTC instance, say, $\mathcal{I}$. The procedure terminates until the cost of the solution found by algorithm $\mathcal{A}$ exceeds αB. Suppose that the value of k is k' in the last iteration of the procedure. Clearly, we need only to prove that $k' \geq OPT$.

Since the optimal solution to rooted MTC spanns OPT vertices using a tree with cost $\leq B$, when running algorithm $\mathcal{A}$ with $k = OPT$, the solution found by $\mathcal{A}$ must have cost $\leq \alpha B$ since $\mathcal{A}$ is an α-approximation algorithm. The search procedure will continue after this point until a tree with cost $> \alpha B$ is found. Therefore, the last value of k must be greater than or equal to OPT. □

The best approximation ratio currently known for rooted k-MST is 2 [12]. Therefore, Theorem 2 shows that there exists a bi-criteria $(1, 2)$-approximation algorithm for the rooted MTC problem, where the first factor one is the approximation ratio, and the second factor 2 is the violation degree of the budget.

3   CMSA Heuristic for Rooted MTC

The CMSA meta-heuristic proposed in [7] is an iterated approach whose each iteration consists of the following four stages.

- Construct: At each iteration, the heuristic generates a certain number of feasible random solutions.
- Merge: The solution components found in these solutions are merged together to form a sub-instance with smaller size than the original instance.

- Solve: Apply an exact solver (e.g., CPLEX) to solve the sub-instances obtained in the merge step.
- Adapt: The heuristic adapts the sub-instances based on the results of each iteration, keeping the solution components that are frequently used and discarding those that are never used after a certain number of iterations.

As for our CMSA heuristic for rooted MTC (see Algorithm 1), the construction stage corresponds to step 5, the merge stage corresponds to steps 6–7, the solving stage corresponds to step 9, and the adaption stage corresponds to steps 11–14.

The secret of success for CMSA is that it repeatedly improves the current solution to the original instance (denoted by $\mathcal{I}$) by solving a generated instance (denoted by $\mathcal{J}$) with smaller size. Instance $\mathcal{J}$ is generated by merging a certain number of random solutions. Consequently, it is structured and easy to be solved. Moreover, instance $\mathcal{J}$ is adapted through iterations, so better solutions may be found in the process.

We solve the smaller instance in our CMSA heuristic by CPLEX. One may wonder why CMSA is compared with CPLEX in the experiments, as CPLEX has already been called as a subroutine in CMSA. The answer is that in our CMSA heuristic CPLEX is only called to solve the generated smaller instance. Since the size of the generated instance is small, CPLEX can terminate within an admissible running time. In the final experiments, CPLEX is again used as a separated algorithm to solve the whole instance of MTC, and is compared with other three algorithms (including CMSA).

3.1 An MILP Model for Rooted MTC

Since in the solving stage of CMSA we need to optimally solve the sub-instances of rooted MTC by using CPLEX, we first introduce an mixed integer linear program (MILP) model for the rooted MTC problem. The model is shown as (MILP).

The key point of the model (MILP) is that its has polynomial size. This is very vital since we will solve this model by CPLEX in the CMSA heuristic. We want to find a tree in the rooted MTC problem. An usual approach to formulate a tree in a mathematical program is to use the cut based constraints. However, this approach would lead to a model of exponential size as there are exponential number of cuts in a graph. Instead, we use the flow approach in our (MILP) model to formulate a tree. By the flow approach, we finally obtain a model for rooted MTC having polynomial size.

$$\max \quad \sum_{v \in V} x_v \qquad\qquad \text{(MILP)}$$

$$\text{s.t.} \quad \sum_{u \in N(r)} f_{ru} - \sum_{u \in N(r)} f_{ur} = \sum_{v \in V,} x_v - 1 \tag{1}$$

$$\sum_{u \in N(v)} f_{uv} - \sum_{u \in N(v)} f_{vu} = x_v, \qquad \forall v \neq r \tag{2}$$

$$\sum_{v \in V} x_v = \sum_{e \in E} y_e + 1 \tag{3}$$

$$\sum_{e \in E} c_e y_e \leq B$$

$$x_v \leq \sum_{e \in \delta(v)} y_e, \qquad \forall v \in V$$

$$x_r = 1$$

$$f_{uv} \leq (n-1)y_e, \qquad \forall e = (u,v)$$

$$f_{vu} \leq (n-1)y_e, \qquad \forall e = (u,v)$$

$$x_v, y_e \in \{0,1\}, \qquad \forall v \in V$$

$$f_{uv}, f_{vu} \geq 0, \qquad \forall e = (u,v)$$

For each $v \in V$, we associate a variable x_v. The equality $x_v = 1$ indicates that vertex v is included in the solution (i.e., the tree). Similarly, we associate a variable y_e with each edge $e \in E$, and $y_e = 1$ indicates that edge e is included in the tree. Besides, for each edge $e = (u,v)$ we set two variables f_{uv} and f_{vu}, where f_{uv} is the flow value from u to v, and f_{vu} is the flow value of the opposite direction.

To force the selected edges form a connected subgraph, our method is to deploy a single source multi-sink flow from the root r to each vertex (with flow value one) in the subgraph. Therefore, constraint (1) guarantees the flow value outgoing from the root r is $\sum_{v \in V} x_v - 1$, i.e., the number of vertices spanned by the tree minus one. If vertex $v \neq r$ is included (i.e., $x_v = 1$), then there is a unit flow from the root to v (constraint (2)). The constraint (3) says that in the subgraph the number of vertices should be equal to the number of edges plus one, guaranteeing that the connected subgraph forms a tree. In (MILP) $N(v)$ is the set of neighbours of vertex v.

3.2 The CMSA Heuristic

The CMSA heuristic for the rooted MTC problem is given in Algorithm 1. Besides the instance as the part of the input, Algorithm 1 requires further three parameters n_a, $age_{\max}$, and d_{rate}, whose meanings are given below.

- n_a: the number of random trees generated at each iteration.
- $age_{\max}$: the maximum allowed age of an edge e.
- d_{rate}: the threshold used when generating a random tree.

At the beginning, Algorithm 1 initializes the best solution currently found (S_{bsf}) and sub-instance E'. The age of every edge is initialized to zero (see line

2 of Algorithm 1). Then, Algorithm 1 repeats the following procedure until the allowed running time is out.

Line 5 of Algorithm 1 corresponds to the construction stage, where a budgeted random tree rooted at r is generated. Lines 6–7 of Algorithm 1 are the merging stage. The algorithm generates n_a random trees and merge them all into E'. Note that E' actually forms a new instance of rooted MTC with smaller size than the original instance G.

Line 9 of Algorithm 1 is the solving stage. We run CPLEX on the MILP model with the instance E' to obtain an optimal tree S_{opt}. If S_{opt} contains more vertices than S_{bsf}, S_{bsf} is replaced by S_{opt}. (See line 10 of Algorithm 1). The last stage adaption is implemented in lines 11–14. For each edge $e \in E'$, if e is in S_{opt}, then its age is reset to zero. Otherwise, the age is increased by one. Then all edges in E' with $age(e) > age_{max}$ are removed from E', which are those never used in any solutions S_{opt}'s in the last age_{max} iterations.

Algorithm 1. CMSA for rooted MTC (Algorithm CMSA)

Input: Graph G, root r, budegt B, n_a, age_{max}, d_{rate}.
Output: A budgeted tree S_{bsf} rooted at r.
1: $S_{bsf} \leftarrow$ NULL, $E' \leftarrow \emptyset$.
2: $\forall e \in E$, $age(e) \leftarrow 0$.
3: **while** CPU time limit not reached **do**
4:     **for** $i \leftarrow 1$ **to** n_a **do**
5:         $S \leftarrow$ GenerateRandomTree(G, d_{rate}, r, B).
6:         $\forall e \in S \setminus E'$, $age(e) \leftarrow 0$.
7:         $E' \leftarrow E' \cup S$.
8:     **end for**
9:     Run CPLEX on the instance E' to obtain a tree S_{opt}.
10:     **if** $|V(S_{opt})| > |V(S_{bsf})|$ **then** $S_{bsf} \leftarrow S_{opt}$.
11:     **for each** $e \in E'$ **do**
12:         **if** $e \in S_{opt}$ **then** $age(e) \leftarrow 0$ **else** $age(e) \leftarrow age(e) + 1$.
13:     **end for**
14:     Remove all edges from E' with $age(e) > age_{max}$.
15: **end while**
16: **return** S_{bsf}.

Since Algorithm 1 calls CPLEX on a sub-instance, each iteration of Algorithm 1 uses exponential time. However, since the solved sub-instance has small size, the running time of each iteration is acceptable in practice. When implementing Algorithm 1, a running time limit is used. The algorithm will terminate when the admissible running time is out.

3.3   Generate a Random Tree

Algorithm 2 is the GenerateRandomTree() algorithm used in Algorithm 1. The algorithm maintains a tree (denoted by S, the set of edges in the tree) rooted at r. Initially, the tree contains only the root. In each iteration, an edge lying in the border of $V(S)$ (denoted by $\delta(V(S))$) is chosen and added to the current tree, until the cost of the tree exceeds the budget B. Here, by border of $V(S)$, we mean the set of edges whose one endpoint is in $V(S)$ and the other endpoint is outside of $V(S)$.

Lines 5–7 of Algorithm 2 describe how we choose an edge from the border A. First, we generate a random number a between 0 and 1. Then we compare a and d_{rate}. If $a \leq d_{rate}$, we choose the edge with the minimum cost from A and add it to S'. Otherwise, we choose an edge uniformly at random from A and add it to S'. Thus, the current tree gets larger, and the border A accordingly changes. At the beginning of each iteration, we save S' in S, so that we can remember the tree before its cost exceeds B.

Algorithm 2. GenerateRandomTree

Input: Graph G, root r, budget B, threshold d_{rate}.
Output: An edge set S that forms a tree.
1: $S' \leftarrow \emptyset$.
2: $A \leftarrow \delta(r)$.
3: **while** $A \neq \emptyset$ **and** $c(S') \leq B$ **do**
4:     $S \leftarrow S'$.
5:     Choose a random number $a \in [0, 1]$.
6:     **if** $a \leq d_{rate}$ **then** let e' be the edge in A with the minimum cost.
7:     **else** let e' be an edge in A chosen uniformly at random.
8:     $S' \leftarrow S' \cup \{e'\}$.
9:     $A \leftarrow \delta(V(S'))$.
10:     $A \leftarrow \text{Reduce}(A)$.
11: **end while**
12: **if** $c(S') > B$ **then return** S **else return** S'.

Algorithm Reduce(A). Note Algorithm 2 calls Reduce(A) as a sub-routine. Algorithm Reduce(A) does the following work. If there are more than one edges in A with the common endpoint outside of $V(S')$, say v, only the edge with the minimum cost remains in A. All the other edges with endpoint v are removed from A.

4 Two Greedy Algorithms

In this section we introduce two greedy algorithms for the rooted MTC problem.

4.1 Greedy Algorithm Based on Prim

Algorithm 3 is a greedy algorithm for rooted MTC which is based on the Prim algorithm for MST. Starting from the root vertex r, each time the algorithm selects the minimum cost edge from the border of the current tree, until the cost of the current tree exceeds the budget B. Algorithm 3 returns the tree just before its cost exceeds B as the solution. This algorithm is essentially identical to Prim, and has the same time complexity as that of Prim. It is known that the Prim algorithm can be implemented in $O(m + n \log n)$ time.

Algorithm 3. Greedy algorithm based on Prim (Algorithm $\mathcal{G}$)

Input: Graph G, root r, budget B.
Output: An edge set S that forms a tree.
1: $S' \leftarrow \emptyset$.
2: $A \leftarrow \delta(r)$.
3: **while** $A \neq \emptyset$ **and** $c(S') \leq B$ **do**
4:     $S \leftarrow S'$.
5:     let e' be the edge in A with the minimum cost.
6:     $S' \leftarrow S' \cup \{e'\}$.
7:     $A \leftarrow \delta(V(S'))$.
8:     $A \leftarrow Reduce(A)$.
9: **end while**
10: **if** $c(S') > B$ **then return** S **else return** S'.

4.2 Greedy Algorithm Based on Average Cost

Recall that an optimal tree to the rooted MTC problem spans as many as possible vertices using edges with total cost at most the budget B. Since in a tree the number of vertices equals to the number of edges plus one, the ratio of total edge cost and the number of edges, called *average cost*, is an intuitively good indicator to find a budgeted tree spanning as many as possible vertices. Inspired by this idea, we design Algorithm 4.

Algorithm repeatedly finds a shortest r-t path which has the length at most the remaining budget and the minimum average cost among $\{t \neq r \mid t \in V\}$, where the average cost of a path is defined as the ratio of its length and the number of edges contained in the path. Once such a path is found, it is united to the current tree S. The vertices in S (note that $r \in S$) are merged into r in G, so that the newly found path does not overlap with S. All self-loops appearing in the merging process are removed. If multiple edges occur, only one edge with the minimum cost is remained.

Each iteration of Algorithm 4 can be implemented in time $O(m + n \log n)$ when using Fibonacci heap to implement Dijkstra's shortest path algorithm. Since at least one vertex is merged in each iteration, the time complexity of Algorithm 4 is $O(mn + n^2 \log n)$.

5 Experimental Evaluation

We do experiments to test the performance of Algorithm CMSA for the rooted MTC problem. In the experiments we also run Algorithm $\mathcal{G}$, Algorithm $\mathcal{P}$, and

---

Algorithm 4. Greedy algorithm using average cost (Algorithm $\mathcal{P}$)

Input: Graph $G = (V, E)$, root r, budget B.
Output: An edge set S that forms a tree.
1: $S \leftarrow \emptyset$.
2: **while** $E(G) \neq \emptyset$ **do**
3:    Find a shortest r-t path P having $c(P) \leq B$ and the minimum average cost among $\{t \neq r \mid t \in V\}$.
4:    **if** not found **then break**.
5:    $B \leftarrow B - c(P)$.
6:    $S \leftarrow S \cup P$.
7:    Merge all vertices in P into r in G.
8: **end while**
9: **return** S.

CPLEX in order to compare their performances with Algorithm CMSA. Note that there are two places we use CPLEX in the experiments. One is the solving stage of Algorithm CMSA. The other is a separate usage where we use CPLEX to solve the whole rooted MTC problem.

We test the above four algorithms on five data sets in the experiments. The first two sets are *ehrfrei* and *blxh* from the benchmark library KCTLIB [4,11]. The last three sets *Gnp*, *Conf*, and *PA* are randomly generated according to three graph models $G[n, p]$, configuration, and preferential attachment, respectively.

Due to space limitation, only experimental findings are stated here. The whole description of the experiments will be given in the full version of the paper. The experimental results show that for the small size instances of rooted MTC, CMSA almost always finds the optimal solutions. For the large size instances, within the same running time limit, CMSA performs best in the sense that it almost always finds solutions better than that of CPLEX, Algorithm $\mathcal{G}$, and Algorithm $\mathcal{P}$. Note that within the running time limit, CPLEX returns the best solution ever found. We also find that in the experiments Algorithm $\mathcal{G}$ performs better than Algorithm $\mathcal{P}$. Our experimental findings show that CMSA can be a good heuristic for the rooted MTC problem in practice.

6   Rooted MTC on Trees

In this section we show that the rooted MTC problem is polynomial-time solvable when the input graph is a tree. Our approach is dynamic programming. When designing a dynamic programm for rooted MTC in trees, one have to consider how to split remaining budget among the subtrees of a node in the input tree. However, this cannot be directly finished in polynomial time since in general the budget is not polynomial bounded. To overcome this difficulty, we use an indirect way to solve the rooted MTC problem. We actually give a dynamic program for its dual, the k-MST problem on trees. This is possible since in our dynamic program for k-MST, we need only consider how to split the number of vertices to be covered in the subtrees of a node. This can be done in polynomial time.

Theorem 3. *The rooted MTC problem is polynomial-time solvable in trees.*

The details of the dynamic programming and the proof of Theorem 3 are omitted due to space limitation, and will appear in the full version of the paper.

7 Conclusions

The rooted MTC problem arises from network design and vehicle routing area. The problem is interesting since it is a dual of the classic k-MST problem. We design a simple heuristic algorithm for rooted MTC by applying the CMSA (construct, merge, solve, and adapt) strategy. Experimental results show that our CMSA heuristic for rooted MTC has very good practical performance.

Acknowledgements. This work is supported by the National Natural Science Foundation of China (62272280 and 61972228), and the Natural Science Foundation of Shandong Province (ZR2021ZD15).

References

1. Arora, S., Karakostas, G.: A $2 + \epsilon$ approximation algorithm for the k-MST problem. In: Proceedings of the 11th Annual ACM-SIAM Symposium on Discrete Algorithms (SODA), pp. 754–759 (2000)
2. Awerbuch, B., Azar, Y., Blum, A., Vempala, S.: Improved approximation guarantees for minimum-weight k-trees and prize-collecting salesmen. SIAM J. Copmput. **28**(1), 254–262 (1998)
3. Bansal, N., Blum, A., Chawla, S., Meyerson, A.: Approximation algorithms for deadline-tsp and vehicle routing with time-windows. In: Proceedings of the 36th Annual ACM Symposium on Theory of Computing (STOC), pp. 166–174 (2004)
4. Blesa, M.J., Xhafa, F.: A C++ implementation of tabu search for k-cardinality tree problem based on generic programming and component reuse. In: Proceedings of Net. ObjectDays Forum, pp. 648–652. Erfurt, Germany (2000)
5. Blum, A., Chawla, S., Karger, D.R., Lane, T., Meyerson, A., Minkoff, M.: Approximation algorithms for orienteering and discounted-reward TSP. SIAM J. Comput. **37**(2), 653–670 (2007)
6. Blum, A., Ravi, R., Vempala, S.: A constant-factor approximation algorithm for the k-MST problem. In: Proceedings of the Annual ACM Symposium on Theory of Computing (STOC), pp. 442–448 (1996)
7. Blum, C., Pinacho, P., López-Ibáñez, M., Lozano, J.: Construct, merge, solve & adapt a new general algorithm for combinatorial optimization. Comput. Oper. Res. **68**, 75–88 (2016)
8. Chaudhuri, K., Godfrey, B., Rao, S., Talwar, K.: Paths, trees, and minimum latency tours. In: Proceedings of the 44th IEEE Annual Symposium on Foundations of Computer Science (FOCS), pp. 36–45 (2003)
9. Chekuri, C., Korula, N., Pál, M.: Improved algorithms for orienteering and related problems. ACM Trans. Algorithms **8**(3), 23:1–23:27 (2012)

10. Chekuri, C., Pal, M.: A recursive greedy algorithm for walks in directed graphs. In: Proceedings of the 46th annual IEEE Smposium on Foundations of Computer Science (FOCS), pp. 245–253 (2005)
11. Ehrgott, M., Freitag, J., Hamacher, H., Maffioli, F.: Heuristics for the k-cardinality tree and subgraph problems. Asia-Pacific J. Oper. Res. **14**(1), 87–114 (1997)
12. Garg, N.: Saving an epsilon: a 2-approximation for the k-MST problem in graphs. In: Proceedings of the 37th Annual ACM Symposium on Theory of Computing (STOC), pp. 302–309 (2005)
13. Johnson, D.S., Minkoff, M., Phillips, S.: The prize collecting steiner tree problem: theory and practice. In: Proceedings of the 11th ACM-SIAM Symposium on Discrete Algorithms (SODA), pp. 760–769 (2000)
14. Lewis, R., Thiruvady, D., Morgan, K.: Finding happiness: an analysis of the maximum happy vertices problem. Comput. Oper. Res. **103**, 265–276 (2019)
15. Pinacho-Davidson, P., Bouamama, S., Blum, C.: Application of CMSA to the minimum capacitated dominating set problem. In: Proceedings of the Genetic and Evolutionary Computation Conference, pp. 321–328 (2019)
16. Ravi, R., Sundaram, R., Marathe, M.V., Rosenkrantz, D.J., Ravi, S.S.: Spanning trees short or small. SIAM J. Discret. Math. **9**(2), 178–200 (1996)
17. Thiruvady, D., Lewis, R.: Recombinative approaches for the maximum happy vertices problem. Swarm Evolution. Comput. **75**, 101188:1–101188:14 (2020)
18. Traub, V., Zenklusen, R.: A better-than-2 approximation for weighted tree augmentation. In: Proceedings of the 62th Annual IEEE Symposium on Foundations of Computer Science (FOCS 2021), pp. 1–12 (2022)

Efficient Algorithms for k-Submodular Function Maximization with p-System and d-Knapsack Constraint

Wenzhe Zhang, Shufang Gong, and Bin Liu[✉][ID]

School of Mathematical Sciences, Ocean University of China, Qingdao 266100, China
binliu@ouc.edu.cn

Abstract. The k-submodular function is a generalization of the submodular function. The k-submodular optimization problems have important applications in influence maximization problems and sensor placement problems with k kinds of sensors. In this paper, we study the problems of maximizing k-submodular functions subject to two kinds of constraints. We set $\alpha = 2$ when f is monotone and $\alpha = 3$ when f is non-monotone. For the p-system constraint, we get a $\frac{1-\epsilon}{p+\alpha}$-approximation ratio. For the intersection of p-system and d-knapsack constraints, we get an approximation ratio of $\frac{1-\epsilon}{p+\alpha+2d}$. And subsequently, we propose an improved algorithm that improves the approximation ratio to $\frac{1-\epsilon}{p+\alpha+\frac{1+\sqrt{5}}{2}d}$.

Keywords: k-submodular maximization · p-system constraint · d-knapsack constraint · Approximation algorithm

1 Introduction

Submodular functions have important applications in many combinatorial optimization problems, such as sensor placement [11,18], influence maximization [10,13], etc. Given a finite set $\mathcal{N}$, a set function $f : \mathcal{N} \to \mathbb{R}_+$ is *submodular* if $f(A \cup \{e\}) - f(A) \geq f(B \cup \{e\}) - f(B)$, for any $A \subseteq B \subseteq \mathcal{N}, e \in \mathcal{N} \setminus B$. An equivalent definition of the submodular function is $f(A) + f(B) \geq f(A \cap B) + f(A \cup B)$ for any subset $A \subseteq \mathcal{N}, B \subseteq \mathcal{N}$.

In recent years, the problem of maximizing k-submodular functions has attracted much attention due to its widely applications in influence maximization with k kinds of topics [17], feature recognition [24] and information coverage maximization [21], etc. Given a finite set $\mathcal{N}$ and an integer k, we define $[k] = \{1, 2 \ldots, k\}$ and the set of k-sets $(k+1)^{\mathcal{N}} = \{(X_1, \ldots, X_k) | X_i \subseteq \mathcal{N}, i \in [k], X_i \cap X_j = \emptyset, i \neq j\}$. A function $f : (k+1)^{\mathcal{N}} \to \mathbb{R}_+$ is called *k-submodular* [7] if

$$f(\mathbf{x}) + f(\mathbf{y}) \geq f(\mathbf{x} \sqcap \mathbf{y}) + f(\mathbf{x} \sqcup \mathbf{y}),$$

This work was supported in part by the National Natural Science Foundation of China (11971447), and the Fundamental Research Funds for the Central Universities (202261097).

where
$$\mathbf{x} \sqcap \mathbf{y} = (X_1 \cap Y_1, \ldots, X_k \cap Y_k),$$

$$\mathbf{x} \sqcup \mathbf{y} = (X_1 \cup Y_1 \setminus (\bigcup_{i \neq 1} X_i \cup Y_i), \ldots, X_k \cup Y_k \setminus (\bigcup_{i \neq k} X_i \cup Y_i)).$$

And a k-submodular function f is *monotone* if

$$f(X_1, \ldots, X_i \cup \{e\}, \ldots, X_k) - f(X_1, \ldots, X_i, \ldots, X_k) \geq 0,$$

for any $\mathbf{x} \in (k+1)^{\mathcal{N}}$, $i \in [k]$, and $e \notin \bigcup_{1 \leq j \leq k} X_j$.

The k-submodular function is an extension of the submodular function, which is the submodular function when $k = 1$. And the k-submodular maximization problems are also NP-hard [29] . In this paper, we consider the problems of maximizing k-submodular functions subject to two kinds of constraints. Given a monotone or non-monotone k-submodular function $f : (k+1)^{\mathcal{N}} \to \mathbb{R}_+$, and $f(\emptyset, \emptyset, \ldots, \emptyset) = 0$, we consider

> Problem (1) : $\{\max f(\mathbf{x}) | \mathbf{x} \in (k+1)^{\mathcal{N}}, \ supp(\mathbf{x}) \in \mathcal{I}\}$;
>
> Problem (2) : $\{\max f(\mathbf{x}) | \mathbf{x} \in (k+1)^{\mathcal{N}}, \ supp(\mathbf{x}) \in \mathcal{I}, c_i(\mathbf{x}) \leq 1, 1 \leq i \leq d\}$.

Let $\mathcal{I} \subseteq 2^{\mathcal{N}}$, we call the pair $(\mathcal{N}, \mathcal{I})$ is a *p-system* if (1) $\emptyset \in \mathcal{I}$; (2) if $B \in \mathcal{I}$ and $A \subseteq B$, we have $A \in \mathcal{I}$; (3) for every $S \subseteq \mathcal{N}$, the ratio between the sizes of the largest base of S and the smallest base of S is at most p. And d-knapsack constraint means that there are d functions $c_1, \ldots, c_d$ and d positive values $B_1, \ldots, B_d$. We assume that $B_i = 1$ for every integer $1 \leq i \leq d$. This can be guaranteed by scaling the cost functions c_i.

Related work
The submodular maximization problem with cardinality constraint can be approximated by a simple greedy algorithm [15] to obtain the ratio of $1 - \frac{1}{e}$, which is the best guarantee one can hope for. Then Badanidiyuru et al. [2] reduced the time complexity from $O(n^2)$ to $O(\frac{n}{\epsilon} \log \frac{n}{\epsilon})$. Due to the phenomenon of big data, the computer cannot store all elements. In this case, Badanidiyuru et al. [1] proposed the streaming algorithm for the cardinality constrained problem and obtained the $(\frac{1}{2} - \epsilon)$-approximation ratio with at most $O(\frac{k}{\epsilon} \log k)$ space. Later on, Kazemi et al. [9] reduced the space complexity to $O(\frac{k}{\epsilon})$. Buchbinder et al. [3] proposed an online $\frac{1}{4}$-approximation algorithm with $O(k)$ memory. For the monotone submodular maximization problem under p-system constraint, Fisher et al. [5] got a $\frac{1}{p+1}$-approximation ratio. For the non-monotone submodular maximization problem under p-system constraint, Gupta et al. [6] proposed a $\frac{p}{(p+1)(3p+3)}$-approximation algorithm. Then Mirzasoleiman et al. [14] increased the approximation ratio to $\frac{p}{(p+1)(2p+1)}$. For the intersection of p-system constraint and d-knapsack constraints, Badanidiyuru and Vondrák [2] used the threshold technique to get a $\frac{1}{p+2d+1}$-approximation ratio. Then Li et al. [12] increased the approximation ratio to $\frac{1+O(\epsilon)}{p+\frac{4}{7}d+1}$.

In recent years, there are extensive research on the k-submodular maximization problems. For the unconstrained problem, Ward and Živný [28] studied this problem and got a $\frac{1}{3}$-approximation ratio. Iwata et al. [8] designed the stochastic algorithm to improve the approximation ratio to $\frac{k}{2k-1}$. Then, Oshima [19] eliminated the random told in Iwata et al. [8] but it increased the time complexity. For the total size constraint, Ohsaka and Yoshida [17] proposed a greedy algorithm for the monotone case and obtained an approximation ratio of $\frac{1}{2}$. Qian et al. [21] proposed a random multi-objective evolutionary algorithm for the monotone k-submodular maximization problem and obtained the same approximate ratio in expectation. Zheng et al. [32] studied the problem of maximizing approximately k-submodular functions subject to the total size constraint and got an approximate ratio of $\frac{(1-\epsilon)^2}{2(1-\epsilon+\epsilon b)(1+\epsilon)}$. Nguyen et al. [16] proposed the streaming algorithm, then obtained a $(\frac{1}{3}-\epsilon)$-approximation ratio when f is monotone and a $(\frac{1}{4}-\epsilon)$-approximation ratio when f is non-monotone. For the individual size constraint, Ohsaka and Yoshida [17] proposed a greedy algorithm for monotone case and obtained an approximation ratio of $\frac{1}{3}$. Then, Alina and Nguyen [4] proposed a streaming algorithm for this problem and got a $\frac{1}{4}$ approximation ratio.

For the knapsack constraint, Zhang et al. [31] studied the k-submodular maximization problem under the individual budget constraint, and got a $\frac{1}{5}(1-\frac{1}{e})$-approximation ratio with $O(n^2 k)$ query complexity. Tang et al. [26] studied the k-submodular maximization problem under the total budget constraint, and got a $\frac{1}{2}(1-\frac{1}{e})$-approximation ratio with $O(n^4 k^3)$ query complexity. Then, Wang and Zhou [27] improved the approximation ratio from $\frac{1}{2}(1-\frac{1}{e})$ to $\frac{1}{2}-\epsilon$. Pham et al. [20] proposed the streaming algorithm, then obtained an approximation ratio of $(\frac{1}{4}-\epsilon)$ when f is monotone and $(\frac{1}{5}-\epsilon)$ approximation ratio when f is non-monotone.

For the matroid constraint, Sakaue et al. [23] proposed a greedy algorithm with an approximation ratio of $\frac{1}{2}$ for the monotone k-submodular function. Rafiey and Yoshida [22] improved the time complexity of Sakaue from $O(nkr)$ to $O(nk \ln r \frac{lnr}{\epsilon})$. For the non-monotone k-submodular maximization problem under the matroid constraint, Li et al. [25] got a $\frac{1}{3}$-approximation algorithm with a high probability. Then, Li et al. [30] also studied monotone and no-monotone k-submodular maximization problem with the intersection of a knapsack and m matroid constraints. And they got a $\frac{1-e^{-(m+2)}}{m+2}$-approximation ratio when f is monotone, a $\frac{1-e^{-(m+3)}}{m+3}$-approximation ratio when f is non-monotone. However, their algorithms may not be polynomial time since the number of greedy swaps cannot be bounded polynomially.

Our Contribution. The main contributions of this paper are as follows.

• In this paper, we propose the greedy algorithm for k-submodular maximization under the p-system constraint. For this constraint, we get a $\frac{1}{p+2}$-approximation ratio when f is monotone and a $\frac{1}{p+3}$-approximation ratio when f is non-monotone.

• We propose a threshold algorithm for k-submodular maximization under the intersection of p-system and d-knapsack constraint to get a $\frac{1-\epsilon}{p+2+2d}$-approximation when f is monotone and a $\frac{1-\epsilon}{p+3+2d}$-approximation ratio when f is non-monotone. We then propose an improved algorithm to improve the approximation ratio to $\frac{1-\epsilon}{p+2+\frac{1+\sqrt{5}}{2}d}$ when f is monotone and $\frac{1-\epsilon}{p+3+\frac{1+\sqrt{5}}{2}d}$ when f is non-monotone. Compared to [30], the problem we study has a wider range of constraints and the algorithmic time is polynomial.

2 Preliminaries

In this section, we will introduce some symbols and basic properties of the k-submodular function. We let $|\mathcal{N}| = n$, $\mathbf{0} = (\emptyset, \emptyset, \ldots, \emptyset)$, and (e, i) be a vector whose i-th position is the set of element e and the rest of the positions are the empty set. For any $\mathbf{x} = (X_1, \ldots, X_k) \in (k+1)^{\mathcal{N}}$, let $supp(\mathbf{x}) = \bigcup_{1 \leq i \leq k} X_i$. For every $e \in supp(\mathbf{s})$, $\mathbf{s}(e)$ denotes the position of element e in $\mathbf{s}$. $\triangle_{e,i} f(\mathbf{x}) = f(\mathbf{x} \sqcup (e, i)) - f(\mathbf{x})$. And $\triangle_{e,i_e} f(\mathbf{s})$ represents benefits from adding element e to the position that produces the greatest marginal benefit. For any $\mathbf{x} = (X_1, \ldots, X_k), \mathbf{y} = (Y_1, \ldots, Y_k) \in (k+1)^{\mathcal{N}}$, we define the partial order relationship $\mathbf{x} \preceq \mathbf{y}$ if $X_i \subseteq Y_i, i \in [k]$. And for any k-sets $\mathbf{x}$ and $\mathbf{y}$, we define $\mathbf{y} \setminus \mathbf{x}$ as follows. Let $\mathbf{y} \setminus \mathbf{x} \preceq \mathbf{y}$ and $supp(\mathbf{y} \setminus \mathbf{x}) = supp(\mathbf{y}) \setminus supp(\mathbf{x})$.

Definition 1. [18] A function $f : (k+1)^{\mathcal{N}} \to \mathbb{R}_+$ is orthant submodular if

$$\triangle_{e,i} f(\mathbf{x}) \geq \triangle_{e,i} f(\mathbf{y})$$

for any $\mathbf{x}, \mathbf{y} \in (k+1)^{\mathcal{N}}$, $i \in [k]$, $\mathbf{x} \preceq \mathbf{y}$, $e \notin supp(\mathbf{y})$.

Definition 2. [18] A function $f : (k+1)^{\mathcal{N}} \to \mathbb{R}_+$ is pairwise monotone if

$$\triangle_{e,i} f(\mathbf{x}) + \triangle_{e,j} f(\mathbf{x}) \geq 0.$$

for any $\mathbf{x} \in (k+1)^{\mathcal{N}}$, $e \notin supp(\mathbf{x})$, $i, j \in [k]$, $i \neq j$.

Proposition 1. *[29] A function f is k-submodular if and only if f is orthant submodular and pairwise monotone.*

Definition 3. We say that the k-sets $\mathbf{z}_1, \ldots, \mathbf{z}_t$ is a partition of a k-set $\mathbf{x}$ if $\mathbf{z}_1, \ldots, \mathbf{z}_t \preceq \mathbf{x}$, and $\bigcup_{i=1}^{t} supp(\mathbf{z}_i) = supp(\mathbf{x})$, $supp(\mathbf{z}_i) \cap supp(\mathbf{z}_j) = \emptyset$, for any $1 \leq i, j \leq t, i \neq j$.

In order to analyze algorithms, we define the following notation. Let (e^j, i^j) be the j-th element added to the output solution of the algorithm. And $\mathbf{s}^j = \{(e^1, i^1), \ldots, (e^j, i^j)\}$ is the solution when adding j elements in the main loop of the algorithm. Let $\mathbf{o}$ be an optimal solution, $\mathbf{o}^j = (\mathbf{o} \sqcup \mathbf{s}^j) \sqcup \mathbf{s}^j$ and $\mathbf{o}^{j-1/2} = (\mathbf{o} \sqcup \mathbf{s}^j) \sqcup \mathbf{s}^{j-1}$. Clearly $\mathbf{o}^0 = \mathbf{o}$. For $\mathbf{s}^{j-1/2}$, we define it in two cases. If $e^j \in supp(\mathbf{o})$, $\mathbf{s}^{j-1/2} = \mathbf{s}^{j-1} \sqcup (e^j, \mathbf{o}(e^j))$. Else, $\mathbf{s}^{j-1/2} = \mathbf{s}^{j-1}$. By the above definitions, we can get $\mathbf{s}^j \preceq \mathbf{o}^{j-1/2} \preceq \mathbf{o}^j$.

3 k-Submodular Maximization Under the p-System Constraint

In this section, we propose the greedy algorithm for Problem (1), i.e. k-submodular maximization under the p-system constraint. The algorithm is as follows.

Algorithm 1. Greedy algorithm

Require: a k-submodular function f, a p-system $(\mathcal{N}, \mathcal{I})$
Ensure: an approximation solution $\mathbf{s}$.
 1: $\mathbf{s}_0 \leftarrow (\emptyset, \emptyset, \ldots, \emptyset)$, $i = 0$
 2: **while** $\mathcal{N} \neq \emptyset$ **do**
 3:      $(e, j) \leftarrow \arg\max_{e \in \mathcal{N}, j \in [k]} \triangle_{e,j} f(\mathbf{s}_i)$
 4:      **if** $supp(\mathbf{s}_i) + e \in \mathcal{I}$ **then**
 5:          $\mathbf{s}_{i+1} \leftarrow \mathbf{s}_i \sqcup (e, j)$, $i = i + 1$
 6:      **end if**
 7:      $\mathcal{N} \leftarrow \mathcal{N} \setminus \{e\}$
 8: **end while**
 9: **return** $\mathbf{s}$

Lemma 1. *For any* $\mathbf{x}, \mathbf{y} \in (k+1)^{\mathcal{N}}$, $\mathbf{x} \preceq \mathbf{y}$, *and* $\mathbf{z}_1, \ldots, \mathbf{z}_t$ *is a partition of* $\mathbf{y} \setminus \mathbf{x}$. *We have*

$$f(\mathbf{y}) - f(\mathbf{x}) \leq \sum_{1 \leq i \leq t} f(\mathbf{x} \sqcup \mathbf{z}_i) - f(\mathbf{x}).$$

Proof.

$$f(\mathbf{y}) - f(\mathbf{x}) = \sum_{1 \leq i \leq t} f(\mathbf{x} \sqcup \mathbf{z}_1 \sqcup \ldots \sqcup \mathbf{z}_i) - f(\mathbf{x} \sqcup \mathbf{z}_1 \sqcup \ldots \sqcup \mathbf{z}_{i-1})$$

$$\leq \sum_{1 \leq i \leq t} f(\mathbf{x} \sqcup \mathbf{z}_i) - f(\mathbf{x}).$$

$\square$

Due to page limitations, we have placed some of the proofs of this section in Appendix A.

Lemma 2. *Let* $\mathbf{o}$ *be an optimal solution of Problem* (1), $\mathbf{s}_l$ *be the output solution of Algorithm 1 and* $\mathbf{c} = \mathbf{o} \setminus \mathbf{s}_l \neq (\emptyset, \emptyset, \ldots, \emptyset)$. *Then there exists a partition of* $\mathbf{c}$: $\mathbf{c}_1, \mathbf{c}_2, \ldots, \mathbf{c}_l$ *such that:*
(1) for all $i \in [k]$, $p_i \leq p$, *where* $p_i = |supp(\mathbf{c}_i)|$;
(2) for all $i \in [k]$, $p_i[f(\mathbf{s}_i) - f(\mathbf{s}_{i-1})] \geq f(\mathbf{s}_l \sqcup \mathbf{c}_i) - f(\mathbf{s}_l)$.

Lemma 3. *Suppose that* $\mathbf{s}^t (1 \leq t \leq l)$ *is the current solution after t iterations of Algorithm 1, and f is a monotone k-submodular function. We have*

$$f(\mathbf{o}) - f(\mathbf{o}^t) \leq f(\mathbf{s}^t).$$

Lemma 4. *Suppose that* $s^t (1 \leq t \leq l)$ *is the current solution after* t *iterations of Algorithm 1, and* f *is a non-monotone* k-*submodular function. We have*

$$f(o) - f(o^t) \leq 2f(s^t).$$

Theorem 1. *Algorithm 1 has* $O(n^2 k)$ *query complexity and outputs a* $\frac{1}{p+2}$-*approximation solution when* f *is monotone, and a* $\frac{1}{p+3}$-*approximation solution when* f *is non-monotone.*

4 k-Submodular Maximization Under the Intersection of p-System and d-Knapsack Constraints

In this section, we propose the threshold greedy algorithm for Problem (2), i.e. k-submodular maximization under the intersection of p-system and d-knapsack constraints. The algorithm is as follows.

Algorithm 2. Threshold Greedy Algorithm With Known ρ

Require: a k-submodular function f, a p-system $(\mathcal{N}, \mathcal{I})$, knapsack-cost functions
$c_i, 1 \leq i \leq d$ and a threshold ρ
Ensure: an approximation solution s.
1: $(e_{max}, j_{max}) \leftarrow \arg\max_{e \in \mathcal{N}, j \in [k]} \triangle_{e,j} f(\mathbf{0})$
2: $s_0 \leftarrow (\emptyset, \emptyset, \ldots, \emptyset), i = 0$
3: **while** $\mathcal{N} \neq \emptyset$ **do**
4:     **if** $\frac{\triangle_{e,j} f(s_i)}{\sum_{1 \leq i \leq d} c_i(e)} \geq \rho$ **then**
5:         **if** $supp(s_i) + e \in \mathcal{I}$, and $\max_{1 \leq t \leq d} c_t(s_i) + c_t(e) \leq 1$
6:             $s_{i+1} \leftarrow s_i \sqcup (e, j)$, $i = i + 1$ **then**
7:         **end if**
8:         **if** $supp(s_i) + e \in \mathcal{I}$ and $\max_{1 \leq t \leq d} c_t(s_i) + c_t(e) \geq 1$ **then**
9:             return $\arg\max\{f(s_i), f((e_{max}, j_{max}))\}$
10:         **end if**
11:     **end if**
12:     $\mathcal{N} \leftarrow \mathcal{N} \setminus \{e\}$
13: **end while**
14: return $\arg\max\{f(s_i), f((e_{max}, j_{max}))\}$

Due to page limitations, we have placed some of the proofs of this section in Appendix B.

Theorem 2. *Algorithm 2 returns a solution* s *that satisfies* $f(s) \geq \min\{\frac{1}{2}\rho, \frac{f(o) - \rho d}{p+2}\}$ *when* f *is monotone and* $f(s) \geq \min\{\frac{1}{2}\rho, \frac{f(o) - \rho d}{p+3}\}$ *when* f *is non-monotone, where* ρ *can be determined later.*

Next, we need to find a suitable threshold value ρ. For a monotone function f, let $\frac{1}{2}\rho = \frac{f(o) - \rho d}{p+2}$. Then, we get a threshold $\rho = \frac{2f(o)}{p+2+2d}$. Similarly, we can get a threshold $\rho = \frac{2f(o)}{p+3+2d}$ when f is non-monotone. Therefore, we propose Algorithm 3 to enumerate the thresholds.

Algorithm 3. Threshold Greedy Algorithm

Require: a k-submodular function f, a p-system $(\mathcal{N}, \mathcal{I})$, knapsack-cost functions
  $c_i, 1 \leq i \leq d$ and a parameter ϵ
Ensure: an approximation solution $\mathbf{s}$.
1: $M \leftarrow \arg\max_{e \in \mathcal{N}, j \in [k]} f((e, j))$
2: **if** f is monotone **then**
3:     $\alpha = 2$
4: **else**
5:     $\alpha = 3$
6: **end if**
7: $\Lambda = \{j \in \mathbb{N}_+ | \frac{M}{1+\epsilon} \leq (1+\epsilon)^j \leq nM\}$
8: **for** $j \in \Lambda$ **do**
9:     $\mathbf{s}_j \leftarrow$ Algorithm 2 $(f, (\mathcal{N}, \mathcal{I}), c_i, \frac{2(1+\epsilon)^j}{p+\alpha+2d})$
10: **end for**
11: **return** $\arg\max\{f(\mathbf{s}_j) | j \in \Lambda\}$

Theorem 3. *Algorithm 3 has $O(n^2 k \frac{\log n}{\epsilon})$ query complexity and provides a $\frac{1-\epsilon}{p+2+2d}$-approximation ratio if f is monotone, and a $\frac{1-\epsilon}{p+3+2d}$-approximation ratio if f is non-monotone.*

Proof. For the monotone case, let $j = \lfloor \log_{1+\epsilon} f(o) \rfloor$. Then we have $\frac{2(1-\epsilon)f(o)}{p+2+2d} \leq \frac{2(1+\epsilon)^j}{p+\alpha+2d} \leq \frac{2f(o)}{p+2+2d}$. By Theorem 2, we have

$$f(\mathbf{s}) \geq \min\{\frac{1}{2}\rho, \frac{f(o) - \rho d}{p+2}\} \geq \frac{1-\epsilon}{p+2+2d}f(o).$$

For the non-monotone case, we have $\frac{2(1-\epsilon)f(o)}{p+3+2d} \leq \frac{2(1+\epsilon)^j}{p+\alpha+2d} \leq \frac{2f(o)}{p+3+2d}$. And by Theorem 2, we have

$$f(\mathbf{s}) \geq \min\{\frac{1}{2}\rho, \frac{f(o) - \rho d}{p+3}\} \geq \frac{1-\epsilon}{p+3+2d}f(o).$$

$\square$

5 Improved Algorithm for k-Submodular Maximization Under the Intersection of p-System and d-Knapsack Constraints

In this section we propose an improved algorithm for k-submodular maximization under the intersection of p-system and d-knapsack constraint. Unlike the previous algorithms, we filter out some "large" set of elements ($\mu \in \mathcal{N}$ is called "large" element if for some $1 \leq i \leq d, c_i(\mu) > \frac{1}{\lambda}$) and output a feasible solution from them. In the next operation, we only need to deal with the "small" set of elements ($\mu \in \mathcal{N}$ is called "small" element if for any $1 \leq i \leq d$, $c_i(\mu) \leq \frac{1}{\lambda}$) and output a better solution among them.

For the set of the "small" element, we proceed as follows. For each element $\mu \in supp(\mathbf{s}_{j+1})$, if it satisfies the d-knapsack constraint, it is added to the set $T_t = \{\mu_1, \ldots, \mu_{t-1}\}(1 \leq t \leq \lambda + 1)$. Until the knapsack constraint is violated when the element μ is added, then μ is noted as μ_t. And at the end, the set with the largest weight $\sum_{1 \leq i \leq d} c_i(T)$ is selected and the position of the corresponding element corresponds with $\mathbf{s}_{j+1}$. The algorithm is as follows.

Algorithm 4. Improved Threshold Greedy Algorithm With Known ρ

Require: a k-submodular function f, a p-system $(\mathcal{N}, \mathcal{I})$, knapsack-cost functions $c_i, 1 \leq i \leq d$ and parameters $\lambda > 1, \epsilon$

Ensure: an approximation solution $\mathbf{s}$.

1: $B \leftarrow \{\mu \in \mathcal{N} | \exists 1 \leq i \leq d, c_i(\mu) > \frac{1}{\lambda}\}$
2: $\mathbf{s}_B \leftarrow \arg\max_{e \in B, i \in [k]} f((e, i))$
3: $\mathbf{s}_0 \leftarrow (\emptyset, \emptyset, \ldots, \emptyset), j \leftarrow 0$
4: **while** there exists an element $\mu \in \mathcal{N} \setminus (supp(\mathbf{s}_j) \cup B)$ such that $supp(\mathbf{s}_j) + \mu \in \mathcal{I}$ and $\Delta_{\mu, i_\mu} f(\mathbf{s}_j) \geq \rho \sum_{1 \leq i \leq d} c_i(\mu)$ **do**
5:     Let v_{j+1} be an element maximizing $\Delta_{\mu, i_\mu} f(\mathbf{s}_j)$ among all the elements
6:     $\mathbf{s}_{j+1} \leftarrow \mathbf{s}_j \cup (v_{j+1}, i_{v_{j+1}})$
7:     **if** $\max_{1 \leq i \leq d} c_i(\mathbf{s}_{j+1}) \leq 1$ **then**
8:         $j \leftarrow j + 1$
9:     **else**
10:         return the output of Extract Algorithm $(\lambda, \mathbf{s}_{j+1})$
11:     **end if**
12: **end while**
13: return $\mathbf{s} = \arg\max\{f(\mathbf{s}_B), f(\mathbf{s}_j)\}$

We place the Extract Algorithm and some of the proofs of this section in Appendix C. According to the SetExtract Algorithm proposed by Li et al. [12], we get the following lemma.

Lemma 5. *[12] If Algorithm 4 returns in Line 10, then* $\max_{1 \leq i \leq d} c_i(\mathbf{s}) \leq 1$, *and* $\sum_{1 \leq i \leq d} c_i(\mathbf{s}) \geq \frac{\lambda}{\lambda+1}$.

Lemma 6. *Let E be the event that Algorithm 4 returns through Line 10. If the event E happens, then Algorithm 5 returns a solution $\mathbf{s}$ that satisfies* $f(\mathbf{s}) \geq \frac{\lambda \rho}{\lambda+1}$.

Lemma 7. *Assume E does not happen, let l be the final value of the variable j of Algorithm 4. Let $\mathbf{o}_l \preceq \mathbf{o}$ and $supp(\mathbf{o}_l) = supp(\mathbf{o}) \setminus (B \cup \{\mu \in supp(\mathbf{o}) \setminus supp(\mathbf{s}_l) | \Delta_{\mu, i_\mu} f(\mathbf{s}_l) < \rho \sum_{1 \leq i \leq d} c_i(\mu)\})$. If $\mathbf{c} = \mathbf{o}_l \setminus \mathbf{s}_l$, we have*

$$f(\mathbf{s}_l) \geq \frac{f(\mathbf{s}_l \sqcup \mathbf{c})}{p + 1}.$$

The proof is similar to Lemma 2.

Lemma 8. *If f is monotone, we have*

$$f(s_l \sqcup c) \geq f(o) - f(b) - \rho d - f(s_l) + \frac{\rho |supp(b)|}{\lambda}.$$

If f is non-monotone, we have

$$f(s_l \sqcup c) \geq f(o) - f(b) - \rho d - 2f(s_l) + \frac{\rho |supp(b)|}{\lambda}.$$

Lemma 9. *When event E does not happen, Algorithm 4 gets a solution s. If f is monotone, we have*

$$f(s) \geq \frac{f(o) - \rho d + \frac{\rho}{\lambda} |supp(b)|}{p + 2 + |supp(b)|}.$$

If f is non-monotone, we have

$$f(s) \geq \frac{f(o) - \rho d + \frac{\rho}{\lambda} |supp(b)|}{p + 3 + |supp(b)|}.$$

Next, we need to find a suitable threshold value ρ. For a monotone function f, let $\frac{f(o) - \rho d + \frac{\rho}{\lambda} |supp(b)|}{p + 2 + |supp(b)|} = \frac{\lambda \rho}{\lambda + 1}$. Then, we get a threshold

$$\rho = \frac{(\lambda + 1) f(o)}{(\lambda + 1)d - \frac{\lambda + 1}{\lambda} |supp(b)| + \lambda(p + 2 + |supp(b)|)}.$$

Similarly, we can get a threshold

$$\rho = \frac{(\lambda + 1) f(o)}{(\lambda + 1)d - \frac{\lambda + 1}{\lambda} |supp(b)| + \lambda(p + 3 + |supp(b)|)}.$$

when f is non-monotone.

Lemma 10. *When f is monotone, let $\rho = \frac{(\lambda+1)f(o)}{(\lambda+1)d - \frac{\lambda+1}{\lambda}|supp(b)| + \lambda(p+2+|supp(b)|)}$. And we set $\lambda = \frac{1+\sqrt{5}}{2}$, Algorithm 4 can get a $\frac{1}{p+2+\frac{1+\sqrt{5}}{2}d}$-approximation ratio. Similarly, Algorithm 4 can get a $\frac{1}{p+3+\frac{1+\sqrt{5}}{2}d}$-approximation ratio when f is non-monotone.*

When $\lambda = \frac{1+\sqrt{5}}{2}$, the value of ρ is only relevant to the optimal solution. Therefore, we can use a technique proposed by Badanidiyuru et al. [1] to estimate the optimal value using the following algorithm.

Algorithm 5. Improved Threshold Greedy Algorithm

Require: a k-submodular function f, a p-system $(\mathcal{N}, \mathcal{I})$, knapsack-cost functions
  $c_i, 1 \le i \le d$ and a parameter ϵ

Ensure: an approximation solution **s**.

1: $M \leftarrow \arg\max_{e \in \mathcal{N}, j \in [k]} f(e, j)$
2: **if** f is monotone **then**
3:    $\alpha = 2$
4: **else**
5:    $\alpha = 3$
6: **end if**
7: $\Lambda = \{ j \in \mathbb{N}_+ | \frac{M}{1+\epsilon} \le (1+\epsilon)^j \le nM \}$
8: **for** $j \in \Lambda$ **do**
9:    $\mathbf{s}_j \leftarrow$ Algorithm 4 $(f, (\mathcal{N}, \mathcal{I}), c_i, \frac{(\sqrt{5}+3)(1+\epsilon)^j}{(\sqrt{5}+3)d+(\sqrt{5}+1)(p+\alpha)})$
10: **end for**
11: **return** $\arg\max\{f(\mathbf{s}_j) | j \in \Lambda\}$

Theorem 4. *Algorithm 6 has $O(\frac{\log n}{\epsilon} nk(n + d))$ time complexity and provides a $\frac{1-\epsilon}{p+2+\frac{1+\sqrt{5}}{2}d}$-approximation ratio if f is monotone, and a $\frac{1-\epsilon}{p+3+\frac{1+\sqrt{5}}{2}d}$-approximation ratio if f is non-monotone.*

Proof. Let $j = \lfloor \log_{1+\epsilon} f(\mathbf{o}) \rfloor$. Then we have

$$\frac{(\sqrt{5}+3)(1-\epsilon)f(\mathbf{o})}{(\sqrt{5}+3)d+(\sqrt{5}+1)(p+\alpha)} \le \frac{(\sqrt{5}+3)(1+\epsilon)^j}{(\sqrt{5}+3)d+(\sqrt{5}+1)(p+\alpha)} \le \frac{(\sqrt{5}+3)f(\mathbf{o})}{(\sqrt{5}+3)d+(\sqrt{5}+1)(p+\alpha)}.$$

By Lemma 10, we have

$$f(\mathbf{s}) \ge \frac{1-\epsilon}{p+\alpha+\frac{1+\sqrt{5}}{2}d} f(\mathbf{o}) \ge \frac{1-\epsilon}{p+\alpha+2d} f(\mathbf{o}).$$

$\square$

6   Conclusion

In this paper, we mainly study maximizing k-submodular functions subject to two kinds of constraints, Problem (1) and Problem (2). Problem (1) is maximizing a k-submodular function with the p-system constraint. And we design a greedy algorithm for this problem. Problem (2) is maximizing a k-submodular function with the intersection of p-system and d-knapsack constraints. We utilize the greedy and threshold ideas to propose a deterministic algorithm. Furthermore, we improve the approximation guarantee of problem (2). In the future, one can investigate other constraints or further improve the algorithm performance.

References

1. Badanidiyuru, A., Mirzasoleiman, B., Karbasi, A., Krause, A.: Streaming submodular maximization: massive data summarization on the fly. In: 20th International Proceedings on KDD, pp. 671–680. ACM, New York, NY, USA (2014)
2. Badanidiyuru, A., Vondrák, J.: Fast algorithms for maximizing submodular functions. In: 25th International Proceedings on SODA, pp. 1497–1514. SIAM, Portland, Oregon, USA (2014)
3. Buchbinder, N., Feldman, M., Schwartz, R.: Online submodular maximization with preemption. ACM Trans. Algorithms **15**(3), 1–31 (2015)
4. Ene, A., Nguyen, H.L.: Streaming algorithm for monotone k-submodular maximization with cardinality constraints. In: 39th International Proceedings on ICML, pp. 5944–5967. PMLR, Baltimore, Maryland, USA (2022)
5. Fisher, M.L., Nemhauser, G.L., Wolsey, L.A.: An analysis of approximations for maximizing submodular set functions - II. In: Balinski, M.L., Hoffman, A.J. (eds.) Polyhedral Combinatorics. Mathematical Programming Studies, vol. 8. Springer, Berlin, Heidelberg (1978). https://doi.org/10.1007/BFb0121195
6. Gupta, A., Roth, A., Schoenebeck, G., Talwar, K.: Constrained non-monotone submodular maximization: offline and secretary algorithms. In: Saberi, A. (ed.) WINE 2010. LNCS, vol. 6484, pp. 246–257. Springer, Heidelberg (2010). https://doi.org/10.1007/978-3-642-17572-5_20
7. Huber, A., Kolmogorov, V.: Towards minimizing k-submodular functions. In: Mahjoub, A.R., Markakis, V., Milis, I., Paschos, V.T. (eds.) ISCO 2012. LNCS, vol. 7422, pp. 451–462. Springer, Heidelberg (2012). https://doi.org/10.1007/978-3-642-32147-4_40
8. Iwata, S., ichi Tanigawa, S., Yoshida, Y.: Improved approximation algorithms for k-submodular function maximization. In: 27th International Proceedings on SODA, pp. 404–413. SIAM, Arlington, VA, USA (2015)
9. Kazemi, E., Mitrovic, M., Zadimoghaddam, M., Lattanzi, S., Karbasi, A.: Submodular streaming in all its glory: Tight approximation, minimum memory and low adaptive complexity. In: 36th International Proceedings on ICML, pp. 3311–3320. PMLR, Long Beach, California (2019)
10. Kempe, D., Kleinberg, J.M., Tardos, É.: Maximizing the spread of influence through a social network. In: 20th International Proceedings on KDD, pp. 137–146. ACM, Washington, DC, USA (2003)
11. Krause, A., McMahan, H.B., Guestrin, C., Gupta, A.: Robust submodular observation selection. J. Mach. Learn. Res. **9**(12), 2761–2801 (2008)
12. Li, W., Feldman, M., Kazemi, E., Karbasi, A.: Submodular maximization in clean linear time. In: 36th International Proceedings on NeurIPS, pp. 7887–7897. New Orleans, LA, USA (2022)
13. Li, Y., Fan, J., Wang, Y., Tan, K.L.: Influence maximization on social graphs: a survey. IEEE Trans. Knowl. Data Eng. **30**(10), 1852–1872 (2018)
14. Mirzasoleiman, B., Badanidiyuru, A., Karbasi, A.: Fast constrained submodular maximization: personalized data summarization. In: 36th International Proceedings on ICML, pp. 1358–1367. JMLR.org, New York City, NY, USA (2016)
15. Nemhauser, G.L., Wolsey, L.A., Fisher, M.L.: An analysis of approximations for maximizing submodular set functions-i. Math. Program. **14**(1), 265–294 (1978)
16. Nguyen, L.N., Thai, M.T.: Streaming k-submodular maximization under noise subject to size constraint. In: 37th International Proceedings on ICML, pp. 7338–7347. PMLR, Virtual Event (2020)

17. Ohsaka, N., Yoshida, Y.: Monotone k-submodular function maximization with size constraints. In: 28th International Proceedings on NIPS, pp. 7694–702. Montreal, Quebec, Canada (2015)
18. Orlin, J.B., Schulz, A.S., Udwani, R.: Robust monotone submodular function maximization. Math. Program. **172**(1–2), 505–537 (2018)
19. Oshima, H.: Derandomization for k-submodular maximization. In: Brankovic, L., Ryan, J., Smyth, W.F. (eds.) IWOCA 2017. LNCS, vol. 10765, pp. 88–99. Springer, Cham (2018). https://doi.org/10.1007/978-3-319-78825-8_8
20. Pham, C.V., Vu, Q.C., Ha, D.K.T., Nguyen, T.T., Le, N.D.: Maximizing k-submodular functions under budget constraint: applications and streaming algorithms. J. Comb. Optim. **44**(1), 723–751 (2022)
21. Qian, C., Shi, J.C., Tang, K., Zhou, Z.H.: Constrained monotone k-submodular function maximization using multiobjective evolutionary algorithms with theoretical guarantee. IEEE Trans. Evol. Comput. **22**(4), 595–608 (2018)
22. Rafiey, A., Yoshida, Y.: Fast and private submodular and k-submodular functions maximization with matroid constraints. In: 28th International Proceedings on ICML, pp. 7887–7897. PMLR, Virtual Event (2020)
23. Sakaue, S.: On maximizing a monotone k-submodular function subject to a matroid constraint. Discret. Optim. **23**, 105–113 (2017)
24. Singh, A.P., Guillory, A., Bilmes, J.A.: On bisubmodular maximization. In: 15th International Proceedings on AISTATS, pp. 1055–1063. JMLR.org, La Palma, Canary Islands, Spain (2012)
25. Sun, Y., Liu, Y., Li, M.: Maximization of k-submodular function with a matroid constraint. In: Du, DZ., Du, D., Wu, C., Xu, D. (eds.) Theory and Applications of Models of Computation. TAMC 2022. LNCS, vol. 13571. Springer, Cham (2022). https://doi.org/10.1007/978-3-031-20350-3_1
26. Tang, Z., Wang, C., Chan, H.: On maximizing a monotone k-submodular function under a knapsack constraint. Oper. Res. Lett. **50**(1), 28–31 (2022)
27. Wang, B., Zhou, H.: Multilinear extension of k-submodular functions. ArXiv abs/2107.07103 (2021)
28. Ward, J., Zivný, S.: Maximizing bisubmodular and k-submodular functions. In: 16th International Proceedings on SODA, pp. 1468–1481. SIAM, Portland, Oregon, USA (2014)
29. Ward, J., Zivný, S.: Maximizing k-submodular functions and beyond. ACM Trans. Algorithms **12**(4), 1–26 (2014)
30. Yu, K., Li, M., Zhou, Y., Liu, Q.: On maximizing monotone or non-monotone k-submodular functions with the intersection of knapsack and matroid constraints. J. Comb. Optim. **45**(3), 1–21 (2023)
31. Zhang, Y., Li, M., Yang, D., Xue, G.: A budget feasible mechanism for k-topic influence maximization in social networks. In: 19th International Proceedings on GLOBECOM, pp. 1–6. IEEE, Waikoloa, HI, USA (2019)
32. Zheng, L., Chan, H., Loukides, G., Li, M.: Maximizing approximately k-submodular functions. In: 15th International Proceedings on SIAM, pp. 414–422. SIAM, Virtual Event (2021)

Data Summarization Beyond Monotonicity: Non-monotone Two-Stage Submodular Maximization

Shaojie Tang[✉][iD]

Naveen Jindal School of Management, University of Texas at Dallas, Richardson, USA
shaojie.tang@utdallas.edu

Abstract. The objective of a two-stage submodular maximization problem is to reduce the ground set using provided training functions that are submodular, with the aim of ensuring that optimizing new objective functions over the reduced ground set yields results comparable to those obtained over the original ground set. This problem has applications in various domains including data summarization. Existing studies often assume the monotonicity of the objective function, whereas our work pioneers the extension of this research to accommodate non-monotone submodular functions. We have introduced the first constant-factor approximation algorithms for this more general case.

1 Introduction

In this paper, we are motivated by the application of data summarization [8,9,16,17] and tackle the two-stage submodular optimization problem. In these applications, we are often faced with multiple user-specific submodular functions, which are used to evaluate the value of a set of items. A typical objective is to select a set of k items to maximize each submodular function [7]. While maximizing a single submodular function has been widely explored in the literature, the feasibility of existing solutions diminishes when confronted with a substantial number of submodular functions and items. Consequently, our objective is to reduce the size of the ground set in a manner that minimizes the loss when optimizing a new submodular function over the reduced ground set, as compared to the original ground set.

The problem at hand can be framed as a two-stage submodular maximization problem, as initially introduced in [1]. While the majority of prior studies in this domain presume that each submodular function exhibits monotone non-decreasing behavior, real-world scenarios often involve objective functions that are non-monotone. These instances include feature selection [4], profit maximization [14], maximum cut [6], and data summarization [9]. A significant contribution presented in our work is the development of the first constant-factor approximation algorithm for the non-monotone two-stage submodular maximization problem, with an approximation ratio of $1/2e$. Remarkably, when the objective function is monotone, our algorithm achieves an improved approximation ratio of $(1 - 1/e^2)/2$, thereby recovering the result presented in [11].

© The Author(s), under exclusive license to Springer Nature Switzerland AG 2024
W. Wu and J. Guo (Eds.): COCOA 2023, LNCS 14461, pp. 277–286, 2024.
https://doi.org/10.1007/978-3-031-49611-0_20

1.1   Related Work

The problem of non-monotone submodular maximization has garnered substantial attention in the literature [3,5,12,13,15]. The current state-of-the-art solution for this problem, especially when accounting for a cardinality constraint, is a 0.385-approximation algorithm [2]. However, it is noteworthy that even though each individual objective function considered in our problem exhibits submodularity, the overall objective function is not submodular in general. As a result, the existing findings on non-monotone submodular maximization do not directly apply to our specific setting.

The most closely related work to our research is the study by [1,10,11]. They have developed constant-factor approximation algorithms, primarily tailored for the monotone case. Our work builds upon and extends their results to address the more general and challenging non-monotone scenario. To achieve this goal, we have integrated the "local-search" approach [11] with "sampling" technique [12] in a non-trivial way, resulting in the creation of a novel sampling-based algorithm. Furthermore, we have incorporated a trimming phase into our algorithm, enabling us to attain the first constant-factor approximation ratio for the non-monotone case.

2   Problem Formulation

The input of our problem is a set of n items Ω. There is a group of m non-monotone submodular functions $f_1, \cdots, f_m : 2^\Omega \to \mathbb{R}_{\geq 0}$. Let $\Delta_i(x, A) = f_i(\{x\} \cup A) - f_i(A)$ denote the marginal gain of adding x to the set A when considering the function f_i. Here we say f_i is submodular if and only if $\Delta_i(x, A) \geq \Delta_i(x, A')$ for any two sets A and A' such that $A \subseteq A' \subseteq \Omega$, and any item $x \in \Omega$ such that $x \notin A'$.

Our objective is to compute a reduced ground set S of size l, where $l \ll n$, such that it yields good performance across all m functions when the choice is limited to items in S. Formally, let

$$F(S) = \sum_{i \in [m]} \max_{A \subseteq S : |A| \leq k} f_i(A) \tag{1}$$

where k is the size constraint of a feasible solution. Our goal is to find an optimal solution $O \subseteq \Omega$ that maximizes F, i.e.,

$$O \in \operatorname*{argmax}_{S \subseteq \Omega : |S| \leq l} F(S). \tag{2}$$

It is worth mentioning that the objective function $F(\cdot)$ is typically non-submodular, as observed in [1]. Consequently, classical algorithms designed for submodular optimization may not provide any approximation guarantees.

3   Algorithm Design and Analysis

Before presenting our algorithm, we need some additional notations. For each $i \in [m]$, we define the gain associated with removing an item y and replacing it with x as $\nabla_i(x, y, A) = f_i(\{x\} \cup A \setminus \{y\}) - f_i(A)$. Then for each $i \in [m]$, we define the largest possible gain brought by x, through local-search, with respect to an existing set A as $\nabla_i(x, A)$. Here the local-search can be realized either by directly adding x to A (while maintaining the cardinality constraint) or by substituting it with an item from A. Formally,

$$\nabla_i(x, A) = \begin{cases} 0 & \text{if } x \in A \\ \max\{0, \max_{y \in A} \nabla_i(x, y, A), \Delta_i(x, A)\} & \text{if } x \notin A \text{ and } |A| < k \\ \max\{0, \max_{y \in A} \nabla_i(x, y, A)\} & \text{if } x \notin A \text{ and } |A| = k \end{cases} \quad (3)$$

Let $\mathsf{Rep}_i(x, A)$ represent the item in A that, when replaced by x, maximizes the incremental gain while maintaining feasibility. Formally,

$$\mathsf{Rep}_i(x, A) = \begin{cases} \emptyset & \text{if } \nabla_i(x, A) = 0 \\ \emptyset & \text{if } \nabla_i(x, A) > 0 \text{ and } |A| < k \\ & \quad \text{and } \max_{y \in A} \nabla_i(x, y, A) < \Delta_i(x, A) \\ \arg\max_{y \in A} \nabla_i(x, y, A) & \text{if } \nabla_i(x, A) > 0 \text{ and } |A| < k \\ & \quad \text{and } \max_{y \in A} \nabla_i(x, y, A) \geq \Delta_i(x, A) \\ \arg\max_{y \in A} \nabla_i(x, y, A) & \text{if } \nabla_i(x, A) > 0 \text{ and } |A| = k \end{cases} \quad (4)$$

Now we are ready to present the design of our algorithm Sampling-Greedy (Algorithm 1). Initially, we introduce a set Φ of l dummy items to the ground set Ω to construct an extended ground set $\Omega' = \Omega \cup \Phi$. Here for all $i \in [m]$, we define $\Delta_i(x, A) = 0$ for any $x \in \Phi$ and $A \subseteq \Omega'$. We introduce these dummy items to prevent negative marginal utility item selection. Notably, these dummy items can be safely removed from the solution without altering its utility.

Throughout the process, Sampling-Greedy maintains a solution set denoted as S, along with a set of feasible solutions $T_i \subseteq S$ for each function f_i (all of which are initially set to empty). In each iteration, it first computes the top l items M from the extended ground set Ω' based on its combined contribution to each f_i, indicated by $\sum_{i=1}^{m} \nabla_i(x, T_i)$. That is,

$$M = \underset{A \subseteq \Omega' : |A| = l}{\arg\max} \sum_{x \in A} \sum_{i=1}^{m} \nabla_i(x, T_i). \quad (5)$$

Then it randomly selects one item, say x^*, from M and adds x^* to S. Sampling-Greedy then verifies if any of the sets T_i can be improved. This can be achieved by either directly adding x^* (while adhering to the cardinality constraint) or substituting it with an item from T_i. For each $i \in [m]$, we update T_i if and only if $\nabla_i(x^*, T_i) > 0$.

Note that there might exist some $i \in [m]$ and $x \in T_i$ such that $f_i(T_i) - f_i(T_i \setminus \{x\}) < 0$. In other words, certain subsets T_i could contain items that provide negative marginal utility to the set T_i. Consequently, we introduce a "trimming" phase (Algorithm 2) to refine each T_i and ensure that no item contributes negative utility to it. This can be achieved through an iterative process of evaluating the marginal utility of each item within T_i and subsequently removing any items with negative marginal utility. By the submodularity of f_i, we can show that after this trimming phase, T_i does not contain any items whose marginal utility if negative. It is also easy to verify that the trimming phase does not decrease the utility of our solution. A formal description of these properties is presented in the following lemma.

Algorithm 1 . Sampling-Greedy

1: $S \leftarrow \emptyset, T_i \leftarrow \emptyset (\forall i \in [m])$
2: **for** $j \in [l]$ **do**
3:    $M = \text{argmax}_{A \subseteq \Omega' : |A| = l} \sum_{x \in A} \sum_{i=1}^{m} \nabla_i(x, T_i).$
4:    randomly pick one item x^* from M, $S \leftarrow S \cup \{x^*\}$
5:    **for** $i \in [m]$ **do**
6:       **if** $\nabla_i(x^*, T_i) > 0$ **then**
7:          $T_i \leftarrow T_i \setminus \text{Rep}_i(x^*, T_i) \cup \{x^*\}$
8:          $T_i \leftarrow \text{Trim}(T_i, f_i)$
9: **return** $S, T_1, T_2, \cdots, T_m$

Algorithm 2 . Trim(B, f_i)

1: $A \leftarrow B$
2: **for** $x \in A$ **do**
3:    **if** $f_i(A) - f_i(A \setminus \{x\}) < 0$ **then**
4:       $A \leftarrow A \setminus \{x\}$
5: **return** A

Lemma 1. *Consider any set of items $B \subseteq \Omega$ and a function f_i. Assume A is returned from Trim(B, f_i), we have $f_i(A) \geq f_i(B)$ and for all $x \in A$, we have $f_i(A) - f_i(A \setminus \{x\}) \geq 0$.*

Proof. The proof that $f_i(A) \geq f_i(B)$ is straightforward, as it follows from the fact that the trimming phase only eliminates items with a negative marginal contribution. We next prove that for all $x \in A$, we have $f_i(A) - f_i(A \setminus \{x\}) \geq 0$. We prove this through contradiction. Suppose there exists an item $y \in A$ such that $f_i(A) - f_i(A \setminus \{y\}) < 0$. Let's denote the solution before considering the inclusion of y as A'. In this case, it must hold that $f_i(A') - f_i(A' \setminus \{y\}) \geq 0$, as otherwise, the trimming phase would eliminate y from the solution. Furthermore, it is straightforward to confirm that $A \subseteq A'$. As a consequence, based on the assumption that f_i is a submodular function, we have $f_i(A) - f_i(A \setminus \{y\}) \geq f_i(A') - f_i(A' \setminus \{y\})$. This, together with $f_i(A') - f_i(A' \setminus \{y\}) \geq 0$, implies that $f_i(A) - f_i(A \setminus \{y\}) \geq f_i(A') - f_i(A' \setminus \{y\}) \geq 0$. This contradicts to the assumption that $f_i(A) - f_i(A \setminus \{y\}) < 0$. $\square$

3.1 Performance Analysis

First, it is easy to verify that Sampling-Greedy requires $O(l(mkl + mn))$ function evaluations. This is because Sampling-Greedy comprises l iterations, where each iteration involves mkl function evaluations in Line 3 of Algorithm 1, along with an additional mn function evaluations in Algorithm 2. In the following theorem, we show that the expected utility of our solution is at least a constant-factor approximation of the optimal solution.

Theorem 1. *Sampling-Greedy returns a random set S of size at most l such that*

$$\mathbb{E}_S[F(S)] \geq \frac{1}{2e} F(O) \tag{6}$$

where O represents the optimal solution.

The rest of this section is devoted to proving this theorem. The basic idea behind the proof is to establish a lower bound on the expected marginal utility achieved by adding x^* to set S after each iteration. We demonstrate that this utility increment is substantial enough to guarantee a $1/2e$ approximation ratio. Consider an arbitrary round $t \in [l]$ of Sampling-Greedy, let S and $T_1, \cdots, T_m$ denote the solution obtained at the end of round t. By the design of Sampling-Greedy, we randomly pick an item x^* from M and add it to S, hence, by the definition of M, the expected marginal utility of adding x^* to S before the "trimming phase" is

$$\mathbb{E}_{x^*}\left[\sum_{i=1}^m \nabla_i(x^*, T_i)\right] = \frac{1}{l} \max_{A \subseteq \Omega' : |A| = l} \sum_{x \in A} \sum_{i=1}^m \nabla_i(x, T_i). \tag{7}$$

Recall that the trimming phase does not decrease utility. Therefore, the ultimate expected utility increment after each iteration is at least $\mathbb{E}_{x^*}[\sum_{i=1}^m \nabla_i(x^*, T_i)]$. Moreover, because F is a monotone function, it is safe to assume that the size of the optimal solution is l, i.e., $|O| = l$. We next provide a lower bound on $\mathbb{E}_{x^*}[\sum_{i=1}^m \nabla_i(x^*, T_i)]$.

Observe that

$$\mathbb{E}_{x^*}\left[\sum_{i=1}^m \nabla_i(x^*, T_i)\right] = \frac{1}{l} \max_{A \subseteq \Omega' : |A| = l} \sum_{x \in A} \sum_{i=1}^m \nabla_i(x, T_i)$$

$$\geq \frac{1}{|O|} \sum_{x \in O} \sum_{i=1}^m \nabla_i(x, T_i) = \frac{1}{l} \sum_{x \in O} \sum_{i=1}^m \nabla_i(x, T_i) \tag{8}$$

Let $O_i \subseteq O$ represent a subset with a maximum size of k items, chosen to maximize f_i, i.e., $O_i = \operatorname{argmax}_{A \subseteq O : |A| \leq k} f_i(A)$. Inequality (8) implies that

$$\mathbb{E}_{x^*}\left[\sum_{i=1}^m \nabla_i(x^*, T_i)\right] \geq \frac{1}{l} \sum_{x \in O} \sum_{i=1}^m \nabla_i(x, T_i) \geq \frac{1}{l} \sum_{i=1}^m \sum_{x \in O_i} \nabla_i(x, T_i). \tag{9}$$

For the sake of simplifying our analysis, let's assume that the size of O_i is fixed at k, i.e., $|O_i| = k$. If O_i does not initially contain exactly k items, we can add additional dummy items to bring its size up to k. Please note that if x is a dummy item, then $\nabla_i(x, T_i) = 0$. This stems from the observation that T_i does not contain any items that would yield negative marginal utility, a consequence of the trimming phase (Lemma 1). Hence, adding dummy items to O_i does not affect inequality (9).

It is easy to verify that there is a mapping π between O_i and T_i such that every item of $O_i \cap T_i$ is mapped to itself, and every item of $O_i \setminus T_i$ is mapped to either the empty set or an item in $T_i \setminus O_i$. We next give a lower bound of $\nabla_i(x, T_i)$.

Lemma 2. *For all $i \in [m]$ and $x \in O_i$, we have*

$$\nabla_i(x, T_i) \geq \Delta_i(x, T_i) - \Delta_i(\pi(x), T_i \setminus \{\pi(x)\}). \tag{10}$$

Proof. We prove this lemma in three cases. We first consider the case when $x \notin T_i$ and $\pi(x) \neq \emptyset$. In this case, the following chain proves this lemma.

$$\nabla_i(x, T_i) \geq f_i(\{x\} \cup T_i \setminus \{\pi(x)\}) - f_i(T_i) \tag{11}$$
$$= \Delta_i(x, T_i) - \Delta_i(\pi(x), T_i \cup \{x\} \setminus \{\pi(x)\}) \tag{12}$$
$$\geq \Delta_i(x, T_i) - \Delta_i(\pi(x), T_i \setminus \{\pi(x)\}) \tag{13}$$

where the first inequality is by the definition of $\nabla_i(x, T_i)$ and the second inequality is by the assumption that f_i is a submodular function.

We next consider the case when $x \notin T_i$ and $\pi(x) = \emptyset$. In this case, because $\pi(x) = \emptyset$, i.e., x is not mapped to any item from T_i, we have $|T_i| < k$. Hence,

$$\nabla_i(x, T_i) = \max\{0, \max_{y \in T_i} \nabla_i(x, y, T_i), \Delta_i(x, T_i)\} \geq \Delta_i(x, T_i). \tag{14}$$

Moreover, $\pi(x) = \emptyset$ implies that

$$\Delta_i(\pi(x), T_i \setminus \{\pi(x)\}) = 0. \tag{15}$$

It follows that

$$\nabla_i(x, T_i) \geq \Delta_i(x, T_i) - 0 = \Delta_i(x, T_i) - \Delta_i(\pi(x), T_i \setminus \{\pi(x)\}), \tag{16}$$

where the inequality is by inequality (14) and the equality is by equality (15).

At last, we consider the case when $x \in T_i$. In this case, we have $\Delta_i(x, T_i) = 0$, and $\Delta_i(\pi(x), T_i \setminus \{\pi(x)\}) \geq 0$, a consequence of the trimming phase (Lemma 1). Hence, $\Delta_i(x, T_i) - \Delta_i(\pi(x), T_i \setminus \{\pi(x)\}) \leq 0$. It follows that

$$\nabla_i(x, T_i) \geq 0 \geq \Delta_i(x, T_i) - \Delta_i(\pi(x), T_i \setminus \{\pi(x)\}). \tag{17}$$

$\square$

Inequality (9) and Lemma 2 imply that

$$\mathbb{E}_{x^*}[\sum_{i=1}^{m} \nabla_i(x^*, T_i)] \geq \frac{1}{l} \sum_{i=1}^{m} \sum_{x \in O_i} \nabla_i(x, T_i) \tag{18}$$

$$\geq \frac{1}{l} \sum_{i=1}^{m} \sum_{x \in O_i} (\Delta_i(x, T_i) - \Delta_i(\pi(x), T_i \setminus \{\pi(x)\})). \tag{19}$$

Because f_i is submodular, we have

$$\sum_{x \in O_i} \Delta_i(x, T_i) \geq f_i(O_i \cup T_i) - f_i(T_i). \tag{20}$$

Moreover, no two items from O_i are mapped to the same item from T_i, we have

$$\sum_{x \in O_i} \Delta_i(\pi(x), T_i \setminus \{\pi(x)\}) = \sum_{y \in T_i} \Delta_i(y, T_i \setminus \{y\}) \leq f_i(T_i) \tag{21}$$

where the inequality is by the assumption that f_i is submodular.

Inequalities (19), (20) and (21) together imply that

$$\mathbb{E}_{x^*}[\sum_{i=1}^{m} \nabla_i(x^*, T_i)] \geq \frac{1}{l} \sum_{i=1}^{m} \sum_{x \in O_i} (\Delta_i(x, T_i) - \Delta_i(\pi(x), T_i \setminus \{\pi(x)\})) \tag{22}$$

$$\geq \frac{1}{l} \sum_{i=1}^{m} (f_i(O_i \cup T_i) - f_i(T_i) - f_i(T_i)) \tag{23}$$

$$= \frac{1}{l} \sum_{i=1}^{m} (f_i(O_i \cup T_i) - 2f_i(T_i)). \tag{24}$$

Taking the expectation over $T_1, \cdots, T_m$ for both the left and right hand sides of (24), we have

$$\mathbb{E}_{T_1, \cdots, T_m}[\mathbb{E}_{x^*}[\sum_{i=1}^{m} \nabla_i(x^*, T_i)]] \tag{25}$$

$$\geq \mathbb{E}_{T_1, \cdots, T_m}[\frac{1}{l} \sum_{i=1}^{m} (f_i(O_i \cup T_i) - 2f_i(T_i))] \tag{26}$$

$$= \mathbb{E}_{T_1, \cdots, T_m}[\frac{1}{l} \sum_{i=1}^{m} (f_i(O_i \cup T_i))] - \mathbb{E}_{T_1, \cdots, T_m}[\sum_{i=1}^{m} \frac{2}{l} f_i(T_i))] \tag{27}$$

$$= \frac{1}{l}\mathbb{E}_{T_1, \cdots, T_m}[\sum_{i=1}^{m} (f_i(O_i \cup T_i))] - \frac{2}{l}\mathbb{E}_{T_1, \cdots, T_m}[\sum_{i=1}^{m} f_i(T_i))] \tag{28}$$

$$\geq \frac{1}{l}(1 - \frac{1}{l})^t \sum_{i=1}^{m} f_i(O_i) - \frac{2}{l}\mathbb{E}_{T_1, \cdots, T_m}[f_i(T_i))] \tag{29}$$

$$= \frac{1}{l}(1 - \frac{1}{l})^t F(O) - \frac{2}{l}\mathbb{E}_{T_1, \cdots, T_m}[f_i(T_i))]. \tag{30}$$

The second inequality is by the observation that $\mathbb{E}_{T_1,\cdots,T_m}[\sum_{i=1}^m (f_i(O_i \cup T_i))] \geq (1-\frac{1}{l})^t \sum_{i=1}^m f_i(O_i)$. To prove this inequality, recall that in each round, Sampling-Greedy randomly picks an item from M to be included in S. Hence, right before entering round t of Sampling-Greedy, each item $x \in \Omega'$ has a probability of at most $p = 1 - (1 - \frac{1}{l})^t$ of being included in S and consequently in T_i for all $i \in [m]$. By Lemma 2.2 of [3], we have $\mathbb{E}_{T_i}[f_i(O_i \cup T_i)] \geq (1-p)f_i(O_i) = (1-\frac{1}{l})^t f_i(O_i)$ for all $i \in [m]$. It follows that $\mathbb{E}_{T_1,\cdots,T_m}[\sum_{i=1}^m (f_i(O_i \cup T_i))] \geq (1 - \frac{1}{l})^t \sum_{i=1}^m f_i(O_i)$.

Let X_t denote the value of $\mathbb{E}_{T_1,\cdots,T_m}[\mathbb{E}_{x^*}[\sum_{i=1}^m \nabla_i(x^*, T_i)]]$ at the end of round t. Inequality (30) implies that

$$X_{t+1} - X_t \geq \frac{1}{l}(1 - \frac{1}{l})^t F(O) - \frac{2}{l}X_t \tag{31}$$

$$\Rightarrow 2(X_{t+1} - X_t) \geq \frac{1}{l}(1 - \frac{1}{l})^t F(O) - \frac{2}{l}X_t \tag{32}$$

$$\Rightarrow 2X_{t+1} - 2X_t \geq \frac{1}{l}(1 - \frac{1}{l})^t F(O) - \frac{2}{l}X_t \tag{33}$$

$$\Rightarrow 2X_{t+1} \geq \frac{1}{l}(1 - \frac{1}{l})^t F(O) + (2 - \frac{2}{l})X_t. \tag{34}$$

Based on the above inequality, we next prove through induction that $2X_t \geq \frac{t}{l}(1 - \frac{1}{l})^{t-1}F(O)$. Note that $X_0 = 0$, meaning that the utility before the start of the algorithm is zero. The induction step is established in the following manner:

$$2X_{t+1} \geq \frac{1}{l}(1 - \frac{1}{l})^t F(O) + (2 - \frac{2}{l})X_t \tag{35}$$

$$\Rightarrow 2X_{t+1} \geq \frac{1}{l}(1 - \frac{1}{l})^t F(O) + (1 - \frac{1}{l})\frac{t}{l}(1 - \frac{1}{l})^{t-1}F(O) \tag{36}$$

$$= \frac{1}{l}(1 - \frac{1}{l})^t F(O) + \frac{t}{l}(1 - \frac{1}{l})^t F(O) \tag{37}$$

$$= \frac{t+1}{l}(1 - \frac{1}{l})^t F(O). \tag{38}$$

It follows that the value of $2X_l$ is at least $(1 - \frac{1}{l})^{l-1}F(O)$, which itself is bounded from below by $(1/e) \cdot F(O)$. Here, X_l represents the expected utility of our algorithm upon completion. Hence, the expected utility of our algorithm is at least $X_l \geq (1/2e) \cdot F(O)$.

3.2   Enhanced Results for Monotone Case

For the case when f_i is both monotone and submodular, we will demonstrate that the approximation ratio of Sampling-Greedy is improved to $(1 - 1/e^2)/2$ which recovers the results presented in [11]. Observe that if f_i is monotone, we have $f_i(O_i \cup T_i) \geq f_i(O_i)$. Hence, inequality (28) implies that

$$\mathbb{E}_{T_1,\cdots,T_m}\left[\mathbb{E}_{x^*}\left[\sum_{i=1}^{m}\nabla_i(x^*,T_i)\right]\right] \tag{39}$$

$$\geq \frac{1}{l}\mathbb{E}_{T_1,\cdots,T_m}\left[\sum_{i=1}^{m}(f_i(O_i\cup T_i))\right] - \frac{2}{l}\mathbb{E}_{T_1,\cdots,T_m}\left[\sum_{i=1}^{m}f_i(T_i))\right] \tag{40}$$

$$\geq \frac{1}{l}\mathbb{E}_{T_1,\cdots,T_m}\left[\sum_{i=1}^{m}(f_i(O_i))\right] - \frac{2}{l}\mathbb{E}_{T_1,\cdots,T_m}\left[\sum_{i=1}^{m}f_i(T_i))\right] \tag{41}$$

$$= \frac{1}{l}\sum_{i=1}^{m}f_i(O_i) - \frac{2}{l}\mathbb{E}_{T_1,\cdots,T_m}\left[\sum_{i=1}^{m}f_i(T_i))\right] \tag{42}$$

$$= \frac{1}{l}F(O) - \frac{2}{l}\mathbb{E}_{T_1,\cdots,T_m}\left[\sum_{i=1}^{m}f_i(T_i))\right] \tag{43}$$

where the first equality is because O_i is a fixed set for all $i \in [m]$. Let X_t denote the value of $\mathbb{E}_{T_1,\cdots,T_m}\left[\mathbb{E}_{x^*}\left[\sum_{i=1}^{m}\nabla_i(x^*,T_i)\right]\right]$ at the end of round t. Inequality (43) implies that

$$X_{t+1} - X_t \geq \frac{1}{l}F(O) - \frac{2}{l}X_t. \tag{44}$$

Previous research [11] has demonstrated that by inductively solving the equation above, we can establish that $X_l \geq ((1 - 1/e^2)/2) \cdot F(O)$.

References

1. Balkanski, E., Mirzasoleiman, B., Krause, A., Singer, Y.: Learning sparse combinatorial representations via two-stage submodular maximization. In: International Conference on Machine Learning, pp. 2207–2216. PMLR (2016)
2. Buchbinder, N., Feldman, M.: Constrained submodular maximization via a non-symmetric technique. Math. Oper. Res. **44**(3), 988–1005 (2019)
3. Buchbinder, N., Feldman, M., Naor, J., Schwartz, R.: Submodular maximization with cardinality constraints. In: Proceedings of the Twenty-Fifth Annual ACM-SIAM Symposium on Discrete Algorithms, pp. 1433–1452. SIAM (2014)
4. Das, A., Kempe, D.: Algorithms for subset selection in linear regression. In: Proceedings of the Fortieth Annual ACM Symposium on Theory of Computing, pp. 45–54 (2008)
5. Gharan, S.O., Vondrák, J.: Submodular maximization by simulated annealing. In: Proceedings of the Twenty-Second Annual ACM-SIAM Symposium on Discrete Algorithms, pp. 1098–1116. SIAM (2011)
6. Gotovos, A., Karbasi, A., Krause, A.: Non-monotone adaptive submodular maximization. In: Twenty-Fourth International Joint Conference on Artificial Intelligence (2015)
7. Krause, A., Golovin, D.: Submodular function maximization. Tractability **3**(71–104), 3 (2014)

8. Lin, H., Bilmes, J.: A class of submodular functions for document summarization. In: Proceedings of the 49th Annual Meeting of the Association for Computational Linguistics: Human Language Technologies, pp. 510–520 (2011)
9. Mirzasoleiman, B., Badanidiyuru, A., Karbasi, A.: Fast constrained submodular maximization: personalized data summarization. In: Proceedings of The 33rd International Conference on Machine Learning (ICML), pp. 1358–1367 (2016)
10. Mitrovic, M., Kazemi, E., Zadimoghaddam, M., Karbasi, A.: Data summarization at scale: a two-stage submodular approach. In: International Conference on Machine Learning, pp. 3596–3605. PMLR (2018)
11. Stan, S., Zadimoghaddam, M., Krause, A., Karbasi, A.: Probabilistic submodular maximization in sub-linear time. In: International Conference on Machine Learning, pp. 3241–3250. PMLR (2017)
12. Tang, S.: Beyond pointwise submodularity: non-monotone adaptive submodular maximization in linear time. Theoret. Comput. Sci. **850**, 249–261 (2021)
13. Tang, S.: Beyond pointwise submodularity: non-monotone adaptive submodular maximization subject to knapsack and k-system constraints. Theoret. Comput. Sci. **936**, 139–147 (2022)
14. Tang, S., Yuan, J.: Adaptive regularized submodular maximization. In: 32nd International Symposium on Algorithms and Computation (ISAAC 2021). Schloss Dagstuhl-Leibniz-Zentrum für Informatik (2021)
15. Tang, S., Yuan, J.: Group equility in adaptive submodular maximization. arXiv preprint arXiv:2207.03364 (2022)
16. Wei, K., Iyer, R., Bilmes, J.: Submodularity in data subset selection and active learning. In: International Conference on Machine Learning, pp. 1954–1963. PMLR (2015)
17. Wei, K., Liu, Y., Kirchhoff, K., Bilmes, J.: Using document summarization techniques for speech data subset selection. In: Proceedings of the 2013 Conference of the North American Chapter of the Association for Computational Linguistics: Human Language Technologies, pp. 721–726 (2013)

GREEDY+MAX: An Efficient Approximation Algorithm for k-Submodular Knapsack Maximization

Zhongzheng Tang[1], Jingwen Chen[2], Chenhao Wang[2,3(✉)], Tian Wang[2,3], and Weijia Jia[2,3]

[1] School of Science, Beijing University of Posts and Telecommunications, Beijing 100876, China
[2] BNU-HKBU United International College, Zhuhai 519087, China
chenhwang@bnu.edu.cn
[3] Beijing Normal University, Zhuhai 519087, China

Abstract. This paper studies the problem of maximizing a k-submodular function under a knapsack constraint. A k-submodular function is a generalization of submodular functions, which takes k disjoint subsets of elements as input and outputs a real value. Many problems in combinatorial optimization and machine leaning can be modeled as k-submodular maximization problems, such as influence maximization, sensor placement, feature selection, etc. In this paper, we propose a novel greedy-based algorithm, called GREEDY+MAX, which augments every partial greedy solution by a feasible item with maximum marginal gain, and returns the best augmented solution. We prove that it achieves a $\frac{1}{3}$-approximation for monotone functions and a $\frac{1}{4}$-approximation for non-monotone functions, with a low query complexity.

Keywords: k-submodularity · knapsack constraint · approximation

1 Introduction

Given a finite nonempty set $V = \{a_1, a_2, \ldots, a_n\}$, a set function $f : 2^V \to \mathbb{R}$ defined on subsets of V is called *submodular* if for all $S, T \subseteq V$,

$$f(S) + f(T) \geq f(S \cap T) + f(S \cup T).$$

Submodular functions play a pivotal role in the realm of operations research [3,10,11,18] and theoretical computer science [1,7]. Examples of these functions encompass matroid rank functions, cut capacity functions and entropy functions.

In this paper, we study a natural generalization of submodularity, the k-submodularity, firstly proposed by Huber and Kolmogorov [6] in 2012. Instead of a single set, k-submodular functions take k disjoint subsets as input and return a real value. Formally, given a finite nonempty set V, the family of k disjoint sets is denoted as $(k+1)^V := \{(X_1, \ldots, X_k) \mid X_i \subseteq V \; \forall i \in [k], X_i \cap X_j = \varnothing \; \forall i \neq j\}$, where $[k] := \{1, \ldots, k\}$.

Definition 1 (k-submodularity [6]). A function $f : (k+1)^V \to \mathbb{R}$ is called k-submodular, if for any $\mathbf{x} = (X_1, \ldots, X_k)$ and $\mathbf{y} = (Y_1, \ldots, Y_k)$ in $(k+1)^V$, we have

$$f(\mathbf{x}) + f(\mathbf{y}) \geq f(\mathbf{x} \sqcup \mathbf{y}) + f(\mathbf{x} \sqcap \mathbf{y}),$$

where

$$\mathbf{x} \sqcup \mathbf{y} := \left(X_1 \cup Y_1 \backslash (\bigcup_{i \neq 1} X_i \cup Y_i), \ldots, X_k \cup Y_k \backslash (\bigcup_{i \neq k} X_i \cup Y_i) \right),$$

$$\mathbf{x} \sqcap \mathbf{y} := (X_1 \cap Y_1, \ldots, X_k \cap Y_k).$$

k-Submodular functions are a generalization of submodular functions. While submodular functions take a single subset of a set V as input and exhibit a *diminishing returns* property, k-submodular functions consider k disjoint subsets, and we must specify not only which element we are adding to the solution, but also which subset it is being added to. This generalization enables the function to capture interactions across multiple dimensions.

k-Submodular functions have found applications in a variety of fields, including influence maximization [24], sensor placement [13], feature selection [21], document summarization [9], and more, with the core of many of these problems being the maximization of a k-submodular function.

- **Sensor Placement:** Various types of sensors are strategically positioned in different locations to gather diverse data, for example, a scenario with k kinds of sensors and N available locations for their placement. With the assumption that each location can host one sensor, we can create a correlation between every k-tuple of disjoint subsets and N locations. The aim is to optimize the information derived from the sensors, hence the assessment of a sensor deployment strategy is performed using k-submodular functions.
- **Influence Maximization:** Assume k topics are available, and social users initially adopt a topic. The objective is to maximize the anticipated number of users ultimately influenced by at least one topic. This mirrors scenarios in viral marketing or product recommendations, wherein a company aims to disseminate an advertising campaign spotlighting k products to users through social networks or the internet.
- **Feature Selection:** Feature selection is paramount in various research domains, such as machine learning and data mining, as it assists in amplifying the analysis of vast datasets by curbing their dimensionality. In multi-class feature selection issues, characterized by a medley of features intertwined with k unrelated prediction variables, the goal is to identify the most informative features and classify them based on the prediction variables. This engenders a k-submodular optimization problem.

The maximization problem of k-submodular functions is NP-hard, being a generalization of the NP-hard submodular maximization problem. Despite this, substantial research has been devoted to developing efficient approximation algorithms with good approximation ratios for this problem under various constraints, e.g., size constraints, matroid constraints, and knapsack constraint, which is the focus of this work.

Our Contributions. In this paper, we investigate the k-submodular maximization problem with a knapsack constraint, referred to as the k-submodular knapsack maximization (kSKM) problem. Each item $a \in V$ has an associated cost $c(a)$, and the goal is to select k disjoint subsets in $(k+1)^V$ that maximizes a k-submodular function, under the constraint that the total cost of the selected items does not exceed a given budget B.

For kSKM, we propose a novel greedy-based algorithm, Greedy+Max. It first runs the Greedy algorithm, which iteratively adds an item into a subset (i.e., an item-dimension pair) that achieves the largest marginal density, until no item fits. At the beginning of each iteration, the current solution maintained by Greedy is called a partial greedy solution. Then, for each partial greedy solution, the algorithm adds one more feasible item into a subset with the maximum marginal gain (this process is Max). Finally, it returns the best one among such augmented solutions. We prove that it achieves a $\frac{1}{3}$-approximation for monotone functions and a $\frac{1}{4}$-approximation for non-monotone functions. The analysis framework follows Yaroslavtsev et al. [27], who propose the Greedy+Max algorithm for the submodular knapsack maximization problem, and prove that it has a $\frac{1}{2}$-approximation for monotone submodular functions.

The time efficiency of algorithms is measured by the query complexity, which is the times that it queries an oracle to find the function value on a given input. The query complexity of Greedy+Max is $O(kn \cdot \min\{n, B\})$. Compared with the best known approximation ratio $(1 - e^{-2})/2 \approx 0.432$ for monotone kSKM and $(1 - e^{-3})/3 \approx 0.316$ for non-monotone kSKM, while their query complexity are $O(n^{10}k^9)$ and $O(n^{23}k^{22})$, respectively, our Greedy+Max algorithm significantly outperforms theirs in terms of time complexity.

Related Work. Huber and Kolmogorov first introduced k-submodular functions a decade ago [6], aiming to express submodularity when selecting k disjoint sets of elements rather than a single set. The concept of k-submodular functions has since evolved into a significant research area [4,5,12,19], with a particular focus on the problem of maximizing k-submodular functions.

Tang *et al.* [23] were the first to explore the k-submodular maximization under a knapsack constraint (i.e., kSKM). They demonstrated that greedily extending all feasible size-2 solutions can achieve $\frac{1}{2}(1 - \frac{1}{e})$-approximation if the function is monotone. Chen *et al.* [2] considered the Greedy+Singleton algorithm, which compares the greedy solution with the best singleton solution and selects the superior one. This algorithm, known for its simplicity and efficiency, has been widely researched in knapsack problems involving linear and submodular objective functions. Chen *et al.* proved an approximation ratio of $\frac{1}{4}(1 - \frac{1}{e})$ for the kSKM.

Pham *et al.* [16] proposed streaming algorithms with approximation ratios of $\frac{1}{4} - \epsilon$ and $\frac{1}{5} - \epsilon$ for monotone and non-monotone cases, respectively. The method requires $O(\frac{n}{\epsilon} \log n)$ queries of the k-submodular function. Wang and Zhou [24] introduced an algorithm based on multilinear extension. It expands a k-submodular function to a continuous space and then rounds the fractional solution, achieving an asymptotically optimal ratio of $\frac{1}{2} - \epsilon$. Additional studies relevant to kSKM include [15, 22, 26, 28].

Researchers have also explored the problem of maximizing k-submodular functions with no constraints or various other types of constraints. Ward and Živný [25] demonstrated that a greedy algorithm is a $\frac{1}{2}$-approximation for unconstrained *monotone* k-submodular maximization. Subsequently, Iwata *et al.* [8] proposed a randomized $\frac{k}{2k-1}$-approximation algorithm, revealing that the ratio is asymptotically tight. Oshima [14] provided a $\frac{k^2+1}{2k^2+1}$-approximation for unconstrained *non-monotone* maximization. For monotone k-submodular maximization under a total size constraint (i.e., at most a given number of items can be selected), Ohsaka and Yoshida [13] proposed a $\frac{1}{2}$-approximation algorithm, while for individual size constraints (i.e., in each dimension at most a given number of items can be selected), they introduced a $\frac{1}{3}$-approximation algorithm. Under a matroid constraint, Sakaue [17] demonstrated that a fully greedy algorithm is a $\frac{1}{2}$-approximation for the monotone case, and Sun *et al.* [20] offered a $\frac{1}{3}$-approximation algorithm for the non-monotone case.

2   Preliminaries

Recall that V is the ground set, and let f be a non-negative k-submodular function defined in Definition 1. For every k-tuple $\mathbf{x} = (X_1, \ldots, X_k) \in (k+1)^V$, there exists a unique set $S = \{(a,d) \mid a \in X_d, \ d \in [k]\}$ consisting of *item-dimension pairs*. In other words, an item-dimension pair (a,d) belongs to S, referred to as a *solution*, if and only if $a \in X_d$ in $\mathbf{x}$. For simplicity, we will use $\mathbf{x}$ and its corresponding solution S interchangeably. For any solution $S \in (k+1)^V$, we define $U(S) := \{a \in V \mid \exists d \in [k] \ (a,d) \in S\}$ as the set of items included, and the *size* is denoted by $|S| = |U(S)|$. For two solutions $S, S' \in (k+1)^V$, $S \subseteq S'$ indicates that all items in $U(S)$ also belong to $U(S')$ and the dimensions in both are consistent. Without loss of generality, we assume that $f(\varnothing) = 0$.

The marginal gain of adding an item-dimension pair (a,d) to S is

$$\Delta_{a,d} f(S) := f(S \cup \{(a,d)\}) - f(S),$$

and the marginal density is $\rho_{a,d}(S) := \frac{\Delta_{a,d} f(S)}{c(a)}$. Ward and Živný [25] prove that a k-submodular function f satisfies the *orthant submodularity*

$$\Delta_{a,d} f(S) \geq \Delta_{a,d} f(S'),$$

$$\forall S, S' \in (k+1)^V \ \text{with} \ S \subseteq S', a \notin U(S'), d \in [k],$$

and the *pairwise monotonicity*

$$\Delta_{a,d_1} f(S) + \Delta_{a,d_2} f(S) \geq 0,$$

$$\forall S \in (k+1)^V \text{ with } a \notin U(S), d_1, d_2 \in [k], d_1 \neq d_2.$$

They further prove that the converse is also true.

Lemma 1 *([25]). A function $f : (k+1)^V \to \mathbb{R}$ is k-submodular if and only if f is orthant submodular and pairwise monotone.*

The following important lemma will be implicitly used in our analysis.

Lemma 2 *([23]). For any solutions S, S' with $S \subseteq S'$, we have*

$$f(S') - f(S) \leq \sum_{(a,d) \in S' \setminus S} \Delta_{a,d} f(S).$$

In the kSKM problem, each item $a \in V$ is associated with a non-negative cost $c(a)$. The cost of a solution S is defined as the sum of the costs of all items included in $U(S)$, denoted as $c(S) = \sum_{a \in U(S)} c(a)$. The goal is to find a solution S that maximizes the value $f(S)$, subject to the constraint that the total cost does not exceed the given budget $B \in \mathbb{R}+$.

3   GREEDY+MAX Algorithm

For the submodular maximization problem under a knapsack constraint, Yaroslavtsev *et al.* [27] prove that GREEDY+MAX gives a $\frac{1}{2}$-approximation in $O(\min\{n, B\} \cdot n)$ queries, where B is the knapsack capacity. Following their work, we describe the GREEDY+MAX algorithm for kSKM.

The GREEDY algorithm starts with an empty set G and in each iteration selects an item-dimension pair (a, d) with the highest marginal density $\rho_{a,d}(G)$ that still fits into the knapsack. We refer to the solution at the beginning of each iteration as a *partial (greedy) solution*, and refer to the resulting solution as the *greedy solution*. The MAX process augments each partial solution with an item-dimension pair of the largest marginal value (as opposed to density). Finally, GREEDY+MAX (see Algorithm 1) returns the best among such *augmented solutions*.

In Algorithm 1, let G_i be the partial greedy solution with i items, and $G_0 = \varnothing$. For each i, it finds an augmenting pair (a_i', d_i') which maximizes $f((a_i', d_i') \cup G_i)$ among all items that still fit, i.e. $c((a_i', d_i') \cup G_i) \leq B$.

Let OPT be the optimal solution f the kSKM. Let $o_1 \in V$ be the item of the largest cost in $U(OPT)$. W.l.o.g. and only for the sake of analysis of approximation we scale the function values and costs so that $f(OPT) = 1$ and $c(OPT) = B = 1$. Let G be the greedy solution computed by GREEDY and let $(g_1, d_1^g), (g_2, d_2^g), \ldots, (g_m, d_m^g)$ be the pairs in G in the order they were added. Then we have $G_i = \{(g_1, d_1^g), \ldots, (g_i, d_i^g)\}$. We introduce a greedy performance function $g(x)$ that continuously tracks the performance of the greedy solution.

Algorithm 1. GREEDY+MAX

Input: $V = \{a_1, \ldots, a_n\}$, non-negative monotone k-submodular function f, cost function $c(\cdot)$, capacity B.

Output: A feasible solution in $(k+1)^V$.

1: $G^0 \leftarrow \varnothing$, $S \leftarrow \varnothing$, $i \leftarrow 1$
2: **while** $V \neq \varnothing$ **do**
3: $\quad (a_i', d_i') \leftarrow \arg\max\limits_{a \in V, d \in [k]} \Delta_{a,d} f(G^{i-1})$: the pair that maximizes the marginal value

4: $\quad$ **if** $f(S) < f(G_{i-1} \cup (a_i', d_i'))$ **then**
5: $\quad\quad S \leftarrow G_{i-1} \cup (a_i', d_i')$
6: $\quad$ **end if**
7: $\quad (g_i, d_i^g) \leftarrow \arg\max\limits_{a \in V, d \in [k]} \frac{\Delta_{a,d} f(G^{i-1})}{c(a)}$: the pair that maximizes the marginal density

8: $\quad G_i \leftarrow G_{i-1} \cup (g_i, d_i^g)$
9: $\quad B \leftarrow B - c(g_i)$
10: $\quad i \leftarrow i + 1$
11: $\quad V \leftarrow V \backslash (\{g_i\} \cup \{a \in V | c(a) > B\})$: remove g_i and all items that no longer fit
12: **end while**
13: **return** S

Definition 2 (Greedy performance function). For $x \in [0, c(G))$, let i be the smallest index such that $c(G_i) > x$. Define $g(x)$ as

$$g(x) = f(G_{i-1}) + (x - c(G_{i-1})) \cdot \rho_{g_i, d_i^g}(G_{i-1}).$$

It is easy to see that g is a continuous and monotone piecewise-linear function such that $g(0) = 0$. The monotonicity of g holds even when f is non-monotone, because f is pairwise-monotone by Lemma 1. Let g' be the right derivative of g, which is well defined on the interval $[0, c(G))$ and non-negative.

Next, we consider adding o_1, the largest item from OPT, to every partial greedy solution, and define a performance function g_1. Note that g_1 is only defined when o_1 still fits. Consider the last item added by the greedy solution before the cost of this solution exceeds $1 - c(o_1)$. Define c^* such that $1 - c(o_1) - c^*$ is the cost of the greedy solution before this item is added.

Consider the case when $c(G) \leq 1 - c(o_1)$. If there is an item $a \in U(OPT)$ in the optimal solution that is not included in G, since $c(a) \leq c(o_1)$ and there is enough capacity for adding it, GREEDY+MAX would continue to add items, giving a contradiction to the fact that G is the output. Thus, all items in OPT must have been included in G. Because GREEDY is a $\frac{1}{2}$-approximation algorithm for maximizing a monotone k-submodular function without constraint [25], we have $2 \cdot f(G) \geq f(OPT)$, and G is already a $\frac{1}{2}$-approximation. Therefore, we only need to consider the case when $c(G) > 1 - c(o_1)$.

Definition 3 (GREEDY+MAX performance lower bound). For any $x \in [0, 1 - c(o_1) - c^*]$, let i be the smallest index so that $c(G_i) > x$. Define $g_1(x)$ as

$$g_1(x) = g(x) + \max_{d \in [k]} \Delta_{o_1, d} f(G_{i-1}).$$

An illustration of functions $g(x)$ and $g_1(x)$ is shown in Fig. 1. Note that g_1 is a lower bound on the performance of GREEDY+MAX, because GREEDY+MAX adds the pair with largest marginal gain to the current partial greedy solution. g_1 is piecewise-linear but not necessarily continuous and monotone.

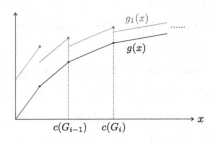

Fig. 1. An illustration of functions $g(x)$ and $g_1(x)$.

4   Approximations

We consider an arbitrary subset of items $V' = \{a_1, a_2, \ldots, a_l\} \subseteq V$. By the pairwise monotonicity, there exists an optimal solution $T = \{(a_1, d_1^*), \ldots, (a_l, d_l^*)\}$ that maximizes the function value over V' without any constraint (note that T may be not feasible for the kSKM). Let $S = \{(a_1, d_1), \ldots, (a_l, d_l)\}$ be the solution returned by the greedy algorithm that aims to maximizing the function value over items V' without any constraint, which considers the items in an order of $a_1, a_2, \ldots, a_l$, and assigns each item to the dimension with largest marginal gain.

For $j = 0, 1, \ldots, l$, define

$$S_j = \{(a_1, d_1), \ldots, (a_j, d_j)\} \text{ and} \tag{1}$$

$$T_j = (T \backslash \{(a_1, d_1^*), \ldots, (a_j, d_j^*)\}) \cup S_j. \tag{2}$$

That is, S_j is the first j item-index pairs in the greedy solution S, and T_j is obtained from the optimal solution T by replacing the first j item-index pairs with S_j. Clearly, $S_0 = \varnothing, S_l = S, T_0 = T$ and $T_l = S$.

The following lemma was noticed by Ward and Živný [25] and Xiao et al. [26], and says that the optimal value $f(T)$ is at most twice the value of any partial greedy solution S_t, plus the total marginal gain of other item-dimension pairs in the optimal solution. For completeness, we write down the proof in our notations.

Lemma 3. For $t = 0, 1, \ldots, l$,

(a) if f is monotone, then $f(T) \leq 2f(S_t) + \sum_{(a,d) \in T_t \backslash S_t} \Delta_{a,d} f(S_t)$;

(b) if f is non-monotone, then $f(T) \le 3f(S_t) + \sum_{(a,d)\in T_t \backslash S_t} \Delta_{a,d} f(S_t)$;

Proof. For $j = 0, \ldots, t-1$, introduce an intermediate $P_j := T_j \backslash (a_{j+1}, d^*_{j+1}) = T_{j+1} \backslash (a_{j+1}, d_{j+1})$. That is, P_j consists of $l-1$ items (excluding a_{j+1}), where the dimensions of items $a_1, \ldots, a_j$ align with those in S, and the dimensions of other items align with those in T. Then

$$f(T_j) = f(P_j) + \Delta_{a_{j+1}, d^*_{j+1}} f(P_j),$$

$$f(T_{j+1}) = f(P_j) + \Delta_{a_{j+1}, d_{j+1}} f(P_j).$$

When f is monotone, the difference of $f(T_j)$ and $f(T_{j+1})$ is

$$f(T_j) - f(T_{j+1}) = \Delta_{a_{j+1}, d^*_{j+1}} f(P_j) - \Delta_{a_{j+1}, d_{j+1}} f(P_j)$$

$$\le \Delta_{a_{j+1}, d^*_{j+1}} f(S_j) \tag{3}$$

$$\le \Delta_{a_{j+1}, d_{j+1}} f(S_j) \tag{4}$$

$$= f(S_{j+1}) - f(S_j).$$

Eq. (3) follows from the fact of $S_j \subseteq P_j$ and the monotonicity of f. Equation (4) follows from the fact that Greedy always assign the index with maximum marginal gain to the item considered, and (a_{j+1}, d_{j+1}) is the $(j+1)$-st pair added by Greedy. Summing up from $j = 0$ to $t - 1$, we have

$$f(T_0) - f(T_t) \le f(S_t) - f(S_0) = f(S_t).$$

Since $S_t \subseteq T_t$ and Lemma 2, we have

$$f(T) \le f(S_t) + f(T_t) \le 2f(S_t) + \sum_{(a,d)\in T_t \backslash S_t} \Delta_{a,d} f(S_t).$$

When f is non-monotone, Eq. (3) no longer holds. Instead, we bound the difference of $f(T_j)$ and $f(T_{j+1})$ by

$$f(T_j) - f(T_{j+1}) = \Delta_{a_{j+1}, d^*_{j+1}} f(P_j) - \Delta_{a_{j+1}, d_{j+1}} f(P_j)$$

$$= 2\Delta_{a_{j+1}, d^*_{j+1}} f(P_j) - [\Delta_{a_{j+1}, d^*_{j+1}} f(P_j) + \Delta_{a_{j+1}, d_{j+1}} f(P_j)]$$

$$\le 2\Delta_{a_{j+1}, d^*_{j+1}} f(P_j) \tag{5}$$

$$\le 2\Delta_{a_{j+1}, d^*_{j+1}} f(S_j)$$

$$\le 2\Delta_{a_{j+1}, d_{j+1}} f(S_j)$$

$$= 2f(S_{j+1}) - 2f(S_j),$$

where Eq. (5) follows from the pairwise monotonicity. Summing up from $j = 0$ to $t - 1$, we have

$$f(T_0) - f(T_t) \le 2f(S_t) - 2f(S_0) = 2f(S_t).$$

Since $S_t \subseteq T_t$ and Lemma 2, we have

$$f(T) \le 2f(S_t) + f(T_t) \le 3f(S_t) + \sum_{(a,d)\in T_t \backslash S_t} \Delta_{a,d} f(S_t).$$

$\square$

Next, we build differential inequalities with respect to g and g_1.

Lemma 4. *For any* $x \in [0, 1 - c(o_1) - c^*]$,

(a) if f is monotone, then $2g_1(x) + (1 - c(o_1)) \cdot g'(x) \geq 1$;
(b) if f is non-monotone, then $3g_1(x) + (1 - c(o_1)) \cdot g'(x) \geq 1$.

Proof. Since the right derivative g' is piecewise-constant, it suffices to prove the inequality only for the points where $x = c(G_{i-1})$ for some $i \geq 1$. We have

$$g_1(x) = g(c(G_{i-1})) + \max_{d \in [k]} \Delta_{o_1, d} f(G_{i-1}) = f(G_{i-1}) + \max_{d \in [k]} \Delta_{o_1, d} f(G_{i-1}). \quad (6)$$

We consider the items in $U(OPT) \cup U(G_{i-1})$. Assume w.l.o.g. that $V' = U(OPT) \cup U(G_{i-1}) = \{a_1, a_2, \ldots, a_l\}$, and $a_1 = g_1$, $a_2 = g_2, \ldots, a_{i-1} = g_{i-1}$. Let T be an optimal solution of the maximization problem over V' without any constraint, and $S = \{(a_1, d_1), \ldots, (a_l, d_l)\}$ be the solution returned by the greedy algorithm, which considers the items in an order of $a_1, a_2, \ldots, a_l$, and assigns each item to the dimension with largest marginal gain. Recall the definitions in (1) and (2), and then we have $S_j = G_j$ for $j = 0, \ldots, i - 1$. We further assume that $o_1 \notin U(G_{i-1})$, and o_1 is the i-th item in V', i.e., $a_i = o_1$. When f is monotone, we apply Lemma 3(a) to $t = i$. (If $o_1 \in U(G_{i-1})$, a similar analysis follows by applying Lemma 3(a) to $t = i - 1$). Since $S_i = S_{i-1} \cup (a_i, d_i) = G_{i-1} \cup (o_1, d_i)$,

$$f(T) \leq 2f(S_i) + \sum_{(a,d) \in T_i \setminus S_i} \Delta_{a,d} f(S_i)$$

$$= 2f(G_{i-1} \cup (o_1, d_i)) + \sum_{(a,d) \in T_i \setminus (G_{i-1} \cup (o_1, d_i))} \Delta_{a,d} f(G_{i-1} \cup (o_1, d_i)). \quad (7)$$

By Eq. (7), we have

$$1 = f(OPT) \leq f(T)$$

$$\leq 2f(G_{i-1} \cup (o_1, d_i)) + \sum_{(a,d) \in T_i \setminus (G_{i-1} \cup (o_1, d_i))} \Delta_{a,d} f(G_{i-1} \cup (o_1, d_i))$$

$$= 2g_1(x) + \sum_{(a,d) \in T_i \setminus (G_{i-1} \cup (o_1, d_i))} \Delta_{a,d} f(G_{i-1} \cup (o_1, d_i)) \quad (8)$$

$$= 2g_1(x) + \sum_{(a,d) \in T_i \setminus (G_{i-1} \cup (o_1, d_i))} c(a) \cdot \rho_{a,d}(G_{i-1} \cup (o_1, d_i)). \quad (9)$$

Eq. (8) follows from Eq. (6), and Eq. (9) follows from the definition of marginal density.

Since $x \leq 1 - c(o_1) - c^*$, all items in $U(T_i \setminus (G_{i-1} \cup (o_1, d_i)))$ still fit at the point of x, as o_1 is the largest item in OPT. Since the greedy algorithm always selects the item with the largest density, then

$$\max_{(a,d)\in T_i\setminus(G_{i-1}\cup(o_1,d_i))} \rho_{a,d}(G_{i-1}\cup(o_1,d_i)) \leq \max_{(a,d)\in T_i\setminus(G_{i-1}\cup(o_1,d_i))} \rho_{a,d}(G_{i-1})$$

$$\leq \rho_{g_i,d_i^g}(G_{i-1}) = g'(x).$$

Hence, we have

$$1 \leq 2g_1(x) + \sum_{(a,d)\in T_i\setminus(G_{i-1}\cup(o_1,d_i))} c(a)\cdot \rho_{a,d}(G_{i-1}\cup(o_1,d_i))$$

$$\leq 2g_1(x) + \sum_{(a,d)\in T_i\setminus(G_{i-1}\cup(o_1,d_i))} c(a)\cdot g'(x)$$

$$= 2g_1(x) + g'(x)\cdot c(T_i\setminus(G_{i-1}\cup(o_1,d_i))) \leq 2g_1(x) + g'(x)\cdot(1-c(o_1)),$$

where the last inequality follows from the fact that $g'(x)$ is always non-negative.

Note that the above analysis also applies to the non-monotone case, since the pairwise monotonicity guarantees the non-negativity of $g'(x)$. When f is non-monotone, applying Lemma 3(b) instead of (a), we have

$$1 \leq 3g_1(x) + g'(x)\cdot(1-c(o_1)).$$

□

Theorem 1. *For the k-submodular maximization problem under a knapsack constraint, GREEDY+MAX is $\frac{1}{3}$-approximation and $\frac{1}{4}$-approximation for monotone and non-monotone functions, respectively, with $O(kn\cdot\min\{n,B\})$ queries.*

Proof. When f is monotone, applying Lemma 4(a) at the point $x = 1-c(o_1)-c^*$, we have

$$2g_1(1-c(o_1)-c^*) + (1-c(o_1))\cdot g'(1-c(o_1)-c^*) \geq 1. \tag{10}$$

If $g_1(1-c(o_1)-c^*) \geq \frac{1}{3}$, then we already have a $\frac{1}{3}$-approximation, because

$$g_1(1-c(o_1)-c^*) = g(1-c(o_1)-c^*) + \max_{d\in[k]} \Delta_{o_1,d}f(G_{i-1})$$

$$= f(G_{i-1}) + \max_{d\in[k]} \Delta_{o_1,d}f(G_{i-1}),$$

which is no more than the function value of the $(i-1)$-th augmented solution. Then we consider the case when $g_1(1-c(o_1)-c^*) < \frac{1}{3}$. By Eq. (10), we have

$$g'(1-c(o_1)-c^*) \geq \frac{1-2g_1(1-c(o_1)-c^*)}{1-c(o_1)} > \frac{1}{3(1-c(o_1))}.$$

Since $g(0) = 0$ and g' is non-increasing by the orthant submodularity, for any $x \in [0,1]$ we have

$$g(x) = \int_0^x g'(y)dy \geq \int_0^x g'(x)dy = g'(x)\int_0^x dy = g'(x)\cdot x.$$

Applying this inequality at the point $x = 1 - c(o_1) - c^*$, then we have

$$g(1 - c(o_1) - c^*) \geq g'(1 - c(o_1) - c^*) \cdot (1 - c(o_1) - c^*) \geq \frac{1 - c(o_1) - c^*}{3(1 - c(o_1))}.$$

Recall that $1 - c(o_1) - c^*$ is the cost of the last partial greedy solution (i.e., G_{i-1}) such that adding o_1 still fits, and thus, the next item that the greedy solution selects has a cost greater than c^*. Therefore, the function value after GREEDY selects this item is at least

$$f(G_{i-1}) + c^* \cdot \rho_{g_i, d_i^g}(G_{i-1}) = g(1 - c(o_1) - c^*) + c^* \cdot g'(1 - c(o_1) - c^*)$$

$$\geq \frac{1 - c(o_1) - c^*}{3(1 - c(o_1))} + \frac{c^*}{3(1 - c(o_1))} = \frac{1}{3}.$$

When f is non-monotone, applying Lemma 4(b) at the point $x = 1 - c(o_1) - c^*$, a similar analysis follows by discussing the two cases when $g_1(1 - c(o_1) - c^*) \geq \frac{1}{4}$ and $g_1(1 - c(o_1) - c^*) < \frac{1}{4}$, which gives a $\frac{1}{4}$-approximation.   □

5   Conclusion

In this work, we explored the monotone and non-monotone k-submodular maximization under a knapsack constraint. We introduced GREEDY+MAX algorithm, which augments all partial greedy solutions by integrating the most advantageous additional item (the one with the highest marginal value). The algorithm provides a $\frac{1}{3}$-approximation for monotone functions and a $\frac{1}{4}$-approximation for non-monotone functions, using $O(kn \cdot \min\{n, B\})$ queries. These results demonstrate that GREEDY+MAX delivers notable approximation ratios while preserving favourable query complexity.

References

1. Buchbinder, N., Feldman, M., Naor, J., Schwartz, R.: Submodular maximization with cardinality constraints. In Proceedings of the Twenty-Fifth Annual ACM-SIAM Symposium on Discrete Algorithms, pp. 1433–1452. SIAM (2014)
2. Chen, J., Tang, Z., Wang, C.: Monotone k-submodular knapsack maximization: an analysis of the Greedy+Singleton algorithm. In: Ni, Q., Wu, W. (eds.) Algorithmic Aspects in Information and Management. AAIM 2022. LNCS, vol. 13513. Springer, Cham (2022). https://doi.org/10.1007/978-3-031-16081-3_13
3. Feige, U.: A threshold of $\ln n$ for approximating set cover. J. ACM **45**(4), 634–652 (1998)
4. Gridchyn, I., Kolmogorov, V.: Potts model, parametric maxflow and k-submodular functions. In: Proceedings of the IEEE International Conference on Computer Vision (ICCV), pp. 2320–2327 (2013)
5. Hirai, H., Iwamasa, Y.: On k-submodular relaxation. SIAM J. Discret. Math. **30**(3), 1726–1736 (2016)

6. Huber, A., Kolmogorov, V.: Towards minimizing k-submodular functions. In: Mahjoub, A.R., Markakis, V., Milis, I., Paschos, V.T. (eds.) ISCO 2012. LNCS, vol. 7422, pp. 451–462. Springer, Heidelberg (2012). https://doi.org/10.1007/978-3-642-32147-4_40

7. Iwata, S., Orlin, J.B.: A simple combinatorial algorithm for submodular function minimization. In: Proceedings of the Twentieth Annual ACM-SIAM Symposium on Discrete Algorithms, pp. 1230–1237. SIAM (2009)

8. Iwata, S., Tanigawa, S., Yoshida, Y.: Improved approximation algorithms for k-submodular function maximization. In: Proceedings of the 27th Annual ACM-SIAM Symposium on Discrete Algorithms (SODA), pp. 404–413 (2016)

9. Lin, H., Bilmes, J.: Multi-document summarization via budgeted maximization of submodular functions. In: Human Language Technologies: The 2010 Annual Conference of the North American Chapter of the Association for Computational Linguistics, pp. 912–920 (2010)

10. Lovász, L.: Submodular functions and convexity. Math. Program. State Art: Bonn **1982**, 235–257 (1983)

11. Nemhauser, G.L., Wolsey, L.A., Fisher, M.L.: An analysis of approximations for maximizing submodular set functions–I. Math. Program. **14**, 265–294 (1978)

12. Nguyen, L., Thai, M.T.: Streaming k-submodular maximization under noise subject to size constraint. In: Proceedings of the 37th International Conference on Machine Learning (ICML), pp. 7338–7347. PMLR (2020)

13. Ohsaka, N., Yoshida, Y.: Monotone k-submodular function maximization with size constraints. In: Proceedings of the 28th International Conference on Neural Information Processing Systems (NeurIPS), vol. 1, pp. 694–702 (2015)

14. Oshima, H.: Improved randomized algorithm for k-submodular function maximization. SIAM J. Discret. Math. **35**(1), 1–22 (2021)

15. Pham, C.V., Ha, D.K.T., Hoang, H.X., Tran, T.D.: Fast streaming algorithms for k-submodular maximization under a knapsack constraint. In Proceedings of the IEEE 9th International Conference on Data Science and Advanced Analytics (DSAA), pp. 1–10. IEEE (2022)

16. Pham, C.V., Vu, Q.C., Ha, D.K.T., Nguyen, T.T., Le, N.D.: Maximizing k-submodular functions under budget constraint: applications and streaming algorithms. J. Comb. Optim. **44**(1), 723–751 (2022)

17. Sakaue, S.: On maximizing a monotone k-submodular function subject to a matroid constraint. Discret. Optim. **23**, 105–113 (2017)

18. Schrijver, A.: Combinatorial optimization: polyhedra and efficiency, volume 24. Springer (2003). https://doi.org/10.1007/s10288-004-0035-9

19. Soma, T.: No-regret algorithms for online k-submodular maximization. In: Proceedings of the 22nd International Conference on Artificial Intelligence and Statistics (AISTATS), pp. 1205–1214. PMLR (2019)

20. Sun, Y., Liu, Y., Li, M.: Maximization of k-submodular function with a matroid constraint. In: Du, DZ., Du, D., Wu, C., Xu, D. (eds.) Theory and Applications of Models of Computation. TAMC 2022. LNCS, vol. 13571. Springer, Cham (2022). https://doi.org/10.1007/978-3-031-20350-3_1

21. Tang, J., Tang, X., Lim, A., Han, K., Li, C., Yuan, J.: Revisiting modified greedy algorithm for monotone submodular maximization with a knapsack constraint. Proc. ACM Measur. Anal. Comput. Syst. **5**(1), 1–22 (2021)

22. Tang, Z., Wang, C., Chan, H.: Monotone k-submodular secretary problems: cardinality and knapsack constraints. Theoret. Comput. Sci. **921**, 86–99 (2022)

23. Tang, Z., Wang, C., Chan, H.: On maximizing a monotone k-submodular function under a knapsack constraint. Oper. Res. Lett. **50**(1), 28–31 (2022)

24. Wang, B., Zhou, H.: Multilinear extension of k-submodular functions. arXiv preprint:2107.07103 (2021)
25. Ward, J., Živný, S.: Maximizing k-submodular functions and beyond. ACM Trans. Algorithms **12**(4), 1–26 (2016)
26. Xiao, H., Liu, Q., Zhou, Y., Li, M.: Approximation algorithms for k-submodular maximization subject to a knapsack constraint. arXiv preprint:2306.14520 (2023)
27. Yaroslavtsev, G., Zhou, S., Avdiukhin, D.: "bring your own greedy"+ max: near-optimal 1/2-approximations for submodular knapsack. In: International Conference on Artificial Intelligence and Statistics, pp. 3263–3274. PMLR (2020)
28. Yu, K., Li, M., Zhou, Y., Liu, Q.: Guarantees for maximization of k-submodular functions with a knapsack and a matroid constraint. In: Ni, Q., Wu, W. (eds.) Algorithmic Aspects in Information and Management. AAIM 2022. LNCS, vol. 13513. Springer, Cham (2022). https://doi.org/10.1007/978-3-031-16081-3_14

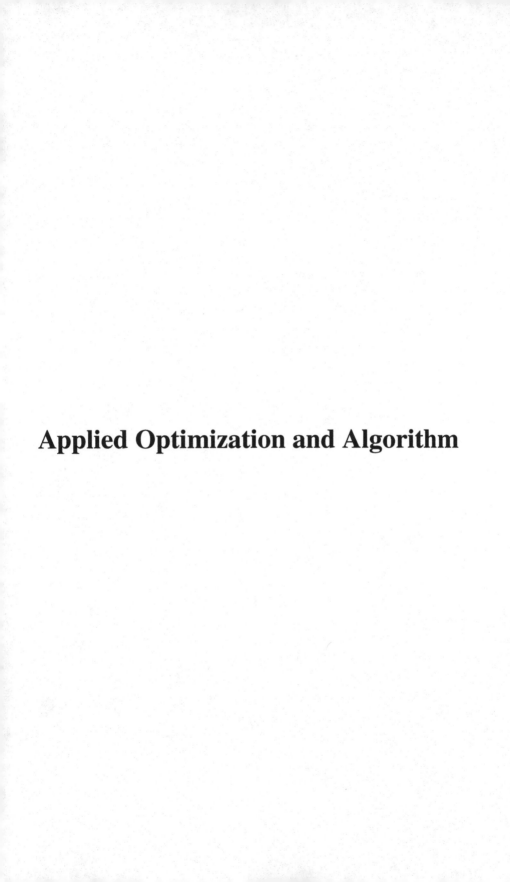

Applied Optimization and Algorithm

Improved Lower Bound for Estimating the Number of Defective Items

Nader H. Bshouty[✉][iD]

Technion - Israel Institute of Technology, Haifa, Israel
bshouty@cs.technion.ac.il

Abstract. Let X be a set of items of size n that contains some defective items, denoted by I, where $I \subseteq X$. In group testing, a *test* refers to a subset of items $Q \subset X$. The outcome of a test is 1 if Q contains at least one defective item, i.e., $Q \cap I \neq \emptyset$, and 0 otherwise.

We give a novel approach to obtaining lower bounds in non-adaptive randomized group testing. The technique produced lower bounds that are within a factor of $1/\log \log \overset{k}{\cdots} \log n$ of the existing upper bounds for any constant k. Employing this new method, we can prove the following result.

For any fixed constants k, any non-adaptive randomized algorithm that, for any set of defective items I, with probability at least $2/3$, returns an estimate of the number of defective items $|I|$ to within a constant factor requires at least

$$\Omega\left(\frac{\log n}{\log \log \overset{k}{\cdots} \log n}\right)$$

tests.

Our result almost matches the upper bound of $O(\log n)$ and solves the open problem posed by Damaschke and Sheikh Muhammad in [8,9]. Additionally, it improves upon the lower bound of $\Omega(\log n / \log \log n)$ previously established by Ron and Tsur [21] and independently by Bshouty [2].

Keywords: Group Testing · Randomized Algorithm · Estimation

1 Introduction

Let X be a set of n items, among which are defective items denoted by $I \subseteq X$. In the context of group testing, a *test* is a subset $Q \subseteq X$ of items, and its result is 1 if Q contains at least one defective item (i.e., $Q \cap I \neq \emptyset$), and 0 otherwise.

Although initially devised as a cost-effective way to conduct mass blood testing [10], group testing has since been shown to have a broad range of applications. These include DNA library screening [20], quality control in product testing [22], file searching in storage systems [16], sequential screening of experimental variables [18], efficient contention resolution algorithms for multiple-access communication [16,26], data compression [14], and computation in the data stream model [7]. Additional information about the history and diverse uses of group testing can be found in [6,11,12,15,19,20] and their respective references.

Adaptive algorithms in group testing employ tests that rely on the outcomes of previous tests, whereas *non-adaptive* algorithms use tests independent of one another, allowing all tests to be conducted simultaneously in a single step. Non-adaptive algorithms are often preferred in various group testing applications [11, 12].

In this paper, we give a novel approach to obtaining lower bounds in non-adaptive group testing. The technique produced lower bounds that are within a factor of $1/\log\log\,.\overset{k}{.}.\,\log n$ of the existing upper bounds for any constant k. Employing this new method, we can prove a new lower bound for non-adaptive estimation of the number of defective items.

Estimating the number of defective items $d := |I|$ to within a constant factor of α is the problem of identifying an integer D that satisfies $d \leq D \leq \alpha d$. This problem is widely utilized in a variety of applications [4,17,23–25].

Estimating the number of defective items in a set X has been extensively studied, with previous works including [3,5,8,9,13,21]. In this paper, we focus specifically on studying this problem in the non-adaptive setting. Bshouty [2] showed that deterministic algorithms require at least $\Omega(n)$ tests to solve this problem. For randomized algorithms, Damaschke and Sheikh Muhammad [9] presented a non-adaptive randomized algorithm that makes $O(\log n)$ tests and, with high probability, returns an integer D such that $D \geq d$ and $\mathbf{E}[D] = O(d)$. Bshouty [2] proposed a polynomial time randomized algorithm that makes $O(\log n)$ tests and, with probability at least $2/3$, returns an estimate of the number of defective items within a constant factor. Damaschke and Sheikh Muhammad [9] gave the lower bound of $\Omega(\log n)$; however, this result holds only for algorithms that select each item in each test uniformly and independently with some fixed probability. They conjectured that any randomized algorithm with a constant failure probability also requires $\Omega(\log n)$ tests. Ron and Tsur [21][1] and independently Bshouty [2] prove this conjecture up to a factor of $\log\log n$. In this paper, we establish a lower bound of

$$\Omega\left(\frac{\log n}{(c\log^* n)^{(\log^* n)+1}}\right)$$

tests, where c is a constant and $\log^* n$ is the smallest integer k such that $\log\log\,.\overset{k}{.}.\,\log n < 2$. It follows that the lower bound is

$$\Omega\left(\frac{\log n}{\log\log\,.\overset{k}{.}.\,\log n}\right)$$

for any constant k.

---

[1] The lower bound in [21] pertains to a different model of non-adaptive algorithms, but their technique implies this lower bound.

An attempt was made to establish this bound in [1]; however, an error was discovered in the proof. As a result, the weaker bound of Bshouty, $\Omega(\log n/\log\log n)$, was proved and published in [2].

The paper is organized as follows: The next subsection introduces the technique used to prove the lower bound. Section 2 defines the notation and terminology used throughout the paper. Section 3 contains the proof of the lower bound.

1.1   Old and New Techniques

In this section, we will explain both the old and new techniques used to prove the lower bounds.

Let $X = [n]$ be the set of items, and let $\mathcal{I} = 2^X$ be the collection of all the possible sets of defective items. The objective is to establish a lower bound for the test complexity of any non-adaptive randomized algorithm that, for any set of defective items $I \in \mathcal{I}$, with probability at least $2/3$, returns an integer $\mathcal{P}(I)$ such that[2] $|I| \leq \mathcal{P}(I) \leq 2|I|$.

Old Technique: The method used by Bshouty in [2] can be described as follows. Suppose we have a non-adaptive randomized algorithm $\mathcal{A}$ that makes s tests, denoted by the random variable set $\mathcal{Q} = \{Q_1, \ldots, Q_s\}$. For every set of defective items $I \in \mathcal{I}$, the algorithm returns an integer $\mathcal{P}(I)$ such that, with probability at least $2/3$, $|I| \leq \mathcal{P}(I) \leq 2|I|$.

First, he defines a partition of the set of tests $\mathcal{Q} = \bigcup_{i=1}^{r} \mathcal{Q}^{(i)}$, where each $\mathcal{Q}^{(i)}$, $i \in [r]$, contains the set of tests of sizes in the interval $[n_i, n_{i+1}]$, with $n_0 = 1$ and $n_{i+1} = poly(\log n) \cdot n_i$. There are $r = \Theta(\log n/\log\log n)$ such intervals. Let c be a small constant. By Markov's bound, there exists j (that depends only on $\mathcal{A}$, not the seed of $\mathcal{A}$) such that, with high probability (at least $1 - 1/c$), $|\mathcal{Q}^{(j)}| \leq cs/r$.

He then identifies an integer d that depends on j (and therefore on $\mathcal{A}$) such that, for every $m \in [d, 4d]$ and for a *uniform random* $I \in \mathcal{I}_m := \{I \in \mathcal{I} : |I| = m\}$, the outcomes of all tests that lie outside $\mathcal{Q}^{(j)}$ can be determined (without having to perform a test) with high probability. This probability is obtained by applying the union bound to the probability that the answer to each test in $\mathcal{Q}\backslash\mathcal{Q}^{(j)}$ can be determined by a randomly selected $I \in \mathcal{I}' := \bigcup_{m \in [d,4d]} \mathcal{I}_m$. The key idea here is that since $\mathcal{Q}^{(j)} = \{Q \in \mathcal{Q} : |Q| \in [n_j, poly(\log n)n_j]\}$, there is d such that for a random uniform set of a defective item of size $m \in [d, 4d]$, with high probability, the answers to all the tests Q that satisfy $|Q| > poly(\log n)n_j$ are 1, and, with high probability, the answers to all the tests Q that satisfy $|Q| < n_j$ are 0.

This proves that the set of tests $\mathcal{Q}^{(j)}$ can, with high probability, estimate the size of a set of defective items of a uniformly random $I \in \mathcal{I}'$, i.e., $|I| \in [d, 4d]$. In particular, it can, with high probability, distinguish[3] between a set of defective items of size d and a set of defective items of size $4d$.

---

[2] The constant 2 can be replaced by any constant.

[3] This is because the algorithm for $|I| = d$ return an integer in the interval $[d, 2d]$ and for $|I| = 4d$ returns an integer in the interval $[4d, 8d]$ and both intervals are disjoint.

If $cs/r < 1$, then $|\mathcal{Q}^{(j)}| < 1$, and therefore, with high probability, $\mathcal{Q}^{(j)} = \emptyset$. This leads to a contradiction. This is because the algorithm cannot, with high probability, distinguish between the case where $|I| = d$ and $|I| = 4d$ without any tests. Therefore, $cs/r \geq 1$, resulting in the lower bound $s > r/c = \Omega(\log n/\log\log n)$ for any non-adaptive randomized algorithm that solves the estimation problem.

New Technique: The union bound required for proving that the outcome of the tests in $\mathcal{Q}\backslash\mathcal{Q}^{(j)}$ can be determined with high probability necessitates a small enough value of r. Consequently, to satisfy the condition $cs/r < 1$, s must also be sufficiently small. This is the bottleneck in getting a better lower bound.

We surmount the bottleneck in this paper by implementing the following technique. Let $\tau = \log^* n$. As in the old technique, we define a partition of the set of tests $\mathcal{Q} = \cup_{i=1}^r \mathcal{Q}^{(i)}$. By Markov's bound there is j such that, with probability at least $1 - 1/\tau$, $|\mathcal{Q}^{(j)}| \leq \tau s/r$. Next, we identify a subset $\mathcal{I}' \subset \mathcal{I}$ such that, for a uniform random $I \in \mathcal{I}'$, the outcomes of all the tests that lie outside $\mathcal{Q}^{(j)}$ can be determined with a probability of at least $1 - 1/\tau$.

We then present the following algorithm $\mathcal{A}'$ that, with high probability, solves the problem for defective sets $I \in \mathcal{I}'$:

Algorithm 1. Algorithm $\mathcal{A}'$ for solving problem $\mathcal{P}$

Require: $I \in \mathcal{I}'$
Ensure: An estimate D of $|I|$
1: Let $\phi : X \to X$ be a uniform random permutation.
2: Let $\mathcal{Q}$ be the set of tests that the algorithm $\mathcal{A}$ makes.
3: For every test that is outside $\mathcal{Q}^{(j)}$, return the determined answer.
4: If $|\mathcal{Q}^{(j)}| \leq \tau s/r$, then make the tests in $\mathcal{Q}_\phi^{(j)} := \{\phi(Q)|Q \in \mathcal{Q}^{(j)}\}$. Otherwise, FAIL.
5: Run the algorithm $\mathcal{A}$ with the above answers to get an estimation D of $|\phi^{-1}(I)|$.
6: Return D.

We then prove that if the algorithm $\mathcal{A}$ solves the estimation problem for any set of defective items I with a success probability of at least $2/3$, then algorithm $\mathcal{A}'$ solves the estimation problem for $I \in \mathcal{I}'$ with a success probability of at least $2/3 - 2/\tau$.

This follows from:

- Making the tests in $\mathcal{Q}_\phi^{(i)}$ with a defective set of item I is equivalent to making the tests in $\mathcal{Q}$ with a defective set of items $\phi^{-1}(I)$, and therefore, with a random uniform defective set of size $|I|$.
- Since $\phi^{-1}(I)$ is a random uniform set of size $|I|$, the answers to the tests outside $\mathcal{Q}^{(j)}$ can be determined with high probability.
- With high probability, $|\mathcal{Q}^{(j)}| \leq \tau s/r$.

Now, as before, if $\tau s/r < 1$, then, with high probability, $\mathcal{Q}^{(j)} = \emptyset$, and the above algorithm does not require any tests to be performed. If, in addition, there are two instances I_1 and I_2 in $\mathcal{I}'$ where $|I_1| < 4|I_2|$ (then, the outcome for I_1 cannot be equal to the outcome of I_2), then we get a contradiction. This is because if the algorithm makes no tests, it cannot distinguish between I_1 and I_2. This contradiction, again, gives the lower bound r/τ for any non-adaptive randomized algorithm that solves the estimation problem.

To obtain a better lower bound, we again take the algorithm $\mathcal{A}'$ that solves $\mathcal{P}$ for $I \in \mathcal{I}'$ with the tests $Q' := Q^{(j)}$ with a success probability $2/3 - 2/\tau$ and, using the same procedure as before, we generate a new non-adaptive algorithm $\mathcal{A}''$ that solves $\mathcal{P}$ for $I \in \mathcal{I}'' \subset \mathcal{I}'$ using the tests in $Q'' \subset Q'$, with success probability $2/3 - 4/\tau$. The test complexity of the algorithm $\mathcal{A}''$ is $\tau^2 s/(rr')$ where $r' = O(\log \log n / \log \log \log n)$ is the number of set partitions of Q'. The lower bound obtained here is now $rr'/\tau^2 = \Omega(\log n /((\log^* n)^2 \log \log \log n))$, which is better than the lower bound r/τ obtained before[4].

If this process is repeated $\ell := \tau/24 - \log^* \tau$ times, we get an algorithm that makes $t := \tau^\ell s/(rr'r'' \cdots)$ tests. If $t < 1$, the algorithm makes no tests and, with a probability of at least[5] $2/3 - 2(\tau/24)/\tau = 7/12 > 1/2$, it can distinguish between two sets of defective items I_1 and I_2 that cannot have the same outcome. This gives the lower bound $rr'r'' \cdots /(\tau^{\tau/24}) = \log n /(c'\tau)^{\tau/24}$ for some constant c'.

Old Attempt: An attempt was made to establish this bound in [1]; however, an error was discovered in the proof. As a result, the weaker bound of $\Omega(\log n / \log \log n)$ was proved and published in [2]. In [1], Bshouty did not use Algorithm 1. Instead, he performs the same analysis on $Q^{(j)}$ (instead of $Q_\phi^{(j)}$), which results in many dependent events in the proof. The key to the success of our analysis is the inclusion of the random permutation ϕ in Algorithm 1. This permutation makes the events independent, allowing us to repeat the same analysis for $Q^{(j)}$.

2   Definitions and Notation

In this section, we introduce some definitions and notation.

We will consider the set of *items* $X = [n] = \{1, 2, \ldots, n\}$ and the set of *defective items* $I \subseteq X$. The algorithm knows n and has access to a test oracle $\mathcal{O}_I$. The algorithm can use the oracle $\mathcal{O}_I$ to make a *test* $Q \subseteq X$, and the oracle answers $\mathcal{O}_I(Q) := 1$ if $Q \cap I \neq \emptyset$, and $\mathcal{O}_I(Q) := 0$ otherwise. We say that an algorithm A, with probability at least $1 - \delta$, α-*estimates* the number of defective items if, for every $I \subseteq X$, A runs in polynomial time in n, makes tests with the oracle $\mathcal{O}_I$, and with probability at least $1 - \delta$, returns an integer D such that $|I| \leq D \leq \alpha|I|$. If α is constant, then we say that the algorithm *estimates the number of defective items to within a constant factor*.

---

[4] We can take τ as a small constant and get the lower bound $\Omega(\log n / \log \log \log n)$.
[5] Unlike in the previous footnote, τ cannot be taken as constant here.

The algorithm is called *non-adaptive* if the tests are independent of the answers of the previous tests and, therefore, can be executed simultaneously in a single step. Our objective is to develop a non-adaptive algorithm that minimizes the number of tests and provides, with a probability of at least $1 - \delta$, an estimation of the number of defective items within a constant factor.

We will denote $\log^{[k]} n = \log \log \overset{k}{\ldots} \log n$, $\log^{[0]} n = n$. Notice that $\log \log^{[i]} n = \log^{[i+1]} n$ and $2^{\log^{[i]} n} = \log^{[i-1]} n$. Let $\mathbb{N} = \{0, 1, \cdots\}$. For two real numbers r_1, r_2, we denote $[r_1, r_2] = \{r \in \mathbb{N} | r_1 \leq r \leq r_2\}$. Random variables and random sets will be in bold.

3   The Lower Bound

In this section, we prove the lower bound for the number of tests in any non-adaptive randomized algorithm that α-estimates the number of defective items for any constant α.

3.1   Lower Bound for Randomized Algorithm

In this section, we prove.

Theorem 1. *Let* $\tau = \log^* n$ *and* α *be any constant. Any non-adaptive randomized algorithm that, with probability at least* $2/3$, α*-estimates the number of defective items must make at least*

$$\Omega \left(\frac{\log n}{(480\tau)^{\tau+1}} \right)$$

tests.

We first prove the following.

Lemma 1. *Let* $n_1 = n$. *Let* $i \geq 1$ *be an integer such that* $\log^{[i]} n \geq \tau := \log^* n$. *Suppose there is an integer* $n_i = n^{\Omega(1)} \leq n$ *and a non-adaptive randomized algorithm* $\mathcal{A}_i$ *that makes*

$$s_i := \frac{\log^{[i]} n}{(480\tau)^{\tau-i+2}} \tag{1}$$

tests and for every set of defective items I *of size*

$$d \in D_i := \left[\frac{n}{n_i}, \frac{n(\log^{[i-1]} n)^{1/4}}{n_i} \right],$$

with probability at least $1 - \delta$, α*-estimates* d. *Then there is an integer* $n_{i+1} = n^{\Omega(1)} \leq n$ *and a non-adaptive randomized algorithm* $\mathcal{A}_{i+1}$ *that makes* s_{i+1} *tests and for every set of defective items* I *of size* $d \in D_{i+1}$, *with probability at least* $1 - \delta - 1/(12\tau)$, α*-estimates* d.

Proof. Let

$$N_i = \left[\frac{n_i}{(\log^{[i-1]} n)^{1/4}}, n_i \right].$$

We will be interested in all the tests Q of the algorithm $\mathcal{A}_i$ that satisfies $|Q| \in N_i$. We now partition N_i into smaller sets. Let

$$N_{i,j} = \left[\frac{n_i}{(\log^{[i]} n)^{4j+4}}, \frac{n_i}{(\log^{[i]} n)^{4j}} \right]$$

where $j = [0, r_i - 1]$ and

$$r_i = \frac{\log^{[i]} n}{16 \log^{[i+1]} n}. \tag{2}$$

Since the lowest endpoint of the interval N_{i,r_i-1} is

$$\frac{n_i}{(\log^{[i]} n)^{4(r_i-1)+4}} = \frac{n_i}{2^{4r_i \log^{[i+1]} n}} = \frac{n_i}{2^{(1/4) \log^{[i]} n}} = \frac{n_i}{(\log^{[i-1]} n)^{1/4}}$$

and the right endpoint of $N_{i,0}$ is n_i, we have, $N_i = \cup_{j=0}^{r_i-1} N_{i,j}$.

Let $\mathcal{Q} = \{Q_1, \dots, Q_{s_i}\}$ be the (random variable) tests that the randomized algorithm $\mathcal{A}_i$ makes. Let $\mathbf{T}_j$ be a random variable representing the number of tests $Q \in \mathcal{Q}$ that satisfies $|Q| \in N_{i,j}$. Since $\mathcal{A}_i$ makes s_i tests, we have $\mathbf{T}_0 + \cdots + \mathbf{T}_{r_i} \le s_i$. Therefore, by (1) and (2), (in the expectation, $\mathbf{E}_j$, j is uniformly at random over $[0, r_i - 1]$ and the other $\mathbf{E}$ is over the random seed of the algorithm $\mathcal{A}_i$)

$$\mathbf{E}_j[\mathbf{E}[\mathbf{T}_j]] = \mathbf{E}[\mathbf{E}_j[\mathbf{T}_j]] \le \frac{s_i}{r_i} = \frac{16 \log^{[i+1]} n}{(480\tau)^{\tau-i+2}}.$$

Therefore, there is $0 \le j_i \le r_i - 1$ that depends only on the algorithm $\mathcal{A}_i$ (not the seed of the algorithm) such that

$$\mathbf{E}[\mathbf{T}_{j_i}] \le \frac{16 \log^{[i+1]} n}{(480\tau)^{\tau-i+2}}.$$

By Markov's bound, with probability at least $1 - 16/(480\tau) = 1 - 1/(30\tau)$,

$$|\{Q \in \mathcal{Q} : |Q| \in N_{i,j_i}\}| = \mathbf{T}_{j_i} \le \frac{\log^{[i+1]} n}{(480\tau)^{\tau-i+1}} = s_{i+1}. \tag{3}$$

Define

$$n_{i+1} = \frac{n_i}{(\log^{[i]} n)^{4j_i+2}}. \tag{4}$$

Since $n_i = n^{\Omega(1)}$ and

$$(\log^{[i]} n)^{4j_i+2} \le (\log^{[i]} n)^{4r_i-2} = \frac{(\log^{[i-1]} n)^{1/4}}{(\log^{[i]} n)^2},$$

we have that $n_{i+1} = n^{\Omega(1)} \leq n$. Notice that this holds even for $i = 1$. This is because $n_1 = n$ and $(\log^{[0]} n)^{1/4} = n^{1/4}$ so $n_2 \geq n^{1/4} \log^2 n = n^{\Omega(1)}$.

Consider the following randomized algorithm $\mathcal{A}'_i$:

1. Let $\mathcal{Q} = \{Q_1, \ldots, Q_{s_i}\}$ be the set of tests of $\mathcal{A}_i$.
2. Choose a uniformly at random permutation $\phi : [n] \to [n]$.
3. Let $\mathcal{Q}' = \{Q'_1, \ldots, Q'_{s_i}\}$ where $Q'_i = \phi(Q_i) := \{\phi(q)|q \in Q_i\}$.
4. Make all the tests in $\mathcal{Q}'$ and give the answer to $\mathcal{A}_i$.
5. Run $\mathcal{A}_i$ with the above answers on $\mathcal{Q}'$ and output what $\mathcal{A}_i$ outputs.

Consider the following algorithm $\mathcal{A}_{i+1}$:

1. Let $\mathcal{Q} = \{Q_1, \ldots, Q_{s_i}\}$ be the set of tests of $\mathcal{A}_i$.
2. Choose a uniformly at random permutation $\phi : [n] \to [n]$.
3. Let $\mathcal{Q}' = \{Q'_1, \ldots, Q'_{s_i}\}$ where $Q'_i = \phi(Q_i) := \{\phi(q)|q \in Q_i\}$.
4. For all the tests in

$$\mathcal{Q}_0 := \left\{ Q'_i \in \mathcal{Q}' : |Q'_i| \leq \frac{n_i}{(\log^{[i]} n)^{4j_i+4}} \right\},$$

answer 0.
5. For all the tests in

$$\mathcal{Q}_1 := \left\{ Q'_i \in \mathcal{Q}' : |Q'_i| \geq \frac{n_i}{(\log^{[i]} n)^{4j_i}} \right\},$$

answer 1.
6. Let

$$\mathcal{Q}'' = \left\{ Q'_i \in \mathcal{Q}' : \frac{n_i}{(\log^{[i]} n)^{4j_i+4}} < |Q'_i| < \frac{n_i}{(\log^{[i]} n)^{4j_i}} \right\}$$
$$= \{Q'_i \in \mathcal{Q}' : |Q'_i| \in N_{i,j_i}\}$$

7. If

$$|\mathcal{Q}''| > s_{i+1} = \frac{\log^{[i+1]} n}{(480\tau)^{\tau-i+1}}$$

return -1 (FAIL) and halt.
8. Make all the tests in $\mathcal{Q}''$.
9. Run $\mathcal{A}_i$ with the above answers on $\mathcal{Q}'$ and output what $\mathcal{A}_i$ outputs.

We now show that for every set of defective items $|I|$ of size

$$d \in D_{i+1} := \left[\frac{n}{n_{i+1}}, \frac{n(\log^{[i]} n)^{1/4}}{n_{i+1}} \right], \tag{5}$$

with probability at least $1 - \delta - 1/(12\tau)$, algorithm $\mathcal{A}_{i+1}$ α-estimates d using s_{i+1} tests.

In algorithm $\mathcal{A}_{i+1}$, Step 8 is the only step that makes tests. Therefore, by step 7 the test complexity of $\mathcal{A}_{i+1}$ is s_{i+1}.

By the definition of D_i, and since, by (4), $n/n_{i+1} > n/n_i$, and, by (4) and (2)

$$\frac{n(\log^{[i]} n)^{1/4}}{n_{i+1}} = \frac{n}{n_i}(\log^{[i]} n)^{4j_i+2.25} \leq \frac{n}{n_i}(\log^{[i]} n)^{4r_i-1.75} \leq \frac{n(\log^{[i-1]} n)^{1/4}}{n_i},$$

we can conclude that $D_{i+1} \subset D_i$.

Consider the following events:

1. Event M_0: For some $\mathbf{Q}' = \phi(\mathbf{Q}) \in \mathbf{\mathcal{Q}}'$ such that $|\mathbf{Q}'| \leq n_i/(\log^{[i]} n)^{4j_i+4}$ (i.e., $\mathbf{Q}' \in \mathbf{\mathcal{Q}}_0$), $\mathbf{Q}' \cap I \neq \emptyset$ (i.e., the answer to the test $\mathbf{Q}$ is 1).
2. Event M_1: For some $\mathbf{Q}' = \phi(\mathbf{Q}) \in \mathbf{\mathcal{Q}}'$ such that $|\mathbf{Q}'| \geq n_i/(\log^{[i]} n)^{4j_i}$, (i.e., $\mathbf{Q}' \in \mathbf{\mathcal{Q}}_1$) $\mathbf{Q}' \cap I = \emptyset$ (i.e., the answer to the test $\mathbf{Q}$ is 0).
3. Event W: $|\mathbf{\mathcal{Q}}''| > s_{i+1} = \frac{\log^{[i+1]} n}{(480\tau)^{\tau-i+1}}$.

The success probability of the algorithm $\mathcal{A}_{i+1}$ on a set I of defective items with $|I| = d \in D_{i+1}$ is (here the probability is over ϕ and the random tests $\mathbf{\mathcal{Q}}$)

$$\mathbf{Pr}[\mathcal{A}_{i+1} \text{ succeeds on } I] = \mathbf{Pr}[(\mathcal{A}'_i \text{ succeeds on } I) \wedge \bar{M}_0 \wedge \bar{M}_1 \wedge \bar{W}]$$
$$\geq \mathbf{Pr}[\mathcal{A}'_i \text{ succeeds on } I] - \mathbf{Pr}[M_0 \vee M_1 \vee W]$$
$$\geq \mathbf{Pr}[\mathcal{A}'_i \text{ succeeds on } I] - \mathbf{Pr}[M_0] - \mathbf{Pr}[M_1] - \mathbf{Pr}[W].$$

Now, since $|\phi^{-1}(I)| = |I|$ and $\mathbf{Q}' \cap I = \phi(\mathbf{Q}) \cap I \neq \emptyset$ if and only if $\mathbf{Q} \cap \phi^{-1}(I) \neq \emptyset$,

$$\mathbf{Pr}_{\phi,\mathbf{\mathcal{Q}}}[\mathcal{A}'_i \text{ succeeds on } I] = \mathbf{Pr}_{\phi,\mathbf{\mathcal{Q}}}[\mathcal{A}_i \text{ succeeds on } \phi^{-1}(I)] \geq 1 - \delta.$$

Therefore, to get the result, it is enough to show that $\mathbf{Pr}[M_0] \leq 1/(300\tau)$, $\mathbf{Pr}[M_1] \leq 1/(300\tau)$ and $\mathbf{Pr}[W] \leq 1/(30\tau)$.

First, since $|\mathbf{Q}'| = |\phi(\mathbf{Q})| = |\mathbf{Q}|$ we have

$$|\mathbf{\mathcal{Q}}''| = |\{\mathbf{Q}'_i : |\mathbf{Q}'_i| \in N_{i,j_i}\}| = |\{\mathbf{Q}_i : |\mathbf{Q}_i| \in N_{i,j_i}\}| = \mathbf{T}_{j_i}.$$

By (3), with probability at most $1/(30\tau)$,

$$|\mathbf{\mathcal{Q}}''| = \mathbf{T}_{j_i} > \frac{\log^{[i+1]} n}{(480\tau)^{\tau-i+1}}.$$

Therefore, $\mathbf{Pr}[W] \leq 1/(30\tau)$.

We now will show that $\mathbf{Pr}[M_0] \leq 1/(300\tau)$. We have, (A detailed explanation of every step can be found below.)

$$\mathbf{Pr}_{\phi,\mathbf{\mathcal{Q}}}[M_0] = \mathbf{Pr}_{\phi,\mathbf{\mathcal{Q}}}[(\exists \mathbf{Q} \in \mathbf{\mathcal{Q}}, \phi(\mathbf{Q}) \in \mathbf{\mathcal{Q}}_0) \ \phi(\mathbf{Q}) \cap I \neq \emptyset] \tag{6}$$
$$= \mathbf{Pr}_{\phi,\mathbf{\mathcal{Q}}}[(\exists \mathbf{Q} \in \mathbf{\mathcal{Q}}, \phi(\mathbf{Q}) \in \mathbf{\mathcal{Q}}_0) \ \mathbf{Q} \cap \phi^{-1}(I) \neq \emptyset] \tag{7}$$

$$\leq s_i \left(1 - \prod_{k=0}^{d-1} \left(1 - \frac{n_i}{(\log^{[i]} n)^{4j_i+4}(n-k)} \right) \right) \tag{8}$$

$$\leq s_i \left(1 - \left(1 - \frac{2n_i}{(\log^{[i]} n)^{4j_i+4}n} \right)^d \right) \tag{9}$$

$$\leq s_i d \frac{2n_i}{(\log^{[i]} n)^{4j_i+4}n} \tag{10}$$

$$\leq \frac{\log^{[i]} n}{(480\tau)^{\tau-i+2}} \cdot \frac{n(\log^{[i]} n)^{1/4}}{n_{i+1}} \cdot \frac{2n_i}{(\log^{[i]} n)^{4j_i+4}n} \tag{11}$$

$$= \frac{\log^{[i]} n}{(480\tau)^{\tau-i+2}} \cdot \frac{n(\log^{[i]} n)^{4j_i+2\frac{1}{4}}}{n_i} \cdot \frac{2n_i}{(\log^{[i]} n)^{4j_i+4}n} \tag{12}$$

$$= \frac{2}{(480\tau)^{\tau-i+2}(\log^{[i]} n)^{3/4}} \leq \frac{1}{300\tau}. \tag{13}$$

(6) follows from the definition of the event M_0. (7) follows from the fact that for any permutation $\phi : [n] \to [n]$ and two sets $X, Y \subseteq [n]$, $\phi(X) \cap Y \neq \emptyset$ is equivalent to $X \cap \Phi^{-1}(Y) \neq \emptyset$. (8) follows from:

1. The union bound and $|\mathcal{Q}_0| \leq |\mathcal{Q}| = s_i$.
2. For a random at uniform ϕ, and a d-subset of $[n]$, $\phi^{-1}(I)$ is a random uniform d-subset of $[n]$.
3. For every $Q \in \mathcal{Q}$ such that $\phi(Q) \in \mathcal{Q}_0$, $|Q| = |\phi(Q)| \leq n_i/(\log^{[i]} n)^{4j_i+4}$.

(9) follows from the fact that since $d \in D_{i+1}$ and $n_{i+1} = n^{\Omega(1)}$ by (5), we have $d \leq n/2$. (10) follows from the inequality $(1-x)^d \geq 1 - dx$. (11) follows from (1) and (5). (12) follows from (4). (13) follows from the fact that since $\log^{[i]} n \geq \tau = \log^* n$, we have $i \leq \tau$, and therefore $(480\tau)^{\tau-i+2} \geq (480\tau)^2 \geq 600\tau$.

We now prove that $\mathbf{Pr}[M_1] \leq 1/(300\tau)$.

$$\mathbf{Pr}_{\phi,\mathcal{Q}}[M_1] = \mathbf{Pr}_{\phi,\mathcal{Q}}[(\exists Q \in \mathcal{Q}, \phi(Q) \in \mathcal{Q}_1) \; \phi(Q) \cap I = \emptyset] \tag{14}$$

$$= \mathbf{Pr}_{\phi,\mathcal{Q}}[(\exists Q \in \mathcal{Q}, \phi(Q) \in \mathcal{Q}_1) \; Q \cap \phi^{-1}(I) = \emptyset] \tag{15}$$

$$\leq s_i \prod_{k=0}^{d-1} \left(1 - \frac{n_i}{(\log^{[i]} n)^{4j_i}(n-k)} \right) \tag{16}$$

$$\leq s_i \left(1 - \frac{n_i}{(\log^{[i]} n)^{4j_i}n} \right)^d \leq s_i \exp\left(-\frac{dn_i}{(\log^{[i]} n)^{4j_i}n} \right) \tag{17}$$

$$\leq \frac{\log^{[i]} n}{(480\tau)^{\tau-i+2}} \exp\left(-\frac{\frac{n}{n_{i+1}} \frac{n_i}{(\log^{[i]} n)^{4j_i}}}{n} \right) \tag{18}$$

$$\leq \frac{\log^{[i]} n}{(480\tau)^{\tau-i+2}} \exp(-(\log^{[i]} n)^2) \tag{19}$$

$$\leq \frac{1}{300\tau}. \tag{20}$$

(14) follows from the definition of the event M_1. (15) follows from the fact that for any permutation $\phi : [n] \rightarrow [n]$ and two sets $X, Y \subseteq [n]$, $\phi(X) \cap Y = \emptyset$ is equivalent to $X \cap \Phi^{-1}(Y) = \emptyset$. In (16), we again use the union bound, $|\mathcal{Q}_1| \leq |\mathcal{Q}| = s_i$, the fact that $\phi^{-1}(I)$ is a random uniform d-subset, and for $Q' \in \mathcal{Q}_1$, $|Q'| \geq n_i/(\log^{[i]} n)^{4j_i}$. (17) follows from the inequalities $(1 - y/(n-k)) \leq (1 - y/n)$ and $1 - x \leq e^{-x}$ for every x, and $y \geq 0$. (18) follows from (1) and (5). (19) follows from (4). (20) follows from the fact that $i \leq \tau$, and therefore $(480\tau)^{\tau-i+2} \geq (480\tau)^2 \geq 300\tau$. $\qquad\square$

We are now ready to prove Theorem 1.

Proof. Suppose, for the contrary, there is a non-adaptive randomized algorithm $\mathcal{A}_1$ that, with probability at least $2/3$, α-estimates the number of defective items and makes

$$m := \frac{\log n}{(480\tau)^{\tau+1}}$$

tests. Recall that $n_1 = n$ and $\log^{[0]} n = n$. We use Lemma 1 with $\delta = 1/3$, $D_1 = [1, n^{1/4}]$ and $s_1 = m$.

Now let ℓ be an integer such that $\log \log^* n < \log^{[\ell]} n \leq \log^* n = \tau$. Then $\log^{[\ell-1]} n > 2^{\log^{[\ell]} n} > 2^{\log \log^* n} = \tau$. Now use Lemma 1 with $i = \ell - 1$ and get

$$s_\ell = \frac{\log^{[\ell]} n}{(480\tau)^{\tau-\ell+2}} \leq \frac{\tau}{(480\tau)^2} < 1.$$

So, algorithm $\mathcal{A}_\ell$ makes no tests and with probability at least $2/3 - \ell/(12\tau) \geq 7/12 > 1/2$ α-estimates the size of defective items I provided that

$$|I| \in D_\ell = \left[\frac{n}{n_\ell}, \frac{n(\log^{[\ell-1]} n)^{1/4}}{n_\ell} \right].$$

In particular, with probability more than $1/2$, we can distinguish between defective sets of size n/n_ℓ and size greater than $\alpha n/n_\ell$ without performing any test, which is impossible. A contradiction. $\qquad\square$

4   Conclusion

In this paper, we have presented a novel approach to obtaining lower bounds in non-adaptive randomized group testing. Our technique has allowed us to establish a lower bound of $\Omega(\log n/((c \log^* n)^{\log^* n}))$, for some constant c, on the test complexity of any randomized non-adaptive algorithm that estimates the number of defective items within a constant factor. This lower bound significantly improves upon the previous bound of $\Omega(\log n/\log \log n)$ that was established in [2, 21].

The key to our success was the introduction of a random permutation ϕ in Algorithm 1, which enabled us to make the events independent and repeat the analysis to the set $\mathcal{Q}^{(j)}$. This crucial step allowed us to overcome the bottleneck in the previous technique and achieve a better lower bound.

A challenging open problem is establishing a lower bound of $\Omega(\log n)$ for the test complexity in non-adaptive randomized algorithms that estimate the number of defective items within a constant factor.

References

1. Bshouty, N.H.: Lower bound for non-adaptive estimate the number of defective items. Electronic Colloquium on Computational Complexity TR18-053 (2018)
2. Bshouty, N.H.: Lower bound for non-adaptive estimation of the number of defective items. In: Lu, P., Zhang, G. (eds.) 30th International Symposium on Algorithms and Computation, ISAAC 2019. LIPIcs, vol. 149, pp. 2:1–2:9. Schloss Dagstuhl - Leibniz-Zentrum für Informatik (2019)
3. Bshouty, N.H., Bshouty-Hurani, V.E., Haddad, G., Hashem, T., Khoury, F., Sharafy, O.: Adaptive group testing algorithms to estimate the number of defectives. ALT (2017). http://arxiv.org/abs/1712.00615
4. Chen, C.L., Swallow, W.H.: Using group testing to estimate a proportion, and to test the binomial model. Biometrics **46**(4), 1035–1046 (1990)
5. Cheng, Y., Xu, Y.: An efficient FPRAS type group testing procedure to approximate the number of defectives. J. Comb. Optim. **27**(2), 302–314 (2014)
6. Cicalese, F.: Fault-Tolerant Search Algorithms - Reliable Computation with Unreliable Information. Monographs in Theoretical Computer Science. An EATCS Series, Springer, Heidelberg (2013). https://doi.org/10.1007/978-3-642-17327-1
7. Cormode, G., Muthukrishnan, S.: What's hot and what's not: tracking most frequent items dynamically. ACM Trans. Database Syst. **30**(1), 249–278 (2005)
8. Damaschke, P., Muhammad, A.S.: Bounds for nonadaptive group tests to estimate the amount of defectives. In: Combinatorial Optimization and Applications - 4th International Conference, COCOA 2010, pp. 117–130 (2010)
9. Damaschke, P., Muhammad, A.S.: Competitive group testing and learning hidden vertex covers with minimum adaptivity. Discrete Math. Alg. Appl. **2**(3), 291–312 (2010)
10. Dorfman, R.: The detection of defective members of large populations. Ann. Math. Statist. 436–440 (1943)
11. Du, D., Hwang, F.K.: Combinatorial Group Testing and Its Applications. World Scientific Publishing Company (2000)
12. Du, D., Hwang, F.K.: Pooling Design and Nonadaptive Group Testing: Important Tools for DNA Sequencing. World Scientific Publishing Company (2006)
13. Falahatgar, M., Jafarpour, A., Orlitsky, A., Pichapati, V., Suresh, A.T.: Estimating the number of defectives with group testing. In: IEEE International Symposium on Information Theory, ISIT 2016, pp. 1376–1380 (2016)
14. Hong, E.S., Ladner, R.E.: Group testing for image compression. IEEE Trans. Image Process. **11**(8), 901–911 (2002). https://doi.org/10.1109/TIP.2002.801124
15. Hwang, F.K.: A method for detecting all defective members in a population by group testing. J. Am. Stat. Assoc. **67**, 605–608 (1972)
16. Kautz, W.H., Singleton, R.C.: Nonrandom binary superimposed codes. IEEE Trans. Inf. Theory **10**(4), 363–377 (1964)

17. Gastwirth, J.L., Hammick, P.A.: Estimation of the prevalence of a rare disease, preserving the anonymity of the subjects by group testing: application to estimating the prevalence of aids antibodies in blood donors. J. Stat. Plann. Inference **22**(1), 15–27 (1989)

18. Li, C.H.: A sequential method for screening experimental variables. J. Amer. Statist. Assoc. **57**, 455–477 (1962)

19. Macula, A.J., Popyack, L.J.: A group testing method for finding patterns in data. Discret. Appl. Math. **144**(1–2), 149–157 (2004)

20. Ngo, H.Q., Du, D.: A survey on combinatorial group testing algorithms with applications to DNA library screening. In: Discrete Mathematical Problems with Medical Applications, Proceedings of a DIMACS Workshop, 8–10 December 1999, pp. 171–182 (1999)

21. Ron, D., Tsur, G.: The power of an example: hidden set size approximation using group queries and conditional sampling. ACM Trans. Comput. Theory **8**(4), 15:1–15:19 (2016)

22. Sobel, M., Groll, P.A.: Group testing to eliminate efficiently all defectives in a binomial sample. Bell Syst. Tech. J. **38**, 1179–1252 (1959)

23. Swallow, W.H.: Group testing for estimating infection rates and probabilities of disease transmission. Phytopathology **75**(8), 882 (1985)

24. Thompson, K.H.: Estimation of the proportion of vectors in a natural population of insects. Biometrics **18**(4), 568–578 (1962)

25. Walter, S.D., Hildreth, S.W., Beaty, B.J.: Estimation of infection rates in population of organisms using pools of variable size. Am. J. Epidemiol. **112**(1), 124–128 (1980)

26. Wolf, J.K.: Born again group testing: multiaccess communications. IEEE Trans. Inf. Theory **31**(2), 185–191 (1985)

Popularity on the Roommate Diversity Problem

Steven Ge$^{(\boxtimes)}$ and Toshiya Itoh

Tokyo Institute of Technology, Meguro, Japan
ge.s.aa@m.titech.ac.jp, titoh@c.titech.ac.jp

Abstract. A recently introduced restricted variant of the multidimensional stable roommates problem is the roommate diversity problem. We study the roommate diversity problem with the notion of popularity.

We show that for the roommate diversity problem with the room size fixed to 2, a popular partitioning of agents is guaranteed to exist and can be computed in polynomial time. By contrast, when there are no restrictions on the room size of a roommate diversity game, a popular partitioning may fail to exist and the problem becomes intractable.

Keywords: Stable roommates problem · Popularity · Dichotomous Trichotomous preferences · Coalition formation · Algorithms · Co-NP-hard

1 Introduction

The formation of stable coalitions in multi-agent systems is a computational problem that has several variations with different conditions on the coalition size. Boehmer and Elkind [5] recently introduced a restricted variant of the multidimensional stable roommates problem called the roommate diversity problem: each agent belongs to one of two types (e.g., red and blue), and the agents' preferences over the rooms solely depend on the fraction of agents of their own type among their roommates. Ties are allowed in the preferences of the agents.

Diversity preferences were originally introduced by Bredereck et al. [7] in the context of hedonic games. Hedonic diversity games have been extensively studied by Bredereck et al. [7], Boehmer and Elkind [4], and Darmann [13]. Ganian et. al. [15] analyzed the parameterized complexity of hedonic diversity games using a combination of the parameters: number of agent types (colors); maximum coalition size; maximum number of coalitions; and number of agent (preference) types.

The roommate diversity model captures important aspects of several real-world coalition formation scenarios, such as seating arrangements at events, and splitting students into teams for group projects. In the latter scenario, a local student may have difficulty communicating in English and an international student may not be able to speak the local language well. Thus, the preference of a student may solely rely on the fraction of international group members, where the ranking of the fractions depend on how much they want to learn the other language and the ease of communication they require.

W. Wu and J. Guo (Eds.): COCOA 2023, LNCS 14461, pp. 316–329, 2024.
https://doi.org/10.1007/978-3-031-49611-0_23

Various notions that define the stability or optimality of a partitioning of agents exist. The notion of popularity was introduced by Gärdenfors [16] in 1975. Popular matchings have been an exciting area of research [20]. Cseh [10] has recently provided a survey on popular matchings.

Computing a popular partition in a stable roommates game is NP-hard, even if the preferences are strict [12,14,19]. One method of obtaining tractability results from intractable coalition formation problems is to put restrictions on the agents' preferences for which the associated computational problems become tractable. This approach has been successfully used in the stable roommates problem [1–3,8,9,11]. Diversity preferences is the same type of approach.

The roommate diversity problem has been studied with the notions envy-freeness, exchange stability, core stability, and Pareto optimality. For the majority of these notions computing a stable/optimal partitioning of a roommate diversity game is NP-hard or co-NP-hard. On the other hand, for some of the stability notions the problem is FPT when using the room size as parameter [5].

1.1 Our Results

We show that popularity on the roommate diversity problem with the room size fixed to 2 is tractable. Particularly, a popular partitioning of agents is guaranteed to exist and can be computed in polynomial time. Additionally, a mixed popular partitioning of agents is always guaranteed to exist in any roommate diversity game. By contrast, when there are no restrictions on the room size of a roommate diversity game, a popular partitioning may fail to exist and the problem becomes intractable. Our intractability results are summarized as follows: Determining the existence of a popular partitioning is co-NP-hard, even if the agents' preferences are trichotomous; Determining the existence of a strictly popular partitioning is co-NP-hard, even if the agents' preferences are dichotomous; A mixed popular outcome for a roommate diversity game is not computable in polynomial time, even if the agents' preferences are dichotomous, unless $P = NP$.

Our hardness proofs are inspired by Theorem 7.3 of the work by Boehmer and Elkind [5] in combination with Sect. 4.2.2 of the work by Brandt and Bullinger [6]. We also use Proposition 1 of the work by Brandt and Bullinger [6] to show that a mixed popular partitioning of agents is always guaranteed to exist in any roommate diversity game.

2 Preliminaries

In this work, we slightly extend the notation and definitions described in Sect. 2 and Theorem 7.3 of the work by Boehmer and Elkind [5], and Sect. 3 of the work by Brandt and Bullinger [6]. For our hardness results, we construct polynomial time reductions from the Exact Cover by 3-Sets (X3C) problem, which is known to be NP-complete [17].

For $s \in \mathbb{N}, t \in \mathbb{N}^+$, we define the integer sets $[t] = \{1, \ldots, t\}$ and $[s,t] = \{s, \ldots, t\}$.

2.1   Roommate Diversity Problem

A roommate diversity game G with room size s and agent set $N = R \cup B$ is a quadruple $(R, B, s, (\succsim_a)_{a \in R \cup B})$, where $|N| = ks$ for some $k \in \mathbb{N}$. The preference relation $\succsim_a$, where $a \in N$, is a complete transitive weak order over $D = \{\frac{i}{s} | j \in [0, s]\}$. We call the agents in R red agents and the agents in B blue agents. An s-sized subset of N is called a room. An outcome $\pi = \{C_1, \ldots, C_k\}$ of G is a partitioning of the agents N into k rooms. Let $\pi(a)$ denote the room in π that contains agent $a \in N$. For a room $C \subseteq N$, let $\theta(C) = \frac{|C \cap R|}{|C|}$ denote the fraction of red agents in C. We say that C has fraction $\theta(C)$ or the fraction of C is $\theta(C)$.

For an agent $a \in N$, the preference relation $\succsim_a$ are the preferences of a over the fraction of red agents in its room. For example, $\frac{2}{s} \succsim_a \frac{3}{s}$ means that agent a likes being in a room with 2 red agents at least as much as being in a room with 3 red agents. Note that a red agent cannot be in a room with fraction $\frac{0}{s}$ and a blue agent cannot be in a room with fraction $\frac{s}{s}$. Thus, discarding these 'impossible' fractions in their preference relation does not impact our results. We write $\theta(S) \succ_a \theta(T)$ and say that agent a strictly prefers room S over room T if $\theta(S) \succsim_a \theta(T)$ and $\theta(T) \not\succsim_a \theta(S)$. Additionally, we say that agent a weakly prefers S over T if $\theta(S) \succsim_a \theta(T)$. If agent a weakly prefers S over T and T over S, we write $\theta(S) \sim_a \theta(T)$ and say that a is indifferent between S and T.

The preference relation of an agent a is trichotomous, if there exists a partitioning of D into three sets D_a^+, D_a^n, D_a^- such that for all $d^+ \in D_a^+, d^n \in D_a^n$, and $d^- \in D_a^-$ it holds that $d^+ \succ_a d^n$ and $d^n \succ_a d^-$. Additionally, for all $d_1^+, d_2^+ \in D_a^+$ it holds that $d_1^+ \sim_a d_2^+$, for all $d_1^n, d_2^n \in D_a^n$ it holds that $d_1^n \sim_a d_2^n$, and for all $d_1^-, d_2^- \in D_a^-$ it holds that $d_1^- \sim_a d_2^-$. We say that agent a approves of, is neutral about, disapproves of the fractions in D_a^+, D_a^n, D_a^- respectively.

We shall use tables to define the tri- and dichotomous preferences. The first column specifies an agent a. The second column denotes the set D_a^+. In the example table below, we have that $D_a^+ = \{f_1, \ldots, f_l\}$. The sets D_a^n and D_a^- are defined analogously in the third and fourth column respectively. In the case of dichotomous preferences, the third column defining the set D_a^n is discarded. The final column may contain additional comments regarding the range of a variable.

Agent	Preference Profile			
	D_a^+	D_a^n	D_a^-	
a	$\{f_1, \ldots, f_l\}$	$\{f_{l+1}, \ldots, f_m\}$	$\{f_{m+1}, \ldots, f_n\}$	"additional comments"

For an outcome π, let D_π^+ denote the agents that approve of the fraction of its room in π, i.e., $D_\pi^+ = \{a \in N | \theta(\pi(a)) \in D_a^+\}$. We define D_π^n and D_π^- analogously.

2.2 Popularity

For outcomes π, π', let $N(\pi, \pi') = \{a \in N | \theta(\pi(a)) \succ_a \theta(\pi'(a))\}$ be the set of agents who prefer π over π' and $\phi(\pi, \pi') = |N(\pi, \pi')| - |N(\pi', \pi)|$. We call $\phi(\pi, \pi')$ the popularity margin of π and π'.

An outcome π is strictly more popular than outcome π' if $\phi(\pi, \pi') > 0$. An outcome π is popular if for any outcome π' we have $\phi(\pi, \pi') \geq 0$. An outcome π is strictly popular if for any other outcome $\pi' \neq \pi$ we have $\phi(\pi, \pi') > 0$. Note that there can be at most one strictly popular outcome.

A mixed outcome $p = \{(\pi_1, p_1), \ldots, (\pi_t, p_t)\}$ is a set of pairs, where for each $i \in [t]$, π_i is an outcome of a roommate diversity game and $(p_1, \ldots, p_t)$ is a probability distribution. For mixed outcomes $p = \{(\pi_1, p_1), \ldots, (\pi_t, p_t)\}$ and $q = \{(\sigma_1, q_1), \ldots, (\sigma_u, q_u)\}$, we define the popularity margin of p and q to be

$$\phi(p, q) = \sum_{i=1}^{t} \sum_{j=1}^{u} p_i q_j \phi(\pi_i, \sigma_j).$$ A mixed outcome p is popular if for any mixed outcome q we have $\phi(p, q) \geq 0$.

2.3 Exact Cover by 3-Sets Problem

Let $X = [m]$, where $m \in \mathbb{N}^+$ and $m \mod 3 = 0$, and let $C = \{A_1, \ldots, A_q\}$ be a collection of 3-element subsets of X. An instance of the X3C problem is a tuple (X, C) and asks: does there exist a subset $C' \subseteq C$ such that C' partitions X? We call such a C' a solution of (X, C).

For $i \in X$, let $J^i = \{j_1^i, \ldots, j_{m_i}^i\}$ be the set of indices of the 3-sets in C that contain i, i.e., $j \in J^i \iff i \in A_j$ (or equivalently $J^i = \{j \in [q] | i \in A_j\}$).

3 Room Size Two

For a roommate diversity game $G = (R, B, 2, (\succsim_a)_{a \in R \cup B})$ with room size 2, a room $S \subseteq R \cup B$, can have exactly one of 3 possible fractions, i.e., $\theta(S) \in \{\frac{0}{2}, \frac{1}{2}, \frac{2}{2}\}$. Let us call a room with fraction $\frac{0}{2}$ or $\frac{2}{2}$ a pure blue or pure red room respectively. A room with fraction $\frac{1}{2}$ is called a mixed room.

As mentioned in Sect. 2, we can discard the 'impossible' fractions from the preference relations of the agents. Thus, the only relevant fractions for a red agent are $\frac{1}{2}$ and $\frac{2}{2}$ and the only relevant fractions for a blue agent are $\frac{0}{2}$ and $\frac{1}{2}$. Let us call an agent that is in a room with one of their most preferred fractions happy. Otherwise, we call the agent sad. That is, an agent $a \in R \cup B$ is happy in outcome π if for each $f \in \{\frac{0}{2}, \frac{1}{2}, \frac{2}{2}\}$ we have $\theta(\pi(a)) \succsim_a f$. An agent $a \in R \cup B$ is sad in outcome π if there exists $f \in \{\frac{0}{2}, \frac{1}{2}, \frac{2}{2}\}$ such that $\theta(\pi(a)) \prec_a f$.

A red agent r can only have one of 3 possible preference relations, namely $\frac{1}{2} \succ_r \frac{2}{2}$, $\frac{1}{2} \prec_r \frac{2}{2}$, or $\frac{1}{2} \sim_r \frac{2}{2}$. We call a red agent r with preference relation $\frac{1}{2} \succ_r \frac{2}{2}$, $\frac{1}{2} \prec_r \frac{2}{2}$, or $\frac{1}{2} \sim_r \frac{2}{2}$ a mixed, pure, or indifferent red agent respectively. We define mixed, pure, and indifferent blue agents using fractions $\frac{0}{2}$ and $\frac{1}{2}$ analogously.

Let us define the set of pure red agents $R^p = \{r \in R | \frac{2}{2} \succ_r \frac{1}{2}\}$, the set of mixed red agents $R^m = \{r \in R | \frac{2}{2} \prec_r \frac{1}{2}\}$, and the set of indifferent red agents

$R^i = \{r \in R | \frac{1}{2} \sim_r \frac{2}{2}\}$. We define the set of pure blue agents B^p, the set of mixed blue agents B^m, and the set of indifferent blue agents B^i analogously. Note that $R = R^p \cup R^m \cup R^i$ and $B = B^p \cup B^m \cup B^i$.

We show that a popular outcome is guaranteed to exist in roommate diversity game G and can be computed in polynomial time by reducing G to the maximum weight perfect matching problem.

Let us define an undirected weighted complete graph $G_m = (R \cup B, E)$ where for each pair of distinct agents $a, b \in R \cup B$, we use $w(a, b)$ to denote the weight of the edge $(a, b) \in E$, where

$$w(a, b) = \begin{cases} 2 & \{a, b\} \text{ is a room with exactly 2 happy agents;} \\ 1 & \{a, b\} \text{ is a room with exactly 1 happy agent;} \\ 0 & \{a, b\} \text{ is a room with exactly 0 happy agents.} \end{cases}$$

Since the room size is 2, we have that $|R \cup B|$ is even. For the weighted graph G_m, let $w(M)$ be the weight of a perfect matching $M \subseteq E$. Given a perfect matching M of G_m, we define the points $p_M(a)$ of an agent $a \in R \cup B$ as follows: For each $(a, b) \in M$,

1. if $w(a, b) = 2$, then $p_M(a) = p_M(b) = 1$;
2. (a) if $w(a, b) = 1$ where a is happy and b is sad, then $p_M(a) = 1$ and $p_M(b) = 0$;
   (b) if $w(a, b) = 1$ where a is sad and b is happy, then $p_M(a) = 0$ and $p_M(b) = 1$;
3. if $w(a, b) = 0$, then $p_M(a) = p_M(b) = 0$.

We have that for any perfect matching M of G_m,

$$w(M) = \sum_{(a,b) \in M} w(a, b) = \sum_{a \in R \cup B} p_M(a). \tag{1}$$

Lemma 1. *For the weighted graph $G_m = (R \cup B, E)$ as defined above, the maximum weight perfect matching M_* of G_m is popular.*

Theorem 1. *Let $G = (R, B, 2, (\succsim_a)_{a \in R \cup B})$ be a roommate diversity game with room size 2. We can find a popular outcome π in polynomial time.*

The proofs of Lemma 1 and Theorem 1, which show the guaranteed existence and polynomial time computability, can be found in [18].

4   Strict Popularity

In this section, we show that determining the existence of a strictly popular outcome is co-NP-hard, even if the preferences are dichotomous. We shall construct a roommate diversity game $G = (R, B, s, (\succsim_a)_{a \in R \cup B})$ with dichotomous preferences from an X3C instance (X, C) such that there exists a solution $C' \subseteq C$ that partitions X if and only if no strictly popular outcome exists for G. The full reduction and proofs can (also) be found in [18].

4.1 Roommate Diversity Game

We set the room size to $s = 5(q+1)+1+m = 5q+6+m$. The agents and their preference profiles are defined in Table 1, Table 2, and Table 3.

4.2 Predefined Outcomes

In this section, we shall define the monolithic outcome π_{mon} and the reduced outcome $\pi_{C'}$, where $C' \subseteq C$ is a solution of (X, C). In these outcomes every agent is assigned a room with a fraction that it approves of.

Table 1. Set of red agents $R = R^{set} \cup R^{mon} \cup R^{red}$.

$$
\begin{aligned}
\text{Set Agents } R^{set} &= \{r_i | i \in X\} = \{r_1, \ldots, r_m\} \\
\text{Redundant Agents } R_j^{red} &= \{r_j^1, \ldots, r_j^{5j-2}\} \qquad \text{for } j \in [q] \\
R^{red} &= R_1^{red} \cup \ldots \cup R_q^{red} \\
\text{Monolith Agents } R^{mon} &= \{r_{mon}^1, \ldots, r_{mon}^{5(q+1)+1}\}
\end{aligned}
$$

Table 2. Set of blue agents $B = B^{even} \cup B^{mon} \cup B^{add} \cup B^{fill}$.

$$
\begin{aligned}
\text{Filling Agents } B_j^{fill} &= \{b_j^1, \ldots, b_j^{s-(5j-2)-3}\} \qquad \text{for } j \in [q] \\
B^{fill} &= B_1^{fill} \cup \ldots \cup B_q^{fill} \\
\text{Additional Agents } B_j^{add} &= \{\tilde{b}_j^1, \tilde{b}_j^2, \tilde{b}_j^3\} \qquad \text{for } j \in [q] \\
B^{add} &= B_1^{add} \cup \ldots \cup B_q^{add} \\
\text{Monolith Agents } B^{mon} &= \{b_{mon}^1, \ldots, b_{mon}^{s-(5(q+1)+1)}\} \\
\text{Evening Agents } B^{even} &= \{b_{even}^1, \ldots, b_{even}^{5(q+1)+1}\}
\end{aligned}
$$

Table 3. Preference profile $(\succsim_a)_a \in R \cup B$.

Agent	Preference Profile			
	D_a^+		D_a^-	
$a = r_i \in R^{set}$	$\left\{\frac{5j_1^i+1}{s}, \ldots, \frac{5j_{m_i}^i+1}{s}\right\} \cup \{1\}$		$D \setminus D_a^+$	
$a = r_j^p \in R_j^{red}$	$\left\{\frac{5j+1}{s}, \frac{5j-2}{s}\right\}$		$D \setminus D_a^+$	$j \in [q]$
$a = r_{mon}^p \in R^{mon}$	$\left\{1, \frac{5(q+1)+1}{s}\right\}$		$D \setminus D_a^+$	
$a = b_j^p \in B_j^{fill}$	$\left\{\frac{5j+1}{s}, \frac{5j-2}{s}\right\}$		$D \setminus D_a^+$	$j \in [q]$
$a = \tilde{b}_j^p \in B_j^{add}$	$\left\{\frac{5j-2}{s}, 0\right\}$		$D \setminus D_a^+$	$j \in [q]$
$a = b_{mon}^p \in B^{mon}$	$\left\{\frac{5(q+1)+1}{s}, 0\right\}$		$D \setminus D_a^+$	
$a = b_{even}^p \in B^{even}$	$\{0\}$		$D \setminus D_a^+$	

Remark 1. Observe that any outcome π, that assigns every agent to a room with a fraction that it approves of, is popular. That is, for any outcome π, such that $\forall_{a \in R \cup B} : \theta(\pi(a)) \in D_a^+$, is popular.

Monolithic Outcome. We define the sets of agents:

$$P_j = B_j^{add} \cup R_j^{red} \cup B_j^{fill} \text{ for } j \in [q];$$
$$R_r = R^{set} \cup R^{mon};$$
$$B_r = B^{mon} \cup B^{even}.$$

Define the monolithic outcome as $\pi_{mon} = \{P_1, \cdots, P_q, R_r, B_r\}$. Note that π_{mon} is a valid partitioning, as every agent in $R \cup B$ is assigned a room and that

$$|P_j| = |R_j^{red}| + |B_j^{fill}| + |B_j^{add}| = (5j - 2) + s - (5j - 2) - 3 + 3 = s,$$
$$|R_r| = |R^{mon}| + |R^{set}| = 5(q + 1) + 1 + m = s,$$
$$|B_r| = |B^{mon}| + |B^{even}| = s - (5(q + 1) + 1) + (5(q + 1) + 1) = s.$$

Every agent is assigned a room by π_{mon} with a fraction that it approves of. The monolithic outcome always exists.

Reduced Outcome. Let $C' \subseteq C$ be a solution that partitions X. We define the following sets of agents:

$$P_j' = \begin{cases} \{r_i \in R^{set} | i \in A_j\} \cup R_j^{red} \cup B_j^{fill}, & \text{if } A_j \in C' \\ B_j^{add} \cup R_j^{red} \cup B_j^{fill}, & \text{if } A_j \notin C' \end{cases} \text{ for } j \in [q];$$
$$R_r' = R^{mon} \cup B^{mon};$$
$$B_r' = B^{even} \cup \bigcup_{A_j \in C'} B_j^{add}.$$

We define the reduced outcome as $\pi_{C'} = \{P_1', \cdots, P_q', R_r', B_r'\}$. Note that $\pi_{C'}$ is a valid outcome, as every agent in $R \cup B$ is assigned a room and that

$$|P_j'| = \begin{cases} |\{r_i \in R^{set} | i \in A_j\}| + |R_j^{red}| + |B_j^{fill}| = 3 + 5j - 2 + s - (5j - 2) - 3 = s \\ |B_j^{add}| + |R_j^{red}| + |B_j^{fill}| = 3 + 5j - 2 + s - (5j - 2) - 3 = s, \end{cases}$$
$$|R_r'| = |R^{mon}| + |B^{mon}| = 5(q + 1) + 1 + s - (5(q + 1) + 1) = s,$$
$$|B_r'| = |B^{even}| + \sum_{A_j \in C'} |B_j^{add}| = 5(q + 1) + 1 + m = s.$$

Every agent is assigned a room by $\pi_{C'}$ with a fraction that it approves of. A reduced outcome exists if and only if (X, C) has a solution.

4.3   Hardness

We demonstrate co-NP-hardness by showing that the monolithic outcome π_{mon} is the only popular outcome and therefore strictly popular, if (X, C) has no solution. Otherwise, we have multiple popular outcomes.

Lemma 2. *Let G be a roommate diversity game constructed as in Sect. 4.1 and π be an outcome of G such that every agent is in a room with a fraction that it approves of. We have that π is either the monolithic outcome π_{mon} or a reduced outcome $\pi_{C'}$, where $C' \subsetneq C$ partitions X.*

Proof. Let $r_i \in R^{set}$ be an arbitrary set agent. As r_i must be assigned a room in π with a fraction that it approves of, we have that $\theta(\pi(r_i)) = 1$ or $\theta(\pi(r_i)) = \frac{5j+1}{s}$, where $j \in J^i$. Let us consider the following cases.

1. $\theta(\pi(r_i)) = 1$.

   Since there are exactly s red agents that approve of fraction 1, namely the agents in $R^{set} \cup R^{mon}$, these agents must be contained in the same room, i.e., $R^{set} \cup R^{mon} \in \pi$. A redundant agent $r_{j'}^p \in R^{red}$ cannot be in a room with fraction $\frac{5j'+1}{s}$ as there are only $s - 3$ remaining agents that approve of that fraction. Thus, any redundant agent $r_{j'}^p$ must be in a room with fraction $\frac{5j'-2}{s}$. The rooms must be $R_{j'}^{red} \cup B_{j'}^{fill} \cup B_{j'}^{add}$ for $j' \in [q]$ as these are exactly the agents that approve of fraction $\frac{5j'-2}{s}$, i.e. for $j' \in [q]$ we have $R_{j'}^{red} \cup B_{j'}^{fill} \cup B_{j'}^{add} \in \pi$. We have s remaining blue agents, namely $B^{even} \cup B^{mon}$, that must belong to the same room. Thus, π must be π_{mon}.

2. $\theta(\pi(r_i)) = \frac{5j+1}{s}$.

   There are exactly s red agents, including r_i, that approve of fraction 1. Since $\theta(\pi(r_i)) \neq 1$ and π is an outcome such that every agent is in a room with a fraction that it approves of, the outcome π cannot contain a room with only red agents. Thus, every set agent $r_{i'} \in R^{set}$ must be in a room with fraction $\frac{5j'+1}{s}$ such that $i' \in A_{j'}$. There are exactly 3 set agents that approve of $\frac{5j''+1}{s}$ for each $j'' \in [q]$. Thus, for some solution $C' \subseteq C$ of (X, C), every set agent must be in a room with the shape $\{r_i | i \in A_j\} \cup R_j^{red} \cup B_j^{fill}$ where $A_j \in C'$, i.e., for each $A_j \in C'$ we have $\{r_i | i \in A_j\} \cup R_j^{red} \cup B_j^{fill} \in \pi$.

   The remaining redundant agents $r_{j'''}^p \in R^{red}$ must be in a room with fraction $\frac{5j'''-2}{s}$ as only $s - 3$ remaining agents approve of fraction $\frac{5j'''+1}{s}$. The rooms must be $R_{j'''}^{red} \cup B_{j'''}^{fill} \cup B_{j'''}^{add}$ for $A_{j'''} \notin C'$, i.e., for each $A_{j'''} \notin C'$ we have $R_{j'''}^{red} \cup B_{j'''}^{fill} \cup B_{j'''}^{add} \in \pi$. Since there is no room with only red agents, the red monolith agents must be in a room with the blue monolith agents, i.e., $R^{mon} \cup B^{mon} \in \pi$. We have s remaining blue agents, namely $B^{even} \cup \bigcup_{A_j \in C'} B_j^{add}$, that must belong to the same room. Thus, π must be $\pi_{C'}$.   □

Theorem 2. *Determining whether a strict popular outcome exists in a roommate diversity game is co-NP-hard, even if the preferences are dichotomous.*

Proof. From Lemma 2 and Remark 1, we have that if (X, C) has no solution, then π_{mon} is strictly popular. Otherwise, we have multiple popular outcomes and therefore no outcome is strictly popular. ☐

5   Mixed Popularity

We show that a mixed popular outcome is guaranteed to exist. However, under the assumption that P≠NP, a mixed popular outcome cannot be computed in polynomial time, even if the preferences are dichotomous.

Theorem 3. *Any roommate diversity game is guaranteed to have a mixed popular outcome.*

Proof. Every roommate diversity game can be viewed as a finite two-player symmetric zero-sum game where the rows and columns of the game matrix are indexed by all possible outcomes $\pi_1, \ldots, \pi_t$ and the entry at (i, j) of the game matrix has value $\phi(\pi_i, \pi_j)$. By the minimax theorem [21], we have that the value of this game is 0. Therefore, any maximin strategy is popular. ☐

To show that a mixed popular outcome for a roommate diversity game cannot be computed in polynomial time, we change the reduction for strict popularity in Sect. 4 by doubling the agents, adding new agents, and changing the room size and preference profiles accordingly. These modifications ensure that the monolithic outcome is strictly popular if (X, C) has no solution and that the monolithic outcome is not popular if (X, C) has a solution. The full reduction and (proofs of) remarks, lemmas, and theorems can be found in [18].

6   Popularity

In this section, we show that a popular outcome is not guaranteed to exist in a roommate diversity game.

Let $R = \{r_1, r_2, r_3\}$ and $B = \{b_1, b_2, b_3, b_4, b_5, b_6\}$. Consider the roommate diversity game $\bar{G} = (R, B, 3, (\succsim_a)_{a \in R \cup B})$ with the following preference profiles:

Agent	Preference Profile		
	D_a^+	D_a^n	D_a^-
$a = r_1$	$\{\frac{1}{3}\}$		$\{\frac{2}{3}, \frac{3}{3}\}$
$a \in \{r_2, r_3\}$	$\{\frac{2}{3}\}$		$\{\frac{1}{3}, \frac{3}{3}\}$
$a \in \{b_1, b_2, b_3, b_4\}$	$\{\frac{1}{3}\}$	$\{\frac{2}{3}\}$	$\{\frac{0}{3}\}$
$a \in \{b_5, b_6\}$	$\{\frac{0}{3}\}$		$\{\frac{1}{3}, \frac{2}{3}\}$

Let us define the rooms $P_1 = \{r_1, \beta_1, \beta_2\}$; $P_2 = \{r_2, r_3, \beta_3\}$; $P_3 = \{b_5, b_6, \beta_4\}$, where $\beta_1, \beta_2, \beta_3, \beta_4 \in \{b_1, b_2, b_3, b_4\}$. An outcome π_{top} is called a top-type outcome if we can write $\pi_{top} = \{P_1, P_2, P_3\}$. We can show that for any non-top-type outcome π, a top-type outcome π' exists such that π' is strictly more popular than π. However, for any top-type outcome π', we can construct another top-type outcome π'' such that π'' is strictly more popular than π'.

Lemma 3. *For any outcome π of $\overline{G}$, we have that there are at least 2 agents in a room with a fraction that it does not approve of.*

Proof. To derive a contradiction, assume that there exists an outcome π of $\overline{G}$ such that $|D_\pi^n \cup D_\pi^-| < 2$. Consider the following cases.

1. $|D_\pi^n \cup D_\pi^-| = 0$.

   Then $b_1, b_2, b_3, b_4 \in D_\pi^+$, therefore each of b_1, b_2, b_3, b_4 must be in a room with fraction $\frac{1}{3}$. We require at least 2 rooms S_1, S_2 with fraction $\frac{1}{3}$ so that each of b_1, b_2, b_3, b_4 is contained in a room with fraction $\frac{1}{3}$. There is exactly 1 red agent that approves of fraction $\frac{1}{3}$, namely r_1. Thus, S_1 or S_2 must contain a red agent r that does not approve of fraction $\frac{1}{3}$. Thus, $|D_\pi^n \cup D_\pi^-| \geq 1$ in this case as $r \in D_\pi^-$. This contradicts $|D_\pi^n \cup D_\pi^-| = 0$.

2. $|D_\pi^n \cup D_\pi^-| = 1$.

   Let us write $D_\pi^n \cup D_\pi^- = \{a\}$. Consider the following cases regarding a.

   2.1 $a = r_1$.

   Then $b_1, b_2, b_3, b_4 \in D_\pi^+$, therefore each of b_1, b_2, b_3, b_4 must be in a room with fraction $\frac{1}{3}$. We require at least 2 rooms S_1, S_2 with fraction $\frac{1}{3}$ so that each of b_1, b_2, b_3, b_4 is contained in a room with fraction $\frac{1}{3}$. There is exactly 1 red agent that approves of fraction $\frac{1}{3}$, namely r_1. Thus, S_1 and S_2 both must contain a red agent $r, r' \in R$ that do not approve of fraction $\frac{1}{3}$. Thus, $|D_\pi^n \cup D_\pi^-| \geq 2$ as $r, r' \in D_\pi^-$. This contradicts $|D_\pi^n \cup D_\pi^-| = 1$.

   2.2 $a \in \{r_2, r_3\}$.

   W.l.o.g. assume that $a = r_2$. We have that $r_3 \in D_\pi^+$, thus the room S that contains r_3 must be of fraction $\frac{2}{3}$. There are exactly 2 agents that approve of fraction $\frac{2}{3}$, namely r_2 and r_3. Since $r_2 \notin S$, as $r_2 \in D_\pi^-$, the room S must contain a red agent $r \neq r_2$ such that $r \in D_\pi^-$. Thus, $|D_\pi^n \cup D_\pi^-| \geq 2$ in this case as $r_2, r \in D_\pi^-$. This contradicts $|D_\pi^n \cup D_\pi^-| = 1$.

   2.3 $a \in \{b_1, b_2, b_3, b_4\}$.

   Analogous to case 2.1.

   2.4 $a \in \{b_5, b_6\}$.

   Analogous to case 2.1.   □

Lemma 4. *For any outcome π of $\overline{G}$, if $|D_\pi^n| = 1$ and $|D_\pi^-| = 1$, then π is a top-type outcome.*

Proof. Let π be an outcome such that $|D_\pi^n| = 1$ and $|D_\pi^-| = 1$. Let us write $D_\pi^n = \{a^n\}$ and $|D_\pi^-| = \{a^-\}$. Since the agents in $\{b_1, b_2, b_3, b_4\}$ are the only agents with a non-empty corresponding D_a^n, we have that $a^n \in \{b_1, b_2, b_3, b_4\}$.

To derive a contradiction, assume that $a^- \notin \{b_1, b_2, b_3, b_4\}$. Then we have that $a^- \in \{r_1, r_2, r_3, b_5, b_6\}$. Let us consider the following cases.

1. $a^- = r_1$.
   Then each agent $b \in \{b_1, b_2, b_3, b_4\} \setminus \{a^n\}$ must be in a room that it approves of, i.e., $b \in D_\pi^+$. We require at least 2 rooms S_1, S_2 with fraction $\frac{1}{3}$ so that each agent in $\{b_1, b_2, b_3, b_4\} \setminus \{a^n\}$ is contained in a room with fraction $\frac{1}{3}$. There is exactly 1 red agent that approves of fraction $\frac{1}{3}$, namely r_1. Thus, S_1 and S_2 both must contain a red agent $r, r' \in R$ that do not approve of fraction $\frac{1}{3}$. Thus, $|D_\pi^-| \geq 2$ as $r, r' \in D_\pi^-$. This contradicts $|D_\pi^-| = 1$.

2. $a^- \in \{r_2, r_3\}$.
   W.l.o.g. assume that $a = r_2$. We have that $r_3 \in D_\pi^+$, thus the room S that contains r_3 must be of fraction $\frac{2}{3}$. There are exactly 2 agents that approve of fraction $\frac{2}{3}$, namely r_2 and r_3. Since $r_2 \notin S$, as $r_2 \in D_\pi^-$, the room S must contain a red agent $r \neq r_2$ such that $r \in D_\pi^-$. Thus, $|D_\pi^-| \geq 2$ as $r_2, r \in D_\pi^-$. This contradicts $|D_\pi^-| = 1$.

3. $a^- \in \{b_5, b_6\}$.
   Analogous to case 1.   □

Remark 2. For any outcome π of $\overline{G}$, we have that $|D_\pi^-| \geq 1$.

Remark 3. For any outcome π of $\overline{G}$, if $|D_\pi^n \cup D_\pi^-| = 2$, then $D_\pi^n \cup D_\pi^- \subseteq \{b_1, b_2, b_3, b_4\}$.

Lemma 5. *Let π be an outcome that is not a top-type outcome. There exists a top-type outcome π' that is strictly more popular than π.*

Proof. By Lemma 3 we have that $|D_\pi^n \cup D_\pi^-| \geq 2$. By Lemma 4 we have that $|D_\pi^n| \neq 1$ or $|D_\pi^-| \neq 1$. Let us consider the following cases.

1. $|D_\pi^-| \neq 1$.
   By Remark 2 and case assumption we have that $|D_\pi^-| \geq 2$.
   1.1. $|D_\pi^-| = 2$.
      1.1.1. $|D_\pi^n| = 0$.
         By Remark 3 we have that $D_\pi^- \subseteq \{b_1, b_2, b_3, b_4\}$. Let us write $D_\pi^- = \{a_1, a_2\}$. Since $a_1, a_2 \in \{b_1, b_2, b_3, b_4\}$, there exists a top-type outcome π' such that $D_{\pi'}^- = \{a_1\}$ and $D_{\pi'}^n = \{a_2\}$. Thus, $\phi(\pi', \pi) = 1$.
      1.1.2. $|D_\pi^n| \geq 1$.
         Let us write $D_\pi^- = \{a_1, a_2\}$ and $a_3 \in D_\pi^n$. We have that $a_3 \in \{b_1, b_2, b_3, b_4\}$. Let π' be a top-type outcome such that $\{a_3\} = D_{\pi'}^n$ and $\{a^-\} = D_{\pi'}^-$. Note that $N(\pi, \pi') \subseteq \{a^-\}$.
         1.1.2.1. $N(\pi, \pi') = \emptyset$.
            Then we have that $a^- \in D_\pi^-$. W.l.o.g. assume that $a_1 = a^-$. We have that $a_2 \in D_{\pi'}^+$ and $a_2 \in N(\pi', \pi)$. Thus, $\phi(\pi', \pi) > 0$.
         1.1.2.2. $N(\pi, \pi') = \{a^-\}$.
            Then we have that $a^- \in D_\pi^n \cup D_\pi^+$ and $a_1, a_2 \in D_{\pi'}^+$. Thus, $a_1, a_2 \in N(\pi', \pi)$. Therefore, $\phi(\pi', \pi) > 0$.
   1.2. $|D_\pi^-| > 2$.
      Let us write $a_1, a_2, a_3 \in D_\pi^-$. Let π' be an arbitrary top-type outcome with $\{a^n\} = D_{\pi'}^n$ and $\{a^-\} = D_{\pi'}^-$. Note that $N(\pi, \pi') \subseteq \{a^n, a^-\}$.

1.2.1. $a^- \in D_\pi^-$.

W.l.o.g. assume that $a_1 = a^-$. In this case, we have that $N(\pi, \pi') \subseteq \{a^n\}$. We have that $a_2, a_3 \in N(\pi', \pi)$. Thus, $\phi(\pi', \pi) > 0$.

1.2.2. $a^- \notin D_\pi^-$.

Since $N(\pi, \pi') \subseteq \{a^-, a^n\}$ and $a_1, a_2, a_3 \in N(\pi', \pi)$, $\phi(\pi', \pi) > 0$.

2. $|D_\pi^n| \neq 1$.

2.1. $|D_\pi^n| = 0$.

By Lemma 3, we have that $|D_\pi^-| \geq 2$. This case is the same as case 1.1.1.

2.2. $|D_\pi^n| \geq 2$.

Let us write $a_1, a_2 \in D_\pi^n$. We have that $\{a_1, a_2\} \subseteq \{b_1, b_2, b_3, b_4\}$. By Remark 2, we have that $|D_\pi^-| \geq 1$.

2.2.1. $a_3 \in \{b_1, b_2, b_3, b_4\}$.

We construct a top-type outcome π' such that $D_{\pi'}^- = \{a_3\}$, $D_{\pi'}^n = \{a_2\}$, and $a_1 \in D_{\pi'}^+$. We have $N(\pi, \pi') = \emptyset$ and $|N(\pi', \pi)| \geq 1$.

2.2.2. $a_3 \notin \{b_1, b_2, b_3, b_4\}$.

We construct a top-type outcome π' s.t. $D_{\pi'}^- = \{a^-\}$, $D_{\pi'}^n = \{a_1\}$, and $a_2, a_3 \in D_{\pi'}^+$. We have $N(\pi, \pi') \subseteq \{a^-\} \wedge a_2, a_3 \in N(\pi', \pi)$. $\square$

Lemma 6. *A top-type outcome π is not popular.*

Proof. Let π be a top-type outcome. We can write

$$\pi = \{P_1, P_2, P_3\} = \{\{r_1, \beta_1, \beta_2\}, \{r_2, r_3, \beta_3\}, \{b_5, b_6, \beta_4\}\},$$

where $\beta_1, \beta_2, \beta_3, \beta_4 \in \{b_1, b_2, b_3, b_4\}$. Let us construct outcome

$$\pi' = \{P_1 \setminus \{\beta_2\} \cup \{\beta_3\}, P_2 \setminus \{\beta_3\} \cup \{\beta_4\}, P_3 \setminus \{\beta_4\} \cup \{\beta_2\}\}.$$

We have that $N(\pi, \pi') = \{\beta_2\}$ and $N(\pi', \pi) = \{\beta_3, \beta_4\}$. Thus, π' is strictly more popular than π. $\square$

Theorem 4. *A popular outcome is not guaranteed to exist in a roommate diversity game G.*

Proof. From Lemma 5 and 6 we have that no popular outcome exists in $\overline{G}$. Thus, not every roommate diversity game permits a popular outcome. $\square$

The idea of the co-NP-hardness reduction is to combine the reduction of Sect. 5 with the proof of a popular outcome not being guaranteed to exist. Particularly, the reduced-type outcome is introduced, which is a combination of the reduced outcome and top-type outcome. In this reduction, we have that a popular outcome exists if and only if (X, C) has no solution. The full reduction and (proofs of) remarks, lemmas, and theorems can be found in [18].

7   Conclusion

We have demonstrated that determining the existence of a (strictly) popular outcome for a roommate diversity game is co-NP-hard and a mixed popular outcome cannot be computed in polynomial time, unless $P = NP$. Even when the preferences are tri- or dichotomous, the problem remains intractable. An avenue for future research would be demonstrating completeness for a certain complexity class. We conjecture that the problem is Π_2^p-complete. The complexity of computing a popular outcome for a roommate diversity game with dichotomous preferences is also unknown.

A popular outcome is guaranteed to exist when the room size is fixed to 2. Additionally, it is possible to compute a popular outcome in polynomial time. As the problem becomes tractable when fixing the room size to 2, it may be possible to construct an FPT algorithm with the room size as parameter.

References

1. Abraham, D.J., Levavi, A., Manlove, D.F., O'Malley, G.: The stable roommates problem with globally-ranked pairs. In: Deng, X., Graham, F.C. (eds.) WINE 2007. LNCS, vol. 4858, pp. 431–444. Springer, Heidelberg (2007). https://doi.org/10.1007/978-3-540-77105-0_48
2. Bartholdi, J., Trick, M.A.: Stable matching with preferences derived from a psychological model. Oper. Res. Lett. 5(4), 165–169 (1986)
3. Biró, P., Irving, R.W., Manlove, D.F.: Popular matchings in the marriage and roommates problems. In: Calamoneri, T., Diaz, J. (eds.) CIAC 2010. LNCS, vol. 6078, pp. 97–108. Springer, Heidelberg (2010). https://doi.org/10.1007/978-3-642-13073-1_10
4. Boehmer, N., Elkind, E.: Individual-based stability in hedonic diversity games. In: Proceedings of the AAAI Conference on Artificial Intelligence, vol. 34, no. 02, pp. 1822–1829 (2020)
5. Boehmer, N., Elkind, E.: Stable roommate problem with diversity preferences. In: Proceedings of the Twenty-Ninth International Joint Conference on Artificial Intelligence. IJCAI'20 (2021). https://doi.org/10.24963/ijcai.2020/14
6. Brandt, F., Bullinger, M.: Finding and recognizing popular coalition structures. J. Artif. Int. Res. **74** (2022). https://doi.org/10.1613/jair.1.13470.https://jair.org/index.php/jair/article/view/13470
7. Bredereck, R., Elkind, E., Igarashi, A.: Hedonic diversity games. In: International Foundation for Autonomous Agents and Multiagent Systems, pp. 565–573 (2019). https://ora.ox.ac.uk/objects/uuid:f99dde29-43d0-4cd8-95a9-6268ae764637
8. Bredereck, R., Chen, J., Finnendahl, U.P., Niedermeier, R.: Stable roommates with narcissistic, single-peaked, and single-crossing preferences. Auton. Agent. Multi-Agent Syst. **34**(2), 53 (2020)
9. Chung, K.S.: On the existence of stable roommate matchings. Games Econ. Behav. **33**(2), 206–230 (2000)
10. Cseh, Á.: Popular matchings, chap. 6, p. 105–122. Lulu. com (2017). https://archive.illc.uva.nl/COST-IC1205/Book/
11. Cseh, A., Juhos, A.: Pairwise preferences in the stable marriage problem. ACM Trans. Econ. Comput. **9**(1) (2021). https://doi.org/10.1145/3434427. https://dl.acm.org/doi/abs/10.1145/3434427

12. Cseh, Á., Kavitha, T.: Popular matchings in complete graphs. Algorithmica **83**(5), 1493–1523 (2021)

13. Darmann, A.: Hedonic diversity games revisited. In: Fotakis, D., Ríos Insua, D. (eds.) ADT 2021. LNCS (LNAI), vol. 13023, pp. 357–372. Springer, Cham (2021). https://doi.org/10.1007/978-3-030-87756-9_23

14. Faenza, Y., Kavitha, T., Powers, V., Zhang, X.: Popular matchings and limits to tractability, pp. 2790–2809. https://doi.org/10.1137/1.9781611975482.173. https://epubs.siam.org/doi/abs/10.1137/1.9781611975482.173

15. Ganian, R., Hamm, T., Knop, D., Schierreich, Š, Suchý, O.: Hedonic diversity games: a complexity picture with more than two colors. In:Proceedings of the AAAI Conference on Artificial Intelligence, vol. 36, no. 5, pp. 5034–5042 (2022). https://doi.org/10.1609/aaai.v36i5.20435. https://ojs.aaai.org/index.php/AAAI/article/view/20435

16. Gärdenfors, P.: Match making: assignments based on bilateral preferences. Syst. Res. Behav. Sci. **20**, 166–173 (1975)

17. Garey, M.R., Johnson, D.S.: Computers and Intractability; A Guide to the Theory of NP-Completeness. W. H. Freeman & Co., USA (1990). https://doi.org/10.1137/1024022. https://epubs.siam.org/doi/10.1137/1024022

18. Ge, S., Itoh, T.: Popularity on the Roommate Diversity Problem. CoRR abs/2210.07911 (2022). https://doi.org/10.48550/arXiv.2210.07911. https://arxiv.org/abs/2210.07911

19. Gupta, S., Misra, P., Saurabh, S., Zehavi, M.: Popular matching in roommates setting is NP-hard. ACM Trans. Comput. Theory **13**(2) (2021). https://doi.org/10.1145/3442354. https://dl.acm.org/doi/10.1145/3442354

20. Manlove, D.: Algorithmics of Matching Under Preferences. In: Bulletin of EATCS (2013)

21. v. Neumann, J.: Zur Theorie der Gesellschaftsspiele. Mathematische Annalen **100**(1), 295–320 (1928). https://doi.org/10.1007/BF01448847. https://link.springer.com/article/10.1007/BF01448847

On Half Guarding Polygons

Erik Krohn[1]([✉]) [iD], Alex Pahlow[2], and Zhongxiu Yang[3] [iD]

[1] University of Wisconsin - Oshkosh, Oshkosh, WI, USA
krohne@uwosh.edu
[2] University of Wisconsin - Madison, Madison, WI, USA
apahlow22@alumni.uwosh.edu
[3] The University of Texas at San Antonio, San Antonio, TX, USA
zhongxiu.yang@utsa.edu

Abstract. Given a polygon P and a set of potential guard locations $G \in P$, the art gallery problem asks for the minimum number of guards needed to guard the polygon. The art gallery problem with different types of polygons has been studied extensively. Variants of the art gallery problem have also been studied including limiting the polygon to be a monotone or an orthogonal polygon. Limitations have also been applied to the guard where a guard cannot see 360° and even limitations on the distance a guard can see.

In this paper, we study the art gallery problem using half guards (guards that see 180°) in various settings. We show that the VC dimension of half guarding a terrain is exactly 2 or 3, depending on certain assumptions, and exactly 4 with half guarding a monotone polygon where all critical points are located on the boundary. We provide so-called art gallery theorems for half guards with different polygon types. Finally, we provide a linear time exact algorithm for half guarding a spiral polygon.

1 Introduction and Related Work

Given a polygon P and a set of potential guard locations $G \in P$, the art gallery problem asks for the minimum number of guards $G' \in G$ needed such that every point in the polygon is seen by at least one point in G'. This problem has been studied extensively [4,5,11]. Two points, p and q, see each other if the $\overline{pq}$ line segment them does not go outside of the polygon. We define a *full guard* as a guard that can see 360°. In our paper, we define a *half guard* as a guard that sees 180° and only sees to the right. Restricting the guard to seeing less than 360° has been studied in [1–3]. VC dimension is a measure of the complexity of some set system. It asks how many guards can be shattered. A set of guards G in P is said to be *shattered* if for every $G_s \subseteq G$, there exists a point that is seen by the guards in G_s and by no guards in $G \setminus G_s$. It has been studied by researchers for many variants of the art gallery problem [6,7,9,10,12,15–17]. Guarding simple polygons with full guards has a VC dimension between 6 and 14 [11]. The terrain guarding problem with full guards has a VC dimension of exactly 4 [14]. The VC dimension of guarding monotone (or simple) polygons with full guards, where key points are limited to being on the boundary of the polygon, was shown to be exactly 6 in both types of polygons [8,9]. The structure half guards add to the art gallery problem is interesting because the difference, as compared to, full guards, is not trivial. For example, convex polygons have a VC dimension of 1 with half guards despite having a VC dimension of 0 with full guards.

1.1 Notation/Definitions

A terrain T is an x-monotone polygonal chain. Let $p < q$ mean that point p is to the left of q, i.e. the x coordinate of $p.x < q.x$. With *half guarding* a polygon (resp. terrain), a point p sees a point q if the line segment connecting p and q does not go outside of the polygon (resp. below the terrain) and $p.x \leq q.x$. Let p and q be two points such that $p.x < q.x$, then $[p, q)$ denotes every point in the polygon (or on the terrain) between p and q, including the vertical line containing p but excluding the vertical line containing q. Let l be the leftmost point and r be the rightmost point of some monotone polygon. A monotone polygon is two x-monotone polygonal chains that do not intersect except at the l and r vertices. The *ceiling (resp. floor)* denotes every boundary point in $[l, r]$ as we travel clockwise (resp. counterclockwise) from l to r. We define *viewpoint* as a point that is exactly seen by a subset of the guards. For example, the viewpoint $vp(AC)$ is a point in the polygon that is seen by guards A and C but is not seen by any other guards.

1.2 Our Results

In Sect. 2.1, we show that the VC dimension of half guarding a terrain is exactly 2 or 3, depending on certain assumptions. In Sect. 2.2, we show that the VC dimension is exactly 4 with half guarding a monotone polygon with guards and viewpoints on the boundary of the monotone polygon. In Sect. 3, we show that $n - 2$ (resp. $n - 1, \frac{n-2}{2}, n - 2$) half guards are sometimes necessary and always sufficient to guard a monotone/simple polygon (resp. terrain, orthogonal polygon, spiral polygon). Finally in Sect. 4, we provide a linear time exact algorithm for half guarding a spiral polygon.

2 VC Dimension

2.1 VC Dimension of Terrains

We start by discussing the VC dimension of terrains with respect to half guards. The VC dimension of a terrain with half guards depends on if a point on the terrain can be considered both a guard and a viewpoint. If guards and viewpoints must be disjoint, then the VC dimension is 2. If a point on the terrain can be both a guard and a viewpoint, then the VC dimension is 3. Figure 1 (resp. 2) shows an example of a terrain being shattered with 2 (resp. 3) guards depending on the assumption. We use the standard *order claim* without proof.

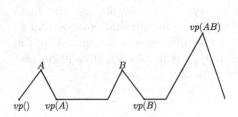

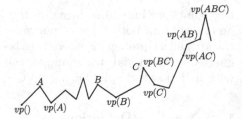

Fig. 1. Terrain shattered by 2 half guards where guards and viewpoints are disjoint

Fig. 2. Terrain shattered by 3 half guards where C and $vp(BC)$ are the same point

Claim. Let A, B, C, D be 4 points on a terrain with $A.x < B.x < C.x < D.x$. If A sees C and B sees D, then A must see D.

Theorem 1. *If a terrain guarding problem does not (resp. does) allow a guard and a viewpoint to be the same point, then the VC dimension of a terrain is exactly 2 (resp. 3).*

Proof. We will first consider the case where guards and viewpoints cannot be located at the same point. Let A, B and C be guards such that $A.x < B.x < C.x$. Assuming that a guard and viewpoint cannot be the same point, the viewpoints that are seen by C must be strictly to the right of C. It follows that $B.x < C.x < vp(BC).x$ and $B.x < C.x < vp(AC).x$. If $vp(BC).x < vp(AC).x$, then we have $B.x < C.x < vp(BC).x < vp(AC).x$. By the order claim, B sees $vp(AC)$, a contradiction. If $vp(AC).x < vp(BC).x$, then $A.x < B.x < vp(AC).x < vp(BC).x$. By the order claim, A sees $vp(BC)$, a contradiction.

Next we consider the VC dimension of terrains where a guard and a viewpoint can be at the same point. In this case, the VC dimension is 3. We achieve a lower bound of 3 by giving an example of a terrain shattering 3 guards in Fig. 2.

We will show that it is impossible for such a terrain to have a VC dimension of 4. Let A, B, C, D be the guards of this terrain with $A.x \le B.x \le C.x \le D.x$:

1. If $vp(AC).x < vp(BD).x$, then $A.x < B.x < C.x \le vp(AC).x < vp(BD).x$.
   By the order claim using $A, B, vp(AC), vp(BD)$, A sees $vp(BD)$.
2. If $vp(BD).x < vp(AC).x$, then $A.x < B.x < C.x < vp(BD).x < vp(AC).x$.
   By the order claim using $B, C, vp(BD), vp(AC)$, B sees $vp(AC)$.

2.2   VC Dimension of Monotone Polygons

We show that the VC dimension of half guarding a monotone polygon is exactly 4. We obtain the lower bound for monotone polygons by giving an example of a monotone polygon being shattered by 4 guards as seen in Fig. 3. We now show that the 5 guards cannot be shattered with a case analysis. A few cases are shown in the paper with the remaining ones moved to the appendix due to lack of space. Proofs for each of these lemmas can be found in the appendix. One can see Figs. 4 and 5 for a visualization of what the proof looks like.

Lemma 1. *Let $s < t < u \leq v$ where s, t, u are on the ceiling (resp. floor), s sees u, t sees v, and s does not see v. The floor (resp. ceiling) must block s from seeing v.*

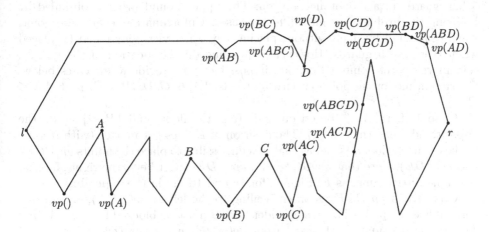

Fig. 3. Monotone polygon shattered by 4 half guards

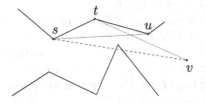

Fig. 4. Visualization of Lemma 1

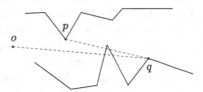

Fig. 5. Visualization of Lemma 2

Lemma 2. *Let p and q be two points on the boundary and $p < q$. If p is on the ceiling (resp. floor) and is blocked from q using the floor (resp. ceiling), then no point in $[l, p]$ can see q.*

Lemma 3. *Let $s < t < u$, where t and u are on opposite sides of the polygon, s sees u, and t does not see u. It must be that t cannot see any point in $[u, r]$.*

Proof. Assume, w.l.o.g., that t is on the floor. Note that t cannot be blocked from u using the ceiling since by Lemma 2, s would not see u. Thus, t must be blocked from u using the floor. Let v denote some point to the right of u. If the $\overrightarrow{tv}$ line crosses above u, then the ceiling will block t from v. If the $\overrightarrow{tv}$ line crosses below u, then the floor will block t from v since the floor is blocking t from u. $\square$

Corollary 1. *Let* $t < u$, *t is on the floor (resp. ceiling), u is on the ceiling (resp. floor), and the floor (resp. ceiling) is blocking t from seeing u. It must be that t cannot see any point in* $[u, r]$.

We obtain an upper bound of 4 by showing that it is impossible to shatter 5 half guards in a monotone polygon. The upper bound proof is obtained by breaking the problem up into different cases. Unfortunately, every viewpoint, when considered by itself without placing any other viewpoints, can be placed when there are 5 guards. However, depending on the location of the guards, certain viewpoint combinations are impossible. We provide a few cases below. Consider a monotone polygon with 5 guards: $\{A, B, C, D, E\}$ such that $A.x \leq B.x \leq C.x \leq D.x \leq E.x$.

Case 1: Let $\{A, C\}$ be on one side (e.g. the floor) and $\{B, D\}$ be on the opposite side (e.g. the ceiling). The position of E does not matter (with respect to the ceiling or floor). We show that it is impossible to place the points $vp(BCE)$ and $vp(ADE)$. Note that $vp(BCE)$ and $vp(ADE)$ must be to the right of, or on the same vertical line, as E. Figures for 1c and 1d are in the appendix.

Case 1a: If $vp(BCE)$ is on the ceiling to the left of $vp(ADE)$, or on same vertical line as $vp(ADE)$, then consider how B must be blocked from $vp(ADE)$. The B guard cannot be blocked from $vp(ADE)$ using the ceiling because of Lemma 1 where $s = B, t = D, u = vp(BCE)$ and $v = vp(ADE)$. The floor must then be used to block B from $vp(ADE)$. By Lemma 2, using $o = A, p = B, q = vp(ADE)$, the A guard would not be able to see $vp(ADE)$. Therefore, B cannot be blocked from $vp(ADE)$. This case is illustrated in Fig. 6.

Case 1b: If $vp(ADE)$ is on the ceiling to the left of $vp(BCE)$, or on same line, then consider how C is blocked from seeing $vp(ADE)$. This case is illustrated in Fig. 7. Similar to the previous argument, if C is blocked from seeing $vp(ADE)$ using the floor, then by Corollary 1 using $t = C, u = vp(ADE)$, C cannot see $vp(BCE)$. If the ceiling blocks C from seeing $vp(ADE)$, then by Lemma 2 using $o = A, p = C, q = vp(ADE)$, A is blocked from seeing $vp(ADE)$.

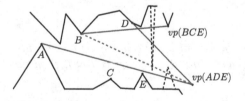

Fig. 6. Visualization of Case 1a

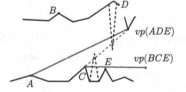

Fig. 7. Visualization of Case 1b

Case 1c: If $vp(BCE)$ is on floor to the left of $vp(ADE)$, or on same line as $vp(ADE)$, then consider how D must be blocked from $vp(BCE)$. If the ceiling blocks D from seeing $vp(BCE)$, then D does not see $vp(ADE)$ by Corollary 1 when $t = D, u = vp(BCE)$. If the floor blocks D from seeing $vp(BCE)$, then by Lemma 2 with $o = C, p = D, q = vp(BCE)$, C cannot see $vp(BCE)$.

Case 1d: If $vp(ADE)$ is on floor to the left of $vp(BCE)$, or on same line as $vp(ADE)$, then consider how B is blocked from $vp(ADE)$. If the floor blocks B from $vp(ADE)$, then by Lemma 2 with $o = A, p = B, q = vp(ADE)$, A does not see $vp(ADE)$. If the ceiling blocks B from $vp(ADE)$, then by Corollary 1 with $t = B, u = vp(ADE)$, B does not see $vp(BCE)$.

Therefore, $\{A, C\}$ and $\{B, D\}$ cannot be on opposite sides of the polygon. These cases are just a few examples of how to show the VC dimension of a monotone polygon with half guards and viewpoints on the boundary is exactly 4. The $2^5 = 32$ cases that we consider, where the remaining ones are in the appendix, are the following: there are any 4 guards that are on the same side (12 cases), $\{A, C\}$ are on the same side and $\{B, D\}$ are on the opposite side (4 cases), $\{A, E\}$ are on the same side and $\{B, C, D\}$ are on the opposite side (2 cases), $\{C, E\}$ are on the same side and $\{A, B, D\}$ are on the opposite side (2 cases), $\{A, B\}$ are on the same side and $\{C, D\}$ are on the opposite side (4 cases), $\{A, D\}$ are on the same side and $\{B, C\}$ are on the opposite side (4 cases), $\{B, E\}$ are on the same side and $\{A, C, D\}$ are on the opposite side (2 cases), and $\{A, B, C\}$ are on the same side and $\{D, E\}$ are on the opposite side (2 cases). Proving these cases are not realizable gives us the following theorem.

Theorem 2. *The VC dimension of half guarding a monotone polygon where guards and viewpoints are on the boundary is exactly 4.*

3   Art Gallery Theorems

In this section, we provide several claims and proofs for half guarding certain types of polygons.

Claim. For half guarding a simple, monotone, or spiral polygon, $n - 2$ half guards are always sufficient and are sometimes necessary to guard the entire polygon.

Proof. We show necessity by constructing the example in Fig. 8. We get sufficiency by triangulating the polygon into $n - 2$ triangles, and placing a point at the leftmost point of each triangle.   □

Claim. For half guarding a terrain, $n - 1$ half guards are always sufficient and are sometimes necessary to guard the entire terrain.

Proof. We show necessity by constructing the example in Fig. 9. We get sufficiency by placing a guard at every vertex in the terrain but the rightmost vertex.   □

Claim. For half guarding an orthogonal polygon, $\frac{n-2}{2}$ half guards are always sufficient and are sometimes necessary to guard the entire polygon.

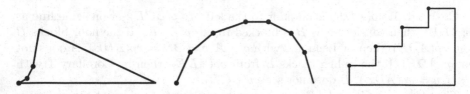

Fig. 8. Polygon with $n - 2$ guards. **Fig. 9.** Terrain with $n - 1$ guards. **Fig. 10.** Orthogonal polygon with $\frac{n-2}{2}$ guards.

Proof. We show necessity by constructing the example in Fig. 10. In this figure, there is a staircase on the "ceiling" that requires 1 guard per "step." Each step consists of 2 vertices. In this figure, the rightmost vertices are excluded since they are seen by the guards being placed for the steps. This gives us $\frac{n-2}{2}$ steps where each step requires a guard. We get sufficiency by first quadrilateralizing the polygon as shown in [13]. We remove a quadrilateral from the polygon such that the remaining polygon is simple and place a guard at the leftmost point of that quadrilateral. In doing so, the quadrilateral that was removed contains 2 vertices that will never be considered again. The guard placed for this quadrilateral is charged to these 2 vertices. The last quadrilateral contains 4 vertices that 1 guard is sufficient for. This gives us as most $\frac{n-2}{2}$ convex quadrilaterals.   □

4   Exact Algorithm for Half Guarding Spiral Polygons

A spiral polygon is a polygon that starts at some vertex s and ends at some vertex t. The outer (inner) boundary are the edges that start at s and are obtained by walking clockwise (or counterclockwise) on the boundary until reaching t. We assume that s and t are vertices on both the outer and inner boundary. The interior angles of the outer (inner) boundary are always less than (greater than) $180°$, see Fig. 11. Let $vp(g)$ be the visibility polygon of some guard g.

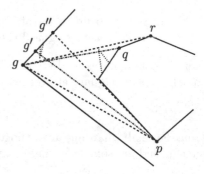

Fig. 11. Blue inner edges and yellow outer edges. (Color figure online)

Fig. 12. Example of Lemmas 4 and 5.

Consider the starting vertex s and let p be some point on the outer (resp. inner) boundary. Let o_{sp} (resp. (i_{sp})) be the distance from s to p as one walks on the outer (resp. inner) boundary from s to p. We first present several lemmas and claims about spiral polygons before introducing the algorithm. Due to lack of space, several proofs have been moved to the appendix. However, quick sketches of each proof are provided.

Lemma 4. *A guard g placed on the outer boundary sees a continuous portion of the inner boundary.*

Lemma 5. *A point p on the inner boundary is seen by a continuous portion of the outer boundary.*

Sketch of Lemmas 4 and 5 Proofs: As shown in Fig. 12, g sees p and r and all of the inner boundary from p to r. No boundary can block g from a point q between p and r on the inner boundary. A similar argument is made for what guards can see p. If g and g'' are outer boundary guards that see p, then an outer boundary guard between g and g'' also sees p.

Lemma 6. *W.l.o.g., assume a guard g is placed on the outer boundary of the polygon such that g can see the inner boundary that is directly below (or above) g. Let p and q be points in P such that p sees q. The line segment connecting p and q cannot cross the boundary of $vp(g)$ more than once.*

Sketch of Lemma 6 Proof: As shown in Fig. 13, no point to the left of g can see through the v_p region. Since the g_e ray crosses at g, no point can see through g_v, cross the g_e ray and then cross the g_e ray again.

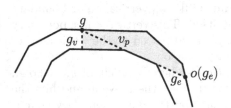

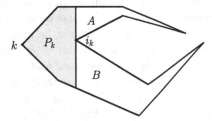

Fig. 13. Visibility polygon with 2 internal edges.

Fig. 14. Sample polygon.

Let k be a vertex in P such that k is only seen by itself, i.e. a guard must be placed at k in order to guard P. Let K be the set of all such vertices. For each vertex $k \in K$, we define i_k to be the leftmost vertex on the inner chain that k sees. Let P_k be the portion of $vp(k)$ that is to the right of k and to the left of i_k, see Fig. 14. Any guard placed at k will see the entire P_k region.

4.1   Vertical Edges

A simple greedy algorithm that places guards from left-to-right is sufficient if vertical outer boundary edges are not allowed. The algorithm and proof can be found in the appendix. To see the issue, in Fig. 15, the optimal solution does not place a guard at either vertex on the leftmost edge. Rather, it must be placed in some exact location. A greedy solution may not be able to guess where to place such a guard on this vertical edge. Therefore, a new algorithm is needed.

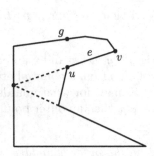

Fig. 15. Optimal solution does not place a guard at a vertex.

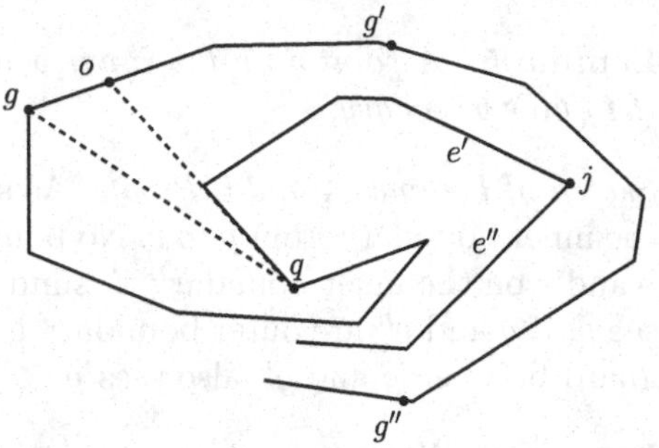

Fig. 16. The guards that see the edges connected to j see the right side of the polygon. Also, an optimal guard o sees more of the inner boundary then g.

Lemma 7. *If the inner boundary and the vertical outer boundary edges are half guarded, and half guards are placed at all $k \in K$, then the polygon is half guarded.*

Proof. If a guard is placed at some $k \in P_k$, then all of P_k is guarded since P_k is a convex polygon. In a similar fashion, one can think of a vertical outer boundary edge as a "point" that must be guarded by itself. This vertical outer boundary edge creates a region similar to P_k regions. A guard placed on the vertical outer boundary edge will see itself and the similar P_k region.

Consider an inner boundary edge $e = (u, v)$ where u is left of v. W.l.o.g., assume that a ray shot upwards from e goes through the polygon and hits the outer boundary, see Fig. 15. Let g be the guard on the outer boundary that is directly above u. The g guard sees u by assumption. The g guard also sees v since there is no inner boundary edge that can block g from v. The outer boundary cannot block g from v otherwise it is not a spiral polygon. By Lemma 4, g sees e. Consider the portion of the polygon directly above e. The g guard will see this entire region since no part of the boundary can block it. Therefore, if some inner boundary edge e is seen, the entire polygon directly above (below) e is seen.

Lastly, let j be a vertex on the inner boundary polygon such that both edges (e' and e'') leaving j are strictly left of j. As shown in Fig. 16 , some guard g' sees the portion of e' just left of j and some guard g'' sees the portion of e'' just left of j. Because the polygon is spiral, the entire portion of the polygon to the right of j is seen by these 2 guards.   □

Lemma 7 says that the algorithm only needs to guard the inner edges and the vertical outer boundary edges after it places a guard at all $k \in K$. The minimum number of guards needed to guard these areas plus the number of guards in K will be the minimum number to guard the entire polygon.

A high level overview of the algorithm is this. Start walking on the inner boundary from s to t and stop when the next point is unseen. The algorithm will find the "best" point on the outer boundary to guard this unseen point. The algorithm presented in [18] gives a linear time algorithm for guarding a spiral polygon with full guards. However, such an algorithm fails with half guarding. An example of the problem is shown in Fig. 16 with a guard being placed at o' when the better solution is to place one at guard g. Therefore, the algorithm must choose a better location for a guard when a vertical outer boundary edge appears. The algorithm repeatedly guards the inner boundary until the entire inner boundary is guarded. We will compare the solution G obtained in Algorithm 1 with some optimal solution $\mathcal{O}$.

Algorithm 1. Half Guarding Spiral Polygon

1: **Input:** Spiral Polygon P
2: $G \leftarrow \emptyset$
3: **for all** $k \in K$ **do**
4:     $G \leftarrow G \cup k$
5: **end for**
6: **while** the entire inner boundary is not yet guarded **do**
7:     Let q be the point on the inner boundary such that the entire inner boundary $[s, q]$ is seen by some guard $g \in G$ and i_{sq} is maximized
8:     **if** q is seen by a vertical edge e on the outer boundary and e is not seen by any $g' \in G$ **then**
9:         Let g be a point on e such that g sees q and o_{sg} is maximized
10:         $G \leftarrow G \cup g$
11:     **else**
12:         Let g be a point on the outer boundary such that g sees q and o_{sg} is maximized
13:         $G \leftarrow G \cup g$
14:     **end if**
15: **end while**
16: **return** G

Our algorithm considers a (possibly disconnected) list of edges that must be guarded. We claim that Algorithm 1 will see at least as much of these edges as some optimal guard set as guards are being placed to guard certain areas of the polygon. The edge list that we wish to guard is the edge list $\mathcal{E}$ that we add to in the following fashion. Start with s and walk on the inner boundary to t. Let $i_0, i_1, \ldots, i_m$ be the vertices of the inner edge boundary where $s = i_0$ and $t = i_m$. When we reach i_j, where $1 \le j \le m$, we add (i_{j-1}, i_j) to $\mathcal{E}$. If (i_{j-1}, i_j) and (i_j, i_{j+1}) are both to the right of j, then we check if j is seen by some vertical outer boundary edge e. If it is, then we add e to $\mathcal{E}$. To illustrate this, in Fig. 17, the yellow inner edges are added to $\mathcal{E}$ first, then the green outer edge is added to $\mathcal{E}$, then the red inner edges are added to $\mathcal{E}$, then the purple outer edge is added to $\mathcal{E}$, and lastly the blue inner edges are added to $\mathcal{E}$. This ordering of edges is how we argue our algorithm stays ahead of the optimal solution.

Let G' (resp $\mathcal{O}'$) be a partial guard covering from G (resp. $\mathcal{O}$). Let $X(\mathcal{E})$ be the maximal continuous guarded section of $\mathcal{E}$ that starts at s for some partial guard placement X. We claim that as the algorithm places guards, $G'(\mathcal{E}) \geq \mathcal{O}'(\mathcal{E})$.

The following lemma was proved as Lemma 3.2 in [18]:

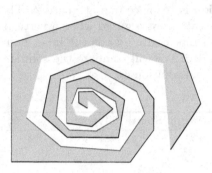

Fig. 17. Example of the order in which edges are added to $\mathcal{E}$. (Color figure online)

Lemma 8. *There exists a minimum stationary posting of guards in a spiral polygon such that all guards are on the convex chain.*

The outer boundary, as described in this paper, is the convex chain since all interior angles of the outer boundary are convex. This gives us the following corollary whose proof is in the appendix due to lack of space.

Corollary 2. *There exists a minimum placement of half guards in a spiral polygon such that all half guards are on the outer boundary.*

Consider the start of the algorithm. Both G' and $\mathcal{O}'$ must contain guards placed at all $k \in K$, else, these points would not be guarded. At this point, $G'(\mathcal{E}) == \mathcal{O}'(\mathcal{E})$. Let q be the point such that $G'(\mathcal{E})$ is maximized. The algorithm will place a guard g on the outer boundary such that g sees q and o_{sg} is maximized. In other words, g is the "furthest" along the outer boundary it can be but still see q. By Corollary 2, we will consider the optimal solution to have had all of its guards pushed to the outer boundary. We will consider 2 cases.

Let $g \in G$ and $o \in \mathcal{O}$ be guards that see q such that o_{sg} and o_{so} are maximal. First we assume that $o_{sg} \geq o_{so}$. If o is to the left of g, then by Lemma 6, g sees more of the unseen inner boundary than o. If o is to the right of g, then since the inner boundary from s to q was already seen, o cannot see an unseen inner point that g, or some guard already in G', sees. Therefore, g dominates o with respect to the unseen portion of $\mathcal{E}$.

Second, assume that $o_{sg} < o_{so}$, see Fig. 16. In this case, an optimal guard o that sees q could see more of the inner boundary than g. This can only happen in step 10 of Algorithm 1. Since a point further along the outer boundary saw q, by Lemma 5, the topmost vertex of that vertical edge sees q and we place a guard at the vertex that has the largest o_{sg} value. An optimal guard o' must also be on the vertical edge in order to see the vertical edge. Therefore, the g guard will dominate the o' guard with respect to the unseen regions of $\mathcal{E}$ up to this point in the algorithm. Now consider the next guard location the algorithm considers. Since g saw q and the optimal guard o sees q. Our subsequent guard will be placed, at minimum, at the o guard location. By Lemma 6, we are then able to charge the subsequent g' guard to the o guard since g' will necessarily dominate the o guard with respect to the unseen regions of $\mathcal{E}$. In this case, g dominates o and g' dominates o' with respect to the unseen regions of $\mathcal{E}$.

The algorithm runs in linear time. The outer boundary points that need to be considered are vertices of the outer boundary and critical locations where visibility of the inner edge changes. An inner edge $e = (u, v)$ can shoot a ray from u through v and v through u to hit the outer boundary. These $O(n)$ locations also need to be considered as potential guard locations. This gives us the following:

Theorem 3. *There is a linear time exact algorithm for half guarding a spiral polygon.*

Appendix

VC dimension Monotone Polygons

The following proofs and cases are provided for completeness.

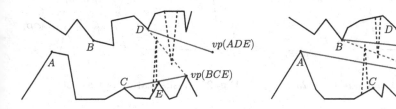

Fig. 18. Visualization of Case 1c   **Fig. 19.** Visualization of Case 1d

Proof. (Proof of Lemma 1) W.l.o.g., assume s, t and u are on the ceiling. If a point p' on the ceiling is used to block s from v such that $s.x < p'.x < u.x$, then s is blocked from u. If a point p' on the ceiling is used to block s from v such that $t.x < p'.x < v.x$, then t is blocked from v. If the ceiling wraps underneath v to block s from v, then the polygon is not monotone. Therefore, if s does not see v, the floor must block it.

Proof. (Proof of Lemma 2) W.l.o.g, assume that p is on the ceiling and the floor is blocking p from q. Let o be some point in $[l, p]$. The $\vec{oq}$ ray lies in between the $\vec{pq}$ ray and the floor. If this were not the case, then p would have blocked o from q. If the floor blocks p from q, the $\vec{oq}$ ray must also go through the floor and therefore, q must also be blocked from o.

Case 2: In this case, $\{A, E\}$ are on the floor (resp. ceiling) and $\{B, C, D\}$ are on the opposite side. In this case, it is impossible to place both $vp(BDE)$ and $vp(ACD)$.

Case 2a: The viewpoint $vp(ACD)$ is on the ceiling to the left of $vp(BDE)$ or on same line as $vp(BDE)$. We consider how C is blocked from $vp(BDE)$. We can't block C from $vp(BDE)$ with the ceiling by Lemma 1, where $s = C, t = D, u = vp(ACD), v = vp(BDE)$. If we try to block C from $vp(BDE)$ using the floor, we end up blocking B from $vp(BDE)$ by Lemma 2 with $o = B, p = C, q = vp(BDE)$).

Case 2b: The viewpoint $vp(BDE)$ is on the ceiling to the left of $vp(ACD)$, or on same line as $vp(ACD)$. By Lemma 1 with $s = B, t = C, u = vp(BDE), v = vp(ACD)$, we must use the floor to block B from $vp(ACD)$. However, if we use the floor to block B from $vp(ACD)$, then by Lemma 2 with $o = A, p = B, q = vp(ACD)$, the A guard is blocked from seeing $vp(ACD)$.

Case 2c: The viewpoint $vp(ACD)$ on floor to the left of $vp(BDE)$. In this case, we consider how B is blocked from $vp(ACD)$. If the ceiling blocks B from $vp(ACD)$, then by Corollary 1 with $t = B, u = vp(ACD)$, B does not see $vp(BDE)$. If the floor blocks B from $vp(ACD)$, then by Lemma 2 with $o = A, p = B, q = vp(ACD)$, the A guard does not see $vp(ACD)$.

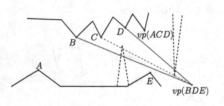

Fig. 20. Visualization of Case 2a

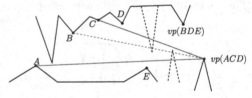

Fig. 21. Visualization of Case 2b

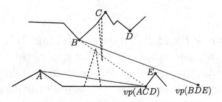

Fig. 22. Visualization of Case 2c

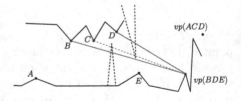

Fig. 23. Visualization of Case 2d

Case 2d: The viewpoint $vp(BDE)$ is on floor to the left of $vp(ACD)$. In this case, consider how C is blocked from seeing $vp(BDE)$. If the floor blocks C from seeing $vp(BDE)$, then by Lemma 2 with $o = B, p = C, q = vp(BDE)$, B would not see $vp(BDE)$. If the ceiling blocks C from $vp(BDE)$, then by Corollary 1 with $t = C, u = vp(BDE)$, C would not see $vp(ACD)$.

Case 3: In this case, $\{C, E\}$ are on the floor (resp. ceiling) and $\{A, B, D\}$ are on the opposite side. In this case, it is impossible to place both $vp(ADE)$ and $vp(BCDE)$.

Case 3a: The viewpoint $vp(ADE)$ on ceiling to the left of $vp(BCDE)$. In this case, we consider how C is blocked from $vp(ADE)$. If the floor blocks C from $vp(ADE)$, then by Corollary 1 with $t = C, u = vp(ADE)$, C does not see $vp(BCDE)$. If the ceiling blocks C from $vp(ADE)$, then by Lemma 2 with $o = A, p = C, q = vp(ADE)$, the A guard does not see $vp(ADE)$.

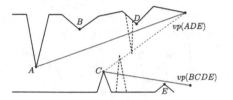

Fig. 24. Visualization of Case 3a

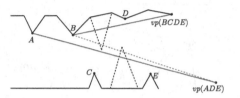

Fig. 25. Visualization of Case 3b

Case 3b: The viewpoint $vp(BCDE)$ is on the ceiling to the left of $vp(ADE)$, or on same line as $vp(BCDE)$. By Lemma 1 with $s = B; t = D; u = vp(BCDE); v = vp(ADE)$, we must use the floor to block B from $vp(ADE)$. However, if we use the floor to block B from $vp(ADE)$, then by Lemma 2 with $o = A, p = B, q = vp(ADE)$, the A guard is blocked from seeing $vp(ADE)$.

Case 3c: The viewpoint $vp(ADE)$ on floor to the left of $vp(BCDE)$. In this case, we consider how B is blocked from $vp(ADE)$. If the floor blocks B from $vp(ADE)$, then by Corollary 1 with $t = B, u = vp(ADE)$, A does not see $vp(ADE)$. If the ceiling blocks B from $vp(ADE)$, then by Lemma 2 with $o = A, p = B, q = vp(ADE)$, the B guard does not see $vp(BCDE)$.

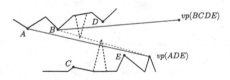

Fig. 26. Visualization of Case 3c

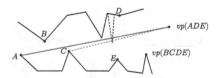

Fig. 27. Visualization of Case 3d

Case 3d: The viewpoint $vp(BCDE)$ is on the floor to the left of $vp(ADE)$, or on same line as $vp(BCDE)$. By Lemma 1 with $s = C; t = E; u = $

$vp(BCDE); v = vp(ADE)$, we must use the ceiling to block C from $vp(ADE)$. However, if we use the ceiling to block C from $vp(ADE)$, then by Lemma 2 with $o = A, p = C, q = vp(ADE)$, the A guard is blocked from seeing $vp(ADE)$.

Case 4: Any four guards are on the same side. For example, $\{A, B, C, D\}$ are on the floor (resp. ceiling). Then it is impossible to place both $vp(ACD)$ and $vp(ABD)$.

Case 4a: The viewpoint $vp(ACD)$ on ceiling to the left of $vp(ABD)$. In this case, we consider how B is blocked from $vp(ACD)$. If the floor blocks B from $vp(ACD)$, then by Corollary 1 with $t = B, u = vp(ACD)$, B does not see $vp(ABD)$. If the ceiling blocks B from $vp(ACD)$, then by Lemma 2 with $o = A, p = B, q = vp(ACD)$, the A guard does not see $vp(ACD)$.

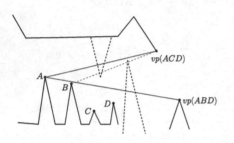

Fig. 28. Visualization of Case 4a

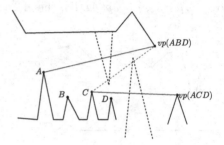

Fig. 29. Visualization of Case 4b

Case 4b: The viewpoint $vp(ABD)$ on ceiling to the left of $vp(ACD)$. In this case, we consider how C is blocked from $vp(ABD)$. If the floor blocks C from $vp(ABD)$, then by Corollary 1 with $t = C, u = vp(ABD)$, C does not see $vp(ACD)$. If the ceiling blocks C from $vp(ABD)$, then by Lemma 2 with $o = A, p = C, q = vp(ABD)$, the A guard does not see $vp(ABD)$.

Case 4c: The viewpoint $vp(ABD)$ is on the floor to the left of $vp(ACD)$, or on same line as $vp(ACD)$. By Lemma 1 with $s = B; t = C; u = vp(ABD); v = vp(ACD)$, we must use the ceiling to block B from $vp(ACD)$. However, if we use the ceiling to block B from $vp(ACD)$, then by Lemma 2 with $o = A, p = B, q = vp(ACD)$, the A guard is blocked from seeing $vp(ACD)$.

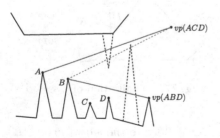

Fig. 30. Visualization of Case 4c

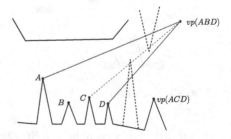

Fig. 31. Visualization of Case 4d

Case 4d: The viewpoint $vp(ACD)$ is on the floor to the left of $vp(ABD)$, or on same line as $vp(ABD)$. By Lemma 1 with $s = C; t = D; u = vp(ACD); v = vp(ABD)$, we must use the ceiling to block C from $vp(ABD)$. However, if we use the ceiling to block C from $vp(ABD)$, then by Lemma 2 with $o = A, p = C, q = vp(ABD)$, the A guard is blocked from seeing $vp(ABD)$.

Case 5: In this case, $\{A, B\}$ are on the floor (resp. ceiling) and $\{C, D\}$ is on the opposite side. In this case, it is impossible to place both $vp(BCE)$ and $vp(ADE)$.

Case 5a: The viewpoint $vp(BCE)$ on ceiling to the left of $vp(ADE)$, or on same line as $vp(ADE)$. By Lemma 1 with $s = C; t = D; u = vp(BCE); v = vp(ADE)$, we must use the ceiling to block C from $vp(ADE)$. However, if we use the floor to block C from $vp(ADE)$, then by Lemma 2 with $o = A; p = C; q = vp(ADE)$, the A guard is blocked from seeing $vp(ADE)$.

Case 5b: The viewpoint $vp(ADE)$ on ceiling to the left of $vp(BCE)$. In this case, we consider how B is blocked from $vp(ADE)$. If the floor blocks B from $vp(ADE)$, then by Corollary 1 with $t = B, u = vp(ADE)$, B does not see $vp(BCE)$. If the ceiling blocks B from $vp(ADE)$, then by Lemma 2 with $o = A, p = B, q = vp(ADE)$, the A guard does not see $vp(ADE)$.

Case 5c: The viewpoint $vp(ADE)$ on floor to the left of $vp(BCE)$. In this case, we consider how C is blocked from $vp(ADE)$. If the ceiling blocks C from $vp(ADE)$, then by Corollary 1 with $t = C, u = vp(ADE)$, C does not see $vp(BCE)$. If the floor blocks C from $vp(ADE)$, then by Lemma 2 with $o = A, p = C, q = vp(ADE)$, the A guard does not see $vp(ADE)$.

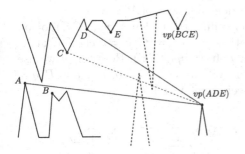

Fig. 32. Visualization of Case 5a

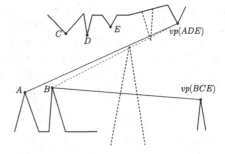

Fig. 33. Visualization of Case 5b

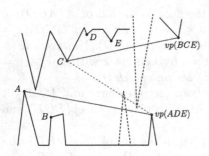
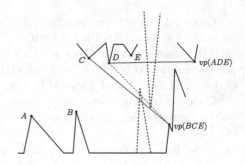

Fig. 34. Visualization of Case 5c   **Fig. 35.** Visualization of Case 5d

Case 5d: The viewpoint $vp(BCE)$ on floor to the left of $vp(ADE)$. In this case, we consider how D is blocked from $vp(BCE)$. If the ceiling blocks D from $vp(BCE)$, then by Corollary 1 with $t = D, u = vp(BCE)$, D does not see $vp(ADE)$. If the floor blocks D from $vp(BCE)$, then by Lemma 2 with $o = C, p = D, q = vp(BCE)$, the A guard does not see $vp(ADE)$.

Case 6: In this case, $\{A, D\}$ are on the floor (resp. ceiling) and $\{B.C\}$ is on the opposite side. In this case, it is impossible to place both $vp(BDE)$ and $vp(ACE)$.

Case 6a: The viewpoint $vp(ACE)$ on ceiling to the left of $vp(BDE)$. In this case, we consider how D is blocked from $vp(ACE)$. If the floor blocks D from $vp(ACE)$, then by Corollary 1 with $t = D, u = vp(ACE)$, D does not see $vp(BDE)$. If the ceiling blocks D from $vp(ACE)$, then by Lemma 2 with $o = A, p = D, q = vp(ACE)$, the A guard does not see $vp(ACE)$.

Case 6b: The viewpoint $vp(BDE)$ on ceiling to the left of $vp(ACE)$, or on same line as $vp(ACE)$. By Lemma 1 with $s = B; t = C; u = vp(BDE); v = vp(ACE)$, we must use the floor to block B from $vp(ACE)$. However, if we use the floor to block B from $vp(ACE)$, then by Lemma 2 with $o = A; p = B; q = vp(ACE)$, the A guard is blocked from seeing $vp(ACE)$.

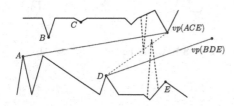
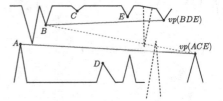

Fig. 36. Visualization of Case 6a   **Fig. 37.** Visualization of Case 6b

Case 6c: The viewpoint $vp(BDE)$ on floor to the left of $vp(ACE)$. In this case, we consider how C is blocked from $vp(BDE)$. If the ceiling blocks C from $vp(BDE)$, then by Corollary 1 with $t = C, u = vp(BDE)$, C does not see $vp(ACE)$. If the ceiling blocks C from $vp(BDE)$, then by Lemma 2 with $o = B, p = C, q = vp(BDE)$, the B guard does not see $vp(BDE)$.

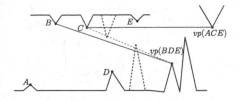

Fig. 38. Visualization of Case 6c

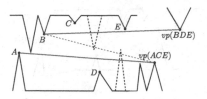

Fig. 39. Visualization of Case 6d

Case 6d: The viewpoint $vp(ACE)$ on floor to the left of $vp(BDE)$. In this case, we consider how B is blocked from $vp(ACE)$. If the ceiling blocks B from $vp(ACE)$, then by Corollary 1 with $t = B, u = vp(ACE)$, B does not see $vp(BDE)$. If the floor blocks B from $vp(ACE)$, then by Lemma 2 with $o = A, p = B, q = vp(ACE)$, the A guard does not see $vp(ACE)$.

Case 7: $\{B, E\}$ are on the floor and $\{A, C, D\}$ are on the ceiling. Then it is impossible to place both $vp(BCE)$ and $vp(ADE)$.

Case 7a: The viewpoint $vp(BCE)$ on ceiling to the left of $vp(ADE)$, or on same line as $vp(ADE)$. By Lemma 1 with $s = C; t = D; u = vp(BCE); v = vp(ADE)$, we must use the floor to block C from $vp(ADE)$. However, if we use the floor to block C from $vp(ADE)$, then by Lemma 2 with $o = A; p = C; q = vp(ADE)$, the A guard is blocked from seeing $vp(ADE)$.

Case 7b: The viewpoint $vp(ADE)$ on ceiling to the left of $vp(BCE)$. In this case, we consider how B is blocked from $vp(ADE)$. If the floor blocks B from $vp(ADE)$, then by Corollary 1 with $t = B, u = vp(ADE)$, B does not see $vp(BCE)$. If the ceiling blocks B from $vp(ADE)$, then by Lemma 2 with $o = A, p = B, q = vp(ADE)$, the A guard does not see $vp(ADE)$.

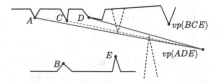

Fig. 40. Visualization of Case 7a

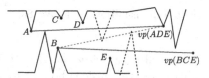

Fig. 41. Visualization of Case 7b

Case 7c: The viewpoint $vp(BCE)$ on floor to the left of $vp(ADE)$, or on same line as $vp(ADE)$. By Lemma 1 with $s = B; t = E; u = vp(BCE); v = vp(ADE)$, we must use the ceiling to block B from $vp(ADE)$. However, if we use the ceiling to block B from $vp(ADE)$, then by Lemma 2 with $o = A; p = B; q = vp(ADE)$, the A guard is blocked from seeing $vp(ADE)$.

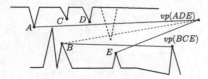

Fig. 42. Visualization of Case 7c

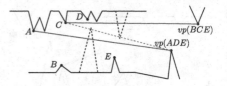

Fig. 43. Visualization of Case 7d

Case 7d: The viewpoint $vp(ADE)$ on floor to the left of $vp(BCE)$. In this case, we consider how C is blocked from $vp(ADE)$. If the ceiling blocks C from $vp(ADE)$, then by Corollary 1 with $t = C, u = vp(ADE)$, C does not see $vp(BCE)$. If the floor blocks C from $vp(ADE)$, then by Lemma 2 with $o = A, p = C, q = vp(ADE)$, the A guard does not see $vp(ADE)$.

Case 8: $\{A, B, C\}$ are on the floor and $\{D, E\}$ are on the ceiling. Then it is impossible to place both $vp(ACE)$ and $vp(BDE)$.

Case 8a: The viewpoint $vp(BDE)$ on the ceiling to the left of $vp(ACE)$. In this case, we consider how C is blocked from $vp(BDE)$. If the floor blocks C from $vp(BDE)$, then by Corollary 1 with $t = C, u = vp(BDE)$, C does not see $vp(ACE)$. If the ceiling blocks C from $vp(BDE)$, then by Lemma 2 with $o = B, p = C, q = vp(BDE)$, the B guard does not see $vp(BDE)$.

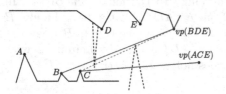

Fig. 44. Visualization of Case 8a

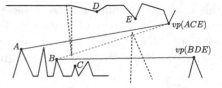

Fig. 45. Visualization of Case 8b

Case 8b: The viewpoint $vp(ACE)$ on the ceiling to the left of $vp(BDE)$. In this case, we consider how B is blocked from $vp(ACE)$. If the floor blocks B from $vp(ACE)$, then by Corollary 1 with $t = B, u = vp(ACE)$, B does not see $vp(BDE)$. If the ceiling blocks B from $vp(ACE)$, then by Lemma 2 with $o = A, p = B, q = vp(ACE)$, the A guard does not see $vp(ACE)$.

Case 8c: The viewpoint $vp(ACE)$ on the floor to the left of $vp(BDE)$. In this case, we consider how D is blocked from $vp(ACE)$. If the ceiling blocks D from $vp(ACE)$, then by Corollary 1 with $t = D, u = vp(ACE)$, D does not see $vp(BDE)$. If the floor blocks D from $vp(ACE)$, then by Lemma 2 with $o = A, p = D, q = vp(ACE)$, the A guard does not see $vp(ACE)$.

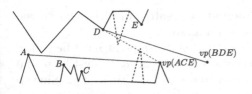

Fig. 46. Case 8c

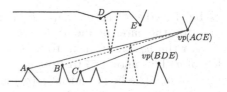

Fig. 47. Case 8d

Case 8d: The viewpoint $vp(BDE)$ on floor to the left of $vp(ACE)$, or on same line as $vp(ACE)$. By Lemma 1 with $s = B; t = C; u = vp(BDE); v = vp(ACE)$, we must use the ceiling to block B from $vp(ACE)$. However, if we use the ceiling to block B from $vp(ACE)$, then by Lemma 2 with $o = A; p = B; q = vp(ACE)$, the A guard is blocked from seeing $vp(ACE)$.

Half Guarding Spiral Polygon

In this section, we provide a simple proof for guarding a spiral polygon that does not contain any vertical edges.

Proof. *(Proof of Lemma 4).* Let p, q and r be 3 points on the inner boundary such that $i_{sp} < i_{sq} < i_{sr}$. Now assume that a guard g on the outer boundary sees p and r but does not see q. The outer boundary cannot block g from q since this would require the outer boundary to pierce the $\overline{gp}$ or the $\overline{gr}$ line segment, see Fig. 12. The inner boundary cannot pierce the $\overline{gq}$ either. If it does, then the inner boundary would require an interior angle less than 180° to reach a spot that g can see again, in this case, the point r. This contradicts the description of a spiral polygon. □

Proof. *(Proof of Lemma 5)* Let g, g' and g'' be 3 guards on the outer boundary such that $o_{sg} < o_{sg'} < o_{sg''}$. Now assume that g and g'' see p but g' does not see p. The inner boundary cannot block g' from p since this would require the inner boundary to pierce the $\overline{gp}$ or the $\overline{g''p}$ line segment, see Fig. 12 (use same figure as last proof but just a different part of the polygon for this part, use a similar dotted line to block g' from p). The outer boundary cannot pierce the $\overline{g'p}$ either. If it does, then the outer boundary would require an interior angle greater than 180° to reach the g'' guard. This contradicts the description of a spiral polygon. □

Proof. Since g is on the outer boundary, there are at most two edges of $vp(g)$ that are interior to P. The vertical line segment, denoted g_v, directly below (above) g is one such edge. If $vp(g)$ has only one such edge, then the claim is trivially true.

If $vp(g)$ has a second edge interior to P, then it must be the case that the inner boundary is blocking g. The outer boundary cannot create such an edge for $vp(g)$ otherwise it would not be a spiral polygon. The inner boundary cannot create more than one such edge, otherwise it would not be a spiral polygon. Let

g_e be the edge of $vp(g)$ that the inner boundary created and let $o(g_e)$ be the point on the outer boundary that g_e hits, see Fig. 13.

For the $\overline{pq}$ line segment to cross both g_v and g_e, the p point must be to the left of g. This allows $\overline{pq}$ to cross g_v. Now consider the point q. It must be to the right of g_v. The line segment $\overline{go(g_e)}$ is, by definition, to the right of g_v. Since p is to the left of g_v, the $\overline{pq}$ line segment must cross the $\overline{go(g_e)}$ line segment twice in order to reach q. Since they are both line segments, this is impossible.   □

Claim. Consider any partial guard placement $\mathcal{G}$. The leftmost unseen point will either be a point in K or a point on the inner boundary.

Proof. If the leftmost unseen point is a point in K, then the claim is trivially true. For each $k \in K$ that has been chosen to be in $\mathcal{G}$, the P_k region for that k is entirely seen. The leftmost unseen point must be to the right of one of these seen P_k regions. Let u be the leftmost unseen point and assume it is not in K nor on the inner boundary.

Since u is not in P_k, in order for u to look up and down and only see the outer boundary, it must be that u is to the right of some inner boundary. In this case, a guard must see some point arbitrarily close to the left of u. Any guard that sees this point arbitrarily close to u must also see u. The outer boundary cannot block this guard from seeing u, else it would not be a spiral polygon. Therefore, when u looks up and down, it must see the inner boundary at least once.

W.l.o.g., assume u looks up to see the inner boundary. Since u is not on the inner boundary, some guard must see the point v that is some ϵ distance above-and-left of u such that the line segment connecting v to u does not cross the inner boundary. Any guard g that sees this point must be blocked from u since u is assumed to be unseen. The inner boundary cannot block g from u since v is seen by g and by assumption, the inner boundary does not cross the $\overline{vu}$ line segment. If the inner boundary does block some guard g from seeing u, then u was not the leftmost unseen point. The only way to block g from u is to use the outer boundary. This is not possible because the interior angles of the outer boundary are less than $180°$. Therefore, if u is unseen, it must be on the inner boundary.   □

We now describe the algorithm and use an exchange argument to prove that the algorithm is no worse than any other solution. Find the leftmost unseen point of the polygon. If the leftmost unseen point is on the outer boundary of the polygon, place a guard at this point. If the leftmost point is not on the outer boundary of the polygon, then shoot a ray up (or down) to hit the outer boundary and place a guard at this location on the outer boundary. Repeat until the entire polygon is seen.

Consider an optimal solution $\mathcal{O}$. Now consider the guards placed by the algorithm. The first guard placed by the algorithm must be a guard in $\mathcal{O}$, else the leftmost point on the outer boundary is unseen. In general, if the leftmost unseen point can only be seen by itself, then it is obvious that the guard we place is identical to the one in $\mathcal{O}$. Now consider an unseen point that can be seen by more than just itself. The optimal solution must contain all guards in K since vertices in K are leftmost and can only be seen by themselves. Each of these points will be the leftmost unseen points at some point and the algorithm will also place all guards at vertices in K. Because of this, for each k, the P_k region associated with it is seen. Let h be the leftmost unseen point that the algorithm considers next and assume h is not a vertex in K. Since all P_k regions are seen and h is on the inner boundary by Claim 4.1, when the h point shoots a ray up (down), it must hit the outer boundary. The algorithm will place a guard at the outer boundary that is directly above (below) h. Consider the optimal guard o' that sees h. By Lemmas 9 and 6, moving o' to this new location will not decrease the visibility of o'. The guard placed by the algorithm will see at least as much as the guards placed in the partial optimal solution. The leftmost unseen point in the greedy algorithm cannot have been seen by a guard in the partial optimal solution since the greedy guards always see at least as much as the optimal guards that have been potentially moved. Because of this, the new guard in the partial optimal solution can be moved to the location specified by the greedy algorithm. Such an optimal guard will never have been considered/moved before.

Proof of Corollary 2

In order to prove the corollary, we use the following lemma.

Lemma 9. *Consider any half-guarding solution for some spiral polygon P. For each guard $k \in K$, no guard is required to be in P_k.*

Proof. For each $k \in K$, the optimal solution must place a guard at k, else k is not seen. Consider a guard g that was placed in P_k. If the P_k region is removed from the polygon, there remains at most 2 polygons that meet at exactly one point, namely, i_k. We call these polygons A and B, see Fig. 14. The g guard cannot see an unseen point in both A and B. Assume that this is possible and let $p \in A$ and $q \in B$ be the two unseen points that g sees. In order to be unseen, neither p nor q can be seen by k. It is not possible to block k from p nor q using the outer boundary since the outer boundary cannot have an interior angle greater than 180°. Therefore, the inner boundary must be used to block k from both p and q. As one looks from g, the $\overline{kp}$ line segment must be closer to the inner boundary than the $\overline{gp}$ line segment. This allows the inner boundary to block k from p, see Fig. 14. In a similar fashion, the line segment $\overline{kq}$ must also be closer to the inner boundary than $\overline{gq}$. Since k is to the left of o, this orientation is impossible. Therefore, g cannot see an unseen point in both A and B.

The g guard can, therefore, only see unseen points in at most one of A or B. W.l.o.g., assume g can see an unseen point in A. Let s be the outer boundary point directly above i_k in polygon A. By Lemma 6, the g guard cannot see an unseen point in A that s does not also see. Therefore, the g guard can be moved to the s location without losing any visibility. Since P_k is, by definition, to the left of i_k, this outer boundary point s is not in P_k. $\square$

Proof. Consider an optimal guard g placed in the interior of the spiral polygon. By Lemma 9, any optimal guard placed in P_k can be moved out of P_k. A simple extension of Lemma 7 says that any optimal guard placed in the polygon must have the inner boundary above (resp. below) it and the outer boundary below (resp. above) it. Let g' be the location on the outer boundary directly above/below g. Since g and g' are on the same vertical line, any point that g sees that g' doesn't see must be to the right of them. Consider Fig. 13, the g guard, by definition, lies on the g'_v edge. In order for g to see a point in the polygon that g' does not see, the g guard must see beyond the g'_e line. The g guard is on one side of this line. We draw a line from g to this point. In order to stay in the polygon, this line must cross the g'_e line. It then must cross the g'_e line again in order to reach a point that g' does not see. This is impossible, therefore, g' dominates g. □

References

1. Abdelkader, A., Saeed, A., Harras, K.A., Mohamed, A.: The inapproximability of illuminating polygons by α-floodlights. In: CCCG, pp. 287–295 (2015)
2. Abello, J., Estivill-Castro, V., Shermer, T.C., Urrutia, J.: Illumination of orthogonal polygons with orthogonal floodlights. Int. J. Comput. Geom. Appl. **8**(1), 25–38 (1998)
3. Bose, P., Guibas, L.J., Lubiw, A., Overmars, M.H., Souvaine, D.L., Urrutia, J.: The floodlight problem. Int. J. Comput. Geom. Appl. **7**(1/2), 153–163 (1997)
4. Ghosh, S.K.: On recognizing and characterizing visibility graphs of simple polygons. In: SWAT, pp. 96–104 (1988)
5. Ghosh, S.K.: On recognizing and characterizing visibility graphs of simple polygons. Discr. Comput. Geomet. **17**(2), 143–162 (1997)
6. Gibson, M., Krohn, E., Rayford, M.: Guarding monotone polygons with half-guards. In: CCCG, pp. 168–173 (2017)
7. Gibson, M., Krohn, E., Wang, Q.: On the VC-dimension of visibility in monotone polygons. In: CCCG (2014)
8. Gibson, M., Krohn, E., Wang, Q.: The VC-dimension of visibility on the boundary of a simple polygon. In: Elbassioni, K., Makino, K. (eds.) ISAAC 2015. LNCS, vol. 9472, pp. 541–551. Springer, Heidelberg (2015). https://doi.org/10.1007/978-3-662-48971-0_46
9. Gibson, M., Krohn, E., Wang, Q.: The VC-dimension of visibility on the boundary of monotone polygons. Comput. Geom. **77**, 62–72 (2019)
10. Gibson-Lopez, M., Yang, Z.: The VC-dimension of limited visibility terrains. In: ISAAC, vol. 212, pp. 1–17. Dagstuhl, Germany (2021)
11. Gilbers, A., Klein, R.: A new upper bound for the VC-dimension of visibility regions. Comput. Geom. **47**(1), 61–74 (2014)
12. Hillberg, H.M., Krohn, E., Pahlow, A.: On the complexity of half-guarding monotone polygons. In: Castaneda, A., Rodríguez-Henríquez, F. (eds.) LATIN 2022: Theoretical Informatics. LATIN 2022. LNCS, vol. 13568. Springer, Cham (2022). https://doi.org/10.1007/978-3-031-20624-5_46
13. Kahn, J., Klawe, M., Kleitman, D.: Traditional galleries require fewer watchmen. SIAM J. Algebr. Discrete Methods **4**(2), 194–206 (1983)
14. King, J.: VC-dimension of visibility on terrains. In: CCCG (2008)

15. Krohn, E., Nilsson, B.J.: The complexity of guarding monotone polygons. In: CCCG, vol. 2012, pp. 167–172 (2012)
16. Krohn, E., Nilsson, B.J.: Approximate guarding of monotone and rectilinear polygons. Algorithmica **66**(3), 564–594 (2013)
17. Krohn, E.A., Nilsson, B.J.: Approximate guarding of monotone and rectilinear polygons. Algorithmica, pp. 1–31 (2012)
18. Nilsson, B., Wood, D.: Optimum watchmen routes in spiral polygons. In: CCCG, pp. 269–272 (1990)

Dynamic Programming for the Fixed Route Hybrid Electric Aircraft Charging Problem

Anthony Deschênes[1]([✉])[iD], Raphaël Boudreault[2][iD], Vanessa Simard[3][iD],
Jonathan Gaudreault[1][iD], and Claude-Guy Quimper[1][iD]

[1] CRISI Research Consortium for Industry 4.0 System Engineering,
Université Laval, Québec, Canada
anthony.deschenes.1@ulaval.ca,
{jonathan.gaudreault,claude-guy.quimper}@ift.ulaval.ca
[2] Thales Digital Solutions, Québec, Canada
raphael.boudreault@thalesgroup.com
[3] NQB.ai, Québec, Canada
vanessa.simard@nqb.ai

Abstract. Air mobility is rapidly moving towards the development and usage of hybrid electric aircraft in multi-flight missions. Aircraft operators must consider numerous infrastructure and operational constraints in their planning, during which predicting energy usage is critical. We introduce this problem as the *Fixed Route Hybrid Electric Aircraft Charging Problem* (FRHACP). Given a fixed route, this problem aims to decide how much to refuel/charge at each terminal as well as the energy types to use during each flight leg (hybridization). The objective is to minimize the total energy-related monetary costs while satisfying scheduling and hybridization constraints. We propose a dynamic programming algorithm to solve this problem and show that it is optimal under assumptions usually satisfied in real-life settings. We then propose a gradient descent post-treatment to relax one of these assumptions while maintaining optimality. Results on realistic instances demonstrate that the developed algorithms outperform greedy heuristics, reaching an average cost reduction of up to 19.4%.

Keywords: Energy Management · Hybrid Electric Aircraft · Air Mobility · Dynamic Programming · Optimization · FRVCP

1 Introduction

Air mobility traditionally involves aircraft powered by combustion engines using carbon-based fuels. In the past years, the interest in alternative propulsion

This work received financial support from the Consortium for Research and Innovation in Aerospace in Quebec (CRIAQ), and the Mitacs Accelerate program.

engines significantly increased with the general aim of reducing aircraft green-house gas emissions. For that purpose, electricity-powered aircraft have been proposed, including hybrid electric aircraft which combines internal combustion engines with electrical power sources. It is envisioned that the future of air mobility will include these aircraft in a significant number of multi-flight missions, even possibly on demand, of varying length and duration [3].

Many challenges arise from the use of electricity as propulsion energy. Not only must one determine the trajectory of the vehicle, but also manage its energy consumption over the whole mission course according to aircraft, infrastructure, security, and schedule specifications. Given a flight route, this management aspect is particularly important in a planning perspective, since charging currently requires a non-negligible and non-linear amount of time [6,15]. Aircraft operators must thus decide how much to refuel and charge at each mission terminal. Furthermore, the consideration of hybrid electric aircraft requires hybridization decisions on the energy types to use (fuel and/or electricity) during each flight leg. These decisions must take into account consumption predictions from non-linear energy models [7,17], as well as mass variations and schedule requirements, to globally minimize energy-related monetary costs.

In this paper, we introduce the above-described optimization problem as the *Fixed Route Hybrid Electric Aircraft Charging Problem* (FRHACP) and propose a Dynamic Programming (DP) algorithm to solve it. Section 2 describes the FRHACP. Section 3 relates this problem to other work in the literature, notably the FRVCP for electric vehicles [7,15]. The DP algorithm is presented in Sect. 4, including details on the assumptions to guarantee its optimality, as well as a post-treatment to relax one of these assumptions. The algorithms are validated and compared to greedy heuristics on realistic instances in Sect. 5, while we conclude in Sect. 6.

2   The Fixed Route Hybrid Electric Aircraft Charging Problem (FRHACP)

The FRHACP considers hybrid electric aircraft in a multi-flight mission setting. A mission is defined as a fixed route $r := (n_1, n_2, \ldots, n_{|N|})$ of subsequent *nodes* $n_i \in N$. Each nodes from the route is either a *terminal* from set T or a *waypoint* from set W ($N := T \cup W$). A terminal is typically an airport, where facilities are available to refuel and charge the aircraft. The route r starts and ends at a terminal, i.e. $n_1, n_{|N|} \in T$, while r induces a natural order $t_1, t_2, \ldots, t_{|T|}$ on the terminals in T. Between consecutive terminals, the route is defined by waypoints, typically reference points in the air that must be part of the aircraft trajectory. We define *legs* as route segments connecting two consecutive nodes such as $L := \{(n_i, n_{i+1}) : i = 1, \ldots, |N|-1\}$.

The FRHACP asks to decide how much to refuel and charge the aircraft at each terminal. Fuel quantity in the aircraft is limited by a minimal security margin f^{min} and the tank capacity f^{max}. Similarly, the aircraft battery State of Charge (SoC) is limited by minimal and maximal security margins, s^{min} and

s^{max}. Each terminal $t \in T$ is also associated with a scheduled departure time d_t^{time} to respect as a hard constraint. The time needed to charge the battery from SoC s_1 to s_2 at terminal $t \in T$ is predicted with $\alpha_t^s(s_1, s_2)$, usually non-linear [6,15]. Refueling duration is given by a constant rate α^f depending on quantity.

Hybridization decisions on the energy types to use (fuel and/or electricity) during each leg are also part of the FRHACP. On that matter, it is known in the literature that the optimal energy management strategy on a leg is to use the fuel first, then the electricity [16]. Furthermore, fuel has a non-negligible mass, here encoded as a constant ratio of m_f depending on volume. This is known to be an important non-linear factor impacting the fuel and electricity consumption [17]. Thus, this problem encodes hybridization on each leg as a percentage of its distance using fuel, while the remaining distance is done using electricity, with fuel used first. Aircraft mass, m_a, and payload mass at terminal t, m_t^p, are also considered. Fuel and electricity consumption prediction models are encoded as functions dependant on the travel distance d and the total mass m, denoted respectively by $\delta^f(d, m)$ and $\delta^s(d, m)$. These functions are usually based on non-linear energy models including numerous other physical parameters [7,17] that are assumed constant on a given leg, but allowed to vary between legs (e.g., speed, altitude, and trajectory angle).

We resume the decisions variables of this problem as follows. For each terminal $t \in T$, $F_t^D \in [f^{min}, f^{max}]$ and $S_t^D \in [s^{min}, s^{max}]$ are respectively the fuel quantity and SoC of the aircraft when departing from terminal t. Then, for each leg $l \in L$, $H_l \in [0, 1]$ is the hybridization on leg l as its percentage traveled using fuel. Intermediate variables F_t^A and S_t^A describe the deduced fuel quantity and SoC upon arrival at terminal $t \in T$.

Finally, each terminal $t \in T$ has a fuel (resp. electricity) cost c_t^f (c_t^s) per refueled (charged) quantity. The FRHACP objective is thus to minimize the mission total cost according to energy decisions, i.e.

$$\min \sum_{t \in T} \left(c_t^f \left(F_t^D - F_t^A \right) + c_t^s \left(S_t^D - S_t^A \right) \right). \tag{1}$$

3   Related Work

The FRHACP is highly related to the *Fixed Route Electric Vehicle Charging Problem* (FRVCP) introduced by Montoya *et al.* [15], which has recently been extended with non-linear energy management in the context of an electric vehicle route planning [7]. In this problem, the objective is to minimize the total route duration including its charging time by considering variable vehicle speed and charging detours, while handling the non-linearity of electricity. The FRVCP has been solved using dynamic programming [5], Mixed Integer Programming [7] and labeling algorithms [10]. The FRHACP can naturally be seen as a variant of the FRVCP adapted to the context of hybrid electric aircraft. The main differences are the hybridization decisions and the objective function.

It is also well known in the literature that the non-linearity of energy models, depending among others on vehicle specifications, speed, mass, and temperature, are essential for energy-related predictions and planning [2,4,6,17]. Prior work in the hybrid electric aircraft domain mainly relates to optimal hybrid management [16], energy architecture [14,18] and fuel/electricity consumption models for different aircraft configurations [13,19]. Notably, *OpenAP* provides open-source aircraft performance and emission models based on open data and accessible for air transport research [17].

4 Dynamic Programming Algorithm

The FRHACP defined in Sect. 2 is more complex than the FRVCP and its variants [5,7] since two energy sources must be simultaneously considered (fuel and electricity). It is thus harder to design a dynamic programming algorithm following the approach of Deschênes *et al.* [5]. At least one state space must be added for the fuel. This would increase the computation time based on the number of sampled fuel quantities, say $\tilde{f}$. When refueling and charging, combinations of fuel and SoC will need to be considered, say $\tilde{f} \cdot \tilde{s}$. Thus we can estimate that the algorithm would be at least $\tilde{f}^2 \cdot \tilde{s}$ times slower, without even considering the additional computation time of 2-dimensional interpolation. With $\tilde{f} = 50$ and $\tilde{s} = 10$, it would be at least 25 000 times slower. Thus, the dynamic programming curse of dimensionality quickly arises.

Nevertheless, under some assumptions, it is possible to design a dynamic programming algorithm that optimally solves the problem. This algorithm is presented in Sect. 4.1. In Sect. 4.2, we develop a gradient descent post-treatment that allows to relax one of these assumptions while maintaining optimality.

4.1 Minimizing Total Cost

The proposed approach looks at the total cost minimization problem from the perspective of minimizing the fuel quantity in a number of subproblems. Each flight between consecutive terminals t_i and t_{i+1} defines a different subproblem, leading to the following question for all nodes n_k between t_i and t_{i+1} inclusively: *Given a current SoC s, what is the minimal fuel quantity $F_{n_k}^*(s)$ needed to reach terminal t_{i+1} from node n_k while satisfying all constraints?* Equation (2) presents the recurrence used to answer this question.

$$F_{n_k}^*(s) = \begin{cases} f^{min} & \text{if } n_k = t_{i+1} \\ \min\limits_{h \in [0,1]} \left[F_{n_{k+1}}^*(s - \Delta_{l_k}^s(h)) + \Delta_{l_k}^f(h) \right] & \text{otherwise} \end{cases} \quad (2)$$

If node n_k is terminal t_{i+1}, the minimal quantity to reach itself is trivially the margin f^{min}. Otherwise, the minimal quantity from n_k depends on the hybridization decision $h \in [0,1]$ on leg $l_k := (n_k, n_{k+1})$. Here, we respectively denote $\Delta_{l_k}^f(h)$ and $\Delta_{l_k}^s(h)$ the fuel and electricity consumption on leg l_k given h. The SoC at node n_{k+1} is thus given by $s - \Delta_{l_k}^s(h)$, while the minimal fuel quantity needed at n_{k+1} is $F^*_{n_{k+1}}(s - \Delta^s_{l_k}(h))$. We must then add the amount of fuel needed on leg l_k as given by $\Delta^f_{l_k}(h)$. Taking the minimal value over all h, $F^*_{n_k}(s)$ returns the minimal fuel quantity from n_k. Proposition 1 directly follows from this inductive reasoning.

Proposition 1. *The problem of minimizing the fuel quantity between consecutive terminals t_i and t_{i+1} admits an optimal substructure. In other words, given a current SoC s, $F^*_{n_k}(s)$ is the minimal fuel quantity to reach t_{i+1} from node n_k for all $n_k \in N$ between t_i and t_{i+1}.*

To solve the recurrence, we compute $\tilde{h}$ values of h and take the minimal computed quantity. $\Delta^f_{l_k}(h)$ is obtained with $\delta^f(h \cdot d_k, m(h))$ where d_k is the total distance of leg l_k and $m(h)$ is the mass, considering m_a, $m^p_{t_i}$, and the fuel mass in the tank depending on h. Similarly, $\Delta^s_{l_k}(h)$ is obtained with $\delta^s((1 - h) \cdot d_k, m(h))$. As in Deschênes *et al.* [5], the state space of s is continuous, thus we use the same techniques to solve the problem. We sample $F^*_{n_k}(s)$ for $\tilde{s}$ different SoC s for each node n_k. Finally, we use Akima interpolation [1] to approximate the overall function $F^*_{n_k}(s)$ for each node n_k.

Constructing the Route Solution. Our proposed DP algorithm solves the problem by constructing decisions for the complete route. It first starts at terminal t_1 with the initial fuel and SoC of the aircraft. Then, it finds $S^D_{t_1}$ satisfying the schedule $d^{time}_{t_1}$ and margin s^{max} using a binary search. From $S^D_{t_1}$, it uses recurrence (2) to compute $F^D_{t_1}$ and follows it until reaching terminal t_2. Note that it always makes sure f^{min}, f^{max}, and s^{min} are satisfied. The hybridization decisions H_{l_k} on legs l_k between t_1 and t_2 are simultaneously computed by the recurrence (as the $arg\,min$). At each following terminal $t_i \in T$, we determine $S^D_{t_i}$ and $F^D_{t_i}$ in the same way. This gives us our final solution. In order to prove this solution is optimal when minimizing the total cost under some assumptions, the following definition is needed.

Definition 1 (Independence of subproblems). *All subproblems are independent if $F^A_t = f^{min}$ and $S^A_t = s^{min}$ $\forall t \in T$ in the optimal solution.*

The independence of subproblems is known to imply *at least* these necessary conditions: (1) Fuel cost c^f_t is the same at each terminal $t \in T$; (2) Between each consecutive terminal, the optimal solution consumes all fuel and electricity. Note that other conditions might be needed to fully ensure independence of subproblems on some instances. For example, it is possible to construct an instance where d^{time}_t constrains the charging time in a way that electricity must be stored from a previous terminal, violating the independence while satisfying the above-mentioned conditions.

Proposition 2. *Suppose that consumption functions $\delta^f(d, m)$ and $\delta^s(d, m)$ are monotonically increasing with respect to d and m, that we have independence of subproblems, and that electricity costs c^s_t are significantly lower than fuel costs c^f_t. Then the DP constructed solution optimally minimizes total cost.*

Proof. With independence of subproblems, the constructed solution is such that all fuel and electricity is consumed between all consecutive terminals. According to Eq. (1) and the assumption about electricity costs, the only way to reduce the total cost would be to reduce the fuel consumption in each subproblem. Since $\delta^f(d, m)$ and $\delta^s(d, m)$ are monotonically increasing (more fuel leads to more mass, increasing the overall consumption), reducing the consumption is only possible by reducing the fuel quantity. However, by Proposition 1, this quantity is already minimal.   □

Complexity Analysis. Suppose the calls to $\delta^f(d, m)$ and $\delta^s(d, m)$ consumption functions are executed in constant time. To solve the problem, we compute the recurrence for $\tilde{s}$ values of SoC for each node $n_k \in N$. Computing $F_{n_k}^*(s)$ has a time complexity of $\Theta(\tilde{h})$ testing $\tilde{h}$ hybridization decisions. Since computing the Akima interpolation is done in a linear time, the overall time complexity is $\Theta(\tilde{s} \cdot \tilde{h} \cdot |N|)$. Thus, the algorithm running time increases *pseudo-linearly* with the number of nodes in the route.

4.2   Gradient Descent Post-treatment

Most of the assumptions of Proposition 2 are usually satisfied in real-life settings, except for the implied condition that fuel cost is the same at each terminal. Algorithm 1 relaxes independence of subproblems by allowing $F_t^A > f^{min}$ at all terminals $t \in T$.

The algorithm starts by computing the DP solution before improving it further with a gradient descent. The problem is encoded as a directed graph $G = (T, E)$, where $E := \{(t_i, t_j) : t_i, t_j \in T, i < j, c_{t_i}^f < c_{t_j}^f\}$. It defines the possibilities of *transferring* fuel between terminals t_i and t_j to save on fuel costs. The action of transferring x quantity of fuel from terminal t_j to t_i, denoted TRANS-FER(x, t_i, t_j), ensures that $F_{t_j}^A$ is increased by x. It is achieved by increasing $F_{t_i}^D$ by *at least* x. The action takes into consideration the non-linearity of $\delta^f(d, m)$ and $\delta^s(d, m)$, i.e. that taking more fuel at terminal t_i will increase the mass and thus the energy consumption until we reach terminal t_j. It takes into account the fact that we may need to add more fuel to compensate for the mass increase or the hybridization correction on the legs to ensure electricity margins. Indeed,

---

Algorithm 1. Gradient Descent Post-Treatment (DP+GD)

1: Compute a solution using the DP algorithm
2: Construct the directed graph $G = (T, E)$, $E := \{(t_i, t_j) : t_i, t_j \in T, i < j, c_{t_i}^f < c_{t_j}^f\}$
3: Initialize gradients $g_e \leftarrow 1, \forall e \in E$
4: **while** $\exists e \in E$ such that $g_e > 0$ **do**
5:     Compute g_e for each edge $e \in E$
6:     Find $(t_i, t_j) \in E$ such that $g_{(t_i, t_j)}$ is maximal
7:       **if** $g_{(t_i, t_j)} > 0$ **then** TRANSFER$(\alpha \cdot g_{(t_i, t_j)}, t_i, t_j)$
8: **return** The updated solution

---

the latter is due to the fact that, because of the schedule, we cannot take more electricity to compensate for the increase in electricity consumption, thus we instead increase the fuel consumption by modifying the hybridization decisions. Finally, the action is only possible if the transfer allows to satisfy f^{max} and $d_{t_i}^{time}$.

The algorithm computes the *gradient* $g_{(t_i,t_j)}$ of each edge $(t_i,t_j) \in E$, i.e. how much a small transfer of fuel from terminal t_j to t_i changes the overall cost of the solution. We then do a gradient descent to transfer fuel on the maximum gradient edge. These transfers are repeated until we reach convergence, i.e. when all gradients are non-positive. If the maximal gradient is $g_{(t_i,t_j)} > 0$, we do $\textsc{Transfer}(\alpha \cdot g_{(t_i,t_j)}, t_i, t_j)$ to transfer the fuel, where α is a strictly positive learning rate. Thus, at each iteration, the solution changes. Since by definition the graph is acyclic and we can only transfer in the direction of the edge (i.e. not backwards), the algorithm terminates in a finite number of steps.

Proposition 3. *Suppose the assumptions of Proposition 2 where we relax the independence of subproblems by allowing $F_t^A > f^{min}$ at all terminals $t \in T$. Then Algorithm 1 converges to the global optimum.*

Proof. Let c_a^* be the cost of solution a returned by Algorithm 1. Suppose the contrary, i.e. that there exists a solution π following the assumptions with total cost $c_\pi^* < c_a^*$. By Proposition 2, we know that the decrease in cost cannot be from using less fuel or using the electricity. Thus, the only option would be by exploiting the relaxed assumption. This implies that solution π takes more fuel at least at one terminal to reduce the total cost, thus that there exist terminals t_i and t_j with $c_{t_i}^f < c_{t_j}^f$ not exploited by solution a. By construction of Algorithm 1, this leads to an edge (t_i, t_j) with gradient $g_{(t_i,t_j)} > 0$. However, this is impossible since the algorithm terminates with all gradients non-positive. □

5   Experiments

The main goal of the experiments is to compare the proposed DP algorithms with heuristics on real-life inspired instances. It aims at showing the benefits of using the electric engine, while doing optimized refueling, charging, and hybridization decisions.

Fuel First Heuristic (FF-H). This heuristic aims at globally maximizing the fuel usage during the flight route. It imposes a hybridization decision of 100% fuel ($H_l := 1$) on each leg. Then, the departure fuel F_t^D is adjusted to minimize the consumption while satisfying the minimal margin f^{min}. Summarized steps of this heuristic are presented in Algorithm 2.

Algorithm 2. Fuel First Heuristic (FF-H)
___
1: Initialize $F_t^D \leftarrow f^{max}$ and $S_t^D \leftarrow s^{min}$ for $t \in T$; $H_l \leftarrow 1$ for $l \in L$
2: Compute F_t^A for $t \in T$
3: For $t_i \in T, i = 1, \ldots, |T|-1$, correct $F_{t_i}^D$ so that $F_{t_{i+1}}^A = f^{min}$
___

Maximize Battery Usage Heuristic (MB-H). This heuristic aims at globally maximizing the electricity usage during the flight route. It tries to impose a hybridization decision of 100% electricity ($H_l := 0$) on each leg. This is often impossible, thus it handles these cases based on a greedy hypothesis of using the fuel first. The minimal quantity of fuel is computed so that the arrival SoC S_t^A reaches the margin s^{min} for $t \in T$. Summarized steps of this heuristic are presented in Algorithm 3.

Algorithm 3 Maximize Battery Usage Heuristic (MB-H)
___
1: Initialize $F_t^D \leftarrow f^{min}$ for $t \in T$
2: Set S_t^D to its maximal value given d_t^{time} for $t \in T$; $H_l \leftarrow 0$ for $l \in L$
3: Compute S_t^A for $t \in T$
4: **for all** $t \in T$ where S_t^A ¡ s^{min} **do**
5:    Set F_t^D to the minimal fuel quantity satisfying s^{min}, with H_l using fuel first
6: For $t_i \in T, i = 1, \ldots, |T|-1$, correct $S_{t_i}^D$ so that $S_{t_{i+1}}^A = s^{min}$
___

5.1 Experimental Setup

We implemented the algorithms described in Sect. 4 and the heuristics in Python. The experiments were performed on an Intel Core i7-8750H CPU @ 2.20 GHz, 6 cores and 8 GB of RAM. The DP algorithm has two different hyperparameters affecting the quality of the solution and its computation time: $\bar{h}$, the number of hybridization values tested on each leg, and $\tilde{s}$, the number of SoC values sampled to determine the minimal fuel. For our experiments, we used $\bar{h} = 40$ and $\tilde{s} = 10$. These values were empirically determined to yield good results in a decent computation time. For the DP+GD algorithm, the gradient is approximated by forward difference. We also used a constant learning rate $\alpha = 500$ empirically determined for converging quickly.

5.2 Instances

Our dataset consists of four real-life inspired instances, created from day-long sequences of commercial flights with the same aircraft in Canada and France. The routes and their schedule are generated using available data in FlightRadar24 [9], while the fuel and electricity costs are directly taken from various credible sources [8,11,12]. For the purpose of comparison, we convert EUR (€) costs in CAD ($) using 1.49 as exchange rate. Table 1 presents the particularities of each instance. In France, PN_{T4W59} describes two round trips between Paris and Nice, while TB_{T5W42} includes a round trip from Toulouse to Lille followed by a round trip from Bordeaux to Marseille. In Canada, MS_{T6W30} includes a flight from Montreal to Quebec City, followed by a round trip to the Magdalen Islands, then a flight to Sept-Îles. OT_{T7W43} describes a flight from Ottawa to Toronto, followed by a round trip to St. John's, Newfoundland.

Table 1. Description of the four real-life inspired instances forming the dataset.

Instance	$|T|$	$|W|$	Duration	Distance (km)	c^s ($/kWh)	c^f ($/L)
PN_{T4W59}	4	59	4 h 30	2740	0.1397	1.46
TB_{T5W42}	5	42	6 h 42	2812	0.1397	1.46
MS_{T6W30}	6	30	7 h 22	2294	0.0533	[1.16, 1.25]
OT_{T7W43}	7	43	9 h 34	4709	[0.0533, 0.1140]	[1.03, 1.28]

All instances suppose a *Cessna S550 Citation II* as the aircraft. Following the approach of Wang *et al.* [18], we suppose a battery of 216 kWh with a mass of 600 kg, giving a total mass m_a of 4256 kg. We also suppose a payload m_t^p varying at terminals $t \in T$ between 400 kg and 800 kg. The Cessna uses Jet-A1 fuel with ratio m_f 0.819 kg/L, a fuel capacity f^{max} of 3260 L, and a refueling rate α^f of 1086 L per minute. In addition, the following security margins are considered: $f^{min} = 163$ L (5%), $s^{min} = 10\%$, and $s^{max} = 95\%$.

We use OpenAP aircraft performance model [17] to predict the fuel consumption as function $\delta^f(d, m)$. To do so, we deduce altitude, distance, speed, and trajectory angle from the instance generated route. We also suppose a cruise phase at an altitude of 10.7 km with a speed of 777 km/h. All other parameters are implicitly encoded in the OpenAP model. For predicting the electricity consumption $\delta^s(d, m)$, we use OpenAP predicted net thrust and convert it to kWh.

For the charging time prediction $\alpha_t^s(s_1, s_2)$, we use for all terminals $t \in T$ the non-linear charging function from Deschênes *et al.* [7]. Although this function is unrealistic given that it has been designed for a 40 kWh battery of an electric vehicle, we envision that charging technology in a near future may allow similar durations.

5.3   Results

Table 2 presents the results of our experiments. For each algorithm (DP, DP+GD, FF-H, MB-H) and each instance, we report the solving time, as well as costs and consumed quantities related to each energy type. We also distinguish the solving time of the algorithms (*internal*) from the calls to OpenAP performance model (*external*). The smallest total costs are in bold. Note that it is possible to check that all instances follow the assumptions discussed in

Table 2. Solving time of each instance in seconds—distinguished between internal (algorithms) and external (OpenAP) time—as well as costs and consumed quantities for fuel and electricity. Results reported for the Dynamic Programming algorithm (DP), DP with the Gradient Descent post-treatment (DP+GD), the Fuel First heuristic (FF-H) and the Maximum Battery heuristic (MB-H).

Instance	Algorithm	Solving time (s)			Costs ($)			Consumption	
		Internal	External	Total	Fuel	Elec.	Total	Fuel (L)	Elec. (kWh)
PN_{T4W59}	DP	0.43	4.65	5.08	4284	148	**4433**	2930	712
	DP+GD	-	-	-	-	-	-	-	-
	FF-H	0.06	0.06	0.12	5316	0	5316	3634	0
	MB-H	0.40	0.55	0.95	4579	128	4707	3131	614
TB_{T5W42}	DP	0.32	3.78	4.10	4335	142	**4477**	2991	805
	DP+GD	-	-	-	-	-	-	-	-
	FF-H	0.05	0.05	0.10	5421	0	5421	3740	0
	MB-H	0.26	0.40	0.66	4630	126	4756	3195	691
MS_{T6W30}	DP	0.23	2.40	2.63	3148	28	3176	2650	899
	DP+GD	0.40	2.56	2.96	3084	28	**3112**	2659	899
	FF-H	0.02	0.02	0.04	4081	0	4081	3429	0
	MB-H	0.14	0.14	0.28	3315	24	3338	2789	798
OT_{T7W43}	DP	0.34	4.32	4.66	6449	95	6543	5408	1210
	DP+GD	1.79	5.94	7.73	6173	95	**6268**	5523	1210
	FF-H	0.05	0.05	0.10	7822	0	7822	6540	0
	MB-H	0.37	0.50	0.87	6836	83	6919	5730	1038

Sect. 4. Since PN_{T4W59} and TB_{T5W42} have no fuel cost variation, DP is optimal for these instances by Proposition 2 and the gradient descent post-treatment is not required. On the other hand, DP+GD is optimal for MS_{T6W30} and OT_{T7W43} by Proposition 3.

As expected, heuristics have the smallest solving times, while all algorithms terminate within 8 s. We remark that the gradient descent can increase the computation time of up to 40.0% (OT_{T7W43}). On average, 92% of DP computation time comes from external calls, i.e. OpenAP. This is reduced to 80 % with DP+GD.

About costs, DP and DP+GD (when applicable) obtain the smallest total cost on all instances. The reduction mainly comes from lower fuel consumption. On instances where the fuel costs vary (MS_{T6W30}, OT_{T7W43}), the gradient descent post-treatment of DP+GD allows an average reduction of 3.1% compared to DP. As expected, the electricity costs remain constant, since the post-treatment does not affect charging decisions. FF-H obtains the highest costs on all instances, with the DP algorithms leading to a reduction of up to 23.7% (average 19.4%). This clearly shows the benefits of using the electric engine. MB-H has smaller costs compared to FF-H, but the DP algorithms can reduce them of up to 9.4% (average 7.0%). This shows that smarter hybridization and refueling decisions can lead to better solutions.

6   Conclusion

In this paper, we introduced the FRHACP, a variant of the FRVCP adapted to the context of hybrid electric aircraft. The problem aims to handle refueling/charging and hybridization decisions given a fixed route, while minimizing energy costs and satisfying various requirements, such as mass and schedule. To solve the problem, we proposed a dynamic programming algorithm that has been shown to be optimal under some assumptions. In order to fit for more real-life settings, we relaxed one of these assumptions and allowed fuel costs to vary between terminals by proposing a gradient descent post-treatment while maintaining optimality. The algorithms were compared to two greedy heuristics on four real-life inspired instances that showed the benefits of considering electric engines and doing smart hybridization decisions. Results demonstrated an average cost reduction of up to 19.4%. The proposed algorithms found the optimal solution within 8 s on all instances.

References

1. Akima, H.: A new method of interpolation and smooth curve fitting based on local procedures. J. ACM **17**(4), 589–602 (1970). https://doi.org/10.1145/321607.321609
2. Ansarey, M., Shariat Panahi, M., Ziarati, H., Mahjoob, M.: Optimal energy management in a dual-storage fuel-cell hybrid vehicle using multi-dimensional dynamic programming. J. Power Sources **250**, 359–371 (2014). https://doi.org/10.1016/j.jpowsour.2013.10.145
3. Ansell, P.J., Haran, K.S.: Electrified airplanes: a path to zero-emission air travel. IEEE Electrification Mag. **8**(2), 18–26 (2020). https://doi.org/10.1109/MELE.2020.2985482
4. De Cauwer, C., Verbeke, W., Coosemans, T., Faid, S., Van Mierlo, J.: A data-driven method for energy consumption prediction and energy-efficient routing of electric vehicles in real-world conditions. Energies **10**(5), 608 (2017). https://doi.org/10.3390/en10050608
5. Deschênes, A., Gaudreault, J., Quimper, C.G.: Dynamic programming for the fixed route electric vehicle charging problem with nonlinear energy management. In: 2022 IEEE 25th International Conference on Intelligent Transportation Systems (ITSC), pp. 3956–3962 (2022). https://doi.org/10.1109/ITSC55140.2022.9922511
6. Deschênes, A., Gaudreault, J., Quimper, C.G.: Predicting real life electric vehicle fast charging session duration using neural networks. In: 2022 IEEE Intelligent Vehicles Symposium (IV), pp. 1327–1332 (2022). https://doi.org/10.1109/IV51971.2022.9827179
7. Deschênes, A., Gaudreault, J., Vignault, L.P., Bernard, F., Quimper, C.G.: The fixed route electric vehicle charging problem with nonlinear energy management and variable vehicle speed. In: 2020 IEEE International Conference on Systems, Man, and Cybernetics (SMC), pp. 1451–1458 (2020). https://doi.org/10.1109/SMC42975.2020.9283062
8. Eurostat: Data - Eurostat. https://ec.europa.eu/eurostat/data
9. FlightRadar24: Live flight tracker - Real-time flight tracker map. https://www.flightradar24.com/

10. Froger, A., Mendoza, J.E., Jabali, O., Laporte, G.: Improved formulations and algorithmic components for the electric vehicle routing problem with nonlinear charging functions. Comput. Oper. Res. **104**, 256–294 (2019). https://doi.org/10. 1016/j.cor.2018.12.013

11. Hydro-Québec: Comparison of electricity prices in major North American cities 2022. Technical report, Hydro-Québec (2022)

12. Jet-A1: Jet A-1 price on fuel. https://jet-a1-fuel.com/

13. Kirschstein, T.: Comparison of energy demands of drone-based and ground-based parcel delivery services. Transp. Res. Part D Transp. Environ. **78**, 102209 (2020). https://doi.org/10.1016/j.trd.2019.102209

14. Lei, T., Min, Z., Gao, Q., Song, L., Zhang, X., Zhang, X.: The architecture optimization and energy management technology of aircraft power systems: a review and future trends. Energies **15**(11), 4109 (2022). https://doi.org/10.3390/ en15114109

15. Montoya, A., Guéret, C., Mendoza, J.E., Villegas, J.G.: The electric vehicle routing problem with nonlinear charging function. Transp. Res. Part B Methodol. **103**, 87–110 (2017). https://doi.org/10.1016/j.trb.2017.02.004

16. Pinto Leite, J.P.S., Voskuijl, M.: Optimal energy management for hybrid-electric aircraft. Aircr. Eng. Aerosp. Technol. **92**(6), 851–861 (2020). https://doi.org/10. 1108/AEAT-03-2019-0046

17. Sun, J., Hoekstra, J.M., Ellerbroek, J.: OpenAP: an open-source aircraft performance model for air transportation studies and simulations. Aerospace **7**(8), 104 (2020). https://doi.org/10.3390/aerospace7080104

18. Wang, M., Mesbahi, M.: To charge in-flight or not: an inquiry into parallel-hybrid electric aircraft configurations via optimal control (2022)

19. Zhang, J., Campbell, J.F., Sweeney, D.C., Hupman, A.C.: Energy consumption models for delivery drones: a comparison and assessment. Transp. Res. Part D Transp. Environ. **90**, 102668 (2021). https://doi.org/10.1016/j.trd.2020.102668

Algorithms for the Ridesharing with Profit Constraint Problem

Qian-Ping Gu and Jiajian Leo Liang(✉)

School of Computing Science, Simon Fraser University, Burnaby, Canada
{qgu,leo_liang}@sfu.ca

Abstract. Mobility-on-demand (MoD) ridesharing is a promising way to improve the occupancy rate of personal vehicles and reduce traffic congestion and emissions. Maximizing the number of passengers served and maximizing a profit target are major optimization goals in MoD ridesharing. We study the ridesharing with profit constraint problem (labeled as RPC) which considers both optimization goals altogether: maximize the total number of passengers subject to an overall drivers' profit target. We give a mathematical formulation for the RPC problem. We present a polynomial-time exact algorithm framework (including two practical implementations of the algorithm) and a $\frac{1}{2}$-approximation algorithm for the case that each vehicle serves at most one passenger. We propose a $\frac{2}{3\lambda}$-approximation algorithm for the case that each vehicle serves at most $\lambda \geq 2$ passengers. Our algorithms revolve around the idea of maximum cardinality matching in bipartite graphs and hypergraphs (set packing) with general edge weight. Based on a real-world ridesharing dataset in Chicago City and price schemes of Uber, we conduct an extensive empirical study on our model and algorithms. Experimental results show that practical price schemes can be incorporated into our model, our exact algorithms are efficient, and our approximation algorithms achieve ∼90% of optimal solutions in the number of passengers served.

Keywords: Ridesharing with profit · exact and approximation algorithms · graph matching · network flow · computational study

1 Introduction

Personal vehicles and mobility-on-demand (MoD) systems are major transportation tools worldwide. MoD systems, such as Uber, Lyft and DiDi, have become popular due to their convenience. MoD system operators and drivers participated in such systems are mostly motivated by profit. Solely focusing on profit and market share from MoD systems and drivers may have increased congestion and CO_2 emissions; the use of MoD has increased the number of single-passenger vehicles on the road significantly [10,19,30]. This, coupled with the saturated personal vehicle usage (with low occupancy rate) in Europe and North America, causes more traffic congestion and emissions [22]. According to studies in [7,11,27], personal vehicles were the main transportation mode in the United

States and Canada in recent years and in more than 200 European cities between 2001 and 2011. In Europe 2017 [11], the transport sector accounted for 27% of total greenhouse gas emissions; and of these 27% gas emissions, 31.55% (8.52% total) were from passenger cars.

On the other hand, there is an urgency to reduce traffic congestion and greenhouse gas emissions. Ridesharing using MoD systems has been proposed and studied in the academia [22,23,28,31]. Major themes from many previous studies include maximizing the total number of passengers served, minimizing the total cost to serve all passengers and maximizing a profit target. Studies, such as those in [2,3,6,29], have shown that ridesharing is a promising effective way to increase the occupancy rate and reduce congestion. An important factor for the adoption of ridesharing in practice is the profit/pricing scheme. Demand-and-pricing of a ridesharing system is important for its adaptability of actual ridesharing in practice [31]. Recently, ridesharing with profit as taxi ridesharing (e.g., [24,25]) and pricing based platform (MoD) equilibrium analysis (e.g., [5, 33]) have received much attention.

This study is motivated by the fact that profits-as-incentives may promote ridesharing in practice for both MoD operators and drivers. The potential of ridesharing has been recognized in the academia, but the potential of ridesharing in profit-maximizing platforms/MoDs is not well understood. The ridesharing problem we study can be summarized in the following (formal definition is given in Sect. 2):

– A centralized system periodically receives a set of ridesharing offer trips (drivers) and a set of ridesharing request trips (passengers). We say a passenger is *served* if the passenger is assigned a driver who can deliver the passenger to his/her destination on time. A driver can serve multiple passengers together, which is a ridesharing *match* consisting of a driver and a group of passengers served by the driver. The system computes a *profit* for each match. An optimizing goal on the profit only, called *Ridesharing with Profit* (**RP**) problem, is to maximize the overall profit obtained from the served matches. In this paper, we study a more complex optimization problem called *Ridesharing with Profit Constraint* (**RPC**) problem which maximizes the number of served passengers subject to satisfying a specified profit goal.

The RPC problem provides a new framework to consider maximizing both the number of passengers served and drivers' profit target. To the best of our knowledge, such an optimization problem has not been studied before. Our model allows a flexible pricing for the MoD system operators and for different pricing schemes (e.g., [21,32]). Although the problem studied by Santos and Xavier [25] is closely related to the RPC problem, their optimization goals differ from ours, and they focus on heuristics. Similarly, only (meta)heuristics are discussed in [20].

Online and offline approaches have been used to handle ridesharing requests. In the online approach, a request trip is processed immediately without the information of later trips. In the offline approach, the system accumulates a set of offer and request trips for each time interval (known as *batching*); and the set of trips is processed at once for that interval. This is a common approach in the literature (e.g., [2, 12, 15, 26]). Under the offline setting, two optimization problems related to ridesharing problems are the Dial-A-Ride problem (DARP) and the Vehicle Routing problem (VRP). There are some major differences between these two problems and the RP/RPC problems. The drivers and passengers in DARP and VRP have less parameters and/or less restricted parameters than that of the drivers and passengers in RPC. A variant of the Vehicle Routing Problems with Profits, called the Team Orienteering Problem (TOP) [17], is related to RPC problem. However, profit calculations in VRPPs and TOP are static compared to the profit calculation in RPC, which is more dynamic since it depends on each different driver-passenger(s) assignment. The most salient difference is that the capacity of a vehicle in DARP and VRP is substantially higher than that in ridesharing. This causes finding an optimal routing for a vehicle in DARP and VRP harder, making many general approaches of DARP and VRP not suitable for the RPC problem since they focus on different fundamentals.

In this paper, we follow the offline setting and our model uses a graph matching approach. All feasible matches between all drivers and passengers are computed first; and then based on some optimization goal/objective, an assignment consisting of a set of disjoint feasible matches is selected. By this approach, the RP problem can be converted to the maximum weight hypergraph matching (or maximum weight set packing problem), which is NP-hard in general [14]. We give a mathematical formulation, an exact and approximate algorithms for the RPC problem. Our algorithms are based on applications of maximum matching in bipartite graphs and hypergraphs. We also conduct empirical studies on our algorithms. One hurdle for empirical studies for the RPC problem is the lack of practical data instances. To clear this hurdle, we incorporate the real-world ridesharing dataset from Chicago City with the driver's profit model of Uber to generate test instances for practical scenarios. Our contributions in this paper are summarized as follows:

1. A new optimization problem (the RPC problem) is studied, and a mathematical formulation of the RPC problem is given. The NP-hardness of the RP problem implies that the RPC problem is NP-hard.
2. We give a polynomial-time exact algorithm framework (including two practical implementations of the algorithm) and a $\frac{1}{2}$-approximation algorithm for a special case of the RPC problem that each match contains $\lambda = 1$ passenger (labeled as RPC1).
3. Another special case of the RPC problem is that only matches with non-negative profit are considered and each match has at most $\lambda \geq 2$ passengers (labeled as RPC+). This case is still NP-hard, and we give a $\frac{2}{3\lambda}$-approximation algorithm for a specific range of profit target in this case.
4. Based on a real-world ridesharing dataset in Chicago City, profit model of Uber and practical scenarios, we create datasets for an extensive computational study on RPC1 and RPC+ problems. Experiment results show that practical profit schemes can be incorporated into our model. The exact algorithm implementations are efficient, and the $\frac{1}{2}$-approximation algo-

rithm achieves 96% to 99% (for different practical scenarios) and the $\frac{2}{3\lambda}$-approximation algorithm achieves 90% of optimal solutions in the number of passengers served.

The rest of the paper is organized as follows. In Sect. 2, we give the preliminaries and formally define the RPC problem. In Sect. 3, we describe the exact algorithms for RPC1. The $\frac{2}{3\lambda}$-approximation algorithm for RPC+ is presented in Sect. 4. We discuss the empirical study and algorithms in Sect. 5.

2   Preliminaries

Let $G(V, E, w)$ be an *edge-weighted* graph with $w : E \to \mathbb{R}$ that assigns each edge $e \in E$ a weight $w(e)$. A path P in G is a sequence of vertices $v_1, v_2, \ldots, v_p$ such that (v_i, v_{i+1}) is an edge of E for $1 \le i \le p - 1$ and denoted by $P = (v_1, \ldots, v_p)$. The *distance* of a path $P = (v_1, \ldots, v_p)$ is defined as $\text{dist}(P) = w(P) = \sum_{i=1}^{p-1} w(v_i, v_{i+1})$. The *length* of a path P is the number of edges in P, denoted by $|P|$.

An MoD system has a road network, modeled as a directed graph $G(V, E, w)$, where V is the set of vertices representing geographical sites, $E \subseteq V \times V$ is the set of edges (each edge represents a connection between two sites), and a distance function $w : E \to \mathbb{R}$ that assigns each edge a weight. The system periodically receives two sets of trips: a set $D = \{\eta_1, \ldots, \eta_k\}$ of k drivers (each operates a vehicle) and a set $R = \{r_1, \ldots, r_l\}$ of l passengers. Each driver $\eta_i \in D$ is represented by a tuple (o_i, d_i, λ_i) of parameters containing an origin location o_i (a vertex $o_i \in V(G)$), a destination location $d_i \in V(G)$, and a passenger capacity $\lambda_i \ge 1$ of η_i's vehicle. Each driver η_i also has an earliest departure time at o_i, a latest arrival time at d_i, a detour time/distance limit and a maximum trip duration. Each passenger $r_i \in R$ is represented by a tuple (o_i, d_i) as defined above, along with an earliest departure time at o_i, a latest arrival time at d_i and a maximum trip duration..

For a driver $\eta_i \in D$ and a group of passengers $R_i \subseteq R$, (η_i, R_i) is a *feasible match* if there exists a feasible path $\text{FP}(\eta_i, R_i)$ in N used by η_i to deliver all of R_i such that travelling along $\text{FP}(\eta_i, R_i)$ satisfies all constraints specified by η_i and every $r_j \in R_i$. These constraints include $|R_i| \le \lambda_i$, detour limit of η_i, and time constraints of η_i and R_i. An *assignment* Π is a set of feasible matches such that for every two feasible matches (η_i, R_i) and (η_j, R_j) in Π, $\eta_i \ne \eta_j$ and $R_i \cap R_j = \emptyset$. For an assignment $\Pi = \{(\eta_i, R_i) \mid \eta_i \in D, R_i \subseteq R\}$, each driver η_i in the assignment follows a *shortest feasible path* $\text{SFP}(\eta_i, R_i)$ to serve trips in R_i. For example, let $R_i = \{r_a, r_q\}$ be the set of passengers in feasible match (η_i, R_i). There are six different visiting orders of R_i in which the passengers of R_i can be picked-up and dropped-off by η_i, which correspond to six paths in road network $G(V, E, w)$. Below are the six visiting orders driver η_i can use:

$$\{(o_i, o_a, o_q, d_a, d_q, d_i), (o_i, o_a, d_a, o_q, d_q, d_i), (o_i, o_a, o_q, d_q, d_a, d_i),$$
$$(o_i, o_q, o_a, d_a, d_q, d_i), (o_i, o_q, d_q, o_a, d_a, d_i), (o_i, o_q, o_a, d_q, d_a, d_i)\}.$$

Path SFP(η_i, R_i) is the path in G corresponds to one of the six visiting orders that is feasible and has the shortest distance. Every feasible match (η_i, R_i) along with SFP(η_i, R_i) can be computed efficiently with small $|R_i|$, as described in [2, 15, 26].

Each feasible match (η_i, R_i) is associated with a revenue $rev(\eta_i, R_i)$, a travel cost $tc(\eta_i, R_i)$ and a profit $w(\eta_i, R_i)$, which are computed by the MoD system. The revenue and cost are decided by several parameters such as SFP(η_i, R_i), travel time, regions, pricing policies, etc. The profit of a feasible match (η_i, R_i) is $w(\eta_i, R_i) = rev(\eta_i, R_i) - tc(\eta_i, R_i)$, which can be negative, and we assume it is expressed in integers (e.g., cents, smallest payable amount). We estimate $rev(\eta_i, R_i)$ and $tc(\eta_i, R_i)$ based on the profit model of Uber and practical scenarios, as described in Sect. 5 (and in [16]).

The RP (ridesharing with profit) problem is to assign passengers of R to drivers D with total profit maximized. In this paper, we focus on a more complex optimization problem, called the *Ridesharing with Profit Constraint* (RPC) problem. In application, MoD may want to serve as many passengers as possible while maintaining a *profit target*. With this in mind, we introduce a profit constraint and a formulation for the RPC problem as follows.

$$\max_{\Pi} \quad \sum_{(\eta_i, R_i) \in \Pi} |R_i| \tag{i}$$

$$\text{subject to} \quad \eta_i \neq \eta_j \wedge R_i \cap R_j = \emptyset, \quad \forall (\eta_i, R_i) \neq (\eta_j, R_j) \in \Pi \tag{ii}$$

$$\sum_{(\eta_i, R_i) \in \Pi} w(\eta_i, R_i) \geq c \tag{iii}$$

The objective function (i) is to maximize the total number of passengers served. Constraint (ii) ensures that each passenger request is assigned to only one driver and each driver serves at most one feasible match (a unique group of passengers). Constraint (iii) ensures the system profit meets a given target. An assignment Π containing any feasible match (η_i, R_i) with negative profit ($w(\eta_i, R_i) < 0$) means that the driver η_i loses money.

We construct an integer-weighted hypergraph $H(V, E, w)$ to represent the formulation (i)-(iii) as follows. Initially, $V(H) = D \cup R$. For each $\eta_i \in D$ and for every subset R_i of R with $1 \leq |R_i| \leq \lambda_i$, create a hyperedge $e = \{\eta_i\} \cup R_i$ in $E(H)$ if (η_i, R_i) is a feasible match. Each edge $e = \{\eta_i\} \cup R_i \in E(H)$ has weight $w(e) = w(\eta_i, R_i)$, the profit of η_i. Remove all isolated vertices from H. Let H^- be the subgraph of H such that H^- contains all edges of H with negative weight and $H^+ = H \setminus H^-$. For an assignment Π, let $w(\Pi) = \sum_{(\eta_i, R_i) \in \Pi} w(\eta_i, R_i)$ be the profit of Π. There are at most $\sum_{1 \leq a \leq \lambda_i} \binom{l}{a}$ edges incident to each η_i in H. Let $\lambda = \max_{\eta_i \in D} \lambda_i$. If λ is a small constant, the size of H is polynomially bounded. In practice, it is reasonable to assume λ is small; however when λ is not small, we may purposely restrict the number of edges incident to each vertex so that $|E(H)|$ becomes reasonable for practice.

For an edge-weighted (hyper)graph $G(V, E)$ and $E' \subseteq E(G)$, the weight of E' is denoted by $w(E') = \sum_{e \in E'} w(e)$, where $w(e)$ is the weight of edge e. A *matching* M in a (hyper)graph G is a set of edges of G such that every pair of edges in M do not have a common vertex. The size $|M|$ of a matching M is the number of edges in M and the weight of M is $w(M)$. The RPC problem is then to find a matching M in H such that $\sum_{\{\eta_i\} \cup R_i \in M} |R_i|$ is maximized and $w(M) > c$. Let c^* be the weight of a maximum weight matching in the above constructed hypergraph H and $c \leq c^*$ be a profit target. Finding a matching M in H^+ with $w(M) = c^*$ is equivalent to finding a maximum weight set packing in H^+, which is NP-hard in general for $\lambda \geq 2$ [14]. This implies Theorem 1 (a proof of this theorem is in [16]).

Theorem 1. *The RPC problem is NP-hard for an arbitrary c and $\lambda \geq 2$.*

Hazan et al. [18] showed that the $(\lambda + 1)$-set packing problem cannot be approximated to within $\Omega(\frac{\ln(\lambda+1)}{\lambda+1})$ in general for $\lambda \geq 2$. There exists a polynomial-time $\frac{2}{\lambda+2}$-approximation algorithm for approximating the maximum profit of Π [4]. However, similar algorithms [4,8] cannot be directly applied to the RPC problem. This is because such algorithms only approximate the maximum profit $w(\Pi)$, and they do not consider the cardinality and the different kinds of elements of each subset/match in Π. Algorithms for the maximum set packing problem (e.g., [13]) cannot apply to the RPC problem either since such algorithms do not consider general integer weight. Due to the NP-hardness of the RPC problem and the inapproximability of the weighted set packing problem, we study two variants of the RPC problem: RPC1 and RPC+. The RPC1 problem variant assumes that for a given instance of the RPC problem, $\lambda_i = 1$ for every driver $\eta_i \in D$ ($\lambda = 1$). To solve the RPC1 variant, we use an approach in solving the maximum matching problem on bipartite graphs. For the RPC+ problem variant, we include one more constraint (called the *non-negative profit constraint*) to formulation (i)-(iii) of the RPC problem:

$$w(\eta_i, R_i) \geq 0, \forall (\eta_i, R_i) \in \Pi. \tag{iv}$$

To solve the RPC+ variant, we use a local search approach similar to the ones in [4,8].

3 RPC1 Variant - Capacity of One

For $\lambda = 1$, the weighted hypergraph $H(V, E, w)$, constructed in Sect. 2, becomes a weighted bipartite graph. A solution to the RPC problem for $\lambda = 1$ (with $c \leq c^*$) is a matching M in H with $w(M) \geq c$ and $|M|$ maximized. We first give a polynomial-time exact algorithm (referred to as **ExactNF**) framework that uses network flow to find an optimal solution and two practical implementations of ExactNF.

ExactNF: Exact Algorithm Framework

1. Construct a flow network $N(V, E)$ from H, where $V(N) = \{s, t\} \cup V(H)$, s is the source, and t is the sink. For each $\eta_i \in V(H)$, create an edge (s, η_i) in

$E(N)$ with cost 0 and capacity 1. For each $(\eta_i, r_j) \in E(H)$, create an edge (η_i, r_j) in $E(N)$ with cost $-w(\eta_i, r_j)$ and capacity 1. For each $r_j \in V(H)$, create an edge (r_j, t) in $E(N)$ with cost 0 and capacity 1. Note that the maximum amount of flow that can be sent from s to t in N is at most $n_{min} = \min\{|V(H) \cap D|, |V(H) \cap R|\}$.

2. For $1 \leq y \leq n_{min}$, find a minimum cost flow f_y of value y (sent from s to t) or conclude that there is no flow of value y in N.

3. For an edge $e \in E(N)$, let $f_y(e)$ be the flow value passing through e in f_y. Let $c(f_y) = \sum_{e \in E(N)|f_y(e)>0} w(e)$ be the cost of flow f_y. If a flow f_y with $c(f_y) \leq -c$ is computed in Step 2, then $y = \arg\max_y -c(f_y) \geq c$, and output the edges $\cup_{e \in E(N)|f_y(e)>0 \wedge e \in E(H)}$ with positive flow value in f_y as solution M; otherwise, conclude there is no matching in H with profit at least c.

Theorem 2. *Algorithm ExactNF finds a matching M with $w(M) \geq c$ and $|M|$ maximized or concludes that there is no matching M with $w(M) \geq c$ in H in polynomial time.*

A proof of Theorem 2 can be found in [16]. The computational time heavily depends on how f_y is computed. We give two implementations (**ExactNF1** and **ExactNF2**) to compute f_y. The first one uses a linear programming (LP) approach to find f_y by a min-cost flow LP formulation, and the second one to find f_y by graph algorithms (details for both algorithms can be found in Sect. 3 of [16]).

ExactNF1: LP Approach for Computing f_y. First, find the maximum flow value y^* of network N. Then for $y = y^*$ to 1, find the min-cost flow f_y of value y in N by a min-cost flow LP formulation. Output a solution M, as described in Step 3 of ExactNF.

Corollary 1. *Algorithm ExactNF1 finds a matching M with $w(M) \geq c$ and $|M|$ maximized or concludes that there is no matching M with $w(M) \geq c$ in H in $O(n_{min} \cdot t(N))$ time, where $t(N)$ is the time to find a min-cost flow f_y by an LP solver.*

ExactNF2: Graph Algorithm Approach for Computing f_y. First, re-weight the edges of N to be non-negative by Johnson's shortest path algorithm [9]. Then, find a min-cost flow f_y of value y in N using the successive shortest path algorithm [1]. Output a solution M, as described in Step 3 of ExactNF.

Theorem 3. *Algorithm ExactNF2 finds a matching M with $w(M) \geq c$ and $|M|$ maximized or concludes that there is no matching M with $w(M) \geq c$ in H in time $O(|V(N)||E(N)| + n_{min} \cdot t(N))$, where $t(N)$ is the time for computing an $s - t$ path in a residual network of N.*

Next, we give a simple greedy $\frac{1}{2}$-approximation algorithm which may have some advantages in practice (referred to as **Greedy**):

1. Compute a maximum weight matching M' in H.

2. Let $M = M'$. For each iteration, select an edge e'' in $H^- = H \setminus H^+$ such that

$$e'' = \text{argmax}_{e \in E(H^-) \setminus M} \mid e \cap e' = \emptyset \; \forall e' \in M \; w(e).$$

If $w(M) + w(e'') \geq c$, then add e'' to M. Repeat this until such an edge e'' does not exist (every edge of H intersects with an edge of M) or $w(M) + w(e'') < c$.

Theorem 4. *Let M be the matching found by the Greedy algorithm and M^* be a matching in H with $w(M^*) \geq c$ and $|M^*|$ maximized. Then, $\frac{|M|}{|M^*|} \geq \frac{1}{2}$, implying Greedy is $\frac{1}{2}$-approximate to RPC1.*

A proof of Theorem 4 can be found in [16]. Algorithm Greedy has a running time of $O(t(H) + m \log m)$, where $t(H)$ is the time to find a maximum weight matching in H and $m = |E(H)|$. Algorithm Greedy runs faster and achieves a higher profit than the exact algorithms in some cases, which provides an alternative choice.

4   RPC+ Variant

Due to the non-negative profit constraint (iv), only edges in H^+ can be selected to solve the RPC+ problem (formulation (i)-(iv)) Inherently, the profit target must be non-negative for the RPC+ problem. In this case, a matching M with $w(M) \geq c$ and $|M|$ maximized may not be an optimal solution to RPC+. For instance, a matching $M_1 = \{e_1, e_2, e_3\}$ with three edges may contain only three passenger vertices of $V(H) \cap R$, whereas a matching $M_2 = \{e_4\}$ can contain four passenger vertices (assuming $\lambda \geq 4$). We need to find a matching M in H such that the number of passenger vertices $V(H) \cap R$ contained in M is maximized and $w(M) \geq c$.

We propose a local search algorithm for $\lambda \geq 2$, called **LS2**. For an edge $e = \{\eta_i\} \cup R_i \in E(H)$, let $R(e) = R_i$ (the passengers of e). For a subset $E' \subseteq E(H)$, let $R(E') = \cup_{e \in E'} R(e)$. For an edge $e \in E(H)$, let $N(e)$ be the set of edges incident to e, and $N^+(e) = N(e) \cap E(H^+)$. By constraint (iv), we only need to consider the subgraph H^+. Although the profit target c is an input parameter of an RPC+ instance, one needs to determine that c should be at most c^* (the weight of a maximum weight matching in H) so that it admits a feasible solution. However, finding c^* is NP-hard as mentioned in the preliminaries. Note that as c gets closer to c^*, the chance of having a lower objective value is higher. We suggest a way to set c to be a reasonable target value. We use a heuristic to compute a weight $\tilde{c}$ to approximate c^* and set $c \leq \tilde{c}$ as a profit target. In fact, our experiment shows that the total profit of solutions with respect to c is not too far way from the total profit of solutions with respect to c^* in practice.

There are two steps in Algorithm LS2. In the first step, LS2 uses the *simple greedy* in [4,8] to find an initial weighted set packing (hypergraph matching) to get $\tilde{c}$. In the second step of LS2, a local search is used to improve the solution computed in the first step. The first step produces a solution with a $\frac{1}{2\lambda}$-approximation ratio, and the second step gives a solution with a

$\frac{2}{3\lambda}$-approximation ratio when a specific condition on the profit target is met. Algorithm LS2 is given in the following, starting with $M' = \emptyset$.

1. In each iteration, select an edge $e'' \in E(H^+)$ that does not intersect with any edge of M' and has maximum weight. That is, find an edge e'' in $E(H^+)$ such that

$$e'' = \mathrm{argmax}_{e \in E(H^+) \setminus M' \mid e \cap e' = \emptyset \ \forall e' \in M'} w(e),$$

   and add e'' to M'. Repeat this until every edge of $E(H^+) \setminus M'$ intersects with M'. Determine c by setting $c \leq \tilde{c} = w(M')$.

2. Let $M = M'$ be the matching obtained after Step 1. Let $A = \{e \in M \mid |R(e)| = 1\}$ and assume $A = \{a_1, \ldots, a_q\}$ with $w(a_i) \leq w(a_j)$ for $1 \leq i < j \leq q$. An *improvement* δ_e of an edge $e \in M$ is a subset of edges in $N^+(e)$ such that

   – $|\delta_e| \leq 2$, all edges of $(M \cup \delta_e) \setminus \{e\}$ are pairwise vertex-disjoint, $|R(\delta_e)| > |R(e)|$ and $w(M) + w(\delta_e) - w(e) \geq c$.

   An improvement δ_e is *maximum* if $|R(\delta_e) \setminus R(M)|$ is maximum among all improvements of e.

   (a) If $\lambda = 2$, execute the following for-loop for each $a_i \in A$.
       – For $i = 1$ to q do, if there is an improvement δ_{a_i} of a_i such that $|R(\delta_{a_i})| = 4$, then perform an *augmentation* as $M = (M \cup \delta_{a_i}) \setminus \{a_i\}$.
   (b) Else if $\lambda \geq 3$, execute the following for-loop for each $a_i \in A$.
       – For $i = 1$ to q do, if there is an improvement of a_i, then find a maximum improvement δ_{a_i} and perform $M = (M \cup \delta_{a_i}) \setminus \{a_i\}$.

   Output M.

The analysis of Algorithm LS2 (including Theorem 5) can be found in [16].

Theorem 5. *Let M' be the matching found by Step 1 of the LS2 algorithm and M be the final matching found by the LS2 algorithm. Let $A = \{e \in M \mid |R(e)| = 1\}$. Let $0 \leq c \leq w(M')$ and M^* be a matching in H^+ such that $|R(M^*)|$ is maximized and $w(M^*) \geq c$. $\frac{|R(M')|}{|R(M^*)|} \geq \frac{1}{2\lambda}$ for $\lambda \geq 1$, and if $c \leq w(M' \setminus A) + \frac{2w(A)}{\lambda+1}$ for $\lambda \geq 2$, then $\frac{|R(M)|}{|R(M^*)|} \geq \frac{2}{3\lambda}$.*

5   Numerical Experiments

We conduct an extensive empirical study to evaluate our model and algorithms for RPC1 and RPC+. To the best of our knowledge, there is no practical test dataset publicly available for the RPC problem. To clear this hurdle, we create a simulation dataset by incorporating a real-world ridesharing dataset from Chicago City with the driver profit model of Uber. A comprehensive and detailed description of the simulation setup, profit estimation and trip generation can be found in Sect. 5 of [16].

5.1   Create Test Instances

The simulated centralized system receives a batch of driver offer trips D and passenger request trips R in a fixed time interval (total of 72 intervals in a day). The drivers and passengers along with their parameters are generated based on a (publicly available) ridesharing dataset in Chicago City. Each data record in the dataset contains the time and location of a completed trip. For each time interval, we first generate a set R of passengers and then a set D of drivers, using the dataset to determine time constraints and the origins and destinations of drivers and passengers.

To estimate revenue $rev(\eta_i, R_i)$, travel cost $tc(\eta_i, R_i)$ and profit $w(\eta_i, R_i)$ of a feasible match (η_i, R_i), we utilize the ridesharing dataset and Uber's cost estimator/price scheme (since the dataset contains both the total amount paid and tips paid by the passengers and its data are reported by rideshare companies in the US). The travel duration and distance of $SFP(\eta_i, R_i)$, vehicle sharing (ridesharing), surge pricing, vehicle type and Uber's commission are all considered in our estimation.

Feasible matches are computed from D and R in each time interval. Shortest paths in our simulation are computed in *real-time*. To speedup the computation for practical reasons, we apply a conditional check to see if a driver $\eta_i \in D$ and a passenger $r_j \in R$ should be considered in a base match by estimating the travel distance without computing any shortest path. A *base match* consists of exactly one driver and one passenger. We also limit the number of matches a driver η_i can have. Any driver η_i can have at most 100 base matches and 500 feasible matches in total; and each passenger can belong to at most 20 base matches.

5.2   Computational Results

A more detailed discussion of the computational results can be found in [16]. All algorithms were implemented in Java, and the experiments were conducted on an Intel Core i7-6700 processor with 2133 MHz of 12 GBs RAM available to JVM. All ILP formulations in our algorithms are solved by CPLEX v12.10.1. We label the algorithm CPLEX uses to solve ILP formulations (i)-(iii) and (i)-(iv) for RPC1 and RPC+ by **Exact**. A passenger $r_j \in R$ is called *served* if $r_j \in R_i$ s.t. match (η_i, R_i) belongs to a solution computed by one of the algorithms.

RPC1 Results. Table 1 shows the overall results for the base case instances with profit targets $c_1 = w(M')$, $c_2 = 0.8 \cdot w(M')$ and $c_3 = 0.6 \cdot w(M')$, where M' is a max-weight matching in H for each interval. As can be seen from Table 1, ExactNF2 has the best running time, and ExactNF1/2 produce optimal solutions with the highest profits. The Greedy solutions serve about 99.76% of passengers served by the optimal solutions. We also tested our model and algorithms on a more practical pricing scheme by considering drivers' day-to-day operation costs, such as gas price, maintenance and depreciation. Such tests paint a more realistic picture, and Greedy has better performances in some cases (please check Sect. 5.4.1 of [16] for more details).

Table 1. Performances of algorithms for RPC1 on base case instances, where $c'_a, 1 \leq a \leq 3$, is the sum of target c_a of all 72 intervals (in dollars).

		Greedy	ExactNF1	ExactNF2	Exact
Total # of passengers served in all intervals	(c'_1 = \$1587436)	109770	109771	109771	109771
	(c'_2 = \$1269949)	109775	110035	110035	110035
	(c'_3 = \$952462)	109775	110035	110035	110035
Total profit of served matches in all intervals	(c'_1 = \$1587436)	1587436	1587436	1587436	1587436
	(c'_2 = \$1269949)	1587432	1586707	1586707	1465676
	(c'_3 = \$952462)	1587432	1586707	1586707	1457338
Avg running time (second) per interval	(c'_1 = \$1587436)	5.573	6.249	4.765	7.030
	(c'_2 = \$1269949)	5.670	6.223	4.484	6.591
	(c'_3 = \$952462)	5.765	6.379	4.565	6.298
Total number of drivers and passengers generated, respectively				124340 and 126625	

RPC+ Results. Recall that the $\frac{2}{3\lambda}$-approximation algorithm that solves RPC+ is labeled as **LS2** and the first step of LS2 is labeled as **SimpleGreedy**. For this variant, the profit target c is upper bounded by the weight $w(M')$ of the matching M' found by SimpleGreedy. Recall that $A = \{e \in M' \mid |R(e)| = 1\}$, as defined in the description of LS2. We set a lower bound $LB = \min\{w(M' \setminus A) + 2w(A)/(\lambda + 1), 0.6w(M')\}$. We tested three profit targets $c_1 = w(M')$, $c_2 = 0.5 \cdot (w(M') - LB) + LB$ and $c_3 = LB$. The overall results are shown in Table 2. The performances of SimpleGreedy and LS2 are about 89.25% and 90.04% of the exact algorithm (Exact), in the total number of passengers served. The running time of LS2 is only 0.031 s longer than that of SimpleGreedy on average. The running time of Exact is 470–630 times longer than that of LS2, depending on the profit target. For very large instances though, Exact may not be suitable for real-time computation.

Table 2. Performances of algorithms for RPC+ on base case instances. For $1 \leq a \leq 3$, $c'_a = \sum_{h=1}^{18} \sum_{h_t=1}^{4} c_a$ (in dollar).

		SimpleGreedy	LS2	Exact
Total # of passengers served in all intervals	(c'_1 = \$845817)	63554	64099	71197
	(c'_2 = \$676653)	63554	64118	71208
	(c'_3 = \$507490)	63554	64118	71208
Total profit of served matches in all intervals	(c'_1 = \$845817)	845817	848677	846893
	(c'_2 = \$676653)	845817	848271	702130
	(c'_3 = \$507490)	845817	848271	681472
Avg running time (second) per interval	(c'_1 = \$845817)	0.0445	0.0761	47.880
	(c'_2 = \$676653)	0.0386	0.0708	33.279
	(c'_3 = \$507490)	0.0397	0.0695	35.664
Total number of drivers and passengers generated, respectively				40573 and 126625

Discussion. We also computed optimal solutions (maximum weight matchings in H) to the RP problem using some RPC1 and RPC+ test instances. The full results can be found in [16]. From Table 3, optimal solutions to RPC1 and RPC+ serve about the same and 8.02% more passengers than the respective RP optimal solutions. The profits of the optimal solutions for RPC1 and RPC+ are reasonably close to that of the RP optimal solutions.

Table 3. Total number # of passengers served and total profit $ of served matches in all intervals. (*) Optimal solutions to RP. ($\diamond$) Optimal solutions to RPC1 (for ExactNF and c_2) and RPC+ (for Exact and c_1).

	RPC1 base	RPC+ base		RPC1 base	RPC+ base
*	#109770	#65913	$\diamond$	#109771	#71197
	$1587436	$893879		$1586707	$846893

Based on the RPC1 and RPC+ results, our algorithms are effective for achieving the optimization goal of RPC in practical scenarios. Although the RPC problem is NP-hard, the exact algorithms are efficient to find optimal solutions (can be practical, depending on applications). Algorithms Greedy and LS2 can achieve 96.1% and 90.04% of the optimal solutions to RPC1 and RPC+, respectively, in the number of passengers served. Furthermore, the average occupancy rate is 1.89 and close to 2 during peak hours for the exact solutions to RPC1. For RPC+, the average occupancy rates for Exact and LS2 are 2.76 and 2.59, respectively (the detailed results can be found in [16]). Both these numbers are better than the reported occupancy rate in the US, which was 1.5 in 2017 [7].

Our model of the RPC problem provides a new framework to incorporate a flexible pricing scheme to maximize the number of passengers served while meeting a profit target. The RPC problem is a more complex variant of the maximum set packing problem, and hence, it is NP-hard. We give a polynomial-time exact algorithm (with two different practical implements) and approximation algorithms for special cases (RPC1 and RPC+) of the problem. Experimental results show that practical profit (price) schemes can be incorporated into our model and suggest that there is potential in profit-maximizing/profit-incentive MoD platforms by utilizing ridesharing.

The RPC+ variant considers only matches with non-negative profit, which may cover the MoD systems' profit-incentive, but it may impose a limit on improving the number of passengers served. It is worth developing algorithms for more general cases where matches with negative profit are also considered although the local search idea of LS2 may not be easily apply to this general case. This is because an improvement in LS2 now has to carefully consider both the weight $w(M)$ and $|R(M)|$ of the current matching M at the same time.

Acknowledgement. The authors thank the reviewers for their constructive comments.

References

1. Ahuja R., Magnanti T., Orlin J.: Network Flows: Theory, Algorithms, and Applications. Prentice Hall, Hoboken (1993)
2. Alonso-Mora, J., Samitha, S., Wallar, A., Frazzoli, E., Rus, D.: On-demand high-capacity ride-sharing via dynamic trip-vehicle assignment. Proc. Nat. Acad. Sci. **114**(3), 462–467 (2017)
3. Amirkiaee, Y., Evangelopoulos, N.: Why do people rideshare? An experimental study. Transport. Res. F: Traffic Psychol. Behav. **55**, 9–24 (2018)
4. Berman, P.: A d/2 approximation for maximum weight independent set in d-claw free graphs. In: SWAT 2000. LNCS, vol. 1851, pp. 214–219. Springer, Heidelberg (2000). https://doi.org/10.1007/3-540-44985-X_19
5. Besbes, O., Castro, F., Lobel, I.: Surge pricing and its spatial supply response. Manage. Sci. **67**(3), 1350–1367 (2021)
6. Caulfield, B.: Estimating the environmental benefits of ride-sharing: a case study of Dublin. Transp. Res. Part D: Transp. Environ. **14**(7), 527–531 (2009)
7. Center for Sustainable Systems, University of Michigan: Personal Transportation Factsheet (2022)
8. Chandra, B., Halldórsson, M.: Greedy local improvement and weighted set packing approximation. J. Algorithms **39**(2), 223–240 (2001)
9. Cormen T.H., Leiserson C.E., Rivest R.L., Stein C.: Introduction to Algorithms, Third Edition. The MIT Press, Cambridge (2009)
10. Diao, M., Kong, H., Jinhua, Z.: Impacts of transportation network companies on urban mobility. Nat. Sustain. **4**, 494–500 (2021)
11. European Environment Agency: Greenhouse gas emissions from transport in Europe (2019)
12. Fielbaum, A., Bai, X., Alonso-Mora, J.: On-demand ridesharing with optimized pick-up and drop-off walking locations. Transp. Res. Part C: Emerg. Technol. **126**, 103061 (2021)
13. Fürer, M., Yu, H.: Approximating the k-set packing problem by local improvements. In: Fouilhoux, P., Gouveia, L.E.N., Mahjoub, A.R., Paschos, V.T. (eds.) ISCO 2014. LNCS, vol. 8596, pp. 408–420. Springer, Cham (2014). https://doi.org/10.1007/978-3-319-09174-7_35
14. Garey, M., Johnson, D.: Computers and Intractability: A Guide to the Theory of NP-Completeness. W.H, Freeman and Company (1979)
15. Gu Q., Liang J.: Multimodal transportation with ridesharing of personal vehicles. In: 32nd International Symposium on Algorithms and Computation, vol. 212, pp. 1–16 (2021)
16. Gu Q., Liang J.: Algorithms for the ridesharing with profit constraint problem. https://doi.org/10.48550/arXiv.2310.04933
17. Gunawan, A., Lau, H.C., Vansteenwegen, P.: Orienteering Problem: a survey of recent variants, solution approaches and applications. Eur. J. Oper. Res. **255**(2), 315–332 (2016)
18. Hazan, E., Safra, S., Schwartz, O.: On the complexity of approximating k-set packing. Comput. Complex. **15**(1), 20–39 (2006)
19. Henao, A., Marshall, W.: The impact of ride-hailing on vehicle miles traveled. Transportation **49**, 2173–2194 (2011)
20. Hsieh, F.-S.: A comparative study of several metaheuristic algorithms to optimize monetary incentive in ridesharing systems. ISPRS Int. J. Geo Inf. **9**(10), 590 (2020)

21. Li, M., Jiang, G., Lo, H.K.: Pricing strategy of ride-sourcing services under travel time variability. Transp. Res. Part E: Logist. Transp. Rev. **159**, 102631 (2022)
22. Martins, L., de la Torre, R., Corlu, C., Juan, A., Masmoudi, M.: Optimizing ridesharing operations in smart sustainable cities: challenges and the need for agile algorithms. Comput. Ind. Eng. **153**, 107080 (2021)
23. Mourad, A., Puchinger, J., Chengbin, C.C.: A survey of models and algorithms for optimizing shared mobility. Transp. Res. Part B: Methodol. **123**, 323–346 (2019)
24. Qian, X., Zhang, W., Ukkusuri, S., Yang, C.: Optimal assignment and incentive design in the taxi group ride problem. Transp. Res. Part B: Methodol. **103**, 208–226 (2017)
25. Santos, D., Xavier, E.: Taxi and ride sharing: a dynamic dial-a-ride problem with money as an incentive. Expert Syst. Appl. **42**(19), 6728–6737 (2015)
26. Simonetto, A., Monteil, J., Gambella, C.: Real-time city-scale ridesharing via linear assignment problems. Transp. Res. Part C: Emerg. Technol. **101**, 208–232 (2019)
27. Statistics Canada, Census: Main mode of commuting (2016)
28. Tafreshian, A., Masoud, N., Yin, Y.: Frontiers in service science: ride matching for peer-to-peer ride sharing: a review and future directions. Serv. Sci. **12**(2–3), 41–60 (2020)
29. Tikoudis, I., Martinez, L., Farrow, K., Bouyssou, C.G., Petrik, O., Oueslati, W.: Ridesharing services and urban transport CO2 emissions: simulation-based evidence from 247 cities. Transp. Res. Part D: Transp. Environ. **97**, 102923 (2021)
30. Tirachini, A., Andres, G.-L.A.: Does ride-hailing increase or decrease vehicle kilometers traveled (VKT)? A simulation approach for Santiago de Chile. Int. J. Sustain. Transp. **14**(3), 187–204 (2020)
31. Wang, H., Yang, H.: Ridesourcing systems: a framework and review. Transp. Res. Part B: Methodol. **129**, 122–155 (2019)
32. Yan, C., Zhu, H., Korolko, N., Woodard, D.: Dynamic pricing and matching in ride-hailing platforms. Nav. Res. Logist. **67**, 705–724 (2020)
33. Zhang, K., Nie, Y.: To pool or not to pool: Equilibrium, pricing and regulation. Transp. Res. Part B: Methodol. **151**, 59–90 (2021)

Multi-Candidate Carpooling Routing Problem and Its Approximation Algorithms

Jiale Zhang[(✉)], Xiuqi Huang, Zifeng Liu, Xiaofeng Gao, and Guihai Chen

Shanghai Jiao Tong University, Shanghai, China
{zhangjiale100,huangxiuqi,liuzifeng}@sjtu.edu.cn,
{gao-xf,gchen}@cs.sjtu.edu.cn

Abstract. Motivated by the carpooling services, we investigate a new and more challenging scenario for carpooling and model it as the Multi-candidate Carpooling Routing Problem (MCRP). The MCRP can be regarded as a new variant of TSP called Generalized Precedence-Constaint Asymmetric Subset Traveling Salesman Path Problem (GPAS-TSPP) and we construct complexity hierarchies for the related problems. We propose a 4-approximation algorithm for its special case Carpooling Routing Problem (CRP), followed by a $(5 + \epsilon)$-approximation algorithm for MCRP on the planar graph. We also design an exact algorithm based on dynamic programming to solve the general MCRP, serving as a benchmark. To the best of our knowledge, we are the first to explore the complexity hierarchy of carpooling problems in the TSP family and give constant-approximation algorithms for these new practical variants.

Keywords: Carpooling · TSP · Approximation · Exact Algorithm

1 Introduction

As a low-carbon way to travel, carpooling services have proliferated in recent years, which provides passengers and drivers with a more economical and efficient travel experience [17]. Route planning is an essential problem for carpooling, requiring determining a sequence with the minimum cost for a car driver to pick up and drop off multiple passengers. It forms complex combinatorial optimization problems different from other routing problems [23]. In this paper, we investigate a more challenging scenario in which the driver can choose from several possible candidate pick-up/drop-off locations for each passenger.

We explain the application scenario of the problem with a toy example shown in Fig. 1. A car, denoted as c, starts from position p_c and picks up passenger 1

This work was supported by the National Key R&D Program of China [2020YFB1707900]; the National Natural Science Foundation of China [62272302, 62172276] and Shanghai Municipal Science and Technology Major Project [2021SHZDZX0102], and DiDi GAIA Research Collaboration Plan [202204]. Xiaofeng Gao is the corresponding author.

W. Wu and J. Guo (Eds.): COCOA 2023, LNCS 14461, pp. 380–391, 2024.
https://doi.org/10.1007/978-3-031-49611-0_27

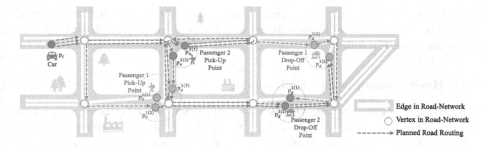

Fig. 1. An Example of Carpooling Routing Problem

and passenger 2 from their respective pick-up points and drops them off at their corresponding drop-off points. Each pick-up point and drop-off point has a set of candidate points for the driver to choose from. We can find the optimal pick-up and drop-off route, as shown by the red path. However, the selection of pick-up points and the sequence of pick-ups and drop-offs greatly affect the length of the route. When we exchange the sequence of candidate pick-ups and drop-offs from $(p_c, p_o^2, p_o^1, p_d^2, p_d^1)$ (red line) to $(p_c, p_o^1, p_o^2, p_d^1, p_d^2)$ (purple line), we find that the total pick-up and drop-off length increases. At the same time, if we select the wrong candidate point, for example, changing passenger 2's pick-up point from $p_o^{2(2)}$ to $p_o^{2(1)}$ (green line), the driver needs to turn around to pick up passenger 2, resulting in the wasted distance.

Therefore, both the selection of candidate pick-up and drop-off positions and their order have a significant impact on the length of the final route. Tiny modifications can reduce the total route weight greatly since the car could avoid the U-turn, detours, and so on [14]. We delve into this multi-candidate scenario, providing theoretical proof and algorithms. The main contributions are summarized as follows:

- We formulate the Multi-Candidate Carpooling Routing Problem (MCRP) problem and conduct it as a new variant of TSP by giving the complexity hierarchy (see Sect. 2).
- We give 4 approximation algorithms for the symmetric version of CRP by shortcut two subgraph solutions based on the 1.5-approximation $s - t$ path TSP algorithm (see Sect. 4).
- We design a $(5 + \epsilon)$-approximation algorithm for MCRP with symmetric and planar constraints, which combines a $(1 + \epsilon)$ approximation algorithm for the group Steiner tree (see Sect. 5).
- We put forward an $O(n^2 \cdot 2^{2l})$ exact algorithm for the general MCRP by improving the dynamic programming method for the TSP problem without increasing the time complexity (see Sect. 6).

2 Problem Formulation and Complexity Hierarchy

We model the road network as a directed graph $G = (V, E, W)$, where $v_i \in V$ represents the road intersection, and $e_{ij} = (v_i, v_j) \in E$ represents the road with a weight $w_{ij} = w(v_i, v_j) \in W$. The weight $w_{ij} \geq 0$ could be defined as road length, travel time, or other customized metrics, and we take the road length as an example for the following discussion. We first introduce some basic concepts.

2.1   Basic Concepts and Definitions

Definition 1 (Carpooling Order). *A carpooling order including a car and* **multiple** *passengers is defined as* $\{l, p_c, p'_c, \mathbb{P}\}$ *(referred as the order in the following). The* p_c *and* p'_c *represent the start position and destination of the car. The* l *represents the number of passengers in the order. The* $\mathbb{P} = \bigcup_{i=1}^{l} \{P_o^i, P_d^i\}$ *represents the set of positions where the car can pick up or drop off passengers. The* P_o^i *and* P_d^i *are the set of positions where the car can pick up and drop off the i-th passenger correspondingly. (All positions are denoted by the vertexes in the graph, i.e.* $p_c, p'_c \in V, \forall i, P_o^i, P_d^i \subset V.)$

Then we define the concept of multiple candidate points below:

Definition 2 (Multiple Candidate Points). *Each passenger may have multiple candidate pick-up and drop-off positions. Denote the number of pick-up and drop-off positions for the i-th passenger in a carpooling order as* k_o^i *and* k_d^i. *The sets of candidate pick-up and drop-off points are* $P_o^i = \{p_o^{i(1)}, p_o^{i(2)}, \cdots, p_o^{i(k_o^i)}\}$ *and* $P_d^i = \{p_d^{i(1)}, p_d^{i(2)}, \cdots, p_d^{i(k_d^i)}\}$, *respectively.*

The carpooling service recommends a sequence of the pick-up and drop-off positions of passengers to guide the driver in completing the carpooling request. We give the definition of a carpooling sequence as follows:

Definition 3 (Carpooling Sequence). *For a given order* $\{l, p_c, p'_c, \mathbb{P}\}$, *where* $\mathbb{P} = \bigcup_{i=1}^{l} \{P_o^i, P_d^i\}$, *a carpooling sequence* $S = \langle p_c, p_1, p_2, \cdots, p_{2l}, p'_c \rangle$ *is a sequence of* $2l + 2$ *positions that satisfies the following constraints:*
Completeness constraint: $\forall i = 1, 2, \cdots, l, \exists p_j \in \{p_1, p_2, \cdots, p_{2l}\}, p_j \in P_o^i$ *and* $\exists p_{j'} \in \{p_1, p_2, \cdots, p_{2l}\}, p_{j'} \in P_d^i$.
Precedence constraint: $\forall i = 1, 2, \cdots, l, p_j \in P_o^i, p_{j'} \in P_d^i$, *there is* $j < j'$.

The completeness constraint is to guarantee all passengers are delivered to their destinations. The precedence constraint is to guarantee the pick-up position p_o^i precedes the drop-off position p_d^i.

Definition 4 (Length of Sequence). *The length* $L(S)$ *for a carpooling sequence* $S = \langle p_c, p_1, p_2, \cdots, p_{2l}, p'_c \rangle$ *is defined as* $L(S) = dist(p_c, p_1) + dist(p_{2l}, p'_c) + \sum_{i=1}^{2l-1} dist(p_i, p_{i+1})$, *where the distance function* $dist(v_i, v_j)$ *denotes the length of the shortest path from* v_i *to* v_j *in the graph* $G = (V, E, W)$.

2.2 Problem Formulation and NP Hardness Proof

Based on the four concepts above, we give the formal statements of the two problems we study in this work.

Definition 5 (Carpooling Routing Problem, CRP). *Given a road network* $G = (V, E, W)$. *For a carpooling order* $(l, p_c, p'_c, \mathbb{P})$, *where* $\mathbb{P} = \bigcup_{i=1}^{l} \{\{p_o^i\}, \{p_d^i\}\}$. *Find a carpooling sequence with the minimum length in* G.

Definition 6 (Multi-Candidate Carpooling Routing Problem, MCRP). *Given a road network* $G = (V, E, W)$. *For a carpooling order* $(l, p_c, p'_c, \mathbb{P},)$, *where* $\mathbb{P} = \bigcup_{i=1}^{l} \{P_o^i, P_d^i\}$, $P_o^i = \{p_o^{i(1)}, p_o^{i(2)}, \cdots, p_o^{i(k_o^i)}\}$, $P_d^i = \{p_d^{i(1)}, p_d^{i(2)}, \cdots, p_d^{i(k_d^i)}\}$, *and the largest number of candidate points is* $k = \max_i \{k_o^i, k_d^i\}$. *Find a carpooling sequence with the minimum length in graph* G.

Theorem 1. *The CRP and MCRP are NP-Hard.*

Proof. Since CRP is a special case of MCRP where $k = 1$, we can just prove CRP is NP-Hard by Karp-reduction from the TSP. To complete the proof, the following discussion considers the decision versions of all mentioned optimization problems, and we skip the derivations regarding thresholds K to provide YES or NO answers for Karp's reduction. For any TSP instance $\mathcal{I}_{TSP}$ defined on a graph $G = (V, E, W)$, where $V = \{v_1, v_2, \cdots, v_n\}$, we can construct a corresponding CRP instance $\mathcal{I}_{CRP}$ with road network G and carpooling order $(n-1, v_1, v_1, \mathbb{P})$, where $\mathbb{P} = \{\{v_2\}, \{v_2\}, \{v_3\}, \{v_3\}, \cdots, \{v_n\}, \{v_n\}\}$. Then $\mathcal{I}_{CRP} \equiv \mathcal{I}_{TSP}$.

2.3 Problem Generalization to Variants of TSP

Motivated by the reduction process from TSP to CRP, we find that CRP and MCRP can be regarded as variants of TSP. We give canonical names to CRP and MCRP for further investigation and analysis. CRP can be regarded as Precedence-Constraints Asymmetric Subset Traveling Salesman Path Problem (PAS-TSPP), and MCRP can be regarded as Generalized Precedence-Constraints Asymmetric Subset Traveling Salesman Path Problem (GPAS-TSPP).

Definition 7 (GPAS-TSPP) *Let* $G = (V, E, W)$ *be a directed graph, with vertex set* $V = \{v_i | i = 1, 2, \cdots, n\}$, *edge set* $E = \{e_{ij} = (v_i, v_j) | v_i, v_j \in V\}$ *(*$|E| = m$*), and weights* $W = \{w_{ij} | e_{ij} \in E\}$. *For a category set* $\mathbb{C} = \{C_1, C_2, \cdots, C_{|\mathbb{C}|}\}$, *with* $|\mathbb{C}|, C_i = \{c_{i,1}, c_{i,2}, \cdots, c_{i,|C_i|}\}, C_i \subset V, 1 \leq i \leq |\mathbb{C}|, C_i \cap C_j = \emptyset$ *for* $\forall(i, j)$ *such that* $i \neq j$, *and a precedence constraints set* $Pr = \{(C_{o1}, C_{d1}), (C_{o2}, C_{d2}), \cdots\}$. *Find a shortest sequence* $S_g = \langle v_1, v_2, \cdots, v_{|\mathbb{C}|} \rangle$ *s.t.* ① $S_g \cap C_i \neq \emptyset$, *for* $1 \leq i \leq |\mathbb{C}|$ ② $\forall(C_{oi}, C_{di}) \in Pr$, *if* $v_j \in C_{oi}, v_{j'} \in C_{di}$, *then* v_j *appears later than* $v_{j'}$ *in* S_g.

The canonical name for this problem implies its differences with the original TSP, which we explain as follows: ① Generalized: There is a category set with disjoint subsets of V as its elements, $\mathbb{C} = \{C_1, C_2, \cdots, C_{|\mathbb{C}|}\}$, with $C_i = \{c_{i,1}, c_{i,2}, \cdots, c_{i,|C_i|}\}, C_i \subset V, 1 \leq i \leq, C_i \cap C_j = \emptyset$ for $\forall(i,j)$ such that $i \neq j$. ② Precedence Constraints: There is a precedence constraint set for the categories $Pr = \{(C_{o1}, C_{d1}), (C_{o2}, C_{d2}), \cdots\}$. Where the vertex in subset C_{oi} is required to appear earlier than the vertex in subset C_{di}. ③ Asymmetric: $w(v_1, v_2)$ may not equals $w(v_2, v_1)$. ④ Subset: $\bigcup_{i=1}^{|\mathbb{C}|} C_i$ is a subset of V that may not equal V. ⑤ Path: Find a path instead of a cycle, with a given start and end.

PAS-TSPP is a special case of GPAS-TSPP where $\forall i, |C_i| = 1$.

2.4   Complexity Hierarchy of Related TSP Variants

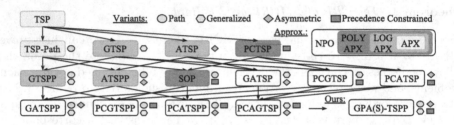

Fig. 2. Complexity Hierarchy of TSP and Its Variants: All Problems are NP-Hard

The GPAS-TSPP problem can be divided into four basic variants of the Traveling Salesman Problem (TSP), including ① Traveling Salesman Path Problem (TSPP), ② Generalized Traveling Salesman Problem (GTSP), ③ Asymmetric Traveling Salesman Problem (ATSP), and ④ Precedence Constrained Traveling Salesman Problem (PCTSP).

Figure 2 provides a complexity hierarchy for TSP variants based on the number of additional variances. The first level is TSP itself. The second level encompasses the four basic variants. The third level includes six variants formed by combining two variances, while the fourth level consists of problems with three or more variances. We review the literature on approximation algorithms for these variants. Note that, we do not take Subset TSP that can be polynomial reduced to TSP as a class of variants. All discussions are based on metric graphs.

For both TSP and TSP-Path, 1.5-approximation algorithms have been developed [13], classifying them as APX problems. The best-known approximation ratios (not special cases) for GTSP [10], ATSP [1], GTSPP, and ATSPP are $O(\log n)$, indicating they fall under the category of LOG-APX. PCTSP and its path version SOP can be approximated within $O(n)$, making them part of POLY-APX. However, for the remaining problems in Fig. 2, there are currently no known approximation algorithms, and we label them as NPO problems. This highlights a gap between the existing approximation algorithms and the problems we are studying.

3   Discussion of Previous Work

Starting from TSP and TSP-Path, in this section we will discuss the approximate algorithms for the GTSP, ATSP, and PCTSP. Table 1 provides an overview of significant research contributions, along with their approximation ratios.

Table 1. A Summary of TSP and Its Variants

Problem	Solution	Constraint	Approximation
TSP	Math Program 1971 [11]	–	2
	DAIO 1991 [13]	–	$3/2$
	IPCO 2023 [15]	–	$3/2 - \epsilon$
TSP-Path	Oper. Res. Lett. 1991 [13]	–	$5/3$
	FOCS 2016 [19]	–	$3/2 + 1/34$
	SODA 2019 [22]	–	$3/2$
GTSP	J. Algorithms 2000 [10]	Geometry	$O(\log^2 n \log \log n \log m)$
	ICALP 2005 [7]	Geometry	$9.1\alpha + 1$
	Proc. Steklov I. Math 2017 [16]	Geometry	$3/2 + 8\sqrt{2} + \epsilon$
ATSP	Networks 1982 [9]	–	$\log(n)$
	SODA 2010 [1]	–	$O(\log n / \log \log n)$
	J. ACM 2020 [21]	Relax Connectivity	$O(1)$
PCTSP	J. Discrete Algorithms 2013 [3]	Linear Order	$2.5 - 2/k$

TSP and TSP-Path. Serdyukov [20] and Christofides [6] give a $3/2$ approximation algorithm for the symmetric TSP with triangle inequality. The Held-Karp relaxation, a well-known LP relaxation of the TSP, is introduced by Held and Karp [11], which provides an approximation ratio of 2. Klein Karlin [15] introduces a maximum entropy algorithm with a deterministic approach, achieving a $3/2 - \epsilon$ approximation ratio. While TSP-Path shares similarities with the TSP, its non-cyclic nature brings difficulty in finding optimal solutions. Hoogeveen [13] proposes a natural variant of Christofides' algorithm for TSP-Path and achieved a $5/3$ approximation ratio. Sebo and Vygen [19] enhance Christofides' trees by removing edges, resulting in a $3/2 + 1/34$ approximation. Zenklusen [22] utilized dynamic programming techniques to obtain a 1.5 approximation for TSP-Path.
GTSP and Group Steiner Tree. Garg et al. [10] propose a randomized algorithm for the group Steiner tree problem, improving the approximation ratio for GTSP and achieving an approximation ratio of $O(\log^2 n \log \log n \log m)$. Elbassioni et al. [7] introduce geometric constraints to the GTSP problem and provide a unified framework that encompasses the TSP with neighborhoods and the group Steiner tree problem, with an approximation ratio of $(9.1\alpha + 1)$. Additionally, Khachai et al. [16]. There is also work that presents an $(1.5 + 8\sqrt{2} + \epsilon)$-approximation algorithm for GTSP under geometric constraints. Solutions to the group Steiner tree problem can be applied for GTSP [10], and previous research

gives appropriate approximations on it. The group Steiner tree problem with geometry constraints can achieve $O\left(\log^2 n \log k \log \log n\right)$ approximation with $O(\log n \log \log n)$ complexity by derandomizing approximation algorithm [4,10]. And via spanner bootstrapping and prize collecting, Bateni et al. [2] gives a PATS algorithm with $(1 + \epsilon)$ approximation with planar constraints, which serves as an inspiration for our problem. Recent research has heavily relied on methods such as Variable Neighborhood Search [8], random global search optimization [18], and Lin-Kernighan adaptation techniques [12].

4   A 4-Approximation Algorithm for Symmetric CRP

Based on the discussion above, we can see that it is quite difficult to get a good approximation for the original CRP and MCRP. In this section, we give 4 approximation algorithm of the symmetric version of CRP. The asymmetry of road networks is low and can be approximated as a symmetric problem.

4.1   Existing 1.5-Approximation Algorithm for the TSP-Path

There is a 1.5-approximation algorithm for the symmetric $s-t$ path TSP [22]. It utilizes a variation of the dynamic programming idea introduced by Traub and Vygen and exploits a seminal result of Karger on the number of near-minimum cuts. For the asymmetric situation, there is a $O(\log n)$- approximation [5], which is based on a combination of local search and dynamic programming. It first constructs a feasible solution using a local search algorithm and then uses dynamic programming to improve the solution and obtain the final approximation. We denote the approximation algorithm for $s - t$ path TSP as $\mathcal{A}_p(G, s, t)$.

- **Input:** $G = (V, E, W)$ is a graph. $s \in V$ is the start point of the path. $t \in V$ is the end point of the path.
- **Output:** A sequence $S_p = \langle v_1, v_2, \cdots, v_n \rangle$, where $n = |V|$, $v_1 = s, v_n = t$ and the length of S_p, denoted as $L(S_p)$.

4.2   Approximation Algorithm for Symmetric CRP

We develop a constant-approximation ratio algorithm for CRP in a metric symmetric graph $G = (V, E, W)$, based on the existing approximation algorithm for $\mathcal{A}_p$. The complexity of the algorithm is $O(n^3) + l \cdot O(\mathcal{A}_p)$.

4.3   Approximation Ratio Analysis for Symmetric CRP

For the convenience of approximation ratio analysis, we use the following symbols in this section. OPT_1 and OPT_2 represent the length of the optimal solution for the TSP-Path on G_1 and G_2 in Algorithm 1 with p_c as its start point and arbitrary destination. OPT represents the length of the optimal solution for the symmetric CRP. For the proof of the approximation ratio, we first present a lemma derived from the triangle inequality (the proof is provided in the appendix).

Algorithm 1: Approximation Algorithm for Symmetric CRP

Input: $G = (V, E, W)$, a carpooling order (l, p_c, p'_c, P).

Output: A Carpooling Sequence S.

1 Run Floyd-Warshall algorithm on G to get pair-wise shortest distances;

2 $G' \leftarrow (V', E', W')$, with $V' = P \cup \{p_c, p'_c\}$ as its vertex set, pair-wise shortest distance as edge weight;

3 $G_1 \leftarrow$ vertex induced subgraph of G' with $\{p_c, p_o^1, \cdots, p_o^l\}$ as vertex set;

4 $G_2 \leftarrow$ vertex induced subgraph of G' with $\{p'_c, p_d^1, \cdots, p_d^l\}$ as vertex set;

5 **for** $i \leftarrow 1$ **to** l **do**

6 $\quad \lfloor \ S_{p1}^{(i)} \leftarrow \mathcal{A}_p(G_1, p_c, p_o^i), \ S_{p2}^{(i)} \leftarrow \mathcal{A}_p(G_2, p_d^i, p'_c);$

7 $S_1 \leftarrow \text{argmin}_i L(S_{p1}^{(i)}), \ S_2 \leftarrow \text{argmin}_i L(S_{p2}^{(i)});$

8 $S \leftarrow$ Concatenate S_1 and S_2;

Lemma 1. *In a metric graph G, the shortest path containing a set of point $P = \{p_1, p_2, \cdots, p_l\}$ is no shorter than the shortest path containing a set $P' = \{p'_1, p'_2, \cdots, p'_{l'}\}$, if $P' \subset P$.*

Proof. Denote the shortest path in G that contains all points in P as $S = \langle p_{i_1}, p_{i_2}, \cdots, p'_{j_1}, \cdots, p'_{j_2}, \cdots, p'_{j_{l'}}, \cdots \rangle$, where $p'_{j_1}, p'_{j_2}, \cdots, p'_{j_{l'}}$ are the point that belong to P'.

By the triangle inequality in metric graph $\forall (v_i, v_j, v_k), w(v_i, v_k) \leq w(v_i, v_j) + w(v_j, v_k)$, we have $L(\langle p'_{j_1}, p'_{j_2} \rangle) \leq L(\langle p'_{j_1}, \cdots, p'_{j_2} \rangle)$, $L(\langle p'_{j_2}, p'_{j_3} \rangle) \leq L(\langle p'_{j_2}, \cdots, p'_{j_3} \rangle)$, $\cdots$, $L(\langle p'_{j_{l'-1}}, p'_{j_{l'}} \rangle) \leq L(\langle p'_{j_{l'-1}}, \cdots, p'_{j_{l'}} \rangle)$.

Let $S_1 = \langle p'_{j_1}, p'_{j_2}, \cdots, p'_{j_{l'}} \rangle$. We have $L(S_1) \leq L(S)$. Let S' be the shortest path containing the set P'. Since S_1 is a feasible path in G containing P', $L(S') \leq L(S_1)$. Therefore, $L(S') \leq L(S_1) \leq L(S)$.

Theorem 2. *The approximation ratio for algorithm 1 is 4.*

Proof. Let $S_1 = \langle p_c, v_1^{(1)}, v_2^{(1)}, v_l^{(1)} \rangle, S_2 = \langle v_1^{(2)}, v_2^{(2)}, v_l^{(2)}, p'_c \rangle$. The length of the sequence output by Algorithm 1 is $L(S) = L(S_1) + L(S_2) + L(\langle v_o^l, v_d^1 \rangle)$.

Since the vertex of G_1, G_2, and $\{v_o^l, v_d^1\}$ are all subsets of P, according to lemma 1, we have $OPT_1, OPT_2, L(\langle v_o^l, v_d^1 \rangle) \leq OPT$. Denote the approximation ratio for algorithm $\mathcal{A}_p$ as γ, we have $L(S_1) \leq \gamma OPT_1$ and $L(S_2) \leq \gamma OPT_2$. Therefore, $L(S_1) + L(S_2) \leq \gamma(OPT_1 + OPT_2) \leq 2\gamma OPT$.

$L(S) \leq (2\gamma + 1)OPT$, since the approximation ratio for $s - t$ path TSP is $\gamma = 1.5$, we have $L(S) < 4 \cdot OPT$. The approximation ratio for algorithm 1 is 4.

5 A $(5 + \epsilon)$-Approximation Algorithm for Planar MCRP

5.1 Existing PTAS for Planar Group Steiner Tree

An approximation scheme for GTSP can be obtained with solutions for the group Steiner tree problem. By applying the classical tree-doubling operation [10], we can create a generalized traveling salesman tour that requires no more than twice the cost of the group Steiner tree. We denote the approximation algorithm for the group Steiner tree as $\mathcal{A}_{gs}(G, \mathbb{C})$.

- **Input:** A symmetric planar graph $G = (V, E, W)$, a category set $\mathbb{C} = \{C_1, C_2, \cdots, C_{|\mathbb{C}|}\}$, for $1 \leq i \leq |\mathbb{C}|, C_i = \{c_{i,1}, c_{i,2}, \cdots, c_{i,|C_i|}\}, C_i \subset V$
- **Output:** A Steiner tree $T = \{V_g, E_g\}$, where $V_g = \{v_1, v_2, \cdots, v_{|\mathbb{C}|}\}$ satisfying $V_g \cap C_i \neq \emptyset$, for $1 \leq i \leq |\mathbb{C}|$, and the length of this tree (sum of the weights of E_g), denoted as $L(E_g)$.

We apply the approximation algorithm in [2] to handle the planar group Steiner tree problem. The algorithm overcomes the barrier of designing a PTAS (Polynomial-Time Approximation Scheme) for this challenging problem by introducing the "spanner bootstrapping" technique. The key steps of the algorithm include constructing a spanner that approximates the optimal solution, as well as a series of carefully designed reward collection processes that balance the cost and benefit of reaching or avoiding specific terminal nodes. The algorithm can find a $(1+\epsilon)$-approximation solution in polynomial time, where ϵ is an arbitrarily small positive constant.

5.2   Constant-Approximation Algorithm Design for MCRP

We develop a $(5 + \epsilon)$-approximation ratio algorithm for MCRP in a metric symmetric planar graph based on the existing $(1 + \epsilon)$-approximation algorithm for the group Steiner tree $\mathcal{A}_{gs}$. The complexity of the algorithm is $O(n) + O(\mathcal{A}_{gs})$.

Algorithm 2: $(5 + \epsilon)$-Approximation Algorithm for Planar MCRP

Input: $G = (V, E, W)$, a carpooling order $(l, p_c, p'_c, \mathbb{P})$.
Output: A Carpooling Sequence S_G.

1 $\mathbb{C} \leftarrow \{\{p_c\}, \{p'_c\}, P_o^1, P_o^2, \cdots, P_o^l, P_d^1, P_d^2, \cdots, P_d^l\}$;
2 $T \leftarrow \mathcal{A}_{gs}(G, \mathbb{C})$;
3 $S_{G1} \leftarrow$ Construct a cycle from T by traversing each edge in T twice;
4 //(Tree doubling to construct a cycle with no more than twice length)
5 $S_G \leftarrow$ Traverse S_{G1} twice from p_c and add p'_c at the end.

5.3   Approxiamtion Ratio Analysis for MCRP

For the convenience of approximation ratio analysis, we use the following symbols in this section. OPT_T represents the length of the optimal solution for the group Steiner tree on $\mathbb{C}$ in Algorithm 2. OPT_G denotes the length of the optimal solution for the symmetric MCRP.

Theorem 3. *The approximation ratio for algorithm 2 is $(5 + \epsilon)$.*

Proof. Firstly, we can see that S_G is a feasible solution. Since S_{G1} contains vertex in $\forall P_o^i, P_d^i \in \mathbb{P}$. S_G contains double S_{G1} so that we can pick up passengers at the vertexes in P_o^i for the first round and drop off them at the vertexes in P_d^i for the second round and go to the destination p_c' at last.

The length of the sequence output by Algorithm 2 is $L(S_G) = 2L(S_{G1}) + L(\langle p_c, p_c' \rangle) = 4L(T) + L(\langle p_c, p_c' \rangle)$. Since all solutions for the GTSP-path on $\mathbb{C}$ are also group Steiner trees on $\mathbb{C}$. We have $OPT_T \leq OPT_G$. By the triangle inequality, $L(\langle p_c, p_c' \rangle) \leq OPT_G$ since the solution for MCRP starts from p_c and ends at p_c' and $\langle p_c, p_c' \rangle$ is a shortcut of it.

Denote the approximation ratio for algorithm $\mathcal{A}_{gs}$ as γ, we have $L(T) \leq \gamma OPT_T$. Therefore, $L(S_G) = 4L(T) + L(\langle p_c, p_c' \rangle) \leq 4\gamma OPT_T + OPT_G \leq 4\gamma OPT_G + OPT_G = (4\gamma + 1)OPT_T$. Since the approximation ratio for the group Steiner tree is $\gamma = 1 + \epsilon$, we have $L(S_G) < (5 + \epsilon)OPT_G$. The approximation ratio for algorithm 2 is $(5 + \epsilon)$.

6 An Exact Algorithm for MCRP

In this section, we design an exact algorithm for MCRP based on dynamic programming, serving as a practical solution for real applications with several passengers. We extend the dynamic programming algorithm for the TSP to suit this problem. To make the algorithm concise, we define the output as the length of the shortest sequence. The sequence itself can be easily obtained by backtracking OPT. The complexity of this exact algorithm is $O(n^2 \cdot 2^{2l})$.

Algorithm 3: Dynamic Programming for MCRP

Input: $G = (V, E, W)$, a carpooling order $(l, p_c, p_c', \mathbb{P})$.
Output: The shortest length of a Carpooling Sequence.

1 $OPT[p_c][0] \leftarrow 0$
2 **for** $s \leftarrow 1$ **to** 2^{2l} **do**
3   Initialize $OPT[v][s] \leftarrow +\infty, \forall v \in V$;
4      $\mathbb{P}_s \leftarrow$ get the s th subset of $\mathbb{P}$
5   **foreach** $C_i \in \mathbb{P}_s$ **do**
6   **if** $C_i = P_o^{i'}$ and $P_d^{i'} \in \mathbb{P}$ **then**
7   continue;   //(Skip invalid states)
8          $i' \leftarrow$ the serial number of subset $\mathbb{P}_s - \{C_i\}$;
9   **foreach** $v \in C_i$ **do**
10   **foreach** $C_j \in \mathbb{P}_s - \{C_i\}$ **do**
11                 $tmp \leftarrow \min_{\forall u \in C_j} \{OPT[u][i']) + w(u, v)\}$;
12                 $OPT[v][s] \leftarrow \min\{OPT[v][s], tmp\}$;

13 **return** $\min_{\forall p \in \bigcup_i P_o^i \cup P_d^i} OPT[p][2^{2l}] + w(p, p_c')$

State function $OPT(v, \mathbb{P}')$ denotes the minimum cost to pass the vertexes of the categories in the subset $\mathbb{P}' \subset \mathbb{P}$ and arrive at v that $\exists C_i \in \mathbb{P}', v \in C_i$, satisfying the precedence constraint: If $\exists P_d^{i'} \in \mathbb{P}'$, then $C_i \neq P_o^{i'}$.

$$OPT(v, \mathbb{P}') = \min_{\forall u \in C_j, C_j \in \mathbb{P}' - \{C_i\}} OPT(u, \mathbb{P}' - \{C_i\}) + w(u, v).$$

This state transition function holds for $\forall v \in C_i$ that satisfies the precedence constraint. To traverse all of the possible legal states, we need to enumerate the subsets of $\mathbb{P}$ in an order that the required states are all obtained when calculating $OPT(v, \mathbb{P}')$. We order the subsets of $\mathbb{P}$ by the binary representations of whether it contains each of the elements. Let $\mathbb{P} = \{P_o^1, P_o^2, \cdots, P_o^l, P_d^1, P_d^2, \cdots, P_d^l\}$, we represent a subset $\mathbb{P}' = \{P_o^{i_1}, P_o^{i_2}, \cdots, P_o^{i_k}, P_d^{i'_1}, P_d^{i'_2}, \cdots, P_d^{i'_{k'}}\}$ with its serial number $\sum_{j=1}^{k} 2^{i_j} + \sum_{j=1}^{k'} 2^{i'_j+l}$. The time complexity to translate a subset to a number is $O(l)$, and vice versa.

7   Conclusion and Future Work

We formulate the MCRP for a new carpooling scenario and classify it into a family of TSP variants. Then we propose a 4-approximation algorithm for its special case CRP, and design a $(5 + \epsilon)$-approximation algorithm for MCRP on the planar graph. We also give an exact dynamic programming algorithm for the general MCRP. To the best of our knowledge, we are the first to explore such carpooling problems from the perspective of the TSP family and give constant-approximation algorithms for these new practical variants. The proposed approximation algorithm for MCRP can be generalized to the general GPAS-TSPP with parameterized constraints on asymmetry and the height of the precedence constraint tree. Hopefully, we can find a polynomial-approximation algorithm for it, which remains as future work.

References

1. Asadpour, A., Goemans, M.X., Mądry, A., Gharan, S.O., Saberi, A.: An $O(\log n/\log \log n)$-approximation algorithm for the asymmetric traveling salesman problem. In: ACM-SIAM Symposium on Discrete Algorithms (SODA), pp. 379–389 (2010)
2. Bateni, M., Demaine, E.D., Hajiaghayi, M., Marx, D.: A PTAS for planar group steiner tree via spanner bootstrapping and prize collecting. In: ACM Symposium on Theory of Computing (STOC), pp. 570–583 (2016)
3. Böckenhauer, H.J., Mömke, T., Steinová, M.: Improved approximations for tsp with simple precedence constraints. J. Discrete Algorithms **21**, 32–40 (2013)
4. Charikar, M., Chekuri, C., Goel, A., Guha, S.: Rounding via trees: deterministic approximation algorithms for group steiner trees and k-median. In: ACM Symposium on Theory of Computing (STOC), pp. 114–123 (1998)
5. Chekuri, C., Pál, M.: An $O(\log n)$ approximation ratio for the asymmetric traveling salesman path problem. Theory Comput. **3**(1), 197–209 (2007)
6. Christofides, N.: Worst-case analysis of a new heuristic for the travelling salesman problem. Carnegie-Mellon University Pittsburgh Pa Management Sciences Research Group, Technical Report (1976)

7. Elbassioni, K., Fishkin, A.V., Mustafa, N.H., Sitters, R.: Approximation algorithms for Euclidean group tsp. In: International Colloquium on Automata, Languages and Programming (ICALP), pp. 1115–1126 (2005)
8. Fatih Tasgetiren, M., Suganthan, P.N., Pan, Q.K.: An ensemble of discrete differential evolution algorithms for solving the generalized traveling salesman problem. Appl. Math. Comput. **215**(9), 3356–3368 (2010). https://doi.org/10.1016/j.amc.2009.10.027
9. Frieze, A.M., Galbiati, G., Maffioli, F.: On the worst-case performance of some algorithms for the asymmetric traveling salesman problem. Networks **12**(1), 23–39 (1982)
10. Garg, N., Konjevod, G., Ravi, R.: A polylogarithmic approximation algorithm for the group steiner tree problem. J. Algorithms **37**(1), 66–84 (2000)
11. Held, M., Karp, R.M.: The traveling-salesman problem and minimum spanning trees. Oper. Res. **18**(6), 1138–1162 (1970)
12. Helsgaun, K.: Solving the equality generalized traveling salesman problem using the Lin–Kernighan–Helsgaun Algorithm. Math. Program. Comput. **7**(3), 269–287 (2015). https://doi.org/10.1007/s12532-015-0080-8
13. Hoogeveen, J.: Analysis of christofides' heuristic: some paths are more difficult than cycles. Oper. Res. Lett. **10**(5), 291–295 (1991)
14. Hu, G., Shao, J., Shen, F., Huang, Z., Shen, H.T.: Unifying multi-source social media data for personalized travel route planning. In: ACM International Conference on Research and Development in Information Retrieval (SIGIR), pp. 893–896 (2017)
15. Karlin, A.R., Klein, N., Oveis Gharan, S.: A deterministic better-than-3/2 approximation algorithm for metric tsp. In: International Conference on Integer Programming and Combinatorial Optimization (IPCO), pp. 261–274 (2023)
16. Khachai, M.Y., Neznakhina, E.: Approximation schemes for the generalized traveling salesman problem. Proc. Steklov Inst. Math. **299**, 97–105 (2017)
17. Liu, H., Luo, K., Xu, Y., Zhang, H.: Car-sharing problem: online scheduling with flexible advance bookings. In: Annual International Conference on Combinatorial Optimization and Applications (COCOA), pp. 340–351 (2019)
18. Schmidt, J., Irnich, S.: New neighborhoods and an iterated local search algorithm for the generalized traveling salesman problem. EURO J. Comput. Optim. **10**, 100029 (2022). https://doi.org/10.1016/j.ejco.2022.100029
19. Sebo, A., Van Zuylen, A.: The salesman's improved paths: A 3/2+ 1/34 approximation. In: IEEE Symposium on Foundations of Computer Science (FOCS), pp. 118–127 (2016)
20. Serdyukov, A.I.: Some extremal bypasses in graphs. Diskretnyi Analiz i Issledovanie Operatsii **17**, 76–79 (1978)
21. Svensson, O., Tarnawski, J., Végh, L.A.: A constant-factor approximation algorithm for the asymmetric traveling salesman problem. J. ACM **67**(6), 1–53 (2020)
22. Zenklusen, R.: A 1.5-approximation for path TSP. In: ACM-SIAM Symposium on Discrete Algorithms (SODA), pp. 1539–1549 (2019)
23. Zheng, T., Jiang, Y.: Driver-rider matching and route optimization in carpooling service for delivering intercity commuters to the high-speed railway station. Expert Syst. Appl. **227**, 120231 (2023)

Maximizing Utilitarian and Egalitarian Welfare of Fractional Hedonic Games on Tree-Like Graphs

Tesshu Hanaka[1]($\boxtimes$) (iD), Airi Ikeyama[2], and Hirotaka Ono[2] (iD)

[1] Kyushu University, Fukuoka, Japan
hanaka@inf.kyushu-u.ac.jp
[2] Nagoya University, Nagoya, Japan
ikeyama.airi.f4@s.mail.nagoya-u.ac.jp, ono@nagoya-u.jp

Abstract. Fractional hedonic games are coalition formation games where the utility of a player is determined by the average value they assign to the members of their coalition. These games are a variation of graph hedonic games, which are a class of coalition formation games that can be succinctly represented. Due to their applicability in network clustering and their relationship to graph hedonic games, fractional hedonic games have been extensively studied from various perspectives. However, finding welfare-maximizing partitions in fractional hedonic games is a challenging task due to the nonlinearity of utilities. In fact, it has been proven to be NP-hard in general and can be solved in polynomial time only for a limited number of graph classes, such as trees. This paper presents (pseudo)polynomial-time algorithms to compute welfare-maximizing partitions in fractional hedonic games on tree-like graphs. We consider two types of social welfare measures: utilitarian and egalitarian. Tree-like graphs refer to graphs with bounded treewidth and block graphs. An NP-hardness result demonstrates that the pseudopolynomial-time solvability is the best possible under the assumption P $\neq$ NP.

Keywords: Fractional hedonic game · Utilitarian welfare · Egalitarian welfare · treewidth · block graphs

1 Introduction

1.1 Definition and Motivation

The hedonic game [11] is a game of modeling coalition formation based on individual preferences. Graphical variants of hedonic games have been considered to express the preferences succinctly. In this work, we deal with a variant of the graphical hedonic games called a *fractional hedonic game* (FHG), a subclass of hedonic games in which each agent's utility is the average of valuations over the

Partially supported by JSPS KAKENHI Grant Numbers JP20H05967, JP21H05852, JP21K17707, JP21K19765, JP22H00513, and JP23H04388.

other agents in the belonging coalition. Similarly to ordinary graphical hedonic games, fractional hedonic games are coalition formation games on a graph, in which vertices represent agents and the weight of edge (i, j) denotes the value that agent i has for agent j. Although fractional hedonic games in the most general setting are defined on weighted directed graphs, simpler versions of fractional hedonic games are well studied [1]. For example, fractional hedonic games are said to be *symmetric* and are represented on undirected graphs when all pairs of two agents are equally friendly. Furthermore, fractional hedonic games are said to be *simple* and are represented on unweighted graphs when all of the edge weights are 1. This work deals with fractional hedonic games on undirected graphs as [1].

In fractional hedonic games, a coalition structure is represented by a partition of the vertices, where each set represents a coalition. Given a coalition structure, the utility of each agent is defined as the average weights of its incident edges, as explained above. This definition implies that if two coalitions contain an identical set of agents (vertices) adjacent to agent v, the sums of the edge weights incident to v are equal, but the smaller coalition is more desirable for v. Such a property is suitable for finding a partition into dense subgraphs, which is why it is used for network clustering.

Under this definition of individual utility, two significant measures are well considered in welfare maximization: utilitarian or egalitarian social welfare. The former measure is the sum of the utilities of all agents. The latter is the minimum utility among the utilities of all agents. Hence, we call the partitions that maximize utilitarian and egalitarian welfare the maximum utilitarian welfare coalition structure (partition) and the maximum egalitarian welfare coalition structure (or partition), respectively.

Although many papers have already studied the computational complexity of finding the maximum utilitarian and egalitarian welfare coalition structure, positive results are few; even for restricted classes of graphs, they are NP-hard, and polynomial-time solvable classes are very restricted as summarized in Table 1. Noteworthy, the complexity of the problem for bounded treewidth remained open [6]. Such unwieldiness might be due to the nonlinearity of the objective function of the problem.

Therefore, this paper tries to enlarge solvable classes of graphs for the problems. In particular, we focus on block graphs and design a polynomial-time algorithm for computing the maximum utilitarian welfare coalition structure. Furthermore, we also focus on graphs with bounded vertex cover numbers and treewidth, for which we can design (pseudo)polynomial-time algorithms for finding the maximum utilitarian and egalitarian welfare coalition structures, which resolves the open problem left in [6]. At the same time, finding the maximum egalitarian welfare partition is shown to be weakly NP-hard even for graphs with vertex cover number 4. The detailed results obtained in this paper are summarized in Our contribution.

1.2   Our Contribution

This paper first shows a polynomial-time algorithm for computing the maximum utilitarian welfare partition on block graphs (Theorem 2). We see how a coalition structure forms in the maximum utilitarian welfare coalition structure on a block graph; by computing average utilities elaborately, we can show that there is a maximum utilitarian welfare coalition structure on a block graph in which every coalition forms a clique or a star. By utilizing the characterization, we design a dynamic programming-based algorithm that runs along a tree structure of a block graph.

Table 1. Complexity of the utilitarian and egalitarian welfare maximization on graph classes.

objective	graph class	complexity
utilitarian	general, unweighted	NP-hard [3,7]
	cubic graphs, unweighted	NP-hard [6]
	$\delta \geq n - 3$, unweighted	P [6]
	bipartite, unweighted	NP-hard [6]
	block, unweighted	P [Theorem 2]
	tree, unweighted	P [7]
egalitarian	general, unweighted	NP-hard [3]

Table 2. Complexity of the utilitarian and egalitarian welfare maximization with graph parameters.

objective	parameter	unweighted	weighted
utilitarian	treewidth	$n^{O(\omega)}$ [Theorem 3]	$(nW)^{O(\omega)}$ [Theorem 3]
	vertex cover number	$n^{O(\tau)}$ [Theorem 4]	$n^{O(\tau)}$ [Theorem 4]
egalitarian	treewidth	$n^{O(\omega)}$ [Theorem 5]	$(nW)^{O(\omega)}$ [Theorem 5]
	vertex cover number	$n^{O(\tau)}$ [Corollary 1]	paraNP-hard [Theorem 6]

We then focus on the complexity of maximizing utilitarian and egalitarian welfare for well-known graph parameters: treewidth and vertex cover number, which indicate how tree-like or star-like a graph is, respectively (Table 2). For the utilitarian welfare maximization, we give an $(nW)^{O(\omega)}$-time algorithm and an $n^{O(\tau)}$-time algorithm where ω is the treewidth of G, τ is the vertex cover number of G, and W is the maximum absolute weight of edges. This resolves the open question left in [6] of whether the utilitarian welfare maximization on unweighted fractional hedonic games (equivalently, DENSE GRAPH PARTITIONING) can be solved in polynomial time on bounded treewidth graphs. We mention that it remains open whether there exists a polynomial-time algorithm on *weighted* bounded treewidth graphs though we can design the one on bounded vertex cover number graphs.

For the egalitarian welfare maximization, we also give an $(nW)^{O(\omega)}$-time algorithm for treewidth ω. We then show that in contrast to the utilitarian case, egalitarian welfare maximization is NP-hard on bounded vertex cover number graphs; it implies that the pseudopolynomial-time solvability is best possible under the assumption P $\neq$ NP.

1.3   Related Work

Hedonic games are coalition formation games, which are well-studied in the fields of Economics, Artificial Intelligence, and Multi-Agent Systems [2,5,8,11]. From the computer science perspective, the computational complexity and parameterized complexity of finding desirable coalition structures in (graphical) hedonic games have garnered significant attention [2,5,14–17]. For further reading for hedonic games, we refer the readers to the relevant chapter of computational social choice book [4].

A fractional hedonic game is a variant of hedonic games. It has been studied from several aspects, including complexity, algorithm, and stability. Here, we only pick up work on algorithms and complexity. From the algorithmic point of view, computing a coalition structure maximizing utilitarian welfare or egalitarian welfare is NP-hard [3,7]. On the other hand, it is computed in polynomial time only for a few graph classes, such as trees [7]. Furthermore, a problem equivalent to computing the maximum utilitarian welfare coalition structure in fractional hedonic games is studied under a different name; DENSE GRAPH PARTITIONING [6,10]. It is the problem of finding a partition with maximum density for a given graph. From the study on DENSE GRAPH PARTITIONING, it is known that computing the maximum utilitarian welfare coalition structure is NP-hard even for cubic graphs [6]. Table 1 shows the complexity of computing the maximum utilitarian welfare coalition structure and computing the maximum egalitarian welfare coalition structure. δ denotes the minimum degree of the input graph.

Most of the proofs are omitted due to space limitations. For the complete proofs, see the full version [13].

2   Preliminaries

2.1   Definitions, Terminologies, and Notation

Let $G = (V, E)$ be an undirected graph. We denote by $G[C] = (C, E(C))$ the subgraph of G induced by $C \subseteq V$ and by $N_G(v)$ the set of neighbors of v in G. The degree $d_G(v)$ of v is defined by $d_G(v) = |N_G(v)|$. We denote by Δ_G the maximum degree of G. The *distance* of u and v is defined by the length of the shortest path between them and denoted by $\text{dist}(u, v)$. The *diameter* of G denote with $\text{diam}(G)$ and $\text{diam}(G) = \max_{u,v \in V} \text{dist}(u, v)$. For a connected graph G, a vertex v is called a *cut vertex* if the graph obtained by deleting v is disconnected. Similarly, for a connected graph G, an edge e is called a *bridge* if the graph obtained by deleting e is disconnected. An *isolated vertex* is a vertex with degree zero. A vertex set K is called a *clique* if $G[K]$ is a complete graph. A vertex set I is called an *independent set* if no two vertices in I are adjacent. For a vertex set $V = \{v_1, \ldots, v_n\}$, a graph whose edge set is $E = \{\{v_1, v_j\} \mid 2 \le j \le n\}$ is called a *star*. The vertex v_1 is called the center of the star, and other vertices are called leaves.

A *block graph* is a graph whose every biconnected component is a clique. A vertex set $S \subseteq V(G)$ is a *vertex cover* if every edge has at least one endpoint in S. The *vertex cover number* $\tau(G)$ is the size of a minimum vertex cover. *Treewidth* is a parameter that represents how close a graph is to a tree and is defined by the minimum *width* among all *tree decompositions* of G, though the definition is omitted here. For the definition of a tree decomposition and its property, see [9].

2.2   Fractional Hedonic Game

A fractional hedonic game is defined on a weighted and directed graph $G = (V, E, w)$, whose weights represent preferences. Without loss of generality, we suppose no edge has weight 0. In this paper, we consider a *symmetric* fractional hedonic game, which is defined on an undirected graph. If all of the edge weight is 1, a fractional hedonic game is called *simple* and it is defined on an unweighted graph.

A partition C of V is called a *coalition structure*. A vertex set $C \in \mathcal{C}$ is called a coalition. The utility $U(v, C)$ of a vertex v that belongs to a coalition C is defined by the sum of edge weights of neighbors of v in C divided by $|C|$, i.e.,

$$U(v, C) = \frac{\sum_{u \in N_{G[C]}(v)} w_{vu}}{|C|}.$$

The *utilitarian welfare* $\mathrm{uw}(C)$ of $C \in \mathcal{C}$ and the one $\mathrm{uw}(\mathcal{C})$ of a coalition structure $\mathcal{C}$ are defined by the sum of the utility of each $v \in C$ and the sum of the utility of each vertex in G, respectively, i.e., $\mathrm{uw}(C) = \sum_{v \in C} U(v, C)$ and $\mathrm{uw}(\mathcal{C}) = \sum_{C \in \mathcal{C}} \mathrm{uw}(C) = \sum_{C \in \mathcal{C}} \sum_{v \in C} U(v, C)$. The *egalitarian welfare* $\mathrm{ew}(C)$ of $C \in \mathcal{C}$ is defined by the minimum utility of $v \in C$, i.e., $\mathrm{ew}(C) = \min_{v \in C} U(v, C)$. Analogously, the egalitarian welfare $\mathrm{ew}(\mathcal{C})$ of a coalition structure $\mathcal{C}$ is defined by the minimum utility of $v \in V$ under $\mathcal{C}$, i.e., $\mathrm{ew}(\mathcal{C}) = \min_{v \in C, C \in \mathcal{C}} U(v, C)$.

In this paper, we consider the problem to find a maximum utilitarian welfare coalition structure and the problem to find a maximum egalitarian welfare coalition structure. We use C^* as an optimal coalition structure.

For fractional hedonic games, the following basic properties hold.

Property 1 ([7]). In a symmetric fractional hedonic game, $\mathrm{uw}(C) = 2|E(C)|/|C|$ holds for any coalition C.

Property 2. In a simple symmetric fractional hedonic game, if a coalition C forms a clique of size k, $\mathrm{uw}(C) = k - 1$.

Property 3. In a simple symmetric fractional hedonic game, if a coalition C forms a star of size k, $\mathrm{uw}(C) = 2(k-1)/k$.

3   Maximizing Utilitarian Welfare on Block Graphs: Characterization

In this section, we characterize an optimal coalition structure on block graphs, and show that there exists a maximum utilitarian welfare coalition structure on the block graph in which each coalition induces a clique or a star. This characterization is used for designing a polynomial-time algorithm on block graphs.

Theorem 1. *There exists a maximum utilitarian welfare coalition structure on the block graph in which each coalition induces a clique or a star.*

To show this, we show the following two lemmas, though the proofs are omitted.

Lemma 1. *Let C be a coalition such that the diameter of $G[C]$ is 2. Then C can be partitioned into cliques or stars without decreasing utilitarian welfare.*

Then we prove that a coalition C with diameter at least 3 can be partitioned into stars and cliques as in Lemma 1. This implies the existence of an optimal coalition structure consisting of stars and cliques.

Lemma 2. *Let C be a coalition such that the diameter of $G[C]$ is at least 3. Then C can be partitioned into coalitions of diameter at most 2 without decreasing utilitarian welfare.*

By combining Lemmas 1 and 2, we can obtain a coalition structure whose coalitions form stars or cliques from any coalition structure. This implies Theorem 1.

4   Maximizing Utilitarian Welfare on Block Graphs: Algorithm

In this section, we show UTILITARIAN WELFARE MAXIMIZATION on block graphs can be computed in polynomial time. We give a polynomial-time algorithm based on dynamic programming using optimal coalition structures shown in Theorem 1.

Theorem 2. UTILITARIAN WELFARE MAXIMIZATION *on unweighted block graphs can be computed in time $O(n\Delta^4)$.*

Our algorithm is a dynamic programming-based algorithm on the block-cut tree.

4.1   Block-Cut Tree

For a block graph $G = (V, E)$, let $\mathcal{B} = \{B_1, \ldots, B_\beta\}$ be the set of maximal cliques, called *blocks*, in G and $C = \{c_1, \ldots, c_\gamma\}$ be the set of cut vertices in G. Then the *block-cut tree* $\mathcal{T}(G) = (\mathcal{X}, \mathcal{E})$ of G is a tree such that $\mathcal{X} = \mathcal{B} \cup C$ and each edge in $\mathcal{E}$ connects a block $B \in \mathcal{B}$ and a cut vertex $c \in C \cap B$. For simplicity, we sometimes write $\mathcal{T}$ instead of $\mathcal{T}(G)$. We call $B \in \mathcal{B}$ a block node and $c \in C$ a cut node of $\mathcal{T}$. For convenience, we consider a block-cut tree rooted by a block node B_r. We denote by $\mathcal{T}_x(G)$ a subtree consisting of a node $x \in \mathcal{X}$ and its descendants on $\mathcal{T}$, by $p_{\mathcal{T}}(x)$ the parent node of $x \in \mathcal{X} \setminus \{B_r\}$, and by $\text{Child}_{\mathcal{T}}(x)$ the set of children nodes of $x \in \mathcal{X}$ in $\mathcal{T}$ (see Figs. 1 and 2). We also denote by V_x the set of vertices in G corresponding to $\mathcal{T}_x$ and by $G[V_x]$ the induced subgraph of G corresponding to $\mathcal{T}_x$. For each block node $B \in \mathcal{B} \setminus \{B_r\}$ and its parent node $p_{\mathcal{T}}(B) \in C$, c_p denotes the cut vertex in $B \cap p_{\mathcal{T}}(B)$. For $B \in \mathcal{B}$, we define $R(B) := B \setminus C$ as the set of non-cut vertices in B. For a rooted block-cut tree, its leaves are block nodes.

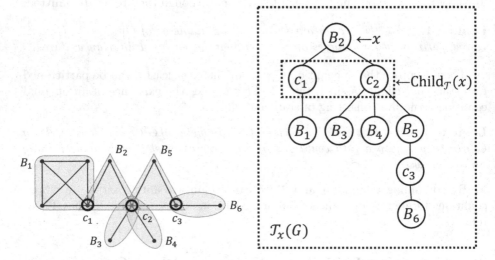

Fig. 1. Example of a block graph.   **Fig. 2.** The corresponding tree structure of block graph in Fig. 1.

4.2   Recurrence Relations: Overviews

By Theorem 1, we design a dynamic programming algorithm on block-cut trees to find an optimal coalition structure such that each coalition is either a clique or a star.

For a block node $B \in \mathcal{B}$, we define $\mathtt{T}_B = \{\mathtt{iso}, \mathtt{cl}, \mathtt{sc}, \mathtt{sl}\}$ as the set of states of the cut vertex $c_p \in B \cap p_{\mathcal{T}}(B)$ that represents the role of c_p within the coalition containing c_p in the intermediate steps of the algorithm for a block node. For a coalition structure $\mathcal{P}_B$ of $G[V_B]$, $\mathtt{iso}$ means that c_p is a singleton in $\mathcal{P}_B$, $\mathtt{cl}$ means that c_p belongs to a coalition that forms a clique of size at least 2 in $\mathcal{P}_B$, $\mathtt{sc}$ means that c_p belongs to a coalition that forms a star (not a singleton) and c_p is its center in $\mathcal{P}_B$, and $\mathtt{sl}$ means that c_p belongs to a coalition that forms a star (not a singleton) and c_p is its leaf in $\mathcal{P}_B$. Figure 3 shows the role of $c_p \in B \cap p_{\mathcal{T}}(B)$ with respect to $\mathtt{T}_B$.

We also define $T_c = \{\texttt{iso}, \texttt{cl}, \texttt{sc}_1, \ldots, \texttt{sc}_{|\text{Child}_{\mathcal{T}}(c)|}, \texttt{scu}, \texttt{sl}\}$ for a cut node $c \in C$ as the set of states of cut vertex c that represents the role of c in the intermediate step of the algorithm. Intuitively, in a cut node, the results of its children's block nodes are integrated. The states $\texttt{iso}$, $\texttt{cl}$, $\texttt{sl}$ of cut vertex c are the same meaning as the ones in T_B. The states $\texttt{sc}_\ell$ denote that the cut vertex c is the center vertex in the coalition that forms a star with ℓ leaves in $G[V_c]$. The state $\texttt{scu}$ denotes that the cut vertex c is the center vertex in the coalition that forms a star with leaves in $G[V_c]$ and the coalition to which $c \in C$ belongs *will* contain one vertex (denote it v_{leaf}) in the parent node $p_{\mathcal{T}}(c) \in \mathcal{B}$ as a leaf. Note that since the coalition forms a star and $p_{\mathcal{T}}(c)$ forms a clique, it can contain at most one vertex in $p_{\mathcal{T}}(c)$. For state $\texttt{scu}$, we need not preserve the number of leaves of the star of c because the maximum utilitarian welfare with respect to $\texttt{scu}$ can be computed from the values with respect to $\texttt{sc}_\ell$ for $1 \le \ell \le |\text{Child}_{\mathcal{T}}(c)|$. Figure 4 shows the role of $c \in C$ with respect to types in T_c.

Now, we define the recurrence relations for block nodes and cut nodes. For block nodes, we define $f^*(B, \texttt{type}_B, k)$ as the maximum utilitarian welfare among coalition structures in $G[V_B]$ that satisfies the following condition:

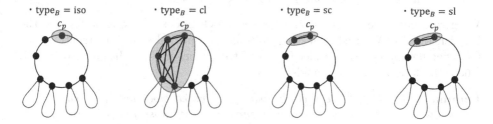

Fig. 3. The role of $c_p \in B \cap p_{\mathcal{T}}(B)$ in B with respect to T_B.

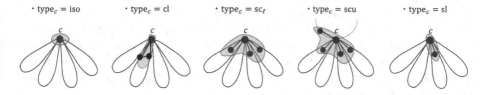

Fig. 4. The role of $c \in C$ in its cut node with respect to T_c.

- c_p is the parent cut node of B and the role of the c_p within the coalition to which c_p belongs is $\texttt{type}_B \in \{\texttt{iso}, \texttt{cl}, \texttt{sc}, \texttt{sl}\}$.
- if $\text{Child}_{\mathcal{T}}(B) \ne \emptyset$, there are exactly k cut vertices $c_1, \ldots c_k \in \text{Child}_{\mathcal{T}}(B)$ such that the coalition that c_i belongs to contain no vertex in $V_{c_i} \setminus \{c_i\}$ for $1 \le i \le k$.

For a cut node $c \in C$ and $\texttt{type}_c \in \{\texttt{iso}, \texttt{cl}, \texttt{sc}_1, \ldots, \texttt{sc}_{|\text{Child}_{\mathcal{T}}(c)|}, \texttt{sl}\}$, let $g^*(c, \texttt{type}_c)$ be the maximum utilitarian welfare in the coalition structures in $G[V_c]$ in which the role of c is $\texttt{type}_c$.

Also, $g^*(c, \mathtt{scu})$ denotes the maximum utilitarian welfare in $G[V_c \cup \{v_{\mathrm{leaf}}\}]$ when the cut node $c \in C$ becomes the center of a star in its coalition and $v_{\mathrm{leaf}} \in p_{\mathcal{T}}(c)$ becomes a leaf of c's star coalition. Then we can compute $g^*(c, \mathtt{scu})$ from $g^*(c, \mathtt{sc}_\ell)$ for $1 \le \ell \le |\mathrm{Child}_{\mathcal{T}}(c)|$ as follows:

$$g^*(c, \mathtt{scu}) = \max_\ell \left\{ g^*(c, \mathtt{sc}_\ell) + \frac{2}{(\ell+1)(\ell+2)} \right\}.$$

Since the utilitarian welfare of a star with ℓ leaves and a star with $\ell+1$ leaves are $2\ell/(\ell+1)$ and $2(\ell+1)/(\ell+2)$, respectively, the increase of the utilitarian welfare when a leaf is added to the coalition that forms a star with ℓ leaves is $2(\ell+1)/(\ell+2) - 2\ell/(\ell+1) = 2/(\ell+1)(\ell+2)$. Thus, the equation holds.

Let $g^*(c, \mathtt{fin}) = \max_{\mathtt{type}_c \in \{\mathtt{cl}, \mathtt{sc}_1, \ldots, \mathtt{sc}_{|\mathrm{Child}_{\mathcal{T}}(c)|}, \mathtt{sl}\}} \{g^*(c, \mathtt{type}_c)\}$ be the maximum utilitarian welfare in $G[V_c]$. If there is no coalition structure that satisfies the conditions, we set $f^*(B, \mathtt{type}_B, k) = -\infty$ and $g^*(c, \mathtt{type}_c) = -\infty$ as invalid cases. In the following, we define the recurrence relations of the dynamic programming algorithm to compute $f^*(B, \mathtt{type}_B, k)$ and $g^*(c, \mathtt{type}_c)$ recursively.

4.3   Recurrence Relations in the Block Nodes

We give recurrence relations only for the block nodes. For cut nodes, see the full version [13]. We first show that for such an optimal coalition structure, the following lemma holds. In $\mathcal{T}$, let $B \in \mathcal{B}$ be the parent node of the cut node $c \in C$. Then, the following lemma holds.

Lemma 3. *Let $B \in \mathcal{B}$ be a block node in a rooted block-cut tree. Then there exists an optimal coalition structure such that at most one cut vertex c in $\mathrm{Child}_{\mathcal{T}}(B)$ that is the center of a star having both vertices in $V_B \setminus B$ and in B as leaves.*

From Lemma 3, there is at most one cut vertex in $\mathrm{Child}_{\mathcal{T}}(B)$ whose role is $\mathtt{scu}$ in the coalition structure that realizes $f^*(B, \mathtt{type}_B, k)$. Therefore, we consider two cases (1) no cut vertex in $\mathrm{Child}_{\mathcal{T}}(B)$ takes the role $\mathtt{scu}$ and (2) exactly one cut vertex in $\mathrm{Child}_{\mathcal{T}}(B)$ takes the role $\mathtt{scu}$. We here see only case (1). For a block node B and $c_1, c_2, \ldots, c_i \in \mathrm{Child}_{\mathcal{T}}(B)$ where $1 \le i \le |\mathrm{Child}_{\mathcal{T}}(B)|$, we define $f(B, \mathtt{type}_B, i, k)$ as the maximum utilitarian welfare in $G[\{c_p\} \cup R(B) \cup \bigcup_{j=1}^i V_{c_j}]$ that satisfies the following conditions:

- c_p is the parent cut node of B and the role of the c_p within the coalition to which c_p belongs is $\mathtt{type}_B \in \{\mathtt{iso}, \mathtt{cl}, \mathtt{sc}, \mathtt{sl}\}$.
- There are exactly k cut vertices among $c_1, c_2, \ldots, c_i$ such that each of k such cut vertices does not form the coalition with vertices in its subtree, and each of exactly $i - k$ other cut vertices among $c_1, c_2, \ldots, c_i$ forms a coalition with only vertices in its subtree.

Note that $0 \le k \le i \le |\mathrm{Child}_{\mathcal{T}}(B)|$. Then, the maximum utilitarian welfare in $G[V_B]$ in Case (1) can be expressed by $f(B, \mathtt{type}_B, |\mathrm{Child}_{\mathcal{T}}(B)|, k)$. We can compute $f(B, \mathtt{type}_B, |\mathrm{Child}_{\mathcal{T}}(B)|, k)$ by the following argument.

As the base case, we consider the maximum utilitarian welfare in $G[\{c_p\} \cup R(B)]$. For convenience, we set this case as $i = 0$. Since $G[\{c_p\} \cup R(B)]$ is a clique, we have

$$
f(B, \mathtt{type}_B, 0, 0) = \begin{cases} 0 & \text{if } R(B) = \emptyset \\ |R(B)| - 1 & \text{if } R(B) \neq \emptyset \,\&\, \mathtt{type}_B = \mathtt{iso} \\ |R(B)| & \text{if } R(B) \neq \emptyset \,\&\, \mathtt{type}_B = \mathtt{cl} \\ \max\{1, |R(B)| - 1\} & \text{otherwise.} \end{cases}
$$

For the case $i > 0$, we need to consider for each $\mathtt{type}_B$. Here, we only show the case of $\mathtt{type}_B = \mathtt{iso}$. For the other cases, see the full version [13].

$$
f(B, \mathtt{iso}, i, k)
$$
$$
= \begin{cases} f(B, \mathtt{iso}, i-1, 0) + g^*(c_i, \mathtt{fin}) & \text{if } k = 0 \\[2mm] \max \left\{ \begin{array}{l} f(B, \mathtt{iso}, i-1, 0) + g^*(c_i, \mathtt{iso}), \\ f(B, \mathtt{iso}, i-1, 1) + g^*(c_i, \mathtt{fin}) \end{array} \right\} & \text{if } k = 1 \,\&\, R(B) = \emptyset \\[4mm] \max \left\{ \begin{array}{l} f(B, \mathtt{iso}, i-1, k-1) + g^*(c_i, \mathtt{iso}) + 1, \\ f(B, \mathtt{iso}, i-1, k) + g^*(c_i, \mathtt{fin}) \end{array} \right\} & \text{otherwise.} \end{cases}
$$

The above equation holds by the following observation: Consider the difference between the coalition structures in $G[\{c_p\} \cup R(B) \bigcup_{j=1}^{i} V_{c_j}]$ and $G[\{c_p\} \cup R(B) \bigcup_{j=1}^{i-1} V_{c_j}]$. When $k = 0$, the cut node c_i in $G[V_{c_i}]$ is not in the same coalition with any vertex in B.

When $k \geq 1$, the cut node c_i in $G[V_{c_i}]$ belongs to either a coalition containing a vertex in B ($g^*(c_i, \mathtt{iso})$) or not ($g^*(c_i, \mathtt{fin})$). Thus, we take the maximum one. Suppose that $k = 1$ and $R(B) = \emptyset$. We have two choices: c_i is a singleton or c_i belongs to a coalition containing only vertices in V_{c_i}. On the other hand, consider the case when $k \geq 2$ or $R(B) \neq \emptyset$. Then if c_i belongs to a coalition with vertices in B, its coalition forms a clique of size at least 2. Thus the utilitarian welfare will increase by 1.

5 Graphs of Bounded Treewidth or Vertex Cover Number

In this section, we design (pseudo)polynomial-time algorithms on graphs of bounded treewidth and bounded vertex cover number for computing a maximum utilitarian welfare coalition structure and a maximum egalitarian welfare coalition structure.

5.1 Maximizing Utilitarian Welfare

For graphs with bounded treewidth, we can design a DP-based algorithm on a nice tree decomposition. We have the theorem below, though we omit the detail.

Theorem 3. UTILITARIAN WELFARE MAXIMIZATION *can be computed in time* $(nW)^{O(\omega)}$ *where ω is the treewidth of an input graph and W is the maximum absolute weight of edges.*

Next, we give a polynomial-time algorithm for computing the maximum utilitarian welfare on graphs of bounded vertex cover numbers.

Theorem 4. UTILITARIAN WELFARE MAXIMIZATION *can be computed in time* $n^{O(\tau)}$ *where τ is the vertex cover number of an input graph.*

To prove this, we first show the key lemma as follows.

Lemma 4. *For any graph G, there exists an optimal coalition structure with at most $\tau + 1$ coalitions.*

Proof. Suppose that there exists an optimal coalition structure with more than $\tau + 1$ coalitions. Let S be a minimum vertex cover. Then at most τ coalitions contain vertices in S. Thus, other coalitions only contain vertices in $V \setminus S$. Since $V \setminus S$ is an independent set, the utilitarian welfare of such a coalition is 0. Therefore, we can merge them into one coalition without decreasing utilitarian welfare. This means that there exists an optimal coalition structure with at most $\tau + 1$ coalitions. □

We then design an algorithm for graphs of bounded vertex cover number. We first guess the number of coalitions in the optimal coalition structure. Let b ($\leq \tau + 1$) be the number of coalitions. We further guess the number of vertices in each coalition. Let $C_1, \ldots, C_b$ be b coalitions and $n_1, \ldots, n_b$ be the number of vertices in them. Note that $n_i \geq 1$ for $1 \leq i \leq b$. The number of possible patterns of $n_1, \ldots, n_b$ is at most n^b.

Let S be a minimum vertex cover of size τ in G. We guess assignments of vertices in S to b coalitions. The number of such possible assignments is at most b^τ. If there exists a coalition C_i such that the number of vertices in it exceeds n_i, then we immediately reject such an assignment.

Finally, we consider assignments of vertices in $V \setminus S$ to coalitions. Suppose that the size of coalition C_j is fixed at n_j. Then if $v \in V \setminus S$ is assigned to C_j, the increase of the utilitarian welfare is computed as $a_{vj} = \sum_{u \in N(v) \cap C_j \cap S} 2w_{vu}/n_j$. Note that $V \setminus S$ is an independent set.

In order to find a maximum utilitarian welfare coalition structure, all we need to do is to find an assignment maximizing the sum of values a_{vj} for $v \in V \setminus S$ under the capacity condition. This can be formulated as the following bin packing problem.

MAX k-BIN PACKING

Input: n items with size 1, k bins with capacity $c_1, \ldots, c_k (\leq n)$, and the value a_{ij} when item $i \in \{1, \ldots, n\}$ is assigned to bin $j \in \{1, \ldots, k\}$

Output: an assignment of items to k bins that maximizes the maximum value when exactly $c_1, \ldots, c_b$ items are assigned to each bottle

MAX k-BIN PACKING can be solved in time $n^{O(k)}$ by a simple dynamic programming algorithm.

Lemma 5. MAX k-BIN PACKING *can be solved in time* $n^{O(k)}$.

Proof. Let $DP[i; d_1, \ldots, d_k]$ be the maximum value obtained by assigning items $1, \ldots, i$ to bins so that the capacity of bin j is d_j for each j. Then, $DP[i; d_1, \ldots, d_k]$ can be computed by dynamic programming. First, we initialize the DP table as $DP[0; 0, \ldots, 0] = 0$. Next, we define the recurrence relation of $DP[i; d_1, \ldots, d_k]$ as follows:

$$DP[i; d_1, \ldots, d_k] = \max \left\{ \max_{j \in \{1, \ldots, k\}} \{DP[i - 1; d_1, \ldots, d_j - 1, \ldots, d_k] + a_{vj}\} \right\}.$$

It is not hard to see the correctness of the recurrence relation. Since the size of DP table is at most n^{k+1} and the recurrence relation can be computed in $O(k)$, the total running time is $O(k \cdot n^{k+1}) = n^{O(k)}$. □

By replacing items by vertices in $V \setminus S$ and bins by coalitions and letting $c_j = n_j - |C_j \cap S|$, we can reduce the above assignment problem to MAX k-BIN PACKING. By Lemma 5, the utilitarian welfare maximization can be computed in time $\sum_{b=1}^{\tau+1} n^b \cdot b^\tau \cdot n^{O(b)} = n^{O(\tau)}$. Therefore, Theorem 4 holds.

5.2   Maximizing Egalitarian Welfare

In this subsection, we consider a coalition structure that maximizes egalitarian welfare. We first design a pseudopolynomial-time algorithm on bounded treewidth graphs. Then we show that the egalitarian welfare maximization is NP-hard on bounded vertex cover number graphs. This is in contrast to the utilitarian welfare maximization.

Theorem 5. EGALITARIAN WELFARE MAXIMIZATION *can be computed in time* $(nW)^{O(\omega)}$ *where* ω *is the treewidth of an input graph and* W *is the maximum absolute weight of edges.*

The algorithm in Theorem 5 is similar to the one in Theorem 3 for computing the maximum utilitarian welfare coalition structure. The different point is preserving in the DP table the maximum value of the minimum utility, i.e., maximum egalitarian welfare, instead of maximum utilitarian welfare. The detail of the proof appears in the full version [13].

Since $\omega \leq \tau$ holds for any graph, we have the following corollary.

Corollary 1. EGALITARIAN WELFARE MAXIMIZATION *can be computed in time* $(nW)^{O(\tau)}$ *where* τ *is the vertex cover number of an input graph and* W *is the maximum absolute weight of edges.*

On the other hand, even restricting $\tau = 4$, we can show that EGALITARIAN WELFARE MAXIMIZATION is NP-hard.

Theorem 6. EGALITARIAN WELFARE MAXIMIZATION *is NP-hard on graphs of bounded vertex cover number.*

Proof. We give a reduction from PARTITION. In the problem, given a set $A = \{a_1, \ldots, a_n\}$ of n integers, the task is to determine whether there exists a subset $A' \subseteq A$ such that $\sum_{a \in A'} = W/2$ where $W = \sum_{a \in A} a$. This problem is NP-hard if we require $|A'| = n/2$ [12]. Thus, we suppose that $|A|$ is even.

From an instance of PARTITION, we construct an instance of EGALITARIAN WELFARE MAXIMIZATION. First, we create four vertices v_1, v_2, w_1, w_2 and n vertices $v_{a_1}, \ldots, v_{a_n}$ corresponding to n integers. Let $V_A = \{v_{a_1}, \ldots, v_{a_n}\}$. Two vertices v_1 and v_2 are connected to each $v_{a_i} \in V_A$ by edges $\{v_1, v_a\}$ and $\{v_2, v_a\}$ with weight $(n/2+2)a_i$, respectively. For w_1 and w_2, we connect them to $v_{a_i} \in V_A$ by edges with weight $(n/2+2)((n+7/2)W - a_i)$. Moreover, we add edges $\{v_1, w_1\}$ and $\{v_2, w_2\}$ with weight $(n/2+2)(n+3)W$. Finally, we connect v_1 and v_2 by a large negative weight edge. The weight $-\infty$ denotes a large negative weight. Let $G = (V, E)$ be the constructed graph. Then G has a vertex cover $\{v_1, v_2, w_1, w_2\}$ of size 4 because the graph obtained by deleting them forms singletons.

We can show that there exists a subset $A' \subseteq A$ such that $\sum_{a \in A'} = W/2$ if and only if there exists a partition $\mathcal{P}$ of V such that the least utility among agents is at least $(n + 7/2)W$. For the details, see the full version [13].   □

References

1. Aziz, H., Brandl, F., Brandt, F., Harrenstein, P., Olsen, M., Peters, D.: Fractional hedonic games. ACM Trans. Econ. Comput. **7**(2), 1–29 (2019)
2. Aziz, H., Brandt, F., Seedig, H.G.: Computing desirable partitions in additively separable hedonic games. Artif. Intell. **195**, 316–334 (2013)
3. Aziz, H., Gaspers, S., Gudmundsson, J., Mestre, J., Täubig, H.: Welfare maximization in fractional hedonic games. In: Proceedings of the Twenty-Fourth International Joint Conference on Artificial Intelligence, IJCAI 2015, pp. 461–467 (2015)
4. Aziz, H., Savani, R.: Hedonic games. In: Brandt, F., Conitzer, V., Endriss, U., Lang, J., Procaccia, A. (eds.) Handbook of Computational Social Choice, pp. 356–376. Cambridge (2016)
5. Ballester, C.: NP-completeness in hedonic games. Games Econ. Behav. **49**(1), 1–30 (2004)
6. Bazgan, C., Casel, K., Cazals, P.: Dense graph partitioning on sparse and dense graphs. In: 18th Scandinavian Symposium and Workshops on Algorithm Theory, SWAT 2022, vol. 227, pp. 13:1–13:15 (2022)
7. Bilò, V., Fanelli, A., Flammini, M., Monaco, G., Moscardelli, L.: Nash stable outcomes in fractional hedonic games: existence, efficiency and computation. J. Artif. Intell. Res. **62**, 315–371 (2018)
8. Bogomolnaia, A., Jackson, M.O.: The stability of hedonic coalition structures. Games Econ. Behav. **38**(2), 201–230 (2002)
9. Cygan, M., et al.: Parameterized Algorithms (2015)
10. Darlay, J., Brauner, N., Moncel, J.: Dense and sparse graph partition. Discret. Appl. Math. **160**(16), 2389–2396 (2012)
11. Drèze, J., Greenberg, J.: Hedonic coalitions: optimality and stability. Econometrica **48**(4), 987–1003 (1980)

12. Garey, M.R., Johnson, D.S.: Computers and Intractability: A Guide to the Theory of NP-Completeness. W. H. Freeman, New York (1979)
13. Hanaka, T., Ikeyama, A., Ono, H.: Maximizing utilitarian and egalitarian welfare of fractional hedonic games on tree-like graphs. CoRR **abs/2310.05139** (2023)
14. Hanaka, T., Kiya, H., Maei, Y., Ono, H.: Computational complexity of hedonic games on sparse graphs. In: PRIMA 2019: Principles and Practice of Multi-Agent Systems - 22nd International Conference, vol. 11873, pp. 576–584 (2019)
15. Hanaka, T., Lampis, M.: Hedonic games and treewidth revisited. In: 30th Annual European Symposium on Algorithms, ESA 2022, vol. 244, pp. 64:1–64:16 (2022)
16. Olsen, M.: Nash stability in additively separable hedonic games and community structures. Theory Comput. Syst. **45**(4), 917–925 (2009)
17. Peters, D.: Graphical hedonic games of bounded treewidth. In: Proceedings of the Thirtieth AAAI Conference on Artificial Intelligence, pp. 586–593 (2016)

The Line-Constrained Maximum Coverage Facility Location Problem

Hiroki Maegawa$^{(\boxtimes)}$, Naoki Katoh, Yuki Tokuni, and Yuya Higashikawa

University of Hyogo, Kobe, Japan
{af21h006,af23v006}@guh.u-hyogo.ac.jp,
{naoki.katoh,higashikawa}@gsis.u-hyogo.ac.jp

Abstract. We consider the maximum coverage facility location problem in the plane. In this paper, we restrict the facilities to be located on the given line, and propose an $O(n^2)$ algorithm for this problem by transforming the problem to the maximum weight k-link path problem in a complete directed acyclic graph, and by proving the concave Monge property inherent to the edge weights of the graph by which the substantial improvement of the running time compared with the straightforward implementation is attained.

Keywords: Maximum Coverage Facility Location Problem · Line-constrained Facility Location Problem · Maximum Weight k-link Path · Concave Monge Property

1 Introduction

This paper considers the maximum coverage facility location problem defined as follows: Given n points with positive weights in the plane, we consider the problem of finding the locations of k facilities that maximize the sum of the weights of the points covered by k disks centered at the facilities. In this paper, we assume that facilities are restricted to be located on the given line.

This problem has been studied by numerous researchers over several decades. See the paper by Farahani et al. (2012) for a comprehensive survey on more than 160 papers on the covering problems in facility location [1]. This paper classified problems in terms of models, algorithms and applications.

de Berg, Cabello and Har-Peled (2009) addressed this problem as NP-hard, and proposed an $O(n \log n)$ time $(1 - \varepsilon)$-approximation algorithm [2].

Megiddo, Zemel and Hakimi (1983) discussed the maximum coverage facility location problem on a network in which vertices have positive weights representing the number of customers and edge weights indicate the distance, and proposed an $O(n^2 k)$ algorithm in case a graph is restricted to a tree [3].

Our paper was motivated by the following practical problem for sensor location problem. Maegawa and Katoh are engaged in joint research with local governments and companies that are conducting research aiming at early detection of road disasters such as slope deformation and falling rocks along roads [4]. As part of this joint research, small sensor devices that can remotely observe data such as slope angle and vibration have been installed along roads, but there are hundreds to thousands of locations where disasters may occur even in one municipality. Considering the problem of optimal placement of sensor devices from the viewpoint of prioritizing the monitoring of high-risk areas with a limited number of devices is practically required. This is our motivation of this paper.

Based on this motivation, we shall consider the maximum coverage problem where the placement of facilities is restricted to a given line [5] and propose a polynomial-time algorithm to find the optimal location of facilities using the algorithm based on the concave Monge property, with the aim of utilizing it in the application field.

The organization of this paper is as follows: Sect. 2 gives preliminaries, a problem formulation and some necessary basic definitions. Section 3 shows that our problem can be reduced to the problem of finding a maximum weight k-link path in a complete directed acyclic graph, and then proposes an $O(n^2)$ algorithm for our problem. Section 4 proves the convex Monge property for our subproblems by which substantial improvement of the running time is attained, and lastly Sect. 5 summarizes our accomplishments and addresses our future challenges.

2   Preliminaries

Suppose that we are given a set of n points $P = \{p_1, p_2, \ldots, p_n\}$ in the plane where a positive weight w_a is associated with each point p_a and a straight line L is also given. Let $F = \{f_1, f_2, \ldots, f_k\}$ be a set of k facilities. When a facility $f_j \in F$ is within a distance d from a point p_a, p_a is said to be *covered* by f_j. In this paper, facilities are assumed to be constrained on the line L. We are asked to locate k facilities on the line L so as to maximize the sum of the weights of the points covered by at least one of these k facilities. Letting $\boldsymbol{y} = (y_1, y_2, \ldots, y_k)$ denote the locations of k facilities $F = \{f_1, f_2, \ldots, f_k\}$ on L, the problem can be formulated as follows:

$$Q(k) : \text{maximize } g_k(\boldsymbol{y}) = \Big\{ \sum w_a \mid p_a \text{ is covered by at least one } f_j \text{ located at } y_j \Big\} \tag{1}$$

The area where a facility can cover the point p_a is defined as a disk D_a with radius d centered at point p_a. The intersection of the disk D_a and the line L is called *coverable interval* $I_a = [l_a, r_a]$ (hereafter referred to as *c-interval* I_a) (i.e., any facility in the c-interval I_a can cover p_a). The weight of I_a is defined to be $w(I_a)$ (which is the weight of the point p_a).

Let E denote the collection of endpoints l_a, r_a of I_a, i.e., $\cup_{a=1}^n \{l_a, r_a\}$. Let $X = \{x_1, x_2, \ldots, x_{2n}\}$ be the set of values in E rearranged in increasing order of their values (see Fig. 1). It is assumed through this paper that all values of x_i's are distinct. An interval $[x_i, x_{i+1})$ defined by two consecutive values of X is called a *primitive interval* (denoted by *p-interval* for short). There are $2n - 1$ p-intervals which are denoted by $J_1, J_2, \ldots, J_{2n-1}$ such that $J_i = [x_i, x_{i+1})$ holds.

Observation 1. *For a point $y \in J_i$ for any i with $1 \le i \le 2n - 1$, the set of points in P that y can cover remains the same irrespective of the position y as long as y belongs to J_i.*

The difficulty of the problem lies in how we enumerate the points without duplication that are covered by at least one of facilities when we are given locations of facilities. In the next section, we will develop an algorithm that overcomes the difficulty.

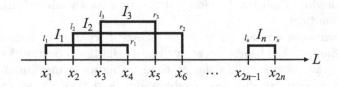

Fig. 1. Line L and c-intervals I_a

3 Algorithm for Our Problem

3.1 Reduction to Maximum k-Link Path Problem

Define the set of c-intervals $\mathcal{I}(i, j)$ as follows:

$$\mathcal{I}(i, j) := \{I_a \mid x_i \le l_a < x_j, 1 \le a \le n\}. \tag{2}$$

$\mathcal{I}(i, j)$ represents the set of c-intervals with the left endpoint being greater than or equal to x_i and less than x_j. For a point y on the line L and for $\mathcal{I}(i, j)$, let

$$cover(y, \mathcal{I}(i, j)) = \{I \mid I \in \mathcal{I}(i, j), y \in I\} \tag{3}$$

which denotes the set of c-intervals in $\mathcal{I}(i, j)$ that contain y, and

$$p\text{-}cover(y) = \{p \in P \mid y \text{ covers } p\}. \tag{4}$$

Let

$$W(y, \mathcal{I}(i, j)) = \Big\{ \sum w(I) \mid I \in cover(y, \mathcal{I}(i, j)) \Big\}. \tag{5}$$

Let $OPT(1, i, j)$ denote the optimal value such that $OPT(1, i, j) = \max\{W(y, \mathcal{I}(i, j) \mid y \in [x_i, x_j]\}$. Let $OPT(p, i, j)$ be the optimal value such that $OPT(p, i, j) = \max\{\sum w_a \mid (y_1, y_2, \ldots, y_p) \text{ with } x_i \le y_1 < y_2 < \cdots < y_p < x_j, p_a \in P \text{ is covered by at least one facility at } y_h \in [x_i, x_j)\}$.

In Sect. 4, the following notation will often be used.

$$value(S, Q) = \max\{W(y, \mathcal{I}(Q)) \mid y \in S\}, \tag{6}$$

where Q is the set of c-intervals defined as $\mathcal{I}(i,j)$ for some interval $[x_i, x_j)$ with $1 \leq i < j \leq 2n$, and S is a subinterval of $[x_i, x_j)$. In particular, when $S = [x_i, x_j)$,

$$value(S, Q) = OPT(1, i, j) \tag{7}$$

holds.

The problem we deal with is to compute $OPT(k, 1, 2n)$. For this purpose, employing the scheme introduced by Higashikawa et al. [6] and by Theorem 1, we reformulate this problem using a k-dimensional vector $\boldsymbol{y}$, and a $(k+2)$-dimensional vector $\boldsymbol{d}$ called *divider vector* given on L. $\boldsymbol{d} = (d_0 = 1, d_1, \ldots, d_{k+1} = 2n)$ such that $d_0 < d_1 < \cdots < d_{k+1}(= 2n)$ such that $x_{d_0}(= x_1) < x_{d_1} < \cdots < x_{d_{k+1}}(= x_{2n})$ holds.

Now define a complete directed acyclic graph (DAG) $G = (N, A)$ such that $N = \{u_1(= 1), u_2, \ldots, u_{2n}(= 2n)\}$ and for every vertex pair (u_i, u_j) with $1 \leq i < j \leq 2n$, there exists an edge which is directed from u_i to u_j and associated with the weight of $OPT(1, i, j)$.

An edge in a path in a complete DAG is called a link of the path. We call a path in the graph a k-link path if the path contains exactly k links. For any two vertices, i and j, we call a path from i to j a maximum k-link path if it contains exactly k links and among all such paths it has the maximum weight.

As will be shown in Theorem 1, solving $Q(k)$ is reduced to finding a maximum weight k-link path from u_1 to u_{2n} in the complete DAG G. Let $\boldsymbol{y}^* = (y_1^*, y_2^*, \ldots, y_k^*)$ with $y_1^* < y_2^* < \ldots < y_k^*$ be an optimal location of $Q(k)$.

Theorem 1. *For an optimal location $\boldsymbol{y}^*$ of $Q(k)$, there exists $(k+1)$-dimensional vector $\boldsymbol{d}^* = (d_0^* = 1, d_1^*, \ldots, d_{k+1}^* = 2n)$ such that the sum of edge weights in the maximum weight k-link path in the complete DAG induced by the pair $(\boldsymbol{y}^*, \boldsymbol{d}^*)$ is equal to the optimal objective value of problem $Q(k)$.*

Proof. $\boldsymbol{y}^*$ is an optimal location of k facilities that maximize the objective function $g_k(\boldsymbol{y})$ of $Q(k)$. We shall show that there exists a divider vector $\boldsymbol{d} = (d_0^* = 1, d_1^*, d_2^*, \ldots, d_{k-1}^*, d_k^* = 2n)$ with $d_0^* < d_1^* < d_2^* < \ldots < d_{k-1}^* < d_k^*$ such that for the maximum weight k-link path induced by $(\boldsymbol{y}^*, \boldsymbol{d}^*)$, the sum of edge weights on the path is equal to the optimal objective value of problem $Q(k)$. We shall prove the existence of such $\boldsymbol{d}^*$ in a constructive manner, namely, in such a way that we construct a sequence of optimal dividers $d_0^*, d_1^*, d_2^*, \ldots, d_{k-1}^*, d_k^*$ in this order.

Let $p\text{-}cover(y_j^*) = \{p_a \mid 1 \leq a \leq n, y_j^* \in I_a\}$, let $d_0^* = 1$.

The first step is to determine d_1^*. Let

$$l_{\min} = \min\{l_a \mid y_2^* \in I_a, y_1^* \notin I_a\} \tag{8}$$

and let $s \in \{1, 2, \ldots, 2n\}$ such that

$$x_s = l_{\min}.$$

Namely, for all left endpoints of c-intervals such that are covered by y_2^* but not by y_1^*, we choose the leftmost point as x_s, we set

$$d_1^* = s. \tag{9}$$

Then, any interval which is covered by y_2^* but not by y_1^* belongs to $\mathcal{I}(d_1^*, d_2^*)$ where $d_2^*(> d_1^*)$ will be determined in the next step.

The second step is to determine d_2^*. This is done in a manner similar to the first step. Let

$$l_{min} = \min\{l_i \mid y_3^* \in I_i, y_h^* \notin I_i (h = 1, 2)\} \tag{10}$$

and let $s \in \{1, 2, \dots, 2n\}$ such that

$$x_s = l_{min}.$$

We set

$$d_2^* = s. \tag{11}$$

Then, any interval which is covered by y_3^* but not by y_1^* or y_2^* belongs to $\mathcal{I}(d_2^*, d_3^*)$ where $d_3^*(> d_2^*)$ will be determined in the next step.

$d_3^*, d_4^*, \dots, d_{k-1}^*$ will be determined essentially in the same manner as the first and second steps (See Fig. 2).

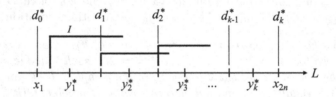

Fig. 2. The optimal solution $\boldsymbol{y}^*$ and dividers $\boldsymbol{d}^*$

We shall prove that $\sum_{1 \le j \le k} W(y_j^*, \mathcal{I}(d_{j-1}^*, d_j^*))$ is equal to the optimal value of $Q(k)$.

We shall define a set $\mathcal{P}^* = \{P_1^*, P_2^*, \dots, P_k^*\}$ as follows:
For $j = 1$,

$$P_1^* = p\text{-}cover(y_1^*). \tag{12}$$

For $j \ge 2$,

$$P_j^* = p\text{-}cover(y_j^*) \setminus p\text{-}cover(\cup_{h=1}^{j-1} y_h^*). \tag{13}$$

This means that P_1^* is a set of points covered by y_1^*, P_2^* is a set of points covered by y_2^* but not y_1^*, and P_3^* is a set of points covered by y_3^* except y_1^* and y_2^*, and so on. The points covered by at least one facility are divided into k disjoint point sets, $\{P_1^*, P_2^*, \dots, P_k^*\}$.

Since y_j^* is a feasible solution for the problem to compute $OPT(1, d_{j-1}^*, d_j^*)$, we have

$$W(y_j^*, \mathcal{I}(d_{j-1}^*, d_j^*)) \le OPT(1, d_{j-1}^*, d_j^*). \tag{14}$$

Therefore,

$$\sum_{j=1}^{k} W(y_j^*, \mathcal{I}(d_{j-1}^*, d_j^*)) \leq \sum_{j=1}^{k} OPT(1, d_{j-1}^*, d_j^*). \tag{15}$$

Since y_j^* covers points of P_j^* which corresponds to the set of c-intervals $\mathcal{I}(d_{j-1}^*, d_j^*)$ and the sum of the point weights of P_j^* is equal to $W(y_j^*, \mathcal{I}(d_{j-1}^*, d_j^*))$,

$$\sum_{j=1}^{k} W(y_j^*, \mathcal{I}(d_{j-1}^*, d_j^*)) \tag{16}$$

equals $g_k(\boldsymbol{y}^*)$. Let z_j^* be the optimal facility location of $OPT(1, d_{j-1}^*, d_j^*)$. Then $\boldsymbol{z}^* = (z_1^*, z_2^*, \dots, z_k^*)$ is a feasible solution of $Q(k)$.

Since z_j^* is an optimal solution of $OPT(1, d_{j-1}^*, d_j^*)$, we have

$$W(y_j^*, \mathcal{I}(d_{j-1}^*, d_j^*)) \leq W(z_j^*, \mathcal{I}(d_{j-1}^*, d_j^*)).$$

Thus,

$$\sum_{j=1}^{k} W(y_j^*, \mathcal{I}(d_{j-1}^*, d_j^*)) \leq \sum_{j=1}^{k} W(z_j^*, \mathcal{I}(d_{j-1}^*, d_j^*)) \tag{17}$$

follows. The right-hand side of (17) is the maximum weight of k-link path in the DAG, and the left-hand side is the optimal value of $Q(k)$. The inequality (15) implies that the value of the right-hand side of (17) must be equal to the optimal value of $Q(k)$. This completes the proof. $\qquad\square$

Schieber [7] showed that minimum k-link path problem can be solved by querying edge weights $n \cdot \min\{O(\sqrt{k \log n} + \log n), 2^{O(\sqrt{\log k \log \log n})}\}$ times if the input DAG satisfies the *concave Monge property*. Since this paper deals with maximization problem, we shall prove instead *convex Monge property*[1], that is, for all integers i, j satisfying $1 \leq i + 1 < j \leq n$,

$$OPT(i, j) + OPT(i + 1, j + 1) \geq OPT(i + 1, j) + OPT(i, j + 1). \tag{18}$$

In case of convex Monge property, the results related to time complexity obtained for concave Monge property can be directly applied to the case for concave Monge property.

3.2 Computing $OPT(1, h, j)$ for $j = h + 1, \dots, 2n$

In order to compute the maximum weight k-link path in a DAG, it is necessary to answer the query for the edge weight $OPT(1, h, j)$ for any h and j.

---

[1] In papers [7,8], the term *convex Monge property* is already used without formal definition. However, its definition is obvious from the context, and thus we used this term.

Algorithm 1. Computation of $OPT(1, h, j)$ for $j = h + 1, \ldots, 2n$

1: **function** OPT(h, events for X)
2:     $W \leftarrow 0$
3:     $W_{max} \leftarrow 0$
4:     **for** $i \leftarrow h, \ldots, 2n - 1$ **do**
5:         **if** event at x_i is Type 1 **then**
6:             $W = W + w(I)$
7:             $W_{max} = \max\{W, W_{max}\}$
8:         **else if** event at x_i is Subtype 2a **then**
9:             $W = W - w(I)$
10:        **end if**
11:        $OPT(1, h, i + 1) = W_{max}$
12:    **end for**
13: **end function**

Lemma 1. *When h is given, Algorithm 1 computes all values of $OPT(1, h, j)$ for $j = h + 1, \ldots, 2n$ in $O(n)$ time.*

Proof. The input for this problem is a set of c-intervals $I_a = [l_a, r_a]$ with weight $w(I_a)(a = 1, 2, \ldots, n)$. Given h and j such that $x_h < x_j$, with $1 \leq h, j \leq 2n$, find a point p with $x_h < p < x_j$ such that for a set of c-intervals $\mathcal{I}$ such that any $I_a \in \mathcal{I}$ satisfies $x_h \leq l_a$ and $p < r_a$, the sum of weights of $I_a \in \mathcal{I}$ is maximum. The algorithm for this problem is as follows:

For each $x_i \in X$ for i with $h \leq i \leq 2n$, we have one of the following events:

Type 1: New c-interval I with $w(I)$ starts from x_i.
Type 2: An existing c-interval I with $w(I)$ terminates at x_i. There are two subtypes:
    Subtype 2a: The left endpoint of I is larger than or equal to x_h.
    Subtype 2b: The left endpoint of I is smaller than x_h.

The algorithm for computing $OPT(1, h, j)$ for $j = h + 1, \ldots, 2n$ is shown Algorithm 1. It consists of single **for** loop for $h, \ldots, 2n$. Therefore, this algorithm computes all values of $OPT(1, h, j)$ for $j = h + 1, \ldots, 2n$ at once in $O(n)$ running time.   □

Lemma 2. *We can query for $OPT(1, h, j)$ in $O(1)$ time if we precompute $OPT(1, h, j)$ for all pairs of values h and j with $1 \leq h < j \leq 2n$ as a preprocessing, using $O(n^2)$ space.*

Proof. Applying Algorithm 1 for all $h = 1, 2, \ldots, 2n$, we can compute $OPT(1, h, j)$ for all pairs of h and j with $1 \leq h < j \leq 2n$ in $O(n^2)$ time, and store them with $O(n^2)$ space. Then we can answer the query for $OPT(1, h, j)$ in $O(1)$.

3.3   Description of Overall Algorithm

As mentioned in Sect. 3.1, our problem is reducible to the problem of finding the maximum weight k-link path in a complete DAG. Our maximization problem can transform into a minimization problem by negating link weights so that the algorithm for finding a minimum weight k-link path in a complete DAG [7] can be directly applied to our problem.

This remark was mentioned in paper [7]. Application IV in [7] deals with the problems of computing the maximum area k-gon and the maximum perimeter k-gon that are contained in a given convex n-gon. The paper [7] mentioned Aggarwal et al. [9] showed that the distance matrix involved in computing the maximum area and the maximum perimeter polygon contained in a given convex polygon has the convex Monge property. It also noted that because finding the maximum weight k-link path in convex DAGs is equivalent to finding the minimum weight k-link path in concave DAGs, the algorithm for computing the minimum k-link path in concave DAGs can be directly applied to compute the maximum k-link path in convex DAGs. Thus we obtain the following.

Proposition 1. *To find the maximum weight k-link path in a complete DAG utilizing the algorithm [6], the edge weights of the graph will be queried $n \cdot \min\{O(\sqrt{k \log n} + \log n), 2^{O(\sqrt{\log k \log \log n})}\}$ times when edge weights possess convex Monge property.*

Theorem 2. *The line-constrained maximum coverage facility location problem can be solved in $O(n^2)$ time with $O(n^2)$ space.*

Proof. From Lemma 2, we can obtain the answer to the query $OPT(1, h, j)$ for $1 \leq h < j < 2n$ in $O(1)$ time with $O(n^2)$ preprocessing time and $O(n^2)$ space. Thus, by Proposition 1, the maximum k-link path in DAG can be found in $O(n \cdot \min\{O(\sqrt{k \log n} + \log n), 2^{O(\sqrt{\log k \log \log n})}\}$ time. Since the time for preprocessing (i.e., $O(n^2)$ time) is the dominant part in the entire computation, the theorem follows.

4   Convex Monge Property of $OPT(1, h, j)$

A function $f : \mathbb{N}^2 \rightarrow \mathbb{R}$ possesses concave Monge property if the following inequality holds for any natural numbers i and j:

$$f(i, j) + f(i + 1, j + 1) \leq f(i + 1, j) + f(i, j + 1). \tag{19}$$

If the following inequality with the inequality sign reversed holds for any natural numbers i and j, f poseses convex Monge property:

$$f(i, j) + f(i + 1, j + 1) \geq f(i + 1, j) + f(i, j + 1). \tag{20}$$

Lemma 3. *Let $f(i, j) = OPT(1, i, j)$. Then, $f(i, j)$ possesses the convex Monge property.*

Proof. We prove that the following inequality holds:

$$OPT(1, i, j) + OPT(1, i+1, j+1) \geq OPT(1, i+1, j) + OPT(1, i, j+1). \quad (21)$$

For the sake of simplicity, we denote $OPT(l, r) = OPT(1, l, r)$. Using this notation, inequality (21) can be rewritten as:

$$OPT(i, j) + OPT(i+1, j+1) \geq OPT(i+1, j) + OPT(i, j+1). \quad (22)$$

Some preparation is required for this proof, and many cases need to be considered afterwards.

The domain intervals of the first and second terms of the left-hand side of inequality (22) are $[i, j)$ and $[i+1, j+1)$, respectively while those of the first and and second terms of the right-hand side of (22) are $[i+1, j)$ and $[i, j+1)$, respectively. We simply denote these four intervals $[i, j)$, $[i+1, j+1)$, $[i+1, j)$ and $[i, j+1)$ as I, II, III, and IV, respectively (see Fig. 3). Using these notations, the inequality of (22) can be rewritten as follows:

$$OPT(\mathrm{I}) + OPT(\mathrm{II}) \geq OPT(\mathrm{III}) + OPT(\mathrm{IV}). \quad (23)$$

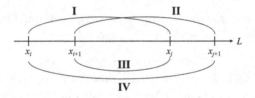

Fig. 3. Intervals referred to by each term in inequality (23)

We abuse the notations $OPT([a, b))$ and $\mathcal{I}([a, b))$ to stand for $OPT(a, b)$ and $\mathcal{I}(a, b)$, respectively. The c-intervals involved in the computation of $OPT(\mathrm{I})$ is $\mathcal{I}(\mathrm{I}) = \mathcal{I}(i, j)$, and that involved in the computation of $OPT(\mathrm{III})$ is $\mathcal{I}(\mathrm{III}) = \mathcal{I}(i+1, j)$. By $\mathcal{I}(\mathrm{I}) \supseteq \mathcal{I}(\mathrm{III})$, the following inequality holds:

$$OPT(\mathrm{I}) \geq OPT(\mathrm{III}). \quad (24)$$

Similarly, the c-intervals that are dealt with by $OPT(\mathrm{II})$ is $\mathcal{I}(\mathrm{II}) = \mathcal{I}(i+1, j+1)$, and since $\mathcal{I}(\mathrm{II}) \supseteq \mathcal{I}(\mathrm{III})$, the following inequality also holds:

$$OPT(\mathrm{II}) \geq OPT(\mathrm{III}). \quad (25)$$

Therefore, if the following inequality holds, convex Monge property (23) follows by (24).

$$OPT(\mathrm{II}) \geq OPT(\mathrm{IV}). \quad (26)$$

Similarly, if the following inequality holds, convex Monge property (23) follows by (25).

$$OPT(\text{I}) \geq OPT(\text{IV}). \tag{27}$$

Note that the domain c-intervals of $OPT(\text{IV})$ is $\mathcal{I}(\text{IV}) = \mathcal{I}(i, j + 1)$, and by $\mathcal{I}(\text{II}) \subseteq \mathcal{I}(\text{IV})$, the following inequality holds:

$$OPT(\text{II}) \leq OPT(\text{IV}). \tag{28}$$

Similarly, by $\mathcal{I}(\text{I}) \subseteq \mathcal{I}(\text{IV})$, the following inequality holds:

$$OPT(\text{I}) \leq OPT(\text{IV}). \tag{29}$$

Inequalities (26) and (28) together lead to:

$$OPT(\text{II}) = OPT(\text{IV}), \tag{30}$$

and inequalities (27) and (29) together lead to:

$$OPT(\text{I}) = OPT(\text{IV}). \tag{31}$$

Property 1. If either (30) or (31) holds, convex Monge property (23) follows.

Now let $optpos(l, r)$ denote the position of the facility that attains $OPT(l, r)$. We define the intervals $[i, i+1)$, $[i+1, j)$, and $[j, j+1)$ as A, B, and C, respectively (as shown in Fig. 4). $optpos(l, r) = B$ means that the optimal location is one of p-$intervals \in [i + 1, j)$, so the $optpos(l, r)$ can then be in either of the intervals A, B, or C. In case there are multiple p-interval that attains $OPT(l, r)$ we use the convention to choose the rightmost one.

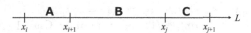

Fig. 4. Definition of intervals A, B, and C

To prove inequality (23), we enumerate all possible combinations of positions of $optpos(\text{I})$, $optpos(\text{II})$, $optpos(\text{III})$, and $optpos(\text{IV})$ (We abuse the notation $optpos(l, r)$ to stand for $optpos([l, r))$). $optpos(\text{I})$ may lie in either A or B, $optpos(\text{II})$ may lie in either B or C, $optpos(\text{III})$ is always B, and $optpos(\text{IV})$ can be A, B, or C. This results in $2 \times 2 \times 1 \times 3 = 12$ possible combinations which are shown in Table 1 (left). $optpos(\text{III})$ is not included in the case name because it always belongs to B.

According to the similarity in proof strategies, the 12 cases are classified into five groups as shown in Table 1 (right). We will prove (23) for each group.

Let *position* denote any of the facility locations A, B, or C, and let *p-interval* be one of the intervals I, II, III, IV. The *value*(*position*, *p-interval*) represents the sum of weights of the c-intervals when the facility is placed at position in *p-interval*. Note that unlike $OPT(l, r)$, it is not always guaranteed to an optimal value.

Table 1. 12 cases for the proof of (23) (left), and their classification into five groups (right)

No.	optpos				Case Name
	(I)	(II)	(III)	(IV)	
1	A	B	B	A	AB-A
2	A	B	B	B	AB-B
3	A	B	B	C	AB-C
4	A	C	B	A	AC-A
5	A	C	B	B	AC-B
6	A	C	B	C	AC-C
7	B	B	B	A	BB-A
8	B	B	B	B	BB-B
9	B	B	B	C	BB-C
10	B	C	B	A	BC-A
11	B	C	B	B	BC-B
12	B	C	B	C	BC-C

No.	Case Name	Group
1	AB-A	Group 1
4	AC-A	
8	BB-B	
11	BC-B	
2	AB-B	Group 2
5	AC-B	
7	BB-A	
10	BC-A	
3	AB-C	Group 3
9	BB-C	
6	AC-C	Group 4
12	BC-C	Group 5

Group 1

Group 1 consists of four cases where $optpos(I) = optpos(IV)$. First, let us consider the cases where $optpos(I) = optpos(IV) = A$: namely, Case 1 (AB-A) and Case 4 (AC-A), and we will then consider the cases where $optpos(I) = optpos(IV) = B$: namely, Case 8 (BB-B) and Case 11 (BC-B).

(1) Case 1 (AB-A) and Case 4 (AC-A)

The difference between $optpos(I)$ and $optpos(IV)$ may occur when there exists a c-interval $I_a = \{[l_a, r_a] \mid l_a = x_j\}$. Specifically, the computation of $optpos(I)$ does not take care of this I_a, while that of $optpos(IV)$ does. However, since $optpos(IV) = A$, such an I_a does not affect the value of $OPT(IV)$ (see Fig. 5). Therefore, $OPT(I) = OPT(IV)$ holds, and thus by Property 1, convex Monge property (23) follows.

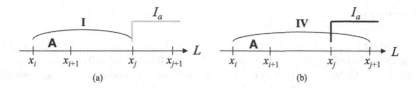

Fig. 5. Illustration of Group 1, Case 1 (AB-A) and Case 4 (AC-A)

(2) **Case 8 (BB-B) and Case 11 (BC-B)**

In this case, the same reasoning as the above applies. If $optpos$(IV) $\in B$, then such I_a does not affect the value of OPT(IV). Therefore, OPT(I) = OPT(IV), and by Property 1, convex Monge property (23) follows.

The proof for Groups 2, 3, 4 and 5 is omitted due to page limitation.

5   Conclusion

In this paper, we proved Theorem 1 by reducing the maximum coverage facility location problem to the problem of finding the maximum weight k-link path in a complete DAG with facility locations being restricted on the given line, taking advantage of the convex Monge property of $OPT(1, i, j)$.

Essential ingredient of this outcome is the $O(n)$ algorithm to solve 1-facility location problem, $OPT(1, i, j)$, and $O(n^2)$ for all pairs of i and j. However, we are sure there is a subquadratic algorithm, i.e., an algorithm faster than $O(n^2)$ time, to solve this subproblem. However, because of the page limit, we could not describe the details. The core idea for this improvement is that we do not precompute the values of $OPT(1, h, j)$ for all possible pairs of h and j, instead for $O(n)$ of (say, $\lceil \sqrt{(n)} \rceil$) values of h, and store these values. For an arbitrary query, say $OPT(1, h, j)$ such that $h = a\lceil \sqrt{(n)} \rceil - b$ with positive integers a and $b(< \lceil \sqrt{(n)} \rceil)$, we use the precomputed values of $OPT(1, a\lceil \sqrt{(n)} \rceil, j)$ to determine $OPT(1, h, j)$ in $O(\log n)$ time by using binary search tree.

Moreover, our problem can have extensions, like a case in which there are two or more lines (roads) in the plane, or the problem with a tree-structured graph being given. These extensions are still our challenges in the future.

Acknowledgments. This work was supported by JSPS KAKENHI Grant Numbers 19H04068, 23H03349. The research conducted in this paper was motivated by our collaborative research with XYMAX Corporation and XYMAX REAL ESTATE INSTITUTE Corporation [4]. We also gratefully acknowledge the financial support from these two companies.

References

1. Farahani, R.Z., Asgari, N., Heidari, N., Hosseininia, M., Goh, M.: Covering problems in facility location: a review. Comput. Ind. Eng. **62**(1), 368–407 (2012)
2. de Berg, M., Cabello, S., Har-Peled, S.: Covering many or few points with unit disks. Theory Comput. Syst. **45**, 446–469 (2009)
3. Megiddo, N., Zemel, E., Hakimi, S.L.: The maximum coverage location problem. SIAM J. Discrete Math. **4**(2), 253–261 (1983)
4. Maegawa, H., Katoh, N., et al.: Proposal for disaster prevention management methods of roadside slopes using IoT sensor devices (in Japanese). J. Soc. Saf. Sci. **41**, 197–207 (2022)
5. Wang, H., Zhang, J.: Line-constrained k-median, k-means, and k-center problems in the plane. Int. J. Comput. Geom. Appl. **26**(03n04), 185–210 (2016)
6. Higashikawa, Y., Golin, M.J., Katoh, K.: Multiple sink location problems in dynamic path networks. Theoret. Comput. Sci. **607**(1), 2–15 (2015)
7. Schieber, B.: Computing a minimum weight k-link path in graphs with the concave Monge property. J. Algorithms **29**(2), 204–222 (1998)
8. Aggarwal, A., Schieber, B., Tokuyama, T.: Finding a minimum-weight k-link path graphs with the concave Monge property and applications. Discrete Comput. Geom. **12**, 263–280 (1994)
9. Aggarwal, A., Klawe, M., Moran, S., Shor, P., Wilber, R.: Geometric applications of a matrix-searching algorithm. Algorithmica **2**, 195–208 (1987)

Graph Planer and Others

On Connectedness of Solutions to Integer Linear Systems

Takasugu Shigenobu[1](✉) and Naoyuki Kamiyama[2]

[1] Graduate School of Mathematics, Kyushu University, Fukuoka, Japan
shigenobu.takasugu.563@s.kyushu-u.ac.jp
[2] Institute of Mathematics for Industry, Kyushu University, Fukuoka, Japan
kamiyama@imi.kyushu-u.ac.jp

Abstract. An integer linear system (ILS) is a linear system with integer constraints. The solution graph of an ILS is defined as an undirected graph defined on the set of feasible solutions to the ILS. A pair of feasible solutions is connected by an edge in the solution graph if the Hamming distance between them is 1. We consider a property of the coefficient matrix of an ILS such that the solution graph is connected for any right-hand side vector. Especially, we focus on the existence of an elimination ordering (EO) of a coefficient matrix, which is known as the sufficient condition for the connectedness of the solution graph for any right-hand side vector. That is, we consider the question whether the existence of an EO of the coefficient matrix is a necessary condition for the connectedness of the solution graph for any right-hand side vector. We first prove that if a coefficient matrix has at least four rows and at least three columns, then the existence of an EO may not be a necessary condition. Next, we prove that if a coefficient matrix has at most three rows or at most two columns, then the existence of an EO is a necessary condition.

Keywords: integer linear system · solution graph · elimination ordering

1 Introduction

An integer linear system (ILS) has an $m \times n$ real coefficient matrix A, an m-dimensional real vector b, and a positive integer d. In this case, a feasible solution of the ILS is an n-dimensional integer vector $x \in \{0, 1, \ldots, d\}^n$ such that $Ax \geq b$. The solution graph of an ILS is defined as an undirected graph defined on the set of feasible solutions to the ILS. A pair of feasible solutions is connected by an edge in the solution graph if the Hamming distance between them is 1.

We consider a property of the coefficient matrix of an ILS such that the solution graph is connected for any right-hand side vector. Especially, we focus on the existence of an elimination ordering (EO) of a coefficient matrix, which is know as the sufficient condition for the connectedness of the solution graph

This work was supported by JSPS KAKENHI Grant Number JP20H05795.

for any right-hand side vector. (See Sect. 2 for the definition of an EO.) That is, we consider the question whether the existence of an EO of the coefficient matrix is a necessary condition for the connectedness of the solution graph for any right-hand side vector.

The results of this paper are summarized as follows. We first prove that if a coefficient matrix has at least four rows and at least three columns, then the existence of an EO may not be a necessary condition (Theorem 1). On the other hand, we also prove that if a coefficient matrix has at most three rows or at most two columns, then the existence of an EO is a necessary condition (Theorem 2). In fact, we prove the contraposition of the statament. That is, we prove that if a coefficient matrix does not have an EO, then there exists an right-hand side vector such that the solution graph is not connected.

Kimura and Suzuki [5] proved that if the coefficient matrix of an ILS has an EO, then the solution graph of the ILS is connected for any right-hand side vector. Precisely speaking, Kimura and Suzuki [5, Theorem 5.1] proved that if the complexity index of the coefficient matrix of an ILS introduced by Kimura and Makino [4] is less than 1, then the solution graph is connected. Furthermore, Kimura and Makino [4, Lemma 3] proved that the complexity index is less than 1 if and only if the coefficient matrix has an EO. The complexity index of the coefficient matrix of an ILS is a generalization of the complexity index for the Boolean satisfiability problem (SAT) introduced by Boros, Crama, Hammer, and Saks [1], and it depends only on the sign of the elements of the matrix. It is known that we can determine whether the coefficient matrix of an ILS has an EO in polynomial time.

The connectedness of the solution graph of an ILS is closely related to a reconfiguration problem of the ILS. A reconfiguration problem is a problem of finding a sequence of feasible solutions from the initial solution to the target solution (see, e.g., [3,6]). For ILS, the standard reconfiguration problems asks whether a given pair of feasible solutions to the ILS belong to the same connected component of the solution graph of the ILS. Therefore, if the solution graph of an ILS is connected, then the answer is always YES. Kimura and Suzuki [5] proved that computational complexity of the reconfiguration problem for the set of feasible solutions of an ILS has trichotomy.

An ILS is closely related to SAT. An instance of SAT can be formulated by an ILS. It is known that computational complexity of the reconfiguration problem of SAT has dichotomy [2,7].

Due to space limitations, we omit the proofs of the statements marked by $\star$.

2   Preliminaries

In this paper, let $\mathbb{R}$ and $\mathbb{Z}_{>0}$ denote the sets of real numbers and positive integers, respectively. For all integers $n \in \mathbb{Z}_{>0}$, we define $[n] := \{1, 2, \ldots, n\}$. First, we formally define an integer linear system and its solution graph. Throughout this paper, we fix a positive integer d. Define $D := \{0, 1, \ldots, d\}$. The set D represents the domain of a variable in an integer linear system.

Definition 1 (Integer linear system). *An integer linear system (ILS) has a coefficient matrix $A = (a_{ij}) \in \mathbb{R}^{[m] \times [n]}$ and a vector $b \in \mathbb{R}^{[m]}$. This ILS is denoted by $I = (A, b)$. A feasible solution to I is a vector $x \in D^{[n]}$ such that $Ax \geq b$. The set of feasible solutions to I is denoted by $R(I)$ or $R(A, b)$.*

Definition 2 (Hamming distance). *Define the function* dist$: \mathbb{R}^{[n]} \times \mathbb{R}^{[n]} \to \mathbb{R}$ *by* dist$(x, y) := |\{j \in \{1, \ldots, n\} : x_j \neq y_j\}|$ *for all vectors $x, y \in \mathbb{R}^{[n]}$. This function is called the Hamming distance on $\mathbb{R}^{[n]}$.*

Definition 3 (Solution graph). *Let R be a subset of $D^{[n]}$. We define the vertex set $V(R) := R$ and the edge set $E(R) := \{\{x, y\} : x, y \in V(R), \text{dist}(x, y) = 1\}$. We define the solution graph $G(R)$ as the undirected graph with the vertex set $V(R)$ and the edge set $E(R)$. Furthermore, for each ILS I, we define $G(I) := G(R(I))$.*

Next, we define elimination and an eliminated matrix. These concepts are used to define an elimination ordering (EO).

Definition 4 (Elimination). *Let $A = (a_{ij})$ be a matrix in $\mathbb{R}^{[m] \times [n]}$. Let j be an integer in $[n]$. We say that A can be eliminated at the column j if it satisfies at least one of the following conditions.*

(i) For all integers $i \in [m]$, if $a_{ij} > 0$, then $a_{ij'} = 0$ for all integers $j' \in [n] \backslash \{j\}$.
(ii) For all integers $i \in [m]$, if $a_{ij} < 0$, then $a_{ij'} = 0$ for all integers $j' \in [n] \backslash \{j\}$.

Definition 5 (Eliminated matrix). *Let A be a matrix in $\mathbb{R}^{[m] \times [n]}$. Let J be a subset of $[n]$. We define the eliminated matrix* elm$(A, J) \in \mathbb{R}^{[m] \times ([n] \backslash J)}$ *as the matrix obtained from A by eliminating the jth column for all integers $j \in J$. We call the matrix* elm(A, J) *the eliminated matrix of A by J.*

Definition 6 (Elimination ordering). *Let A be a matrix in $\mathbb{R}^{[m] \times [n]}$. Let $S = (j_1, j_2, \ldots, j_n)$ be a sequence of integers in $[n]$. Then S is called an elimination ordering (EO) of A if, for all integers $t \in [n]$,* elm$(A, \{j_1, j_2, \ldots, j_{t-1}\})$ *can be eliminated at j_t.*

Finally, we define the sign function as follows.

Definition 7 (Sign function). *For all real numbers $x \in \mathbb{R}$, the sign function* sgn$: \mathbb{R} \to \{-1, 0, 1\}$ *is defined as follows. If $x < 0$, then we define* sgn$(x) := -1$. *If $x = 0$, then we define* sgn$(x) := 0$. *If $x > 0$, then we define* sgn$(x) := 1$.

2.1 Our Contribution

In this paper, we prove following theorems.

Theorem 1. *Suppose that $m \geq 4$ and $n \geq 3$. Then there exists a matrix $A \in \mathbb{R}^{[m] \times [n]}$ satsisfying the following conditions. (i) A does not have an EO. (ii) For all vectors $b \in \mathbb{R}^{[m]}$, the solution graph $G(R(A, b))$ is connected.*

Theorem 2. *Let A be a matrix in $\mathbb{R}^{[m] \times [n]}$. Suppose that, for all vectors $b \in \mathbb{R}^{[m]}$, the solution graph $G(R(A, b))$ is connected. Then if at least one of $m \leq 3$ and $n \leq 2$ holds, then A has an EO.*

3   Proof of Theorem 1

First, we prove following lemma. This lemma plays an important role in the proof of Theorem 1.

Lemma 1. *There exists a matrix $A \in \mathbb{R}^{[4] \times [3]}$ satsisfying the following conditions. (i) The matrix A does not have an EO. (ii) For all vectors $b \in \mathbb{R}^{[4]}$, the solution graph $G(R(A, b))$ is connected.*

Proof. We define the matrix A as follows.

$$A = \begin{pmatrix} 1 & 1 & 0 \\ 1 & -1 & 0 \\ -1 & 0 & 1 \\ -1 & 0 & -1 \end{pmatrix}. \tag{1}$$

It is not difficult to see that A does not have an EO.

Suppose that $R(I)$ is not empty. We take arbitrary vectors $b \in \mathbb{R}^{[4]}$ and $s, t \in R(I)$. Without loss of generality, we suppose that $s_1 \geq t_1$. We take the path P from s to t defined by

$$s = \begin{pmatrix} s_1 \\ s_2 \\ s_3 \end{pmatrix} \to u^1 = \begin{pmatrix} s_1 \\ t_2 \\ s_3 \end{pmatrix} \to u^2 = \begin{pmatrix} t_1 \\ t_2 \\ s_3 \end{pmatrix} \to t = \begin{pmatrix} t_1 \\ t_2 \\ t_3 \end{pmatrix}.$$

We prove that $u^1 \in R(I)$ because

$$s_1 + t_2 \geq t_1 + t_2 \geq b_1, \quad s_1 - t_2 \geq t_1 - t_2 \geq b_2,$$
$$-s_1 + s_3 \geq b_3, \quad -s_1 - s_3 \geq b_4.$$

We prove that $u^2 \in R(I)$ because

$$t_1 + t_2 \geq b_1, \quad t_1 - t_2 \geq b_2,$$
$$-t_1 + s_3 \geq -s_1 + s_3 \geq b_3, \quad -t_1 - s_3 \geq -s_1 - s_3 \geq b_4.$$

These imply that every vertex of P is contained in $R(I)$. Thus, $G(I)$ is connected. This completes the proof. □

Proof (Theorem 1). Let A be the matrix defined in (1). If $m > 4$ or $n > 3$, then we add rows and columns whose all elements are 0 to A until A becomes an $m \times n$ matrix. Lemma 1 completes the proof. □

4   Proof of Theorem 2

First, we prove Lemma 2, which we call Expansion Lemma. Expansion Lemma means that the columns which can be eliminated have nothing to do with the connectedness of the solution graph.

4.1 Expansion Lemma

Let $A = (a_{ij})$ be a matrix in $\mathbb{R}^{[m] \times [n]}$. Suppose that A does not have an EO. We define the subsets $\Delta, E \subseteq [n]$ as the output of Algorithm 1. We define the matrix $A^r = (a_{ij}^r)$ as the submatrix of A whose index set of columns is Δ.

Algorithm 1. Algorithm for defining Δ and E.

1: $\Delta \leftarrow \emptyset, E \leftarrow \emptyset$
2: **while** $\mathrm{elm}(A, E)$ can be eliminated at some column **do**
3:     Find an index $j \in E$ at which the matrix $\mathrm{elm}(A, E)$ can be eliminated.
4:     $E \leftarrow E \cup \{j\}$
5: **end while**
6: $\Delta \leftarrow [n] \setminus E$
7: Output Δ, E

Lemma 2 (Expansion Lemma). *Suppose that there exists a vector $b^r \in \mathbb{R}^{[m]}$ such that the solution graph $G(I^r)$ of the ILS $I^r = (A^r, b^r)$ is not connected. Then there exists a vector $b \in \mathbb{R}^{[m]}$ such that the solution graph $G(I)$ of the ILS $I = (A, b)$ is not connected.*

Proof. We define the vector $x^e \in D^E$ as follows.

$$x_k^e := \begin{cases} 1 & \text{if the column vector } A_k \text{ is eliminated by the rule (i) in Definition 4} \\ 0 & \text{otherwise.} \end{cases}$$

With this vector x^e, we define the vector $b^e \in D^{[m]}$ by

$$b_i^e := \sum_{k \in E} a_{ik}^e d(1 - x_k^e) \quad (i \in [m]).$$

With the vectors b^e and b^r, we define the vector b by $b := b^r + b^e$. We prove that the solution graph $G(I)$ of the ILS $I = (A, b)$ is not connected. For each vector $z \in D^\Delta$ and each vector $\zeta \in D^E$, we define the vector $(z, \zeta) \in D^{[n]}$ by

$$(z, \zeta)_k := \begin{cases} z_k & (k \in \Delta) \\ \zeta_k & (k \in E). \end{cases}$$

Proposition 1. *There exists a vector $\zeta' \in D^E$ such that $(z, \zeta') \in R(I)$ for all feasible solutions $z \in R(I^r)$.*

Proof. We define the vector $\zeta' \in D^E$ by $\zeta_k' := d(1 - x_k^e)$ for all integers $k \in E$.

Since $z \in R(I^r)$, for all integers $i \in [m]$, we have $\sum_{k \in \Delta} a_{ik}^r z_k \geq b_i^r$. Thus, for all integers $i \in [m]$, we have

$$\sum_{k \in [n]} a_{ik}(z, \zeta')_k - b_i = \sum_{k \in \Delta} a_{ik}^r z_k - b_i^r + \sum_{k \in E} a_{ik}^e \zeta_k' - b_i^e$$

$$\geq \sum_{k \in E} a_{ik}^e \zeta_k' - \sum_{k \in E} a_{ik}^e d(1 - x_k^e) = 0.$$

Thus, we have $(z, \zeta') \in R(I)$. This completes the proof. $\square$

Proposition 2. *For all vectors* $z \in D^\Delta \backslash R(I^r)$ *and* $\zeta \in D^E$, *we have* $(z, \zeta) \notin R(I)$.

Proof. Since z is not a feasible solution in $R(I^r)$, there exists an integer $i \in [m]$ such that $\sum_{k \in \Delta} a_{ik}^r z_k < b_i^r$.

We prove that there exists an integer $j \in \Delta$ such that $a_{ij}^r \neq 0$. Suppose that $a_{ij}^r = 0$ for all integers $j \in \Delta$. For all vectors $z' \in R(I^r)$, $0 = \sum_{k \in \Delta} a_{ik}^r z_k' \geq b_i^r$. Therefore, we have $\sum_{k \in \Delta} a_{ik}^r z_k = 0 \geq b_i^r$. It contradicts $\sum_{k \in \Delta} a_{ik}^r z_k < b_i^r$. There exists an integer $j \in \Delta$ such that $a_{ij}^r \neq 0$. We fix such an integer $j \in \Delta$.

We prove that, for all integers $k \in E$, $x_k^e = 1$ (resp. $x_k^e = 0$) implies $a_{ik}^e \leq 0$ (resp. $a_{ik}^e \geq 0$). Suppose that there exist an integer $k \in E$ such that $x_k^e = 1$ (resp. $x_k^e = 0$) and $a_{ik}^e > 0$ (resp. $a_{ik}^e < 0$). Since $x_k^e = 1$ (resp. $x_k^e = 0$) and $k \in E$, the column vector A_k is eliminated by the rule (i) (resp. (ii)) in Definition 4. Therefore, since $a_{ik}^e > 0$ (resp. $a_{ik}^e < 0$), for all integers $j' \in [n] \backslash \{j\}$, we have $a_{ij'} = 0$. However, we have $a_{ij}^r \neq 0$ and $j \in \Delta \subseteq [n] \backslash \{j\}$. This is a contradiction. Thus, for all integers $k \in E$, $x_k^e = 1$ (resp. $x_k^e = 0$) implies $a_{ik}^e \leq 0$ (resp. $a_{ik}^e \geq 0$).

We define E^1 (resp. E^0) as the set of integers $k \in E$ such that $x_k^e = 1$ (resp. $x_k^e = 0$). For all integers $k \in E^1$, since $a_{ik}^e \leq 0$, we have $a_{ik}^e \zeta_k \leq a_{ik}^e 0 = a_{ik}^e d(1 - x_k^e)$. Similarly, for all integers $k \in E^0$, since $a_{ik}^e \geq 0$, we have $a_{ik}^e \zeta_k \leq a_{ik}^e d = a_{ik}^e d(1 - x_k^e)$. Thus, for all integers $k \in E$, we have $a_{ik}^e \zeta_k \leq a_{ik}^e d(1 - x_k^e)$. We have

$$\sum_{k \in [n]} a_{ik}(z, \zeta)_k - b_i = \sum_{k \in \Delta} a_{ik}^r z_k - b_i^r + \sum_{k \in E} a_{ik}^e \zeta_k - b_i^e < \sum_{k \in E} a_{ik}^e \zeta_k - b_i^e$$

$$\leq \sum_{k \in E} a_{ik}^e d(1 - x_k^e) - \sum_{k \in E} a_{ik}^e d(1 - x_k^e) = 0,$$

where the strict inequality follows from $\sum_{k \in \Delta} a_{ik}^r z_k < b_i^r$. This completes the proof. $\square$

We take vectors $p, q \in R(I^r)$ that are not connected on $G(I^r)$. We take a vector $\zeta' \in D^E$ satisfying the condition in Proposition 1. We obtain $(p, \zeta'), (q, \zeta') \in R(I)$. We take an arbitrary path P from (p, ζ') to (q, ζ') on $G(D^{[n]})$. Let $(p, \zeta') = (u^{(0)}, v^{(0)}) \to (u^{(1)}, v^{(1)}) \to \cdots \to (u^{(\ell)}, v^{(\ell)}) = (q, \zeta')$ denote P.

Define the map $F^r : D^{[n]} \to D^\Delta$ by $F^r((z, \zeta)) := z$ for all vectors $(z, \zeta) \in D^{[n]}$. Define the path P^r as $F^r((u^{(0)}, v^{(0)})) \to \cdots \to F^r((u^{(\ell)}, v^{(\ell)}))$ on $G(D^\Delta)$

(P^r may contain some duplicate vertices). Since p, q are not connected on $G(I^r)$, there exists a positive integer $k < \ell$ such that $F^r((u^{(k)}, v^{(k)})) \notin R(I^r)$.

By Proposition 2, $F^r((u^{(k)}, v^{(k)})) \notin R(I^r)$ implies that, for any vector $\zeta \in D^E$, $(F^r((u^{(k)}, v^{(k)})), \zeta) \notin R(I)$. If we take $v^{(k)}$ as ζ, then we have $(u^{(k)}, v^{(k)}) = (F^r((u^{(k)}, v^{(k)})), v^{(k)}) \notin R(I)$. This implies that P is not a path in $G(I)$. Thus, the solution graph $G(I)$ is not connected. This completes the proof. □

4.2 Two Rows

In this subsection, we consider the case where the coefficient matrix of an ILS consists of two rows.

Proposition 3 ($\star$). *Let $A = (a_{ij})$ be a matrix in $\mathbb{R}^{[2] \times [n]}$. Suppose that A cannot be eliminated at any column. Then for all integers $j \in [n]$, $\mathrm{sgn}(a_{1j}) = -\mathrm{sgn}(a_{2j})$, $\mathrm{sgn}(a_{1j}) \neq 0$, and $\mathrm{sgn}(a_{2j}) \neq 0$.*

Lemma 3. *Let $A = (a_{ij})$ be a matrix in $\mathbb{R}^{[2] \times [n]}$. Suppose that A cannot be eliminated at any column. Then there exists a vector $b \in \mathbb{R}^{[2]}$ such that the solution graph $G(I)$ of the ILS $I = (A, b)$ is not connected.*

Proof. We define the set of integer $\{j_1^1, \ldots, j_n^1\} = [n]$ (resp. $\{j_1^2, \ldots, j_n^2\} = [n]$) by $|a_{1j_k^1}| \leq |a_{1j_{k+1}^1}|$ (resp. $|a_{2j_k^2}| \leq |a_{2j_{k+1}^2}|$) for all integers $k \in [n-1]$. That is, we arrange the elements in each row in non-decreasing order.

Define the vector $x \in \{0, 1\}^{[n]}$ by

$$x_k := \begin{cases} 0 & (a_{1k} > 0) \\ 1 & (a_{1k} < 0). \end{cases}$$

Notice that Proposition 3 implies that $a_k \neq 0$.

We define the vector $b \in \mathbb{R}^{[2]}$ by

$$\begin{pmatrix} b_1 \\ b_2 \end{pmatrix} := \begin{pmatrix} \sum_{k \in [n] \setminus \{j_2^1\}} a_{1k} d(1 - x_k) + a_{1j_2^1}((d-1)(1 - x_{j_2^1}) + x_{j_2^1}) \\ \sum_{k \in [n] \setminus \{j_1^2\}} a_{2k} d(1 - x_k) + a_{2j_1^2}((d-1)(1 - x_{j_1^2}) + x_{j_1^2}) \end{pmatrix}.$$

Then we consider the ILS $I = (A, b)$. We define the vectors $p, q \in D^{[n]}$ as follows.

$$p_k := \begin{cases} d(1 - x_k) & (k \neq j_1^1) \\ (d-1)(1 - x_k) + x_k & (k = j_1^1), \end{cases} \quad q_k := \begin{cases} d(1 - x_k) & (k \neq j_2^1) \\ (d-1)(1 - x_k) + x_k & (k = j_2^1). \end{cases}$$

We prove that the vectors p, q belong to $R(I)$ and they are not connected on the solution graph $G(I)$.

Proposition 4 ($\star$). *The vectors p, q belong to $R(I)$.*

Proposition 5 ($\star$). *For all integers $j \in [n]$, $a_{1j}(1 - 2x_j) = |a_{1j}|$ and $a_{2j}(1 - 2x_j) = -|a_{2j}|$.*

We define Y as the set of vectors $y \in D^{[n]}$ such that $\mathrm{dist}(q, y) = 1$. In other words, the subset Y is the set of neighborhood vertices of q on $G(D^{[n]})$. Then we prove that $y \notin R(I)$ for all vectors $y \in Y$.

We arbitrarily take a vector $y \in Y$. From the definition, the following equation is obtained for the vector y.

$$y_k = \begin{cases} q_k & (k \neq j) \\ \xi & (k = j), \end{cases}$$

where j is an integer in $[n]$ and ξ is an integer in D such that $\xi \neq q_j$.

Case 1 $(j \neq j_2^1)$. If $j \neq j_2^1$, then we have

$$\sum_{k \in [n]} a_{1k}y_k - b_1 = \sum_{k \in [n] \setminus \{j\}} a_{1k}q_k + a_{1j}\xi - b_1$$

$$= \sum_{k \in [n] \setminus \{j, j_2^1\}} a_{1k}d(1 - x_k) + a_{1j_2^1}((d-1)(1 - x_{j_2^1}) + x_{j_2^1}) + a_{1j}\xi$$

$$- \left(\sum_{k \in [n] \setminus \{j_2^1\}} a_{1k}d(1 - x_k) + a_{1j_2^1}((d-1)(1 - x_{j_2^1}) + x_{j_2^1}) \right)$$

$$= a_{1j}(\xi - d(1 - x_j)).$$

Case 1.1 $(a_{1j} > 0)$. If $a_{1j} > 0$, then $x_j = 0$. Since $q_j = d(1 - x_j) = d$, the inequality $0 \leq \xi \leq d - 1$ is obtained. We have

$$\sum_{k \in [n]} a_{1k}y_k - b_1 = a_{1j}(\xi - d(1 - x_j)) \leq a_{1j}((d-1) - d) = -a_{1j} < 0.$$

Case 1.2 $(a_{1j} < 0)$. If $a_{1j} < 0$, then $x_j = 1$. Since $q_j = d(1 - x_j) = 0$, the inequality $1 \leq \xi \leq d$ is obtained. We have

$$\sum_{k \in [n]} a_{1k}y_k - b_1 = a_{1j}(\xi - d(1 - x_j)) \leq a_{1j}(1 - 0) = a_{1j} < 0.$$

Case 2 $(j = j_2^1)$. If $j = j_2^1$, then we have

$$\sum_{k \in [n]} a_{1k}y_k - b_1 = \sum_{k \in [n] \setminus \{j\}} a_{1k}q_k + a_{1j}\xi$$

$$- \left(\sum_{k \in [n] \setminus \{j\}} a_{1k}d(1 - x_k) + a_{1j}((d-1)(1 - x_j) + x_j) \right)$$

$$= a_{1j}(\xi - ((d-1)(1 - x_j) + x_j)).$$

Case 2.1 $(a_{1j} > 0)$. If $a_{1j} > 0$, then $x_j = 0$. Thus, $q_j = (d-1)(1-x_j) + x_j = d - 1$. Therefore, either $0 \leq \xi \leq d - 2$ or $\xi = d$ is satisfied. We have

$$\sum_{k \in [n]} a_{1k}y_k - b_1 = a_{1j}(\xi - ((d-1)(1-x_j) + x_j)) = a_{1j}(\xi - (d-1)).$$

Case 2.1.1 $(0 \leq \xi \leq d - 2)$. If $0 \leq \xi \leq d - 2$, then we have

$$\sum_{k \in [n]} a_{1k}y_k - b_1 = a_{1j}(\xi - (d-1)) \leq a_{1j}((d-2) - (d-1)) = -a_{1j} < 0.$$

Case 2.1.2 $(\xi = d)$. If $\xi = d$, then we consider b_2.
Case 2.1.2.1 $(j \neq j_1^2)$. By Proposition 5, if $j \neq j_1^2$, then we have

$$\sum_{k \in [n]} a_{2k}y_k - b_2$$

$$= \sum_{k \in [n]} a_{2k}y_k - \left(\sum_{k \in [n] \setminus \{j_1^2\}} a_{2k}d(1 - x_k) + a_{2j_1^2}((d-1)(1 - x_{j_1^2}) + x_{j_1^2}) \right)$$

$$= a_{2j}(\xi - d(1 - x_j)) + a_{2j_1^2}(y_{j_1^2} - ((d-1)(1 - x_{j_1^2}) + x_{j_1^2}))$$

$$= a_{2j}(d - d(1 - 0)) + a_{2j_1^2}(q_{j_1^2} - ((d-1)(1 - x_{j_1^2}) + x_{j_1^2}))$$

$$= a_{2j_1^2}(d(1 - x_{j_1^2}) - ((d-1)(1 - x_{j_1^2}) + x_{j_1^2}))$$

$$= a_{2j_1^2}(1 - 2x_{j_1^2}) = -|a_{2j_1^2}| < 0.$$

Case 2.1.2.2 $(j = j_1^2)$. If $j = j_1^2$, then $a_{2j} < 0$ follows from $a_{1j} > 0$ and Proposition 3.

$$\sum_{k \in [n]} a_{2k}y_k - b_2$$

$$= \sum_{k \in [n]} a_{2k}y_k - \left(\sum_{k \in [n] \setminus \{j\}} a_{2k}d(1 - x_k) + a_{2j}((d-1)(1 - x_j) + x_j) \right)$$

$$= a_{2j}(\xi - ((d-1)(1 - x_j) + x_j))$$

$$= a_{2j}(d - ((d-1)(1 - 0) + 0)) = a_{2j} < 0.$$

Due to space limitations, we omit the remaining part of the proof.

We obtain $y \notin R(I)$ for all the cases. Therefore, any neighborhood vertex of q on $G(D^{[n]})$ is not a feasible solution. This completes the proof. $\quad\square$

Lemma 4. *Let $A = (a_{ij})$ be a matrix in $\mathbb{R}^{[2] \times [n]}$. Suppose that A does not have an EO. Then there exists a vector $b \in \mathbb{R}^{[2]}$ such that the solution graph $G(I)$ of the ILS $I = (A, b)$ is not connected.*

Proof. We define A^r in the same way as in Sect. 4.1. Then A^r cannot be eliminated at any column. By Lemma 3, there exists a vector b^r such that the solution graph $G(R(A^r, b^r))$ is not connected. Lemma 2 completes the proof. $\quad\square$

4.3   Two Columns

In this subsection, we consider the case where the coefficient matrix of an ILS consists of two columns. We prove the following lemma.

Lemma 5 ($\star$). *Let $A = (a_{ij})$ be a matrix in $\mathbb{R}^{[m] \times [2]}$. Suppose that A does not have an EO. Then there exists a vector $b \in \mathbb{R}^{[m]}$ such that the solution graph $G(I)$ of the ILS $I = (A, b)$ is not connected.*

4.4   Three Rows

In this subsection, we consider the case where the coefficient matrix of an ILS consists of three rows. We prove the following lemma. At the end of this section, we prove Theorem 2.

Lemma 6. *Let $A = (a_{ij})$ be a matrix in $\mathbb{R}^{[3] \times [n]}$. Suppose that A cannot be eliminated at any column. Then there exists a vector $b \in \mathbb{R}^{[3]}$ such that the solution graph $G(I)$ of the ILS $I = (A, b)$ is not connected.*

Proof. For all integers $i_1, i_2 \in [3]$, we define

$$\Lambda_{i_1, i_2} := \{j \in [n] : \mathrm{sgn}(a_{i_1 j}) = -\mathrm{sgn}(a_{i_2 j}) \neq 0\}.$$

Proposition 6 ($\star$). *If there exist integers $i_1, i_2 \in [3]$ such that $|\Lambda_{i_1, i_2}| \geq 2$, then for all integers $d \in \mathbb{Z}_{>0}$, there exists a vector $b \in \mathbb{R}^{[3]}$ such that the solution graph $G(I)$ of the ILS $I = (A, b)$ is not connected.*

Proposition 7 ($\star$). *If $n \neq 3$, then there exist integers $i_1, i_2 \in [3]$ such that $|\Lambda_{i_1, i_2}| \geq 2$.*

Proposition 8 ($\star$). *Suppose that $n = 3$ and there are distinct integers $i_1', i_2', i_3' \in [3]$ such that $\Lambda_{i_1', i_2'} \cap \Lambda_{i_2', i_3'} \neq 0$. Then there exist integers $i_1, i_2 \in [3]$ such that $|\Lambda_{i_1, i_2}| \geq 2$.*

Proposition 6 implies that if there exist integers $i_1, i_2 \in [3]$ such that $|\Lambda_{i_1, i_2}| \geq 2$, then the proof is done. Suppose that, for all integers $i_1, i_2 \in [3]$, $|\Lambda_{i_1, i_2}| = 1$. By Proposition 7, we have $n = 3$. By Proposition 8, $\Lambda_{1,2}$, $\Lambda_{2,3}$, and $\Lambda_{1,3}$ are pairwise disjoint. Notice that, for all integers $j \in [3]$, there exist integers $i_1, i_2 \in [3]$ such that $j \in \Lambda_{i_1, i_2}$. Without loss of generality $\Lambda_{1,2} = \{1\}$, $\Lambda_{2,3} = \{2\}$, and $\Lambda_{1,3} = \{3\}$. We have $a_{12} = a_{23} = a_{31} = 0$. For example, if $a_{12} \neq 0$, then $2 \in \Lambda_{1,2}$ or $2 \in \Lambda_{1,3}$.

We define the vector $x \in \{0, 1\}^{[3]}$ by

$$x_k := \begin{cases} 0 & \text{if } a_{kk} > 0 \\ 1 & \text{if } a_{kk} < 0 \end{cases} \qquad (k \in [3]).$$

For all integers $i \in [3]$, we assume that $\{j_1^i, j_2^i, j_3^i\} = [3]$ and $|a_{ij_1^i}| \leq |a_{ij_2^i}| \leq |a_{ij_3^i}|$. By the definition, for all integers $i \in [3]$, we have $a_{ij_1^i} = 0$. We define the vector $b \in \mathbb{R}^{[3]}$ by

$$
b_i := \begin{cases}
\displaystyle\sum_{k \in [n]} a_{ik} d(1 - x_k) & \text{if } j_2^i = i \\[4mm]
\displaystyle\sum_{k \in [n]} a_{ik}((d-1)(1-x_k) + x_k) & \text{if } j_2^i \neq i
\end{cases}
\qquad (i \in [3]).
$$

Then we consider the ILS $I = (A, b)$.

We define the vectors $p, q \in D^{[3]}$ as follows.

$$
\begin{aligned}
p_i &:= d(1 - x_i) & (i \in [3]), \\
q_i &:= (d-1)(1 - x_i) + x_i & (i \in [3]).
\end{aligned}
$$

We prove that the vectors p, q belong to $R(I)$ and they are not connected on the solution graph $G(I)$.

Proposition 9 ($\star$). *The vectors p, q belong to $R(I)$.*

Proposition 10 ($\star$). *For all integers $j \in [3]$, we have $a_{jj}(1 - 2x_j) = |a_{jj}|$. For all integers $i \in [3]$ and all integers $s \in \{2, 3\}$, if $j_s^i \neq i$, then we have $a_{ij_s^i}(1 - 2x_{j_s^i}) = -|a_{ij_s^i}|$.*

We define Y as the set of vectors $y \in D^{[n]}$ such that $\text{dist}(p, y) = 1$. In other words, the subset Y is the set of neighborhood vertices of p on $G(D^{[n]})$. Then we prove that $y \notin R(I)$ for all vectors $y \in Y$.

We arbitrarily take a vector $y \in Y$. From the definition, the following equation is obtained for the vector y.

$$
y_k = \begin{cases}
p_k & (k \neq \ell) \\
\xi & (k = \ell).
\end{cases}
$$

where ℓ is an integer in $[3]$ and ξ is an integer in D such that $\xi \neq p_\ell$.

Suppose that $j_2^\ell = \ell$. For all vectors $y \in Y$, we have

$$
\sum_{k \in [3]} a_{\ell k} y_k - b_\ell = \sum_{k \in [3]} a_{\ell k} y_k - \sum_{k \in [3]} a_{\ell k} d(1 - x_\ell) = a_{\ell\ell}(\xi - d(1 - x_\ell)).
$$

If $a_{\ell\ell} > 0$, then $x_\ell = 0$. Since $p_\ell = d(1 - x_\ell) = d$, the inequality $0 \leq \xi \leq d - 1$ is obtained. We have

$$
\sum_{k \in [3]} a_{\ell k} y_k - b_\ell = a_{\ell\ell}(\xi - d(1 - x_\ell)) \leq a_{\ell\ell}(d - 1 - d) = -a_{\ell\ell} < 0.
$$

If $a_{\ell\ell} < 0$, then $x_\ell = 1$. Since $p_\ell = d(1 - x_\ell) = 0$, the inequality $1 \leq \xi \leq d$ is obtained. We have

$$
\sum_{k \in [3]} a_{\ell k} y_k - b_\ell = a_{\ell\ell}(\xi - d(1 - x_\ell)) \leq a_{\ell\ell} < 0.
$$

Suppose that $j_2^\ell \neq \ell$. By Proposition 10, for all vectors $y \in Y$, we have

$$\sum_{k \in [3]} a_{\ell k} y_k - b_\ell = a_{\ell j_2^\ell} d(1 - x_{j_2^\ell}) + a_{\ell\ell} \xi - b_\ell$$

$$= a_{\ell j_2^\ell}(d(1 - x_{j_2^\ell}) - ((d-1)(1 - x_{j_2^\ell}) + x_{j_2^\ell}))$$
$$+ a_{\ell\ell}(\xi - ((d-1)(1 - x_\ell) + x_\ell))$$
$$= a_{\ell j_2^\ell}(1 - 2x_{j_2^\ell}) + a_{\ell\ell}(\xi - ((d-1)(1 - x_\ell) + x_\ell))$$
$$= -|a_{\ell j_2^\ell}| + a_{\ell\ell}(\xi - ((d-1)(1 - x_\ell) + x_\ell)).$$

If $a_{\ell\ell} > 0$, then $x_\ell = 0$. Since $p_\ell = d(1 - x_\ell) = d$, the inequality $0 \leq \xi \leq d - 1$ is obtained.

$$\sum_{k \in [3]} a_{\ell k} y_k - b_\ell = -|a_{\ell j_2^\ell}| + a_{\ell\ell}(\xi - ((d-1)(1 - x_\ell) + x_\ell))$$

$$\leq -|a_{\ell j_2^\ell}| + a_{\ell\ell}(d - 1 - (d-1)) = -|a_{\ell j_2^\ell}| < 0.$$

If $a_{\ell\ell} < 0$, then $x_\ell = 1$. Since $p_\ell = d(1 - x_\ell) = 0$, the inequality $1 \leq \xi \leq d$ is obtained. We have

$$\sum_{k \in [3]} a_{\ell k} y_k - b_\ell = -|a_{\ell j_2^\ell}| + a_{\ell\ell}(\xi - ((d-1)(1 - x_\ell) + x_\ell))$$

$$\leq -|a_{\ell j_2^\ell}| + a_{\ell\ell}(1 - 1) = -|a_{\ell j_2^\ell}| < 0.$$

For all vectors $y \in Y$, we obtain $y \in R(I)$. Therefore, any neighborhood vertex of p on $G(D^{[n]})$ is not a feasible solution. This completes the proof. □

Lemma 7. *Let $A = (a_{ij})$ be a matrix in $\mathbb{R}^{[3] \times [n]}$. Suppose that A does not have an EO. Then there exists a vector $b \in \mathbb{R}^{[3]}$ such that the solution graph $G(I)$ of the ILS $I = (A, b)$ is not connected.*

Proof. We define A^r in the same way as in Sect. 4.1. Then A^r cannot be eliminated at any column. By Lemma 6, we have the vector b^r such that the solution graph $G(R(A^r, b^r))$ is not connected. Lemma 2 completes the proof. □

Proof (Theorem 2). We consider the contraposition of the statement in Theorem 2. If $m = 2$ (resp. $n = 2$, $m = 3$), Lemma 4 (resp. Lemma 5, Lemma 7) completes this proof. □

References

1. Boros, E., Crama, Y., Hammer, P.L., Saks, M.: A complexity index for satisfiability problems. SIAM J. Comput. **23**(1), 45–49 (1994). https://doi.org/10.1137/S0097539792228629
2. Gopalan, P., Kolaitis, P.G., Maneva, E., Papadimitriou, C.H.: The connectivity of Boolean satisfiability: computational and structural dichotomies. SIAM J. Comput. **38**(6), 2330–2355 (2009). https://doi.org/10.1137/07070440X

3. Ito, T., et al.: On the complexity of reconfiguration problems. Theoret. Comput. Sci. **412**(12–14), 1054–1065 (2011). https://doi.org/10.1016/j.tcs.2010.12.005
4. Kimura, K., Makino, K.: Trichotomy for integer linear systems based on their sign patterns. Discret. Appl. Math. **200**, 67–78 (2016). https://doi.org/10.1016/j.dam.2015.07.004
5. Kimura, K., Suzuki, A.: Trichotomy for the reconfiguration problem of integer linear systems. Theoret. Comput. Sci. **856**, 88–109 (2021). https://doi.org/10.1016/j.tcs.2020.12.025
6. Nishimura, N.: Introduction to reconfiguration. Algorithms **11**(4), 52 (2018). https://doi.org/10.3390/a11040052
7. Schwerdtfeger, K.W.: A computational trichotomy for connectivity of Boolean satisfiability. J. Satisfiability, Boolean Model. Comput. **8**(3–4), 173–195 (2014). https://doi.org/10.3233/sat190097

An Exact Algorithm
for the Line-Constrained Bottleneck
k-Steiner Tree Problem

Jianping Li[1(✉)], Suding Liu[1], and Junran Lichen[2]

[1] School of Mathematics and Statistics, Yunnan University,
East Outer Ring South Road, Kunming 650504, People's Republic of China
jianping@ynu.edu.cn, suding2020@163.com
[2] School of Mathematics and Physics, Beijing University of Chemical Technology,
No.15, North Third Ring East Road, Beijing 100190, People's Republic of China
J.R.Lichen@buct.edu.cn

Abstract. In this paper, we address the line-constrained bottleneck k-Steiner tree (LcBkStT) problem. Specifically, given an input line l, a set P of n points in $\mathbb{R}^2$ and a positive integer k, we are asked to find at most k Steiner points located on this line l and additionally a spanning tree T_l on these $n + k$ points, the objective is to minimize the length of the longest edge in T_l, where the edges in T_l are not allowed to cross this line l and the length of each edge in T_l is equal 0 if the two endpoints of that edge are located on the aforementioned line l. Using a technique of oriented Voronoi diagram, we design an exact algorithm for the LcBkStT problem in $O(n \log n + f(k) \cdot n^k)$ time, where $f(k)$ is a function dependent only on the positive integer k. This algorithm is an exact algorithm for the LcB1StT problem (for $k = 1$) in $O(n \log n)$ time.

Keywords: Line-constrained bottleneck Steiner tree · Steiner points · Computational geometry · Oriented Voronoi diagram · Exact algorithms

1 Introduction

The minimum spanning tree (MST) problem is one of classic and well-known combinatorial optimization problems [16,23]. The Euclidean minimum spanning tree (EMST) problem is a special version of the MST problem, and it is defined as follows. Given a set P of n points in $\mathbb{R}^2$, it is asked to find an interconnecting tree of minimum total length, whose vertices are the n points in P. The EMST

This paper is supported by the National Natural Science Foundation of China [Nos. 12361066, 12101593]. Junran Lichen is also supported by Fundamental Research Funds for the Central Universities [No.buctrc202219], Suding Liu is supported by the China Scholarship Council [No. 202107030013], and Jianping Li is also supported by Project of Yunling Scholars Training of Yunnan Province [No. K264202011820].

W. Wu and J. Guo (Eds.): COCOA 2023, LNCS 14461, pp. 434–445, 2024.
https://doi.org/10.1007/978-3-031-49611-0_31

problem is solved by using many well-known exact algorithms [9,16,23] to solve the MST problem in at least $O(n^2)$ time. Using techniques to construct the Voronoi diagram of n points in P, Shamos and Hoey [24] in 1975 presented an exact algorithm to resolve the EMST problem in $O(n \log n)$ time.

The minimum Steiner tree (MStT) problem [16] is one of the most fundamental *NP*-hard problems, which is a generalization of the MST problem and the shortest path problem. Over the past five decades, the MStT problem has been studied extensively, and many approximation algorithms to solve this problem can be found in those references [1,6,21]. The Euclidean minimum Steiner tree (EMStT) problem is a new form of the MStT problem. Specifically, given a set P of n points in $\mathbb{R}^2$, called terminals, it asks to find a shortest possible tree interconnecting these n points in $\mathbb{R}^2$, allowing the addition of auxiliary points to the set (Steiner points). Graham and Johnson [13] in 1977 showed that the EMStT problem is still *NP*-hard. There are many approximation algorithms to solve this problem [2,19,20,27].

Wang and Du [25] in 2002 addressed the bottleneck Steiner tree (BStT) problem. Specifically, given a set P of n terminals in $\mathbb{R}^2$ and a positive integer k, it is asked to find a Steiner tree, with at most k Steiner points, the objective is to minimize the length of a longest edge in such a Steiner tree. The same authors [25] proved that the BStT problem remains *NP*-hard even to approximate within ratio $\sqrt{2}$, unless $\mathcal{P} = \mathcal{NP}$, then they presented a 2-approximation algorithm to solve the BStT problem. Wang and Li [26] in 2002 presented the approximation algorithm that achieves an approximation ratio of 1.866 for the BStT problem. Cardei et al. [7] in 2006 provided by far the best-known approximation algorithm with approximation ratio of $\sqrt{3} + \epsilon$, where ϵ is an arbitrary positive number.

Bae et al. [4] in 2010 focused on finding exact solutions to the BStT problem for a small constant k. Having based on geometric properties of optimal location of Steiner points, Bae et al. [4] presented an optimal $\Theta(n \log n)$-time exact algorithm for the case $k = 1$ and an $O(n^2)$-time algorithm for the case $k = 2$ of the BStT problem, respectively. The same authors [4] also presented an optimal $\Theta(n \log n)$-time exact algorithm for any constant k for a special case where there is no edge between Steiner points. In addition, Bae et al. [3] in 2011 studied this problem in the L_p metric for any $1 \leqslant p \leqslant \infty$ and then presented a fixed-parameter tractable algorithm, running in $f(k) \cdot n \log^2 n$ time, for the L_1 and the L_∞ metrics and an exact algorithm, running in $f(k) \cdot (n^k + n \log n)$ time, for the L_p metric for any fixed rational p with $1 < p < \infty$, where $f(k)$ is a function dependent only on this constant k.

Georgakopoulos and Papadimitriou [14] in 1987 considered the 1-Steiner tree problem. Specifically, given a set P of n points in $\mathbb{R}^2$, it is asked to find a new point $s \in \mathbb{R}^2$ such that the total length of the minimum spanning tree on the set $P \cup \{s\}$ is as short as possible. Using a technique of oriented Voronoi diagram (OVD) [8], the same authors [14] presented an exact algorithm to solve the 1-Steiner tree problem in $O(n^2)$ time.

Holby [15] in 2017 discussed a variation of the EMStT problem by introducing a free Steiner line and attempting to construct a minimum Steiner network using this line, and the author [15] proposed a heuristic algorithm for this variation on larger sets. Li et al. [17] in 2020 reconsidered this Holby's problem and restated that one as the 1-line Euclidean minimum Steiner tree (1L-EMStT) problem. Specifically, given an input line l and a set P of n terminals in $\mathbb{R}^2$, it is asked to find a Steiner tree T_l to interconnect this line l and the n terminals with the shortest possible length, where the length of each edge in T_l is equal 0 if the two endpoints of that edge are located on the line l. When Steiner points added are all located on such a line l, the authors [17] called this version as the constrained Euclidean minimum Steiner tree (CEMStT) problem. Li et al. [17] designed a polynomial-time exact algorithm for the CEMStT problem and then a 1.214-approximation algorithm for the 1L-EMStT problem in $O(n \log n)$ time.

Chen and Zhang [10] in 2000 considered the constrained Euclidean minimum spanning tree problem, which is a variation of the EMST problem. Given an input line l and a set P of n points located at the same side of this line l in $\mathbb{R}^2$, it is asked to find one point s on this line l such that the total length of the minimum spanning tree on the set $P \cup \{s\}$ is as short as possible. The same authors [10] applied the divide-and-conquer technique to design an exact algorithm to solve this problem in $O(n^2)$ time.

Bose et al. [5] in 2022 addressed the restricted k-Steiner tree problem. Specifically, given an input line l and a set P of n points in $\mathbb{R}^2$, it is asked to find a Steiner tree T such that the number of Steiner points in T are no more than k, which are all located on the line l, the objective is to minimize the total length of that Steiner tree T. When $k = 1$, the same authors [5] refer to this problem as the restricted 1-Steiner tree problem. Bose et al. [5] provided an exact algorithm to solve the restricted k-Steiner tree problem in $O(n^k)$ time for any constant $k > 1$ and an exact algorithm for the restricted 1-Steiner tree problem in $\Theta(n \log n)$ time, respectively.

Motivated by these aforementioned interesting problems, we address the line-constrained bottleneck k-Steiner tree (LcBkStT) problem. Specifically, given an input line l, a set $P = \{p_1, p_2, \ldots, p_n\}$ of n points (called as terminals), located at the outside of this line l in $\mathbb{R}^2$, and a positive integer k, it is asked to find at most k Steiner points located on this line l and additionally a spanning tree T_l on these $n + k$ points, the objective is to minimize the length of the longest edge in such a Steiner tree T_l, i.e., $\min_{T_l} \max\{|\overline{pq}| \mid e = \overline{pq} \in T_l\}$, where the edges in T_l are not allowed to cross the line l, the length of each edge in T_l is equal 0 if the two endpoints of that edge are located on the line l, and otherwise the length of edge is the Euclidean distance between two endpoints. In particular, we are interested in the version $k = 1$ of the LcBkStT problem, and we refer this problem as the line-constrained bottleneck 1-Steiner tree (LcB1StT) problem.

The bottleneck Steiner tree problems and their variations have many known applications in VLSI layout [11], multi-facility location, and wireless communication network design [18]. The LcBkStT problem has an immediate application in the design of wireless networks. A security and protection system includes many sensors capable of collecting the changes of environment and sending signals to data collection node through the sensor network. To ensure the transmission efficiency of the sensor network, it is common practice to utilize wired networks at a global scale, while employing wireless sensor networks at a local level. Consequently, the supplementary sensors onto the wired network become essential for relaying the signals collected by the wireless sensors into the wired network.

Generally, sensors use batteries to provide power. In the radio-frequency wireless sensor networks, the power required to transfer a signal is related to the distance between the source and destination sensors. In general, shorten distance among nodes means longer network lifetime.

2   Properties of Bottleneck Steiner Trees

In this section, we present some properties of bottleneck Steiner trees to ensure the correctness of our algorithms.

Given any two points p and q in $\mathbb{R}^2$, we denote by $\overline{pq}$ the line segment between two points p and q, and we denote by $|\overline{pq}|$ the Euclidean length of this segment $\overline{pq}$. Given a point p and a straight line l in $\mathbb{R}^2$, we denote by $|\overline{pl}|$ the Euclidean length of perpendicular segment $\overline{pl}$ from the point p to the fixed line l, and we denote by l_p the vertical foot from the point p to the line l, i.e. $\overline{pl_p} \perp l$.

Lemma 1. [24] *The Euclidean minimum spanning tree problem can be optimally solved by the Shamos-Hoey algorithm in $O(n \log n)$ time.*

Given a set P of n points in $\mathbb{R}^2$, a (Euclidean) bottleneck spanning tree of P is a spanning tree of P such that the length of the longest edge is minimized. We call the length of the longest edge in a bottleneck spanning tree T of P as the bottleneck of that tree T, denoted by $B(T)$.

Lemma 2. [4] *Each Euclidean minimum spanning tree of P is a bottleneck spanning tree of P.*

Given a set P of n points, an input line l in $\mathbb{R}^2$ and a positive integer k as an instance $(P, l; k)$ of the LcBkStT problem, we denote by T_l^* the shortest optimal solution that is an optimal Steiner tree for the LcBkStT problem with the minimum total length.

Lemma 3. *Each shortest optimal solution T_l^* must have the following properties (1) No two edges, in T_l^*, cross each other; (2) If two edges, in T_l^*, on the same side of the line l meeting at a point form an angle, then this angle is of at least $60°$; (3) If two edges, in T_l^*, on the same side of the line l form an angle of exactly $60°$, then these two edges have the same length.*

Proof. (1) We may suppose, to the contrary, that two edges $\overline{ac}$ and $\overline{bd}$ in T_l^* cross at point o. Note that quadrangle $abcd$ must have an inner angle of at least $90°$. Without loss of generality, we may assume that $\angle abc \geqslant 90°$. Then, we have $|\overline{ab}| < |\overline{ac}|$ and $|\overline{bc}| < |\overline{ac}|$. When an edge $\overline{ac}$ is removed from T_l^*, T_l^* would be broken into two parts, the one containing vertex a and the other containing vertex c, respectively. One of the two parts, which contains the vertex a, must contain vertex b. Adding an edge $\overline{bc}$ results in a shorter tree to be still optimal for the CBkStT problem. This contradicts the fact that T_l^* is the shortest optimal solution. Thus, (1) holds.

(2) We may suppose, to the contrary, that two edges $\overline{ab}$ and $\overline{bc}$ meet at point b and that these two edges form an angle $\angle abc < 60°$. Then, we have $|\overline{ac}| <$ $\max\{|\overline{ab}|, |\overline{bc}|\}$. Using an edge ac with length $|\overline{ac}|$ to substitute for either an edge ab or an edge bc with length $\max\{|\overline{ab}|, |\overline{bc}|\}$, we would reduce the total length of that tree T_l^*, contradicting the length-minimality of T_l^* among optimal solutions. Thus, (2) holds. The statement (3) can be proved by a similar arguments.   □

Without loss of generality, we may treat the input line l as the x-axis in $\mathbb{R}^2$. This line can be parameterized by x-coordinates. Let an interval on l be the set of points on l in between and including two fixed x-coordinates, called the endpoints of the interval. The coordinates of the leftmost point and the rightmost point in P are 0 and r, respectively. Then, the Steiner points we want to determine are in the interval $[0, r]$.

We consider the positive x-axis to be the basis for measuring angles, so that 0 radians is the positive x-axis, $\pi/3$ radians is a counterclockwise rotation of the positive x-axis about the origin by $\pi/3$ radians, and so on. Given two radians θ_1 and θ_2, we denote by $C(p, \theta_1, \theta_2)$ the cone with apex p and limiting angles θ_1 and θ_2. We denote by C_p^i the cone $C(p, (i-1)\pi/3, i\pi/3)$, where p is a point in $[0, r]$ and $i = 1, 2, \ldots, 6$. In other words, the cones C_p^i are obtained by dividing the plane up into six interior-disjoint cones of angle $\pi/3$ all apexed on p.

Lemma 4. *There exists a shortest optimal Steiner tree T_l^* for the LcBkStT problem such that the Steiner points s^* in T_l^* have degree at most six. These neighbors of s^* are located in different cones $C_{s^*}^i$. If the point p_i in $C_{s^*}^i$ is the neighbor of s^* in T_l^*, then p_i is the closest point of P to s^* in $C_{s^*}^i$.*

Proof. By using the property (2) of Lemma 3, if two edges on the same side of this line l meeting at a Steiner point s^* may form angle, then this angle is at least 60°. If there are four or more edges that are incident to s^* on the same side of the line l (see Fig. 1), then we obtain $\alpha_1 + \alpha_2 + \alpha_3 + \alpha_4 + \alpha_5 > 180°$, where $\alpha_2, \alpha_3, \alpha_4 \geq 60°$. This contradicts that a straight angle is 180°. Then, there are at most three edges that are incident to s^* on the same side of the line l. Thus, the Steiner point s^* has degree at most six.

Let point p_i be the neighbor of s^* in $C_{s^*}^i$. We may suppose, to the contrary, that q_i is the closest point of P to s^* in $C_{s^*}^i$, i.e., $|\overline{s^*q_i}| < |\overline{s^*p_i}|$. We have the facts $\angle p_i s^* q_i < 60°$ and $|\overline{p_i q_i}| < |\overline{s^*p_i}|$. When the edge $\overline{s^*p_i}$ is removed from T_l^*, then T_l^* would be broken into two parts. Adding either edge $\overline{s^*p_i}$ or edge $\overline{p_i q_i}$ results in a shorter tree to be still optimal for the LcBkStT problem (see Fig. 2). This contradicts that T_l^* is the shortest optimal solution.   □

Lemma 5. [5] *Given a set P of n points and a straight line l in $\mathbb{R}^2$, using the OVD algorithm [8], the interval $[0, r]$ on the line l can be divided into $O(n)$ interior-disjoint intervals such that each interval I has the property that, for every pair of points $p, q \in I$, the closest point in $C_p^i \cap P$ to p is the same as in $C_q^i \cap P$ to q (see Fig. 2), and it runs in $\Theta(n \log n)$ time using $O(n)$ space.*

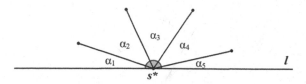

Fig. 1. Four edges that are incident to s^* on the same side of the line l

Given a set P of n points in $\mathbb{R}^2$, we denote by $MST(P)$ any fixed minimum spanning tree of P. Denote $e_1, e_2, \ldots, e_{n-1}$ to be the edges of $MST(P)$ in the order that their lengths are not increasing. Given an instance $(P, l; k)$ of the LcB*k*StT problem, where we treat the input line l as the x-axis in $\mathbb{R}^2$, we denote by P_1 the n_1 points in P to be located on the upper side of l and by P_2 the n_2 points in P to be located on the lower side of l, respectively. We have $n = n_1 + n_2$.

Lemma 6. *For any two minimum spanning trees $MST(P_1)$ on the set P_1 and $MST(P_2)$ on the set P_2, there exists a shortest optimal Steiner tree T_l^* for the LcB*k*StT problem such that each edge of T_l^* either belongs to $MST(P_1) \cup MST(P_2)$ or is incident to a Steiner point s^*.*

Proof. Suppose that there are two points $p, q \in P_1$ such that the edge $\overline{pq}$ belongs to T_l^*, but not to $MST(P_1)$. Let τ be the unique path in $MST(P_1)$ between p and q, and P_1' and P_1'' be the bipartition of P_1 obtained by removing $\overline{pq}$ from T_l^*. Note that τ excludes the edge $\overline{pq}$. Then, there must exist an edge $e \neq \overline{pq}$ on τ such that e connects P_1' and P_1'', i.e., this edge e is between one point in P_1' and another in P_1''. Because $MST(P_1)$ is a minimum spanning tree on the set P_1 and T_l^* is a shortest optimal Steiner tree, we have $|e| = |\overline{pq}|$. Then, we remove the edge $\overline{pq}$ from T_l^* and add e to obtain a new Steiner tree, denote by T_l'. Hence, T_l' is another optimal bottleneck Steiner tree. We repeat the above operations to obtain a shortest optimal Steiner tree with the claimed property after performing the operation a finite number of times. □

Lemma 7. *Let $e_1', e_2', \ldots, e_{n_1-1}'$ be the edges of $MST(P_1)$ in the edge-length non-increasing order and $e_1'', e_2'', \ldots, e_{n_2-1}''$ the edges of $MST(P_2)$ in the edge-length non-increasing order. Then, the optimal Steiner tree T_l^* for the LcB*k*StT problem must satisfy $B(T_l^*) \geqslant \max\{|e_{3k}'|, |e_{3k}''|\}$.*

Proof. We may assume, to the contrary, that $B(T_l^*) < |e_{3k}'|$. There are at least $3k + 1$ connected components when we remove the first $3k$ edges $e_1', e_2', \ldots e_{3k}'$ from $MST(P_1)$. Using Lemma 6, we obtain that there is at least one Steiner point s in T_l^* with degree four. By Lemma 4, this contradicts that there are at most three edges that are incident to s on the same side of the line l. □

Lemma 8. *Given an instance $(P, l; k)$ of the LcB*k*StT problem, there exists a optimal Steiner tree T_l^* such that, for any Steiner point s in T_l^*, the edge $\overline{sp_1}$ either is perpendicular to the line l, i.e., $(\overline{sp_1} \perp l)$, or has the same length as $\overline{sp_2}$, i.e., $|\overline{sp_1}| = |\overline{sp_2}|$, where $\overline{sp_i}$ ($i \leqslant 6$) are the edges in T_l^* that are incident to the Steiner point s and $|\overline{sp_1}| \geqslant |\overline{sp_2}| \geqslant \cdots$.*

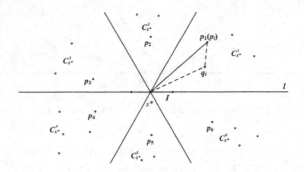

Fig. 2. The plane is divided into six cones of $\pi/3$ radians. The blue points labelled as p_i are the closest point of P to s^* in $C^i_{s^*}$, where $1 \leqslant i \leqslant 6$. If s^* connects to a point in the cone $C^i_{s^*}$, s^* must connect to p_i. The green interval I has the property that this is true everywhere we slide s^* and its cones in I. Every point along the green interval I of l has the same potential neighbors (the blue points) in the same cone. (Color figure online)

Proof. Suppose that T_l be an optimal solution of the LcBkStT problem that the edge $\overline{sp_1}$ in T_l is not perpendicular to the line l and the length of $\overline{sp_1}$ is longer than $\overline{sp_2}$, where $\overline{sp_i}$ ($i \leqslant 6$) are the edges in T_l that are incident to the Steiner point s, satisfying that $|\overline{sp_1}| > |\overline{sp_2}| \geqslant \cdots$. Let l_{p_1} be the vertical foot from the point p_1 to the line l, $i.e.$, $\overline{p_1 l_{p_1}} \perp l$. Now, we slightly move point s towards point l_{p_1} until either $\overline{sp_1} \perp l$ or $|\overline{sp_1}| = |\overline{sp_2}|$ (see Fig. 3). Thus, we can obtain another optimal Steiner tree with the claimed property. $\qquad\square$

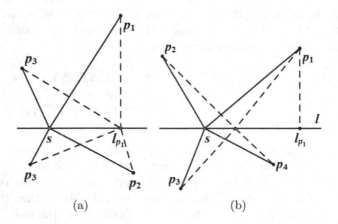

(a)   (b)

Fig. 3. (a) Sliding point s towards point l_{p_1} until the edge $\overline{sp_1}$ is perpendicular to the line l; (b) Sliding point s towards point l_{p_1} until $|\overline{sp_1}| = |\overline{sp_2}|$.

3 The LcB1StT Problem

In this section, we present an exact algorithm for the line-constrained bottleneck 1-Steiner tree (LcB1StT) problem. In details, the LcB1StT problem is to find just one Steiner point s^* on the line l and additionally a spanning tree T_s^* on the set $P \cup \{s^*\}$, the objective is minimize the longest edge in T_s^*, i.e., $\min_{s^*} \max\{|\overline{pq}| \mid e = \overline{pq} \in T_s^*\}$.

Given an instance (P, l) of the LcB1StT problem, where we treat the input line l as the x-axis in $\mathbb{R}^2$, we may choose the $n_1 - 1$ edges of $MST(P_1)$ to satisfy $|e_1'| \geqslant |e_2'| \geqslant \cdots \geqslant |e_{n_1-1}'|$ and the $n_2 - 1$ edges of $MST(P_2)$ to satisfy $|e_1''| \geqslant |e_2''| \geqslant \cdots \geqslant |e_{n_2-1}''|$ (see Lemma 7), respectively. For convenience, we may denote $e_1, \ldots, e_{n-2}$ to be the $n - 2$ edges of $MST(P_1) \cup MST(P_2)$ in the edge-length non-increasing order. We denote by e_b the edge in the order, satisfying $|e_b| > \max\{|e_3'|, |e_3''|\}$ and $|e_{b+1}| \leqslant \max\{|e_3'|, |e_3''|\}$. Then, we have that $b \leqslant 4$.

By using Lemma 7, the optimal Steiner tree T_s^* must have $B(T_s^*) \geqslant |e_{b+1}|$. We can remove at most four longest edges from $MST(P_1) \cup MST(P_2)$. Thus, new edges added are all incident to Steiner point s^* on the line l. Our exact algorithm to solve the LcB1StT problem is described in details as follows.

Algorithm: $\mathcal{A}_1$
Input: a straight line l and a set $P = \{p_1, p_2, \ldots, p_n\}$ of n points in $\mathbb{R}^2$;
Output: a line-constrained bottleneck 1-Steiner tree T_A.
Begin
Step 1 Set $T_A = \emptyset$, $B(T_A) = +\infty$;
Step 2 Use the Shamos-Hoey algorithm [24] to construct two Euclidean minimum spanning trees $MST(P_1)$ and $MST(P_2)$, respectively;
Step 3 Use the OVD algorithm [8] to divide the input line l into $O(n)$ interior-disjoint intervals;
Step 4 **For** each interval I along l **do**:
    For $i = 0$ to b **do**:
        Remove i longest edge(s) from $MST(P_1) \cup MST(P_2)$;
        Find the optimal Steiner point s_i for the interval I_i;
        If $(B(T_{s_i}) < B(T_A))$ **then**
            $T_A = T_{s_i}$, $B(T_A) = B(T_{s_i})$;
Step 5 Output T_A.
End

Theorem 1. *The algorithm $\mathcal{A}_1$ is an exact algorithm for solving the LcB1StT problem, and it runs in $O(n \log n)$ time, where n is the number of points in P.*

Proof. Given an instance (P, l) of the LcB1StT problem, denote by T_s^* the optimal Steiner tree for the LcB1StT problem to satisfy the following

(1) The Steiner point s^* in T_s^* has degree at most six;
(2) The edge $\overline{s^*p_1}$ either is perpendicular to line l ($\overline{s^*p_1} \perp l$) or has the same length as $\overline{s^*p_2}$ ($|\overline{s^*p_1}| = |\overline{s^*p_2}|$), where $\overline{s^*p_i}$ ($i \leqslant 6$) are the edges in T_s^* that are incident to the Steiner point s^* and $|\overline{s^*p_1}| \geqslant |\overline{s^*p_2}| \geqslant \cdots$;

(3) Each edge of T_s^* either belongs to $MST(P_1)$ and $MST(P_2)$ or is incident to the Steiner point s^*.

By Lemma 5, the interval $[0, r]$ on the line l can be divided into $O(n)$ interior-disjoint intervals such that each interval I has the property that, for every pair of points $p, q \in I$, the closest point in $C_p^i \cap P$ to p is the same as in $C_q^i \cap P$ to q. Now, we show how to find the optimal Steiner point s for the interval I.

As shown in Fig. 2, each interval I has the property that, for any point $s \in I$, the six potential neighbors are determined. Consider an interval I and its set of potential neighbors $P_I \subset P$ of size at most six. For each subset P_i of P_I, using Lemma 8, we can compute a constant number of candidate optimal Steiner points in I. In each interval, we solve an optimization problem to find the optimal placement for a Steiner point s_i in that interval, $i.e.$, minimize the longest distances of potential neighbors to the Steiner point, which takes $O(1)$ time since each of these $O(1)$ subproblems has $O(1)$ size.

We execute repeat steps in each interval to look for the solution by finding the optimal placement of a Steiner point in that interval. Once we have computed an optimal placement for a Steiner point in each computed interval, we want to compute one Steiner point, which is the one of these $O(n)$ candidates to produce the minimum bottleneck Steiner tree, $i.e.$, the candidate s_A that produces the smallest length of the $B(T_A)$. Once we determine the candidate optimal Steiner points s_i for the interval I, we need to compare s_i against the current best solution s_A. In other words, for each candidate s_i, we need to compare $B(T_{s_i})$ with $B(T_A)$.

We may assume that the Steiner point s^* in the optimal Steiner tree T_s^* is located in the interval I. Denote by s_i as the candidate optimal Steiner points in I. We can obtain that $B(T_{s_i}) \leqslant B(T_{s^*})$. Because s_A is the optimal Steiner points among all the candidates, we have $B(T_A) \leqslant B(T_{s_i})$. Thus, we have $B(T_A) \leqslant B(T_{s_i}) \leqslant B(T_{s^*})$. Because T_s^* is the optimal Steiner tree for the LcB1StT problem, we can obtain that $B(T_A) = B(T_{s^*})$. Further, the algorithm $\mathcal{A}_1$ is an exact algorithm for the LcB1StT problem.

The complexity of the Algorithm $\mathcal{A}_1$ can be determined in the following ways. (1) By Lemma 1, the Euclidean minimum spanning trees $MST(P_1)$ and $MST(P_2)$ can be computed by Shamos-Hoey algorithm [24] in $O(n \log n)$ time. (2) By Lemma 5, the OVD algorithm [8] runs in $\Theta(n \log n)$ time using $O(n)$ space. (3) The interval $[0, r]$ on the line l can be divided into $O(n)$ interior-disjoint intervals. For each interval, we find the optimal placement for a Steiner point s_i which takes $O(1)$ time. Thus, the whole algorithm $\mathcal{A}_1$ can be implemented in $O(n \log n)$ time. $\qquad\square$

4   The LcBkStT Problem

In this section, we extend the arguments for a single Steiner point to multiple k (> 1) Steiner points for the line-constrained bottleneck k-Steiner tree problem.

As similar arguments in an instance (P, l) of the LcB1StT problem, we may choose $e_1', \ldots, e_{n_1-1}'$ to be the $n_1 - 1$ edges of $MST(P_1)$ and $e_1'', \ldots, e_{n_2-1}''$ to be the $n_2 - 1$ edges of $MST(P_2)$, in the edge-length non-increasing order. Denote by $e_1, e_2, \ldots, e_{n-2}$ as the $n - 2$ edges of $MST(P_1) \cup MST(P_2)$ in the edge-length non-increasing order. We denote by e_d the edge in the order, satisfying $|e_d| > \max\{|e'_{3k}|, |e''_{3k}|\}$ and $|e_{d+1}| \leqslant \max\{|e'_{3k}|, |e''_{3k}|\}$. We obtain the fact $d \leqslant 6k - 2$. By using Lemma 7, the optimal Steiner tree T^*_l must have $B(T^*_l) \geqslant |e_{d+1}|$.

We can use the idea of the single Steiner point location to solve multiple Steiner points of the LcBkStT problem. For a fixed integer c, where $1 \leqslant c \leqslant d$, we remove $e_1, e_2, \ldots, e_c$ from $MST(P_1) \cup MST(P_2)$ to obtain some subtrees $T^c_1, T^c_2, \cdots$. We then find at most k Steiner points on the line l to reconnect the T^c_i with new edges incident to the Steiner points, minimizing the longest edge length. Finally, we choose the best solution among all integers c as an optimal solution for the original problem.

This motivates another variation of the LcBkStT problem, called the bottleneck Steiner tree problem with a fixed topology on subtrees. As shown in earlier works [12,22], the term "topology" is used to denote a tree whose vertices are labeled by a single or multiple terminals in P or symbolically by each Steiner point without its location. Then, a topology can be seen as a combinatorial structure that consists of information only about how the terminals and Steiner points to be connected, which can be obtained from a Steiner tree. The topology $\mathcal{T}_p$ appears in this problem consists of a set of vertices $V = \{s_1, \ldots, s_k, t_1, t_2, \ldots\}$, where t_i represents subtree T^c_i consisting of terminals in P and s_j represents a Steiner point. Thus, the strategy for the LcBkStT problem is summarized in the following

(1) Use the Shamos-Hoey algorithm [24] to construct two Euclidean minimum spanning trees $MST(P_1)$ and $MST(P_2)$, respectively.
(2) Use the OVD algorithm [8] to divide the input line l into $O(n)$ interior-disjoint intervals.
(3) Enumerate all possible ways to put k Steiner points into $O(n)$ different intervals; There are at most $O(n^k)$ possible to locate this k Steiner points.
(4) Do a binary search on the collected critical values $\{|e_1|, \ldots, |e_d|\}$ to find the optimal value e_c; Remove edges $e_1, e_2, \ldots, e_c$ from $MST(P_1) \cup MST(P_2)$ to obtain some subtrees $T^c_1, T^c_2, \cdots$.
(5) Enumerate all possible (combinatorially distinct) topology trees. For a fixed integer c, we have at most $k^{\sum_{i=0}^{6} C_6^i} (= k^{O(1)})$ possible topology trees.
(6) For a fixed topology $\mathcal{T}_p$ on all subtrees T^c_i and k Steiner points as vertices, using Lemma 8, we can find an optimal location of k Steiner points in $O(k)$ time.

Theorem 2. *Given an instance $(P, l; k)$ of the LcBkStT problem, an optimal line-constrained bottleneck k-Steiner tree can be produced in $O(n \log n + f(k) \cdot n^k)$ time, where $f(k) = k^{O(1)}$.*

Proof. From the aforementioned strategy, we can find an optimal locations of k Steiner points and then an optimal line-constrained bottleneck k-Steiner tree, as shown in the theorem. We divide the input line l into $O(n)$ intervals, and then enumerate all possible ways to put k Steiner points into $O(n)$ different intervals. For fixed c with $1 \leq c \leq d$, we remove $e_1, e_2, \ldots, e_c$ from $MST(P_1) \cup MST(P_2)$ to obtain some subtrees $T_1^c, T_2^c, \cdots$. We enumerate all possible topology trees T_p that consist of a set of vertices $V = \{s_1, \ldots, s_k, t_1, t_2, \ldots\}$, where t_i represents subtree T_i^c consisting of terminals in P and s_j represents a Steiner point. We solve an optimization problem to find the optimal placement for the Steiner points with a fixed topology T_p, *i.e.*, minimize the longest distances of potential neighbors to the Steiner points. Enumerating all such topologies gives us an exact solution to the LcBkStT problem.

The complexity of the algorithm for the LcBkStT problem can be determined as follows. (1) By Lemma 1, the Euclidean minimum spanning trees $MST(P_1)$ and $MST(P_2)$ can be computed by Shamos-Hoey algorithm [24] in $O(n \log n)$ time. (2) By using Lemma 5, the OVD algorithm [8] runs in $\Theta(n \log n)$ time using $O(n)$ space. There are at most $n^k \cdot \log(6k - 2) \cdot k^{\sum_{i=0}^{6} C_6^i}$ possible topology trees. For each fixed topology tree, we can find an optimal location of k Steiner points in $O(k)$ time. Let $f(k) = \log(6k - 2) \cdot k^{\sum_{i=0}^{6} C_6^i} \cdot O(k) = k^{O(1)}$. Thus, the whole algorithm can be implemented in $O(n \log n + f(k) \cdot n^k)$ time. $\qquad \square$

5 Conclusion

In this paper, we consider the line-constrained bottleneck k-Steiner tree problem. It is asked to find at most k Steiner points on the line l such that the length of the longest edge in the spanning tree on these $n + k$ points is minimized.

(1) We design an exact algorithm for the LcB1StT problem, and this algorithm runs in $O(n \log n)$ time.
(2) We present an exact algorithm for the LcBkStT problem, and it runs in $O(n \log n + f(k) \cdot n^k)$ time, where $f(k)$ is a function dependent only on k.

For our further research, a challenging task is to solve the LcBkStT problem with lower running time.

References

1. Aazami, A., Cheriyan, J., Jampani, K.R.: Approximation algorithms and hardness results for packing element-disjoint Steiner trees in planar graphs. Algorithmica **63**(1–2), 425–456 (2012)
2. Arora, S.: Polynomial time approximation schemes for Euclidean traveling salesman and other geometric problems. J. ACM **45**(5), 753–782 (1998)
3. Bae, S.W., Choi, S., Lee, C., Tanigawa, S.: Exact algorithms for the bottleneck Steiner tree problem. Algorithmica **61**(4), 924–948 (2011)
4. Bae, S.W., Lee, C., Choi, S.: On exact solutions to the Euclidean bottleneck Steiner tree problem. Inform. Process. Lett. **110**(16), 672–678 (2010)
5. Bose, P., D'Angelo, A., Durocher, S.: On the restricted k-Steiner tree problem. J. Comb. Optim. **44**(4), 2893–2918 (2022)
6. Byrka, J., Grandoni, F., Rothvoß, T., Sanità, L.: Steiner tree approximation via iterative randomized rounding. J. ACM **60**(1), 1–33 (2013)

7. Cardei, I., Cardei, M., Wang, L., Xu, B., Du, D.: Optimal relay location for resource-limited energy-efficient wireless communication. J. Global Optim. **36**(3), 391–399 (2006)
8. Chang, M.S., Huang, N.F., Tang, C.Y.: An optimal algorithm for constructing oriented Voronoi diagrams and geographic neighborhood graphs. Inform. Process. Lett. **35**(5), 255–260 (1990)
9. Chazelle, B.: A minimum spanning tree algorithm with inverse-Ackermann type complexity. J. ACM **47**(6), 1028–1047 (2000)
10. Chen, G., Zhang, G.: A constrained minimum spanning tree problem. Comput. Oper. Res. **27**(9), 867–875 (2000)
11. Chiang, C., Sarrafzadeh, M., Wong, C.K.: A powerful global router: based on Steiner min-max trees. In: Proceedings of the IEEE International Conference on Computer-Aided Design, pp. 2–5. Santa Clara (1989)
12. Ganley, J.L., Salowe, J.S.: Optimal and approximate bottleneck Steiner trees. Oper. Res. Lett. **19**(5), 217–224 (1996)
13. Garey, M.R., Graham, R.L., Johnson, D.S.: The complexity of computing Steiner minimal trees. SIAM J. Appl. Math. **32**(4), 835–859 (1977)
14. Georgakopoulos, G., Papadimitriou, C.H.: The 1-Steiner tree problem. J. Algorithms **8**(1), 122–130 (1987)
15. Holby, J.: Variations on the Euclidean Steiner tree problem and algorithms. Rose-Hulman Undergraduate Math. J. **18**(1), 123–155 (2017)
16. Korte, B., Vygen, J.: Combinatorial Optimization: Theory and Algorithms, 3rd edn. Springer-Verlag, Berlin (2008)
17. Li, J.P., Liu, S.D., Lichen, J.R., Wang, W.C., Zheng, Y.J.: Approximation algorithms for solving the 1-line Euclidean minimum Steiner tree problem. J. Comb. Optim. **39**(2), 492–508 (2020)
18. Li, Z., Xiao, W.: Determining sensor locations in wireless sensor networks. Int. J. Distrib. Sens. Netw. **11**(8), 914625 (2015)
19. Mitchell, J.S.B.: Guillotine subdivisions approximate polygonal subdivisions: a simple polynomial-time approximation scheme for geometric TSP, k-MST, and related problems. SIAM J. Comput. **28**(4), 1298–1309 (1999)
20. Rao, S.B., Smith, W.D.: Approximating geometrical graphs via "panners" and "banyans". In: Proceedings of the Thirtieth Annual ACM Symposium on Theory of Computing, pp. 540–550. New York (1999)
21. Robins, G., Zelikovsky, A.: Tighter bounds for graph Steiner tree approximation. SIAM J. Discrete Math. **19**(1), 122–134 (2005)
22. Sarrafzadeh, M., Wong, C.K.: Bottleneck Steiner trees in the plane. IEEE Trans. Comput. **41**(3), 370–374 (1992)
23. Schrijver, A.: Combinatorial Optimization: Polyhedra and Efficiency, 1st edn. Springer-Verlag, Berlin (2003)
24. Shamos, M.I., Hoey, D.: Closest-point problems. In: 16th Annual Symposium on Foundations of Computer Science, pp. 151–162. IEEE Computer Society, Long Beach, Calif (1975)
25. Wang, L.S., Du, D.Z.: Approximations for a bottleneck Steiner tree problem. Algorithmica **32**(4), 554–561 (2002)
26. Wang, L.S., Li, Z.M.: An approximation algorithm for a bottleneck k-Steiner tree problem in the Euclidean plane. Inform. Process. Lett. **81**(3), 151–156 (2002)
27. Williamson, D.P., Shmoys, D.B.: The Design of Approximation Algorithms, 1st edn. Cambridge University Press, Cambridge (2011)

The Longest Subsequence-Repeated Subsequence Problem

Manuel Lafond[1] ⓘ, Wenfeng Lai[2], Adiesha Liyanage[3], and Binhai Zhu[3](✉) ⓘ

[1] Department of Computer Science, Université de Sherbrooke,
Quebec J1K 2R1, Canada
`manuel.lafond@usherbrooke.ca`
[2] College of Computer Science and Technology, Shandong University, Qingdao, China
[3] Gianforte School of Computing, Montana State University, Bozeman, MT 59717,
USA
`bhz@montana.edu`

Abstract. Motivated by computing duplication patterns in sequences, a new fundamental problem called the longest subsequence-repeated subsequence (LSRS) is proposed. Given a sequence S of length n, a letter-repeated subsequence is a subsequence of S in the form of $x_1^{d_1} x_2^{d_2} \cdots x_k^{d_k}$ with x_i a subsequence of S, $x_j \neq x_{j+1}$ and $d_i \geq 2$ for all i in $[k]$ and j in $[k-1]$. We first present an $O(n^6)$ time algorithm to compute the longest cubic subsequences of all the $O(n^2)$ substrings of S, improving the trivial $O(n^7)$ bound. Then, an $O(n^6)$ time algorithm for computing the longest subsequence-repeated subsequence (LSRS) of S is obtained. Finally we focus on two variants of this problem. We first consider the constrained version when Σ is unbounded, each letter appears in S at most d times and all the letters in Σ must appear in the solution. We show that the problem is NP-hard for $d = 4$, via a reduction from a special version of SAT (which is obtained from 3-COLORING). We then show that when each letter appears in S at most $d = 3$ times, then the problem is solvable in $O(n^4)$ time.

Keywords: Tandem duplications · Longest common subsequence · Longest letter-duplicated subsequence · NP-completeness · Dynamic programming

1 Introduction

Finding patterns in long sequences is a fundamental problem in string algorithms, combinatorial pattern matching and computational biology. In this paper we are interested in long patterns occurring at a global level, which has also been considered previously. One prominent example is to compute the longest square subsequence of a string S of length n, which was solved by Kosowski in $O(n^2)$ time in 2004 [6]. The bound is conditionally optimal as any $o(n^{2-\varepsilon})$ solution would lead to a subquadratic bound for the traditional Longest Common Subsequence (LCS) problem, which is not possible unless the SETH conjecture fails

[2]. Nonetheless, a slight improvement was presented by Tiskin [18]; and Inoue et al. recently tried to solve the problem by introducing the parameter M (which is the number of matched pairs in S) and r (which is the length of the solution) [5].

In biology, it was found by Szostak and Wu as early as in 1980 that gene duplication is the driving force of evolution [17]. There are two kinds of duplications: arbitrary segmental duplications (i.e., an arbitrary segment is selected and pasted at somewhere else) and tandem duplications (i.e., in the form of $X \rightarrow XX$, where X is any segment of the input sequence). It is known that the former duplications occur frequently in cancer genomes [4,13,16]. On the other hand, the latter are common under different scenarios; for example, it is known that the tandem duplication of 3 nucleotides CAG is closely related to the Huntington disease [12]. In addition, tandem duplications can occur at the genome level (acrossing different genes) for certain types of cancer [14].

As duplication is common in biology, it was not a surprise that in the first sequenced human genome around 3% of the genetic contents are in the form of tandem repeats [10]. In 2004, Leupold et al. posed a fundamental question regarding tandem duplications: what is the complexity to compute the minimum tandem duplication distance between two sequences A and B (i.e., the minimum number of tandem duplications to convert A to B). In 2020, Lafond et al. answered this open question by proving that this problem is NP-hard for an unbounded alphabet [7]. Later in [8], Lafond et al. proved that the problem is NP-hard even if $|\Sigma| \geq 4$ by encoding each letter in the unbounded alphabet proof with a square-free string over a new alphabet of size 4 (modified from Leech's construction [11]), which covers the case most relevant with biology, i.e., when $\Sigma = \{A, C, G, T\}$ or $\Sigma = \{A, C, G, U\}$ [8]. Independently, Cicalese and Pilati showed that the problem is NP-hard for $|\Sigma| = 5$ using a different encoding method [3].

Besides duplication, another driving force in evolution is certainly mutation. As a simple example, suppose we have a toy singleton genome ACGT (note that a real genome certainly would have a much larger alphabet) and it evolves through two tandem duplications ACGT·ACGT·ACGT then another one on the second GTA to have $H =$ ACGT·AC·GTA·GTA·CGT. If in H some mutation occurs, e.g., the first G is deleted and the second G is changed to T to have $H' =$ ACT·AC·TTA·GTA·CGT, then it is difficult to retrieve the tandem duplications from H'. Motivated by the above applications, Lai et al. [9] recently proposed the following problem called the *Longest Letter-Duplicated Subsequence*: Given a sequence S of length n, compute a longest letter-duplicated subsequence (LLDS) of S, i.e., a subsequence of S in the form $x_1^{d_1} x_2^{d_2} \cdots x_k^{d_k}$ with $x_i \in \Sigma$, where $x_j \neq x_{j+1}$ and $d_i \geq 2$ for all i in $[k]$, j in $[k-1]$ and $\sum_{i \in [k]} d_i$ is maximized. A simple linear time algorithm can be obtained to solve LLDS. But some constrained variation, i.e., all letters in Σ must appear in the solution, is shown to be NP-hard.

In this paper, we extend the work by Lai et al. by looking at a more general version of LLDS, namely, the Longest Subsequence-repeated Subsequence (LSRS) problem of S, which follows very much the same definition as above except that each x_i is a subsequence of S (instead of a letter). As a comparison, for the sequence H', one of the optimal LLDS solutions is AATTTT $= A^2 T^4$ while the LSRS solution is ACAC·TAGTAG = $(AC)^2(TAG)^2$ which clearly gives more information about the duplication histories. This motivates us studying LSRS and related problems in this paper. Let d be the maximum occurrence of any letter in the input string S, with $|S| = n$. Let $LSDS+(d)$ be the constrained version that all letters in Σ must appear in the solution, and the maximum occurrence of any letter in S is at most d. We summarize the results of this paper as follows.

1. We show that the longest cubic subsequences of all substrings of S can be solved in $O(n^6)$ time, improving the trivial $O(n^7)$ bound.
2. We show that LSRS can be solved in $O(n^6)$ time.
3. When $d \geq 4$, $LSRS+(d)$ is NP-complete.
4. When $d = 3$, $LSRS+(3)$ can be solved in $O(n^4)$ time.

Note that the parameter d, i.e., the maximum duplication number, is practically meaningful in bioinformatics, since whole genome duplication is a rare event in many genomes and the number of duplicates is usually small. For example, it is known that plants have undergone up to three rounds of whole genome duplications, resulting in a number of duplicates bounded by 8 [20].

It should also be noted that our LSRS and LSRS+ problems seem to be related to the recently studied problems Longest Run Subsequence (LRS) [15], which is NP-hard; and Longest (Sub-)Periodic Subsequence [1], which is polynomially solvable. But these two problems are different from our LSRS and LSRS+ problems. For instance, in an LRS solution a letter can appear in at most one run while in our LSRS and LSRS+ solutions, say ACAC · TAGTAG for the input string H', a substring (e.g., AC) can appear many times, hence a letter (e.g., A) could appear many times but non-consecutively in LSRS and LSRS+ solutions. On the other hand, in the Longest (Sub-)Periodic Subsequence problem one is very much only looking for the repetition of a single subsequence of the input string, while obviously in our LSRS and LSRS+ problems we need to find the repetitions of multiple subsequences of the input string (e.g., AC and TAG).

This paper is organized as follows. In Sect. 2 we give necessary definitions. In Sect. 3 we give an $O(n^6)$ time algorithm for computing the longest cubic subsequences of all substrings of S, as well as the solution for LSRS. In Sect. 4 we prove that $LSRS+(4)$ is NP-hard and then we show that $LSRS+(3)$ can be solved in polynomial time. We conclude the paper in Sect. 5.

2   Preliminaries

Let $\mathbb{N}$ be the set of natural numbers. For $q \in \mathbb{N}$, we use $[q]$ to represent the set $\{1, 2, ..., q\}$ and we define $[i, j] = \{i, i+1, ..., j\}$. Throughout this paper, a sequence S is over a finite alphabet Σ. We use $S[i]$ to denote the i-th letter in S and $S[i..j]$ to denote the substring of S starting and ending with indices i and j respectively. (Sometimes we also use $(S[i], S[j])$ as an interval representing the substring $S[i..j]$.) With the standard run-length representation, S can be represented as $y_1^{a_1} y_2^{a_2} \cdots y_q^{a_q}$, with $y_i \in \Sigma, y_j \neq y_{j+1}$ and $a_i \geq 1$, for $i \in [q], j \in [q-1]$. Finally, a *subsequence* of S is a string obtained by deleting some letters in S. Specifically, a *square subsequence* of S is a subsequence of S in the form of X^2, where X is also a subsequence of S; and a *cubic subsequence* of S is a subsequence of S in the form of X^3, where X is a also subsequence of S. One is certainly interested in the longest ones in both cases.

A subsequence S' of S is a subsequence-repeated subsequence (SRS) of S if it is in the form of $x_1^{d_1} x_2^{d_2} \cdots x_k^{d_k}$, with x_i being a subsequence of S, $x_j \neq x_{j+1}$ and $d_i \geq 2$, for $i \in [k], j \in [k-1]$. We call each $x_i^{d_i}$ in S' a *subsequence-repeated block* (SR-block, for short). For instance, let $S = \texttt{ACGAGCGCAGCGA}$, then $S_1 = \texttt{AGAG} \cdot \texttt{CGCGCG}$, $S_2 = \texttt{ACGACG} \cdot \texttt{CGCG}$ and $S_3 = \texttt{ACGACG} \cdot \texttt{CACA}$ are multiple solutions for the longest subsequence-duplicated subsequence of S, where any maximal substring in S_i separated by $\cdot$ forms a SR-block. As a separate note, given this S, the longest square subsequence is $\texttt{CAGCG} \cdot \texttt{CAGCG} = (\texttt{CAGCG})^2$ and the longest cubic subsequence is $\texttt{CGACGACGA} = (\texttt{CGA})^3$.

3 A Polynomial-Time Solution for LSRS

In this section we proceed to solve the LSRS problem. Firstly, as a subroutine, we need to compute the longest cubic subsequences of all $O(n^2)$ substrings of S in $O(n^6)$ time. Assuming that is the case, we have a way to solve LSRS as follows.

3.1 Solution for the LSRS Problem

With Kosowski's quadratic solution for the longest square subsequence (even though we could achieve our goal without using it, see Sect. 3.2) and our $O(n^6)$ time solution for the longest cubic subsequence (details to be given in Sect. 3.2), we solve the LSRS problem by dynamic programming. We first have the following observation.

Observation 1. *Suppose that there is an optimal LSRS solution for a given sequence S of length n, in the form of $x_1^{d_1} x_2^{d_2} \ldots x_k^{d_k}$. Then it is possible to decompose it into a generalized SR-subsequence in the form of $y_1^{e_1} y_2^{e_2} \ldots y_p^{e_p}$, where*

- *$2 \leq e_i \leq 3$, for $i \in [p]$,*
- *$p \geq k$,*
- *y_j does not have to be different from y_{j+1}, for $j \in [p-1]$.*

The proof is straightforward: For any natural number $\ell \geq 2$, we can decompose it as $\ell = \ell_1 + \ell_2 + \ldots + \ell_z \geq 2$, such that $2 \leq \ell_j \leq 3$ for $1 \leq j \leq z$. Consequently, for every $d_i > 3$, we could decompose it into a sum of 2's and 3's. Then, clearly, given a generalized SR-subsequence, we could easily obtain the corresponding SR-subsequence by combining $y_i^{e_i} y_{i+1}^{e_{i+1}}$ when $y_i = y_{i+1}$.

We now design a dynamic programming algorithm for LSRS. Let $L(i)$ be the length of the optimal LSRS solution for $S[1..i]$. Let $Q2[i,j]$ and $Q3[i,j]$ store the longest square and cubic subsequences of $S[i..j]$ respectively. The recurrence for $L(i)$ is as follows.

$$L(0) = 0,$$
$$L(1) = 0,$$
$$L(i) = \max \begin{cases} L(j) + Q2[j+1, i], & j < i-1 \\ L(j) + Q3[j+1, i], & j < i-2 \end{cases}$$

Computing all the cells $Q2[j, k]$ takes $O(n^4)$ time as there are $O(n^2)$ cells and each can be computed using Kosowski's algorithm in quadratic time. (As we will show right after Theorem 1, the $O(n^4)$ time bound can also be obtained without using Kosowski's algorithm.) Computing all $Q3[j, k]$ takes $O(n^7)$ time: there are $O(n^2)$ cells, each can be computed in $O(n^5)$ time using the only known brute-force solution. However, in the next subsection we show that the longest cubic subsequences of all substrings of S, i.e., all $Q3[j, k]$ can be computed in $O(n^6)$ time. Therefore, after $Q2[-, -]$ and $Q3[-, -]$ are all computed, it takes $O(n^2)$ time to update and fill the whole table $L(-)$. The value of the optimal LSRS solution for S can be found in $L(n)$. Consequently, we have a running time of $O(n^6)$. To make the solution complete, we next show the algorithm for computing the longest cubic subsequences of all substrings of S.

3.2   An $O(n^6)$ Time Bound for the Longest Cubic Subsequences of All Substrings of the Input String

First of all, notice that an $O(n^5)$ time brute-force solution for the longest cubic subsequence problem is trivial: just enumerate in $O(n^2)$ time all the cuts cutting S into three substrings, and then compute the longest common subsequence over this triple of substrings in $O(n^3)$ time. The longest of all would give us the solution. Then, to compute all $Q3[j, k]$ it takes $O(n^7)$ time since there are $O(n^2)$ cells. To improve this bound, a different idea is needed.

The idea is that when one computes the longest common subsequence of three sequences A, B and C, one would use dynamic programming to compute, for each triple of i, j, k, the longest common subsequence of $A[1..i], B[1..j]$ and $C[1..k]$. When i, j are fixed this dynamic programming algorithm can in fact compute the longest common subsequences of $A[1..i], B[1..j]$ and all $C[1..k']$, with $1 \leq k' \leq k$. Therefore, by enumerating i and j, in $O(n^2 \cdot n^3) = O(n^5)$ time, we can compute all longest cubic subsequences of a prefix $A \cdot B \cdot C$ of S. To compute the longest cubic subsequences of all substrings of S, it suffices to run the above algorithm on every suffix of S. Hence, in $O(n) \cdot O(n^5) = O(n^6)$ time we can compute the longest cubic subsequences of all substrings of S.

Theorem 1. *The longest cubic subsequences of all substrings of an input string of length n can be computed in $O(n^6)$ time and $O(n^3)$ space.*

Note that we can use this idea to compute the longest square subsequences for all substrings of the input string S in $O(n^4)$ time, without using Kosowski's algorithm at all. In this case, using the standard dynamic programming for computing the longest common subsequence of A and B, we compute all longest square subsequences of a prefix $A \cdot B$ of the input sequence S in $O(n^3)$ time. Then we run this algorithm on all suffix of S, giving a total running time of $O(n^4)$. In this process, there is no need to use Kosowski's algorithm.

Finally, together with the algorithm in Sect. 3.1, we have the following theorem.

Theorem 2. *The longest subsequence-repeated subsequence problem can be solved in $O(n^6)$ time.*

4   The Variants of LSRS

In this section, we focus on the following variations of the LSRS problem.

Definition 1. Constrained Longest Subsequence-Repeated Subsequence *(LSRS+ for short)*
   Input: *A sequence S with length n over an alphabet Σ and an integer ℓ.*
   Question: *Does S contain a subsequence-repeated subsequence S' with length at least ℓ such that all letters in Σ appear in S'?*

Definition 2. Feasibility Testing *(FT for short)*
   Input: *A sequence S with length n over an alphabet Σ.*
   Question: *Does S contain a subsequence-repeated subsequence S'' such that all letters in Σ appear in S''?*

For LSRS+ we are really interested in the optimization version, i.e., to maximize ℓ. Note that, though looking similar, FT and the decision version of LSRS+ are different: if there is no feasible solution for FT, certainly there is no solution for LSRS+; but even if there is a feasible solution for FT, computing an optimal solution for LSRS+ could still be non-trivial.

Finally, let d be the maximum number of times a letter in Σ appears in S. Then, we can represent the corresponding versions for LSRS+ and FT as $LSRS+(d)$ and $FT(d)$ respectively.

It turns out that (the decision version of) $LSRS+(d)$ and $FT(d)$ are both NP-complete when $d \geq 4$, while when $d = 3$ both $LSRS+(3)$ and $FT(3)$ can be solved in $O(n^4)$ time. We present the details below.

4.1   LSRS+(4) is NP-Hard

We first show that $(3^+, 1, 2^-)$-SAT is NP-complete; in this version of SAT all variables appear positively in 3-CNF clauses (i.e., clauses containing exactly 3 positive literals) and each variable appears exactly once in total in these 3-CNF clauses; moreover, the negation of the variables appear in 2-CNF clauses (i.e., clauses containing 2 negative literals), possibly many times. A *valid* truth assignment for an $(3^+, 1, 2^-)$-SAT instance ϕ is one which makes ϕ true; moreover, each 3-CNF clause has exactly one true literal.

A folklore reduction was discussed in the internet at some point; here we give a formal sketch of the proof.

Theorem 3. $(3^+, 1, 2^-)$-*SAT is NP-complete.*

Proof. As the problem is easily seen to be in NP, let us focus more on the reduction from 3-COLORING. In 3-COLORING, given a graph $G = (V, E)$, one needs to assign one of the 3 colors to each of the vertex $u \in V$ such that for any edge $(u, v) \in E$, u and v are given different colors.

For each vertex u, we use u_1, u_2 and u_3 to denote the 3 colors, then, obviously, we have the 3-CNF clause $(u_1 \vee u_2 \vee u_3)$. Therefore, the positive 3-CNF formulae are

$$C^+ = \bigwedge_{u \in V} (u_1 \vee u_2 \vee u_3).$$

We have 2 kinds of 2-CNF clauses. First, for each $u \in V$, we have a type-1 2-CNF clause which demands that one cannot select two colors i and j for u at the same time:

$$\overline{u_i \wedge u_j} = (\bar{u}_i \vee \bar{u}_j),$$

for $1 \leq i \neq j \leq 3$. Then, for each edge $(u, v) \in E$, we have a type-2 2-CNF clause which demands that u and v cannot have the same color i:

$$\overline{u_i \wedge v_i} = (\bar{u}_i \vee \bar{v}_i),$$

for $i = 1, 2, 3$.

Let C^- be the conjunction of these 2-CNF clauses. Then $\phi = C^+ \wedge C^-$, and it is clear that G has a 3-coloring if and only if ϕ has a valid truth assignment. The reduction obviously takes linear time. Hence the theorem is proven. □

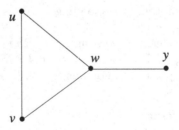

Fig. 1. An illustration of the proof of Theorem 3. In this case, $C^+ = (u_1 \vee u_2 \vee u_3) \wedge (v_1 \vee v_2 \vee v_3) \wedge (w_1 \vee w_2 \vee w_3) \wedge (y_1 \vee y_2 \vee y_3)$.

In Fig. 1, we show an example for the proof of Theorem 3. The example will be used in the following paragraphs.

We next reduce $(3^+, 1, 2^-)$-SAT to FT(4). Let the input ϕ for $(3^+, 1, 2^-)$-SAT be constructed directly from a 3-COLORING instance; moreover, let ϕ have $3n$ variables $x_1, x_2, \cdots, x_{3n}$ and m 2-CNF clauses. We label its 3-CNF clauses as $F_1^+, F_2^+, \cdots, F_n^+$ and its 2-CNF clauses as $F_1^-, F_2^-, \cdots, F_m^-$. (For the example in Fig. 1, we can take as $u_1, u_2, u_3, \cdots, y_1, y_2, y_3$ alphabetically as $x_1, x_2, \cdots, x_{12}$.)

In the example in Fig. 1, the type-1 2-CNF clauses on u are

$$(\bar{u}_1 \vee \bar{u}_2) \wedge (\bar{u}_1 \vee \bar{u}_3) \wedge (\bar{u}_2 \vee \bar{u}_3) = F_1^- \wedge F_2^- \wedge F_3^-,$$

the type-2 2-CNF clauses on edge (u, v) are

$$(\bar{u}_1 \vee \bar{v}_1) \wedge (\bar{u}_2 \vee \bar{v}_2) \wedge (\bar{u}_3 \vee \bar{v}_3) = F_4^- \wedge F_5^- \wedge F_6^-,$$

and the type-2 clauses on edge (u, w) are

$$(\bar{u}_1 \vee \bar{w}_1) \wedge (\bar{u}_2 \vee \bar{w}_2) \wedge (\bar{u}_3 \vee \bar{w}_3) = F_7^- \wedge F_8^- \wedge F_9^-.$$

For each variable x_i, let $L(x_i)$ be the list of type-1 2-CNF clauses containing $\bar{x}_i$ followed with the list of type-2 2-CNF clauses containing $\bar{x}_i$, each repeating twice consecutively. (For the example in Fig. 1, $L(x_1) = L(u_1) = F_1^- F_1^- F_2^- F_2^- \cdot F_4^- F_4^- F_7^- F_7^-$.) For each F_i^+ we also define three unique letters $1_i, 2_i$ and 3_i. Hence, the alphabet we use to construct the final sequence H is

$$\Sigma = \{F_j^- | j \in [m]\} \cup \{1_i, 2_i, 3_i | i \in [n]\} \cup \{g_k, g_k' | k \in [n-1]\},$$

where g_k and g_k' are used as separators.

Let $F_i^+ = (x_{i,1} \vee x_{i,2} \vee x_{i,3})$. We construct

$$H_i = 2_i \cdot L(x_{i,1}) \cdot 1_i 2_i 1_i \cdot L(x_{i,2}) \cdot 2_i \cdot L(x_{i,3}) \cdot 3_i 2_i 3_i,$$

where 2_i and all the 2-CNF clauses each appear 4 times in H_i, while 1_i and 3_i each appears twice. Finally we construct a sequence H as

$$H = H_1 \cdot g_1 g_1' g_1 g_1' \cdot H_2 \cdot g_2 g_2' g_2 g_2' \cdot H_3 \cdots g_{n-1} g_{n-1}' g_{n-1} g_{n-1}' \cdot H_n,$$

where g_j and g_j' each appears twice. We claim that ϕ has a valid truth assignment if and only if H induces a feasible SRS which contains all $1_i, 2_i, 3_i$ ($i = 1..n$), all F_j^- ($j = 1..m$) and all $g_k g_k' g_k g_k'$ ($k = 1..n-1$).

The forward direction, i.e., when ϕ has a valid truth assignment, is straightforward. In this case, suppose exactly one of $x_{i,1}$, $x_{i,2}$ and $x_{i,3}$ (say $x_{i,j}$, $1 \leq j \leq 3$) is assigned TRUE, then we delete $L(x_{i,j})$ for $j = 1, 2$ or 3 in H_i. Finally we delete some $1_i, 2_i$ and 3_i to obtain a feasible solution H_i' as follows.

1. If $j = 1$, we have $H_i' = 2_i 1_i 2_i 1_i \cdot L(x_{i,2}) \cdot L(x_{i,3}) \cdot 3_i 3_i$.
2. If $j = 2$, we have $H_i' = L(x_{i,1}) \cdot 1_i 2_i 1_i 2_i \cdot L(x_{i,3}) \cdot 3_i 3_i$.
3. If $j = 3$, we have $H_i' = L(x_{i,1}) \cdot 1_i 1_i \cdot L(x_{i,2}) \cdot 2_i 3_i 2_i 3_i$.

It is noted that exactly one $L(x_{i,j})$ is deleted in all three cases. (We focus on $j = 1$ next.) Hence, the deleted letters (2CNF clauses in the form of F_k^-) in $L(x_{i,1})$

would still appear in the claimed feasible solution, even after the deletion of $L(x_{i,1})$. For example, if F_k^- is type-1 which contains $\bar{x}_{i,1}$ and $\bar{x}_{i,2}$, then $F_k^- F_k^-$ must appear in $L(x_{i,2})$, which is not deleted. Similarly, if F_k^- is type-2 which contains $\bar{x}_{i,1}$ and $\bar{x}_{\ell,1}$, where (x_i, x_ℓ) is an edge in the graph G, then $F_k^- F_k^-$ appears in $L(x_{\ell,1})$ which must be kept — if $L(x_{\ell,1})$ were also deleted, it would imply that both x_i and x_ℓ are colored with color-1. Therefore, all the 2CNF clauses appear in the claimed feasible solution.

The reverse direction is slightly more tricky. We first show the following lemma.

Lemma 1. *If H admits a feasible (SRS) solution, then exactly two (non-empty subsequences) of $L(x_{i,1}), L(x_{i,2})$ and $L(x_{i,3})$ appear in the feasible solution H' (or, exactly one of the three is completely deleted from H).*

Proof. See full version.   □

Let $L'(x_{i,j})$ be a non-empty subsequence of $L(x_{i,j})$. With the above lemma, the reverse direction can be proved as follows.

1. If $H_i' = 2_i 1_i 2_i 1_i \cdot L'(x_{i,2}) \cdot L'(x_{i,3}) \cdot 3_i 3_i$, then assign $x_{i,1} \leftarrow TRUE$, $x_{i,2} \leftarrow FALSE$, $x_{i,3} \leftarrow FALSE$.
2. If $H_i' = L'(x_{i,1}) \cdot 1_i 2_i 1_i 2_i \cdot L'(x_{i,3}) \cdot 3_i 3_i$, then assign $x_{i,1} \leftarrow FALSE$, $x_{i,2} \leftarrow TRUE$, $x_{i,3} \leftarrow FALSE$.
3. If $H_i' = L'(x_{i,1}) \cdot 1_i 1_i \cdot L'(x_{i,2}) \cdot 2_i 3_i 2_i 3_i$, then assign $x_{i,1} \leftarrow FALSE$, $x_{i,2} \leftarrow FALSE$, $x_{i,3} \leftarrow TRUE$.

Clearly, this gives a valid truth assignment for ϕ — as all the 2-CNF clauses $(\bar{x}_{i,j} \vee \bar{x}_{k,\ell})$ must appear in $L'(x_{i,j})$ or $L'(x_{k,\ell})$ in H', and at least one of $x_{i,j}$ and $x_{k,\ell}$ is assigned FALSE. We thus have the following theorem.

Theorem 4. *FT(4) is NP-complete.*

Since FT(4) is NP-complete, the optimization problem $LSRS+(4)$ is certainly NP-hard.

Corollary 1. *The optimization version of* LSRS+(4) *is NP-hard.*

Note that the above proof implies that the optimization version of $LSRS+(4)$ does not admit any polynomial-time approximation (regardless of the approximation) as any such approximate solution would form a feasible solution for FT(4). In fact, using a similar argument as in [9], even finding a good bi-criteria approximation, i.e., approximating the optimal length as well as the maximum number of letters covered, for $LSRS+(4)$ is not possible (unless P=NP). On the other hand, we show next that $LSRS+(3)$ is polynomially solvable.

4.2 LSRS+(3) is Polynomially Solvable

We now try to solve LSRS+(3), where the input is a sequence S of length n where each letter appears at most three times and at least twice. As a matter of fact, an optimal solution must be in the form of $x_1^{d_1} \cdots x_k^{d_k}$, where x_i is a subsequence of S and $d_i \in \{2, 3\}$ for $i \in [k]$. Throughout this subsection we assume that all letters in S appear at least twice and at most three times — if a letter appears only once in S then there is no solution for the corresponding LSRS+(3) instance.

Our idea is again dynamic programming, based on the above observation that in an optimal solution each SR-block is either a square or a cube. Define 6 tables, $S_2[i, j], C_2[i, j], S_3[i, j], C_3[i, j], L[i, j]$ and $C[i, j]$, with $1 \le i < j \le n$. The first 4 are only used to initialize $L[i, j]$'s.

- $C_2[i, j]$ is the set of letters that all appear at least twice in $S[i..j]$ if at least one letter appears exactly twice in $S[i, j]$; otherwise $C_2[i, j]$ is empty.
- $S_2[i, j]$ is the length of a longest square subsequence in $S[i, j]$ containing all the letters in $C_2[i, j]$. If such a local feasible solution does not exist, set $S_2[i, j] \leftarrow -1$; otherwise, we say that this local feasible solution is *2-feasible*.
- $C_3[i, j]$ is the set of letters that all appear three times in $S[i..j]$ if no letter in $S[i, j]$ appears exactly twice; otherwise $C_3[i, j]$ is empty.
- $S_3[i, j]$ is the length of a longest cubic subsequence (if exists) in $S[i, j]$ containing all the letters in $C_3[i, j]$; otherwise $S_3[i, j]$ is the length of a longest square subsequence containing all the letters in $C_3[i, j]$. If such a local feasible solution (cube or square) does not exist, set $S_3[i, j] \leftarrow -1$; otherwise, we say that this local feasible solution is *3-feasible* or *2-feasible* respectively (depending on whether the local solution is cubic or square).
- $C[i, j]$ is the set of letters appearing at least twice in $S[i..j]$. The $C[i, j]$'s can be computed, each with a linear scan, in a total of $O(n^3)$ time. $C[i, j]$'s are only used to enforce the coverage condition.
- $L[i, j]$ is the length of a feasible solution of $S[i..j]$ which covers all the letters in $C[i, j]$.

The initial values of $L[i, j]$, for $i < j$, can be set as follows.

$$
L[i, j] = \begin{cases} S_3[i, j] & \text{if } S_3[i, j] > 0 \\ S_2[i, j] & \text{else if } S_2[i, j] > 0 \\ -1 & \text{otherwise.} \end{cases}
$$

Note that the initialized solution might not be final. For example, $S[1..9] = ababbcacc$, then $S_3[1, 9] = S_2[1, 9] = -1$ and $L[1, 9] = -1$, i.e., there is no local 3-feasible (cubic) or 2-feasible (square) solution that covers all the letters in $C[1..9]$. But obviously an optimal solution, i.e., $(ab)^2 c^3$, exists. Hence we need to proceed to update $L[i, j]$.

Then we update the general case for $L[i, j]$ recursively as follows. This is done bottom-up, ordered by the ascending length of $S[i..j]$.

$$L[i,j] = \begin{cases} \max\{\max_{i<k<j}\{L[i,k]+L[k+1,j]\}, L[i,j]\} & \text{if } L[i,k] > 0, L[k+1,j] > 0 \\ & \text{and} \\ & C[i,j] == C[i,k] \cup C[k+1,j] \\ -1 & \text{otherwise.} \end{cases}$$

Two examples can be used to illustrate the update step. In the first example, $S[1..9] = abacbabcc$, $L[1,9]$ is initially assigned with $S_3[1,9] = 6$ (which corresponds to $(abc)^2$). After the update step $L[1,9] = 8$ (i.e., corresponds to $(ab)^3(cc)$). In the second example, $S[1..9] = abacabccb$, $L[1,9]$ is also initialized with $S_3[1,9] = 6$ (which corresponds $(abc)^2$). After the update step $L[1,9] = 6$ (which corresponds to $(ab)^2(cc)$ or $(abc)^2$ — in the actual implementation, there is no need to perform an explicit update for this example; but such a piece of information cannot be known before the update step, as shown in the first example).

Note that the condition $C[i,j] == C[i,k] \cup C[k+1,j]$, is to ensure that $L[i,j]$ is updated only when all the letters in $C[i,j]$ are covered. Then, the maximum of $L[i,k] + L[k+1,j]$, if greater than $L[i,j]$, replaces (the previous) $L[i,j]$.

An optimal solution is computed if $L[1,n] > 0$, and its solution value is stored in $L[1,n]$. Clearly, with an additional table, one can easily retrieve such an optimal solution, if exists.

Regarding the correctness of our algorithm, we have several simple lemmas.

Lemma 2. $C_2[i,j] \cap C_3[i,j] = \emptyset$.

Proof. This is obvious, as, by definition, $C_2[i,j]$ is non-empty only when there is a letter appearing exactly twice in $S[i,j]$. On the other hand, $C_3[i,j]$ is non-empty when there is no letter appearing exactly twice in $S[i,j]$. The two conditions are complementary.   □

Regarding $S_3[i,j]$, the following lemma says that if a 3-feasible solution does not exist then any of those 2-feasible solutions could be stored.

Lemma 3. *If a 3-feasible solution for $S_3[i,j]$ does not exist, then any 2-feasible solution for $S[i,j]$ can be stored (without changing the optimal solution value).*

Proof. By definition, $C_3[i,j]$ contains all the letters in $S[i,j]$ which appear exactly three times; moreover, there is no letter x which appears exactly twice in $S[i,j]$. Hence, if a letter y appears exactly once in $S[i,j]$ it would never appear as a local feasible solution for $S_3[i,j]$.

Therefore, if a 3-feasible solution does not exist, by definition, we would consider only a 2-feasible solution for $S[i..j]$ which covers all the letters in $C_3[i,j]$. The length of such a 2-feasible solution is exactly $2 \cdot |C_3[i,j]|$.   □

An example for this lemma is $S[1..6] = $ baabab. There is no 3-feasible solution. On the other hand, either abab or baba would make a valid 2-feasible solution, to be stored in $S_3[1,6]$. On the other hand, aabb could also make a final solution via the update of $L[i, j]$, made of two 2-feasible solutions aa and bb, for $S_2[1, 3]$ and $S_2[4, 6]$ respectively. But aabb itself is not considered as a 2-feasible solution in, say, $S[1..6]$.

Theorem 5. *Given a string S of length n, where each letter appears at most three times, the problem of LSRS+(3) can be solved in $O(n^4)$ time.*

Proof. See full version. □

Clearly, LSRS+(3) has a solution (i.e., $L[1, n] > 0$) if and only if FT(3) has a feasible solution.

5   Concluding Remarks

Obviously, the most prominent open problem is to decide if it is possible to compute the longest cubic subsequence in $o(n^5)$ time. Recently, Wang and Zhu gave an improved $O(k^3n^2)$ time algorithm (where k is the minimum number of letters deleted to have a feasible solution), but the worst case running time is still $O(n^5)$ when $k = \Theta(n)$ [19].

Acknowledgments. We thank anonymous reviewers for some insightful comments.

References

1. Bannai, H., Tomohiro, I., Köppl, D.: Longest (sub-)periodic subsequence. CoRR abs/2202.07189 (2022)
2. Bringmann, K., Künnemann, M.: Quadratic conditional lower bounds for string problems and dynamic time warping. In: Guruswami, V. (ed.) IEEE 56th Annual Symposium on Foundations of Computer Science, FOCS 2015, Berkeley, CA, USA, 17–20 October, 2015, pp. 79–97. IEEE Computer Society (2015)
3. Cicalese, F., Pilati, N.: The tandem duplication distance problem is hard over bounded alphabets. In: Flocchini, P., Moura, L. (eds.) IWOCA 2021. LNCS, vol. 12757, pp. 179–193. Springer, Cham (2021). https://doi.org/10.1007/978-3-030-79987-8_13
4. Ciriello, G., Miller, M.L., Aksoy, B.A., Senbabaoglu, Y., Schultz, N., Sander, C.: Emerging landscape of oncogenic signatures across human cancers. Nat. Genet. **45**, 1127–1133 (2013)
5. Inoue, T., Inenaga, S., Bannai, H.: Longest square subsequence problem revisited. In: Boucher, C., Thankachan, S.V. (eds.) SPIRE 2020. LNCS, vol. 12303, pp. 147–154. Springer, Cham (2020). https://doi.org/10.1007/978-3-030-59212-7_11
6. Kosowski, A.: An efficient algorithm for the longest tandem scattered subsequence problem. In: Apostolico, A., Melucci, M. (eds.) SPIRE 2004. LNCS, vol. 3246, pp. 93–100. Springer, Heidelberg (2004). https://doi.org/10.1007/978-3-540-30213-1_13
7. Lafond, M., Zhu, B., Zou, P.: The tandem duplication distance is NP-hard. In: Paul, C., Bläser, M. (eds.) 37th International Symposium on Theoretical Aspects of Computer Science, STACS 2020, 10–13 March 2020, Montpellier, France. LIPIcs, vol. 154, pp. 15:1–15:15. Schloss Dagstuhl - Leibniz-Zentrum für Informatik (2020)

8. Lafond, M., Zhu, B., Zou, P.: Computing the tandem duplication distance is NP-hard. SIAM J. Discrete Math. **36**(1), 64–91 (2022)

9. Lai, W., Liyanage, A., Zhu, B., Zou, P.: Beyond the longest letter-duplicated subsequence problem. In: Bannai, H., Holub, J. (eds.) 33rd Annual Symposium on Combinatorial Pattern Matching, CPM 2022, 27–29 June 2022, Prague, Czech Republic. LIPIcs, vol. 223, pp. 7:1–7:12. Schloss Dagstuhl - Leibniz-Zentrum für Informatik (2022)

10. Lander, E., et al.: International human genome sequencing consortium: initial sequencing and analysis of the human genome. Nature **409**(6822), 860–921 (2001)

11. Leech, J.: A problem on strings of beads. Math. Gaz. **41**(338), 277–278 (1957)

12. Macdonald, M., et al.: A novel gene containing a trinucleotide repeat that is expanded and unstable on Huntington's disease. Cell **72**(6), 971–983 (1993)

13. Cancer Genome Atlas Research Network: Integrated genomic analyses of ovarian carcinoma. Nature **474**, 609–615 (2011)

14. Oesper, L., Ritz, A.M., Aerni, S.J., Drebin, R., Raphael, B.J.: Reconstructing cancer genomes from paired-end sequencing data. BMC Bioinform. **13**(Suppl 6), S10 (2012)

15. Schrinner, S., Goel, M., Wulfert, M., Spohr, P., Schneeberger, K., Klau, G.W.: Using the longest run subsequence problem within homology-based scaffolding. Algorithms Mol. Biol. **16**(1), 11 (2021)

16. Sharp, A.J., et al.: Segmental duplications and copy-number variation in the human genome. Am. J. Hum. Genet. **77**(1), 78–88 (2005)

17. Szostak, J.W., Wu, R.: Unequal crossing over in the ribosomal DNA of saccharomyces cerevisiae. Nature **284**, 426–430 (1980)

18. Tiskin, A.: Semi-local string comparison: algorithmic techniques and applications (2013). arXiv:0707.3619

19. Wang, L., Zhu, B.: Algorithms and hardness for the longest common subsequence of three strings and related problems. In: Nardini, F.M., Pisanti, N., Venturini, R. (eds.) SPIRE 2023. LNCS, vol. 14240, pp. 367–380. Springer, Cham (2023). https://doi.org/10.1007/978-3-031-43980-3_30

20. Zheng, C., Kerr Wall, P., Leebens-Mack, J., de Pamphilis, C., Albert, V.A., Sankoff, D.: Gene loss under neighborhood selection following whole genome duplication and the reconstruction of the ancestral Populus genome. J. Bioinform. Comput. Biol. **7**(03), 499–520 (2009)

An Approximation Algorithm
for Covering Vertices by 4^+-Paths

Mingyang Gong[1], Zhi-Zhong Chen[2]([✉]), Guohui Lin[1], and Lusheng Wang[3]

[1] University of Alberta, Edmonton, Canada
{mgong4,guohui}@ualberta.ca
[2] Tokyo Denki University, Saitama, Japan
zzchen@mail.dendai.ac.jp
[3] City University of Hong Kong, Hong Kong SAR, China
lusheng.wang@cityu.edu.hk

Abstract. This paper deals with the problem of finding a collection of vertex-disjoint paths in a given graph $G = (V, E)$ such that each path has at least four vertices and the total number of vertices in these paths is maximized. The problem is NP-hard and admits an approximation algorithm which achieves a ratio of 2 and runs in $O(|V|^8)$ time. The known algorithm is based on time-consuming local search, and its authors ask whether one can design a better approximation algorithm by a completely different approach. In this paper, we answer their question in the affirmative by presenting a new approximation algorithm for the problem. Our algorithm achieves a ratio of 1.874 and runs in $O(\min\{|E|^2|V|^2, |V|^5\})$ time. Unlike the previously best algorithm, ours starts with a maximum matching M of G and then tries to transform M into a solution by utilizing a maximum-weight path-cycle cover in a suitably constructed graph.

1 Introduction

Throughout this paper, a graph always means a simple undirected graph without parallel edges or self-loops, and an approximation algorithm always means one running in polynomial time. Let k be a positive integer. Given a graph $G = (V, E)$, MPC_v^{k+} is the problem of finding a collection of vertex-disjoint paths each with at least k vertices in G so that the total number of vertices in these paths is maximized. Note that we can assume that each path in the output collection has at most $2k - 1$ vertices. This is because we can split a path having $2k$ or more vertices into two or more paths each having at least k and at most $2k - 1$ vertices. MPC_v^{k+} has numerous real-life applications such as transportation networks [9]. In this paper, we mainly focus on MPC_v^{4+}.

On the one hand, MPC_v^{k+} is related to many important optimization problems. For example, Berman and Karpinski [3] consider *the maximum path cover problem*, which is the problem of finding a collection of vertex-disjoint paths in a given graph so that the total number of edges in the paths is maximized. For other related path cover problems with different objectives, the reader is referred

© The Author(s), under exclusive license to Springer Nature Switzerland AG 2024
W. Wu and J. Guo (Eds.): COCOA 2023, LNCS 14461, pp. 459–470, 2024.
https://doi.org/10.1007/978-3-031-49611-0_33

to [1–5,8,15,16] for more details. On the other hand, MPC_v^{k+} can be viewed as a special case of *the maximum-weight $(2k - 1)$-set packing problem* [10,14] because the former can be easily reduced to the latter as follows. Recall that an instance of the latter problem is a collection $\mathcal{C}$ of sets each having a non-negative weight and at most $2k - 1$ elements. The objective is to select a collection of pairwise-disjoint sets in $\mathcal{C}$ so that the total weight of the selected sets is maximized. To reduce MPC_v^{k+} to the maximum-weight $(2k-1)$-set packing problem, it suffices to construct an instance $\mathcal{C}$ of the latter problem from a given instance graph G of MPC_v^{k+}, where $\mathcal{C}$ is the collection of all paths of G with at least k and at most $2k - 1$ vertices and the weight of each path P in $\mathcal{C}$ is the number of vertices in P. This reduction leads to an approximation algorithm for MPC_v^{k+} achieving a ratio of k because the maximum-weight $(2k-1)$-set packing problem can be approximated within a ratio of k [10] or within a slightly better ratio of $k - \frac{1}{63,700,992} + \epsilon$ [14] for any $\epsilon > 0$.

MPC_v^{k+} can be solved in polynomial time if $k \leq 3$ [5], but is NP-hard otherwise [11]. Kobayashi *et al.* [11] design an approximation algorithm for MPC_v^{4+} achieving a ratio of 4. Afterwards, Gong et al. [9] give the formal definition of MPC_v^{k+} and present an approximation algorithm for MPC_v^{k+} which achieves a ratio of $\rho(k) \leq 0.4394k + 0.6576$ and runs in $O(|V|^{k+1})$ time. The core of their algorithm is three local improvement operations, each of which increases the number of vertices in the current solution by at least 1 if it is applicable. The algorithm stops when none of the three operations is applicable. They employ an amortization scheme to analyze the approximation ratio of their algorithm by assigning the vertices in the optimal solution to the vertices of the solution outputted by their algorithm. For the special case where $k = 4$, they design two more local improvement operations to increase the number of vertices or the number of paths with exactly 4 vertices in the current solution, and then use a more careful amortization scheme to prove that the approximation ratio of their algorithm is bounded by 2 although the running time jumps to $O(|V|^8)$. As an open question, they ask whether one can design better approximation algorithms for the problem by completely different approaches.

1.1   Our Contribution and Design Highlights

In this paper, we answer the open question in the affirmative for the case where $k = 4$. Motivated by the approaches in [5,6,12] for similar problems, one may want to design an approximation algorithm for MPC_v^{k+} by first computing a maximum path-cycle cover $\mathcal{C}$ of the input graph G and then transforming $\mathcal{C}$ into a solution for G. Unfortunately, this approach to maximizing the number of edges does not seem to work. Our new idea for designing a better approximation algorithm for MPC_v^{4+} is to let the algorithm start by computing a maximum matching M in the input graph G. The intuition behind this idea is that the paths in an optimal solution for G can cover at most $\frac{5}{2}|M|$ vertices. So, it suffices to find a solution for G of which the paths cover a large fraction of the endpoints of the edges in M. To this purpose, our algorithm then constructs a maximum-weight path-cycle cover C in an auxiliary graph suitably constructed from M

and G. Our algorithm further tries to use the edges in C to connect a large fraction of the edges of M into paths with at least four vertices. If the algorithm fails to do so, then it will be able to reduce the problem to a smaller problem and in turn uses recursion to get a good solution.

Due to lack of space, the proofs of most lemmas are omitted here and will be shown in the journal version.

2   Basic Definitions

Throughout the remainder of this paper, we fix an instance G of MPC_v^{4+} for discussion. Let $n = |V(G)|$ and $m = |E(G)|$. For the graph G, let $V(G)$ and $E(G)$ be the vertex and edge set of G.

For a subset F of $E(G)$, we use $V(F)$ to denote the set $\{v \in V(G) \mid v$ is an endpoint of an edge in $F\}$. A *spanning subgraph* of G is a subgraph H of G with $V(H) = V(G)$. For a set F of edges in G, $G - F$ denotes the spanning subgraph $(V(G), E(G) \setminus F)$. In contrast, for a set F of edges with $V(F) \subseteq V(G)$ and $F \cap E(G) = \emptyset$, $G + F$ denotes the graph $(V(G), E(G) \cup F)$. The *degree* of a vertex v in G, denoted by $d_G(v)$, is the number of edges incident to v in G. A vertex v of G is *isolated* in G if $d_G(v) = 0$. The *subgraph induced by* a subset U of $V(G)$, denoted by $G[U]$, is the graph (U, E_U), where $E_U = \{\{u, v\} \in E(G) \mid u, v \in U\}$. Two vertex-disjoint subgraphs of G are *adjacent* in G if G has an edge between them.

A *cycle* in G is a connected subgraph of G in which each vertex is of degree 2. A *path* in G is either a single vertex of G or a connected subgraph of G in which exactly two vertices (called the *endpoints*) are of degree 1 and the others (called the *internal vertices*) are of degree 2. A *path component* of G is a connected component of G that is a path. If a path component is an edge, then it is called an *edge component*. The *order* of a cycle or path C, denoted by $|C|$, is the number of vertices in C. A *triangle* of G is a cycle of order 3 in G. A *k-path* of G is a path of order k in G, while a *k^+-path* of G is a path of order k or more in G. A *matching* of G is a (possibly empty) set of edges of G in which no two edges share an endpoint. A *maximum matching* of G is a matching of G whose size is maximized over all matchings of G. A *path-cycle cover* of G is a set F of edges in G such that in the spanning subgraph $(V(G), F)$, the degree of each vertex is at most 2. A *star* is a connected graph in which exactly one vertex is of degree ≥ 2 and each of the remaining vertices is of degree 1. The vertex of degree ≥ 2 is called the *center*, while the other vertices are the *satellites* of the star.

Notation 1. *For a graph G,*

- *$OPT(G)$ denotes an optimal solution for the instance graph G of MPC_v^{4+}, and $opt(G)$ denotes the total number of vertices in $OPT(G)$;*
- *$ALG(G)$ denotes the solution for G outputted by a specific algorithm, and $alg(G)$ denotes the total number of vertices in $ALG(G)$.*

3   The Algorithm for MPC_v^{4+}

Our algorithm for MPC_v^{4+} consists of multiple phases. In the first phase, it computes a maximum matching M in G in $O(\sqrt{n}m)$ time [13], initializes a subgraph $H = (V(M), M)$, and then repeatedly modifies H and M as described in Sect. 3.1. With a small loss of vertices, OPT can be transferred to a matching (by moving edges) and thus we have the following lemma.

Lemma 1. $|V(M)| \geq \frac{4}{5} opt(G)$.

3.1   Modifying H and M

We here describe a process for modifying H and M iteratively. The process consists of two steps. During these two steps, the following will be an invariant, which will be proved in Lemma 2.

Invariant 1. M is both a maximum matching of G and a subset of $E(H)$. Each connected component K of H is an edge, a triangle, a star, or a 5-path. Moreover, if K is a 5-path, then the two edges of $E(K)$ incident to the endpoints of K are in M; otherwise, exactly one edge of K is in M.

Initially, Invariant 1 clearly holds. Since M is a maximum matching, any two vertices of $V(G) \setminus V(H)$ cannot be adjacent to each other. Moreover, for any vertex $u_0 \in V(G) \setminus V(H)$, either it is incident to two different edge components e_0, e_1; incident to an unique edge component e_0 or not incident to any edge components of H. Generally speaking, for the first case, we present an operation to generate a 5-path by connecting u_0 with e_0, e_1. For the second case, u_0 and e_0 form a triangle or a star with other vertices of $V(G) \setminus V(H)$. Lastly, if no vertex of $V(G) \setminus V(H)$ is incident to e_0, then e_0 remains an edge component of H.

Definition 1. An augmenting triple with respect to H is a triple $(u_0, e_0 = \{v_0, w_0\}, e_1 = \{v_1, w_1\})$ such that $u_0 \in V(G) \setminus V(H)$, both e_0 and e_1 are edge components of H. We modify H and M as follows:

C1. If $\{u_0, v_0\}, \{u_0, v_1\} \in E(G)$, then add u_0 and the edges $\{u_0, v_0\}, \{u_0, v_1\}$ to H.

C2. If $\{u_0, v_0\}, \{w_0, v_1\} \in E(G)$, then add u_0 and the edges $\{u_0, v_0\}, \{w_0, v_1\}$ to H and then modify M by replacing e_0 with $\{u_0, v_0\}$.

Clearly, once the above modification is executed, a 5-path is generated. We next give the first two steps for H and M as follows.

Step 1.1 Repeatedly modify H and M with an augmenting triple until it is not applicable;

Step 1.2 Add all those edges $\{u, v\} \in E(G)$ such that $u \in V(G) \setminus V(H)$ and v is an endpoint of an edge component of H, as well as their endpoints u, to H.

We have the next lemma on H and M at the end of Step 1.2.

Lemma 2. The above Steps 1.1–1.2 can be done in $O(\min\{m^2 n, n^4\})$ time such that Invariant 1 always holds.

3.2 Bad Components and Rescuing Them

We consider the subgraph H and the maximum matching M at the end of Step 1.2. In the sequel, a component always means a *connected* component. Note that a 5-path of H can be contained in the solution, but we can not form a 4^+-path from any other components of H, which are defined as *bad*.

Definition 2. *A* bad component *of H is a connected component that is not a 5-path. By Invariant 1, a bad component is an edge, a triangle or a star.*

Clearly, moving all bad components of H will lead to a large loss of the vertices in $V(M)$. So, in this subsection, we construct a maximum-weighted path-cycle cover to connect a bad component of H to another bad component or a 5-path as many as possible such that we are able to form more 4^+-paths from bad components. We call it a *rescue* process of bad components.

Step 2.1 Construct a spanning subgraph G_1 of G of which the edge set consists of all the edges $\{v_1, v_2\}$ of G such that v_1 and v_2 appear in different components of H and at least one of the components is bad.

Definition 3. *A set F of edges in G_1 saturates a bad component K of H if at least one edge in F is incident to a vertex of K. The weight of F is the number of bad components saturated by F.*

Lemma 3. *A maximum-weighted path-cycle cover in G_1 can be found in $O(mn \log n)$ time.*

Step 2.2. Compute a maximum-weighted path-cycle cover C of G_1 (as in the proof of Lemma 3).
Step 2.3. As long as C contains an edge e such that $C \setminus e$ has the same weight as C, repeatedly remove e from C, that is, C is updated to $C \setminus e$.

Notation 2. $-$ C denotes the maximum-weighted path-cycle cover of G_1 computed at the end of Step 2.3.
$-$ M_C denotes the subset of the maximum matching M containing those edges in 5-paths of H or in bad components of H saturated by C.

Intuitively, the maximum-weighted path-cycle cover C connects as many bad components as possible with each other or 5-paths. So, the vertices of M_C is relatively larger than $opt(G)$ since it only removes the vertices (exactly two vertices in $V(M)$ by Invariant 1) of each bad component not saturated by C. The following lemma shows this fact.

Lemma 4. $|V(M_C)| \geq \frac{4}{5}opt(G)$.

3.3   Structure of Composite Components of $H + C$

By Lemma 4, in order to obtain a good approximate solution for G, it suffices to focus on M_C instead of its superset M. By Step 2.3, we remove the edges of C such that the weight of C is unchanged. So, it gives us simpler structures of each connected component of $H + C$.

Notation 3. – $H+C$ denotes the spanning subgraph $(V(G), E(H) \cup C)$. In the sequel, we use K to refer to a component in $H + C$.
– $(H+C)_m$ denotes the graph obtained from $H+C$ by contracting each component of H into a single node. In other words, the nodes of $(H+C)_m$ one-to-one correspond to the components of H and two nodes are adjacent in $(H + C)_m$ if and only if C contains an edge between the two corresponding components. We use $(K)_m$ to refer to the component of $(H + C)_m$ corresponding to the component K in $H + C$.

We next show the structures of each component $(K)_m$.

Lemma 5. For each component $(K)_m$ of $(H + C)_m$ (see Notation 3), the following statements hold:

1. $(K)_m$ is an isolated node, an edge, or a star.
2. If $(K)_m$ is an edge, then at least one endpoint of $(K)_m$ corresponds to a bad component of H.
3. If $(K)_m$ is a star, then each satellite of $(K)_m$ corresponds to a bad component of H.

If $(K)_m$ is isolated, then K is defined as *isolated* as well. Otherwise, K is a *composite component* and it contains two or more components of H, which are connected through the edges of C. If $(K)_m$ is isolated, K is a 5-path or a bad component of H, not saturated by C. Recall that we only focus on the vertices of M_C. So, we can assume $(K)_m$ is a 5-path if $(K)_m$ is isolated. We next discuss the case that $(K)_m$ is an edge or a star. By the second statement in Lemma 5, when $(K)_m$ is an edge, we choose an endpoint corresponding to a bad component of H as the satellite, while the other endpoint as the center.

Definition 4. For each composite component K of $H + C$, its center element is the component of H corresponding to the center of $(K)_m$, and it is denoted as K_c in the sequel; the other components of H contained in K are the satellite elements of K.
   Every vertex v of K_c is defined as an *anchor*. The edge connecting v to a satellite element S in C is called the *rescue-edge* for S and v is called the supporting anchor for S. For a nonnegative integer j, an anchor v is a j-*anchor* if v is the supporting anchor for exactly j satellite elements of $H + C$.

Since C is a path-cycle cover of G_1, each anchor is a 0-, 1-, or 2-anchor. When $(K)_m$ is isolated, then K is a 5-path and thus $opt(K) = 5$. One might ask whether the solution $OPT(K)$ can be easily computed for a composite component K. The following lemma answers this question in the affirmative.

Lemma 6. *For each component K of $H + C$, an $OPT(K)$ can be computed in $O(1)$ time.*

Definition 5. *For each composite component K of $H + C$, let $s(K) = |V(K) \cap V(M_C)|$. A critical component of $H + C$ is a component K with $\frac{s(K)}{opt(K)} \geq \frac{14}{11}$.*

Generally speaking, by computing an $OPT(K)$ for every K of $H + C$ and outputting their union as an approximate solution for G, we obtain an approximation algorithm for MPC_v^{4+} achieving a ratio of $\frac{5}{4} \max_K \frac{s(K)}{opt(K)}$ because of Lemma 4, unless K is *critical* (Definition 4) and *responsible* (Definition 8). If K is an isolated 5-path, then by Invariant 1, we have $\frac{s(K)}{opt(K)} = \frac{4}{5}$. But if K is a composite component, $\frac{s(K)}{opt(K)}$ is not necessarily small (smaller than our target value which is about 1.4992). Next, we show the possible structures of a critical component in Fact 1.

Fact 1. *A critical component K of $H + C$ has one 2-anchor or two 2-anchors. Moreover, K_c is an edge or a 5-path. If K_c is an edge, then $s(K) = 8$; if K_c is a 5-path, then $s(K) \in \{8, 10, 14, 16, 18\}$. Figure 1 shows all possible structures of a critical component.*

Proof. We can prove the fact by discussing the number of 2-anchors in K_c.

Remark 1. Even if a satellite element S of K can be a star or triangle, we almost always draw only one edge of S in Fig. 1 for simplicity.

Definition 6. *A 2-anchor of $H + C$ is critical if it appears in a critical component of $H + C$. A satellite element of $H + C$ is critical if its rescue-anchor is critical in $H + C$.*

By Fact 1, every critical component must have one or two critical 2-anchor.

Definition 7. *Suppose that v is a 0- or 1-anchor in $H + C$ and S is a satellite element in $H + C$ such that S has a vertex w with $\{v, w\} \in E(G)$. Then, moving S to v in $H + C$ is the operation of modifying C by replacing the rescue-edge of S with the edge $\{v, w\}$.*

Suppose a critical component K has exactly one 2-anchor. Then if we move one of critical satellite element to an isolated 5-path (if possible), K and the 5-path will be both not critical since they do not have 2-anchor. So, such moving decreases the number of critical components by one. However, we cannot guarantee that every moving will reduce critical components. In Fact 2, we discuss the different movings and their effects.

Fact 2. *For each critical component K of $H + C$ and its critical satellite element S, the following statements hold:*

1. *If we move S to another component (not K), then K is no longer critical and will not become isolated.*

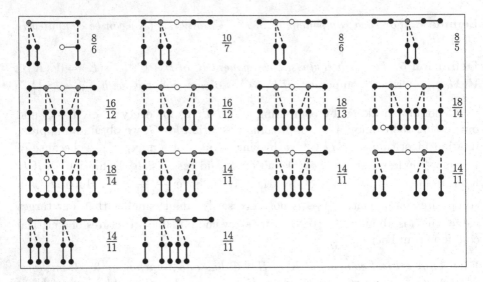

Fig. 1. The possible structures for a critical component K of $H + C$, where thick (respectively, dashed) edges are in the matching M (respectively, the path-cycle cover C), thin edges are not in $M \cup C$, the filled (respectively, blank) vertices are in (respectively, not in) $V(M)$, gray vertices are 2-anchors, and the fraction on the right side of each structure is $\frac{s(K)}{opt(K)}$.

2. *If v is a 0-anchor of K such that v is adjacent to S in G, then moving S to v in $H + C$ makes K no longer critical.*

3. *If v is a 1-anchor in K, then moving S to v makes K remain critical only if K has the first structure of Fig. 1, or K has the last or the second last structure in the bottom row of Fig. 1 and the rescue-anchor of S is the leftmost 2-anchor in K.*

Roughly speaking, we might need to pay more attention to 1-anchors since if a critical satellite element is moved to a 1-anchor v, then v is possible to become a critical 2-anchor. So, we introduce the following definition of 1-anchors.

Definition 8. *Let K be a composite component of $H + C$. If K has a 1-anchor v such that G has an edge between v and some critical satellite-element S of $H + C$ in G and moving S to v in $H + C$ makes K critical in $H + C$, then we call K a responsible component of $H + C$ and call v a responsible 1-anchor of $H + C$.*

By the third statement in Fact 2, a component of $H + C$ can be both critical and responsible only if it has the first or one of the last two structures in Fig. 1.

Lemma 7. *Suppose that a component K of $H + C$ is both critical and responsible. If K has the first structure in Fig. 1, then $s(K) = 8$ and we find a feasible solution with at least 7 vertices in $O(1)$ time; otherwise, $s(K) = 14$ and we find a feasible solution with at least 12 vertices in $O(1)$ time.*

By the above lemma, we know for each critical and responsible component K, we can find a feasible solution for K in constant time, which is still denoted as $OPT(K)$ for ease of presentation, with $\frac{s(K)}{opt(K)} < \frac{14}{11}$. Now, we can regard each critical and responsible component K as a non-critical component. So, any critical component cannot be responsible or vice versa. Hereafter, a critical component always refers to a critical but not responsible component and a responsible component always refers to a responsible but not critical component.

By Definition 8, the structure for a responsible component of $H + C$ can only be obtained by deleting a critical satellite-element from one of the structures in Fig. 1. So, by Fig. 1, we can easily list all possible structures for responsible components of $H + C$.

3.4 Operations for Modifying Critical Components

In this subsection, we define three operations for modifying C (and accordingly one or more critical components of $H + C$) so that after the modification, $H + C$ will hopefully have fewer critical components. Let v be a vertex of a satellite element S in a critical component K and v' be a vertex of K' in $H + C$. We remark that K and K' may be the same. Suppose $\{v, v'\} \in E(G) \setminus C$ and we design the following three operations.

Operation 1. *Suppose v' is a 0-anchor of K or v' is a non-responsible 1-anchor. Then, the operation modifies C by replacing the rescue-edges of S with $\{v, v'\}$.*

Clearly, Operation 1 does not change the weight of C by the first statement in Fact 2. Suppose v' is a 0-anchor. If $K = K'$, then after Operation 1, K is no longer critical by the second statement of Fact 2. Then, we suppose $K \neq K'$. Obviously, K is no longer critical but K' may become critical after Operation 1. So, Operation 1 may not necessarily decrease but does not increase the number of critical components in $H + C$. Fortunately, Operation 1 changes v' from a 0-anchor to a 1-anchor. Similarly, it is not hard to check if v' is a non-responsible 1-anchor, then Operation 1 makes K, K' both not critical. So, Operation 1 decreases the number of 0-anchors in $H + C$ by 1 or the number of critical components in $H + C$ by 1. Obviously, Operation 1 does not change the number of components in $H + C$.

Operation 2. *Suppose v' is in a satellite-element S' of K' and the center element K'_c of K' is an edge or a star to which no satellite element other than S' is adjacent in $H + C$. Then, the operation modifies C by replacing the rescue-edge of S with $\{v, v'\}$.*

Obviously, Operation 2 does not change the weight of C by the first statement of Fact 2. Note that K' has no 2-anchor and hence K' is not critical by Fact 1. So, $K \neq K'$ since K is critical. Moreover, after Operation 2, S' becomes the center element of K' and hence Lemma 5 still holds. Furthermore, by the first statement in Fact 2 and Fact 1, K, K' are not critical after Operation 2 and thus Operation 2 decreases the number of critical components in $H + C$ by 1. Clearly, Operation 2 does not change the number of components in $H + C$. Before Operation 2, K' may have one 0-anchor x. After Operation 2, x will be in a satellite element of $H + C$ and hence will not be a 0-anchor, but S' will become a center element with two satellite elements adjacent to it in $H + C$, implying that one vertex of S' may become a 0-anchor in $H + C$ (or not an anchor, if S' is a star). In summary, Operation 2 does not increase the number of 0-anchors in $H + C$.

Operation 3. *Suppose v' appears in a satellite-element S' of K' and K'_c is a 5-path , or K'_c is an edge or a star to which at least one more satellite element other than S' is adjacent in $H + C$. Then, the operation modifies C by replacing the rescue-edges of S and S' with the edge $\{v, v'\}$.*

By the first statement of Fact 2, Operation 3 does not change the weight of C since K, K' will not be an isolated bad component of H. Operation 3 uses the edge $\{v, v'\}$ to connect S and S' into a new composite component K_{new} of $H + C$. By Fact 1, K_{new} is not critical. If $K = K'$, then clearly Operation 3 does not increase the number of critical components in $H + C$. Otherwise, Operation 3 makes K not critical because of the first statement in Fact 2, but it is possible that Operation 3 makes K' critical. In any case, Operation 3 does not increase the number of critical components in $H + C$. Luckily, Operation 3 always increases the number of components in $H + C$ by 1.

Lemma 8. *Operations 1–3 can be repeatedly performed at most $O(n^2)$ times.*
Suppose that G has an edge $\{v, v'\}$ such that v is in a critical satellite-element S of $H + C$ and $v' \notin V(S)$ when none of Operations 1–3 is applicable. Then v' is a 2-anchor or a responsible 1-anchor.

3.5 Bounding $opt(G)$

Let R denote the set of vertices $v \in V(H)$ such that v is a 2-anchor or a responsible 1-anchor in $H + C$. By Lemma 8, once none of Operations 1–3 is applicable, any critical satellite element can be only incident to the vertices of R. Clearly, $|R \cap V(K)|$ is bounded by the total number of 1- and 2-anchors in K_c (K_c is not a triangle). Thus, if K_c is an edge, then $|R \cap V(K)| \in \{0, 1, 2\}$; if K_c is a star, then $|R \cap V(K)| \in \{0, 1\}$; if K_c is a 5-path, then $|R \cap V(K)| \in \{0, 1, 2, 3, 4, 5\}$. By Fact 1 and Lemma 7, each critical component K has one or two 2-anchors and no responsible 1-anchor. That is, $|R \cap V(K)| \in \{1, 2\}$.

Notation 4. *For the components in $H + C$, we define the notations as follows.*

- *Let $\mathcal{K}$ be the set of composite components or isolated 5-paths of $H + C$.*
- *For each $i \in \{0, 1, 2, 3, 4, 5\}$, let $\mathcal{K}_i \subseteq \mathcal{K}$ be a subset of $\mathcal{K}$ such that $|R \cap V(K)| = i$.*
- *For each $i \in \{1, 2\}$, let $\mathcal{K}_{i,c}$ be the set of critical components in $\mathcal{K}_i$.*
- *Let R_c be the set of 2-anchors in the critical components of $H + C$.*

- $U_c = \bigcup_{v \in R_c} \{w \in V(H) \mid w$ is in a critical satellite-element whose rescue-anchor is $v\}$.
- Let $G_c = G[V(G) \setminus (R_c \cup U_c)]$.

Lemma 9. $opt(G) \leq opt(G_c) + 7 \sum_{i=1}^{5} i|\mathcal{K}_i|$.

The above lemma indicates that though there are many critical components in $H+C$, after we "destroyed" all the critical components, the problem is reduced to a smaller problem on G_c and $opt(G_c)$ is not far away from $opt(G)$. So, it is possible to get a good solution of G by recursively solving the problem on G_c.

4   Summary of the Algorithm

Let $r = \frac{15+\sqrt{505}}{20} \approx 1.874$ be the positive root to the quadratic equation $10r^2 - 15r - 7 = 0$. Our algorithm proceeds as follows.

0. If $|V(G)| \leq 4$, find an optimal solution by brute-force search, output it, and then halt.
1. Construct the graph H as follows:
   (a) Compute a maximum matching M in G and initialize $H = (V(M), M)$.
   (b) Modify M and H by performing Steps 1.1 and 1.2 in Sect. 3.1.
2. Compute a maximum path-cycle cover C and modify it as follows:
   (a) Perform Steps 2.1, 2.2, and 2.3 in Sect. 3.2 to compute a maximum path-cycle cover C in an auxiliary graph G_1.
3. Repeatedly perform Operations 1, 2, and 3 in Sect. 3.4 to modify C, until none of them is applicable.
4. If no component of $H + C$ is critical, or $\frac{\sum_{i=1}^{5} i|\mathcal{K}_i|}{|\mathcal{K}_{1,c}|+2|\mathcal{K}_{2,c}|} > \frac{5}{7}r$, then
   (a) compute $OPT(K)$ for each component K of $H+C$ that is not an isolated bad component of H by Lemma 6;
   (b) output their union as a solution for G, and then halt.
5. Otherwise, there is at least one critical component and $\frac{\sum_{i=1}^{5} i|\mathcal{K}_i|}{|\mathcal{K}_{1,c}|+2|\mathcal{K}_{2,c}|} \leq \frac{5}{7}r$.
   (a) Recursively call the algorithm on the graph G_c to obtain a solution $ALG(G_c)$.
   (b) For each $v \in R_c$, compute a 5^+-path P_v since v is a 2-anchor.
   (c) Output the union of $ALG(G_c)$ and $\cup_{v \in R_c} P_v$, and halt.

Theorem 1. *The   running   time   of   the algorithm is bounded by $O(\min\{m^2n^2, n^5\})$ and the approximation ratio is at most $r = \frac{15+\sqrt{505}}{20} < 1.874$.*

Due to the page limitation, the proof of Theorem 1 is omitted here.

References

1. Asdre, K., Nikolopoulos, S.D.: A linear-time algorithm for the k-fixed-endpoint path cover problem on cographs. Networks **50**, 231–240 (2007)
2. Asdre, K., Nikolopoulos, S.D.: A polynomial solution to the k-fixed-endpoint path cover problem on proper interval graphs. Theoret. Comput. Sci. **411**, 967–975 (2010)
3. Berman, P., Karpinski, M.: 8/7-approximation algorithm for (1,2)-TSP. In: Proceedings of ACM-SIAM SODA 2006, pp. 641–648 (2006)
4. Cai, Y., et al.: Approximation algorithms for two-machine flow-shop scheduling with a conflict graph. In: Wang, L., Zhu, D. (eds.) COCOON 2018. LNCS, vol. 10976, pp. 205–217. Springer, Cham (2018). https://doi.org/10.1007/978-3-319-94776-1_18
5. Chen, Y., et al.: Path cover with minimum nontrivial paths and its application in two-machine flow-shop scheduling with a conflict graph. J. Comb. Optim. **43**, 571–588 (2022)
6. Chen, Z.-Z., Konno, S., Matsushita, Y.: Approximating maximum edge 2-coloring in simple graphs. Discret. Appl. Math. **158**, 1894–1901 (2010)
7. Gabow, H.N.: An efficient reduction technique for degree-constrained subgraph and bidirected network flow problems. In: Proceedings of ACM STOC 1983, pp. 448–456 (1983)
8. Gomez, R., Wakabayashi, Y.: Nontrivial path covers of graphs: existence, minimization and maximization. J. Comb. Optim. **39**, 437–456 (2020)
9. Gong, M., Fan, J., Lin, G., Miyano, E.: Approximation algorithms for covering vertices by long paths. In: Proceedings of MFCS 2022. LIPIcs, vol. 241, pp. 53:1–53:14 (2022)
10. Hochbaum, D.S.: Efficient bounds for the stable set, vertex cover and set packing problems. Discret. Appl. Math. **6**, 243–254 (1983)
11. Kobayashi, K., et al.: Path cover problems with length cost. In: Mutzel, P., Rahman, M.S., Slamin (eds.) WALCOM 2022. LNCS, vol. 13174, pp. 396–408. Springer, Cham (2022). https://doi.org/10.1007/978-3-030-96731-4_32
12. Kosowski, A.: Approximating the maximum 2- and 3-edge-colorable subgraph problems. Discret. Appl. Math. **157**, 3593–3600 (2009)
13. Micali, S., Vazirani, V.V.: An $O(\sqrt{|V|}|E|)$ algorithm for finding maximum matching in general graphs. In: Proceedings of IEEE FOCS 1980, pp. 17–27 (1980)
14. Neuwohner, M.: An improved approximation algorithm for the maximum weight independent set problem in d-claw free graphs. In: Proceedings of STACS 2021, pp. 53:1–53:20 (2021)
15. Pao, L.L., Hong, C.H.: The two-equal-disjoint path cover problem of matching composition network. Inf. Process. Lett. **107**, 18–23 (2008)
16. Rizzi, R., Tomescu, A.I., Mäkinen, V.: On the complexity of minimum path cover with subpath constraints for multi-assembly. BMC Bioinform. **15**, S5 (2014)

V-Words, Lyndon Words and Substring circ-UMFFs

Jacqueline W. Daykin[1,2,3], Neerja Mhaskar[4(✉)], and W. F. Smyth[4]

[1] Department of Computer Science, Aberystwyth University, Aberystwyth, Wales
jwd6@aber.ac.uk
[2] Department of Information Science, Stellenbosch University, Stellenbosch,
South Africa
[3] Normandie University, UNIROUEN, LITIS, 76000 Rouen, France
[4] Department of Computing and Software, McMaster University, Hamilton, Canada
{pophlin,smyth}@mcmaster.ca

Abstract. We say that a family $\mathcal{W}$ of strings over Σ^+ forms a Unique
Maximal Factorization Family if and only if for every $w \in \mathcal{W}$, w has a
unique maximal factorization. Then an UMFF $\mathcal{W}$ is a circ-UMFF whenever it contains exactly one rotation of every primitive string $x \in \Sigma^+$.
V-order is a non-lexicographical total ordering on strings that determines
a circ-UMFF. In this paper we propose a generalization of circ-UMFF
called the substring circ-UMFF and extend the combinatorial research on
V-order by investigating connections to Lyndon words. Then we extend
concepts to considering any total order. Applications of this research
arise in efficient text indexing, compression, and search tasks.

Keywords: circ-UMFF · Combinatorics · Factorization · Lyndon
word · Substring circ-UMFF · Total order · UMFF · V-order · V-word

1 Introduction

V-order [8] (Definition 3) is a non-lexicographic global order on strings that was
introduced more than 25 years ago [8,9]. Similar to conventional lexicographical
order (lexorder), V-order string comparison can be performed using a simple linear time, constant space algorithm [3–5,12,13], further improved in [1,2]. Much
theoretical research has been done on this ordering [12,13,17], including efficient construction of the so-called V-BWT or V-transform [17], a variant of the
lexicographic Burrows-Wheeler transform (BWT).

In this paper we extend combinatorial research on V-order and circ-UMFFs.
We first show that there are infinitely more V-words (Definition 9) than Lyndon
words (Definition 1). Then we study instances of circ-UMFFs having similar
properties to V-words and/or Lyndon words. Finally, we propose a generalization
of the circ-UMFF (Definition 17) called the substring circ-UMFF (Definition 28)
and show that for a generalized order $\mathcal{T}$, with order relation $\ll$, classes of border-
free words exist that form circ-UMFFs and substring circ-UMFFs, respectively.

© The Author(s), under exclusive license to Springer Nature Switzerland AG 2024
W. Wu and J. Guo (Eds.): COCOA 2023, LNCS 14461, pp. 471–484, 2024.
https://doi.org/10.1007/978-3-031-49611-0_34

2 Preliminaries

A *string* (or *word*) is an ordered sequence of elements drawn from a finite totally ordered set Σ of cardinality $\sigma = |\Sigma|$, called the *alphabet*. The elements of Σ are referred to a *characters* (*letters*). We refer to strings using mathbold: x, w instead of x, w. The length of a string $w[1..n]$ is $|w| = n$. The *empty string* of length zero is denoted by ε. The set of all nonempty strings over the alphabet Σ is denoted by Σ^+, with $\Sigma^* = \Sigma^+ \cup \varepsilon$. If $x = uwv$ for (possibly empty) strings $u, w, v \in \Sigma^*$, then u is a *prefix*, w a *substring* or *factor*, and v a *suffix* of x. A substring u of w is said to be *proper* if $|u| < |w|$. A string w has a *border* u if u is both a proper prefix and a proper suffix of w. If w has only the empty border ε then it is called *border-free*.

For $x = x[1..n]$ and an integer sequence $0 < i_1 < i_2 < \cdots < i_k \leq n$, the string $y = x[i_1]x[i_2]\cdots x[i_k]$ is said to be a *subsequence* of x, *proper* if $|y| < n$. If $x = u^k$ (a concatenation of k copies of u) for some nonempty string u and some integer $k > 1$, then x is said to be a *repetition*; otherwise, x is *primitive*. A string $y = R_i(x)$ is the ith *conjugate* (or *rotation*) of $x = x[1..n]$ if $y = x[i+1..n]x[1..i]$ for some $0 \leq i < n$ (so that $R_0(x) = x$).

Definition 1. *A* **Lyndon word** *[6] is a primitive string that is minimum in lexorder $<$ over its conjugacy class.*

The following Lyndon factorization (*LF*) theorem is fundamental in stringology and underpins the wide-ranging applications of Lyndon words:

Theorem 2. *[6] Any nonempty string x can be written uniquely as a product $LF_x = x = u_1 u_2 \cdots u_k$ of $k \geq 1$ Lyndon words, with $(u_1 \geq u_2 \geq \cdots \geq u_k)$.*

For further stringological definitions and theory, see [7, 25].

3 V-order

In this section we start by defining V-order and describing some of its important properties used later in the paper.

Let $x = x_1 x_2 \cdots x_n$ be a string over Σ. Define $h \in \{1, \ldots, n\}$ by $h = 1$ if $x_1 \leq x_2 \leq \cdots \leq x_n$; otherwise, by the unique value such that $x_{h-1} > x_h \leq x_{h+1} \leq x_{h+2} \leq \cdots \leq x_n$. Let $x^* = x_1 x_2 \cdots x_{h-1} x_{h+1} \cdots x_n$, where the star $*$ indicates deletion of x_h. Write $x^{s*} = (...(x^*)^*...)^*$ with $s \geq 0$ stars. Let $g = \max\{x_1, x_2, \ldots, x_n\}$, and let k be the number of occurrences of g in x. Then the sequence $x, x^*, x^{2*}, \ldots$ ends $g^k, \ldots, g^1, g^0 = \varepsilon$. From all strings x over Σ we form the *star tree* (see Example 5), where each string x labels a vertex and there is a directed edge upward from x to x^*, with the empty string ε as the root.

Definition 3 (*V*-order [8]). *We define V-order $\prec$ for distinct strings x, y.*

First $x \prec y$ if in the star tree x is in the path $y, y^*, y^{2*}, \ldots, \varepsilon$. If x, y are not in a path, there exist smallest s, t such that $x^{(s+1)*} = y^{(t+1)*}$. Let $s = x^{s*}$ and $t = y^{t*}$; then $s \neq t$ but $|s| = |t| = m$ say. Let $j \in [1..m]$ be the greatest integer such that $s[j] \neq t[j]$. If $s[j] < t[j]$ in Σ then $x \prec y$; otherwise, $y \prec x$. Clearly $\prec$ is a total order on all strings in Σ^*.

See the star tree path and star tree examples in Examples 4 and 5, respectively.

Example 4. *[Star tree path] Figure 1 illustrates the star tree for the case $x \prec y$ if in the star tree x is in the path $y, y^*, y^{2*}, \ldots, \varepsilon$. Consider the V-order comparison of the strings $x = 929$ and $y = 922911$. The subscript h indicates the V letter to be deleted (defined above as $x_{h-1} > x_h \leq x_{h+1} \leq x_{h+2} \leq \cdots \leq x_n$). Since 929 is in the path of star deletions of 922911, therefore $929 \prec 922911$.*

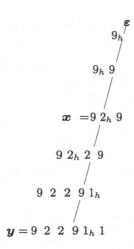

Fig. 1. $929 \prec 922911$

Example 5. *[Star tree] Figure 2 illustrates the star tree for the non-path case using the V-order comparison of the words $x = unique$ and $y = equitant$. As in the previous example, the subscript h indicates the V letter to be deleted (defined above as $x_{h-1} > x_h \leq x_{h+1} \leq x_{h+2} \leq \cdots \leq x_n$). The circled letters are those compared in alphabetic order (defined above as $s[j] \neq t[j]$).*

Definition 6 (V-form [8,9,12,13]). *The V-form of any given string x is*

$$V_k(x) = x = x_0 g x_1 g \cdots x_{k-1} g x_k,$$

where g is the largest letter in x—thus we suppose that g occurs exactly k times. Note that any x_i may be the empty string ε and we let $\mathcal{L}_x = g$, $\mathcal{C}_x = k$.

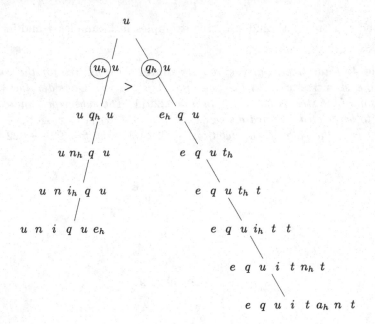

Fig. 2. *unique* ≻ *equitant*

Lemma 7. *[8, 9, 12, 13] Suppose we are given distinct strings x and y with corresponding V-forms*

$$x = x_0 \mathcal{L}_x x_1 \mathcal{L}_x x_2 \cdots x_{j-1} \mathcal{L}_x x_j,$$
$$y = y_0 \mathcal{L}_y y_1 \mathcal{L}_y y_2 \cdots y_{k-1} \mathcal{L}_y y_k,$$

where $j = \mathcal{C}_x$, $k = \mathcal{C}_y$. Let $h \in 0..\max(j, k)$ be the least integer such that $x_h \neq y_h$. Then $x \prec y$ if and only if one of the following conditions holds:

(C1) $\mathcal{L}_x < \mathcal{L}_y$
(C2) $\mathcal{L}_x = \mathcal{L}_y$ *and* $\mathcal{C}_x < \mathcal{C}_y$
(C3) $\mathcal{L}_x = \mathcal{L}_y$, $\mathcal{C}_x = \mathcal{C}_y$ *and* $x_h \prec y_h$.

Lemma 8. *[12, 13] For given strings x and y, if y is a proper subsequence of x, then $y \prec x$.*

By Lemma 8, suffix sorting becomes a trivial problem in V-order, which otherwise requires a non-trivial linear algorithm in lexorder [16]. For instance, for $x = 7547223$ we have $3 \prec 23 \prec 223 \prec 7223 \prec 47223 \prec 547223 \prec 7547223$.

We now introduce the V-order equivalent of the lexorder Lyndon word:

Definition 9 (*V-word* [9]). *A string x over an ordered alphabet Σ is a **V-word** if it is the unique minimum in V-order $\prec$ in the conjugacy class of x.*

Thus, like a Lyndon word, a V-word is necessarily primitive.

Example 10. $[\prec]$ *We can apply Definition 3, equivalently the methodology of Lemma 7, to conclude that*

$$6263 \prec 6362 \prec 2636 \prec 3626,$$

so that 6263 is a V-word, while on the other hand 2636 is a Lyndon word. Similarly, 62626263 and 929493 are V-words, while conjugates 26262636 and 294939 are Lyndon words.

We now define another important ordering below:

Definition 11 (V-word order). *Suppose x and y are V-words on an ordered alphabet Σ. If xy is also a V-word, then we write $x <_V y$; if not, then $x \geq_V y$.*

Thus, corresponding to the Lyndon factorization into Lyndon words using $\geq$ (Theorem 2), we arrive at a *V-order factorization* expressed in terms of V-word order $\geq_V$:

Lemma 12. *[12, 13, 16] (V-order Factorization) Using only linear time and space (see Algorithm VF in [13]), a string x can be factored uniquely, using V-word order, into V-words $x = x_1 x_2 \cdots x_m$, where $x_1 \geq_V x_2 \geq_V \cdots \geq_V x_m$.*

Example 13. *For $x = 33132421$, the Lyndon decomposition (computed using lexorder) is $3 \geq 3 \geq 13242 \geq 1$, while the V-order factorization identifies nonextendible V-words 33132 and 421 with $33132 \geq_V 421$. (Note however that $33132 \prec 421$! See [11] for more background on this phenomenon.) Similarly, from Example 10, the string*

$$x = uvw = (6263)(62626264)(929493)$$

has the unique V-order factorization $u \geq_V v \geq_V w$, even though $u \prec v \prec w$.

It will also be useful to order strings x, y based on a lexicographic approach to their factorizations into identified substrings; this will be applied in Sect. 5 to handle string factorization based not on letters but substrings. We call this ordering, denoted $\prec_{LEX(F)}$, *lex-extension* order, expressed here with respect to substring ordering using $\prec$—but note that other substring ordering methodologies could instead be applied.

Definition 14 (Lex-extension order [9, 13, 17]). *Suppose that according to some factorization F, two strings $x, y \in \Sigma^+$ are expressed in terms of nonempty factors:*

$$x = x_1 x_2 \cdots x_m, \quad y = y_1 y_2 \cdots y_n.$$

Then $x \prec_{LEX(F)} y$ if and only if one of the following holds:

(1) x is a proper prefix of y (that is, $x_i = y_i$ for $1 \leq i \leq m < n$); or
(2) for some least $i \in 1..\min(m,n)$, $x_j = y_j$ for $j = 1, 2, \ldots, i1$, and $x_i \prec y_i$.

4 UMFF and Circ-UMFF Theory

Motivated by classical Lyndon words, investigations into combinatorial aspects of the factoring and concatenation of strings led to the concepts of UMFF and circ-UMFF [9–11] whose properties we overview here and apply in Sect. 5.

For given $x = x[1..n] \in \Sigma^+$, if $x = w_1 w_2 \cdots w_k$, $1 \leq k \leq n$, then $w_1 w_2 \cdots w_k$ is said to be a **factorization** of x; moreover, if every factor w_j, $1 \leq j \leq k$, belongs to a specified set W, then $w_1 w_2 \cdots w_k$ is said to be a **factorization of x over W**, denoted by $F_W(x)$. A subset $W \subseteq \Sigma^+$ is a **factorization family** (FF) of Σ if for every nonempty string x on Σ there exists a factorization $F_W(x)$. If, for every $j = 1, 2, ..., k$, every factor w_j is of maximum length, then the factorization $F_W(x)$ is unique and said to be **maximal**.

Definition 15. *Let W be an FF on an alphabet Σ. Then W is a* **unique maximal factorization family** *(UMFF) if and only if there exists a maximal factorization $F_W(x)$ for every string $x \in \Sigma^+$.*

Factors cannot overlap in a unique maximal factorization of a string:

Lemma 16. *(The* **xyz** *Lemma [9]) An FF W is an* **UMFF** *if and only if whenever $xy, yz \in W$ for some nonempty y, then $xyz \in W$.*

An important class of UMFFs can now be specified:

Definition 17. *An UMFF W over Σ^+ is a* **circ-UMFF** *if and only if it contains exactly one rotation of every primitive string $x \in \Sigma^+$.*

A circ-UMFF W expresses a concatenation order:

Definition 18. *([11]) If a circ-UMFF W contains strings u, v and uv, we write $u <_W v$ (called W-order).*

Observe that V-word order (Definition 11) is a particular instance of the W-order.

Structural properties of circ-UMFFs are summarized as follows:

Theorem 19. *([10]) Let W be a circ-UMFF.*
(1) If $u \in W$ then u is border-free.
(2) If $u, v \in W$ and $u \neq v$ then uv is primitive.
(3) If $u, v \in W$ and $u \neq v$ then $uv \in W$ or $vu \in W$ (but not both).
(4) If $u, v, uv \in W$ then $u <_W v$ where $<_W$ is a total order on W.
(5) If $w \in W$ and $|w| \geq 2$ then there exist $u, v \in W$ with $w = uv$.

The first known circ-UMFF is believed to be the set of Lyndon words, whose specific W-order is lexorder; that is, the same ordering of the strings of Σ^+ is used to obtain Lyndon words:

Theorem 20. *([20]) Let $\mathcal{L}$ be the set of Lyndon words, and suppose $u, v \in \mathcal{L}$. Then $uv \in \mathcal{L}$ if and only if u precedes v in lexorder.*

Note that, from Definition 15, V-order factorization determines an UMFF, which, by Definitions 9 and 17, is a circ-UMFF.

5   V-words, Lyndon Words and Circ-UMFF

In this section we further investigate the relationship and differences between Lyndon and V-words and introduce generalized words over any total order.

We begin with the following observation, stated in [11], which follows immediately from Duval's fundamental Theorem 20 [20]:

Observation 21. *Let Σ^*_{lex} denote the lexicographic total ordering of Σ^*. Then the lexordered set $\mathcal{L}$ of Lyndon words is a suborder of Σ^*_{lex}.*

However, observe that there is no corresponding architecture for V-words. In V-ordered Σ^*, for $x = 21, y = 31$, we have $x \prec y$ by Lemma 7 (C1), while in the class of V-words we have $x \geq_V y$ by Definition 11 of V-word order. For further details on the distinction between $\prec$ and $\geq_V$ see Lemma 3.16 in [13].

Lyndon words and V-words are generally distinct [13]. For instance, the integer string 1236465123111 factors into Lyndon words $(12316465)(123)(1)(1)(1)$ and into V-words $(1)(2)(3)(6465123111)$—no correspondence whatever. Nevertheless, as the following result tells us, when substrings are restricted to a single letter, a rather remarkable result holds, which is a newly observed special case of Theorem 4.1 in [9] and leads to the concept of V-Lyndons:

Lemma 22. *Suppose x is a string whose V-form is $x = \mathcal{L}_x x_1 \mathcal{L}_x x_2 \cdots x_{j-1} \mathcal{L}_x x_j$, where $x_0 = \varepsilon$ and $|x_h| = 1$ for $1 \leq h \leq j$. Let $x' = x_1 x_2 \cdots x_{j-1} x_j$. Then x is a V-word if and only if x' is a Lyndon word.*

To see that the requirement $|x_h| = 1$ is necessary, consider $x = 321312$ with $\mathcal{L}_x = 3$, $|x_1| = |x_2| = 2$. Certainly $x' = x_1 x_2 = 2112$ is *not* a Lyndon word, but since $x \prec 312321$, x is a V-word. Thus Lemma 22 does not generalize to V-form substrings with $|x_h| > 1$. Nonetheless, there does exist a kind of reciprocity between infinite classes of Lyndon words and V-words:

Observation 23. *For any Lyndon word $x[1..n]$, $n \geq 2$, on ordered alphabet Σ:*

(1) If $\mathcal{L}_x$ is the largest letter in x, then $(\mathcal{L}')^k x$ is a V-word for $\mathcal{L}' > \mathcal{L}_x$ and every integer $k > 0$.
(2) If ℓ_x is the smallest letter in x, then $(\ell_x)^k x$ is also a Lyndon word for every integer $k > 0$.

We shall call V-words having the form specified by Lemma 22 V-***Lyndons***. Then, building on Lemma 22 and Observation 23(1), we can show that there are infinitely more V-words than there are Lyndon words:

Theorem 24. *Suppose that $\ell[1..n]$ is a Lyndon word over an ordered alphabet Σ and further that there exists $\mathcal{L}_\ell \in \Sigma$ such that $\mathcal{L}_\ell > \ell[i]$ for $i \in 1..n$. Then we can construct infinitely many V-words from ℓ over Σ.*

Proof. For the first V-word $\boldsymbol{v}_1$, applying Lemma 22 we write $\boldsymbol{\ell}$ as $\boldsymbol{v}_1[1..2n]$ where for $i \in 1..2n$, if i is odd, $\boldsymbol{v}_1[i] = \mathcal{L}_\ell$, while if i is even, $\boldsymbol{v}_1[i] = \boldsymbol{\ell}[i/2]$; that is, $\boldsymbol{v}_1 = \mathcal{L}_\ell\boldsymbol{\ell}[1]\mathcal{L}_\ell\boldsymbol{\ell}[2]..\mathcal{L}_\ell\boldsymbol{\ell}[n]$.

For V-words $\boldsymbol{v}_h$, $h > 1$, rewrite $\boldsymbol{\ell}$ as $\boldsymbol{v}_h = \mathcal{L}_\ell\boldsymbol{\ell}[1]^h\mathcal{L}_\ell\boldsymbol{\ell}[2]^h..\mathcal{L}_\ell\boldsymbol{\ell}[n]^h$. Lemma 7 (C1) shows that if $a \prec b$ for letters a, b (that is, $a < b$ in Σ), then $a^h \prec b^h$ and hence the Lyndon property (Lemma 22) of $\boldsymbol{\ell}$ is preserved for $\boldsymbol{v}_h$ using Definition 14 for lex-extension order of strings. $\square$

It might then be natural to suppose that V-words exhibit the same structural properties as Lyndon words and support equivalent string operations. For instance, a defining property of Lyndon words is that they are strictly less in lexorder than any of their proper suffixes; that is, for a Lyndon word $\boldsymbol{\ell} = \boldsymbol{p}_\ell\boldsymbol{s}_\ell$, with $\boldsymbol{p}_\ell, \boldsymbol{s}_\ell \neq \epsilon$, we have

$$\boldsymbol{\ell} < \boldsymbol{s}_\ell < \boldsymbol{s}_\ell\boldsymbol{p}_\ell.$$

This central Lyndon property relates to two important operations on strings: ordering and concatenation. For Lyndon words, these operations are consistent with respect to lexorder: that is, for every proper suffix $\boldsymbol{s}_\ell$ of $\boldsymbol{\ell}$, by virtue of the ordering $\boldsymbol{\ell} < \boldsymbol{s}_\ell$, we can construct a Lyndon word $\boldsymbol{\ell}\boldsymbol{s}_\ell{}^h$ by concatenation for every $h \geq 1$.

In contrast, for V-order, these operations are not necessarily consistent. First, by Lemma 8, a proper suffix $\boldsymbol{u}$ of a string $\boldsymbol{x}$ is less than $\boldsymbol{x}$ in V-order; thus, for example, given the V-word $\boldsymbol{v} = 43214123$, even though substrings $23 \prec \boldsymbol{v}$ and $4123 \prec \boldsymbol{v}$, on the other hand, by definition of a V-word, $\boldsymbol{v} \prec 41234321$. So for a V-word $\boldsymbol{v} = \boldsymbol{p}_v\boldsymbol{s}_v$, with $\boldsymbol{p}_v, \boldsymbol{s}_v \neq \epsilon$, we have

$$\boldsymbol{s}_v \prec \boldsymbol{v} \prec \boldsymbol{s}_v\boldsymbol{p}_v.$$

Nevertheless, like a Lyndon word, a V-word can be concatenated with any of its proper suffixes (although they are less in V-order) to form a larger V-word (Lemma 3.21 in [13]). Hence we are interested in those combinatorial properties related to operations like concatenation and indexing in conjugacy classes which hold both for Lyndon words and V-words. Examples include border-freeness, existence of $\boldsymbol{uv}$ and $\boldsymbol{vu}$ in the conjugacy class where $\boldsymbol{u}$ and $\boldsymbol{v}$ are Lyndon words, and the FM-index Last First mapping property [16, 21].

We proceed to introduce a general form of order, $\mathcal{T}$:

Definition 25. *Let $\mathcal{T}$ be any total ordering of Σ^* with order relation $\ll$ so that given distinct strings $\boldsymbol{x}, \boldsymbol{y}$ they can be ordered deterministically with the* **relation** $\ll$: *either $\boldsymbol{x} \ll \boldsymbol{y}$ or $\boldsymbol{y} \ll \boldsymbol{x}$.*

So for Lyndon words (V-words) the ordering $\mathcal{T}$ is lexorder (V-order) and the corresponding order relation $\ll$ is $<$ ($\prec$). Using the general order $\mathcal{T}$, we can extend Definition 1 (9) from Lyndon words (V-words) to $\mathcal{T}_\ll$-words:

Definition 26. *A string $\boldsymbol{x}$ over an ordered alphabet Σ is said to be a $\mathcal{T}_\ll$-**word** if it is the unique minimum in $\mathcal{T}$-order $\ll$ in the conjugacy class of $\boldsymbol{x}$.*

Similarly, Definition 14 can be generalized by replacing the order $\prec$ by $\mathcal{T}$-order $\ll$: $x \ll_{LEX(F)} y$ if and only if $x_i \ll y_i$ and $x_j = y_j$, $1 \leq j < i$. We call this $\mathcal{T}_{lex}$-**order**.

Applications of circ-UMFFs in the literature arise in linear-time variants of the Burrows-Wheeler transform: the V-order based transform $V - BWT$ [17]; the binary Rouen transform $B - BWT$ derived from binary block order which generated twin transforms [15]; the degenerate transform $D - BWT$ for indeterminate strings implemented with lex-extension order [18] which supports backward search [14]. These instances stimulate the quest for new circ-UMFFs and we pave the way for this by next introducing a generalization of circ-UMFFs.

6   Substring circ-UMFF: Generalization of circ-UMFF

Definitions 25 and 26 encourage considering conjugacy classes for substrings $\mathcal{L}_x x_i$ rather than individual letters:

Definition 27. *Given a string* $x = \mathcal{L}_x x_1 \mathcal{L}_x x_2 \cdots x_{j-1} \mathcal{L}_x x_j$ *in V-form over an ordered alphabet* Σ, *with maximal letter* $\mathcal{L}_x$, *then a string* $y = \mathcal{R}_t(x) = \mathcal{L}_x x_t \mathcal{L}_x x_{t+1} \cdots x_{j-1} \mathcal{L}_x x_j \mathcal{L}_x x_1 \cdots \mathcal{L}_x x_{t-1}$ *is the* t^{th} **substring conjugate** *(or* **substring rotation***) of* $x = x[1..n]$.

We then obtain a form of the **xyz** Lemma 16 for substrings:

Corollary 1. *An FF* $\mathcal{W}$ *is an UMFF if and only if whenever* $xy, yz \in \mathcal{W}$ *for some nonempty* y, *where each of* xy, yz *and* y *are in V-form, then* $xyz \in \mathcal{W}$.

Then a natural generalization of circ-UMFFs is the *substring circ-UMFF* where a conjugate is selected from the conjugacy class of substrings of a string rather than the usual rotation of letters.

Definition 28. *An UMFF* $\mathcal{W}$ *over* Σ^+ *is a* **substring circ-UMFF** *if and only if it contains exactly one substring rotation of every primitive string* $x \in \Sigma^+$ *expressed in V-form.*

Definition 29. *A string* $x = \mathcal{L}_x x_1 \mathcal{L}_x x_2 \cdots x_{j-1} \mathcal{L}_x x_j$ *in V-form over an ordered alphabet* Σ, *with maximal letter* $\mathcal{L}_x$, *is said to be a* $\mathcal{T}_{lex}$-**word** *if it is the unique minimum in* $\mathcal{T}_{lex}$-*order in the conjugacy class of the* $\mathcal{L}_x x_i$.

To clarify, consider the primitive integer string 431412 where, with reference to V-form, $\mathcal{L} = 4$, then the conjugate 412431 is least in co-lexorder (lexorder of reversed strings) for the letter-based conjugates, while the substring conjugate 431412 is least in Lex-Ext co-lexorder in the comparison of 431412 and 412431. Note that depending on the particular context, the substrings may be of the form $\mathcal{L}_x x_i$ such as when using co-lexorder and $x_0 = \varepsilon$, or in the case of V-order, Lemma 7 Part (C3) shows that it suffices for the substrings to simply have the form x_i. The following then holds for circ-UMFFs over an ordered alphabet Σ and a letter $\mathcal{L}$ such that for $\sigma_i \in \Sigma$, $\sigma_i < \mathcal{L}$:

Theorem 30. *Suppose $\mathcal{T}$-order $\ll$ is a total order over Σ^*. (i) The class of border-free $\mathcal{T}_\ll$-words forms a circ-UMFF $\mathcal{T}$ over the conjugacy class of letters. (ii) The class of border-free $\mathcal{T}_{lex}$-words forms a substring circ-UMFF over the conjugacy class of substrings.*

Proof. For this we apply Lemma 16 and Theorem 3.1 in [10], that is Theorem 19, which describes the structure of a circ-UMFF.

Part (i). Let $\mathcal{T}$ consist of precisely the set of border-free $\mathcal{T}_\ll$-words over Σ^*, and we will show that $\mathcal{T}$ is a circ-UMFF. First, by construction of $\mathcal{T}$ every letter in Σ is in $\mathcal{T}$ so we proceed to consider strings of non-unit length.

Suppose that xy and yz, with x, y, z nonempty, are both border-free $\mathcal{T}_\ll$-words in $\mathcal{T}$, where by Definition 26 they must be primitive. Consider the string xyz and suppose that it is the repetition u^k, $k > 1$. If u is a proper prefix of xy then xy is bordered, whereas if xy is a prefix of u then yz is bordered, and so we conclude that xyz is primitive. Since xyz is primitive then it must have a border-free conjugate - the classic argument for this is that the conjugate in the conjugacy class which is a Lyndon word must be border-free. So we next consider a border-free conjugate $c^{\mathcal{T}}$ of xyz which is in $\mathcal{T}$.

So suppose that we do not have xyz minimal in $\mathcal{T}$-order in its conjugacy class and border-free. Let $x = x_1 x_2 \cdots x_r$, $y = y_1 y_2 \cdots y_s$ and $z = z_1 z_2 \cdots z_t$, $r, s, t \geq 1$. Assume then that a conjugate $c = x_{c+1} \cdots x_r yz x_1 \cdots x_c$, $c + 1 > 1$ and $r > 1$, is minimal, border-free, and belongs to $\mathcal{T}$. Then applying Lemma 16 to distinct c and xy would imply that the bordered word $x_{c+1} \cdots x_r yz xy$ is in $\mathcal{T}$. So assume that a conjugate $c' = y_{d+1} \cdots y_s zx y_1 \cdots y_d$, $d + 1 \geq 0$, is minimal and belongs to $\mathcal{T}$. Again applying Lemma 16 to distinct c' and xy would imply that the bordered word $y_{d+1} \cdots y_s zxy$ is in $\mathcal{T}$. Finally, assume that $c'' = z_{e+1} \cdots z_t xyz_1 \cdots z_e$, $e + 1 > 1$, and a similar argument for c'' and yz would imply that the bordered word $z_{e+1} \cdots z_t xyz$ is in $\mathcal{T}$. Hence the primitive conjugate xyz is border-free and so must be the one, $c^{\mathcal{T}}$, in $\mathcal{T}$, and by construction of $\mathcal{T}$, xyz must be least in $\mathcal{T}$-order $\ll$ in its conjugacy class. Applying Lemma 16, since xy, yz and xyz with non-empty y all belong to $\mathcal{T}$ we have that $\mathcal{T}$ is an UMFF. From each conjugacy class of a primitive string we have selected a border-free word for $\mathcal{T}$ thus satisfying Definition 17 and moreover circ-UMFFs are necessarily border-free (Theorem 19 Part (1)). Hence we conclude that $\mathcal{T}$ is a circ-UMFF.

Part (ii). This follows similarly to Part (i) and by substituting Definition 26 for $\mathcal{T}_\ll$-words with Definition 29 for $\mathcal{T}_{lex}$-words and applying Corollary 1 for substrings. Here we let $\mathcal{T}^{\mathcal{S}}$ consist of precisely the set of border-free $\mathcal{T}_{lex}$-words over Σ^*. Then, by construction of $\mathcal{T}^{\mathcal{S}}$, the strings of length one have the form $\mathcal{L}w$ where $w \in \Sigma^*$. So if w is the empty string we get the word $\mathcal{L}$. Then strings of length ℓ will have ℓ occurrences of $\mathcal{L}$. In the analysis of $\mathcal{T}^{\mathcal{S}}$, for the existence of a border-free substring conjugate, we refer to the circ-UMFF IL of Indeterminate Lyndon Words [18] whose words are necessarily border-free (an indeterminate string on an alphabet Σ is a sequence of nonempty subsets of Σ). Then there exists a border-free indeterminate Lyndon word in the substring conjugacy class. The rest of the argument follows as in Part (i).   □

Observe that Theorem 30 shows that if a circ-UMFF or a substring circ-UMFF is defined using a total order (which is not necessary) then every element of the (substring) circ-UMFF is obtained using the same total order and no other ordering technique. Observe further that the proof did not depend on any particular method of totally ordering Σ^*, however the method must be total for border-free primitive strings in Σ^*.

We illustrate concepts from Theorem 30 with the following:

Example 31. *Consider the primitive and border-free integer string* $x = 3177412$. *Then the (unordered) conjugacy class of* x *is given by*

$$
\begin{array}{ccccccc}
3 & 1 & 7 & 7 & 4 & 1 & 2 \\
2 & 3 & 1 & 7 & 7 & 4 & 1 \\
1 & 2 & 3 & 1 & 7 & 7 & 4 \\
4 & 1 & 2 & 3 & 1 & 7 & 7 \\
7 & 4 & 1 & 2 & 3 & 1 & 7 \\
7 & 7 & 4 & 1 & 2 & 3 & 1 \\
1 & 7 & 7 & 4 & 1 & 2 & 3 \\
\end{array}
$$

The third conjugate, 1231774, is the Lyndon word as it is least in lexorder. The sixth conjugate, 7741231, is the V-word as it is least in V-order; it is also a co-lexorder word as it is least in co-lexorder; furthermore, it is least in relex order (reverse lexorder). The fifth conjugate, 7412317, is the sixth largest in lexorder but is a bordered word. And the seventh conjugate, 1774123, is least in alternating lexorder (indexing strings from 1, odd indexed letters are compared with < and even with >).

Implementing the FM-Index in V-order was considered in [16] leading to V-order substring pattern matching using backward search whereby computing only on the k conjugates starting with the greatest letter, essentially a substring circ-UMFF, reduced the BWT matrix to $O(nk)$ space. Hence, the substring circ-UMFF concept promises future optimization opportunities in particular related to indexing and pattern matching applications.

7   Galois Words

Finally, we show the necessity of the border-free requirement in Theorem 30 with Galois words [24], which are based on alternating lexicographic order (alternating lexorder, denoted $\prec_{alt}$), defined informally for comparing conjugates as: starting with <, string letters are compared in alternating < and > order. The Alternating Burrows-Wheeler Transform (ABWT) is an analogous transform to the

BWT which applies alternating lexorder [22]; an algorithmic perspective of the ABWT is given in [23].

It is straightforward to show that $\prec_{alt}$ is a total order and so a candidate for Theorem 30, so consider an analogous concept to Lyndon and V-words:

Definition 32. *[24] A primitive word w is a **Galois word** if for each non-trivial factorization $w = uv$, $w \prec_{alt} vu$.*

Examples of Galois words: ab, aba, abb, $abba$, $ababa$, $ababaa$, $ababba$. Observe that these words are not necessarily border-free and can be palindromic. Applying Lemma 16 to the Galois words $xy = ababa$ and $yz = ab$ with $y = a$ then $xyz = ababab$, namely a repetition, while Galois words are necessarily primitive. Hence, Galois words do not form an UMFF or therefore circ-UMFF. A related sentiment (Example 44 in [19]) shows that, while it follows from Theorem 20 that the unique maximal Lyndon factorization of a word has the least number of Lyndon factors, this is not necessarily the case with Galois words.

8   Concluding Comments

The concept of circ-UMFFs for uniquely factoring strings is a generalization of Lyndon words, which are known to be border-free string conjugates. The literature includes instances of circ-UMFFs for both regular and indeterminate (degenerate) strings. In this paper we have extended current knowledge on circ-UMFF theory including a further generalization to substring circ-UMFFs. Known instances of circ-UMFFs have been defined using a total order over Σ^*, such as V-order for arbitrary alphabets generating V-words, binary B-order generating B-words, and lex-extension order generating indeterminate Lyndon words.

We establish here that given any total ordering methodology $\mathcal{T}$ over Σ^*, and a subset of Σ^* consisting of border-free conjugates minimal in $\mathcal{T}$-order, the subset defines a circ-UMFF. An analogous result is established for substring circ-UMFFs. The border-free requirement is illustrated using Galois words by showing that they do not necessarily yield unique maximal string factorization – Galois words are thus worthy of deeper investigation in this context.

We have also delved further into the relationship between Lyndon and V-words, in particular showing that there are infinitely more V-words than Lyndon words. Novel concepts are illustrated throughout.

Acknowledgements. Funding: The third author was funded by the Natural Sciences & Engineering Research Council of Canada [Grant Number 10536797].

References

1. Alatabbi, A., Daykin, J.W., Kärkkäinen, J., Rahman, M.S., Smyth, W.F.: V-Order: new combinatorial properties & a simple comparison algorithm. Discrete Appl. Math. **215**, 41–46 (2016)
2. Alatabbi, A., Daykin, J.W., Mhaskar, N., Rahman, M.S., Smyth, W.F.: A faster V-order string comparison algorithm. In: Proceedings of Prague Stringology Conference, pp. 38–49 (2018)
3. Alatabbi, A., Daykin, J., Rahman, M.S., Smyth, W.F.: Simple linear comparison of strings in *V*-order. In: Pal, S.P., Sadakane, K. (eds.) WALCOM 2014. LNCS, vol. 8344, pp. 80–89. Springer, Cham (2014). https://doi.org/10.1007/978-3-319-04657-0_10
4. Alatabbi, A., Daykin, J.W., Rahman, M.S., Smyth, W.F.: String comparison in V-order: new lexicographic properties & on-line applications. arXiv:1507.07038 (2015)
5. Alatabbi, A., Daykin, J.W., Rahman, M.S., Smyth, W.F.: Simple linear comparison of strings in V-order. Fundam. Inform. **139**(2), 115–126 (2015). https://doi.org/10.3233/FI-2015-1228
6. Chen, K.T., Fox, R.H., Lyndon, R.C.: Free differential calculus, IV - the quotient groups of the lower central series. Ann. Math. **68**, 81–95 (1958)
7. Crochemore, M., Hancart, C., Lecroq, T.: Algorithms on Strings. Cambridge University Press, New York (2007)
8. Danh, T.N., Daykin, D.E.: The structure of V-order for integer vectors. In: Hilton, A.J.W. (ed.) Congressus Numerantium, vol. 113, pp. 43–53. Utilitas Mathematica Publishing Inc., Winnipeg (1996)
9. Daykin, D.E., Daykin, J.W.: Lyndon-like and V-order factorizations of strings. J. Discrete Algorithms **1**(3–4), 357–365 (2003)
10. Daykin, D.E., Daykin, J.W.: Properties and construction of unique maximal factorization families for strings. Internat. J. Found. Comput. Sci. **19**(4), 1073–1084 (2008)
11. Daykin, D.E., Daykin, J.W., Smyth, W.F.: Combinatorics of unique maximal factorization families (UMFFs). Fund. Inform. **97**(3), 295–309 (2009)
12. Daykin, D.E., Daykin, J.W., Smyth, W.F.: String comparison and Lyndon-like factorization using V-order in linear time. In: Symposium on Combinatorial Pattern Matching, vol. 6661, pp. 65–76 (2011)
13. Daykin, D.E., Daykin, J.W., Smyth, W.F.: A linear partitioning algorithm for hybrid Lyndons using V-order. Theoret. Comput. Sci. **483**, 149–161 (2013)
14. Daykin, J.W., et al.: Efficient pattern matching in degenerate strings with the Burrows-Wheeler transform. Inf. Process. Lett. **147**, 82–87 (2019). https://doi.org/10.1016/j.ipl.2019.03.003
15. Daykin, J.W., et al.: Binary block order Rouen transform. Theor. Comput. Sci. **656**, 118–134 (2016). https://doi.org/10.1016/j.tcs.2016.05.028
16. Daykin, J.W., Mhaskar, N., Smyth, W.F.: Computation of the suffix array, Burrows-Wheeler transform and FM-index in V-order. Theor. Comput. Sci. **880**, 82–96 (2021). https://doi.org/10.1016/j.tcs.2021.06.004
17. Daykin, J.W., Smyth, W.F.: A bijective variant of the Burrows-Wheeler transform using V-order. Theoret. Comput. Sci. **531**, 77–89 (2014)
18. Daykin, J.W., Watson, B.W.: Indeterminate string factorizations and degenerate text transformations. Math. Comput. Sci. **11**(2), 209–218 (2017). https://doi.org/10.1007/s11786-016-0285-x

19. Dolce, F., Restivo, A., Reutenauer, C.: On generalized Lyndon words. Theor. Comput. Sci. **777**, 232–242 (2019). https://doi.org/10.1016/j.tcs.2018.12.015
20. Duval, J.P.: Factorizing words over an ordered alphabet. J. Algorithms **4**(4), 363–381 (1983)
21. Ferragina, P., Manzini, G.: Opportunistic data structures with applications. In: Proceedings of 41st Annual Symposium on Foundations of Computer Science, (FOCS 2000), pp. 390–398 (2000)
22. Gessel, I.M., Restivo, A., Reutenauer, C.: A bijection between words and multisets of necklaces. Eur. J. Comb. **33**(7), 1537–1546 (2012). https://doi.org/10.1016/j.ejc.2012.03.016
23. Giancarlo, R., Manzini, G., Restivo, A., Rosone, G., Sciortino, M.: The alternating BWT: an algorithmic perspective. Theor. Comput. Sci. **812**, 230–243 (2020). https://doi.org/10.1016/j.tcs.2019.11.002
24. Reutenauer, C.: Mots de Lyndon généralisés 54. Sém Lothar. Combin. (B54h), pp 16 (2006)
25. Smyth, B.: Computing Patterns in Strings. Pearson/Addison-Wesley (2003)

The Two-Center Problem of Uncertain Points on Trees

Haitao Xu$^{(\boxtimes)}$ and Jingru Zhang

Cleveland State University, Cleveland, OH 44115, USA
h.xu12@vikes.csuohio.edu, j.zhang40@csuohio.edu

Abstract. In this paper, we consider the (weighted) two-center problem of uncertain points on a tree. Given are a tree T and a set $\mathcal{P}$ of n (weighted) uncertain points each of which has m possible locations on T associated with probabilities. The goal is to compute two points on T, i.e., two centers with respect to $\mathcal{P}$, so that the maximum (weighted) expected distance of n uncertain points to their own expected closest center is minimized. This problem can be solved in $O(|T| + n^2 \log n \log mn + mn \log^2 mn \log n)$ time by the algorithm for the general k-center problem. In this paper, we give a more efficient and simple algorithm that solves this problem in $O(|T| + mn \log mn)$ time.

Keywords: Algorithms · Two-center · Trees · Uncertain points

1 Introduction

Facility locations play a significant role in operations research due to its wide applications in transportation, sensor deployments, circuit design, etc. Consider the inherent uncertainty of collected data caused by measurement errors, sampling discrepancy, and object mobility. It is natural to consider facility location problems on uncertain points. The locational model for uncertain points has been considered a lot in facility locations [10, 14, 16], where the location of an uncertain point is represented by a probability density function (pdf). In this paper, we study the two-center problem, one of classical facility location problems, for uncertain points on a tree under the locational model.

Let T be a tree. We consider each edge e of T as a line segment of a positive length so that we can talk about "points" on e. Formally, we specify a point x of T by an edge e that contains x and the distance between x and an incident vertex of e. For any two points p and q on T, the distance $d(p, q)$ is the sum of all edges on the simple path from p to q. Let $\mathcal{P}$ be a set of n uncertain points $P_1, \cdots, P_n$. Each $P_i \in \mathcal{P}$ is associated with m locations $p_{i1}, p_{i2}, \cdots, p_{im}$ each being a point on T, and each location p_{ij} has a probability $f_{ij} \geq 0$ for P_i appearing at location p_{ij}. Additionally, each P_i has a non-negative weight w_i.

For any (deterministic) point x on T, the distance of any uncertain point P_i to x is the *expected* version and defined as $\sum_{j=1}^{m} f_{ij} \cdot d(p_{ij}, x)$. We use $\mathsf{Ed}(P_i, x)$ to denote the expected distance of P_i to x. Let x_1 and x_2 be any points on T. We say that an uncertain point P_i is *expectedly closer* to x_1 if $\mathsf{Ed}(P_i, x_1) \leq \mathsf{Ed}(P_i, x_2)$.

W. Wu and J. Guo (Eds.): COCOA 2023, LNCS 14461, pp. 485–497, 2024.
https://doi.org/10.1007/978-3-031-49611-0_35

Define $\phi(x_1, x_2) = \max_{1 \le i \le n}\{w_i \cdot \min(\mathsf{Ed}(P_i, x_1), \mathsf{Ed}(P_i, x_2))\}$. The *two-center* problem aims to compute two points on T so as to minimize $\phi(x_1, x_2)$ and the two optimal points, denoted by q_1^* and q_2^*, are called centers with respect to $\mathcal{P}$ on T. We say that P_i is *covered* by q_1^* if P_i is expectedly closer to q_1^*.

The algorithm [17] for the general k-center problem can address our problem in $O(|T| + n^2 \log n \log mn + mn \log^2 mn \log n)$ time. In this paper, however, we present an $O(|T| + mn \log mn)$-time algorithm with the assistance of our proposed linear-time approach for the decision two-center problem. Note that the time complexity of our algorithm almost matches the $O(|T| + mn \log m + n \log n)$ result [18] for the case of T being a path.

1.1   Related Work

If every P_i has exactly one location then the problem falls into the deterministic case. Ben-Moshe et al. [3] adapted Megiddo's prune-and-search technique [13] to solve in $O(n)$ time the deterministic two-center on a tree where each vertex is a demand point. On a cactus graph, the two-center problem was addressed in their another work [2], and an $O(n \log^3 n)$-time algorithm was proposed. On a general graph, Bhattacharya and Shi [5] reduced the decision problem into the two-dimensional Klee's measure problem [7] so that the problem can be solved in polynomial time. The planar version was studied in several works [6,9,15]. The state-of-the-art result is an $O(n \log^2 n)$ deterministic algorithm given by Wang [15].

In general, every P_i has more than $m > 1$ locations on T. As mentioned above, Wang and Zhang [17] considered the general k-center problem so that the two-center problem can be addressed in $O(|T| + n^2 \log n \log mn + mn \log^2 mn \log n)$ time. If T is a path, the two-center was solved in $O(|T| + mn \log m + n \log n)$ time in our previous work [18]. One of the most related problems is the one-center problem. Wang and Zhang [16] generalized Megiddo's prune-and-search technique to solve the one-center of $\mathcal{P}$ on a tree in linear time. Hu and Zhang [10] studied the uncertain one-center problem on a cactus graph and proposed an $O(|T| + mn \log mn)$-time algorithm. Moreover, Li and Huang [11] considered the planar Euclidean uncertain k-center and gave an approximation algorithm. Later, Alipour and Jafari [1] improved their result to an $O(3 + \epsilon)$-approximation and proposed a 10-approximation algorithm for any metric space.

1.2   Our Approach

The locations of the uncertain points of $\mathcal{P}$ may be in the interior of edges of T. A vertex-constrained case happens when all locations are at vertices of T and each vertex of T contains locations. As shown in [17], any general case can be reduced to a vertex-constrained case. In the following, we focus on discussing the vertex-constrained case.

Let λ^* be the minimized objective value. The median of each $P_i \in \mathcal{P}$ is the point where $\mathsf{Ed}(P_i, x)$ reaches its minimum. We first solve the decision problem that determines whether $\lambda \ge \lambda^*$ for any given λ, i.e., decide if λ is *feasible*.

This can be addressed in $O(mn \log^2 mn)$ time by the algorithm [17], which relies on several dynamic data structures requiring $O(mn \log^2 mn)$-time constructions. Since the convexity of each $\mathsf{Ed}(P_i, x)$ on any path, we develop a much simpler algorithm that is free of any data structures and runs faster in $O(mn)$ time.

Regarding our decision problem, there is an observation that there exists an edge e on T such that a center must be placed on e to cover all uncertain points whose medians are on one of the two subtrees generated by removing e from T. Such an edge is called a *peripheral-center* edge. Our decision algorithm first computes a peripheral-center edge in $O(mn)$ time. With this edge, the feasibility of λ can be known in $O(mn)$ time.

To compute centers q_1^* and q_2^*, we first compute the two *critical* edges that respectively contain q_1^* and q_2^*. Because an essential lemma can decide in $O(mn)$ time which split subtree of any given point contains a critical edge. Our algorithm finds each critical edge on T recursively with the assistance of this lemma. Once the two edges are found, q_1^* and q_2^* can be computed in $O(mn \log n)$ time.

2 Preliminaries

Let u and v be any two vertices on T. Denote by $e(u, v)$ the edge incident to both u and v. For any two points p and q on T, we let $\pi(p, q)$ be the simple path between p and q. As in [17], the lowest common ancestor data structure [4] can be applied to T so that with an $O(mn)$ preprocessing work, the path length of $\pi(p, q)$, i.e., the distance $d(p, q)$, can be known in constant time.

Let π be any simple path on T and x be any point on π. For any $P_i \in \mathcal{P}$, as analyzed in [16], $\mathsf{Ed}(P_i, x)$ is a convex piece-wise linear function in $x \in \pi$, and it monotonically increases or decreases as x moves on any edge of π from one ending vertex to the other.

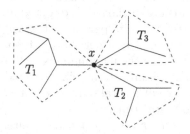

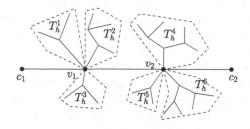

Fig. 1. The point x has three split subtrees T_1, T_2 and T_3.

Fig. 2. Illustrating the tree T_h for the case $C = 2$ where $V = \{v_1, v_2\}$ and $\Gamma(V) = \{T_h^1, \cdots, T_h^6\}$.

Let x be any point on T. Removing x from T generates several disjoint subtrees. Consider x as an "open vertex" on each subtree that is free of any locations. We call each obtained subtree a *split subtree* of x. See Fig. 1 for an example. For any $P_i \in \mathcal{P}$ and any subtree T' of T, we refer to the sum of probabilities of P_i's all locations in T' as the *probability sum* of P_i in T'. Denote by p_i^* the median of P_i that minimizes $\mathsf{Ed}(P_i, x)$ among all points on T. The following lemma was given in [16].

Lemma 1. *[16] Consider any point x on T and any uncertain point P_i of $\mathcal{P}$.*

1. *If x has a split subtree whose probability sum of P_i is greater than 0.5, then p_i^* must be in that split subtree.*
2. *The point x is p_i^* if the probability sum of P_i in each of x's split subtrees is less than 0.5.*
3. *The point x is p_i^* if x has a split subtree in which the probability sum of P_i is equal to 0.5.*

Note that the median of P_i may not be unique. But all points on T minimizing $\mathsf{Ed}(P_i, x)$ induce a connected subtree of T. Let p_i^* be any of these points. Due to the convexity of $\mathsf{Ed}(P_i, x)$, we have that $\mathsf{Ed}(P_i, x)$ monotonically increases as x moves along any path away from p_i^*.

3   The Decision Algorithm

Given any value $\lambda > 0$, the decision problem is to determine whether there exist no more than two points, i.e., centers, on T so that the expected distance of each $P_i \in \mathcal{P}$ to at least one of them is at most λ. If yes, then $\lambda \geq \lambda^*$ and so λ is *feasible*. Otherwise, $\lambda < \lambda^*$ and λ is *infeasible*. Clearly, λ^* is the smallest feasible value.

To establish a context for our work, we first introduce the algorithm [17] for the decision k-center problem. Let T_m be the minimum subtree of T spanning the medians of n uncertain points. We say that an uncertain point is covered by a center if their expected distance is at most λ. The convexity of $\mathsf{Ed}(P_i, x)$ leads the greedy algorithm that places minimum centers in the bottom-up manner whenever it "has to." During the post-order traversal, for each vertex v, considering all 'active' uncertain points whose medians are in the subtree rooted at v, a center must be placed on the edge incident to v and its parent vertex u if one of these uncertain points has its expected distance at u larger than λ. If so, then the center is placed on this edge at the point where the maximum expected distance of these uncertain points equals λ, and next all uncertain points covered by this center are "deactivated." With the assistance of several data structures, minimum centers can be placed on T in $O(mn \log^2 mn)$ time.

We say that an edge e of T is a *peripheral-center* edge if a center must be placed on e so that this center is determined by all uncertain points whose medians are on one of the two subtrees generated by removing e from T (but keeping its two incident vertices), and the center on e is a *peripheral* center.

Regarding our problem, the goal is to determine whether two peripheral centers can be found on T so that the expected distance of every uncertain point to at least one of them is no more than λ. Our algorithm thus first finds a peripheral-center edge and then applies Lemma 2 to decide the feasibility of the given λ. If it is feasible then the two centers are returned. The proof of Lemma 2 is in the full paper.

Lemma 2. *Given a peripheral-center edge e on T, we can determine in $O(mn)$ time whether $\lambda \geq \lambda^*$, and if yes, the two centers can be computed in $O(mn)$ time.*

To find a peripheral edge on T, each round of our algorithm consists of two pruning steps: The first pruning step is a recursive procedure that "shrinks" T to find a peripheral-center edge recursively. After at most $\log m + 1$ recursive steps, we obtain a subtree of at most $|\mathcal{P}|/2$ vertices. It follows that at least a quarter of uncertain points are pruned from $\mathcal{P}$ in the second step. After at most $\log n$ rounds, only $O(1)$ uncertain points remain, or a subtree of T with $O(1)$ vertices is obtained. At this point, a peripheral-center edge can be found in $O(mn)$ time.

3.1   The First Pruning Step

We first compute the *centroid* c of T, which is a vertex so that every split subtree of c has no more than $|T|/2$ vertices, in $O(|T|)$ time by traversing the tree [12,13]. We then decide whether a center is at c, and if yes, then Lemma 2 is used to decide the feasibility of λ in $O(mn)$ time. Otherwise, we determine which split subtree of c contains a peripheral-center edge. Lemma 6 can be utilized to solve this problem in $O(|T|)$ time. Note that if c has more than one such split subtree then c is associated with a flag equal to true in Lemma 6.

Next, we set c as a *connector* on the obtained split subtree T' in $O(|T|)$ time as [16] by traversing the *connector subtree* $\{c\} \cup (T/T')$, denoted by $T'(c)$, to compute the location information of each P_i in $T'(c)$. Specifically, we first create two *information arrays* $F_c[1 \cdots n]$ and $D_c[1 \cdots n]$. Initialize them as zero. We then traverse $T'(c)$ and during the traversal, for each location p_{ij}, we add f_{ij} to $F_c[i]$ and add value $w_i f_{ij} \cdot d(p_{ij}, c)$ to $D_c[i]$. Last, we associate c on T' with the two information arrays. Clearly, for each $1 \leq i \leq n$, $F_c[i]$ is the probability sum of P_i in $T'(c)$, and $D_c[i]$ is called the *distance sum* of P_i's locations in $T'(c)$ to c.

In $O(|T|)$ time, we obtain a subtree of at most $|T|/2$ vertices that must contain a peripheral-center edge. We continue to perform the above procedure recursively on T' to find a peripheral-center edge. Suppose we are about to perform the h-th recursive step. Denote by T_{h-1} the obtained subtree after the $(h-1)$-th recursive step. T_{h-1} consists of at most $|T|/2^{h-1}$ vertices and at most $h - 1$ connectors.

Similarly, we first compute its centroid c in $O(|T_{h-1}|)$ time. During the traversal, we also count the number of vertices with true flags. If more than one such vertices exist, then at least three centers must be placed on T to cover $\mathcal{P}$ under λ since two (pruned) connector subtrees and T_h each must contain center(s) and so we return $\lambda < \lambda^*$. Otherwise, we decide if one center must be placed at c. If yes, Lemma 2 is applied to decide λ's feasibility immediately in $O(mn)$ time. Otherwise, we determine which split subtree of c contains a peripheral-center edge. Lemma 6 can address the problem in $O(|T_{h-1}| + n \cdot (h - 1))$ time.

In general, we obtain a subtree T_h of at most $|T|/2^h$ vertices, and it must contain a peripheral-center edge. Note that T_h has h connectors. We then set c as a connector on T_h by computing its information arrays $F_c[1 \cdots n]$ and $D_c[1 \cdots n]$.

This can be done in the above way except that when we visit a connector u on T_h, we scan its information arrays to add $F_u[i]$ to $F_c[i]$ and add value $D_u[i] + w_i F_u[i] d(u, c)$ to $D_c[i]$ for each $1 \leq i \leq n$. Thus, the time complexity of the h-th recursive step is $O(|T_{h-1}| + n \cdot (h - 1))$.

We perform the above procedure for $h = 1 + \log m$ recursive steps. At this moment, by the definition of the centroid, we have $|T_h| \leq |T|/2^h = mn/2^h = n/2$. The running time of the h recursive steps, i.e., the first pruning step, is $O(\sum_{i=0}^{h-1}(T_i + i \cdot n))$, which is $O(mn)$ due to $T_0 = T$ and $h = 1 + \log m$.

Due to $|T_h| \leq n/2$, we can see that there are at least $n/2$ uncertain points in $\mathcal{P}$ so that they have no locations in T_h. Let $\mathcal{P}'$ be the subset of these uncertain points. To compute this subset, we create an array $I[1 \cdots n]$ so that $P_i \in \mathcal{P}'$ if $I[i]$ is true. It is easy to see that $I[1 \cdots n]$, i.e., $\mathcal{P}'$, can be computed in $O(|T_h|)$ time by traversing T_h. Below we will show in the second pruning step that at least half of $\mathcal{P}'$, i.e., a quarter of uncertain points in $\mathcal{P}$, can be pruned.

3.2   Pruning Uncertain Points

We first traverse T_h to count the number of vertices with true flags. If more than one such vertices exist, then we return $\lambda < \lambda^*$ in that at least three centers are needed to be placed on T. Otherwise, denote by C the number of connectors on T_h. Depending on the value of C, our algorithm will proceed accordingly for three cases: $C = 1$, $C = 2$ and $C > 2$.

The Case $C = 1$: Let c be the connector on T_h. If the flag of c is true, then the connector subtree $T_h(c)$ must contain centers. By Lemma 6, we also have that peripheral-center edge(s) on T_h must be decided by uncertain points with medians on T_h, i.e., whose probability sums in T_h are at least 0.5. Because Lemma 6 always returns a subtree without any true-flag vertices unless no such subtrees exist. Keeping c's flag being true leads that $\mathcal{P}'$ can be pruned as computing a peripheral-center edge on T_h further. So, we keep entries of $I[1 \cdots n]$ for $\mathcal{P}'$ being true.

Otherwise, the connector subtree $T_h(c)$ does not contain any centers. Denote by x_t any point on T_h at distance t to c. For each $P_i \in \mathcal{P}'$, since all its locations are in $T_h(c)$, we have $\mathsf{Ed}(P_i, x_t) = \mathsf{Ed}(P_i, c) + w_i \cdot t$ where $\mathsf{Ed}(P_i, c) = D_c[i]$. Denote by $P_{i'}$ the uncertain point of $\mathcal{P}'$ that determines the furthest point to c on T_h to cover $\mathcal{P}'$, i.e., that has the smallest t-value by resolving $\mathsf{Ed}(P_i, c) + w_i \cdot t = \lambda$ for all $P_i \in \mathcal{P}'$. Clearly, any point on T_h that covers $P_{i'}$ must cover all in $\mathcal{P}'$. Thus, $\mathcal{P}' - \{P_{i'}\}$ can be pruned in the further step to compute a peripheral-center edge on T_h. To find $P_{i'}$, we scan $I[1 \cdots n]$ to compute the t-value of each $P_i \in \mathcal{P}'$: If $I[i]$ is true, we compute in $O(1)$ time $(\lambda - \mathsf{Ed}(P_i, c))/w_i$, i.e., the t-value. The larger t-value is always maintained during the scan so that $P_{i'}$ can be found in $O(n)$ time. Last, we set $I[i']$ as false. Hence, it takes $O(n)$ time to find all uncertain points of $\mathcal{P}'$ that will be pruned.

It can be seen that for either case, at least a half of $\mathcal{P}'$, that is, a quarter of uncertain points in $\mathcal{P}$, can be pruned as we further search for peripheral-center edges on T_h. Additionally, if $I[i]$ for any $1 \leq i \leq n$ is true then P_i can be pruned. Last, we reconstruct a tree to prune these uncertain points. Traverse the connector subtree $T_h(c)$, i.e., $\{c\} \cup T/T_h$. For each location p_{ij} on $T_h(c)$ with $I[i]$ being false, we create a *dummy* vertex v, set v as an incident vertex of c on T_h by a *dummy* edge of length $d(c, v)$, and reassign p_{ij} to v. These can be carried out totally in $O(|T|)$ time.

A tree T^+ is thus obtained: T^+ contains at most $3n/4$ uncertain points and hence its size is $3mn/4$. T^+ has dummy vertices and each of them is a leaf. Additionally, the subtree generated by removing all dummy vertices from T^+ is exactly T_h which contains a peripheral-center edge. As the flag of c on T^+ is maintained, computing a peripheral-center edge on T_h is equivalent to computing a non-dummy peripheral-center edge on T^+.

The Case $C = 2$: Let c_1 and c_2 be the two connectors on T_h. Consider the path $\pi(c_1, c_2)$ between c_1 and c_2. Let V be the set of vertices on $\pi(c_1, c_2)$ except for c_1 and c_2. For any vertex $v \in V$, denote by $\Gamma(v)$ the set of all split subtrees of v in T_h excluding the two containing c_1 and c_2. Let $\Gamma(V) = \cup_{v \in V} \Gamma(v)$. V and $\Gamma(V)$ can be computed in $O(T_h)$ time. See Fig. 2 for an example.

We proceed with determining whether $\Gamma(V)$ contains peripheral-center edges. This is an instance of the *center-edge detecting* problem, defined below in Sect. 3.4, and it can be solved in $O(|T|)$ time by Lemma 7. If one split subtree T' of $\Gamma(V)$ is returned then we can reduce this case to the case $C = 1$ by setting the vertex v' of T' in V as a connector on T' in $O(|T_h| + n)$ time. Note that the flag of v' is set properly in Lemma 7. Otherwise, if no subtrees are returned then $\pi(c_1, c_2)$ must contain a peripheral-center edge. Depending on whether $T_h(c_1)$ and $T_h(c_2)$ contain centers, we have different pruning approaches.

Suppose c_1's flag is equal to true, i.e., $T_h(c_1)$ contains centers. Let $\mathcal{P}'_{c_1}$ be the subset of uncertain points in $\mathcal{P}'$ whose medians are in $T_h(c_1)$ and $\mathcal{P}'_{c_2}$ be the subset of the remaining uncertain points in $\mathcal{P}'$ whose medians are in $T_h(c_2)$. Lemma 6 implies that the peripheral-center edge on $\pi(c_1, c_2)$, i.e., that center, is decided by uncertain points whose medians are in $T/T_h(c_1)$. Keeping c_1's flag being true allows us to prune $\mathcal{P}'_{c_1}$ in the further rounds of computing a peripheral-center edge on $\pi(c_1, c_2)$. So, we keep entries of $I[1 \cdots n]$ for $\mathcal{P}'_{c_1}$ being true.

Next, we find uncertain points of $\mathcal{P}'_{c_2}$ that can be pruned in the further rounds. Let x_t be any point on $\pi(c_1, c_2)$ at distance t to c_1. Because all locations of uncertain points in $\mathcal{P}'$ are in the two connector subtrees. Each $P_i \in \mathcal{P}'$ has $\mathsf{Ed}(P_i, x_t) = \mathsf{Ed}(P_i, c_1) + w_i F_{c_1}[i] \cdot t - w_i F_{c_2}[i] \cdot t$, which is $D_{c_1}[i] + D_{c_2}[i] + w_i(F_{c_1}[i] - F_{c_2}[i])t + w_i F_{c_2}[i]d(c_1, c_2)$. Clearly, $\mathsf{Ed}(P_i, x_t)$ changes linearly as x_t moves from c_1 to c_2. We resolve $\mathsf{Ed}(P_i, x_t) = \lambda$ in $O(1)$ time for each $P_i \in \mathcal{P}'_{c_2}$. (For any $1 \leq i \leq n$, if $I[i]$ is true and $F_{c_2}[i] \geq 0.5$ then $P_i \in \mathcal{P}'_{c_2}$.) Let t' be the largest value among all obtained, and t' defines the furthest point to c_2 on $\pi(c_1, c_2)$ that covers $\mathcal{P}'_{c_2}$.

If $t' < 0$ then $\mathsf{Ed}(P_i, x) < \lambda$ for each $P_i \in \mathcal{P}'_{c_2}$ and any $x \in \pi(c_1, c_2)$. This means that any center on $\pi(c_1, c_2)$ is irrelevant to $\mathcal{P}'_{c_2}$. We thus are allowed to prune $\mathcal{P}'_{c_2}$ when we shall compute a peripheral-center edge on $\pi(c_1, c_2)$. Otherwise, $t' \geq 0$ and let $x_{t'}$ be the point at distance t' to c_1 on $\pi(c_1, c_2)$. We then call

Lemma 6 on $x_{t'}$ to decide in $O(|T_h| + n)$ time whether $x_{t'}$ must contain a center. If yes, then the feasibility of λ can be decided in $O(mn)$ time by Lemma 2, and otherwise, a peripheral-center edge must be on $\pi(x_{t'}, c_2)$ in that the split subtrees in $\Gamma(V)$ of each v on $\pi(x_{t'}, c_2)$ do not contain centers. It follows that $\mathcal{P}'_{c_2}$ can be pruned during the further searching for the peripheral-center edge on $\pi(x_{t'}, c_2)$. Hence, we join a vertex v' for $x_{t'}$ into the path if $x_{t'}$ is in the interior of an edge, and set the flag of v' as true in $O(1)$ time to find a peripheral-center edge on $\pi(v', c_2)$ in further steps. For either $t' < 0$ or $t' \geq 0$, $\mathcal{P}'_{c_2}$ can be pruned, and so we keep entries of $I[1 \cdots n]$ being true for $\mathcal{P}'_{c_2}$.

It is clear to see that if one of c_1 and c_2 has a true flag, we find in $O(|T|)$ time a subpath π of $\pi(c_1, c_2)$ that contains a peripheral-center edge and at least a half of uncertain points in $\mathcal{P}'$ that can be pruned in the further rounds of computing a peripheral-center edge on π. Notice that the flags of vertices are maintained in the above procedure.

On the other hand, both c_1 and c_2 have false flags. We first compute the furthest point x_1 (resp., x_2) to c_1 (resp., c_2) on $\pi(c_1, c_2)$ that covers $\mathcal{P}'_{c_1}$ (resp., $\mathcal{P}'_{c_2}$), which can be computed in $O(n)$ time as the above. We then find irrelevant uncertain points for each below case.

In the first case, x_1 is to the left of x_2. There must be a peripheral-center edge on $\pi(x_2, c_2)$ (resp., $\pi(c_1, x_1)$) if split subtrees in $\Gamma(V)$ of vertices on $\pi(x_2, c_2)$ (resp., $\pi(c_1, x_1)$) do not contain any vertex with a true flag, which can be determined in $O(|T_h|)$ time. Suppose it is on $\pi(x_2, c_2)$. Since any point on $\pi(x_2, c_2)$ cannot cover $\mathcal{P}'_{c_1}$ under λ, $\mathcal{P}'_{c_1}$ can be pruned by setting the flag of x_2 as true in our further searching on $\pi(x_2, c_2)$. Additionally, all in $\mathcal{P}'_{c_2}$ except for the one determining x_2 can be pruned, which can be found in $O(n)$ time. So, we join a vertex v' of a true flag for x_2 if necessary, and set the entry in $I[1 \cdots n]$ as false for the uncertain point determining x_2. The case where $\pi(c_1, c_1)$ does contain a peripheral-center edge can be processed similarly. It is not hard to see that in $O(|T_h| + n)$ time we obtain a subpath π of $\pi(c_1, c_2)$ that must contain a peripheral-center edge, and all irrelevant uncertain points can be found in $O(n)$ time.

Another case is that x_1 is to the right of x_2. Similarly, if split subtrees in $\Gamma(V)$ of vertices on $\pi(c_1, x_2)$ (resp., $\pi(x_1, c_2)$) contain any vertex of a true flag then there must be a peripheral-center edge on $\pi(x_2, c_2)$ (resp., $\pi(c_1, x_1)$), which can be known in $O(|T_h|)$ time. If $\pi(x_2, c_2)$ (resp., $\pi(c_1, x_1)$) contains a peripheral-center edge, then $\mathcal{P}'$ can be pruned except for the one in $\mathcal{P}'_{c_2}$ (resp., $\mathcal{P}'_{c_1}$) determining x_2 (resp., x_1). For the situation where a peripheral-center edge is on $\pi(x_2, c_2)$ (resp., $\pi(c_1, x_1)$), we join a vertex for x_2 if necessary and set its flag as true, and then reset the entry of $I[1 \cdots i]$ as false for that uncertain point determines x_2 (resp., x_1). Clearly, all these operations can be done in $O(|T_h| + n)$ time.

Otherwise, only split subtrees in $\Gamma(V)$ of vertices on $\pi(x_2, x_1)/\{x_1, x_2\}$ may contain a true-flag vertex. If such a split subtree exists, supposing it intersects $\pi(c_1, c_2)$ at vertex v', then we apply Lemma 6 to v' to decide which split subtree of v' contains a peripheral-center edge. Because $\pi(c_1, c_2)$ must contain a peripheral-center edge. A split subtree of v' containing c_1 or c_2 must be returned. If it includes c_2 (resp., c_1), then $\mathcal{P}'$ can be pruned as we further compute a peripheral-center edge on $\pi(v', c_2)$ (resp., $\pi(c_1, v')$). For either case, we maintain the path containing a peripheral-center edge and set the flag of v' as true. Clearly, this case can be handled in $O(|T_h| + n)$ time as well.

If no true-flag vertices are found in split subtrees of $\Gamma(v)$ of vertices on $\pi(x_2, x_1)/\{x_1, x_2\}$, then solving the decision problem is equivalent to solving the *path-constrained* version that is to decide whether two points can be found on $\pi(c_1, c_2)$ to cover $\mathcal{P}$ by the given λ. This is because no centers are necessary to be placed on split subtrees of $\Gamma(V)$ and the two connectors subtrees. Moreover, the path-constrained decision version can be solved in $O(mn)$ time by Lemma 3, whose proof is in the full paper.

Lemma 3. *The path-constrained decision problem can be solved in $O(mn)$ time.*

It follows that for the case where x_1 is to the right of x_2, we spend $O(|T_h| + n)$ time on computing a subpath π of $\pi(c_1, c_2)$ that contains a peripheral-center edge, and at least a quarter of uncertain points of $\mathcal{P}$ that can be pruned in our further rounds on π.

Last, x_1 is exactly x_2. We first decide whether a center must be placed at x_1 by Lemma 6 in $O(|T_h| + n)$ time. If yes, then the feasibility of λ can be decided by Lemma 2 in $O(mn)$ time. Otherwise, one split subtree of x_1 is returned and it must contain either c_1 or c_2. Assume it contains c_2. We shall compute a peripheral-center edge on $\pi(x_2, c_2)$ so that all uncertain points in $\mathcal{P}'$ except for the one determining x_2 can be pruned. Similar to the above, we join a vertex v' for x_2, set its flag as true since there must be a center on the connector subtree $T_h(v')$, and reset the entry in $I[1 \cdots n]$ of that uncertain point leading x_2 to be true.

As a consequence, for the case $C = 2$, generally, in $O(|T|)$ time, either we reduce the problem into the case $C = 1$, or we obtain a subpath π on $\pi(c_1, c_2)$ that is known to contain a peripheral-center edge. For the former case, the above approach for the case $C = 1$ is applied to prune at least a quarter of uncertain points of $\mathcal{P}$ in $O(|T|)$ time. For the other case, supposing the two ending vertices of π are u and u', we then prune these irrelevant uncertain points as follows. Traverse the connector subtrees $T_h(u)$ and $T_h(u')$, which can be obtained in $O(|T|)$ time. For each location p_{ij} in $T(u)$ (resp., $T(u')$), we create a dummy vertex incident to u (resp., u') by a dummy edge of length $d(p_{ij}, u)$ (resp., $d(p_{ij}, u')$) if $I[i]$ is false. We perform the same procedure on each split subtree in $\Gamma(V)$ of each vertex v on $\pi(u, u')/\{u, u'\}$ and additionally, we set v's flag as true if its split subtrees in $\Gamma(V)$ have a true-flag vertex. It is not hard to see that in $O(|T|)$ time we obtain a tree T^+ so that T^+ contains at most $3n/4$ uncertain points and $T^+/\pi(u, u')$ are induced by dummy vertices. Clearly, computing a peripheral-center edge on T is equivalent to computing a non-dummy peripheral-center edge on T^+.

The Case $C > 2$: In general, T_h has more than two connectors. We utilize an approach similar to [16] to "shrink" T_h until the problem is reduced to one of the previous two cases.

A vertex z of T_h is called a *connector-centroid* if each split subtree of z has no more than $C/2$ connectors, which can be found in linear time [16]. We first compute the connector-centroid z of T_h and then determine whether a center must be at z, and if not, which split subtree of z contains a peripheral-center edge. These can be decided in $O(|T_h| + h \cdot n)$ time by applying Lemma 6 to z. Generally, a split subtree is returned. Set z as a connector on this subtree in $O(|T_h| + h \cdot n)$ time. Let T_{h+1} be the obtained tree. Clearly, the size of T_{h+1} is at most $|T_h|$ but it has no more than $C/2$ connectors. We perform the above procedure recursively on T_{h+1}. After at most $\log C$ steps, we obtain a subtree T' with at most two connectors. The total time complexity is $O(\sum_{i=1}^{\log C}(|T_h| + n \cdot C/2^i))$, which is $O(|T|)$ due to $C \le 1 + \log m$ and $T_h \le \frac{n}{2}$. It follows that the above $O(|T|)$ pruning approach is applied to T' accordingly to prune at least a quarter of uncertain points from $\mathcal{P}$. Consequently, a tree T^+ containing at most $3n/4$ uncertain points is achieved in $O(|T|)$ time so that computing a peripheral-center edge on T_h is equivalent to computing a non-dummy peripheral-center edge on T^+.

3.3   Wrapping Things Up

The above procedure gives an $O(|T|)$-time algorithm that computes a tree T^+ of at most $3n/4$ uncertain points and at most $3mn/4$ vertices, such that computing a peripheral-center edge on T is equivalent to computing a non-dummy peripheral-center edge on T^+. Note that all dummy vertices on T^+ are leaves and flags of vertices are maintained. Because our Lemma 6 and Lemma 7 are defined in Sect. 3.4 consider the situation where the given tree may contain dummy vertices and connectors. We thus continue the same procedure recursively on T^+ to compute a non-dummy peripheral-center edge.

After $h - 1$ rounds, the obtained tree T_{h-1}^+ consists of at most $(\frac{3}{4})^{h-1} \cdot n$ uncertain points and at most $(\frac{3}{4})^{h-1} \cdot mn$ vertices. It is not hard to see that performing the h-th round on T_{h-1}^+ takes $O(|T_{h-1}^+|)$ time. We stop until we obtain a tree T' that contains $O(1)$ uncertain points or T' consists of $O(1)$ non-dummy vertices, that is, after at most $\log n$ rounds. At this moment, a peripheral-center edge of T can be computed in $O(mn)$ time by Lemma 4, whose proof is in the full paper. Thus, the total time complexity is $O(\sum_{i=1}^{\log n}(\frac{3}{4})^i \cdot mn)$ time, which is $O(mn)$.

Lemma 4. *The non-dummy peripheral-center edge on T' can be computed in $O(mn)$ time.*

Once a peripheral-center edge is found, we call Lemma 2 to decide the feasibility of λ in an additional $O(mn)$ time. We thus have the following result.

Lemma 5. *The decision two-center problem can be solved in $O(mn)$ time.*

3.4   Lemma 6 and Lemma 7

Let T_h be a tree obtained after several recursive steps of the first pruning step in a round of our algorithm. Suppose T_h contains n_h uncertain points and t connectors. Additionally, T' may contain dummy vertics but all of them are leaves, and at most one vertex of T' has a true flag. We have the following lemma and its proof is in the full paper.

Lemma 6. *Given any point x on T_h, we can decide in $O(|T_h| + t \cdot n_h)$ time whether a center must be placed at x, and if not, which split subtree of x on T_h contains a peripheral-center edge.*

Furthermore, we have the following result for solving the peripheral-center detecting problem on the obtained tree T^+ after several rounds of the above algorithm. So, T^+ has n^+ uncertain points and $|T^+| = mn^+$. All dummy vertices of T^+ are leaves, and T^+ may contain at most one true-flag vertex. Note that no connectors are on T^+ (Fig. 3).

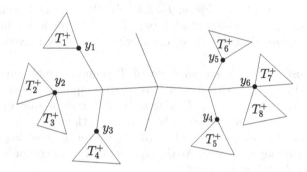

Fig. 3. Illustrating an example for the peripheral-center detecting problem: $Y = \{y_1, y_2, \cdots, y_6\}$ and $\Gamma = \{T_1^+, T_2^+, \cdots, T_8^+\}$ shown with triangles.

Given are a set Y of points $y_1, \cdots, y_t$ on T^+ and a set Γ of split subtrees $T_1^+, \cdots, T_s^+$ of points in Y on T^+. The peripheral-center detecting problem is to decide which split subtree of Γ contains a peripheral-center edge. We have the following result and its proof is in the full paper.

Lemma 7. *The peripheral-center detecting problem can be solved in $O(|T^+|)$ time.*

4   Computing Centers q_1^* and q_2^*

This section presents our algorithm that computes centers q_1^* and q_2^* in $O(mn \log mn)$ time. We say that an edge of T containing a center is a *critical edge*. Similar to the decision algorithm, our algorithm recursively computes each critical edge on T with the assistance of the following key lemma. See its proof in the full paper.

Lemma 8. *Given any point x on T, we can decide whether x is a center, and if not, which split subtree of x contains a critical edge; further, if one center is at x then centers q_1^* and q_2^* can be computed in $O(mn)$ time.*

At the beginning, T is known to contain a critical edge and so we let $T_0 = T$. We first compute the centroid c of T_0 in $O(|T_0|)$ time and apply Lemma 8 to c in $O(mn)$ time. Either we obtain two adjacent vertices of c so that the two corresponding split subtrees of c each must have a critical edge, or only one adjacent vertex of c is returned so that both critical edges are in that split subtree of x containing this vertex. For the former case, we set T_1 as either subtree, which can be done in $O(|T_0|)$ time, and let the flag of c on T_1 as true. In the other case, we let T_1 be the only split subtree. Clearly, $|T_1| \leq |T_0|/2$. We continue to search in T_1 for a critical edge. First, compute the centroid c of T_1 in $O(|T_1|)$ time, and then apply Lemma 8 to c on T_1, which always takes $O(mn)$ time. If two split subtrees are obtained, then the one without any true-flag vertex must contain a critical edge. We thus let T_2 be this subtree and set c's flag as true on T_2. Otherwise, let T_2 be the only split subtree returned. Clearly, $|T_2| \leq |T_1|/2$. We recursively find a critical edge on T_2. Clearly, the obtained subtree T_h of the $h - 1$-th recursive step consists of at most $|T^0|/2^{h-1}$ vertices. Since Lemma 8 always runs in $O(mn)$ time, the time complexity of each recursive step is $O(mn)$ time.

After at most $\log mn$ steps, an edge of T remains and it must contain a center. Denote by e_1^* this critical edge. We then adapt the above procedure to find the other critical edge e_2^* on T. The only difference is that every recursive step, if two split subtrees are returned by Lemma 8, then we always consider the one excluding e_1^*. Therefore, e_2^* can be obtained after at most $\log mn$ recursive steps, i.e., in $O(mn \log mn)$ time. At this moment, we can compute q_1^* and q_2^* in $O(mn)$ time by the following lemma.

Lemma 9. *Given two critical edges e_1^* and e_2^* on T, we can find centers q_1^* and q_2^* in $O(mn \log n)$ time.*

As mentioned in Sect. 2, any given general case can be reduced into a vertex-constrained case in $O(|T| + mn \log mn)$ time. We have the following result.

Theorem 1. *The two-center problem of n uncertain points on a tree T can be solved in $O(|T| + mn \log mn)$ time.*

References

1. Alipour, S.: Improvements on approximation algorithms for clustering probabilistic data. Knowl. Inf. Syst. **63**, 2719–2740 (2021)
2. Ben-Moshe, B., Bhattacharya, B., Shi, Q., Tamir, A.: Efficient algorithms for center problems in cactus networks. Theoret. Comput. Sci. **378**(3), 237–252 (2007)
3. Ben-Moshe, B., Bhattacharya, B., Shi, Q.: An optimal algorithm for the continuous/discrete weighted 2-center problem in trees. In: Correa, J.R., Hevia, A., Kiwi, M. (eds.) LATIN 2006. LNCS, vol. 3887, pp. 166–177. Springer, Heidelberg (2006). https://doi.org/10.1007/11682462_19

4. Bender, M., Farach-Colton, M.: The LCA problem revisited. In: Proceedings of the 4th Latin American Symposium on Theoretical Informatics, pp. 88–94 (2000)
5. Bhattacharya, B., Shi, Q.: Improved algorithms to network p-center location problems. Comput. Geom. **47**(2), 307–315 (2014)
6. Chan, T.: Dynamic planar convex hull operations in near-logarithmic amortized time. In: Proceedings of the 40th IEEE Symposium on Foundations of Computer Science (FOCS), pp. 92–99 (1999)
7. Chan, T.: Klee's measure problem made easy. In: Proceedings of the 2013 IEEE 54th Annual Symposium on Foundations of Computer Science, pp. 410–419 (2013)
8. Chen, D., Wang, H.: A note on searching line arrangements and applications. Inf. Process. Lett. **113**, 518–521 (2013)
9. Eppstein, D.: Faster construction of planar two-centers. In: Proceedings of the 8th Annual ACM-SIAM Symposium on Discrete Algorithms, pp. 131–138 (1997)
10. Hu, R., Kanani, D., Zhang, J.: Computing the center of uncertain points on cactus graphs. In: Proceedings of the 34th International Workshop on Combinatorial Algorithms, pp. 233–245 (2023)
11. Huang, L., Li, J.: Stochastic k-center and j-flat-center problems. In: Proceedings of the 28th ACM-SIAM Symposium on Discrete Algorithms (SODA), pp. 110–129 (2017)
12. Kariv, O., Hakimi, S.: An algorithmic approach to network location problems. I: the p-centers. SIAM J. Appl. Math. **37**(3), 513–538 (1979)
13. Megiddo, N.: Linear-time algorithms for linear programming in R^3 and related problems. SIAM J. Comput. **12**(4), 759–776 (1983)
14. Nguyen, Q., Zhang, J.: Line-constrained l_∞ one-center problem on uncertain points. In: Proceedings of the 3rd International Conference on Advanced Information Science and System, vol. 71, pp. 1–5 (2021)
15. Wang, H.: On the planar two-center problem and circular hulls. Discrete Comput. Geom. **68**(4), 1175–1226 (2022)
16. Wang, H., Zhang, J.: Computing the center of uncertain points on tree networks. Algorithmica **78**(1), 232–254 (2017)
17. Wang, H., Zhang, J.: Covering uncertain points on a tree. Algorithmica **81**, 2346–2376 (2019)
18. Xu, H., Zhang, J.: The two-center problem of uncertain points on a real line. J. Comb. Optim. **45**(68) (2023)

Space-Time Graph Planner
for Unsignalized Intersections with CAVs

Caner Mutlu, Ionut Cardei[(⊠)], and Mihaela Cardei

Department of Electrical Engineering and Computer Science, Florida Atlantic
University, Boca Raton, FL 33431, USA
{cmutlu,icardei,mcardei}@fau.edu
http://www.eecs.fau.edu

Abstract. Emerging autonomous intersection management systems
control the entry order and trajectory for connected and autonomous
vehicles ready to traverse a road intersection. They aim to compute trajectories that are safe and optimal in order to reduce congestion, environmental impact, and to cut travel time. We propose a novel approach for
computing the fastest waypoint trajectory using search in a discretized
space-time graph that produces collision-free paths with variable vehicle speeds complying with traffic rules and vehicle dynamics constraints.
The resulting trajectories allow high levels of intersection sharing, high
evacuation rate, with a low algorithm runtime even with large scenarios
with 1200 vehicles (5.5 s on a laptop).

Keywords: space-time graph · intersection management · autonomous
cars

1 Introduction

There are about 15 million road intersections in the continental US alone,
and 44% of the road incidents occur at intersections [1]. Time waste and
fuel consumption [2] at intersections also have negative societal impact. In
contrast to traditional signaled intersections, signal-free intersections con-
trolled by Autonomous Intersection Management Systems (AIMS) do not
employ semaphores and provide conflict-free intersection transit for Connected
Autonomous Vehicles (CAVs). CAVs talk to AIMS on a network, such as cellular,
vehicle-to-vehicle (V2V), vehicle-to-instructure (V2I), and others [3].

A CAV contacts the nearby AIMS and sends an admission request with its
predicted arrival time at the intersection entry and its intended intersection exit
lane. The AIMS considers new requests, the trajectories of CAVs already in transit, and computes control commands for each new vehicle. The computation can
be centralized in the infrastructure or it can be distributed among coordinating
CAVs. Possible AIMS goals include minimizing exit time for each vehicle in a
fair first-in/first-out (FIFO) way or maximizing the global exit flow, across all
vehicles.

W. Wu and J. Guo (Eds.): COCOA 2023, LNCS 14461, pp. 498–511, 2024.
https://doi.org/10.1007/978-3-031-49611-0_36

Management of CAVs in an unsignaled AIMS environment can be categorized under two broad problem classes, "vehicular scheduling" and "vehicular control" [4]. Both can be addressed separately, but a true optimal trajectory would require solving both concurrently. A vehicle control system formulates a solution describing commands for vehicle actuators, such as steering, throttle, braking. Some examples use model predictive control [5] or optimal control [6]. On the other hand, a vehicle scheduler is a trajectory planner that gives a sequence of waypoints (location + time) that must be traversed by the CAV. At a minimum, it just provides an entry sequence to the intersection, enough for vehicles to avoid collisions if they stick to their desired lane. Reservation-based systems [6,7] schedule vehicles to leave the intersection in FIFO order, underutilizing shared intersection space. Solutions that solve a discrete optimization problem [8] are limited by the exponential growth of the search space - unfeasible for realistic large scenarios with hundreds of vehicles. Graph search methods model the intersection as a graph in 2D. Depth first spanning tree (DFST) methods [4,9] do conflict analysis and determine an entry order that increases parallel access to the intersection.

In this paper we are concerned with the problem of finding the shortest collision-free space-time trajectory through an intersection, constrained by traffic rules and vehicle limitations. Such a trajectory can then be passed to the CAV's navigation unit to generate actuator commands.

We propose a solution – the Fastest Trajectory Planner algorithm – that a) models the intersection road map as a discretized graph G_u; b) expands G_u's vertices and edges to the time dimension into a space-time graph G_t so that a G_u vertex or an edge used at a particular time corresponds to vertices and edges in G_t that will be removed from G_t for subsequent vehicles; c) finds the fastest variable-speed trajectory complying to constraints using a shortest path algorithm in the space-time graph. Our algorithm was inspired by our earlier work on drone traffic management [10–13], with addition of the variable speed capability and vehicle dynamic constraints. The algorithm has a low runtime complexity and scales well: scenarios with 1200 vehicles at a 4-lane 4-way intersection are solved in about 5.5 s on a typical laptop, with code in Python.

This paper continues with related work in Sect. 2, the problem statement in Sect. 3, the proposed algorithm in Sect. 4.4, a performance evaluation in Sect. 5, and conclusions (Sect. 6).

2   Related Work

Papers [4,14] study global optimality for vehicular scheduling problems in an AIMS. Their method models vehicles with vertices and they build a Conflict Directed Graph where edges map from pair-wise path conflicts. An Improved Depth First Search Spanning Tree is used to design a conflict-free passing order through the intersection. A second algorithm uses a complementary Coexisting Undirected Graph built from nonconflicting vehicle pairs to compute the Minimum Clique Cover. That gives an optimal passing order with the minimum evacuation time.

The conflict-duration approach in [15] builds a Gantt-chart inspired conflict-duration diagram. Its axes are conflict locations and timing stamps. The conflict-duration diagram registers double or triple conflicts between vehicles. By considering the physical size (L x W) of each vehicle, the total duration where a physical conflict persists between two or three vehicles is represented on the conflict-duration diagram as overlapping time duration at a particular conflict point. By removal of the overlapping time region, through rescheduling speed profile of one or more vehicle(s), a conflict can be prevented between any pair of vehicles.

In our prior work on drone traffic management, we developed the concept of shortest path search in a space-time graph for vehicle trajectory planning. We initially formulated the point-to-point trajectory planner for drone package delivery in [10] using multi-source/multi-destination BFS on the space-time graph. The planner computes shortest space-time paths with edges traversed in one time unit and no constraints on vehicle dynamics. We improved that approach in [12] with a batch scheduling method that has a lower complexity and better results. We addressed in [11] the problem of energy-constrained drone package delivery with multiple warehouses and customers using a multi-source A* algorithm running on the space-time graph. More recently, [13] presents a multi-source/multi-destination search algorithm for the fastest trajectory between two disjoint groups of vertices in the space-time graph. This is suitable for drone planning when the operator has multiple drones available stationed through the network and has to deliver packages to multiple customers.

Our contribution in this paper differs from prior work with space-time graphs by complying to vehicle dynamic constraints and by allowing multiple possible times for space-time edge traversal, necessary for supporting variable average edge velocities. The collision constraints and resource sharing rules are different from drone scenarios.

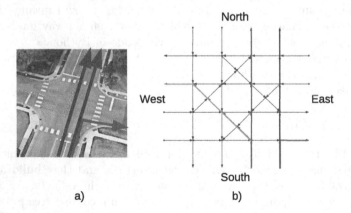

a)   b)

Fig. 1. (a) a 4-lane, 4-way intersection. (b) the graph for the intersection traffic road network. Paths possible from the two South entry vertices are highlighted with different colors. (Color figure online)

3   Problem Statement

Figure 1a illustrates a typical 4-lane, 4-way intersection, with legal trajectories highlighted. Vehicles moving in perpendicular directions have paths that intersect at shared points. The shared space forces vehicles to serialize their passage in order to avoid collisions. An AIMS runs the algorithms presented in Sect. 4.4. It uses the graph representation for the road network described below.

We model a road map as a directed graph with tuple $G = (V, E, pos)$, with edge set E and vertex set V, as in Fig. 1b. A segment corresponds to a directed edge $(u, v) \in E$. If a road segment (u, v) is bidirectional, then $(v, u) \in E$, too. Function $pos : V \to \mathbb{R}^2$ defines the position of a vertex.

A collision between two vehicles occurs when their trajectories overlap in space and time. Vehicles can safely occupy the same space at different times sufficiently separated.

A vehicle's admission request is as an object $Request(src, dst, t_s, T_{max})$, where src, $dst \in V$ are the source and destination vertices, $t_s \in \mathbb{R}_{\geq 0}$ is the vehicle arrival time at the intersection, and $T_{max} \in \mathbb{R}^+$ is the maximum allowed trajectory duration. $T_{max} = \infty$ if a vehicle may take indefinitely to exit the intersection. The intersection manager accepts requests over a finite time interval $[0, t_{end}]$.

Vehicle movement on the road is limited by the maximum legal speed s_{max}, maximum acceleration a^+_{max}, maximum deceleration a^-_{max}, and vehicle length, packaged all in an object $Cons(s_{max}, a^+_{max}, a^-_{max}, L)$.

The computed trajectory between the $req.src$ and $req.dst$ is defined by a list of space-time waypoints that the vehicle must reach: $Tr(times, positions, velocities)$ indicating the time at each waypoint in $times = (t_0, t_1, ..., t_{m-1})$, the position of each waypoint $positions = (p_0, p_1, ..., p_{m-1})$, and a velocity vector for each waypoint.

A Tr object for which a solution cannot be found has no waypoints: $Tr((), (), ())$, where $()$ is the empty sequence. Otherwise, that is a *valid* trajectory.

We define the problem of finding trajectories for vehicles on a traffic map as follows:

Problem Definition. *Given a road network graph $G = (V, E, pos)$, vehicle constraints $Cons$, and a list of vehicle admission objects $(Request)_i$ over a time interval $[0, t_{end}]$, the **Fastest Trajectory Planning problem** is finding a trajectory with m_i waypoints for each vehicle i through the road network that has the earliest arrival time t_{m_i-1}, subject to these conditions:*

1. *vehicles move on edges in E,*
2. *there are no collisions between any two vehicles on the road network,*
3. *vehicle constraints as defined by $Cons$ are satisfied at all times.*

The problem objective is locally greedy. An algorithm that globally minimizes the maximum delay is NP-complete because of the combinatorial explosion of the number of ways in which vehicle moves can be sequenced over time edges and space-time edges.

Performance Metrics

The intersection trajectory planner accepts a sequence of N requests $reqs$ and produces a sequence of N Tr objects, from which n are valid: $trj_i = Tr(times_i, positions_i, velocities_i)$, $i = 0, ..., N-1$, and with attribute $times_i = (t_0, t_1, ..., t_{m-1})_i$, for m_i waypoints. We define the following performance metrics for a planning solution:

Definition 1. *The **trajectory delay** for a valid trajectory i is the difference between trajectory arrival time at destination and the request start time: $delay = t_{m-1} - t_0$.*

The following metrics apply to a batch $reqs$ of N Requests resulting in n valid trajectories that complete in $t_{evac} = \max_i t_{m_i-1}$.

Definition 2. *The **average trajectory delay** is $delay_{avg} = \frac{1}{n} \sum_i delay_i$ over valid trajectories. The **maximum trajectory delay** is $delay_{max} = \max_i delay_i$*

Definition 3. *The **request admission ratio** is the fraction of valid trajectories vs. the total number of requests submitted, $adm = n/N$.*

Definition 4. *The **traffic flow rate** is the number of vehicles that reach their destination vertex per time unit. This is the exit rate from the intersection. The traffic flow rate over a time period of duration t_{evac} is $flow_T = n/t_{evac}$ $[s^{-1}]$.*

We make the following assumptions to design our algorithm:

1. The waypoint trajectory is converted by the CAV's own control systems to commands for actuators (throttle control, braking, steering) to maintain a trajectory with high fidelity.
2. Without loss of generality, all vehicles have the same dynamic constraints.
3. The optimization objective for the planning algorithm is to minimize the delay of each request while preserving the *first in - first out* order at intersection entry lanes. Minimizing the travel time reduces the overall utilization of shared intersection resources, such as graph edges and vertices, contributing to increased traffic flow.

4 The Space Time Graph Methodology for Trajectory Search

The proposed solution, Fastest Trajectory Planner (FTP), is inspired from the Space-Time graph planner for the drone delivery problems introduced in articles [10–13]. In contrast to our earlier work, the new algorithm works for autonomous cars carrying people and goods, supports multiple average speeds on graph edges, enforces dynamic vehicle constraints (e.g. min/max acceleration), and applies collision avoidance rules specific to road vehicles.

4.1   Collision Avoidance and Graph Representation

Figure 2 illustrates several collision scenarios on graph G. Figure 2a shows two vehicles moving on different edges in G_u towards the same vertex. Figure 2b shows the red vehicle moving on an edge towards a vertex v occupied by the blue vehicle standing still. Figure 2c shows the red vehicle on edge (u, v) and the blue vehicle on edge (v, u) moving towards each other. A vehicle (red) can stand still in the middle of an edge (u, v) while the blue vehicle comes barreling towards it from vertex u, Fig. 2d.

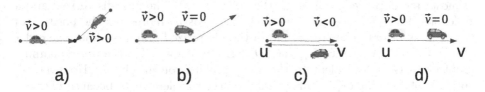

Fig. 2. Examples of collisions scenarios.

At the same time, two vehicles moving on the same long edge in the same direction, with similar speeds should be perfectly fine, with no collision.

Three salient observations are apparent:

(a) Graph edges and vertices occupied by a vehicle at a time are resources that must be allocated to vehicles with mutual exclusion on that time.

(b) Time-dependent allocation of resources for one request controls allocation for other vehicles, hence their movement, collision avoidance, and performance metrics.

(c) The original graph G derived from the road map has insufficient space and time resolution for an adequate fine-grained granularity to achieve effective resource reuse.

The basic approach of our proposed algorithm is summarized here:

1. Discretize the original road map graph G to a spatial *unit graph* G_u that has fine-grained "granularity", e.g. 5 m.

2. From G_u, build a *space-time graph* G_t, with *time edges* for each vertex for a vehicle that stands there still for a time unit, and *space-time* edges for a vehicle that moves from one vertex to another during one or more time units.

3. Time variable availability of edges and vertices in G_u is modeled by existence of edges in the space-time graph.

4. A vehicle trajectory is expressed by a path in this space-time graph. Edges and vertices in G_t that form a trajectory and their adjacent neighbors are "allocated" exclusively to a vehicle's trajectory and removed from G_t. A trajectory (vehicle) cannot use vertices and edges in G_t already allocated to other trajectories.

5. The fastest ending (shortest) path in the space-time graph is a good approximation to the fastest trajectory in the original graph G.

Next are the key ideas underlining the Fastest Trajectory Planner's algorithm.

4.2 Discretized Graph and Discrete Time

The original road map graph G is discretized with unit length D (e.g. 5 m) so that each original edge $(u, v) \in E$ is split into smaller edges of length D and at most one shorter edge (w, v) at the end of (u, v). The new discretized graph is denoted by $G_u(V_u, E_u, pos)$, with $V \subseteq V_u$. $pos(u)$ represents the position of vertex u, as before. The size bound of the discretized graph is about the order of $|V_u| = \Theta(D^{-1}|V|)$, and $|E_u| = \Theta(D^{-1}|E|)$. Since $V \subseteq V_u$, the route planning problem on G is equivalent to the same problem on graph G_u. However, an optimal solution for G_u is suboptimal for the problem in G because of space discretization error.

Discretizing the graph allows one to run the planner in discrete time, with time units of δ_t (e.g. 1 s) expressed in time *ticks*.

We allow an edge to be traversed in discrete multiples of δ_t: $\{\delta_t, 2\delta_t, ..., p\delta_t\}$. This is called a *slow fragment* and p is called the *edge time multiplier*. We also allow for up to q consecutive edges to be traversed in just one δ_t time interval. We call this a *fast fragment* and q is the *edge speed multiplier*. The *unit edge speed* is $s_u = \frac{D}{\delta_t}$. The set of possible average speeds on edges in E_u of length D is $\{s_u, 2s_u, 3s_u, ..., q\,s_u\} \cup \{\frac{s_u}{2}, \frac{s_u}{3}, ..., \frac{s_u}{p}\}$, where constants $p, q \in \mathbb{N}^+$ are selected such that $qs_u \leq s_{max}$ and $\frac{s_u}{p}$ exceeds the minimum possible vehicle speed allowed. In our simulations $p = 2$ and $q = 4$. These parameters also affect runtime complexity, as discussed later.

4.3 The Space-Time Graph

We assume the planner computes trajectories for a sequence of requests over a finite time horizon $H > 0$, with $H = \min\{\{req_i.tf\}_{i=0..n-1} \cup \{T_{max}\}\}$. The discrete time horizon is defined as $K = \lfloor \frac{H}{\delta_t} \rfloor$, where T_{max} is the maximum simulation time.

The **space-time graph** G_t is built from the discretized unit graph G_u as follows. Each vertex $u \in V_u$ converts to K space-time vertices $(k, u) \in V_t$. Time edges in E_t are defined as $((k, u), (k + 1, u))$ for all $0 \leq k < K - 1$ and $u \in V_u$. Space-time edges are defined as $((k, u), (k + 1, v))$ for all $0 \leq k < K - 1$ and $(u, v) \in E_u$. A space-time edge is added for each edge in the discretized graph G_u and each time unit. The size of G_t is given by $|V_t| \in \Theta(K\delta_t^{-1}|V|)$, and $|E_t| \in \Theta(K\delta_t^{-1}|E|)$. The space-time graph for a very simple G_u is shown in Fig. 3a.

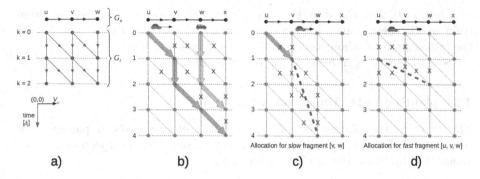

Fig. 3. a) Space-time graph G_t is derived from G_u by "extending" (u, v) edges in time (red) and by adding time edges (blue) for each vertex. b) space-time paths for two vehicles (red and blue wide arrows). Edges crossed with an **X** are pruned from G_t after admission to prevent collisions. c) a *slow fragment* (green dashed line) of one G_u edge traversed in more than one single tick at half the normal edge speed. d) a *fast fragment* (green dashed line) of more than 2 edges in one tick leads to higher speeds. (Color figure online)

4.4   The Fastest Trajectory Planner Algorithm

The key ideas behind the use of the space-time graph are:

1. A shortest path for $Request(src, dst, t_s, T_{max})$ in the discretized graph G_u between vertices src and dst can be found by computing the "earliest ending" path in the space-time graph G_t between any space-time vertices (k_s, src) and (k_f, dst), with $t_s \leq k_s\delta_t \leq k_f\delta_t \leq t_s + T_{max}$, such that k_f is the minimum such value that is possible. This is the same as the multi-source, multi-destination shortest path problem in the space-time graph.
2. A space-time edge $e = ((k_u, u), (k_u + 1, v))$ is traversable at time k_u from u to v only if $e \in E_t$. We remove (prune) space-time edges from G_t to prevent collisions when ulterior requests (with later start times) are computed. Figure 3b shows shortest space-time paths for two vehicles. Blue's path $u \rightarrow x$ is computed first. The algorithm finds a path with the edges covered by the wide blue arrow and, then, it prunes from G_t the edges marked with **X** signs from G_t. The path computation for the red vehicle will not *see* those deleted edges and it will find the space-time path drawn with the wide red arrows. The edges marked with X signs will be pruned from E_t. Space-time edges in G_t used by a solution path are removed from G_t
3. Support for multiple speeds (and dynamic constraints) is added by modifying the Dijkstra algorithm to consider during the "vertex expansion step" space-time path fragments (i.e. subpaths) corresponding to multiple traversal times and multiple space-time edges.

No Optimality with Constraints: since our planner enforces dynamic constraints (min/max acceleration) involving successive edges, we cannot prove that it finds the fastest path. Without constraints on acceleration, it does.

Several functions implement the planning algorithm.

4.5   planRequests: Top Level Algorithm

The entry to the planner is a function called *planRequests*, implemented in Algorithm 1. Function *planRequests(G_u, reqs)* computes a trajectory for each *Request* in list *reqs*, in the given input order.

Algorithm 1 : compute N admssion from list *requests*.

1: **function** planRequests(G_t, *requests*)   ▷ Process a list of requests
2:     *trjs* = [ ]   ▷ empty list
3:     **for all** *reqinrequests* **do**
4:         *path* ← computePath(G_u, *req*)   ▷ plan one path: list of V_t vertices
5:         *trj* ← convertToTrajectory(G_u, *req*, *path*)   ▷ convert path to *Tr*
6:         *trjs.append(trj)*
7:     **return** *trjs*   ▷ all trajectories, including $Tr((),(),())$ for failed ones

---

Line 4 calls function *computePath(G_t, req)* to compute the space-time path for the current request *req*. That is the main part of the planning algorithm. It returns in variable *path* a list of vertices (waypoints) in G_t that defines a space-time trajectory in format $[(k_0, v_0), (k_1, v_1), ...]$, indicating that the vehicle must be at vertex v_0 at tick k_0, at v_1 at tick k_1, etc. This works also when multiple space-time edges are traversed in one tick.

4.6   computePath: The Shortest Space-Time Path Algorithm

The *computePath(G_t, req)* function call (Algorithm 2) computes the fastest space-time path from vertex *req.src* to vertex *req.dst* using only available edges in the space-time graph G_t. This algorithm runs a multi-source/multi-destination version of Dijkstra's shortest path algorithm. Once a space-time vertex (k_u, u) is reached our algorithm explores all feasible path fragments that start from time tick k_u like this:

- *slow fragments* (u, v) with exactly two space vertices that can be traversed in $1, 2, ..., q$ ticks; the average speed on this fragment does not exceed s_u (Fig. 3c).
- *fast fragments* $(u, ..., v)$ with three or more space vertices that can be traversed in exactly 1 tick; the average speed on this fragment may exceed D/δ_t (Fig. 3d).

A path fragment is *feasible* if it satisfies intersection/traffic lane constraints, its space-time edges are available in G_t, and if speed, acceleration/deceleration constraints (including vs. the previous fragment) are satisfied.

The priority queue that orders space-time vertex expansion holds objects of type $QueueEntry(k1, k0, priorVelocity, fragment, priorQe)$. $fragment$ is a list of V_u vertices $[u, ..., v]$ that forms a subpath in G_t traversable from tick $k0$ at u, arriving at v on tick $k1$. The fragment is constrained by available space-time edges in G_t and vehicle constraints. $priorQe$ is the currently expanding $QueueEntry$ object. $priorVelocity$ is the average 2D velocity vector across the $priorQe.fragment$ fragment.

QueueEntry objects in the priority queue created on line 5 are ordered by their $k1$ attribute, the arrival at their fragment's end vertex.

Algorithm 2. computes the shortest space-time path for one *Request*.

1: **function** computePath(G_t, req)
2:     $ks \leftarrow \lfloor req.ts/\delta_t \rfloor$ and $kf \leftarrow \lfloor req.tf/\delta_t \rfloor$ ▷ arrival ticks; last allowed exit time in ticks
3:     $explored \leftarrow \{(ks, req.src)\}$
4:     $queue \leftarrow$ new $PriorityQueue()$
5:     $queue.enqueue($new $QueueEntry(ks, ks, (0.0, 0.0), [req.src], None))$
6:     $ktime \leftarrow ks$ ▷ ktime keeps the current exploration time tick
7:     $path \leftarrow [\,]$
8:     **while** $path == [\,]$ and $queue.size > 0$ and $ktime \leq kf$ **do**
9:         $qe \leftarrow queue.dequeue()$ ▷ ordered by fragment end time, qe.k1
10:         **continue** if impossible to reach $req.dst$ from $qe.fragment[0]$ by tick tf
11:         update $ktime$ from $qe.k1$
12:         $nextQes \leftarrow discoverFragments(G_t, req, explored, qe)$
13:         **if** $nextQes.size > 0$ **then**   ▷ if fragments were found
14:             **for all** nqe in $nextQes$ with $nqe.k1 \leq kf$ **do**
15:                 **if** $req.dst \in nqe.fragment$ **then**   ▷ reached destination?
16:                     $path \leftarrow$ extract path from nqe and its predecessors
17:                     prune space time edges in $path$ from G_t
18:                     **break**
19:                 **else**
20:                     $queue.enqueue(nqe)$
21:     **return** $path$

Function $discoverFragments$ computes the feasible fragments consisting of *feasible* edges in the space-time graph: available in G_t and that satisfy the constraints on vehicle dynamics and intersection lanes (line 12). Line 16 checks if the destination vertex was reached. If so, it computes $path$ from the chain of queue entries, going backwards in time towards the root queue entry. In case of failure to find a path, the function returns the empty list $[\,]$. The call to $discoverFragments(G_t, req, explored, qe)$ explores the current space-time vertex from $(qe.k1, qe.fragment.last)$ and returns new $QueueEntry$ objects for the shortest *feasible* fragments with duration between 1 and maximum q ticks. Its algorithm runs a Breadth-First Search starting from space-time vertex $(qe.k1, qe.fragment.last)$ in G_t with a search radius limited to q ticks.

Each shortest feasible fragment $[u, ..., v]$ returned by function *discoverFragments*, with u at $k0$ and v and $k1$ has these properties:

– it has no cycles if it is longer than 2 vertices;
– it forms no cycles going back on the queue entry chain history;
– has only space-time edges available in G_t and that comply with constraints;
– there is no other faster fragment with the same space endpoints u and v.

Function *discoverFragments* uses the space-time vertices in parameter *explored* and updates it during search with each encountered space-time vertex.

Runtime Complexity Analysis

In the following, f is the maximum number of exits reachable from any entry vertex, typically 2-4. The constrained *effective* average out-degree for exploration in V_u is $b \gtrsim 1$. In the 4-lane 4-way example from Fig. 1, $b = 1.0833$ and $f = 1$. Search in the space-time graph search will now branch only on time edges.

Other parameters for runtime complexity include (with typical values): the total number of requests to consider N: $10^1 - 10^3$, the discrete time horizon for trajectory computation $K = \Theta\left(H\delta_t^{-1}\right)$: $10^1 - 10^3$, the edge time multiplier p: 4 - 8, and the edge average speed multiplier q: 1, 2.

The runtime of *convertToTrajectory* is $O(K)$. The time complexity of the top-level *planRequests* algorithm is $O(NfK(p + b^{q+1} + \log_2 fK))$, with a heap priority queue. For intersections parameters f, p, q have moderate values and can be considered constant. In that case, the runtime is $O(NfK \log_2 fK)$ and does not seem to depend directly on the road map's graph topology, but on the discretization granularity D.

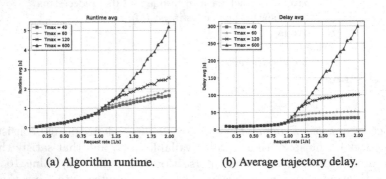

(a) Algorithm runtime.   (b) Average trajectory delay.

Fig. 4. Algorithm runtime and average trajectory delay.

5   Performance Evaluation

We evaluate the performance of the Fastest Trajectory Planner for the 4-lane, 4-way intersection in Fig. 1b. $|V_u| = 36$, $|E_u| = 56$ edges, with average out-degree 1.55. Legal lanes restrict the *effective* edge out-degree during search to $b = 1.0833$. The space edge discretization length $D = 10\,\mathrm{m}$ and the time tick unit is $\delta_t = 1\,\mathrm{s}$. The speed multiplier $q = 2$ and the edge time multiplier $p = 4$. The top acceleration/deceleration is $2m/s^2$, consistent with a comfortable ride with enough braking ability.

Vehicle admission requests (random source/destination, no U-turns) are generated from time 0 to $t_{end} = 600\,\mathrm{s}$ with a rate that varies from 0.1/s ($N = 60$) to 2/s ($N = 1200$) in 0.05 s ($\delta N = 30$) increments. All charts have on the horizontal axis this independent variable. The maximum allowed trajectory delay $T_{max} \in \{40, 60, 120, 600\}$s.

Saturation starts when the request rate approaches 1/s, $N = 600$ requests. When resource availability drops, vehicles experience higher waiting time, and metrics start deteriorating. This congestion behavior, with an inflection point, is common in scheduling with resource sharing. After congestion starts, requests will be rejected, bringing relief for resource contention. Most relief is seen for scenarios with $T_{max} < 600$.

The average runtime is shown in Fig. 4a. It is proportional to the trajectory duration and it has the $N \log_2 N$ asymptotic trend. The worst running time was for a request rate of 2/s ($N = 1200$), with 4.323 ms/request, and a 5.5 s total.

Figure 4b shows the average trajectory delay. It has a very slow growth under 1 request/s, followed by a sharper growth above 1/s, when congestion begins that tapers off, converging to T_{max}. Note the a lower T_{max} value causes more requests to be dropped. This is evident in the request admission ratio chart from Fig. 5a. The admission ratio stays at 100% for all scenarios before congestion begins (at 1/s–1.1/s). After that, the admission ratio starts a linear drop, delayed by a higher value for T_{max}.

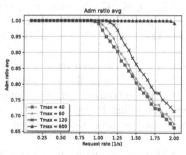
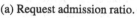

(a) Request admission ratio.

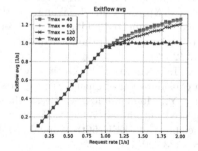

(b) Average intersection exit flow rate (vehicle exits/second).

Fig. 5. Request admission ratio and intersection exit flow rate.

Figure 5b shows the evolution of the intersection exit flow rate vs. request rate. It follows the identity function before the congestion threshold, for all T_{max} values. It is constant for $T_{max} = 600$ s after that since the admission ratio is 100% up to the end and no requests are dropped. However, for $T_{max} < 600$ s the admission ratio is less than 100%, allowing only shorter trajectories. That causes a higher exit flow rate.

Finally, in the maximum traffic flow regime, we counted on average 14 vehicles present at the same time in the intersection. This high resource utilization should lead to superior traffic flow rates compared to alternatives.

6   Conclusions

This paper proposes a novel algorithm for the Fastest Trajectory Planning problem for intersections with CAVs. The algorithm uses a shortest path search in a space-time discretized graph derived from the original road network graph. The algorithm enforces vehicle constraints (e.g. acceleration/deceleration) and it has a low runtime compared to that reported for state of art algorithms using Conflict Detection Graphs [4,14]. The algorithm scales well with the number of admission requests and with the traffic graph size, the main limitation being the maximum path duration parameter T_{max}.

Future research directions include improving the search algorithm with A* and local search heuristics that reorder vehicle advance at each search step.

References

1. Choi, E.-H.: Crash Factors in Intersection-Related Crashes: An On-Scene Perspective, HS-811 366 (2010)
2. Wang, J., Guo, X., Yang, X.: Efficient and safe strategies for intersection management: a review. Sensors **21**(9), 3096 (2021)
3. Kiela, K., et al.: Review of V2X-IoT standards and frameworks for ITS applications. Appl. Sci. **10**(12), 4314 (2020)
4. Chen, C., Xu, Q., Cai, M., Wang, J., Wang, J., Li, K.: Conflict-free cooperation method for connected and automated vehicles at unsignalized intersections: graph-based modeling and optimality analysis. IEEE Trans. Intell. Transp. Syst. **23**(11), 21897 (2022)
5. He, X., Liu, X., Liu, H.X.: Optimal vehicle speed trajectory on a signalized arterial with consideration of queue. Transp. Res. C Emerg. Technol. **61**, 106–120 (2015)
6. Zhang, Y., Malikopoulos, A., Cassandras, C.G.: Decentralized optimal control for connected automated vehicles at intersections including left and right turns. In: Proceedings of IEEE 56th Annual Conference on Decision Control (CDC), pp. 4428–4433 (2017)
7. Xu, B., Ban, X.J., Bian, Y., Wang, J., Li, K.: V2I based cooperation between traffic signal and approaching automated vehicles. In: Proceedings of IEEE Intelligent Vehicles Symposium (IV) (2017)
8. Xu, X., Zhang, Y., Li, L., Li, W.: Cooperative driving at unsignalized intersections using tree search. IEEE Trans. Intell. Transp. Syst. **21**(11), 4563–4571 (2019)

9. Xu, B., et al.: Distributed conflict-free cooperation for multiple connected vehicles at unsignalized intersections. Transp. Res. C Emerg. Technol. **93**, 322–334 (2018)

10. Steinberg, A., Cardei, M., Cardei, I.: UAS path planning using a space-time graph. In: IEEE SysCon (2020)

11. Papa, R., Cardei, I., Cardei, M.: Energy-constrained drone delivery scheduling. In: Wu, W., Zhang, Z. (eds.) COCOA 2020. LNCS, vol. 12577, pp. 125–139. Springer, Cham (2020). https://doi.org/10.1007/978-3-030-64843-5_9

12. Steinberg, A., Cardei, M., Cardei, I.: UAS batch path planning with a space-time graph. IEEE Open J. Intell. Transp. Syst. **2**, 60–72 (2021). https://doi.org/10.1109/OJITS.2021.3070415

13. Papa, R., Cardei, I., Cardei, M.: Generalized path planning for UTM systems with a space-time graph. IEEE Open J. Intell. Transp. Syst. **3**, 351–368 (2022). https://doi.org/10.1109/OJITS.2022.3171502

14. Chen, C., et al.: A graph-based conflict-free cooperation method for intelligent electric vehicles at unsignalized intersections. In: IEEE International Intelligent Transportation Systems Conference (2021)

15. Deng, Z., Shi, Y., Han, Q., Lu, L., Shen, W.: A conflict duration graph-based coordination method for connected and automated vehicles at signal-free intersections. Appl. Sci. **10**(18), 6223 (2020)

The Two Sheriffs Problem: Cryptographic Formalization and Generalization

Kota Sugimoto[1]([✉]), Takeshi Nakai[2], Yohei Watanabe[1,3],
and Mitsugu Iwamoto[1]

[1] The University of Electro-Communications, Tokyo, Japan
{sugimoto-k,watanabe,mitsugu}@uec.ac.jp
[2] Toyohashi University of Technology, Aichi, Japan
nakai@cs.tut.ac.jp
[3] National Institute of Advanced Industrial Science and Technology, Tokyo, Japan

Abstract. The two sheriffs problem is the following problem. There are two sheriffs, and each of them has their own list of suspects. Assuming that these lists are the result of a proper investigation, we can say that a culprit is the intersection of them even if the sheriffs do not know who the culprit is. Now, they wish to identify the culprit through an open channel, i.e., to compute the intersection of two lists, without letting an eavesdropper know the culprit who observed all communications. This cryptographic problem was proposed by Beaver et al., and a combinatorial solution using a bipartite graph was proposed. In this paper, we propose a formulation of the two sheriffs problem by introducing a secrecy evaluation based on the eavesdropper's attack success probability. Furthermore, we propose an improved version of Beaver et al.'s protocol that an arbitrary number of players can execute and has less attack success probability.

Keywords: Two sheriffs problem · communication protocol · shared secret

1 Introduction

1.1 Background

We say two parties, Alice and Bob, share a secret if there is a question Q that only they know its answer. Using the shared secret between the two parties, they can share a bit x with public communication without any other secrets. For example, suppose that Q is a yes-or-no question, and only Alice and Bob know its answer. With a public channel, Alice sends Bob a message "The bit x is 0 if the answer of Q is YES, 1 otherwise," and Bob determines the bit x is either 0 or 1 since he knows the answer of Q. As a result, Alice and Bob share a bit x. In contrast, no one except Alice and Bob cannot determine the bit x.

This work was supported by JSPS KAKENHI Grant Numbers JP23H00468, JP23H00479, JP23K17455, JP23K16880, JP22H03590, JP21H03395, JP21H03441, JP18H05289, and MEXT Leading Initiative for Excellent Young Researchers.

W. Wu and J. Guo (Eds.): COCOA 2023, LNCS 14461, pp. 512–523, 2024.
https://doi.org/10.1007/978-3-031-49611-0_37

We can also consider a situation where Alice and Bob do not know the answer to the question Q; however, each has some non-independent knowledge about it. For example, when Alice and Bob each make a narrowed-down list of possible answers to question Q, they are somehow convinced that both lists contain the answer to question Q. In this setting, they do not know the answer; however, they know that any of the elements in their list is the answer. Under these circumstances, can Alice and Bob share the answer using communication with public communications (without anyone determining the answer), and can they share a bit using it? In [1], Beaver et al. performed a combinatorial analysis of the above setting in a mathematical model using bipartite graphs and proposed a communication protocol based on it. As an example of such situations, they introduced the two sheriffs problem.

Two Sheriffs Problem: The problem settings are the following. There is a list of suspects in a case. Two sheriffs, Alice and Bob, narrowed down the list each other by conducting investigations independently. If the two sheriffs investigate in proper ways, it is natural to assume that the culprit is the intersection of their lists. Unfortunately, they are now separated and it is not possible to meet to identify the culprit. Hence, they want to identify the culprit by only using public communications, where Eve eavesdrops on their communications. The goal of the two sheriffs problem is to establish secure communications under the above situations.

Formally, the above problem can be described as follows. For a set of whole suspects S, which is public information, two players, Alice and Bob, have $A \subseteq S$ and $B \subseteq S$, respectively. Assume that they do not know each other's set, but they are convinced that the intersection exists and it is the culprit they want to know. Now, the two players are going to start communications for specifying the set intersection of A and B, namely, $A \cap B$. However, an eavesdropper, Eve, taps all communications between Alice and Bob to know $A \cap B$. To sum up, the two sheriffs problem asks one to construct a communication protocol that allows the two players who share no secrets to obtain $A \cap B$ while not allowing Eve, who taps all communications, to identify it.

Related Work. Related work of two sheriffs problem is not so many. The two sheriffs problem is an application of the problem called *isolation of common secret* [1]. Related to this problem, the problem called *cryptogenography problem* was proposed in [2], and it is recently discussed in [3,4]. Cryptogenography aims to design the protocols to share a secret held by one of the players keeping who had the secret hidden from the eavesdropper who monitors the transcripts. The upper and lower bounds of the success probabilities of attacks. On the other hand, the success probability in the two sheriffs problem is not discussed in [1].

1.2   Motivation and Our Contribution

Beaver et al. [1] proposed three protocols for solving the two sheriff problem, one is deterministic and the other two are probabilistic. We summarize the compar-

Table 1. Comparison of BHW protocols [1] and Ours

	Number of players	Number of Steps	Execution condition	Attack success probability
BHW-1	2	4	$\|\mathcal{S}\| \geq 2\|\mathcal{A}\|\|\mathcal{B}\|$	$1/2$
BHW-2	2	4	$\|\mathcal{S}\| \geq 2\|\mathcal{A}\|\|\mathcal{B}\|$	$1/2$
BHW-3	2	2	$\|\mathcal{S}\| \geq \|\mathcal{B}\|(1 + \sqrt{\|\mathcal{A}\| - 1})^2$	$(\|\mathcal{A}\| - 1)/\|\mathcal{A}\|$
Ours	$n\ (\geq 2)$	$2n$	$\|\mathcal{S}\| \geq 2\prod_{\ell=1}^{n} \|\mathcal{S}_\ell\|$	$\lfloor \|\mathcal{S}\| / \prod_{\ell=1}^{n} \|\mathcal{S}_\ell\| \rfloor^{-1}$

ison among the three protocols in Table 1. The first protocol, called BHW-1, is the deterministic one that works only when $|\mathcal{S}| \geq 2|\mathcal{A}||\mathcal{B}|$ with the success probability $1/2$ assuming that the dummy set in the protocol is chosen uniformly at random. The same performance can be obtained by the second non-deterministic protocol, called BHW-2. The third protocol, BHW-3, is also non-deterministic, which attains a smaller number of steps at the expense of attack success probability. Namely, BHW-2 and BHW-3 have a trade-off between the number of steps and attack success probability.

On the other hand, there are several problems in BHW protocols. First, the definition of security against the eavesdropper is weak. Actually, in Beaver et al.'s definition, the communication is secure if the suspect is not uniquely determined by the eavesdropper, and further, it is not formalized mathematically. Second, the problem only considered two-party protocols, and only the intersection of two lists is computed.

This paper formalizes a framework of two sheriffs problem where the security is evaluated by success probability by the eavesdropper when the dummy set is selected uniformly at random. The formulation involves the *multi-sheriff* problem as a generalization of the two sheriffs problem regarding the number of players. In addition, we propose a protocol for the multi-sheriff problem. Our protocol is superior to the BHW protocols in terms of success probability and the number of players. Extension of our protocol to set operations other than the set intersection is possible, but it will be reported in the full version due to space limitations.

The comparison among BHW protocols and ours is shown in Table 1. Our protocol can be executed by any $n\ (\geq 2)$ players, the attack success probability is the same or smaller than that of protocols Beaver et al. proposed (BHW protocol), and the number of Steps becomes the same as BHW protocols when the number of players is 2. In addition, BHW protocol can be executed when the execution condition, $|\mathcal{S}| \geq 2|\mathcal{A}||\mathcal{B}|$, is satisfied, and our protocol can be executed when $|\mathcal{S}| \geq 2\prod_{\ell=1}^{n} |\mathcal{S}_\ell|$, the generalized version of the execution condition of BHW protocol, is satisfied.

1.3   Organization

The rest of this paper is organized as follows. We first formalize the two sheriffs problem in Sect. 2, which extends the original problem to a multi-shriff problem. After describing the idea of BHW-2 protocol in Sect. 3, we propose our protocol for multi-sheriff problem in Sect. 4, and the security proof is sketched. In Sect. 4, the extension of our protocol to the other set operations is briefly discussed.

2   Formalization of the Multi-Sheriff Problem

For a finite set $\mathcal{X}$, we denote its cardinality by $|\mathcal{X}|$. We use also $[n]$ to denote $\{1, 2, \ldots, n\}$ for any positive interger $n \in \mathbb{N}$.

In this section, we introduce the multi-sheriff problem, the generalized version of the two sheriffs problem, and formalize it. We denote n players $P_1, P_2, \ldots, P_n$, instead of Alice and Bob in the two sheriffs problem, and for all $k \in [n]$, $\mathcal{S}_k$ is the set the player P_k has. The multi-sheriff problem is equivalent to the two sheriffs problem where $n = 2$.

2.1   Settings of the Multi-Sheriff Problem

There are a set $\mathcal{S}$, which is public information, and $n \ (\geq 2)$ players $P_1, P_2, \ldots, P_n$ who have $\mathcal{S}_1, \mathcal{S}_2, \ldots, \mathcal{S}_n$, subsets of $\mathcal{S}$, respectively. Each player only knows their own set and does not share any secrets with other players. They aim to share the set intersection $\bigcap_{\ell=1}^{n} \mathcal{S}_\ell \ (\neq \emptyset)$ after communicating with each other via authenticated channels. However, an eavesdropper, Eve, taps all communications among the n players to determine $\bigcap_{\ell=1}^{n} \mathcal{S}_\ell$.

2.2   Multi-Sheriff Problem Protocol Construction

We define the multi-sheriff problem protocol π as a pair of two phases, the communication and output phases.

Communication Phase: Each player creates a transcript from previous communications and sends it to other players. For simplicity, we assume that all transcripts are broadcast to all other players and transcripts are subsets of the universe $\mathcal{S}$ since players' information is only subsets of $\mathcal{S}$. Let $\mathcal{M}_i \subseteq 2^{\mathcal{S}}$ be an i-th transcript created by some player, say P_k, and Tran_i be the *next message function* that outputs $\mathcal{M}_i$. Tran_i is a probabilistic algorithm that takes as input all up to $(i-1)$-th transcripts $\mathcal{M}_1, \ldots, \mathcal{M}_{i-1}$ and P_k's set $\mathcal{S}_k$, and outputs the i-th transcript $\mathcal{M}_i$. For every $i = 1, 2, \ldots$, a player $P_{j_i} \in \{P_1, P_2, \ldots, P_n\}$ runs

$$\mathcal{M}_i \leftarrow \mathsf{Tran}_i(\mathcal{M}_1, \mathcal{M}_2, \ldots, \mathcal{M}_{i-1}, \mathcal{S}_{j_i}),$$

as an i-th transcript and broadcasts it.

Output Phase: Suppose that the players run τ-round communications. Each player P_k runs the *output function* Output_{P_k} that takes as input the universe, all transcripts, and their own set and outputs a set most likely to be $\bigcap_{i=1}^{n} \mathcal{S}_i$, denoted by $\mathcal{M}_{P_k}$. Eve also has their output function Output_E that takes as input the universe and all transcripts and outputs $\mathcal{M}_E$. Formally, the output functions are probabilistic algorithms written as follows:

$$\mathcal{M}_{P_k} \leftarrow \mathsf{Output}_{P_k}(\mathcal{S}, \mathcal{M}_1, \ldots, \mathcal{M}_\tau, \mathcal{S}_k),$$
$$\mathcal{M}_E \leftarrow \mathsf{Output}_E(\mathcal{S}, \mathcal{M}_1, \ldots, \mathcal{M}_\tau),$$

where $k \in [n]$.

2.3 Requirements of the Multi-Sheriff Problem

We define the correctness property and security notion for the multi-sheriff problem. First, we define the correctness; every party gets the correct output, i.e., $\bigcap_{k=1}^{n} \mathcal{S}_k$.

Definition 1 (Correctness). *We say that a τ-round protocol π satisfies the correctness if π satisfies that:*

$$\Pr\left(\mathcal{M}_{P_1} = \mathcal{M}_{P_2} = \cdots = \mathcal{M}_{P_n} = \bigcap_{k=1}^{n} \mathcal{S}_k\right) = 1,$$

where $\mathcal{M}_i \leftarrow \mathsf{Tran}_i(\mathcal{M}_1, \mathcal{M}_2, \ldots, \mathcal{M}_{i-1}, \mathcal{S}_{j_i})$ and $\mathcal{M}_{P_k} \leftarrow \mathsf{Output}_{P_k}(\mathcal{S}, \mathcal{M}_1, \ldots, \mathcal{M}_\tau, \mathcal{S}_k)$ for $i \in [\tau]$ and $k \in [n]$.

We next formalize a security notion. To do so, we define the success probability of Eve's attack, i.e., the probability that Eve succeeds in guessing $\bigcap_{k=1}^{n} \mathcal{S}_k$.

Definition 2 (Attack Success Probability). *Let π be a τ-round protocol. For any (even computationally unbounded) eavesdropper Eve, the attack success probability is defined as:*

$$\epsilon(\mathsf{Output}_E) := \Pr\left(\mathcal{M}_E = \bigcap_{k=1}^{n} \mathcal{S}_k\right), \tag{1}$$

where the probability is taken over the possible outcomes of the internal coin tosses of $\mathsf{Tran}_1, \ldots, \mathsf{Tran}_\tau$, and Output_E.

Definition 3 (p-Secrecy). *We say that a τ-round protocol π satisfies p-secrecy if Eve's attack success probability satisfies:*

$$\max_{\mathsf{output}_E} \epsilon(\mathsf{output}_E) \leq p. \tag{2}$$

We say that a protocol π solves the multi-sheriff problem with a probability p if it satisfies Definitions 1 and 3.

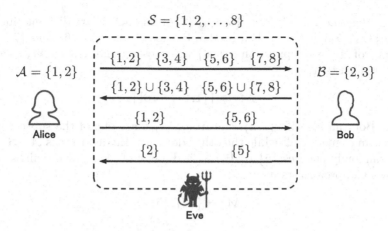

Fig. 1. An example of BHW protocol execution

3 BHW Protocols for Two Sheriffs Problem

Beaver et al. [1] proposed three protocols for the two sheriffs problem, and we explain the simplest one among them, which we refer to as the BHW protocol. Based on the formalization of the Sect. 2, we describe how the BHW protocol works (Fig. 1).

3.1 Overview of the BHW Protocol

We describe how the BHW protocol works with a concrete example as introduced in [1]. For simplicity, we denote P_1 as Alice, P_2 as Bob, and the sets each of them has as $\mathcal{A}$ and $\mathcal{B}$. We assume $|\mathcal{S}| \geq 2|\mathcal{A}||\mathcal{B}|$ and $\mathcal{A} \cap \mathcal{B} \neq \emptyset$. In this example, it is assumed that $\mathcal{S} = [8]$, $\mathcal{A} = \{1, 2\}$ and $\mathcal{B} = \{2, 3\}$.

Communication Phase

Step 1. Alice partitions $\mathcal{S}$ into $|\mathcal{S}| / |\mathcal{A}| = 4$ sets so that one of them is $\mathcal{A} = \{1, 2\}$, and the elements of the other three sets, where the size of each is two, are randomly chosen. For example, Alice randomly chooses $\{3, 4\}$, $\{5, 6\}$, and $\{7, 8\}$, and she defines $\mathcal{M}_1$ below and broadcasts it:

$$\mathcal{M}_1 = \{\{1, 2\}, \{3, 4\}, \{5, 6\}, \{7, 8\}\}.$$

Step 2. Bob calculates the intersections of $\mathcal{B} = \{2, 3\}$ and each element of $\mathcal{M}_1 = \{\mathcal{X}_1^{(1)}, \ldots, \mathcal{X}_4^{(1)}\}$, then he makes two sets according to the rule whether or not $\mathcal{X}_i^{(1)}$ shares elements of $\mathcal{B} = \{2, 3\}$; one is the union of $\{1, 2\}$ and $\{3, 4\}$ and the other is the union of $\{5, 6\}$ and $\{7, 8\}$. He defines $\mathcal{M}_2$ as follows and broadcasts it:

$$\mathcal{M}_2 = \{\{1, 2\} \cup \{3, 4\}, \{5, 6\} \cup \{7, 8\}\}.$$

Step 3. Alice can identify that $\{5,6\} \cup \{7,8\}$ and $\mathcal{A} \cup \mathcal{B}$ are disjoint since $\mathcal{A} \cap (\{5,6\} \cup \{7,8\}) = \emptyset$. She then randomly chooses one of $\{5,6\}$ and $\{7,8\}$ as a dummy of $\mathcal{A}$ (here suppose that $\{5,6\}$ was chosen), and defines $\mathcal{M}_3$ as follows and broadcasts it.

$$\mathcal{M}_3 = \{\{1,2\},\{5,6\}\}.$$

Step 4. Bob can identify $\{1,2\} = \mathcal{A}$ from $\mathcal{M}_3$, since one of the subsets in $\mathcal{M}_3$ shares an element of $\mathcal{B}$ while the other does not. He determines $\mathcal{A} \cap \mathcal{B} = \{2\}$, and randomly chooses $\{5\}$ or $\{6\}$ as a dummy of $\mathcal{A} \cap \mathcal{B}$. He defines $\mathcal{M}_4$ as follows and broadcasts it.

$$\mathcal{M}_4 = \{\{2\},\{5\}\}.$$

Output Phase

Alice: Alice can determine $\mathcal{A} \cap \mathcal{B} = \{2\}$ since only one of the subsets of $\mathcal{M}_4$ shares an element of $\mathcal{B}$, i.e., $\mathcal{A} \cap \mathcal{B}$. Therefore, she outputs $\{2\}$ as $\mathcal{M}_{\text{Alice}}$.
Bob: Bob already determined $\mathcal{A} \cap \mathcal{B} = \{2\}$ in Step 4, and so he outputs $\{2\}$ as $\mathcal{M}_{\text{Bob}}$.

Correctness and Secrecy: In this example, BHW protocol obviously satisfies the correctness defined in Definition 1 since it holds that

$$\Pr\left(\mathcal{M}_{\text{Alice}} = \mathcal{M}_{\text{Bob}} = \mathcal{A} \cap \mathcal{B}\right) = 1.$$

It is obvious that Eve cannot identify which set is $\mathcal{A}$ in Step 1. In Step 2, two sets are computed from randomly chosen two subsets from $\mathcal{M}_1$ from Eve's view. In Step 3, Eve now knows $\mathcal{A}$ is either $\{1,2\}$ or $\{5,6\}$. Finally, after Step 4, she knows that either $\{2\}$ or $\{5\}$ is $\mathcal{A} \cap \mathcal{B}$. However, regardless of her computational power, she cannot guess $\mathcal{A} \cap \mathcal{B}$ with more than probability $1/2$ since all the same transcripts might appear even if Therefore, the optimal attack strategy is to randomly choose either one of $\{2\}$ or $\{5\}$ as $\mathcal{M}_{\text{E}}$. Therefore, the attack success probability $\epsilon(\text{output}_{\text{E}}) = \Pr(\mathcal{M}_{\text{E}} = \{2\}) = 1/2$, and this holds for any attacker Eve. Hence, the BHW protocol satisfies the $1/2$-secrecy defined in Definition 3 since it holds that

$$\max_{\text{output}_{\text{E}}} \epsilon(\text{output}_{\text{E}}) = 1/2.$$

4   Our Solution for Multi-Sheriffs Problem

In this section, we construct a protocol for the multi-sheriff problem based on the above BWH protocol. The technical challenge to construct such a protocol is two-fold. First, all protocols shown by Beaver et al. are two-player protocols, and therefore it is unclear how we generalize it to an n-player protocol. Second, from the point of cryptologic view, the attack success probability should be as small as possible. However, the success probability of each Beaver et al.'s protocol is at least $1/2$ regardless of the sizes of $\mathcal{S}$, $\mathcal{A}$, and $\mathcal{B}$.[1] Therefore, we aim to realize a protocol that provides smaller success probability than $1/2$.

4.1   Observation and Our Construction Idea

Generalizing the Number of Players. The core construction idea of the BHW protocols is the partitioning of the set $\mathcal{S}$ into two sets; one includes $\mathcal{A} \cup \mathcal{B}$ and the other, called the dummy, is disjoint to $\mathcal{A} \cup \mathcal{B}$. Steps 1 and 2 of the BHW protocol provide this partitioning properly.

This idea can be applied when generalizing the number of players. In particular, the number of players does not affect the number of partitions. For example, suppose a three-player setting where Alice, Bob, and Carol have $\mathcal{A}$, $\mathcal{B}$, and $\mathcal{C}$, respectively. Then, as in the BHW protocol, partitioning into two sets is sufficient to give the meaningful upper bound of the attack success probability; if $\mathcal{S}$ can be partitioned into a set including $\mathcal{A} \cup \mathcal{B} \cup \mathcal{C}$ and that disjoint to $\mathcal{A} \cup \mathcal{B} \cup \mathcal{C}$, we can construct a protocol that has the success probability at most $1/2$ as in the BHW protocol. The same holds even for n players. Therefore, we aim to show how we realize such partition of $\mathcal{S}$ into two sets, one includes $\bigcup_{i=1}^{n} \mathcal{S}_i$ and the other is disjoint to $\bigcup_{i=1}^{n} \mathcal{S}_i$.

Improving Secrecy. The original two sheriffs problem [1] asks whether there exists a protocol that Alice and Bob can share the intersection of their sets while Eve cannot determine it from the transcripts. In this sense, all protocols proposed by Beaver et al. are sufficient since the attack success probability is less than one. In particular, in the BHW protocol in the previous section, a single dummy is sufficient to provide the success probability $1/2$.

In this work, we aim to achieve better success probability than the BHW protocol by increasing the number of dummies. Observing Step 4 of the BHW protocol, the number of subsets included in the last transcript $\mathcal{M}_4$ directly leads to the attack success probability, since the partitioning technique provides the two partitioned sets such that each of them is equally likely to include the intersection. Therefore, we extend the partitioning technique to provide as many partitioned sets as possible while each is equally likely to include an intersection of all players' sets.

4.2   Construction of Our Protocol

Execution Conditions. We assume that it holds $|\mathcal{S}| \geq 2 \prod_{\ell=1}^{n} |\mathcal{S}_\ell|$ and $\bigcap_{\ell=1}^{n} \mathcal{S}_\ell \neq \emptyset$. These execution conditions are generalized ones of the BHW protocol (also see Sect. 3.1). Indeed, the conditions are equivalent to those of the BHW protocol when $n = 2$.

---

[1] To be precise, the attack success probability of the third protocol is at most $(|\mathcal{A}| - 1)/|\mathcal{A}|$, and it means that it depends on the size of $|\mathcal{A}|$ (also see Table 1). Nonetheless, it holds $(|\mathcal{A}| - 1)/|\mathcal{A}| \geq 1/2$ for any $\mathcal{A}$ s.t. $|\mathcal{A}| \geq 2$.

Overview of Our Protocol. As described above, the attack success probability depends on the number of partitions. Therefore, for notational simplicity, let

$$\alpha_k := \left\lceil \frac{|\mathcal{S}|}{\prod_{\ell=1}^{k} |\mathcal{S}_\ell|} \right\rceil,$$

be the number of partitions that a player P_k initially creates. Note that it holds $\alpha_1 \geq \cdots \geq \alpha_n$. The communication phase of our protocol consists of $2n$ steps. Specifically, for all $k \in [n]$, a player P_k creates two transcripts in Steps k and $n + k$. In Step k, a player P_k partitions $\mathcal{S}$ into α_k sets. At the end of Step n, i.e., after all players created their first transcripts, they share α_n equal-sized partitioned sets, $T^{(n)}, \mathcal{D}_1^{(n)}, \mathcal{D}_2^{(n)}, \ldots, \mathcal{D}_{\alpha_n-1}^{(n)}$, where $(\bigcup_{k=1}^{n} \mathcal{S}_k) \subset T^{(n)}$ and $\mathcal{D}_\ell^{(n)} \cap (\bigcup_{k=1}^{n} \mathcal{S}_k) = \emptyset$ for all $\ell \in [\alpha_n - 1]$. In the second round, the players try to identify the intersection with their second transcripts. At the end of Step $2n$, all players share α_n equal-sized partitioned sets, $T^{(2n)}, \mathcal{D}_1^{(2n)}, \mathcal{D}_2^{(2n)}, \ldots, \mathcal{D}_{\alpha_n-1}^{(2n)}$, where $T^{(2n)} = \bigcap_{k=1}^{n} \mathcal{S}_k$ and $\mathcal{D}_\ell^{(2n)} \subset \mathcal{D}_\ell^{(n)}$ for all $\ell \in [\alpha_n - 1]$. In the output phase, each player P_k just outputs a set $\mathcal{X}$ such that it holds $\mathcal{X} \cap \mathcal{S}_k \neq \emptyset$. Note that there exists only one set satisfying the condition in $\mathcal{M}_{2n}$.

Communication Phase

Step 1. Let $\widetilde{\mathcal{S}}$ be an arbitrary subset of $\mathcal{S}$ such that $\widetilde{\mathcal{S}}$ includes $\mathcal{S}_1$ and $|\widetilde{\mathcal{S}}| = \alpha_1 |\mathcal{S}_1|$. P_1 sets $T^{(1)} = \mathcal{S}_1$ and divides $\widetilde{\mathcal{S}} \setminus \mathcal{S}_1$ into $\alpha_1 - 1$ sets $\mathcal{D}_1^{(1)}, \mathcal{D}_2^{(1)}, \ldots,$ $\mathcal{D}_{\alpha_1-1}^{(1)}$ such that it holds $\left| \mathcal{D}_\ell^{(1)} \right| = |\mathcal{S}_1|$ for every $\ell \in [\alpha_1 - 1]$.[2] P_1 broadcasts $\mathcal{M}_1 := \left\{ T^{(1)}, \mathcal{D}_1^{(1)}, \mathcal{D}_2^{(1)}, \ldots, \mathcal{D}_{\alpha_1-1}^{(1)} \right\}$.

Step k $(= 2, 3, \ldots, n)$. P_k computes and broadcasts $\mathcal{M}_{k-1} = \{T^{(k)}, \mathcal{D}_1^{(k)}, \mathcal{D}_2^{(k)}, \ldots, \mathcal{D}_{\alpha_k-1}^{(k)}\}$ as follows.

(I) Initially, let $\widetilde{\mathcal{M}}_{k-1} = \mathcal{M}_{k-1}$. For $\ell = 1, 2, \ldots, \alpha_k - 1$, P_k makes an ℓ-th dummy $\mathcal{D}_\ell^{(k)}$ as follows. Roughly speaking, P_k randomly chooses (unchosen) $|\mathcal{S}_k|$ sets from the last transcript $\mathcal{M}_{k-1}$, and sets the union of the sets as the ℓ-th dummy. Namely, P_k randomly chooses $|\mathcal{S}_k|$ distinct sets

$$\mathcal{X}_1, \mathcal{X}_2, \ldots, \mathcal{X}_{|\mathcal{S}_k|} \in \widetilde{\mathcal{M}}_{k-1},$$

such that it holds $\mathcal{X}_i \cap \mathcal{S}_k = \emptyset$ for every $i \in [|\mathcal{S}_k|]$, and sets

$$\mathcal{D}_\ell^{(k)} = \bigcup_{i=1}^{|\mathcal{S}_k|} \mathcal{X}_i, \text{ and } \widetilde{\mathcal{M}}_{k-1} = \widetilde{\mathcal{M}}_{k-1} \setminus \{\mathcal{X}_1, \mathcal{X}_2, \ldots, \mathcal{X}_{|\mathcal{S}_k|}\}.$$

---

[2] Even if it holds that $\mathcal{S}_k \not\subset \widetilde{\mathcal{S}}$ for some $k \in \{2, \ldots, n\}$, it does not matter; the protocol works well since $\widetilde{\mathcal{S}}$ includes $\mathcal{S}_1$, which also includes $\bigcap_{k=1}^{n} \mathcal{S}_k$.

(II) P_k computes $T^{(k)}$ as follows. P_k randomly chooses $|S_k|$ distict sets

$$\mathcal{Y}_1, \mathcal{Y}_2, \ldots, \mathcal{Y}_{|S_k|} \in \widetilde{\mathcal{M}}_{k-1},$$

such that it holds $\mathcal{Y}_i \cap S_k \neq \emptyset$ for every $i \in [|S_k|]$, and sets

$$T^{(k)} = \bigcup_{i=1}^{|S_k|} \mathcal{Y}_i.$$

Note that although there might be unused elements in $\widetilde{\mathcal{M}}_{k-1} \cup \{\mathcal{Y}_1, \ldots, \mathcal{Y}_{|S_k|}\}$, it does not matter and the protocol works well.

Step $n+1$. P_1 finds $T^{(n)}$ such that $S_1 \subset T^{(n)}$ from $\mathcal{M}_n$, and sets $T^{(n+1)} = S_1$. Let $\mathcal{D}_1^{(n)}, \mathcal{D}_2^{(n)}, \ldots, \mathcal{D}_{\alpha_n-1}^{(n)}$ be other $\alpha_n - 1$ dummy elements of $\mathcal{M}_n$. For all $\ell \in [\alpha_n - 1]$, P_1 randomly chooses $\mathcal{D}_\ell^{(n+1)} \subseteq \mathcal{D}_\ell^{(n)}$ such that $\left|\mathcal{D}_\ell^{(n+1)}\right| = \left|T^{(n+1)}\right|$. P_1 broadcasts $\mathcal{M}_{n+1} = \left\{T^{(n+1)}, \mathcal{D}_1^{(n+1)}, \mathcal{D}_2^{(n+1)}, \ldots, \mathcal{D}_{\alpha_n-1}^{(n+1)}\right\}$.

Step $n+k$ (for $k = 2, 3, \ldots, n$). P_k finds $T^{(n+k-1)}$ such that $S_k \cap T^{(n+k-1)} \neq \emptyset$ from $\mathcal{M}_{n+k-1}$, and sets

$$T^{(n+k)} = S_k \cap T^{(n+k-1)}.$$

Let $\mathcal{D}_1^{(n+k-1)}, \mathcal{D}_2^{(n+k-1)}, \ldots, \mathcal{D}_{\alpha_n-1}^{(n+k-1)}$ be other $\alpha_n - 1$ dummy elements of $\mathcal{M}_{n+k-1}$. For all $\ell \in [\alpha_n - 1]$, P_k randomly chooses

$$\mathcal{D}_\ell^{(n+k)} \subseteq \mathcal{D}_\ell^{(n+k-1)} \text{ such that } \left|\mathcal{D}_\ell^{(n+k)}\right| = \left|T^{(n+k)}\right|.$$

Finally, P_k broadcasts $\mathcal{M}_{n+k} := \left\{T^{(n+k)}, \mathcal{D}_1^{(n+k)}, \mathcal{D}_2^{(n+k)}, \ldots, \mathcal{D}_{\alpha_n-1}^{(n+k)}\right\}$.

Output Phase

Local operations. For any $k \in [n]$, since there exists only one set $\mathcal{X} \in M_{2n}$ such that it holds $\mathcal{X} \cap S_k \neq \emptyset$, P_k can determine and output $\bigcap_{\ell=1}^n S_\ell = \mathcal{X}$.

4.3   Proof for Correctness and Secrecy

We have the following theorems.

Theorem 1. *Our protocol satisfies the correctness in Definition 1 under the execution condition $|S| \geq 2 \prod_{\ell=1}^n |S_\ell|$.*

Proof. In Step k for any $k \in \{2, 3, \ldots, n\}$, P_k can make $T^{(k)}$ that includes $\bigcup_{\ell=1}^k S_\ell$ since $T^{(1)} = S_1$ and $\bigcap_{\ell=1}^n S_k \neq \emptyset$ is assumed. Specifically, in Step k, $T^{(k-1)}$ contains at least one element of S_k since it holds $\bigcap_{\ell=1}^n S_\ell \neq \emptyset$, and therefore, P_k can correctly compute $T^{(k)}$. At the end of Step n, all players have

$\mathcal{M}_n = \left\{ \mathcal{T}^{(n)}, \mathcal{D}_1^{(n)}, \mathcal{D}_2^{(n)}, \ldots, \mathcal{D}_{\alpha_n-1}^{(n)} \right\}$ and they can determine $\mathcal{T}^{(n)}$ since only $\mathcal{T}^{(n)}$ contains (at least one) elements of $\mathcal{S}_k$ for all $k \in [n]$; all other dummy sets do not include any element of them. Hence, from Steps $n+1$ to $2n$, each player P_k can determine $\mathcal{T}^{(n+k-1)}$ and compute its intersection with $\mathcal{S}_k$. Therefore, our protocol satisfies the correctness since all players can compute and output $\bigcap_{k=1}^{n} \mathcal{S}_k$ from $\mathcal{M}_{2n}$. □

Theorem 2. *Our protocol satisfies p-secrecy in Definition 3 under the execution condition* $|\mathcal{S}| \geq 2 \prod_{\ell=1}^{n} |\mathcal{S}_\ell|$, *where* $p = \lfloor |\mathcal{S}| / \prod_{\ell=1}^{n} |\mathcal{S}_\ell| \rfloor^{-1}$.

Proof. It is obvious that Eve cannot identify which element of $\mathcal{M}_1$ is $\mathcal{S}_1$ in Step 1. After Step k $(k = 2, 3, \ldots, n)$, α_k sets are computed from randomly chosen $|\mathcal{S}_k|$ element of $\mathcal{M}_k$ from Eve's view. After Step $n+1$, Eve knows that one of the α_n elements of $\mathcal{M}_{n+1}$ must be $\mathcal{S}_1$, however, she cannot determine which element it is since each element in $\mathcal{M}_{n+1}$ is equally likely to be $\mathcal{S}_1$. Similarly, after Step $n+k$ $(k = 2, 3, \ldots, n)$, Eve knows that one of the α_n elements of $\mathcal{M}_{n+k}$ must be $\bigcap_{\ell=1}^{k} \mathcal{S}_\ell$. However, since each element in $\mathcal{M}_{n+k}$ is equally likely to be $\bigcap_{\ell=1}^{k} \mathcal{S}_\ell$, Eve cannot determine it from $\mathcal{M}_{n+k}$. Hence, regardless of her computational power, she cannot guess $\bigcap_{\ell=1}^{n} \mathcal{S}_\ell$ with more than probability α_n^{-1}. Therefore, the optimal attack strategy for any attacker Eve is to randomly choose one of the α_n elements of $\mathcal{M}_{2n}$ as $\mathcal{M}_E$. Therefore, the attack success probability is $\epsilon(\text{output}_E) = \Pr(\mathcal{M}_E = \bigcap_{\ell=1}^{n} \mathcal{S}_\ell) = \alpha_n^{-1} = \lfloor |\mathcal{S}| / \prod_{\ell=1}^{n} |\mathcal{S}_\ell| \rfloor^{-1}$. Thus, our protocol satisfies the p-secrecy where $p = \lfloor |\mathcal{S}| / \prod_{\ell=1}^{n} |\mathcal{S}_\ell| \rfloor^{-1}$ since it holds that

$$\max_{\text{output}_E} \epsilon(\text{output}_E) = \left\lfloor \frac{|\mathcal{S}|}{\prod_{\ell=1}^{k} |\mathcal{S}_\ell|} \right\rfloor^{-1}.$$

It completes the proof. □

4.4   Extension of Our Protocol

As can be seen above, we generalized the BHW protocol and showed our protocol that enables n players to share the intersection of their sets $\mathcal{S}_1, \mathcal{S}_2, \ldots, \mathcal{S}_n$. In this section, we show that this protocol can be extended to a protocol that allows n players to share the result of arbitrary set operations on $\mathcal{S}_1, \mathcal{S}_2, \ldots, \mathcal{S}_n$.

We describe below the construction idea of the generalized protocol, and will give the formal description in the full version of this paper. From Steps $n+1$ to $2n$ of our protocol, P_k receives $\bigcap_{\ell=1}^{k-1} \mathcal{S}_\ell$ (included in $\mathcal{M}_{n+k-1}$) and updates it to $\bigcap_{\ell=1}^{k} \mathcal{S}_\ell$. Finally, P_n calculates $\bigcap_{\ell=1}^{n} \mathcal{S}_\ell$ and shares it among all players. The procedures are obviously specific to the set intersection. Roughly speaking, we can extend the procedures so that each player has all sets of $\mathcal{S}_1, \mathcal{S}_2, \ldots, \mathcal{S}_n$ at the last step of the communication phase. To do so, we modify the protocols so that the transcripts always contain all elements of $\mathcal{S}_1, \ldots, \mathcal{S}_n$, whereas the proposed protocol allows players to discard some elements in Steps 1 to n, which we call the discarded elements the remaining set. For the proposed protocol, we do not care what elements the remaining set contains since we are only interested in the intersection of the players' sets; the remaining set never contains any element in the intersection. On the other hand, e.g., if we consider a protocol for the set union, the remaining set is crucial to construct the protocol since it might contain some elements in some player's set. Therefore, we modify the steps so that they do not produce any remaining set.

References

1. Beaver, D., Haber, S., Winkler, P.: On the isolation of a common secret. Math. Paul Erdös II 121–135 (1997)
2. Brody, J., Jakobsen, S.K., Scheder, D., Winkler, P.: Cryptogenography. In: ITCS, pp. 13–22 (2014)
3. Doerr, B., Kunnemann, M.: Improved protocols and hardness results for the two-player cryptogenography problem. IEEE Trans. Inf. Theory **66**(9), 5729–5741 (2020). https://doi.org/10.1109/tit.2020.2978385. https://ieeexplore.ieee.org/document/9115678/
4. Jakobsen, S.K.: Information theoretical cryptogenography. J. Cryptol. **30**(4), 1067–1115 (2016). https://doi.org/10.1007/s00145-016-9242-8.pdf

Author Index

W. Wu and J. Guo (Eds.): COCOA 2023, LNCS 14461, pp. 525–527, 2024.
https://doi.org/10.1007/978-3-031-49611-0

Printed in the United States
by Baker & Taylor Publisher Services